Algorithms for Intelligent Syst

CW01390560

Series Editors

Jagdish Chand Bansal, Department of Mathematics, South Asian University,
New Delhi, Delhi, India

Kusum Deep, Department of Mathematics, Indian Institute of Technology Roorkee,
Roorkee, Uttarakhand, India

Atulya K. Nagar, School of Mathematics, Computer Science and Engineering,
Liverpool Hope University, Liverpool, UK

This book series publishes research on the analysis and development of algorithms for intelligent systems with their applications to various real world problems. It covers research related to autonomous agents, multi-agent systems, behavioral modeling, reinforcement learning, game theory, mechanism design, machine learning, meta-heuristic search, optimization, planning and scheduling, artificial neural networks, evolutionary computation, swarm intelligence and other algorithms for intelligent systems.

The book series includes recent advancements, modification and applications of the artificial neural networks, evolutionary computation, swarm intelligence, artificial immune systems, fuzzy system, autonomous and multi agent systems, machine learning and other intelligent systems related areas. The material will be beneficial for the graduate students, post-graduate students as well as the researchers who want a broader view of advances in algorithms for intelligent systems. The contents will also be useful to the researchers from other fields who have no knowledge of the power of intelligent systems, e.g. the researchers in the field of bioinformatics, biochemists, mechanical and chemical engineers, economists, musicians and medical practitioners.

The series publishes monographs, edited volumes, advanced textbooks and selected proceedings.

More information about this series at http://www.springer.com/series/16171

Xiao-Zhi Gao · Rajesh Kumar · Sumit Srivastava ·
Bhanu Pratap Soni
Editors

Applications of Artificial Intelligence in Engineering

Proceedings of First Global Conference on
Artificial Intelligence and Applications
(GCAIA 2020)

🦄 Springer

Editors
Xiao-Zhi Gao 🆔
School of Computing
University of Eastern Finland
Kuopio, Finland

Rajesh Kumar
Department of Electrical Engineering
Malaviya National Institute of Technology
Jaipur, Rajasthan, India

Sumit Srivastava
Department of Information Technology
Manipal University Jaipur
Jaipur, Rajasthan, India

Bhanu Pratap Soni
Department of Electrical Engineering
University of Engineering and Management
Jaipur, Rajasthan, India

ISSN 2524-7565   ISSN 2524-7573 (electronic)
Algorithms for Intelligent Systems
ISBN 978-981-33-4606-2   ISBN 978-981-33-4604-8 (eBook)
https://doi.org/10.1007/978-981-33-4604-8

This Springer imprint is published by the registered company Springer Nature Singapore Pte Ltd.
The registered company address is: 152 Beach Road, #21-01/04 Gateway East, Singapore 189721, Singapore

Preface

This 1st Global Conference on Artificial Intelligence and Applications (GCAIA 2020) was expected to be held in city of royals, Jaipur, but due to this pandemic situation of COVID-19 we have conducted this conference in a virtual mode. A visionary educationalist Prof. Dr. Satyajit Chakrabarti established the University of Engineering and Management, Jaipur, in 2011. He is presently the chancellor of the University of Engineering and Management, Jaipur. This university has Lush Green Campus of 32 acres situated in Sikar Road (NH-11), Udaipuria Mod, Jaipur 303807, Rajasthan. University of Engineering and Management, Jaipur, is one of the leading educational institutes in Rajasthan under the umbrella of Institute of Engineering and Management (IEM), Kolkata, West Bengal. Global Conference on Artificial Intelligence and Applications (GCAIA 2020) gives a premier forum to the researcher, engineers, corporates and scholar students for sharing their ideas of research in the field of artificial intelligence. The platform of GCAIA actually mitigates the gaps between academia, industry and governmental ethics, and the academic researchers would expose results and findings of laboratory-based experiments. The platform of GCAIA will also bring regional and international concerns for the discussions of recent advances in modern computation intelligence.

A tremendous amount of efforts was made by the reviewers to improve the quality of the research paper by making constructive and critical comments. The authors have implemented the modifications as suggested by the reviewers of this conference. For this aspect, author's efforts are really been appreciated. We are very much grateful to the session chairs, national and international committee members, student volunteers, administrative assistant and of course our faculty members for making this conference a grand success. We also thank the authors from different parts of the world for their good quality research paper submission for this conference. In this conference, a total of 336 research papers were submitted, and out of 336 papers 75 papers were accepted and presented. It was the quality of their presentation and their passion for research in the field of artificial intelligence which makes this conference a grand success.

Kuopio, Finland   Dr. Xiao-Zhi Gao
Jaipur, India   Prof. Dr. Rajesh Kumar
Jaipur, India   Prof. Dr. Sumit Srivastava
Jaipur, India   Prof. Bhanu Pratap Soni

Contents

About the Editors

Xiao-Zhi Gao received his B.Sc. and M.Sc. degrees from the Harbin Institute of Technology, China, in 1993 and 1996, respectively. He obtained his D.Sc. (Tech.) degree from the Helsinki University of Technology (now Aalto University), Finland, in 1999. He has been working as Professor at the University of Eastern Finland, Finland, since 2018. He is also Guest Professor at the Harbin Institute of Technology, Beijing Normal University and Shanghai Maritime University, China. Prof. Gao has published more than 400 technical papers in refereed journals and international conferences. His current Google Scholar H-index is 30. His research interests are nature-inspired computing methods with their applications in optimization, data mining, machine learning, control, signal processing and industrial electronics.

Dr. Rajesh Kumar received his B.Tech. degree from NIT Kurukshetra, India, the M. Tech. degree in Power System and the Ph.D. degree in Intelligent Systems MNIT, Jaipur. He was Postdoctorate Research Fellow in the Department of Electrical and Computer Engineering at the National University of Singapore (NUS), Singapore, from 2009 to 2011. Currently, he has been working as Professor with the Department of Electrical Engineering, MNIT, Jaipur. Dr. Kumar research interests focus on intelligent systems, machine intelligence, power management, smart grid and robotics. Dr. Kumar has published over 450 research articles, has supervised 20 Ph.D. and more than 30 M.Tech. thesis. He has 12 patents to his name. He received 03 academic awards, 12 best paper awards, 06 best thesis awards, 04 professional awards and 25 student awards. He has received the Career Award for Young Teachers in 2002 from the Government of India. He is on 12 Journal Editorial Boards. He is Associate Editor of IEEE Access, IEEE ITeN (Industrial Electronics technology News), Associate Editor, Swarm and Evolutionary Computation, Elsevier, Associate Editor, IET Renewable and Power Generation, Associate Editor, IET Power Electronics, Associate Editor, International Journal of Bio Inspired Computing, and Deputy Editor-in-Chief, CAAI Transactions on Intelligence Technology, IET. Dr. Kumar is Senior Member of IEEE (USA), Fellow of IET (UK), Fellow of IE (INDIA), Fellow of IETE, Life Member of CSI, Senior Member of IEANG and Life Member of ISTE. He was Honorary Secretary of IETE Rajasthan Chapter and Excom IEEE Delhi PES-IAS. Currently, he is Vice Chairman of IEEE Rajasthan Subsection.

Sumit Srivastava is currently Professor in the Department of Information Technology, Manipal University Jaipur (MUJ). He was holding the post of Head of the Department of Information Technology and recently promoted to Controller of Examination at MUJ. He has done his MCA from BITS, Mesra, Ranchi, and Ph.D. in Data Mining from the University of Rajasthan. He has 17 years of Teaching and nearly 12 years of Research Experience. Dr. Sumit has also been an invited speaker in the area of algorithms, machine learning, and information & knowledge discovery at various short-term courses. He has published around 85+ research papers in SCI, Scopus & peer-reviewed journals and conferences of international repute. He is currently Senior Member of IEEE, CSI & ACM and has been associated with various scientific societies as Member. He is also serving as Editorial Member in Springer Journal of Image Processing & Computer Vision, Guest Editor in Wiley Journal of Computer Vision, Reviewer in Journal of Imaging & Sensing and Journal of Computing and Communication and also Convener for the two international conferences of Smart System, Innovation and Computing (SSIC) in collaboration with Springer.

Bhanu Pratap Soni received the Bachelor of Technology with honors in Electrical Engineering and Master of Technology degree with honors in Power System Engineering from the Rajasthan Technical University, Kota, India, in the years 2011 and 2014, respectively, and submitted his thesis in Intelligent Power Systems for Ph.D. degree at the Department of Electrical Engineering, Malaviya National Institute of Technology (MNIT) Jaipur, India, in 2019. He is currently Associate Professor at the Department of Electrical Engineering at the University of Engineering and Management, Jaipur. He has received Young Scientist Award (under ITS, SERB) in 2018 from the Department of Science Technology, Govt. of India. He has also worked as Research Associate from 2015 to 2018 at the Department of Electrical Engineering, Malaviya National Institute of Technology, Jaipur. He is currently Member of IEEE and IACSIT and has been associated with various scientific societies as Member. He has published around 40+ research papers in SCI, Scopus & peer-reviewed journals and conferences of national and international repute. He is also serving as Reviewer in Applied Soft Computing, Cogent Engineering, International Transactions on Electrical Energy Systems and Scientia Iranica. He has also worked as Associate Chief Editor of Proceedings of 2nd International Conference on Advance Trends in Engineering & Technology (ICATET-2014), ACEIT, Jaipur, India. His background is in the field of artificial intelligence and machine learning, optimization, power system stability, smart grid, optimization and deep learning.

Chapter 1
Application of Supervised Learning for Voltage Stability Assessment

Asna Akhtar, Ankit Kumar Sharma, Abhishek Kumar Gupta, and Vivek Kumar Jain

1 Introduction

The voltage stability is more and more turning into a proscribing considering the trendy energy systems because of the assorted adjustments continuously brought to satisfy steadily increasing load demand without revolutionizing transmission and generation requirements. Due to this which has necessitated employment, the techniques assessed and resolved the crisis of voltage stability. Voltage stability is defined as the potential of the facility system to stay up acceptable and consistent voltage level at all buses in the system under normal conditions and after being subjected to the disturbance. Therefore, voltage stability analysis is essential to spot the critical buses in an exceeding power grid, i.e., buses which are closed to their voltage instability or just about voltage collapse point and to assist the planning engineers and operators to take suitable moves to keep away from voltage instability problems [1, 2]. The usual procedures available for the calculation of voltage stability of several mechanisms are supported by the load flow solution practicability, the distinctiveness of Jacobian, optimal power flow, bifurcation technique, etc. In this paper, the efforts are made to assess the voltage stability in terms of network equivalence to urge a global scenario or picture of voltage stability. The particular system is reduced into the same two-bus system, i.e., approach-related at the only line, by using all constraints regarding the

A. Akhtar (✉) · A. K. Gupta (✉) · V. K. Jain
Department of EE, Jaipur National University, Jaipur, India
e-mail: asnaakhtar9@gmail.com

A. K. Gupta
e-mail: abhi.kr.gupta@gmail.com

V. K. Jain
e-mail: Vivekkumar.jn@gmail.com

A. Kumar Sharma (✉)
Department of EE, University of Engineering and Management, Jaipur, India
e-mail: ankit.krishnaa@gmail.com

line and, subsequently, the global voltage stability indices for signifying the state of the particular system. The load flow solution provides all the parameters of the equivalent system of the first system. In this equivalent system, at the receiving end, a load having series equal impedance is kept, but the sending end voltage is kept at the reference voltage. The concept of a single line equivalent is further used to determine the voltage collapse proximity. The determination of an accurate global voltage stability index is feasible if a similar two-bus system is represented by the entire power grid accurately and faithfully.

In this equivalent model for stability of voltage, an influence system is obtained through lumping shunt admittances and series impedances of transmission lines altogether inside the series equivalent impedance obtained from any load flow study achieved on the particular dimension system [3–7]. G. C. Swetha et al. used an artificial neural network (ANN) for assessment of the power system with voltage stability.

This method applies to ANN with Newton–Raphson (N-R) load flow analysis. This paper result achieves from the ANN technique which is compared with the consequence of offline N-R load flow analysis in terms of accuracy to forecast the condition of the power system [8]. Bhavik Suthar et al. used an outline program to get actual output for all input model with including the Q limits of the generators [9]. N. Mittal et al. provide the different application of layer recurrent neural network model for load forecasting and use 3n in layer recurrent model of ANN. Soni B. P. et al. discussed the SVM approach for contingency analysis of the power system [10–12]. Basics about the different types of ANN topologies are understood by using [13–15].

In this part of ANN techniques for both real (P) and reactive load (Q), each smallest bus is randomly varied and observed the effect on the performance of the global voltage stability margin. In this paper, voltage stability assessment has done by different ANN network models. Comparisons are applied for various network topologies with a conventional N-R method.

2   Two-Bus Equivalent Pi Network

The equivalent two-bus pi-network model is developed and projected as follows: Suppose two-bus systems which are sending end bus and a load bus receiving end bus as shown in Fig. 1.

The behavior and properties of the projected two-bus equivalent model ought to be alike as a result of the multi-bus network and modify the estimation of voltage stability [3–5].

Therefore, the power equations for this equivalent network area unit are written as:

$$S_g = P_g + j Q_g = \vec{V}_s \vec{I}_s^* = (S_{se} + S_{sh}) + S_{load} \tag{1}$$

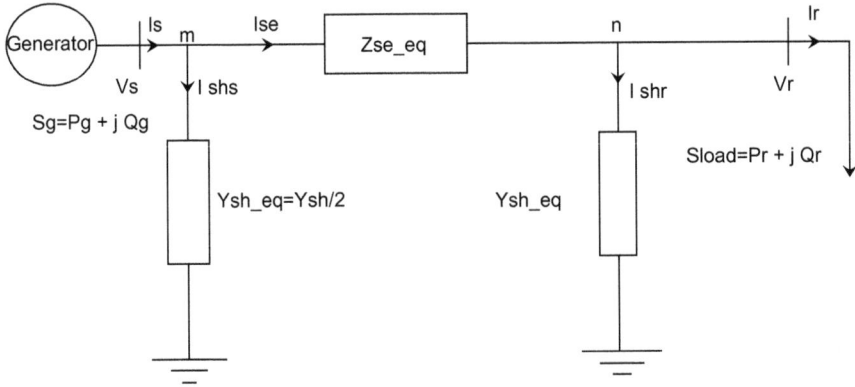

Fig. 1 Two-bus equivalent pi- network

$$S_{\text{se}} = \left(\vec{V}_{\text{s}} - \vec{V}_{\text{r}} \right) \vec{I}_{\text{se}}^{*} \tag{2}$$

$$S_{\text{sh}} = \vec{V}_{\text{s}} \vec{I}_{\text{shs}}^{*} + \vec{V}_{\text{r}} \vec{I}_{\text{shr}}^{*} \tag{3}$$

Applying KCL at node m and we get:

$$\vec{I}_{\text{se}}^{*} = \frac{S_{\text{g}}}{\vec{V}_{\text{s}}} - S_{\text{sh}} \left(\frac{\vec{V}_{\text{s}}^{*}}{\left| \vec{V}_{\text{s}} \right|^{2} + \left| \vec{V}_{\text{r}} \right|^{2}} \right) \tag{4}$$

At node n

$$\vec{I}_{\text{se}}^{*} = S_{\text{sh}} \left(\frac{\vec{V}_{\text{s}}^{*}}{\left| \vec{V}_{\text{s}} \right|^{2} + \left| \vec{V}_{\text{r}} \right|^{2}} \right) + \frac{S_{\text{load}}}{\vec{V}_{\text{r}}} \tag{5}$$

where

$I_{\text{shs}}, I_{\text{shr}}$ shunt branch currents at receiving and sending end.
$V_{\text{s}}, V_{\text{r}}$ sending and receiving end voltage.
$I_{\text{s}}, I_{\text{r}}$ sending and receiving end current.
I_{se} series equivalent impedance current.

For this experimental calculation, we get find out the equivalent shunt admittance and equivalent series impedance.

$$Z_{\text{se\_eq}} = \frac{\left(\vec{V}_{\text{s}} - \vec{V}_{\text{r}} \right)}{\vec{I}_{\text{se}}} \tag{6}$$

$$Y_{\text{sh\_eq}} = \frac{\vec{I}_{\text{shr}}}{\vec{V}_r} = \frac{\vec{I}_{\text{shs}}}{\vec{V}_s} \tag{7}$$

Thus, the equivalent two-bus pi-network model is obtained.

3 Global Voltage Stability Analysis of Multi-Bus Power System

For this part two-bus network, the same of a multi-bus power grid is taken for universal voltage stability index with different parameters of the global network as follows:

$$\begin{bmatrix} V_s \\ I_s \end{bmatrix} = \begin{bmatrix} A & B \\ C & D \end{bmatrix} \begin{bmatrix} V_r \\ I_r \end{bmatrix} \tag{8}$$

where
$A = D = 1 + \frac{YZ}{2}; \ B = Z; \ C = Y\left(1 + \frac{YZ}{4}\right)$
Assuming
$\left[Z = Z_{\text{se\_eq}} \text{ and } \frac{Y}{2} = Y_{\text{sh\_eq}} \right]$
Let as assume

$$A = |A|\angle\alpha \ ; \ B = |B|\angle\beta; \ \vec{V}_s = \left|\vec{V}_s\right|\angle\theta; \ \vec{V}_r = \left|\vec{V}_r\right|\angle\delta; \ \delta < \theta$$

Solving for the receiving end current:

$$I_r = \frac{\left|\vec{V}_s\right|}{|B|}\angle\theta - \beta - \frac{|A|\left|\vec{V}_r\right|}{|B|}\angle\alpha - \beta + \delta \tag{9}$$

Complex power of receiving end given by:

$$S_r = \vec{V}_r\vec{I}_r^* = \left|\vec{V}_r\right|\angle\delta\left[\frac{\left|\vec{V}_s\right|}{|B|}\angle - \theta + \beta - \frac{|A|\left|\vec{V}_r\right|}{|B|}\angle - \alpha + \beta - \delta\right] \tag{10}$$

The active power and reactive power at the receiving end which sending end voltage is constant is given by:

$$P_r = \frac{\left|\vec{V}_r\right|}{|B|}\cos(\beta + \delta) - \frac{|A|\left|\vec{V}_r\right|^2}{|B|}\cos(\beta - \alpha)$$

$$Q_r = \frac{\left|\vec{V_r}\right|}{|B|} \sin(\beta + \delta) - \frac{|A|\left|\vec{V_r}\right|^2}{|B|} \sin(\beta - \alpha)$$

The Jacobian matrix is given by:

$$J = \begin{matrix} \frac{\partial P_r}{\partial \delta} & \frac{\partial P_r}{\partial V_r} \\ \frac{\partial Q_r}{\partial \delta} & \frac{\partial Q_r}{\partial V_r} \end{matrix} = \frac{1}{|B|} \left[\begin{matrix} -\left|\vec{V_r}\right| \sin(\beta + \delta) \cos(\beta + \delta) - 2|A|\left|\vec{V_r}\right| \cos(\beta - \alpha) \\ \left|\vec{V_r}\right| \cos(\beta + \delta) \sin(\beta + \delta) - 2|A|\left|\vec{V_r}\right| \sin(\beta - \alpha) \end{matrix} \right]$$

$$(11)$$

The determinant of Jacobian matrix is given as:

$$\Delta[J] = \frac{1}{|B|^2} \left[2|A|\left|\vec{V_r}\right|^2 \cos(\delta + \alpha) - \left|\vec{V_r}\right| \right] \tag{12}$$

Equation (13) for critical point of voltage stability, $\Delta[J] = 0 \; \Delta[J] = 0$

$$\left|\vec{V_r}\right| = V_{cr} = \frac{1}{2|A| \cos(\delta + \alpha)} \tag{13}$$

Voltage stability limits are measured by critical value (V_{cr}) at the receiving end voltage. To maintain the global voltage stability, $\Delta[J] = 0$ the low value of V_{cr} indicates a better voltage profile along with side higher load catering capability results in better voltage stability. The value of GVSM (the global voltage stability margin, secure global voltage stability) should be equal to $\Delta[J]$, from Eq. (12). Present operating conditions for GVSM points on the global voltage security status or the global system voltage collapse [1] (Fig. 2).

4 Computational Procedure

The artificial neural network is discovered in the year 1958 by psychologist Frank Rosenblatt [16]. Artificial neural network (ANN) operates by creating a load of processing elements, and at some point in a biological brain, each aspect is analogous to individual neuron. The training of ANNs is done through adopting a network and comparing the output received with the input training and target data. Mainly the preparation is administered to match the network output to focus on data. In this paper, voltage stability assessment is conducted by using a conventional N-R method and RBFN and FFBP. These are two network models of ANN. We found that the outcomes from RBFN are more accurate when we compare the results from the ANN networks. So, we emphasized only on RBFN.

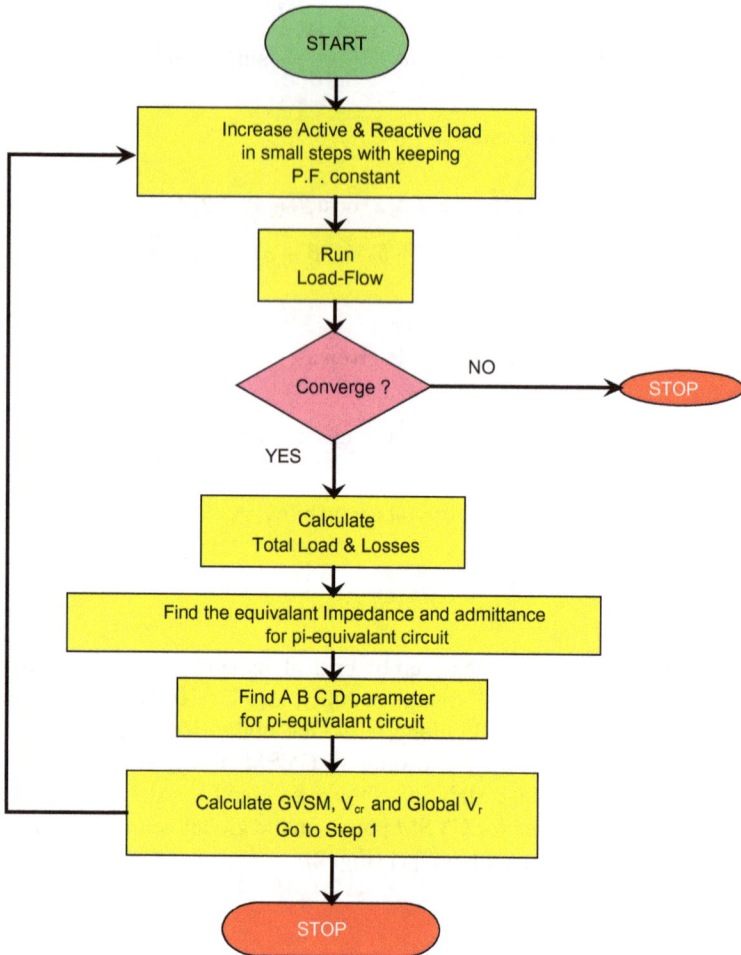

Fig. 2 Flow chart for V_{cr}, global V_r, and GVSM [13]

4.1 *RBF Neural Network Architecture*

The RBFNN is a radial basis function neural network; it is an artificial neural network that uses radial basis functions as activation functions. It is a feedforward neural network; it consists of three layers: an input layer, one hidden layer, and one output layer. All inputs are connected to each hidden neuron, a hidden layer with a nonlinear RBF activation function, and a linear output layer. For the RBF neural network, a learning process is made by combining the initiating center, width for RBF units, and computing weights for connectors [17]. According to the theory of function approximation, the Gaussian feature is used in it, i.e., there are two-layer feedforward network, and a collection of radial basis functions is implemented by

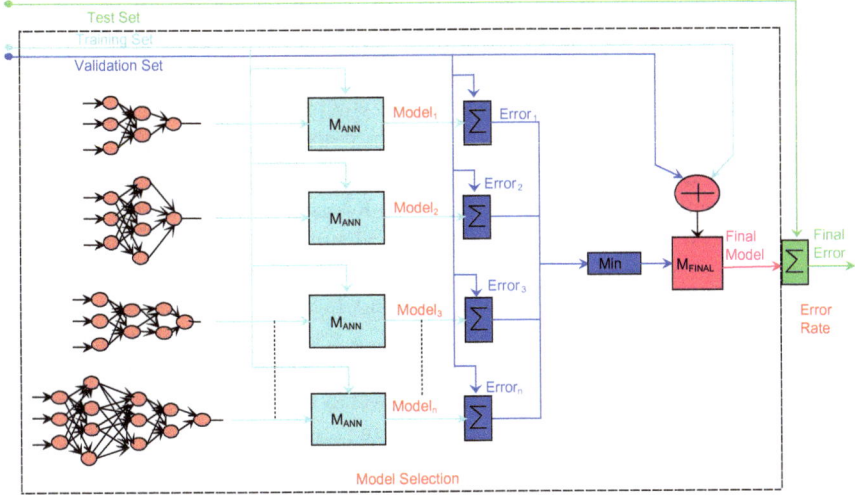

Fig. 3 Architecture for proposed ANN

hidden nodes. The linear summation function features as in a multi-layer perceptron (MLP) is implemented via output nodes. Weights are determined from the input layer to the hidden layer and from the hidden layer to the output layer in the first and second stages, respectively. Thus, the network training is divided into two phases; this makes interpolation very good for the developed architecture of network given in Fig. 3. Outlines for RBFN networks are given by [18, 19].

The equation of exact interpolation is

$$f(X) = \sum_{p=1}^{N} W_p \varphi \left(\left\| X - X^p \right\| \right) \tag{14}$$

The equation of determining the weights is

$$f\left(X^q\right) = \sum_{p=1}^{N} W_p \varphi \left(\left\| X^q - X^p \right\| \right) = t^p \tag{15}$$

And in the equation of improved RBFN network for the Gaussian basis function, we have:

$$\varphi_j(X) = \exp \left[-\frac{\left\| X - \mu_j \right\|^2}{2\sigma_j^2} \right] \tag{17}$$

For the training of artificial neural network, base case values of both actual and reactive loads at all the buses are varied randomly, and input data sets are generated from offline Newton–Raphson load flow analysis. In data collection, the input data is split into three categories named train data, validation data, and test data. N-R load flow analysis has conducted all steps remotely, and corresponding global voltage stability margin GVSM is calculated. The MATLAB is employed as a computing tool [8]. A total of 60 inputs and 82 outputs for IEEE 30 bus systems and 28 inputs and 40 outputs for IEEE 14 bus systems, whole 100 load samples are generated, and a total of 100 load samples are produced by offline N-R load flow analysis approach. The training process of neural networks helps us to identify the topology of an artificial neural network. The speed factor, the transfer function of neurons, or the method of network initialization is the essential factors on which the training speed depends. For the training process of the neural network, a set of network inputs and target outputs are required, and enough information about the network is also required in order to simulate the right prediction of the power system. To train the feedforward back propagation neural network, different training algorithms are being used. In this paper, Levenberg–Marquardt (LM) [18, 19] training technique is employed because this algorithm is suitable for medium-sized neural network, and of faster training and good convergence.

5 Simulation Result

A computer software program has been developed in the MATLAB 2015a upbringing to perform the simulations. An IEEE 14 bus test system and a typical IEEE 30 bus test system are used under ideal testing conditions. The IEEE 14 bus system represents some of the American electrical power system that is found in the Midwestern US as of February 1962. This 14-bus system has 14 buses, 5 generators, and 9 load buses. A portion of the American Electric Power System (in the Midwestern US) as of December 1961 is represented for the IEEE 30 bus system. This 30-bus system has 30 buses, 6 generators, and 24 load buses [20].

5.1 For IEEE 14 Test Bus System

For validation of the proposed approach, three operating scenarios which are considered here are as follows:

Case-I near the base case or at minimum loading (where the values of actual and reactive power loading of load buses are approximate values).

Case-II At medium loading (where the system operating load is increased by (2.24 p.u.) from base case value), i.e., it is half of the value of high system operating loading.

Case-III At high system operating load (where the system operating load is increased by (3.2 p.u.) from base case value).

The values of IEEE 14 test bus system for the global voltage stability margin (GVSM) and critical voltage V_{cr} for all these three cases are given in Tables 1 and 2. The following points emerge from this analysis.

a. Near the base case or at minimum loading, it is observed that values of GVSM and critical voltage V_{cr} are in a secure range which is employed and predicted by feedforward back propagation (FFBP) and radial basis function neural network (RBFN), i.e., for the system, the value of GVSM is higher and for V_{cr} it is lower.

Table 1 Different global voltage stability margin (GVSM) values for IEEE 14 test bus system at different operating loadings

GVSM				
At minimum loading				
N-R	3.32008	3.30638	3.21975	3.19575
RBFN	3.32008	3.30885	3.29754	3.21335
FFBP	3.31946	3.29425	3.24002	3.16152
At medium loading				
N-R	2.34058	2.32092	2.301070	2.281019
RBFN	2.30434	2.28793	2.271394	2.254718
FFBP	2.30472	2.28808	2.271274	2.254305
At high operating loading				
N-R	1.40571	1.37306	1.33969	1.30554
RBFN	1.38129	1.37821	1.32645	1.29823
FFBP	1.35412	1.35411	1.33399	1.31889

Table 2 Different critical voltage (V_{cr}) values for IEEE 14 test bus system at different operating loadings

V_{cr}				
1. At minimum loading				
N-R	0.5130	0.5134	0.51382	0.5142
RBFN	0.5132	0.5135	0.5139	0.51428
FFBP	0.51454	0.51471	0.514906	0.51159
2. At medium loading				
N-R	0.54652	0.547666	0.54882	0.55001
RBFN	0.54652	0.547666	0.54882	0.55001
FFBP	0.54649	0.54765	0.54882	0.55003
3. At high operating loading				
N-R	0.65240	0.65679	0.66139	0.66619
RBFN	0.65240	0.65679	0.66139	0.66619
FFBP	0.65375	0.65677	0.66094	0.66322

The analysis of GVSM and V_{cr} is shown in tables. These values are validated by using the offline N-R method. For RBFNN, it is also observed that the prediction accuracy is advanced or we can say higher as compared to diverse networks such as FFBP.

b. In the condition of case 2 or at medium loading, by all prediction methods, a considerable decrease or increase in the numerical values of GVSM and V_{cr} is observed, respectively, which is likewise given in Tables and, also RBFN is most accurate here, in this case.

c. For case 3, after continuous growth in the system operating load, the system has reached close to the collapse point. The values of GVSM tend to 0, and V_{cr} becomes better and dominant. The prediction capability of all the networks is verified with the aid of the offline N-R method. From this evaluation, it is miles concluded that RBFNN is an appropriate topology to spot the crucial buses in IEEE 14 bus test system.

d. When we increase the system operating load, the value of the global voltage stability margin decreases, and the power system goes toward instability. The GVSM turns into a decline or could also be zero at point of voltage collapse or voltage fall apart. Due to the profile of GVSM, the effect of local voltage collapse phenomena in the global scenario is quite reliable.

e. The pattern of global receiving end voltage is reducing while increasing the system operating load. For series equivalent, the receiving end voltage is excessive that shows the improved voltage stability limit; however, while the load will increase, its value reached in the region of voltage collapse.

f. Figures 4 and 5 show the prediction errors of RBFNN for GVSM and V_{cr}, respectively. From this analysis, it is concluded that RBFNN will be an appropriate topology for assessing system voltage stability in the IEEE 14 bus test system.

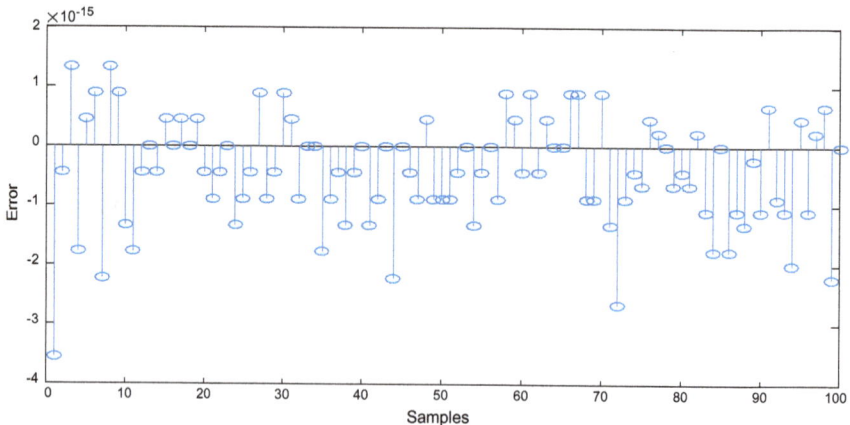

Fig. 4   For IEEE 14 bus system GVSM error plot of RBFNN topology

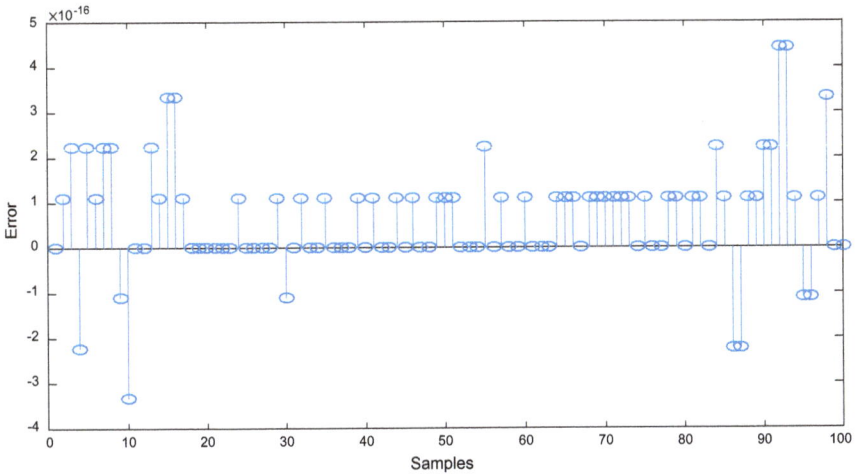

Fig. 5 For IEEE 14 bus system V_{cr} error plot of RBFNN topology

5.2 Case Study of IEEE 30 Test Bus System

For validation of the proposed approach, three operating scenarios which are considered here are as follows:

Case-I Near the base case or at minimum loading (where the values of actual and reactive power loading of load buses are approximate values).

Case-II At medium loading (where the system operating load is increased by (1.89 p.u.) from base case value), i.e., it is half of the value of high system operating loading.

Case-III At high system operating load (where the system operating load is increased by (2.7 p.u.) from base case value.

The values of IEEE 30 test bus system for the global voltage stability margin (GVSM) and critical voltage V_{cr} for all these three cases are given in Tables 3 and 4. The following points emerge from this analysis.

a. Near the base case or at minimum loading, it is observed that values of GVSM and critical voltage V_{cr} are in a secure range which is employed and predicted by feedforward back propagation (FFBP) and radial basis function neural network (RBFN), i.e., for the system, the value of GVSM is higher and for V_{cr} it is lower. The analysis of GVSM and V_{cr} is shown in tables. These values are validated by using the offline N-R method. For RBFNN, it is also observed that the prediction accuracy is advanced or we can say higher as compared to diverse networks such as FFBP.

b. In the condition of case 2, at medium loading, considerable amount of decrease or increase within the numerical values of GVSM and V_{cr}, respectively, are observed by all prediction methods, which is in addition given in Tables.

Table 3 Different global voltage stability margin (GVSM) values for IEEE 30 test bus system at different operating loadings

GVSM				
At minimum loading				
N-R	2.0263	2.0161	2.00581	1.9954
RBFN	2.0263	2.0161	2.00581	1.9954
FFBP	1.976	1.9677	1.9575	1.9458
At medium loading				
N-R	1.2083970	1.1948587	1.1812201	1.1674790
RBFN	1.2083970	1.1948587	1.1812201	1.1674790
FFBP	1.204047	1.1906726	1.1772599	1.1638012
At high operating loading				
N-R	0.38603	0.3595	0.3318	0.3029
RBFN	0.38603	0.3595	0.3318	0.3029
FFBP	0.37652	0.3570	0.3387	0.3217

Table 4 Critical voltage (V_{cr}) values for IEEE 30 test bus system at different operating loadings

V_{cr}				
At minimum loading				
N-R	0.510053	0.51031	0.51057	0.510849
RBFN	0.510053	0.51031	0.51057	0.510849
FFBP	0.510592	0.51074	0.51092	0.511130
At medium loading				
N-R	0.532629	0.533288	0.5339579	0.5346382
RBFN	0.532629	0.533288	0.5339579	0.5346382
FFBP	0.5326338	0.5332992	0.5339750	0.5346614
At high operating loading				
N-R	0.58461361	0.58630905	0.588058	0.58986
RBFN	0.58461361	0.58630905	0.588058	0.58986
FFBP	0.58504366	0.58643431	0.587569	0.588424

c. For case 3, after continuous growth in the system operating load, the system has reached close to the collapse point. The values of GVSM tend to 0, and V_{cr} becomes better and dominant. The prediction capability of all the networks is verified with the aid of the offline N-R method. From this evaluation, its miles concluded that RBFNN is the best topology to spot the crucial buses in IEEE 30 bus test system.

d. When we increase the system operating load, then the value of the global voltage stability margin decreases, and the power system goes toward instability. At the point of voltage collapse, the GVSM becomes lower or could also be zero.

e. The profile of the global receiving end voltage is reducing while increasing the system operating load. For series equivalent, the receiving end voltage is excessive, which shows the improved voltage stability limit; however, while the load increases, its value reached close to the region voltage collapse.

f. Figures 6 and 7 show the prediction errors of RBFNN for GVSM and V_{cr}, respectively. From this analysis, it is concluded that RBFNN will be an appropriate topology for assessing system voltage stability in the IEEE 30 bus test system.

Series impedance and shunt admittances of a given system are lumped within the series impedance for the series equivalent two-bus model. The profile shows voltage collapse at a higher level. In a series equivalent network, with an increase in system

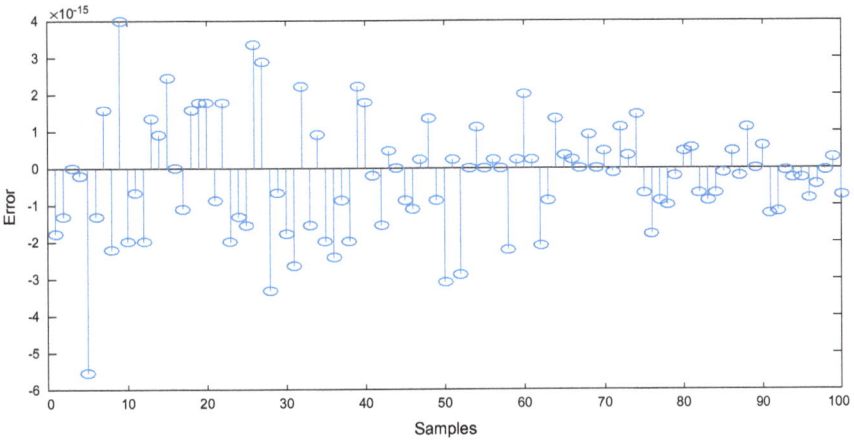

Fig. 6 For IEEE 30 bus system GVSM error plot of RBFNN topology

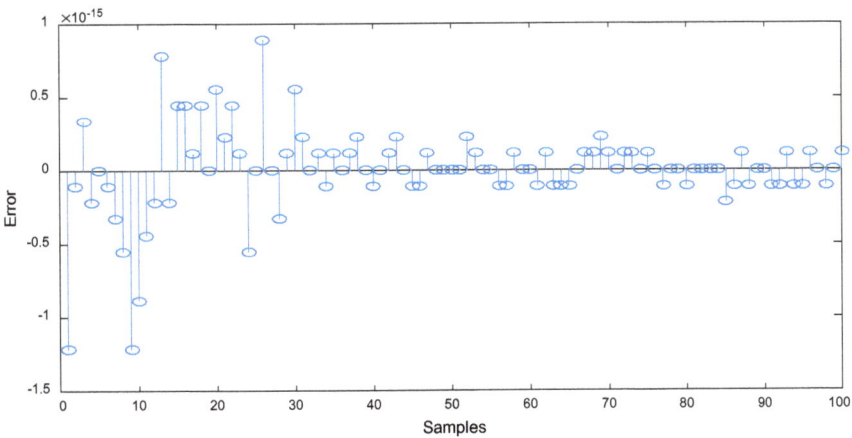

Fig. 7 For IEEE 30 bus system V_{cr} error plot of RBFNN topology

load, there are no visible changes in V_{cr}. But V_{cr} is far vulnerable to change in system operating load in the case of the pi-equivalent network. In the pi-equivalent model, V_{cr} increases with an increase in system working load. It shows more vital operating circumstances, and sooner or later, voltage collapse occurs at a higher load. In the case of the series model, the value of critical voltage (V_{cr}) depends on the power factor of the load, and in this model, there is no effect of system parameters. The simulation results of the voltage stability analysis using the proposed technique give higher accuracy and reliability.

6 Conclusion

In this paper, for assessment of voltage stability of a multi-bus power system, an alternative methodology is proposed to evaluate the same two-bus pi-network model where series and shunt parameters of transmission lines are lumped separately within the sort of series and shunt equivalent.

The equivalent network parameters such as GVSM, V_{cr}, and global receiving end voltage, are geared up to experience any change in the system accurately and efficiently as compared to two-bus series equivalent methodology.

An innovative approach named GVSM is employed to assess the voltage instability or in different words to determine the proximity of the current system state from voltage falls apart.

This paper provides a GVSM based voltage stability assessment using ANN for IEEE 14 and IEEE 30 test bus systems. Some significant highlights of this work are given below.

a. To decide the health of the power system, GVSM is calculated in the manner of stability for the given power system network. Prediction of GVSM and V_{cr} are validated via different neural network topologies.
b. The proposed method's main gain is that it suggests a good understanding between target data (N-R) and RBFN output. By comparing the results with other topologies of ANN, we found that prediction accuracy of RBFN is best, and the results obtained with RBFN are much better and accurate. This fact is validated by the prediction error indices.
c. Fast computation of the global voltage stability margin, GVSM, is also provided by this approach. By the usage of this supervised learning, an approaching operator can examine any unknown load pattern. We designed an efficient neural network by varying active and reactive loads and found out the GVSM, V_{cr}, and global receiving end voltage in the offline N-R method and ANN method.
d. The future scope of this paper is in the application of the proposed approach in the wind dominated system or micro grid, where the system observes uncertainty in generation lies in.

References

1. Nagendra P, Dey SHN, Paul S (2011) An innovative technique to evaluate network equivalent for voltage stability assessment in a widespread sub-grid system. Int J Electr Power Energy Syst 33(3):737–744
2. Kundur P (1994) Power system stability and control. McGraw-Hill, NewYork, USA. ISBN-13:9780070635159
3. Gubina F, Strmcnik B (1995) Voltage collapse proximity index determination using voltage phasors approach. IEEE Trans Power System 10(2):788–794
4. Gubina F, Strmcnik B (1997) A simple approach to voltage stability assessment in radial networks. IEEE Trans Power System 12(3):1121–1128
5. Moghavvemi M, Faruque MO (2001) Technique for assessment of voltage stability in ill-conditioned radial distribution network. IEEE Power Eng Rev 21(1):58–60
6. Pantoš M, Verbič G, Gubina F (2006) An improved method for assessing voltage stability based on network decomposition. Int J Electr Power Energy Syst 28(5):324–330
7. Kashem MA, Ganapathy V, Jasmon GB (2000) Network reconfiguration for enhancement of voltage stability in distribution networks. IEE Proc-Gener Trans Distrib 147(3):171–175
8. Swetha GC, Sudarshana Reddy HR (2014) Voltage stability assessment in power network using artificial neural network. Int J Adv Res Electr Electron Instrum Eng 3(3)
9. Suthar B, Balasubramanian R (2007) A novel ANN based method for online voltage stability assessment. In: 2007 International conference on intelligent systems applications to power systems. IEEE, pp 1–6
10. Soni BP, Saxena A, Gupta V (2016) Application of support vector machines for fast and accurate contingency ranking in large power system. Information systems design and intelligent applications. Springer, India, pp 327–335
11. Soni BP, Saxena A, Gupta V (2015) Support vector machine based approach for accurate contingency ranking in power system. In: Annual IEEE India conference (INDICON), New Delhi, India, pp 1–5. https://doi.org/10.1109/INDICON.2015.7443689
12. Soni, BP, Saxena A, Gupta V (2016) A least square support vector machine-based approach for contingency classification and ranking in a large power system. Cogent Eng 3(1):1137201
13. Sharma AK, Saxena A (2017) Prediction of line voltage stability index using supervised learning. J Electr Syst 13(4):696–708
14. Sharma AK et al (2018) Voltage stability assessment using artificial neural network. In: 2018 IEEMA engineer infinite conference (eTechNxT). IEEE
15. Saxena A, Sharma AK (2016) Assessment of global voltage stability margin through radial basis function neural network. Adv Electr Eng 2016:1–11
16. Rosenblatt F (1958) The perceptron: a probabilistic model for information storage and organization in the brain. Psychol Rev 65(6):386–408
17. Gupta S, Saxena A, Soni BP (2015) Optimal placement strategy of distributed generators based on radial basis function neural network in distribution networks. Proc Comput Sci 57:249–257
18. Bishop CM (1995) Neural networks for pattern recognition. Clarendon Press, Oxford
19. Hagan MT et al (1995) Neural network design, 2nd edn.
20. https://labs.ece.uw.edu/pstca/pf30/pg_tca30bus.htm

Chapter 2
Breast DCE-MRI Segmentation for Lesion Detection Using Clustering with Fireworks Algorithm

Tapas Si and Amit Mukhopadhyay

1 Introduction

1.1 Background

According to World Health Organization (WHO)'s report [1], it is estimated that 6,27,000 women in the world have been died from breast cancer and it is approximately 15% of all types of cancer deaths among women in 2018. Breast cancer is the most common cancer in women in India and accounts for 14% of all cancers in women [2, 3]. Organized and opportunistic screening programs in the developed countries result in a significant decrease in mortality caused due to breast cancer [4]. Recently, dynamic contrast- enhanced magnetic resonance imaging (DCE-MRI) is widely used for breast cancer detection, diagnosis and treatment planning or surgery.

1.2 Related Works

Several methods have been developed for lesion detection and its characterization in breast DCE-MRI in the recent past. Among them, methods based on Artificial Intelligence (AI), Machine Learning (ML) and Soft Computing (SC) are more popular. The statistical learning method namely Markov Random Field (MRF) is used

T. Si (✉)
Department of Computer Science and Engineering, Bankura Unnayani Institute of Engineering, Bankura, West Bengal 722146, India
e-mail: c2.tapas@gmail.com

A. Mukhopadhyay
Department of Electronics & Communication Engineering, Bankura Unnayani Institute of Engineering, Bankura, West Bengal 722146, India
e-mail: amitaecece@gmail.com

in breast lesion segmentation in studies [5–8]. Fuzzy c-means (FCM)-based breast lesion segmentation method is proposed in the study [9]. Shokouhi [10] used a region growing method based on FCM and vesselness filter to segment the breast lesion. An Artificial Neural Network (ANN)-based diagnostic prediction of breast lesion using morphology and texture features was proposed by Nie et al. in [11]. Yao et al. [12] developed a breast lesion segmentation method based on a committee of support vector machine (SVM) using wavelet features. Wang et al. [13] proposed a hierarchical SVM with coil sensitivity correction to segment the breast lesion. Gubern-Mérida et al. [14] developed a fibroglandular tissue segmentation method in the breast area using the expectation–maximization (EM) method. For further study, a review on applications of ML techniques for breast lesion segmentation can be obtained from the study [15]. Apart from the ML techniques, classical methods were also used in breast lesion segmentation. Wei et al. [16] combined the classic moment preserving and a thresholding adjustment method to segment the fibroglandular tissue in the breast. McClymont [17] developed a lesion segmentation method using a mean-shift and graph cut algorithm. Wang et al. [18] proposed a tumor segmentation and characterization method using both kinetic information and morphological features of breast DCE-MRI. Sim et al. [19] presented a computer-aided detection auto-probing (CADAP) system for breast lesion using a spatial-based discrete Fourier transform. Gubern-Mérida et al. [20] developed a probabilistic atlas-based segmentation method for breast lesions. Khalvati et al. [21] proposed a robust atlas-based automated segmentation method in breast DCE-MRI.

1.3   Objectives

After making the survey of the literature on breast DCE-MRI segmentation, it is found that no hard-clustering technique with meta-heuristic algorithms is used in breast lesion segmentation in DCE-MRI till date. That's why modified hard-clustering techniques with FWA [22, 23] algorithm is proposed to segment the lesion in breast DCE-MRI in this paper. This paper also demonstrates the novel application of FWA algorithm in medicine. The experimental results of the proposed method are compared with the PSO-based hard-clustering technique and K-means algorithm. The experimental results demonstrate that the proposed method statistically outperforms other methods used in the comparative study.

1.4   Organization of This Paper

The remaining of this paper is organized in the following way: Sect. 2 gives the description of DCE-MRI dataset and proposed methods. In this section, FWA algorithm and clustering techniques with it are described. The experimental setup is given in Sect. 3. Experimental results and discussion are given in Sect. 4. Finally, a conclusion with future works is given in Sect. 5.

2 Materials and Methods

2.1 DCE-MRI Dataset

Total 10 Sagittal T2-Weighted DCE-MRI slices of two patients have been taken from Cancer Genome Atlas Breast Invasive Carcinoma (TCGA-BRCA) [24, 25]. All MRI slices having a size greater than 256×256 are resized to 256×256. All the breast DCE-MRI slices are given in Fig. 1.

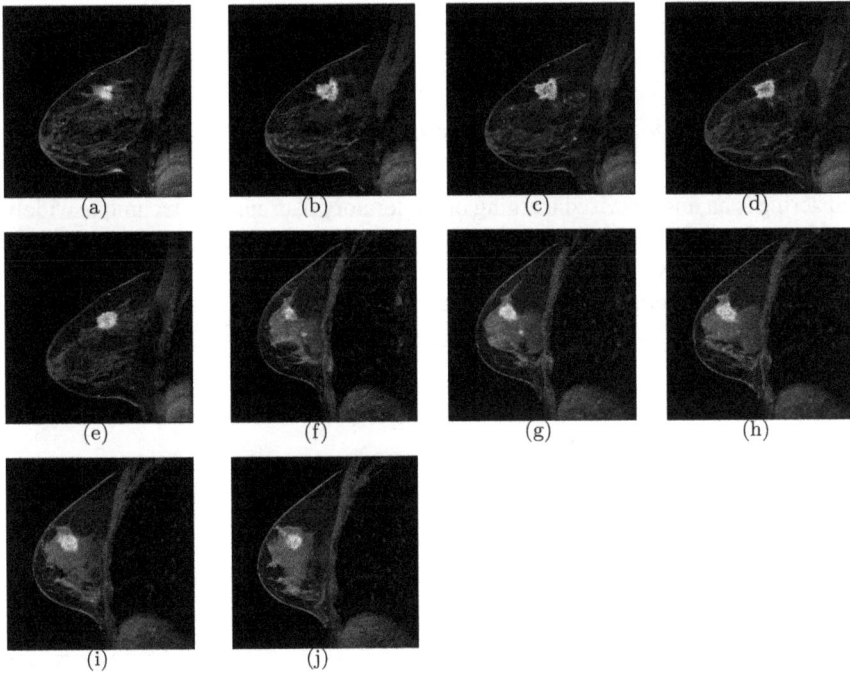

Fig. 1 Breast DCE-MR images. MRI slices of patient–1: **a–e** and MRI slices of patient–2: **f–j**

2.2 Proposed Method

The proposed methods have three steps as the following:

1. Preprocessing
2. Segmentation
3. Postprocessing.

The outline of the proposed FWA-based segmentation method is given in Fig. 2.

2.3 Preprocessing

The segmentation process faces difficulties due to the presence of noise and intensity inhomogeneitis (IIH) in the MR images. Therefore, noise is removed using anisotropic diffusion filter [26]. Then IIH is corrected using max filter-based method [27].

2.4 Segmentation Using Clustering with FWA

Clustering is an unsupervised learning or exploratory data analysis technique widely used in biomedical applications like MRI image analysis [28]. Let $\mathbb{X} = \{\vec{X}_1, \vec{X}_2, \ldots, \vec{X}_\mathcal{N}\}$ is a set of input patterns where $\vec{X}_i = (x_{i1}, x_{i2}, \ldots, x_{id}) \in \mathcal{R}^d$ and each x_{ij} is the attribute of input patterns. Hard or partitional clustering attempts to seek a $\mathcal{K}$ partition of $\mathbb{X}$, $\mathbb{C} = \{\mathcal{C}_1, \mathcal{C}_2, \ldots, \mathcal{C}_\mathcal{K}\}(\mathcal{K} \leq \mathcal{N})$ such that

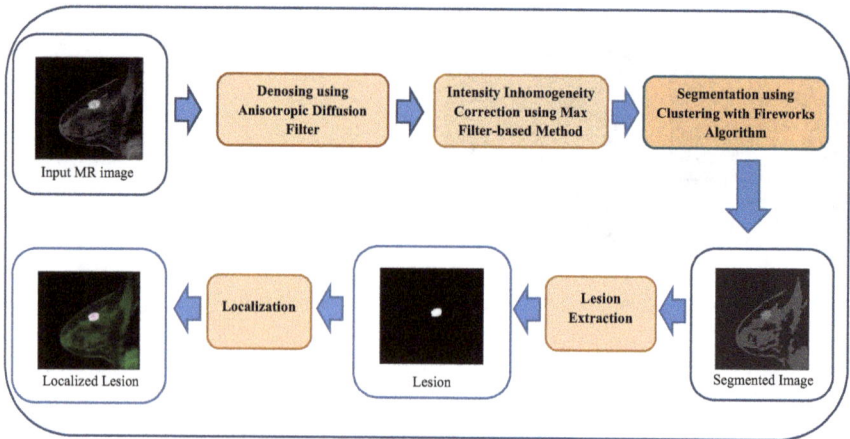

Fig. 2 Outline of the proposed method

1. $C_k \neq \phi, \quad k = 1, 2, \ldots, \mathcal{K}$
2. $\cup_{k=1}^{\mathcal{K}} C_k = \mathbb{X}$
3. $C_k \cap C_l = \phi, \quad k, l = 1, 2, \ldots, \mathcal{K}$ and $k \neq l$.

In this paper, a modified clustering technique with FWA (CFWA) is used in the segmentation. A clustering solution is represented by ith candidate solution in FWA as $\mathcal{Y}_i = \{\vec{m}_1, \vec{m}_2, \ldots, \vec{m}_{\mathcal{K}}\}$ where $\vec{m}_j$ is the jth cluster center. The two-dimensional MR images are converted into one-dimensional data points. The size of the data points is $\mathcal{N} = \mathcal{P} \times \mathcal{Q}$ where $\mathcal{P}$ and $\mathcal{Q}$ are the numbers of rows and columns of the images respectively. For $\mathcal{N}$ data points of d dimension and for a user-specified cluster number $\mathcal{K}$, candidate solutions in FWA are the real vector of dimension $\mathcal{K} \times d$. In this work, clustering is performed on gray values of MR images and therefore, $d = 1$. Hence, the resultant dimension of the candidate solutions is $\mathcal{K}$.

The intra-cluster spread (ICS) was used as an objective function defined as:

$$ICS = \sum_{k=1}^{\mathcal{K}} \sum_{\vec{x}_i \in C_k} \| \vec{x}_i - \vec{m}_k \|^2 \tag{1}$$

In a good clustering solution, a good *trade-off* between compactness (i.e., intra-cluster distance) and separation (i.e., inter-cluster distance) is needed. In ICS, only intra-cluster distance in all cluster is considered and it is minimized using FWA algorithm in the study [23]. Separation is also needed to be maximized for good clustering. On the other hand, the hard-clustering with basic FWA is already applied in document clustering [23]. Here, ICS is used as an objective function. In this paper, FWA-based clustering technique is modified by incorporating *DB-index* [29] as objective function in it in place of ICS. *DB-index* is the ratio of intra-cluster compactness and inter-cluster separation. It is discussed in Sect. 2.5.

2.4.1 FWA Algorithm

FWA is a swarm intelligence algorithm mimicking the explosion of fireworks in the night at the sky. There are two types of spark namely 'explosion sparks' and 'Gaussian sparks' created in FWA. Exploration (i.e., global search) and exploitation (i.e., local search) are two important properties of the meta-heuristic algorithm. For a good search technique, there should be a good *trade-off* between exploration and exploitation. In FWA, the exploration is done using the bad fireworks having a higher amplitude of explosion whereas exploitation is done using the good fireworks having a lower amplitude of explosion. The number of 'explosion sparks' is higher for good fireworks whereas it is lower for bad fireworks. Let $\mathcal{X}_i(x_{i1}, x_{i2}, \cdots, x_{id})$ denotes the ith firework having fitness, i.e., objective function value f_i where d is the dimension of the search space. The number of sparks (m_i) is computed by the following equation:

$$m_i = \mathcal{M} \times \frac{f_{\max} - f_i + \xi}{\sum_{i=1}^{n}(f_{\max} - f_i) + \xi} \tag{2}$$

where $f_{max} = \max(f_i), i = 1, 2, 3, \ldots, n$, $\mathcal{M}$ is the maximum number of 'explosion sparks' and ξ is the computer-generated very small number to avoid 'division by zero' error. The amplitude (a_i) of the 'explosion sparks' for ith firework is computed by the following equation:

$$a_i = \mathcal{A} \times \frac{f_i - f_{min} + \xi}{\sum_{i=1}^{n} (f_i - f_{min}) + \xi} \tag{3}$$

where $f_{min} = \min(f_i)$, $\mathcal{A}$ is the maximum amplitude of explosion and $\mathcal{A} = (\mathcal{X}_{max} - \mathcal{X}_{min})$. The explosion sparks are generated using Algorithm 1 and Gaussian sparks are generated using Algorithm 2. The flowchart of FWA is given in Fig. 3.

Algorithm 1: Explosion Spark Generation

Input: Firework $\mathcal{X} = (x_1, x_2, \ldots, x_d)$ of d dimensions
Output: Explosion spark $\mathcal{X}_e = (x_{e1}, x_{e2}, \ldots, x_{ed})$
1 Initialize the spark: $\mathcal{X}_e \leftarrow \mathcal{X}$
2 Select the number of position randomly: $l \leftarrow [d \times rand(0, 1)]$
3 Calculate the purturbation: $\delta x \leftarrow a_i \times rand(-1, 1)$
4 **for** $j \leftarrow 1$ **to** l **do**
5   Select the index randomly: $k \in [1, d]$
6   $x_{ek} \leftarrow x_{ek} + \delta x$
7   **if** $x_{ek} < \mathcal{X}_{min}$ *or* $x_{ek} > \mathcal{X}_{min}$ **then**
8     $x_{ek} \leftarrow \mathcal{X}_{min} + |x_{ek}| \% (\mathcal{X}_{max} - \mathcal{X}_{min})$
9   **end if**

10 **return** $\mathcal{X}_e$

Algorithm 2: Gaussian Spark Generation

Input: Firework $\mathcal{X} = (x_1, x_2, \ldots, x_d)$ of d dimensions
Output: Gaussian spark $\mathcal{X}_g = (x_{g1}, x_{g2}, \ldots, x_{gd})$
1 Initialize the spark: $\mathcal{X}_g \leftarrow \mathcal{X}$
2 Select the number of position randomly: $l \leftarrow [d \times rand(0, 1)]$
3 **for** $j \leftarrow 1$ **to** l **do**
4   Select the index randomly: $k \in [1, d]$
5   $x_{gk} \leftarrow \mathcal{X}_{min} + (\mathcal{X}_{max} - \mathcal{X}_{min}) \times rand(0, 1)$

6 **return** $\mathcal{X}_e$

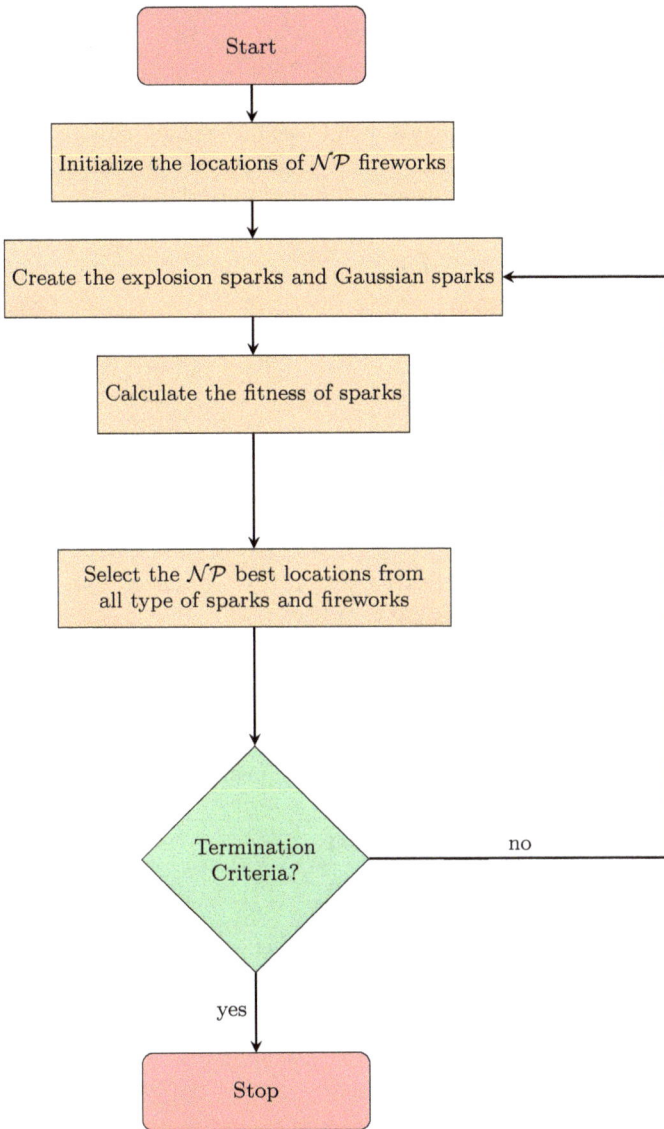

Fig. 3 Flowchart of FWA

2.5 Objective Function

In this work, *DB-index* is used as an objective function in the FWA-based clustering technique. Previously, *DB-index* was used to validate the clustering performance in the segmentation of brain MRI [28, 30, 31]. *DB-index* is the ratio of intra-cluster compactness and inter-cluster separation. ith intra-cluster compactness is defined as:

$$S_{i,t} = \left[\frac{1}{N_i} \sum_{\vec{X} \in C_i} \| \vec{X} - \vec{m}_i \|^t \right]^{\frac{1}{t}} \tag{4}$$

Inter–cluster distance, i.e., separation between ith and jth cluster is defined as:

$$d_{ij,p} = \left[\sum_{i=1}^{d} |m_{i,k} - m_{j,k}|^p \right]^{\frac{1}{p}} = \| \vec{m}_i - \vec{m}_j \|^p \tag{5}$$

where $\vec{m}_i$ is the centroid of ith cluster,t, $p \geq 1$, t, p are integer, N_i is the number of elements in the ith cluster C_i. $R_{i,qt}$ is expressed as

$$R_{i,tp} = \max_{j \in K, j \neq i} \left\{ \frac{S_{i,t} + S_{j,t}}{d_{ij,p}} \right\} \tag{6}$$

Finally, *DB–index* is measure by the following equation:

$$DB(K) = \frac{1}{K} \sum_{i=1}^{K} R_{i,qt} \tag{7}$$

The smallest *DB–index* indicates a valid optimal clustering. In the current work, *DB–index* is used as fitness for fireworks in FWA.

2.6 Postprocessing

After obtaining the clustered image from the clustering process of DCE-MRI, the lesions are extracted. In DCE-MRI, the pixels belong to the regions of lesions having the hyper-intensities. Therefore, the cluster labels of the regions of lesions having the highest values and cluster images are thresholded using the highest cluster labels. Lesions are extracted from the original MR images using the thresholded binary images and localized in the original MR images.

3 Experimental Setup

3.1 Parameters Settings

The parameters of FWA are set as follows: number of fireworks = 10, maximum number of explosion sparks = 40, bound constraints on number of explosion sparks = [8, 32].

3.2 Termination Criteria

1. Maximum number of function evaluation = 3000, or
2. There is no improvement in the best solution in successive 20 iterations.

3.3 PC Configuration

1. CPU: Intel(R) Core(TM) i7-4770 @3.40 GHz
2. RAM: 8 GB
3. Operating System: Windows 7 Professional 64-bit
4. Software: MATLAB 2016b.

4 Results and Discussion

The MR images are first de-noised using anisotropic diffusion filter and IIH are corrected using Max filter-based method. After that, CFWA is used to cluster the preprocessed MR images. After clustering the MR images, lesions are extracted from the segmented images and lesions are localized in the original MR images in the postprocessing step. The PSO, and K-means algorithms are used in the same framework of the proposed method to cluster the MR images. The PSO-based clustering techniques are termed as CPSO in this work. Each method is run on 10 MR images for 51 independent runs. The mean and standard deviation of the DB-index values over 51 independent runs are tabulated in Table 1. The bold-faced fonts in this table indicate better results. It is observed that the CFWA method achieved the lowest mean DB–index values among all the methods. That indicates that CFWA performs better clustering for breast MR images. To test the statistical significance of the quantitative results, Wilcoxon Signed Ranks Test [32], a non-parametric test, has been conducted using mean *DB-index* over 51 independent runs. The test results are tabulated in Table 2. From this table, it is observed that CFWA statistically outperforms CPSO, and K-means methods in clustering, i.e., segmentation of breast

Table 1 Mean and standard deviation (in parentheses) of DB-index values over 51 independent runs

MRI#	CPSO	K-means	CFWA
1	0.08695 (0.0012)	0.14457 (9.41E−05)	**0.08548 (0.0002)**
2	0.09124 (0.0013)	0.14794 (0.0005)	**0.08976 (0.0004)**
3	0.09273 (0.0009)	0.14884 (0.0011)	**0.09139 (0.0002)**
4	0.09133 (0.0014)	0.15167 (1.40E−16)	**0.08974 (0.0002)**
5	0.08831 (0.0009)	0.14858 (1.05E−06)	**0.08737 (8.39E−05)**
6	0.08820 (0.0007)	0.14251 (2.06E−05)	**0.08740 (0.0001)**
7	0.08840 (0.0008)	0.14458 (0.0001)	**0.08736 (9.14E−05)**
8	0.08824 (0.0010)	0.14704 (9.28E−06)	**0.08738 (9.62E−05)**
9	0.08801 (0.0006)	0.14348 (0.0002)	**0.08740 (0.0002)**
10	0.08833 (0.0014)	0.13812 (2.26E−06)	**0.08739 (0.0001)**

Table 2 Wilcoxon Signed Ranks Test Statistics on mean *DB-index* over 51 independent runs. R^+: sum of positive ranks, R^-: sum of negative ranks

S. No.	Comparison	R^+	R^-	Z	$p(2-tailed)$	Winner
1	CPSO versus CFWA	55	0	−2.803060	0.005062 < 0.01	CFWA
2	K-means versus CFWA	55	0	−2.803060	0.005062 < 0.01	CFWA
3	K-means versus CPSO	55	0	−2.803060	0.005062 < 0.01	CPSO

DCE-MRI. CPSO statistically outperforms the K-means method. The qualitative, i.e., visual results of clustering using different methods are given in Figs. 4 and 5 for patients 1 and 2 respectively. The lesion images are given in Figs. 6 and 7 for patients 1 and 2 respectively. The localized lesion images are given in Figs. 8 and 9 for patients 1 and 2 respectively. If the quantitative (i.e., *DB-index*) and qualitative (i.e., visual) results of the K-means algorithm are compared with other methods, it can be observed that K-means provides poor clustering solutions because it suffers from premature convergence in local optima. PSO also can not provide better solutions than FWA because it also suffers from premature convergence and gets stuck into local optima.

Robustness is another important aspect of meta-heuristic algorithms. It is measured in terms of the standard deviation of best results over 51 independent runs. The lower standard deviation indicates higher robustness of the meta-heuristics. From Table 1, it is observed that the standard deviations of *DB-index* values over 51

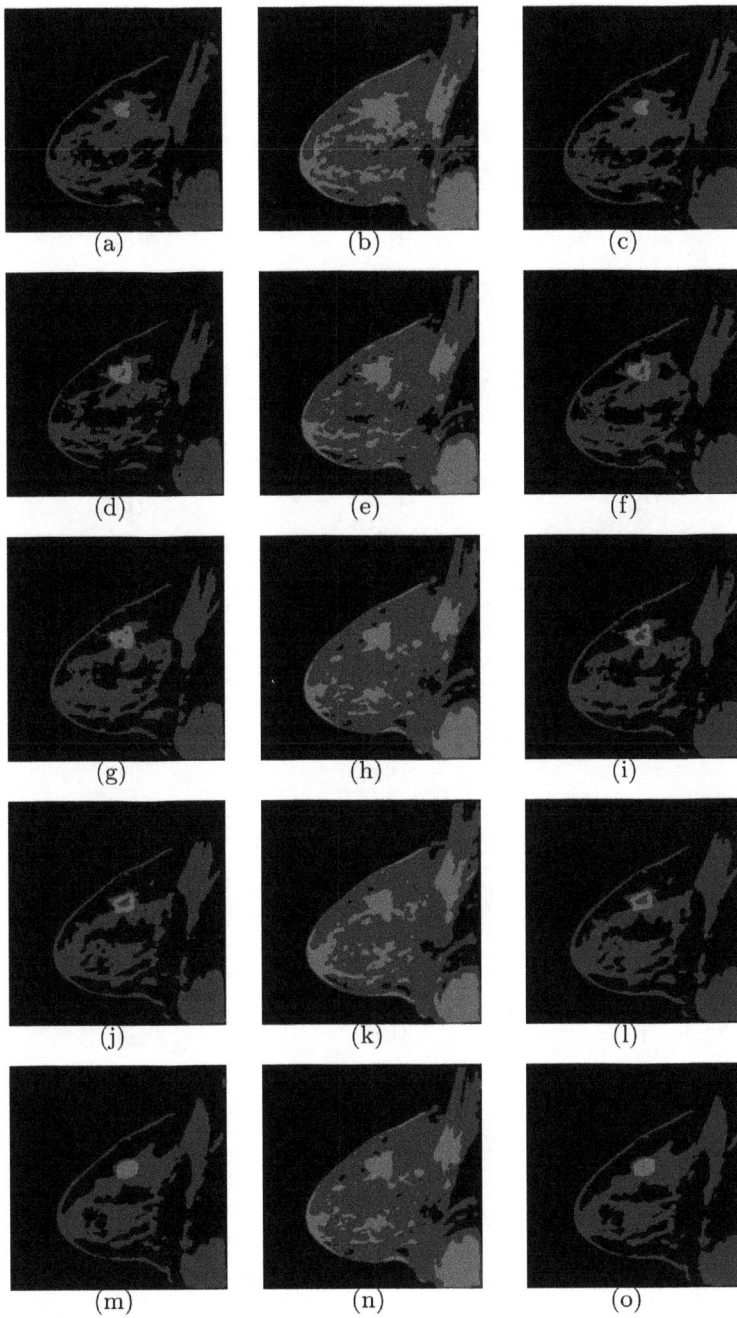

Fig. 4 Clustered images generated using different methods for patient–1. 1st column: CPSO, 2nd column: K-means, 3rd column: CFWA

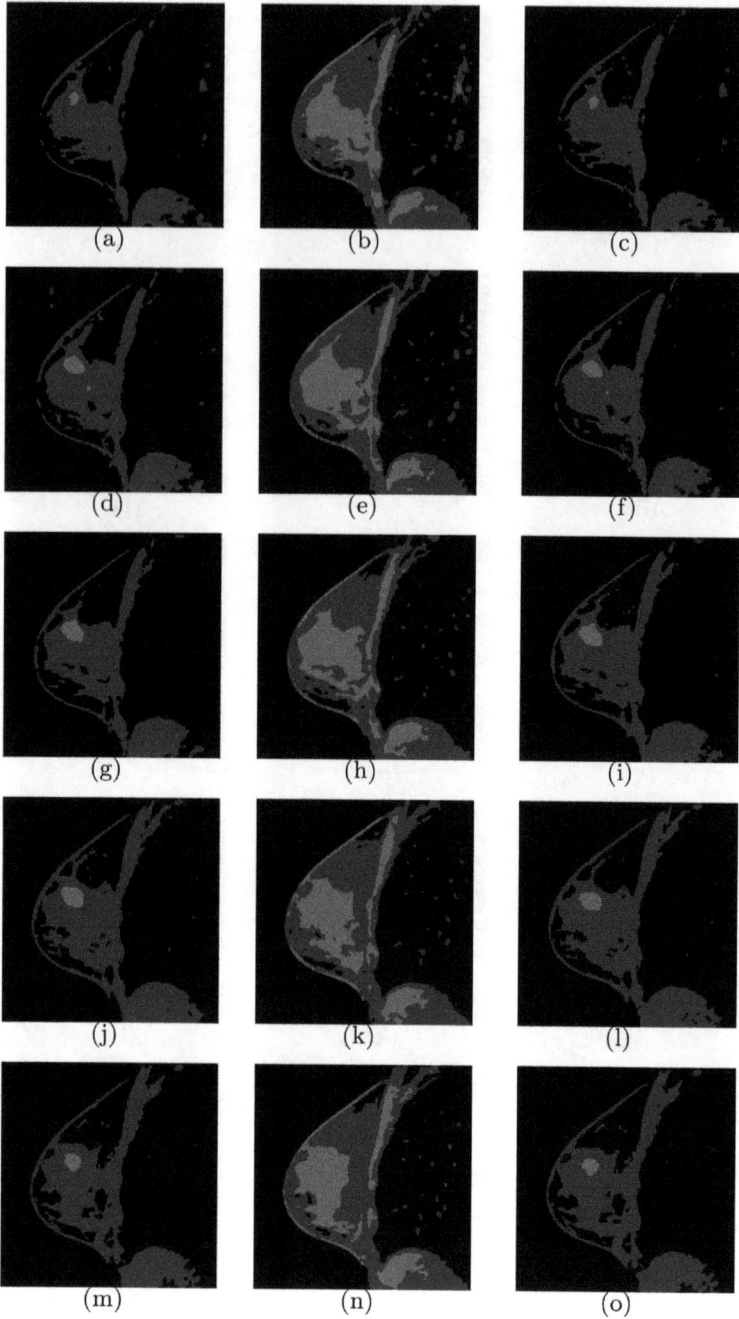

Fig. 5 Clustered images generated using different methods for patient–2. 1st column: CPSO, 2nd column: K-means, 3rd column: CFWA

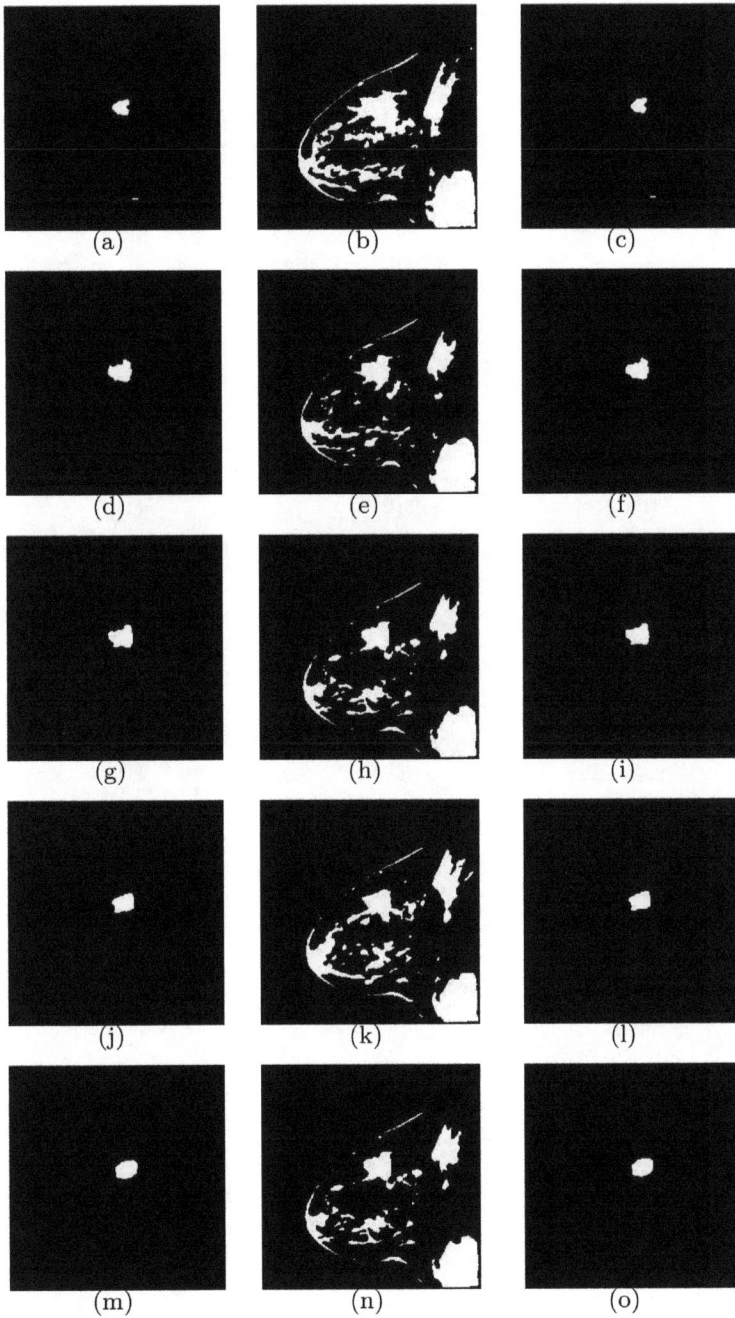

Fig. 6 Extracted lesions from segmented images generated using different methods for patient–1. 1st column: CPSO, 2nd column: K-means, 3rd column: CFWA

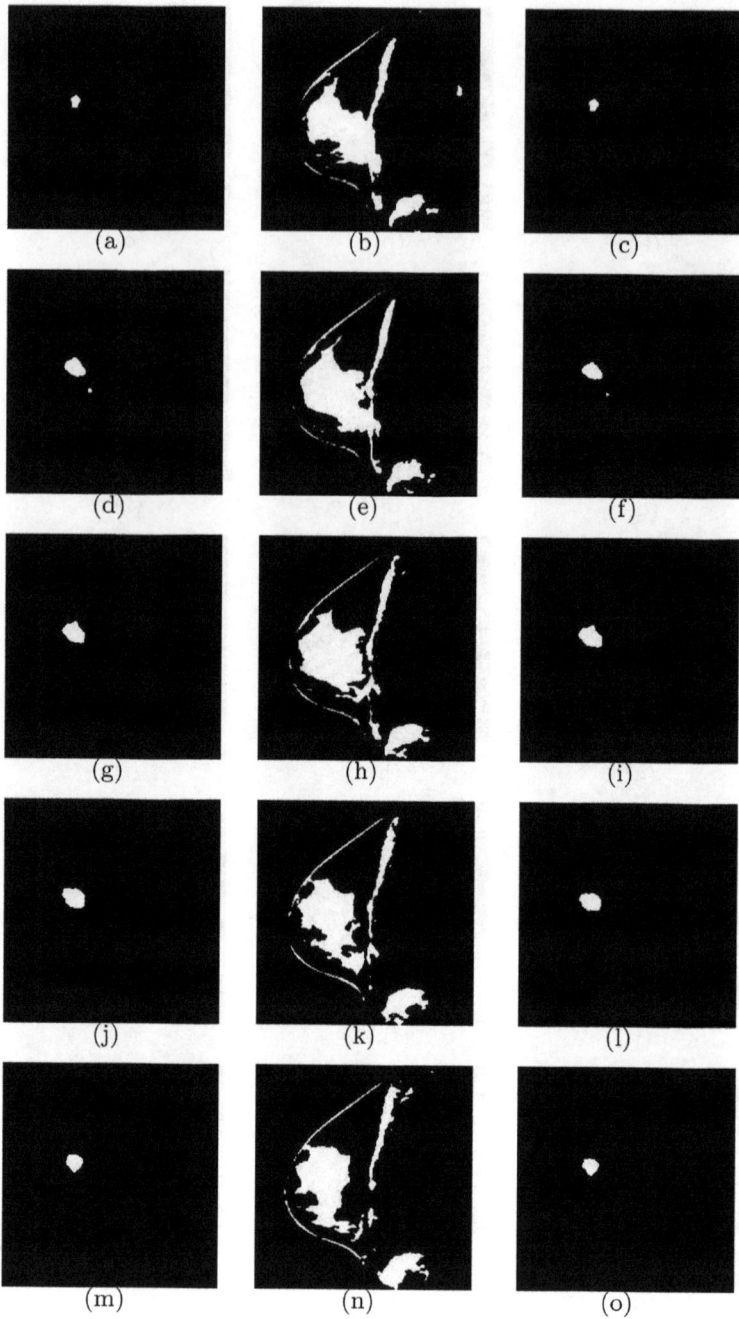

Fig. 7 Extracted lesions from segmented images generated using different methods for patient–2. 1st column: CPSO, 2nd column: K-means, 3rd column: CFWA

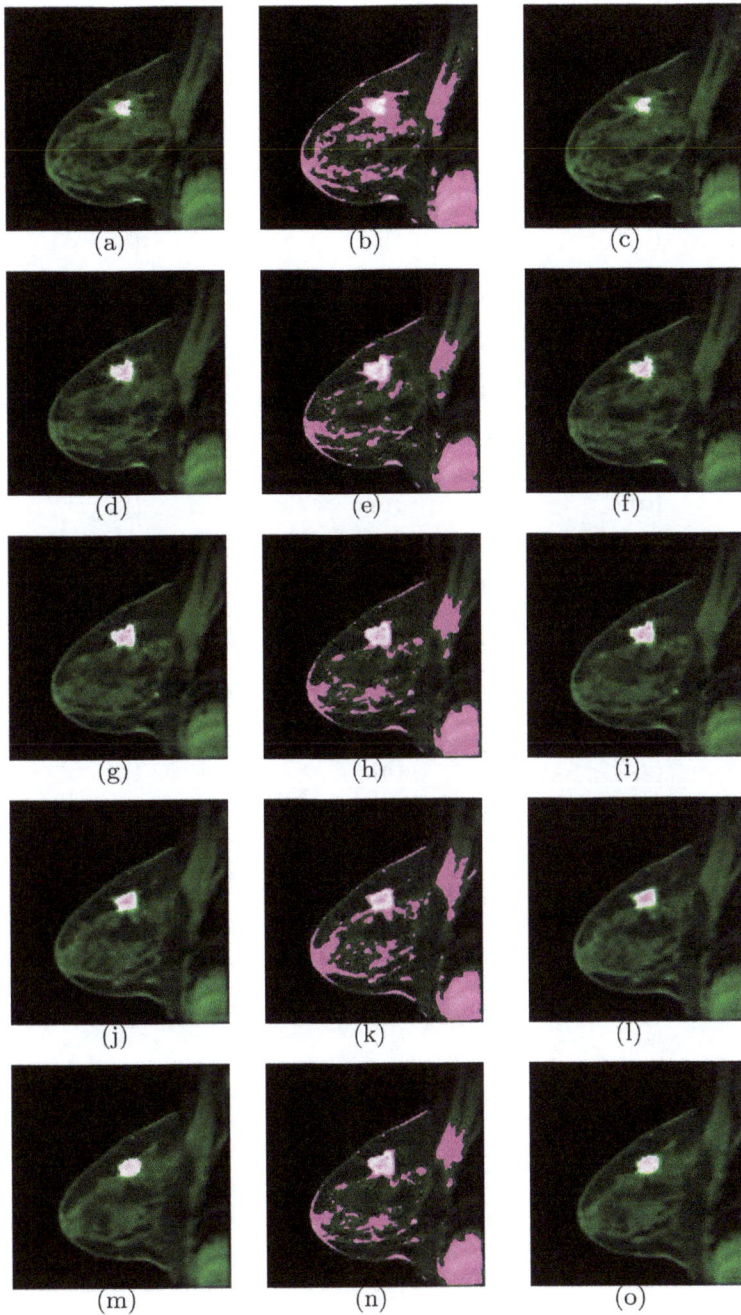

Fig. 8 Localized lesions (bright colored spot) in MR images for patient–1. 1st column: CPSO, 2nd column: K-means, 3rd column: CFWA

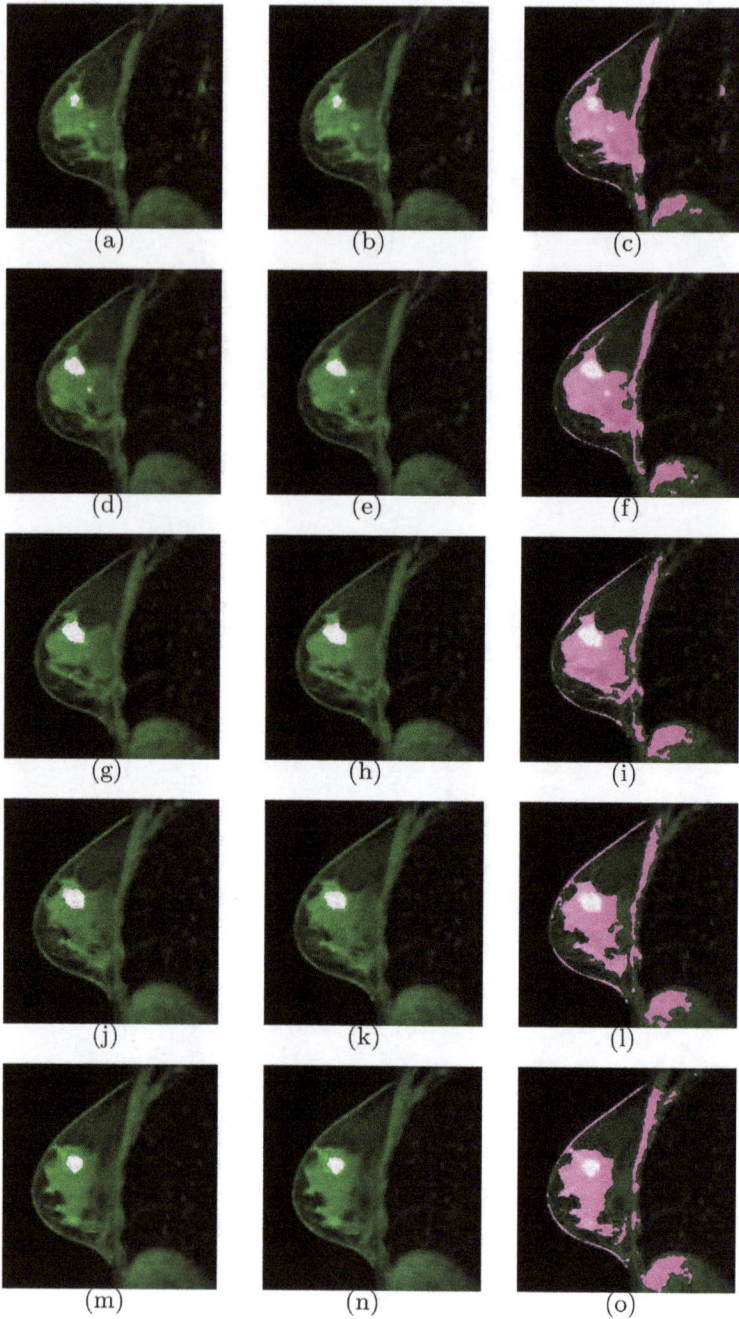

Fig. 9 Localized lesions (bright colored spot) in MR images for patient–2. 1st column: CPSO, 2nd column: K-means, 3rd column: CFWA

independent runs obtained by CFWA are smaller than CPSO for all the MR slices. Therefore, CFWA is more robust than CPSO. Though K-means provides a smaller standard deviation than CFWA for 7 MR slices, its segmentation results are not promising.

The average computational time costs for CFWA, CPSO, and K-means are 145.43, 63.32, and 0.1139 s respectively. Though the average computational time of K-means is lower than other methods, it is unable to provide better solutions. Though CFWA takes higher computational time than CPSO, it provides a better solution than CPSO.

From the above analysis and discussion of experimental results, it may be concluded that the proposed FWA-based segmentation method is both efficient and effective in lesion segmentation of breast DCE-MRI.

5 Conclusion

In this paper, a segmentation method has been presented for breast lesion segmentation in DCE-MRI. In the proposed method, the FWA-based hard-clustering technique is used in the segmentation. The performance of the proposed method is compared with the PSO-based clustering technique and K-means algorithm. Both quantitative and qualitative results are compared. The proposed method statistically outperforms other methods. In this paper, single objective clustering technique is used. In the future, multi-objective clustering technique can be used to segment the lesion in breast DCE-MRI. In this paper, clustering techniques are applied on MR images using only gray-level features. In the future, wavelet features can be used in clustering to further improve the segmentation performance.

References

1. https://www.who.int/cancer/prevention/diagnosis-screening/breast-cancer/en/. Accessed online: 5 Dec 2019
2. Ferlay J, Soerjomataram I, Ervik M et al (2013) GLOBOCAN 2012 v1.0, Cancer incidence and mortality worldwide: IARC CancerBase No. 11 [Internet]. International Agency for Research on Cancer, F. Lyon
3. Bray JS, Ren JS, Masuyer E et al (2013) Estimates of global cancer prevalence for 27 sites in the adult population in 2008. Int J Cancer 132(5):1133–45
4. Shetty MK (Ed) (2015) Breast cancer screening and diagnosis—a synopsis. Springer Science+Business Media, New York
5. Wu Q, Salganicoff M, Krishnan A, Fussell DS, Markey MK.: (2006) Interactive lesion segmentation on dynamic contrast enhanced breast MR using a Markov Model. In: Reinhardt JM, Pluim JPW (Eds) Medical imaging 2006: image processing, Proc. of SPIE, vol 6144, 61444M
6. Azmi R, Norozi N (2011) A New Markov random field segmentation method for breast lesion segmentation in MR images. J. Med. Sig. Sens. 1(3):156
7. Jayender J, Gombos E, Chikarmane S, Dabydeen D, Jolesz FA, Vosburgh KG (2013) Statistical learning algorithm for in situ and invasive breast carcinoma segmentation. Comput Med Imaging Graph 37:281–292

8. Ribes S, Didierlaurent D, Decoster N, Gonneau E, Risser L, Feillel V, Caselles O (2014) Automatic segmentation of breast MR images through a Markov random field statistical model. IEEE Trans Med Imaging 33(10):1986–1996 Oct

9. Chen W, Giger ML, Bick U (2006) A Fuzzy C-Means (FCM)-based approach for computerized segmentation of breast lesions in dynamic contrast-enhanced MR images. Acad Radiol 13:63–72

10. Shokouhi SB, Fooladivanda A, Ahmadinejad N (2017) Computer-aided detection of breast lesions in DCE-MRI using region growing based on fuzzy C-means clustering and vesselness filter. EURASIP J Adv Signal Process 39

11. Nie K, Chen JH, Yu HJ, Chu Y, Nalcioglu O, Su MY (2008) Quantitative analysis of lesion morphology and texture features for diagnostic prediction in breast MRI. Acad Radiol 15:1513–1525

12. Yao J, Chen J, Chow C (2009) Breast tumor analysis in dynamic contrast enhanced MRI using texture features and wavelet transform. IEEE J Sel Top Signal Process 3(1):94–100 Feb

13. Wang Y, Morrell G, Heibrun ME, Payne A, Parker DL (2013) 3D multi-parametric breast MRI segmentation using hierarchical support vector machine with coil sensitivity correction. Acad Radiol 20:137–147

14. Gubern-Mérida A, Kallenberg M, Mann RM, Martí R, Karssemeijer N (2015) Breast segmentation and density estimation in breast MRI: a fully automatic framework. IEEE J Biomed Health Inform. 19(1):349–357

15. Yassin NIR, Omran S, Houby EMF El, Allam H (2017) Machine learning techniques for breast cancer computer aided diagnosis using different image modalities: a systematic review. Comput Methods Programs Biomed. https://doi.org/10.1016/j.cmpb.2017.12.012

16. Wei CH, Li Y, Huangc PJ, Gwoa CY, Harmse SE (2012) Estimation of breast density: an adaptive moment preserving method for segmentation of fibroglandular tissue in breast magnetic resonance images. Euro J Radiol 81:e618–e624

17. McClymont D, Mehnert A, Trakic A, Kennedy D, Crozier S (2014) Fully automatic lesion segmentation in breast MRI using mean-shift and graph-cuts on a region adjacency graph. J Magn Reson Imaging 39:795–804

18. Wang TC, Huang YH, Huang CS, Chen JH, Huang GY, Chang YC, Chang RF (2014) Computer-aided diagnosis of breast DCE-MRI using pharmacokinetic model and 3-D morphology analysis. Magn Reson Imaging 32:197–205

19. Sim KS, Chia FK, Nia ME, Tso CP, Chong AK, Abbas SF, Chong SS (2014) Breast cancer detection from MR images through an auto-probing discrete Fourier transform system. Comput Biol Med 49:46–59

20. Gubern-Mérida A, Martí R, Melendez J, Hauth JL, Mann RM, Karssemeijer N, Platel B (2014) Automated localization of breast cancer in DCE-MRI. Med. Image Anal. https://doi.org/10.1016/j.media.2014.12.001

21. Khalvati F, Ortiz CG, Balasingham S, Martel AL (2015) Automated segmentation of breast in 3D MR images using a robust atlas. IEEE Trans Med Imaging 34(1):116–125

22. Tan Y, Zhu Y (2010) Fireworks algorithm for optimization. In: Tan Y et al (eds) ICSI 2010, Part I, LNCS, vol 6145. Springer, Heidelberg, pp 355–364

23. Tan Y (2015) Fireworks algorithm. Springer GmbH

24. Lingle W, Erickson BJ, Zuley ML, Jarosz R, Bonaccio E, Filippini J, Gruszauskas N (2016) Radiology data from the cancer genome atlas breast invasive carcinoma [TCGA-BRCA] collection. Cancer Imaging Arch. https://doi.org/10.7937/K9/TCIA.2016.AB2NAZRP

25. Clark K, Vendt B, Smith K, Freymann J, Kirby J, Koppel P, Moore S, Phillips S, Maffitt D, Pringle M, Tarbox L, Prior F (2013) The cancer imaging archive (TCIA): maintaining and operating a public information repository. J Digit Imaging 26(6):1045–1057

26. Mohana J, Krishnavenib V, Guo Y (2014) A survey on the magnetic resonance image denoising methods. Biomed Signal Process Control 9:56–69

27. Balafar MA, Ramli AR, Mashohor S (2010) A new method for MR grayscale inhomogeneity correction. Artif Intell Rev 34:195–204

28. Si T, De A, Bhattacharjee AK (2016) MRI brain lesion segmentation using generalized opposition-based glowworm swarm optimization. Int J Wavelets Multiresolut Inf Process 14(5):1650041 (29 pages)
29. Davies DL, Bouldin DW (1979) A cluster separation measure. IEEE Trans Pattern Anal Mach Intell 1(2):224–227 Apr
30. Si T, De A, Bhattacharjee AK (2014) Brain MRI segmentation for tumor detection using Grammatical Swarm based clustering algorithm. In: Proceedings of IEEE international conference on circuits, power and computing technologies (ICCPCT-2014), Nagercoil, India
31. Si T, De A, Bhattacharjee AK (2015) Grammatical swarm based segmentation methodology for lesion segmentation in brain MRI. Int J Comput Appl 121(4):1–8
32. Derrac J, Garcla S, Molina D, Herrera F (2011) A practical tutorial on the use of nonparametric statistical tests as a methodology for comparing evolutionary and swarm intelligence algorithms. Swarm Evol Comput 1:3–18

Chapter 3
Swarm Programming Using Moth-Flame Optimization and Whale Optimization Algorithms

Tapas Si

1 Introduction

Automatic programming [1] is a machine learning technique by which computer programs are generated automatically in any arbitrary language. SP [2] is an automatic programming technique which uses SI algorithms as search engine or learning algorithms. The grammar-based SP is a type of SP in which *Context-free Grammar* (CFG) is used to generate computer programs in a target language. Genetic Programming (GP) [3] is an evolutionary algorithm in which tree-structured genome is used to represent a computer program and Genetic algorithm (GA) is used as a learning algorithm. Grammatical Evolution (GE) [4, 5] is a variant of grammar-based GP (GGP) [6] in which linear genome i.e., array of integer codons is used to represent a genotype and Backus-Naur Form (BNF) of CFG is used to generate the computer programs (i.e., phenotype) from the genotype. Generally, variable-length GA is used as a learning algorithm in GE. SP uses GP-like tree-structure genome which represents a computer program and SI algorithms are used as learning algorithms. Roux and Fonlupt [7] proposed ant programming in which ant colony optimizer (ACO) [8] was used to generate the computer programs. Karaboga et al. [9] proposed artificial bee colony programming (ABCP) for symbolic regression and artificial bee colony (ABC) algorithm [10] was used as learning algorithm. Mahanipour and Nezamabadi-pour [11] proposed Gravitation Search Programming (GSP) in which Gravitation Search Algorithm (GSA) [12] was used as a learning algorithm. The grammar-based SP are the variants of GE where SI algorithms are used as search engines or learning algorithms through genotype-to-phenotype mapping using BNF of CFG. Grammatical Swarm (GS) [13–15], Grammatical Bee Colony (GBC) [16], and Grammatical Fireworks algorithm (GFWA) [17] are grammar-based SP. Particle

T. Si (✉)
Department of Computer Science and Engineering, Bankura Unnayani Institute of Engineering, Bankura, West Bengal 722146, India
e-mail: c2.tapas@gmail.com

© The Author(s), under exclusive license to Springer Nature Singapore Pte Ltd. 2021
X.-Z. Gao et al. (eds.), *Applications of Artificial Intelligence in Engineering*, Algorithms for Intelligent Systems, https://doi.org/10.1007/978-981-33-4604-8_3

Swarm Optimizer (PSO) [18] was used as search engine in GS. ABC algorithm was used as search engine in GBC and Fireworks algorithm (FWA) [19] was used as search engine in GFWA. Till date, as per author's best knowledge, Moth-Flame Optimization (MFO) [20] and Whale Optimization algorithm (WOA) [21] are not used in automatic programming. Therefore, in the current work, two grammar-based SP algorithms GMFO and GWO are proposed. In GMFO algorithm, MFO is used as a search engine to generate computer programs automatically through genotype-to-phenotype mapping using BNF of CFG. Similar to GMFO, WOA is used as a search engine in GWO for automatic computer program generation. The proposed two methods are applied to Santa Fe Ant Trail, symbolic regression and 3-input multiplexer problem. The results of GMFO and GWO are compared to the results of GFWA and GBC. The experimental results demonstrates that the proposed two methods can be used in automatic computer program generation in any arbitrary language.

2 Materials and Methods

In this work, two grammar-based SP such as GMFO and GWO are proposed. In GMFO, MFO algorithm is used as a search engine or learning algorithm to generate computer programs automatically through genotype-to-phenotype mapping. In GMFO, each individual is an array of integer codons in the range [0, 255] and it represents the genotype and derived computer program from genotype using BNF of CFG is known as phenotype. First of all, it is primary concern to define problem specific CFG in BNF. An example of CFG in BNF is given below:

```
1. <expr> := (<expr><op><expr>) (0) | <var> (1)
2. <op> := + (0) | - (1) | * (2) | / (3)
3. <var> := x1 (0) | x2 (1)
```

The ith individual of the search engine, i.e., set of d integer codons $X_i(x_1, x_2, \ldots, x_d)$ are initialized as follows:

$$x_j = \text{round}(255 \times \text{rand}(0, 1)) \tag{1}$$

A *mapping process* maps the rule numbers from the codons in the derivation process of computer programs in the following way: rule = (codon integer value) MOD (number of rules for the current non-terminal)

The representation of genotype, phenotype and genotype-to-phenotype mapping are given in Fig. 1. Similar to GFMO, the positions of whale are the set of integer codons in the range [0, 255]. The same genotype-to-phenotype mapping process as in GMFO is used to generate the computer programs. During the derivation process, if the derivation is run out of codons, then the process restarts from the beginning.

Fig. 1 Genotype-to-phenotype mapping process

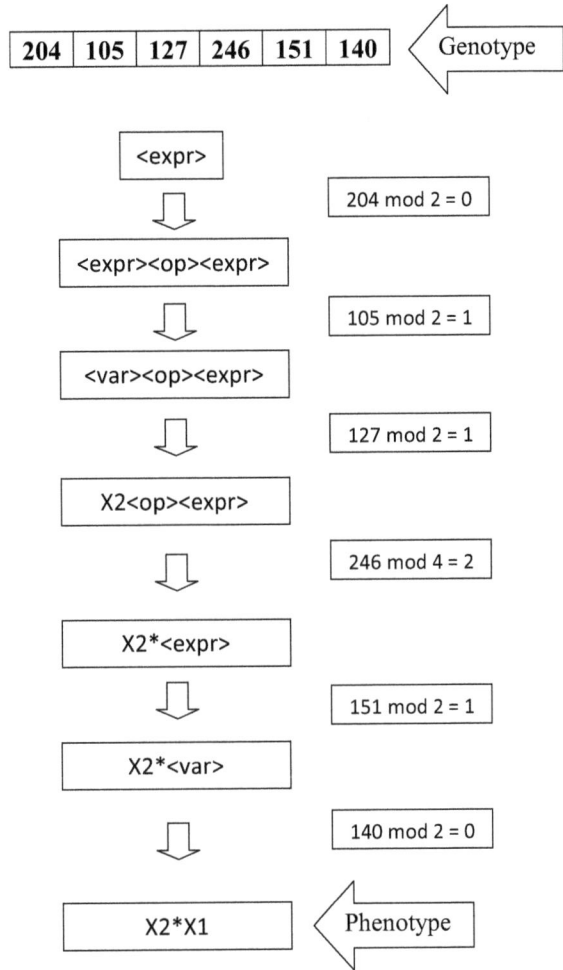

| 204 | 105 | 127 | 246 | 151 | 140 | ◁ Genotype |

<expr>

⬇ 204 mod 2 = 0

<expr><op><expr>

⬇ 105 mod 2 = 1

<var><op><expr>

⬇ 127 mod 2 = 1

X2<op><expr>

⬇ 246 mod 4 = 2

X2*<expr>

⬇ 151 mod 2 = 1

X2*<var>

⬇ 140 mod 2 = 0

X2*X1   ◁ Phenotype

This process is called as *wrapping*. After a certain number of wrapping, if there is any non-terminal remain in the derived program, then the corresponding individual is denoted as invalid. The invalid individual is replaced by the valid individual later on during the search process.

Search engines or learning algorithms are another important part of GMFO and GWO. As already mentioned before, MFO and WOA are used as search engines in GMFO and GWO respectively. MFO algorithm is based on the transverse orientation, i.e., the navigation method of moths in nature. Moths fly in the night by maintaining a fixed angle with respect to the moon for travelling in a straight line for long distance. But they are trapped in a useless or deadly spiral around the artificial light source.

This phenomena is modelled in MFO algorithm. The detail description of MFO algorithm can be obtained from [20]. WOA is another nature-inspired meta-heuristic algorithm which mimics the social behaviour of humpback whales. The humpback whales live alone or in group and their favourite prey is krill and small fish herds. Their foraging behaviour, i.e., bubble-net feeding method is done by creating bubbles along a circle. The Bubble-net attacking method (i.e., exploitation phase) has two steps namely shrinking encircling mechanism and spiral updating position. The searching mechanism for prey is used to create exploration of the search space. The detail of WOA can be obtained from [21].

3   Experimental Setup

3.1   Benchmark Problems

Three benchmark problems such as Santa Fe Ant Trail (SFAT), symbolic regression, and 3-input multiplexer are chosen for the experiment. The objective of the SFAT problem is to find out the program by which an ant can eat 89 piece of food placed in 32×32 grid in 600 time steps. The target function for symbolic regression problem is $f(x) = x + x^2 + x^3 + x^4$ and 100 fitness cases are generated randomly in the range $[-1,1]$. 3-input multiplexer problem has 8 fitness cases. The detail descriptions and problem specific defined grammar can be obtained from [16, 17].

3.2   Parameters Settings

The parameters of GMFO are set as the following: number of search agents ($\mathcal{N}$) = 30, dimension = 100. The parameters of GWO are set as the following: number of search agents ($\mathcal{N}$) = 30, dimension = 100. The other parameters in GMFO and GWO are dynamically controlled as in [20, 21]. As the functions are not evaluated for invalid individual in the above algorithms, the maximum number generations or iterations are not fixed for the comparative study. Therefore, each of GFW, GBC, GMFO, and GWO algorithms is allowed to run for maximum 30,000 number of function evaluations (FEs) in a single run. All algorithms are terminated when they reach the maximum FEs or target error. The target errors are set to 0, 0.01 and 0 for ant trail, symbolic regression and 3-multiplexer problems respectively. The numbers of wrapping are set to 3, 2 and 1 for ant trail, symbolic regression and 3-multiplexer problems respectively.

3.3 PC Configuration

- Operating System: Windows 10 Pro
- CPU: Intel(R) Core(TM) i7-9700K @3.6 GHz
- RAM: 64 GB
- Software: Matlab 2018a.

4 Results and Discussion

The proposed GMFO and GWO algorithms are applied to Santa Fe Ant Trail (SFAT), symbolic regression, and 3-input multiplexer problems. The experiments are repeated for 30 times independently for each algorithm. The mean and standard deviation of best-run-errors over 30 independent runs are given in Table 1. The number of successful runs and success rates (in %) over 30 independent runs are given in Table 2. The success rate (SR) is calculated as follows:

$$SR = \frac{\text{number of achieving the target}}{\text{number of total runs}}$$

The number of successful runs and success rates (in %) over 30 independent runs are given in Table 3. The results of GFWA and GBC are obtained from study [17] as the current work is the part of the same project.

From Table 1, it is observed that the GWO performs better than other algorithms for SFAT problem. GFWA performs better than other algorithms for symbolic regression problem. GBC performs better than others for 3-multiplexer problem. If the results

Table 1 Mean and standard deviation of best-run-errors over 30 independent runs

Algorithms	Santa Fe ant trail	Symbolic regression	3-Multiplexer
GFWA	24.57(16.9516)	**6.65 (7.246)**	0.93(0.2537)
GBC	32.57(17.8609)	10.35(7.1404)	**0.7(0.4661)**
GMFO	35.53(22.8212)	10.35(7.814)	0.8(0.407)
GWO	**23.57(17.2880)**	10.15(7.4338)	1.00(0.00)

Table 2 Number of successful runs and success rates (in %) over 30 independent runs

Algorithms	Santa Fe ant trail	Symbolic regression	3-Multiplexer
GFWA	1(3.33%)	**15(50.00%)**	2(6.67%)
GBC	0(0.00%)	7(23.33%)	**9(30.00%)**
GMFO	**2(6.67%)**	9(30.00%)	6(20.00%)
GWO	1(3.33%)	9(30.00%)	0(0.00%)

Table 3 Mean and standard deviation of FEs over 30 independent runs

Algorithms	Santa Fe ant trail	Symbolic regression	3-Multiplexer
GFWA	29, 917(453.88)	23943(8657.80)	29062(4415.30)
GBC	30, 000(0.00)	27, 076(6935.20)	**23549(10,420.00)**
GMFO	**28224.9(6812.3395)**	**21,586.9(13,097.6074)**	26, 200(9020)
GWO	29553.8(2524.6001)	22, 432.07(12, 023.5602)	30, 000(0.00)

of GMFO are compared with GWO, it can be observed that GWO provides a higher accuracy than GMFO for SFAT and symbolic regression problems whereas GMFO provides higher accuracy than GWO only for 3-multiplexer problem.

From Table 2, it is observed that the success rate of GMFO is higher than all others algorithms for SFAT problem. GFWA provides higher success rate than others for regression problem whereas GBC provides higher success rate than others for multiplexer problem. If the success rates of GMFO and GWO are compared, then it can be observed that GMFO has higher success rate than GWO for SFAT and multiplexer problems and there is a tie for regression problem.

From Table 3, it is observed that the mean FEs taken by GMFO is lower than other algorithms for SFAT and regression problems whereas the mean FEs taken by GBC is lower than others for multiplexer problem. The computer programs evolved by GMFO and GWO are given below:

The successful ant program evolved by GMFO (ant eats all 89 pieces of food):

```
if(foodahead()) if(foodahead()) if(foodahead())
if(foodahead())    if(foodahead()) move(); else
if(foodahead()) if(foodahead())  left();  else
if(foodahead()) if(foodahead()) left(); else
left(); end; else   if(foodahead()) right();
else if(foodahead())  if(foodahead()) if(foodahead())
left(); else left(); end;   else move(); end; else
move(); end; end; end; end;    else    if(foodahead())
if(foodahead()) if(foodahead()) if(foodahead())
if(foodahead()) right(); else left(); end; else left();
end; else    move(); end;  else left(); end; else move();
end; end; end; else    right(); end; else    if(foodahead())
move(); else move(); end; end; else left(); end; else
left();   end; move(); left(); if(foodahead()) move();
else if(foodahead()) right(); else right(); end; end; right();
```

The ant program evolved by GWO (ant eats 88 out of 89 pieces of food):

```
if(foodahead()) left(); else right(); end; right();
if(foodahead()) move(); else left(); end; move(); left();
```

A successful program evolved by GMFO for symbolic regression problem (absolute error $= 1.7837e - 15$):

```
plus(times(x,plus(times(x,plus(x,times(x,times(pdivide(x,x),x))))),
x)),x)
```

A successful program evolved by GWO for symbolic regression problem (absolute error $= 4.6668e - 15$):

```
times(plus(plus(times(minus(times(x,x),pdivide(x,x)),x),x),x),
plus(x,pdivide(x,x)))
```

A successful program evolved by GMFO for 3-multiplexer problem (absolute error $= 0$):

```
nor(nor(nor(x3,x1),nor(nor(x1,nor(x1,x1)),x2)),nor(nand(nor(x2,x1),
nor(x1,x1)),x3))
```

A program evolved by GWO for 3-multiplexer problem (absolute error $= 1$):

```
nand(or(x3,x1),x2)
```

From the discussion of the results, it is found that no single algorithm performs better than all other algorithms for all problems in this study. The above presented results are the experimental evidence of the fact that the proposed GMFO and GWO algorithms can be used in automatic computer program generation in any arbitrary language.

5 Conclusion

This paper presents two grammar-based swarm programming methods namely GMFO and GWO. The proposed methods are applied to solve SFAT, symbolic regression, and 3-input multiplexer problems. The experimental results demonstrate that the proposed methods can be used to generate computer programs automatically in any arbitrary language. In this study, the basic version of MFO and WOA are utilized as search engines or learning algorithms in genotype-to-phenotype mapping. The update version of these algorithms can be used to obtain better performance in automatic computer program generation. In the future, the GMFO and GWO can be applied to real-world problems such as data classification and regression analysis.

References

1. Rich C, Waters RC (1998) Automatic programming: myths and prospects. IEEE Comput 21(8):40–51
2. Olmo JL, Romero JR, Ventura S (2014) Swarm-based metaheuristics in automatic programming: a survey. WIREs Data Mining Knowl Discov. https://doi.org/10.1002/widm.1138
3. Koza JR (1992) Genetic programming: on the programming of computers by means of natural selection. MIT Press
4. Ryan C, Collins J.J, O'Neill M (1998) Grammatical evolution: evolving programs for an arbitrary language. In: Banzhaf W, Poli R, Schoenauer M, Fogarty TC (eds) EuroGP 1998. LNCS, vol 1391, Springer, Heidelberg, pp 83–95
5. O'Neill M, Ryan C (2001) Grammatical evolution. IEEE Trans Evol Comput 5(4):349–358
6. Mckay RI, Hoai NX, Whigham PA, Shan Y, O'Neill M (2010) Grammar-based genetic programming: a survey. Genet Program Evolvable Mach 11:365–396
7. Roux O, Fonlupt C (2000) Ant programming: or how to use ants for automatic programming. In: International conference on swarm intelligence (ANTS), pp 121–129
8. Dorigo M, Maniezzo V, Colorni A (1996) Ant system: optimization by a colony of cooperating agents. IEEE Trans Syst Man Cybern B Cybern 26:29–41. https://doi.org/10.1109/3477.484436
9. Karaboga D, Ozturk C, Karaboga N, Gorkemli B (2012) Artificial bee colony programming for symbolic regression. Inform Sci 209:1–15. https://doi.org/10.1016/j.ins.2012.05.002
10. Karaboga D (2005) An idea based on honey bee swarm for numerical optimization. In: Technical Report-TR06, Erciyes University, Engineering Faculty, Computer Engineering Department
11. Mahanipour A, Nezamabadi-pour H (2019) GSP: an automatic programming technique with gravitational search algorithm. Appl Intell 49:1502–1516. https://doi.org/10.1007/s10489-018-1327-7
12. Rashedi E, Nezamabadi-Pour H, Saryazdi S (2009) GSA: a gravitational search algorithm. Inf Sci 179:2232–2248
13. O'Neill M, Brabazon A (2004) Grammatical swarm. In: Genetic and evolutionary computation conference (GECCO), pp 163–174
14. O'Neill M, Brabazon A (2006) Grammatical swarm: the generation of programs by social programming. Nat Comput 5(4):443–462
15. O'Neill M, Leahy F, Brabazon A (2006) Grammatical swarm: a variable-length particle swarm algorithm. In: Swarm intelligent systems, studies in computational intelligence. Springer, pp 59–74
16. Si T, De A, Bhattacharjee AK (2013) Grammatical bee colony. In: Panigrahi BK et al. (eds) SEMCCO 2013, Part I, LNCS vol 8297, pp 436–445
17. Si T (2016) Grammatical evolution using fireworks algorithm. In: Pant M et al (eds) Proceedings of fifth international conference on soft computing for problem solving, advances in intelligent systems and computing, vol 436. https://doi.org/10.1007/978-981-10-0448-34
18. Kennedy J, Eberhart R (1995) Particle swarm optimization. In: IEEE international conference on neural networks, Perth, Australia
19. Tan Y, Zhu Y (2010) Firework Algorithm for Optimization. In: Tan Y et al (eds) ICSI 2010, Part I, LNCS 6145. Springer, Berlin Heidelberg, pp 355–364
20. Mirjalili S (2015) Moth-flame optimization algorithm: a novel nature-inspired heuristic paradigm. Knowl-Based Syst (2015). https://doi.org/10.1016/j.knosys.2015.07.006
21. Mirjalili S, Lewis A (2016) The whale optimization algorithm. Adv Eng Softw 95:51–67

Chapter 4
Nonlinear Regression Analysis Using Multi-verse Optimizer

Jayri Bagchi and Tapas Si

1 Introduction

Regression analysis is a statistical method to explain the relationship between independent and dependent variables or parameters and predict the coefficients of the function. Linear regression analysis involves those functions that are linear combination of the independent parameters, whereas nonlinear regression analysis is a type of regression analysis in which given data is modeled with a function that is a nonlinear combination of multiple independent variables. Some examples on nonlinear regression functions are exponential, logarithmic, trigonometric, power functions. Regression is the most important and widely used statistical technique with many applications in business and economics.

Ozsoy et al. [1] performed an estimation of nonlinear regression model parameters using PSO. This study compares the optimal and estimated parameters, and its results hence show that the estimation of the coefficients using PSO yield reliable results. Mohanty [2] applied PSO to astronomical data analysis. The results show that PSO requires tuning of few parameters compared to GA but is found to be slightly worse than GA. A case study of PSO in regression analysis by Cheng et al. [3] utilized PSO to solve a regression problem in the dielectric relaxation field. The results show that PSO with ring structure has a good mean solution than PSO with a star structure. Erdogmus and Ekiz [4] proposed a nonlinear regression analysis using PSO and GA for some test problems show that GA shows better performance in estimating the values of the coefficients. Their work further shows that such heuristic optimization algorithms can be an alternative to classic optimization methods. Lu et al. [5] performed a selection of most important descriptors to build QSAR models using modified PSO (PSO-MLR) and compared the results with GA (GA-MLR). The results reveal that PSO-MLR performed better than GA-MLR for the prediction

J. Bagchi (✉) · T. Si
Department of Computer Science & Engineering, Bankura Unnayani Institute of Engineering,
Bankura, West Bengal 722146, India
e-mail: jayribagchi@gmail.com

T. Si
e-mail: c2.tapas@gmail.com

© The Author(s), under exclusive license to Springer Nature Singapore Pte Ltd. 2021
X.-Z. Gao et al. (eds.), *Applications of Artificial Intelligence in Engineering*, Algorithms
for Intelligent Systems, https://doi.org/10.1007/978-981-33-4604-8_4

set. Barmpalexis et al. [6] applied multi-linear regression, PSO, and artificial neural networks in the pre-formulation phase of mini-tablet preparation to establish an acceptable processing window and identify product design space. Their results show that DoE-MLR regression equations gave good fitting results for 5 out of 8 responses, whereas GP gave the best results for the other 3 responses. PSO-ANNs were only to fit all selected responses simultaneously. Cerny et al. [7] proposed a new type of genotype for prefix gene expression programming (PGEP). PGEP, improved from gene expression programming (GEP), is used for signomial regression (SR). The method was called differential evolution-prefix gene expression programming (DE-PGEP) which allows for expression and constants to coexist in the same vector spaced representation and be evolved simultaneously. Park et al. [8] proposed PSO-based signomial regression (PSR) to solve nonlinear regression problems. Their work attempted to solve the signomial function by estimating the parameters using PSO. Mishra [9] evaluates the performance of differential evolution at nonlinear curve fitting. Results show that DE has been successful to obtain optimum results even if parameter domains were wide but it could not reach near-optimal results for the CPC-X problems which are the challenge problems for any nonlinear least-squares algorithm. Gilli and Schumann [10] used DE, PSO, and threshold accepting methods to estimate the parameters of linear regression. Yang et al. [11] constructed the linear regression models for the symbolic interval values data using PSO.

The objective of this paper is to perform a nonlinear regression analysis using the MVO algorithm [15]. The proposed method is applied to 10 well-known benchmark nonlinear regression problems. A comparative study is conducted with PSO [12]. The experimental results with statistical analysis demonstrate that the proposed method outperforms PSO.

1.1 Organization of This Paper

The remaining of the paper is organized as follows: The proposed method is discussed in Sect. 2. The experimental setup including the regression models and dataset description is given in Sect. 3. The results and discussion are given in Sect. 4. Finally, the conclusion with future works is given in Sect. 5.

2 Materials and Methods

2.1 Regression Analysis

Regression analysis is a statistical technique to estimate the relationships among the variables of a function. It is a commonly used method for obtaining the prediction function for predicting the values of the response variable using predictor variables [14]. There are three types of variable in regression such as

- The unknown coefficients or parameters, denoted as β, may represent a scalar or a vector
- The independent variable or predictor variable, i.e., input vector $X = (x_1, x_2, \ldots, x_n)$
- The dependent variable or response variable, i.e., output y.

The regression model in basic form can be defined as:

$$y \approx f(x, \beta) \tag{1}$$

where $\beta = (\beta_0, \beta_1, \beta_2, \ldots, \beta_m)$.

A linear regression model is a model of which output variable is the linear combination of coefficients and input variables, and it is defined as [13]:

$$y = \beta_0 + \beta_1 x_1 + \beta_2 x_2 + \cdots + \beta_n x_n + \xi \tag{2}$$

where ξ is a random variable, a disturbance that perturbs the output y. A nonlinear regression model is a model of which output variable is the nonlinear combination of coefficients and input variables. The nonlinear regression model is defined as follows [13]:

$$y = f(x, \beta) + \xi \tag{3}$$

where f is the nonlinear function.

In regression analysis, an optimizer is used to search the coefficients, i.e., parameters so that the model fits well the data. In the current work, the unknown parameters of different nonlinear regression models are searched using MVO algorithm. The MVO algorithm is discussed next.

2.2 MVO Algorithm

Multi-verse optimizer is an optimization algorithm whose design is inspired by the multi-verse theory in Physics [15]. Multi-verse theory in Physics states that there exist multiple universes and each universe possesses its own inflation rate which is responsible for the creation of stars, planets, asteroids, meteoroids, black holes, white holes, wormholes, physical laws for that universe.

For a universe to be stable, it must have a minimum inflation rate. So the goal of the MVO algorithm is to find the best solution by reducing the inflation rate of the universes which is also the fitness value. Now, observations from multi-verse theory show that universes with higher inflation rate have more white holes and universes with a low inflation rates have more black holes. So to have a stable situation, objects from white holes have to travel to black holes. Also, the objects in each universe may travel randomly to the best universe through wormholes.

In MVO, each solution represents a universe and each variable to be an object in the universe. Further, the inflation rate is assigned to each universe which is proportional to the fitness value of each universe. MVO uses the concept of black holes and white holes for exploring search spaces and wormholes to exploit search spaces. When a tunnel is established between two universes, it is assumed that the universe with a higher inflation rate has more white holes and the universe with a lower inflation rate has more black holes. So universes exchange objects from white holes to black holes which improves the average inflation rates of all universes over the iterations. In order to mathematically model the above idea, the roulette wheel mechanism is used that selects one of the universes with a high inflation rate to contain a white hole and allows objects from that universe to move into the universe containing a black hole and relatively low inflation rate. At every iteration, universes are sorted and one of them is selected by the roulette wheel to have a white hole. Assuming that U is the matrix of universes with d parameters and n candidate solutions.

Roulette wheel selection mechanism based on the normalized inflation rate is illustrated as below:

$$x_i(j) = \begin{cases} x_k(j) & r_1 < \mathrm{NI}(U_i) \\ x_i(j) & r_1 >= \mathrm{NI}(U_i) \end{cases} \tag{4}$$

Here $x_i(j)$ indicates jth parameter of ith universe, U_i shows ith universe, $\mathrm{NI}(U_i)$ is the normalized inflation rate of ith universe, r_1 is a random number in the interval $[0, 1]$, and $x_k(j)$ is the jth parameter of the kth universe selected by roulette wheel mechanism.

As this is done with the sorted universes, so the universes with low inflation rates have a higher probability if sending objects through white/black holes. Now in order to perform exploitation, it is considered each universe has wormholes to transport its objects randomly through space. In order to improve average inflation rates, it is assumed that wormhole tunnels are established between a universe and the best universe obtained so far. The formulation of the mechanism is:

$$x_i^j = \begin{cases} \begin{cases} x_j + \mathrm{TDR} * ((\mathrm{ub}_j - \mathrm{lb}_j) * r_4 + \mathrm{lb}_j) & r_3 < 0.5 \\ x_j - \mathrm{TDR} * ((\mathrm{ub}_j - \mathrm{lb}_j) * r_4 + \mathrm{lb}_j) & r_3 >= 0.5 \end{cases} & r_2 < \mathrm{WEP} \\ x_i^j & r_2 >= \mathrm{WEP} \end{cases} \tag{5}$$

Here x_j indicates the jth parameter of best universe formed so far, traveling distance rate (TDR) and wormhole existence probability (WEP) are coefficients, lb_j is the lower bound of jth variable, ub_j is the upper bound of jth variable, x_i^j indicates the jth parameter of ith universe, and $r_2, r3, r_4$ are random numbers in $[0, 1]$.

WEP is for defining the probability of the existence of wormhole. It is to be increased linearly over the iterations for better exploitation results. TDR defines the distance rate that an object can travel through the wormhole to the best universe obtained so far. TDR is decreased over the iterations to increase the accuracy of local search by the following rule:

$$\text{TDR} = 1 - \frac{l^{\frac{1}{p}}}{L^{\frac{1}{p}}} \qquad (6)$$

where l is the current iteration and L is the maximum number of iterations. p is the exploitation accuracy over iterations, and generally, it is set to 6. The update rule of WEP is as follows:

$$\text{WEP} = \text{WEP}_{\text{min}} + l * \left(\frac{\text{WEP}_{\text{max}} - \text{WEP}_{\text{min}}}{L} \right) \qquad (7)$$

where WEP_{min} and WEP_{max} indicate the minimum and maximum range of WEP. The flowchart of MVO is given in Fig. 1.

2.3 Regression Analysis Using MVO

In this work, MVO is used to search the unknown parameters (β) of the nonlinear regression model. Let us assume a model has m number of parameters and then the dimension of the universe in MVO is m. The ith universe is represented by $X_i = (x_1, x_2, \ldots, x_m)$. The inflation rate of the universe is the objective function value. The mean square error (MSE) is used as an objective function in this work. The MSE is calculated as follows:

$$\text{MSE} = \frac{1}{N} \sum_{i=1}^{N} (y_i - y_i')^2 \qquad (8)$$

where y_i and y_i' are the target and predicted output of the ith input data, respectively. MVO algorithm minimizes MSE to fit the data. N is the number of samples in the dataset.

3 Experimental Setup

3.1 Regression Model and Dataset Description

In this work, 10 regression model is analyzed and the datasets for the models have been collected from [16]. The description of different regression models and their dataset is given in Table 1.

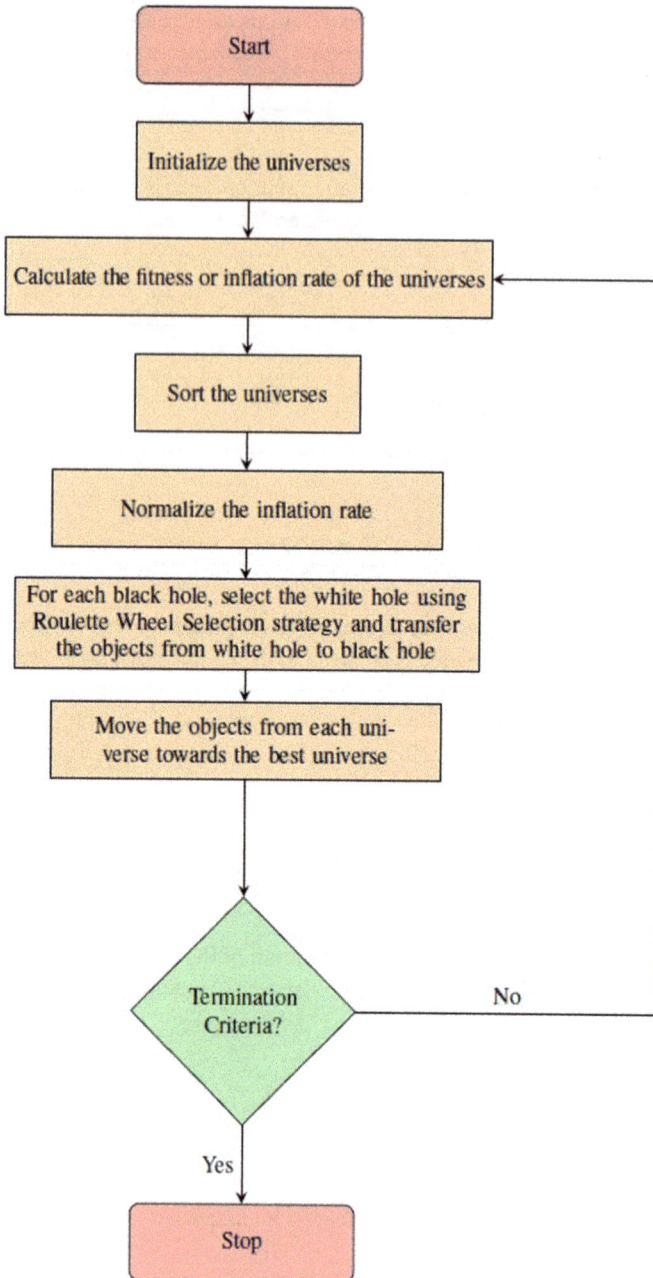

Fig. 1 Flowchart of MVO

Table 1 Regression model and dataset description

S. No.	Name	Model	No. of coefficients	No. of samples
1	Misra1a	$\beta_1(1 - \exp(-\beta_2 x))$	2	14
2	Gauss1	$\beta_1 \exp(-\beta_2 x) + \beta_3 \frac{-(x-\beta_4)^2}{\beta_5^2} + \beta_6 \frac{-(x-\beta_7)^2}{\beta_8^2}$	8	250
3	DanWood	$\beta_1 x^{\beta_2}$	2	6
4	Nelson	$\exp(\beta_1 - \beta_2 x_1 \exp(-\beta_3 x_2))$	3	128
5	Lanczos2	$\beta_1 \exp(-\beta_2 x) + \beta_3 \exp(-\beta_4 x) + \beta_5 \exp(-\beta_6 x)$	6	24
6	Roszman1	$\beta_1 - \beta_2 x - \frac{\arctan \frac{\beta_3}{x - \beta_4}}{\pi}$	4	25
7	ENSO	$\beta_1 + \beta_2 \cos \frac{2\pi x}{12} + \beta_3 \sin \frac{2\pi x}{12} + \beta_5 \cos \frac{2\pi x}{\beta_4} + \beta_6 \sin \frac{2\pi x}{\beta_4} + \beta_8 \cos \frac{2\pi x}{\beta_7} + \beta_9 \sin \frac{2\pi x}{\beta_7}$	9	168
8	MGH09	$\frac{\beta_1(x^2 + x\beta_2)}{x^2 + x\beta_3 + \beta_4}$	4	11
9	Thurber	$\frac{\beta_1 + \beta_2 x + \beta_3 x^2 + \beta_4 x^3}{1 + \beta_5 x + \beta_6 x^2 + \beta_7 x^3}$	7	37
10	Rat42	$\frac{\beta_1}{1 + \exp(\beta_2 - \beta_3 x)}$	3	9

3.2 Parameters Setting

The parameters of MVO are set as the following: Number of universe = 30, $WEP_{max} = 1$, $WEP_{min} = 0.2$, exploitation accuracy $(p) = 6$, the maximum number of iterations = 100.

The parameters of PSO are set as the following: population size=30, $w_{max} = 0.9$, $w_{min} = 0.4$, $c_1 = 2.05$, $c_2 = 2.05$, the maximum number of iterations = 100.

3.3 PC Configuration

- CPU: Intel i3-4005U 1.70 GHz
- RAM: 4 GB
- Operating System: Windows 7
- Software Tool: MATLAB R2018a.

4 Results and Discussion

In this work, nonlinear regression analysis has been performed using the MVO algorithm for 10 regression models. 'Hold-out' cross-validation method is used. 80% of the dataset is used in training of the model, and the remaining 20% of the dataset is used as test data for the model. The experiment is repeated 31 times for each model. The same experiment is conducted using PSO for the comparative study. The quality of the results has been measured in terms of training and testing MSE errors over 31 independent runs. The mean and standard deviation of MSE values in training over 31 runs are given in Table 2. The mean and standard deviation of MSE values in testing over 31 runs are given in Table 3.

From Table 2, it is observed that MVO performs better in training than PSO for most of the models. The standard deviations of training MSEs of MVO are also lower than that of PSO. It is observed from Table 3 that the mean testing MSEs are better than that of PSO for most of the models. To test the significance in the difference of performance of MVO and PSO, a nonparametric statistical test, Wilcoxon's Signed Ranked Test [12] has been carried out with significance level (α) = 0.05. The p-values and null hypothesis values (h) are given in Table 3. The p-values less than 0.05 indicate statistically significant difference in the performance, whereas p-values greater than or equal to 0.05 depict no significant difference in the performance of the algorithms. In Table 3, $h = 1$ indicates MVO statistically outperforms PSO, $h = -1$ indicates PSO statistically outperforms MVO, and $h = 0$ indicates no significant difference in the performance. From Table 3, it is observed that the h-values come out to be 1 for 8 out of 10 models and it signifies the statistically better performance of MVO over PSO. PSO statistically outperforms MVO for the Roszman1 model. There is no significant difference in the performance of MVO and PSO for ENSO model. The robustness of metaheuristic algorithms is measured in terms of standard deviations. From Tables 2 and 3, it can be observed that the standard deviation of PSO is lower than PSO that indicates that MVO is more robust than PSO in nonlinear

Table 2 Mean and standard deviation of training residual errors over 31 independent runs

Model	PSO	MVO
Misra1a	3.389 (0.1696)	**0.2638 (0.1246)**
Gauss1	80.4833 (130.1000)	**5.5966 (0.4977)**
Danwood	0.0076 (0.0288)	**6.05E−04 (1.42E−05)**
Nelson	0.0499 (0.0441)	**0.0271 (1.37E−04)**
Lanczos2	5.96E−04 (8.37E−04)	**2.76E−06 (5.01E−06)**
Roszman1	2.62E−05 (2.20E−05)	**1.56E−05 (4.42E−07)**
ENSO	11.2482 (1.48E−06)	11.2482 (1.82E−05)
MGH09	**2.58E−05 (1.01E−06)**	2.65E−05 (2.86E−07)
Thurber	7.15E+02 (8.01E+02)	**5.46E+02 (4.03E+02)**
Rat42	1.3186 (0.4807)	**0.9483 (0.0043)**

Table 3 Mean and standard deviation of testing residual errors over 31 independent runs

Model	PSO	MVO	p-Value	h
Misra1a	1.9869 (0.1696)	**0.1381 (0.0677)**	1.17E$-$06 $\ll$ 0.5	1
Gauss1	76.8515 (133.73033)	**6.2735 (0.7319)**	1.17E$-$06 $\ll$ 0.5	1
DanWood	0.0061 (0.0061)	**0.0013 (7.04E$-$05)**	5.99E$-$06 $\ll$ 0.5	1
Nelson	0.0707 (0.0567)	**0.0403 (5.54E$-$04)**	2.12E$-$04 $\ll$ 0.5	1
Lanczos2	8.39E$-$04 (0.0012)	**2.41E$-$06 (4.26E$-$06)**	2.56E$-$06 $\ll$ 0.5	1
Roszman1	**2.96E$-$05 (5.61E$-$06)**	3.78E$-$05 (2.55E$-$06)	3.10E$-$06 $\ll$ 0.5	-1
ENSO	13.4705 (6.06E$-$06)	13.4705 (7.45E$-$05)	0.9064 $>$ 0.5	0
MGH09	3.91E$-$05 (7.19E$-$06)	**3.48E$-$05 (1.18E$-$06)**	0.004 $<$ 0.5	1
Thurber	9.16E+02 (1.05E+03)	**3.57E+02 (2.10E+02)**	3.70E$-$03 $\ll$ 0.5	1
Rat42	2.617 (0.3246)	**0.4734 (0.0349)**	1.17E$-$06 $\ll$ 0.5	1

$h = 1$ indicates MVO statistically outperforms PSO, $h = -1$ indicates PSO statistically outperforms MVO, and $h = 0$ indicates no significant difference in the performance

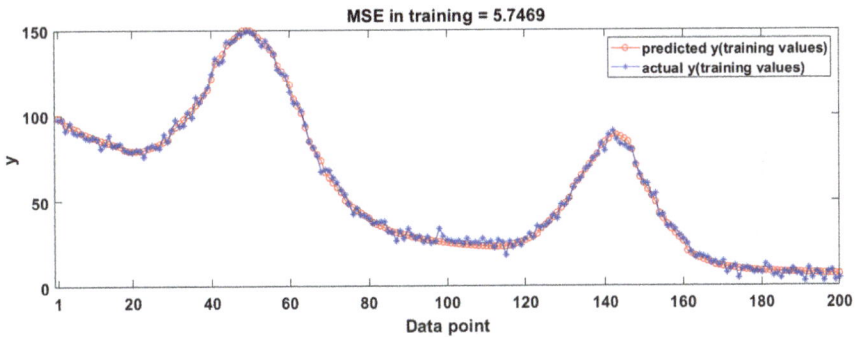

Fig. 2 Model prediction results using MVO for training data in regression analysis of Gauss1 model

regression. Model prediction results using MVO for training and testing data in regression analysis of the Gauss1 model are given in Figs. 2 and 3, respectively. From these graphs, it is observed that MVO almost fits the training and testing curves for Gauss1 model. The convergence graph of MVO and PSO is given Fig. 4. From this graph, it is observed that MVO has better convergence behavior than PSO.

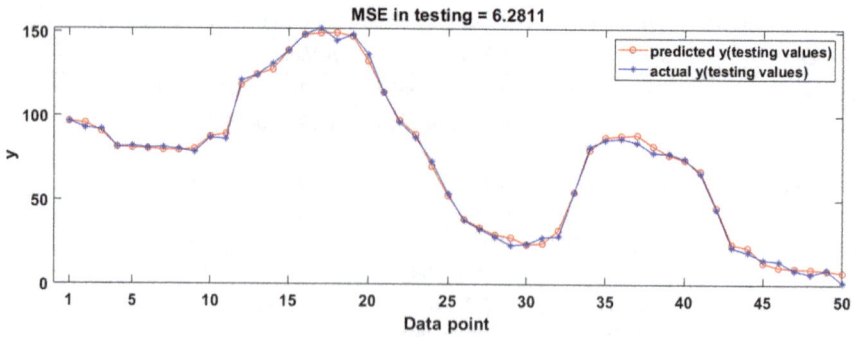

Fig. 3 Model prediction results using MVO for testing data in regression analysis of Gauss1 model

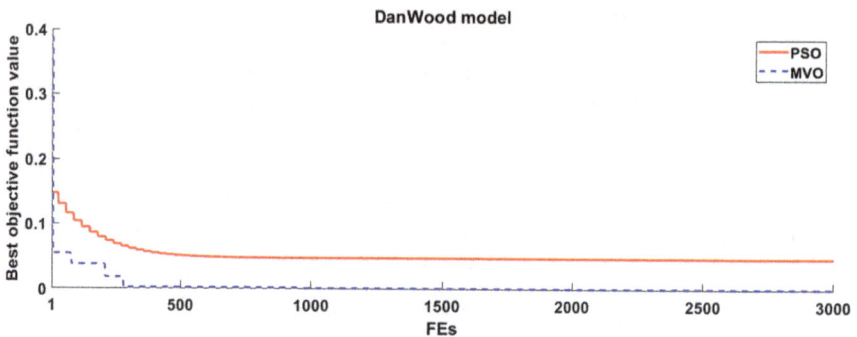

Fig. 4 Convergence graph of MVO for Danwood model

5 Conclusion

In this paper, MVO is used for nonlinear regression analysis. MVO is applied to search the parameters of different regression models. For the experiment, 10 well-known benchmark regression models are used. A comparative study has been carried out with PSO. The experimental results demonstrate that the proposed method statistically outperforms PSO in nonlinear regression analysis. In the future, different metaheuristic algorithms will be studied in nonlinear regression analysis.

References

1. Ozsoy VS, Orkcu HH (2016) Estimating the parameters of nonlinear regression models through Particle Swarm Optimization. Gazi Univ J Sci 29(1):187–199
2. Mohanty SD (2015) Particle Swarm Optimization and regression analysis–I. Astron Rev 7(2):29–35

3. Cheng S, Zhao C, Wu J, Shi Y (2013) Particle Swarm Optimization in regression analysis: a case study. In: International conference on swarm intelligence, Part I. LNCS, vol 7928, pp 55–63
4. Erdogmus P, Ekiz S (2016) Nonlinear regression using Particle Swarm Optimization and genetic algorithm. Int J Comput Appl (0975-8887) 153(6)
5. Lu J, Shen Q, Jiang J, Shen G, Yu R (2004) QSAR analysis of cyclooxygenase inhibitor using particle swarm optimization and multiple linear regression. J Pharm Biomed Anal 35:679–687
6. Barmpalexis P, Karagianni A, Karasavvaides G, Kachrimanis K (2018) Comparison of multi-linear regression, particle swarm optimization artificial neural networks and genetic programming in the development of mini-tablets. Int J Pharm 551(1–2):166–176
7. Cerny B, Nelson P, Zhou C (2008) Using differential evolution for symbolic regression and numerical constant creation. In: Proceedings of the 10th annual conference on genetic and evolutionary computation (GECCO '08), pp 1195–1202
8. Park S, Song N, Yu W, Kim W: PSR: PSO-based signomial regression model. Int J Fuzzy Logic Intell Syst 19(4):307–314
9. Mishra SK (2007) Performance of differential evolution method in least squares fitting of some typical nonlinear curves. SSRN-Elsevier, MPRA Paper No. 4656
10. Gilli M, Schumann E (2010) Robust regression with optimisation heuristics. In: Brabazon A et al (eds) Natural computing in computational finance, vol 3. SCI 293, pp 9–30
11. Yang C, Chuang C, Jeng J, Tao C (2011) Constructing the linear regression models for the symbolic interval-values data using PSO algorithm. In: IEEE proceedings of 2011 international conference on system science and engineering
12. Derrac J, Garcla S, Molina D, Herrera F (2011) A practical tutorial on the use of nonparametric statistical tests as a methodology for comparing evolutionary and swarm intelligence algorithms. Swarm Evol Comput 1:3–18
13. Bates D, Watts D (1988) Nonlinear regression analysis and its applications. Wiley, New York
14. Franklin AG, Iyer HK (1994) Regression analysis: concepts and application. Duxbury Press, Belmont
15. Mirjalili S, Mirjalili SM, Hatamlou A (2016) Multi-Verse Optimizer: a nature inspired algorithm for global optimization. Neural Comput Appl 27:495–513
16. https://www.itl.nist.gov/div898/strd/nls/nls_main.shtml

Chapter 5
An Extended ACO Based Routing Algorithm for Cognitive Radio Network

Vishal Raj and Mahendra Kumar Murmu

1 Introduction

Cognitive radio (CR) is a cutting edge technology to solve the spectrum shortage problem in wireless communication environments [1–3]. Users of cognitive radio can be viewed as secondary user (SU) and primary user (PU). The secondary user has the potential to identify the unused spectrum called *holes* and finds the most suitable frequency band (or holes) using dynamic spectrum accessing policy [2]. The SU node opportunistically identifies the spectrum holes and use it for communication while ensuring that the primary user of the spectrum remain unaffected. Further, the secondary users must switch the band once the primary user is detected. This new research in the development of the cognitive radio network foresees a further improvement in spectrum efficiency. Each spectrum opportunities (SOP) acts upon one or multichannel, which are available to use by the secondary user, dynamically. The set of available channels to secondary nodes in the radio environment varies with space and time. For a larger network, a routing algorithm may be helpful and suit the dynamic aspect of a multi-hop CRN. Only some of the research work has been done in this direction. Most of them are based on to modify the traditional routing algorithm, for example, applying and mapping AODV in a multi-hop cognitive radio network [4, 5]. The protocol broadcasts the route request packet to all the channels, which increases the routing overhead and the discovery time. Thus, the overhead increases if so the number of SU nodes in the networks.

Therefore, the present works intent to propose an ant-based algorithm which could find the global most favorable path with minimum delay and maximum rate of delivery. The contribution introduces improved ACO for routing [6–15].

V. Raj (✉) · M. K. Murmu
Department of Computer Engineering, National Institute of Technology, Kurukshetra, Haryana 136119, India
e-mail: vishal_31803208@nitkkr.ac.in

57
X.-Z. Gao et al. (eds.), *Applications of Artificial Intelligence in Engineering*, Algorithms for Intelligent Systems, https://doi.org/10.1007/978-981-33-4604-8_5

1.1 Ant Colony System

Ant colony system [16] was introduced by Dorigo. Further, Dorigo and Gambardella proposes improved colony optimization-based approach in [17]. This approach focuses on a centralized or global update for pheromone, so that the search space concentrates close to the best solution found so far in order to improve the overall convergence speed.

Ant colony system has broadly four modifications over ant system:

- Different Transition Rule: The transition rule is modified to further improve on exploration by tuning parameters. Earlier algorithm concentrates on the best solution rather than exploring constantly for new solution within search space.

$$j = \begin{cases} \arg \max_{j \in \text{neighbour}} \left([\tau_{ij}(t)]^{\alpha} [n_{ij}]^{\beta} \right) & \text{if } q \leq q_0 \\ J & \text{otherwise} \end{cases}$$

- where q is a random variable distributed over [0,1], and q_0 is an adjustable parameter between 0 and 1. This makes the solution search space concentrate close to the best possible solution in order to improve the overall convergence speed, instead of exploring constantly.

$$P_{ij}^k(t) = \begin{cases} \dfrac{\left([\tau_{ij}(t)]^{\alpha} [n_{ijl}]^{\beta} \right)}{\sum_{i \in \text{neighbour}} [\tau_{is}]^{\alpha} [n_{is}]^{\beta}} \end{cases}$$

- The agent (ant) that generates the best tour is only allowed to update the pheromone. The pheromone is updated only on the edges of the best tour found so far.

$$\tau_{ij}(t) = (1 - \rho)\tau_{ij}(t) + \rho \Delta \tau_o$$

- The local pheromone is updated to diminish the level of pheromone on the visited edge. This is performed to avoid the stagnation of algorithm and allow agent to explore more.
- Use of tabu list to restrict the choice of the next city to visit. The tabu list consists a list of visited nodes sorted in increasing order. The list is scanned by the agent (ant) to make sure that the transaction over an already visited node will not happen.

Ant colony system has some enhanced feature which is:

- There is a balance between the new edge exploration and the exploitation of prior accumulated knowledge.
- The best or optimal length tour has the global pheromone update.
- During the building of a solution by an ant, the local pheromone is updated.

1.2   State Transaction Phases with ACS:

- *Graph formulation phase*: The process formulates the graph $G(V, E)$.
- *Initialization phase*: The pheromone deposition and ant initialization taken place.
- *Transaction phase*: The movement of ants in ACS is accomplished using pheromone and heuristics information of the path.
- *Global pheromone updating*: At each iteration, the best path is selected. The updated pheromone performs as information passed from the previous iterations and use it in the next iteration.
- *Termination phase*: The algorithm terminates when an optimum solution is reached or stopping criteria meets.

The rest of the paper is arranged as follows. Sect. 2 describes the assumed network model. Then, we describe various routing matrices based on mathematical expression, given in Sect. 3. In next, we illustrate routing tables in Sect. 4 and then included algorithm in Sect. 5. Finally, we conclude our work in section.

2   Network Model

Suppose a CRN consist of n SU nodes and k channels. The secondary user node is designated with their unique identification number. In CRN, any SU node may initialize to transmit a routing request. The node in the cognitive radio network is aware of their available channel and position (or neighboring nodes) in the network. Thus, each node in CRN can select spectrum holes and exchange the information with their neighbor nodes, periodically. Similarly, the k channels are viewed as sum of common control channel (CCC), local channel set (LCS), and global channel set (GCS). The SU needs to constantly sense the spectrum and vacate the channel in the presence of a PU, and to achieve this, the CR user usually coordinates together using control message on a common medium called the common control channel [18, 19]. Here, common control channel is the subset of local channel set, and LCS is the subset of global channel set. Using Fig. 1, we illustrate our network model as follows.

The SU nodes have the broad frequency range, which can be further labeled into many channels. The SU senses each working channel, periodically. The basic challenge of multi-hop routing in CRN is the availability of dynamic heterogeneous channels. The above-shown Fig. 1 represents a typical scenario of CRN, which consist of three primary user networks. The network topology shown in Fig. 1 is updated when a licensed user interrupts.

If the SU broadcasts "HELLO" packet on its licensed channel, the node detects the packet if it is within the transmission range. Let us suppose that the CRN has "n" number of the unused channel, and therefore, the node monitors every channel one by one for the "HELLO" message for its uses. If the channel has no "HELLO"

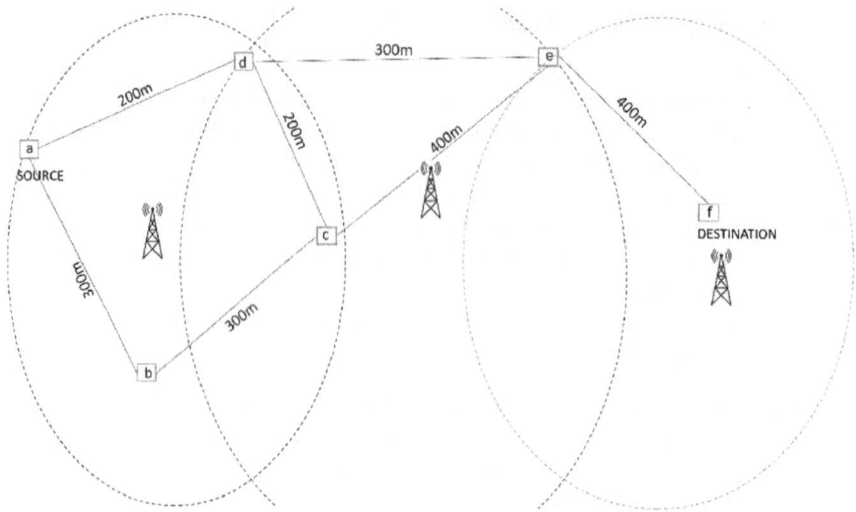

Fig. 1 Typical scenario of CRN

message, the SU adds that channel in its available or vacant channel list. The SU node broadcasts information, periodically.

3 Routing Metrics

The proposed algorithm is an on-demand based routing algorithm [20]. The algorithm finds a valid and optimal path from source to destination. For the working of this algorithm, all nodes must have knowledge about their available channels and their position information including their neighboring nodes.

There is some assumption made for efficient working of the proposed algorithm:

- A CCC channel present allows the CR nodes to send control packets.
- The graph G must be constructed, which is based on the spectrum sensing result of CR's physical layer before starting the routing algorithm.
- The SU nodes can able to adjust the working frequency range for transmitting data.
- Each node maintains a list of local available channel that is not in the use of the PU. Thus, each on its own detect SOP (the frequency which is being unoccupied and is presently available).

Whenever a node wants to send data, route discovery process begins, and the node is labeled as source node. The source node will send the forward ant, which based on pheromone level of the path will select the route. Thus, the forward ant will route between source and destination node. The backward ant will do the pheromone update. Whenever a forward ant on node i has to select the next node, it uses following measures (1) or (2):

Forward ant

$$
s = \begin{cases} \arg\max_{j \in \text{neighbour}} \left([\tau_{ijl}]^\alpha [n_{ijl}]^\beta \right) & \text{if } q \leq q_0 \\ S & \text{otherwise} \end{cases} \tag{1}
$$

The next node j is being selected according to the probability distribution:

$$
S = \begin{cases} \dfrac{\left([\tau_{ijl}]^\alpha [n_{ijl}]^\beta \right)}{\sum_{i \in \text{neighbour}} [\tau_{is}]^\alpha [n_{is}]^\beta} & j \in \text{neighbour} \\ 0, & \text{otherwise} \end{cases} \tag{2}
$$

where q is a randomly generated number by the forward ants within uniform distribution $[0,1]$. q_0 is a predefined threshold parameter within $(0,1)$. If $q \leq q_0$, then the ant uses the previous knowledge to choose further the nested node in the channel based on pheromone distribution τ_{ijl} and heuristic visibility n_{ijl}.

τ_{ijl} is the pheromone distribution between node i and j through channel l. n_{ijl} is the heuristic visibility between node i and j through channel l:

$$
n_{ijl}(t) = \frac{k_1}{t_{sij}} + \frac{k_2}{t_{pj}} + \frac{k_3}{t_{\text{backoff}}} \tag{3}
$$

"t_{si}" is the switching delay for neighbor node i when it transmits data in different channel

$$
t_{si} = \frac{10\,\text{ms}(\text{band}_i - \text{band}_j)}{10\,\text{mHz}} \tag{4}
$$

t_{qj} is the propagation delay to node j.

$$
t_{qj} = \frac{d}{v} \tag{5}
$$

"d" is the distance between pair of nodes i and node j, and v is the velocity of signal in the medium.

If there are more than one secondary node in the same channel, then they will contend for the spectrum resource. This leads to back off delay which can be calculated as follows:

$$
t_{\text{backoff}}(\text{num}_j) = \frac{1}{(1 - P_c)\left(1 - (1 - P_c)^{\frac{1}{\text{num}_i - 1}}\right)} w_0 \tag{6}
$$

where P_c is the probability of a contending node experiencing collision and w_0 is the dimension of minimum contention window. The back off delay increased when many node routes in the same channel.

When the forward ant reached the destination, the backward ants are generated. The backward ants contain the address of all the intermediate nodes and thus follow the same path as developed by the forward ant. The backward ant also updates the local pheromone concentration. The local pheromone concentration is based on:

$$\tau_{ijl} = (1 - \rho)\tau_{ijl}(t) + \rho \Delta \tau_{ijl}(t, t + 1) \tag{7}$$

where ρ is the coefficient of pheromone updating and it lies between $(0 < \rho < 1)$.

$$\Delta \tau_{ijl} = \begin{cases} \frac{1}{d_n} \\ 0 \end{cases} \tag{8}$$

Here, $\Delta \tau_{ijl}$ is the pheromone update on a path between node i to node j for ant l.

When each the backward ants reaches the source node, a search scheme starts which further accelerates the convergence of the algorithm and reduces path delay. Let us assume that the forward ant chooses a set of vertices' $= \{v_1, v_2, v_3 \ldots v_i\}$. The search scheme for ant's k's path which is called the first path consists of following steps:

- Choose a path already covered by another ant called the j's path called the second path.
- For each path, search their corresponding vertices.
- If two path having same vertex, then the two paths are split up further into two segments. The initial part which contains the same vertex and the ending part is obtained by the second selected tour. At the same time, the second new path can be identified by inversing the preceding operation.
- Compute the path latency. If the path has with few vertex's or better latency, then the path is accepted.

Finally, the global optimization rule is updated on the global best selected path. The global pheromone updating rule is based on:

$$\tau(r, s) = (1 - \rho)\tau(r, s) + \rho \Delta \tau(r, s) \tag{9}$$

$$\Delta \tau(r, s) = \begin{cases} \frac{1}{L_g}, & \text{if}(r, s) \in \text{global optimized path} \\ 0 \end{cases} \tag{10}$$

where ρ is the coefficient of pheromone updating and it lies between $(0 < \rho < 1)$ and L_g is the global optimized path.

After the global update is made, the pheromone concentration in node is cleared.

Table 1 Information in immediate node

c	nn_id	τ	n	d	c	l

n_id the SU node id; nn_id the neighbor node id; τ the pheromone; n the heuristic desirability; d delay; c the available channel list; l the link

Table 2 Information in ant

s_add	d_add	A_id	[]	b/f	ttl

s_add the source address; d_add the destination address; A_id the ant id; [] the visited node list; b/f the backward or forward ant; ttl time to leave

4 Routing Algorithm

The following list of information is maintained by the SU node, as given in Tables 1 and 2 [21].

The SU node which intends to send the data dispatches M ants.

Now, our routing algorithm works as follows:

Input: α, β, Δ, q_0, k_1, k_2, k_3, m, G, s, t.

Output: A path from source to destination based on metrics constraint.

(1) Initialize the number of iteration t. Then, s sends M ants with ant $id\ k$ whose ttl is T. Whenever an ant reaches at any node, the ttl value is reduced by 1. If ttl becomes zero, the ant is assumed to be killed.

(2) When the node j receives ant

    (a) Current node Id is checked in the visited node list. If node id matches, the ant is assumed to be killed to avoid path loop, else continue.

    (b) Compute the heuristic probability for the next node using Eqs. (1) and (2).

    (c) Update all information.

(3) When source nodes receive the backward ant from t, apply the search process.

(4) When all the backward ant reaches to the source node or the timer is timeout, s extracts the paths in temp path list and selects the path which has minimal total delay. The selected path is the final path after iteration \.

(5) If the iteration time is lower than the limit, return to the initial step. Otherwise, extract all paths in the available path list in order to select the final path.

5 Proposed Algorithm

The routing algorithm starts from any SU node connected. The next node selection algorithm works as follows (given in Fig. 2). The algorithm initializes with initial parameters such as maximum number of iterations, number of ants, initial pheromone amount, and constant evaporation value.

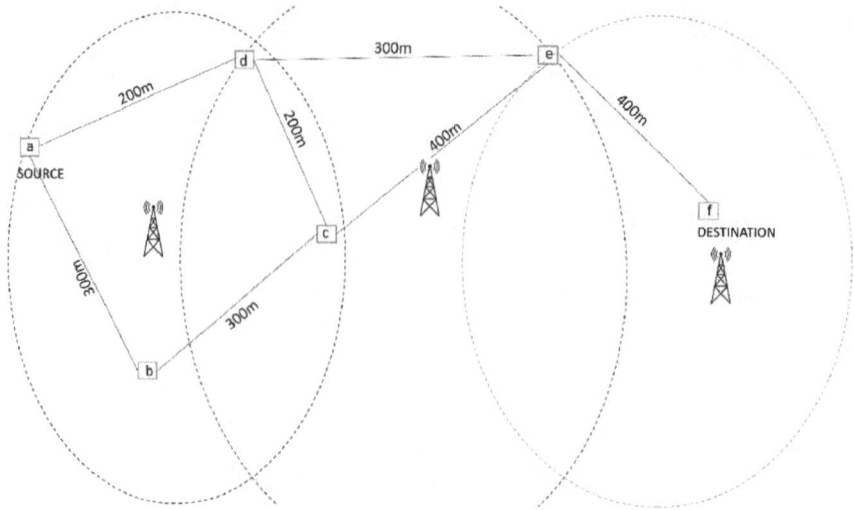

Fig. 2 Working scenario of CRN

We created a typical scenario as shown above in Fig. 1 where three primary networks are present. The protection range of each primary network is 600 m, and SU node's transmission range is 400 m. We have assumed that the channel capacity is 1. At least, a common control channel is available every time for the SU user. We assume that SU node could simultaneously transmit signal on multiple available data channels if multiple channel are available. This is useful to increase data rate.

We assume that the number of ants in our case is 5. The value of ant is 0.5, and available channels are set 1 and 2, respectively. The evaporation rate is set 0.5. The pheromone count at each node is same and assumed 0.01 (Fig. 3).

During first iteration, suppose $ant\_id = 1$ starts network exploration (Fig. 4). We have a q value at node A which we assumed as larger one. Then, it selects the path based on Eq. (2) (Fig. 5).

Now, in next, q is smaller, comparatively. Therefore, based on Eq. (1), node f is selected, after which the backward ants are generated which follows the same route and updates the pheromone of the path based on Eq. (7), shown in Fig. 6.

The ants are generated after every three data packets transmission. Table 3 represents the algorithm at first iteration and second iteration.

6 Conclusion

In our present work, we have designed and proposed an ACS based routing protocol for CRNs. It is a further extension of the work proposed in [6]. It is a biological inspired approach based adaptive routing algorithm for CRN. We have utilized

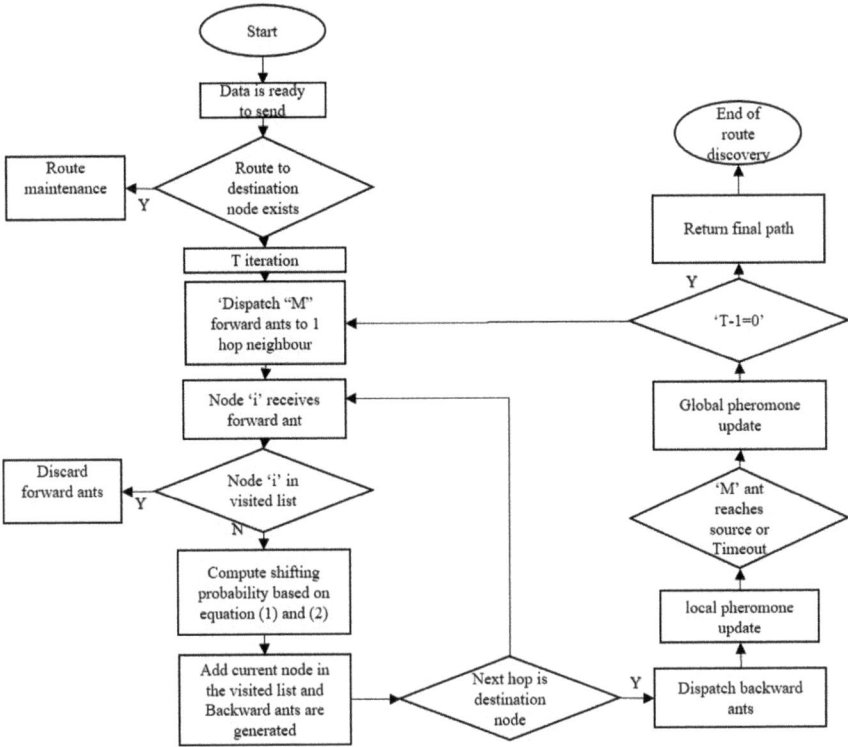

Fig. 3 Flowchart of proposed algorithm

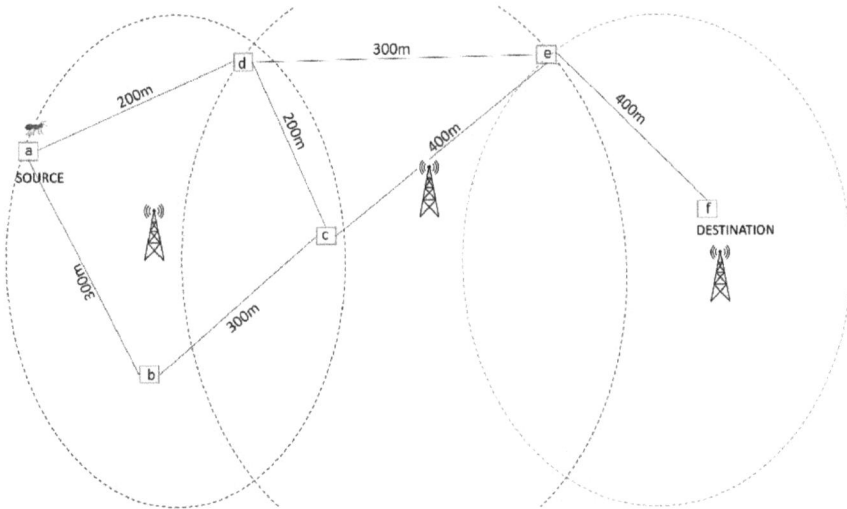

Fig. 4 Initial position of forward ant

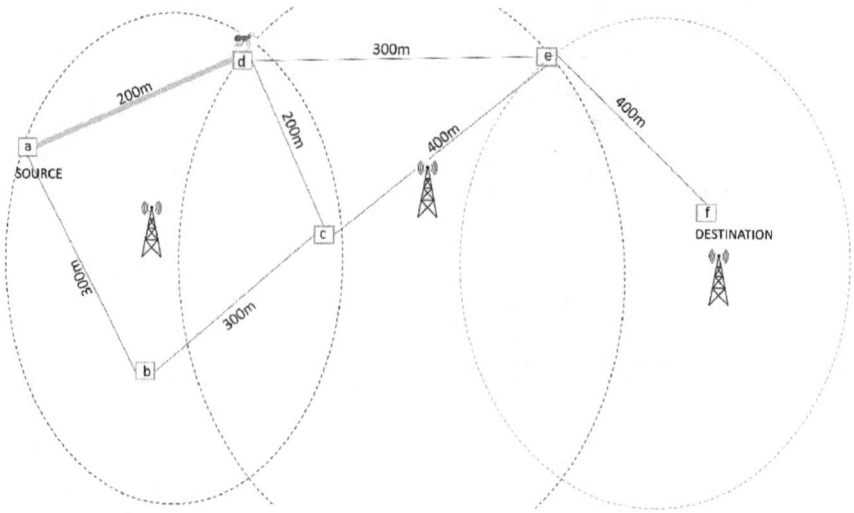

Fig. 5 Position of forward ant

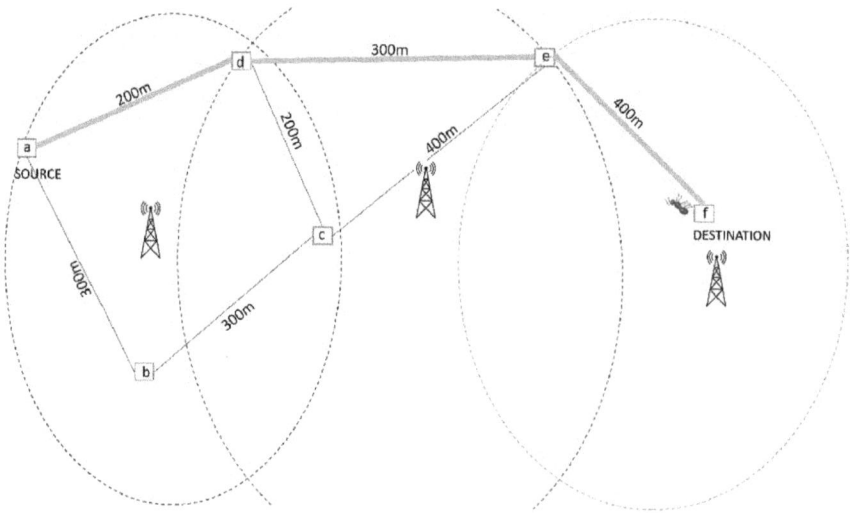

Fig. 6 Backward ant generated

forward and backward ants to fit our algorithm to the dynamic environment of CRN. The forward ants discover new path based on transaction rule to the destination. The backward ant updates information of routing table of each node. The illustration of the algorithm has been included using suitable example. The present method improves reliability of the routing protocol.

Table 3 Example of working of algorithm

Source node	Destination node	Ant ID	Visited node list	TTL	B/F
End of first iteration selected path [D, C, F] pheromone updated					
a	f	1	[D, E, F]		
a	f	2	[B, C, F]		
a	f	3	[D, E, F]		
a	f	4	[D, C, F]		
a	f	5	[D, E, F]		
a	f	6	[B, C, E, F]		
End of second iteration selected path [D, C, F] pheromone updated					
a	f	1	[D, E, F]		
a	f	2	[D, C, F]		
a	f	3	[D, E, F]		
a	f	4	[D, E, F]		
a	f	5	[B, C, F]		
a	f	6	[D, C, E, F]		

References

1. Mitola J (2006) Cognitive radio architecture: the engineering foundations of radio XML. Wiley
2. Akyildiz IF, Won-Yeol L, Vuran MC, Mohanty S (2006) NeXt generation/dynamic spectrum access/cognitive radio wireless networks: A survey. J Comput Netw 50(13):2127–2159
3. Akyildiz IF, Won-Yeol L, Chowdhury KR (2009) CRAHNs: cognitive radio ad hoc networks. J Ad Hoc Netw 7(5):810–836
4. Al-Rawi HAA, Yau K-LA (2013) Routing in distributed cognitive radio networks: a survey. Wirel Personal Commun 69:1983–2020
5. Elrhareg H, Ridouani M, Hayar A (2019) Routing protocols on cognitive radio networks: survey. In: IEEE international smart cities conference (ISC2), pp 296–302
6. Huang P, Zhou J (2013) ACO-based routing algorithm for cognitive radio networks. Theor Appl Complex Netw 264504 (2013). https://doi.org/10.1155/2013/264504
7. Thakare AN, Latesh B, Thomas A (2018) Analysis of bio-inspired scalable routing protocol for cognitive radio sensor networks, project: analysis of bio-inspired scalable routing protocol for cognitive radio sensor networks
8. Qadir J (2015) Artificial intelligence based cognitive routing for cognitive radio networks. Artif Intell Rev:25–96
9. Çavdar T, Guler E (2018) HyMPRo: a hybrid multi-path routing algorithm for cognitive radio ad hoc networks. Telecommun Syst 69. https://doi.org/10.1007/s11235-018-0426-4
10. Khalife H, Ahuja S, Malouch N, Krunz M (2008) Probabilistic path selection in opportunistic cognitive radio networks. In: Proceedings of the IEEE Globecom. https://doi.org/10.1109/GLOCOM.2008.ECP
11. Dutta N, Deva Sarma HK (2017) A probability based stable routing for cognitive radio adhoc networks. Wirel Netw 23:65–78. https://doi.org/10.1007/s11276-015-1138-2
12. Xin C, Ma L, Shen C-C (2008) A path-centric channel assignment framework for cognitive radio wireless networks. Mobile Netw Appl (Kluwer) 13(5):463–476
13. Hoque MA, Hong X (2012) BioStaR: a bio-inspired stable routing for cognitive radio networks. In: 2012 international conference on computing, networking and communications (ICNC), pp 402–406

14. Xu Y-H, Wu Y, Song J (2017) Joint channel assignment and routing protocol for cognitive radio wireless sensor networks. Wirel Personal Commun 97:1–22 (2017). https://doi.org/10.1007/s11277-017-4491-x

15. Gangambika RG, Sri Pavan B (2017) PSO and OLSR based protocol for cognitive radio sensor network, pp 1–5. https://doi.org/10.1109/CSITSS.2017.8447725

16. Mitola J (2000) Cognitive radio: an integrated agent architecture for software defined radio, Thesis (PhD), Department of Teleinformatics, Royal Institute of Technology (KTH), Stockholm, Sweden

17. Dorigo M, Maniezzo V, Colorni A (1996) The ant system: optimization by a colony of cooperating agent. IEEE Trans Syst Man Cybern Part B 26(1):1–13

18. Lo BF (2011) A survey of common control channel design in cognitive radio networks. Phys Commun 4(1):26–39

19. Pyo CW, Hasegawa M (2007) Minimum weight routing based on a common link control radio for cognitive wireless ad hoc networks. In: Proceedings of the international conference on wireless communications and mobile computing (IWCMC), pp 399–404. https://doi.org/10.1145/1280940.1281026

20. Salim S, Moh S (2013) On-demand routing protocols for cognitive radio ad hoc networks. EURASIP J Wirel Commun Netw. https://doi.org/10.1186/1687-1499-2013-102

21. Youssef M, Ibrahim M, Abdelatif M, Chen L, Vasilakos AV (2013) Routing metrics of cognitive radio networks: a survey. Commun Surveys Tutor 16(1)

Chapter 6
A Review: Image Classification and Object Detection with Deep Learning

Aditi and Aman Dureja

1 Introduction

Deep learning has come a long way in object detection. Leading to the progress and rapid growth o deep learning (DL) [1], significant results have been achieved in a variety of areas including robotics, health care, biology, commerce, and so on. Such advancements help open a new domain for machine learning. Deep neural network (DNN) [2, 3] is used in a variety of fields where artificial intelligence is very beneficial. DL and the deep convolution neural network (DCNN) [4] have significantly increased the efficiency furthermore than the state of the art in the fields mentioned above which relates to traditional machine learning (ML) methods, such as supporting vector machine (SVM) [5] and Naïve Bayes [6]. The fundamental concept of DL comes from artificial neural network (ANN) [7]. This part introduces several of those instances where DNNs are currently having a significant effect: human activity recognition [8, 9], facial expression recognition [10], image classification [11, 12], feature modeling [13], object detection [14], etc. There have been big contributions in the computer vision field by the experts which have helped shape this domain as the world have it today. Several reasons are included in the literature that why a deep framework has benefits for the portrayal of attributes, including the statement that by recombining weights along various paths around the channel, attributes may be replicated exponentially [15, 16]. For this review paper, various models and other research papers have been researched and drawn important conclusions from them.

Aditi (✉)
Department of Computer Science and Engineering, Faculty of Engineering and Technology, PDM University, Bahadurgarh, India
e-mail: aditi958205@gmail.com

A. Dureja
Department of Computer Science and Engineering, Faculty of Engineering and Technology, PDM University, Bahadurgarh, India
e-mail: amandureja@gmail.com

© The Author(s), under exclusive license to Springer Nature Singapore Pte Ltd. 2021 69
X.-Z. Gao et al. (eds.), *Applications of Artificial Intelligence in Engineering*, Algorithms for Intelligent Systems, https://doi.org/10.1007/978-981-33-4604-8_6

Each of the models that have been discussed in this paper has some or the other revolutionized computer vision. All the research discussed in this paper has created an impact by some or the other way by convolutional neural networks. Convolutionary neural network (CNN) first was demonstrated in 1998 by Fukushima [17], which have extensive applications in behavior identification, sentence classification [18], text recognition [19], face recognition [20], object detection and localization [21, 22], image character development [23], etc. The term "convolutional neural network" implies that the device requires a logical linear method which is named as convolution rather than a standard matrix multiplication in at least one of its layers. CNN includes at least one convolutionary layer, preceded by at least one completely linked layer, as in the typical multilayer neural network. The term "convolutionary neural network" suggests that the system uses a linear mathematical method called convolution, instead of multiplying the general matrix in at least one of its layers. A CNN included at least one convolution layer and then followed in a regular multilayer neural network by at least one fully connected layer. They are one of the most important techniques used in image recognition, machine vision systems, and in data science where they are demonstrably impressive. One of the most prevalent strategies used in the paradigm of image recognition and computer vision is factually remarkable in the arena of data science. The remainder of this paper is organized as follows: In Sect. 2, a detailed review summary of image classification and object detection models with their configuration and implementation results are discussed with its architecture. A detailed analysis and important comparison are made in Sect. 3. And the last section describes the conclusion of the paper.

2 Various Successful Models in Computer Vision

In this section, a detailed discussion and results have been drawn for the various models that have shaped today's computer vision domain.

2.1 Deep CNN

The following model and its architecture were published in "ImageNet Classification with Deep convolutional neural networks" [24]. For this, the dataset includes LabelMe, that majorly contains hundreds of thousands of separate images, and ImageNet, which contains high-resolution images in over 22,000 categories. To do such heavy learning, a model with great learning capabilities is needed. The convolutional networks (CNNs) constitute a single phase of the model. Its volume can be controlled by its wide range and width, and it also makes a very strong and accurate image views. Apart from CNN's attractive features, it is efficient in its spatial design, but they are still very expensive to apply in high-resolution images.

Table 1 Performance comparison on test dataset of ILSVRC-2010

Model name	Top-1 error (%)	Top-5 error (%)
Sparse coding	47.1	28.2
SIFT + FVs	45.7	25.7
CNN	37.5	17.0

The architecture of the network consists of eight depths of layers, five of which are convolutional and three are fully connected. It turns out that about one million teaching instances are more than required to fit on a single GPU. So, the network is spread to two GPUs. The simulation system used puts a portion of neurons in both the GPU's, in a specific bundle: GPUs share knowledge only in certain parts. The training of models was done by using batch size of 128 instances, a momentum of 0.9, and a weight decline of 0. 0005. The same training rate was applied across all the layers, which was adjusted from time to time. The heuristic followed was ten when the validation rate of the suspension stopped improving. The reading level was started at 0.01, and it was lowered three times before ending. Table 1 draws the comparison between three different models, namely Sparse coding, SIFT, and the proposed model using ILSVRC-2010. The proposed model achieved the least error percentage of 37.5 in top 1 and 17 in top 5.

The model was entered at the ILSVRC-2012 competitions and CNN formulated reaches an error rate of 18.2% in Top-5. The results indicate that a huge, deep neural network with limited access can reach coding results in highly challenging data using supervised learning.

2.2 Maxout Network

This section describes a new simple model called maxout designed to optimize dropout performance and improve the accuracy of the fastest model estimation [25]. We consider the question of constructing models to exploit the newly implemented estimated model-averaging technique named dropout. Maxout networks have a performance of maximum number of inputs even though its dropout is a natural companion built both for promoting dropout optimization which increases precision for the quickly approximate model-averaging technique [26]. Dropout provides an inexpensive and easy way to train both types of models that share parameters and almost combine the predictions of these models. Dropout is commonly viewed as an indiscriminate operating tool that provides greater performance for improvement when it is used for most of the models [27]. Here is a simple model called maxout with the beneficial features of both optimization and stop-motion modeling technique [28]. The model has a dynamic architecture, like a deep neural network or multilayer perceptron, using a new type of startup function: Maxout leaves many elements for traditional rendering functions [29, 30] as the model was tested against a variety of

Table 2 Misclassification rates on the MNIST data

Model	Test error (%)
2-layer CNN + 2-layer NN	0.53
Stochastic pooling [28]	0.47
Conv. maxout + dropout	0.45

Table 3 Misclassification rates on CIFAR dataset

Model name	Test error (%)
Stochastic pooling [28]	15.13
CNN + spearmint	14.98
Conv. maxout + dropout	11.68

data. The first data used was MNIST. MNIST contains 28 × 28-pixel black-and-white images for digitized digits 0–9, with training examples reaching 60,000 and test examples reaching 10,000. For the work, the model was trained with two layers that are connected via a maxout and softmax layer. The result of maxout compared with other popular models. Table 2 draws the comparison between the proposed and two different state of the approaches [31, 32]. The misclassification rate on MNIST dataset is the least on the proposed approach.

The second data used was the CIFAR-10. The CIFAR-10 database contains 32 × 32 colorful images drawn in 10 classes separated by 50,000 learning and 10,000 testing images. This data was improved by ZCA's standard and white comparisons. The result of maxout compared with other popular models is Table 3 draws the comparison between the proposed and two different state of the approaches. The misclassification rate on CIFAR dataset is the least on the proposed approach.

The main reason maxout works so well is because of a dropout style training class. To behave differently in SGD by de-signing, it will require a gradient to change visual such as choosing which units are used to minimize the change. Strong evidence is shown above that the dropout achieves a good fit for high-resolution models in deep models.

2.3   Network in Network

In this section, a deep network structure called "network on network" (NIN) is introduced to improve representation discrimination of localized overlap in receiving medium [33]. Small neural networks having very complex structures were created to extract information inside the receiving field [34, 35]. The emphasis is on a small neural network with multilayer perceptron, feature approximate functionality. Feature maps are obtained by sliding small networks on the same input as CNN and then given in the following section. By filtering through a sequence followed by offline operations [36], sliding a smaller network over the input has been proposed

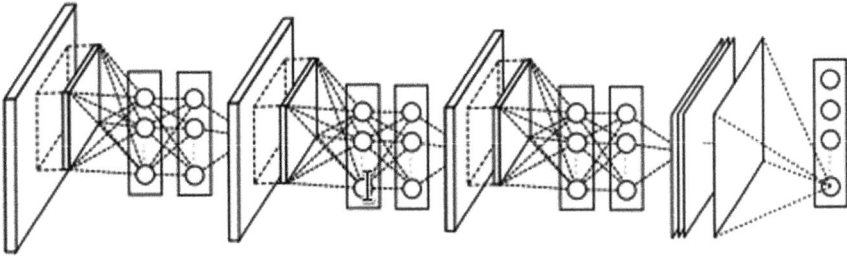

Fig. 1 Complete architecture of network in network. NINs include the overlapping of three mlpconv layers and one global average pooling layer

Table 4 Error rates of test data of MNIST dataset

Model name	Test error (%)
2-layer CNN + 2-layer NN	0.53
NIN + dropout	0.47
Conv. maxout + dropout	0.45

in the previous work. Network in network is proposed for a standard view, a small network integrated with the CNN structure for best use at all signal levels. Figure 1 visualizes the architecture of the proposed model that is network in network with details about each layer of the model.

A multilayer perceptron is advised for this task because it is compatible with the formation of validated networks and it can be a deep learning model, coupled with the motive to reuse different features. This new type of layer is called mlpconv. MNIST is a simpler data compared to CIFAR-10 because only a few parameters are required, and hence, it is used first to test the model. Table 4 compares the result of the proposed model to previous works that have used similar schemes.

The case of CIFAR-100 hyperparameters was not set, but the same setup for CIFAR-10 data was used [37, 38]. The only difference is that the last mlpconv produces 100 map results. The 35.68% diagnostic error of the CIFAR-100 exceeds the optimum current results without increasing data by more than a percent. So, the proposed deep network of novels called "network in network" (NIN) has a structure which contains mlpconv layers that use perceptrons in multilayer to convince the input and the global standard configuration layer.

2.4 OverFeat

An attached architecture using convolutional networks for classification, localization, and adoption was proposed [39]. It is shown that a multiscale and sliding windows approach can work well in ConvNet, which is a new method of deep learning in local practice by learning to predict object boundaries [40]. The main point is to

Table 5 Classification test on validation data

Approach	Top 1 error (%)	Top 5 error (%)
Krizhevsky et al.	40.7	18.2
OverFeat—1 fast model, scale 1, coarse stride	39.28	17.12
OverFeat—1 fast model, scale 1, fine stride	39.01	17.12
OverFeat—1 model, 4 corners + center + flip	35.60	14.71
OverFeat—7 fast models, 4 scales, fine stride	35.10	13.86
OverFeat—7 accurate models, 4 scales, fine stride	33.96	13.24

show that training the image acquisition network, finding and detecting images can increase the classification accuracy as well as gain local accuracy across functions [41]. Some authors have proposed doing part-time localization based on ConvNet. The simplest method involves training ConvNet to separate the center pixel from its viewing window. But where regions are to be divided into sections, it is best to make certain fractions [42, 43]. The thought is to train ConvNet to separate the center pixel of the view window by categorizing the wrong item, using the window as the context for the decision [44, 45]. Local names and their availability are used in such a manner that is consistent with use in the ImageNet 2013 competition [46]. Table 5 compares different variations of the proposed model with itself and other standard models and provides error percentage during classification task.

The improvements are proceeded on network building and signaling step. From the classification network, the classifier is replaced by the compression network and trained to predict the specific binding boxes for each level and dimension [47, 48]. Then the prediction of clustering is combined with the results of segmentation at each location. Multiple image fields can be trained at the same time. As the model is a modification of convolutional, the weights are shared in every location [49, 50]. Negative training is performed on the fly. It is reported that the results of the ILSVRC 2013 competition ranked 3rd with 19.4% mean median accuracy (mAP).

2.5 R-CNN

A simple and intelligent acquisition algorithm is introduced that upgrades mean average accuracy (mAP) by over 30% in relation to the excellent result of VOC 2012 [51]. The proposition unites two crucial insights:

- one can use high-capacity neural networks of higher order in lower frequency circuits to generate household items.
- when labeled data are labeled as scarce, monitored prior to training for auxiliary work, followed by swearing-in domain, increasing the intensity of the operation.

As circuit suggestions are combined with CNNs, the approach is called R-CNN: Regions with CNN system. This unique system:

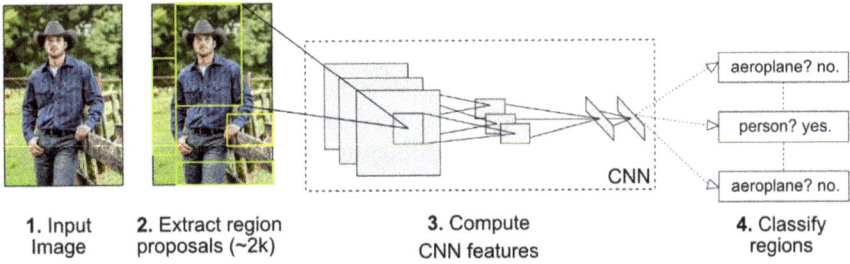

1. Input Image   **2.** Extract region proposals (~2k)   **3.** Compute CNN features   **4.** Classify regions

Fig. 2 1 Receives an input image, 2 draw out close to two thousand bottom-up region propositions, 3 enumerate features of every proposition using a "CNN," and then 4 categories every region using SVMs [6]

- Takes the input image,
- Outputs around the low-frequency circuit suggestions in 2000,
- Every proposal shows different features by the usage of large neural network and
- Divides every region using SVMs of a specific category.

Figure 2 describes the different regions of the proposed method with CNN features and their working.

The proposed model achieves a mAP of "53.7% in PASCAL VOC 2010." In contrast, it passes on a 35.1% mAP using the identical regional proposals but accompanied by a local pyramid. It solves "CNN localization problem" which works inside "recognition using district" standard, that achieved the discovery and the semantic component [52]. During the experiment, the method generated 2000 propositions for the individual components of the input image, extracting feature furniture of the exact length from each with the help of CNN, and then subdividing every region into a specific section of the SVMs. Table 6 compares the average precision of various versions of the model attained using VOC 2007 dataset.

The model consists of three modules. The first presents the proposals of the independent regions of the category. These propositions describe the set of nominations for the recipient [53]. The second section is a large-scale visual network that releases the feature vectors of every region. Lastly, the third module is a collection of special SVMs in the class [54]. Without a 4096 feature-builder from each application using Caffe, the CNNR-CNN implementation achieves a mAP of 31.4%, far more than the second-best result of 24.3% from on OverFeat.

Table 6 Average precision test on VOC 2007 dataset

Model name	mAP (%)
R-CNN T-Net	54.2
R-CNN T-Net BB	58.5
R-CNN O-Net	62.2
R-CNN O-Net BB	66

2.6 Deeper ConvNets

This section investigates the effect of the depth of the network for its precision on the huge image identification arrangement [55]. The major benefaction is a comprehensive examination of increasing depth of networks using a structure with very compact (3 × 3) filters, indicating that noteworthy improvements in preprocessing that can be accomplished by compressing abyss to 16–19 layers. To measure the improvements brought by ConvNet's depth to optimal depth, all the ConvNet layer settings were arranged using the same concept [56]. The composite layer stack (with different depths for different structures) is accompanied through" three fully connected (FC) layers": the first two in the beginning contains 4096 channels each, the third assemble a 1000 ILSVRC split—which accommodates 1000 channels (per class). Softmax is the last layer. Optimization of completely integrated levels is identical across networks. Such ConvNet configuration is quite disparate from that used by the top-level installation of the ILSVRC-2012. Instead of using large reception fields in the first concert layers, the smallest 3 × 3 acceptor fields were used in the entire net, which are available for insertion in all pixels. For immediate reasons, large-scale models were trained by refining all levels of a single-dimensional model in the identical order, before training with $S = 384$. The training of multi-GPU uses similarity of data and is done through dividing every set of "training images into various GPU pieces," working in parallel GPU per unit. In the ILSVRC-2014 challenge classification task, the "VGG" team secured 2nd place with a "7.3% test error" by a combination of 7 different models. After installation, error rate was reduced to 6.8% using an ensemble of 2 models. Convents' is much deeper than previous generation models, with excellent result in ILSVRC-2012 and ILSVRC-2013 design competitions. Better result (7.0% test error), using one GoogLeNet at 0.9%. The models performed well on a variety of tasks and data sources, by comparing or using complex recognition pipelines through built-in deep image representations.

2.7 Deep ConvNets with Hashing

The idea in this section is that binary codes can be read by availing a subtle layer to introduce the concepts used to store class labels [50]. CNN can also be availed to read photo presentations. Some supervised techniques require binary input of binary code reading. However, this technique can be used to read hash codes and image representations in a point-based manner and is therefore suited to high-quality data [57]. The detail of highest level is about pixel-level data from human-level insights. Although several aspects of the manual have been suggested to represent the images, the performance of these visual descriptions is still mediocre [58]. Deployment of deep structures also brings practical learning to feature retrieval. To compare with conventional methods, this approach is ideal for big data. Road features are as follows:

- An uncomplicated yet fruitful learning framework that introduces rapid image presentation.
- The deep CNN learned concurrently image presentations and hashing functions for quickly retrieving images with minimal tempering to the model.

The proposed method is the best of the most sophisticated functions in the MNIST and CIFAR-10 database. The model boosts the leading performance of CIFAR-10 data with 30% accuracy and MNIST data with 1% accuracy. Compared to the conventional two-way twins, the method reads binary hashing codes in a quantitative way and is effortlessly identified by data size. This approach incorporates three key elements:

- The first element is pre-monitored training in the enormous ImageNet dataset.
- The second element prepares the network with a one-time learning layer for a feature representation of a domain and a set of hash-like functions.
- The last element returns the same images and one query with a separate search.

Separation is done by a suggestion to image representations and similar hash functions at the same time. The proposed network is adjusted to the domain's structured data by moving backwards to monitor domain synchronization. The model was tested against some of the popular datasets like:

- MNIST Dataset—compiled with handwritten images of numbers 0–9. It contains 60,000 training and 10,000 testing images.
- CIFAR-10 Dataset—compiled of 10 item categories with each category containing 6,000 images. The data is divided into fifty thousand training and ten thousand testing images.
- NUS-WIDE Dataset—combination of twenty-seven thousand betting in total and eighty-one categories. Data is collected by capturing images from Web sites.

A comprehensive CNN feature layer is added for reading special image presentations and a collection of hash-like functions which does not depend on duplicate data matching. For MNIST data, the system with 48 latent connections learns at 0.50% error rate and superior than the other methods. The method is more efficient (98.5% recapture accuracy) regardless of the count of images detected. Also, the method improves the precision from 97.5% obtained by CNNH+ [31] to 98.5%, which studies hashing functions by decoding the details of pairs. Then the network model is adjusted to the CIFAR-10 data and finally obtain an average accuracy of 89.5% after 50,000 repetitions. Finally, the testing is proceeded in big data for NUS-WIDE to show the increase and efficiency of the method. After 150,000 double training, the model provides 83.75% accuracy. Table 7 lists various models with different approaches and compares it with our proposed model. The data used for measuring accuracy is MNIST and the model achieved leave test error percentage.

Table 7 (Error %) model accuracy comparison for classification on MNIST dataset

Methods	Test error (%)
2-layer CNN + 2-layer NN	0.53
Stochastic pooling	0.47
NIN + dropout	0.47
Conv. maxout + dropout	0.45
Proposed method without 48 nodes latent layer	0.50

2.8 Inception

A deep neural network architecture is proposed in this section of secret name "inception" which achieved a futuristic result of differentiation [53]. The basic feature of this design is the enhanced use of resources. With a vigilant designed model, the network's depth is increased while maintaining a constant budget. Inception deployment to ILSVRC 2014 uses 12 less variables than the triumphant model structure of Krizhevsky et al., while being more precise and detailed. It is assumed that each portion from the preceding layer communicate to a region of input image, and these portions are organized into filter layers. In the bottommost layers, the corresponding units focus on the local regions. Therefore, it would end up with multiple combinations of groups embedded in one region, as suggested. To avoid packet alignment issues, current text encoding is restricted to size filters 1×1, 3×3 and 5×5; this decision was based on an easy compliment rather than necessary. A designation network comprising of portions of the above type mounted on top of each other, with infrequent layers of stride 2 to stop the grid cutout [37]. Figure 3 is a flowchart describing the working of naïve version of inception model. Figure 4 is a flowchart describing the working of Inception model using dimensionality reduction.

The arrangement follows the logical idea that perceptible knowledge should be processed at separate levels and then integrated so that the succeeding section can extract features at separate levels simultaneously.

The arrangement follows the logical idea that perceptible knowledge should be processed at separate levels and then integrated so that the succeeding section can extract features at separate levels simultaneously. The total number of layers used for the establishment of the model is approximately a hundred. It was established that moving from fully connected parts to the central network ameliorate the accuracy by 0.6%, although the dropout usage persisted significant even after abolishing the fully connected layers. The exceptional achievement of the deep networks in this work advocate that the features generated by the layers within the network should have a greater prejudice. For training, their mislaying is added to the aggregate weighted network. During submission, supplementary networks are discarded.

Szegedy [25] suggested a computer vision Creation concept. On single frame measurement, they get 21.2 top-1 and 5.6 top-5 mistake. Xia [26] propounded the flower classification model Inception-v3. The research is performed on the data collection for the Oxford-17 and the Oxford-102 flower. They get mAP 95 on

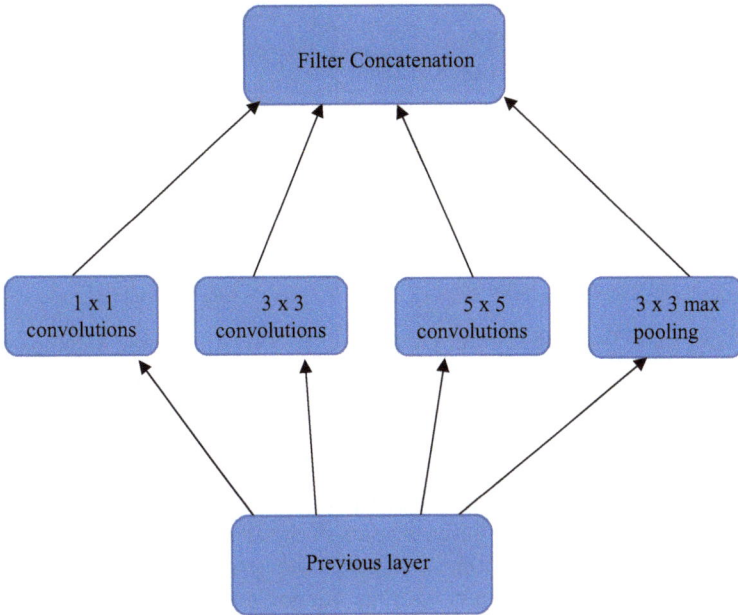

Fig. 3 Inception module, naive version

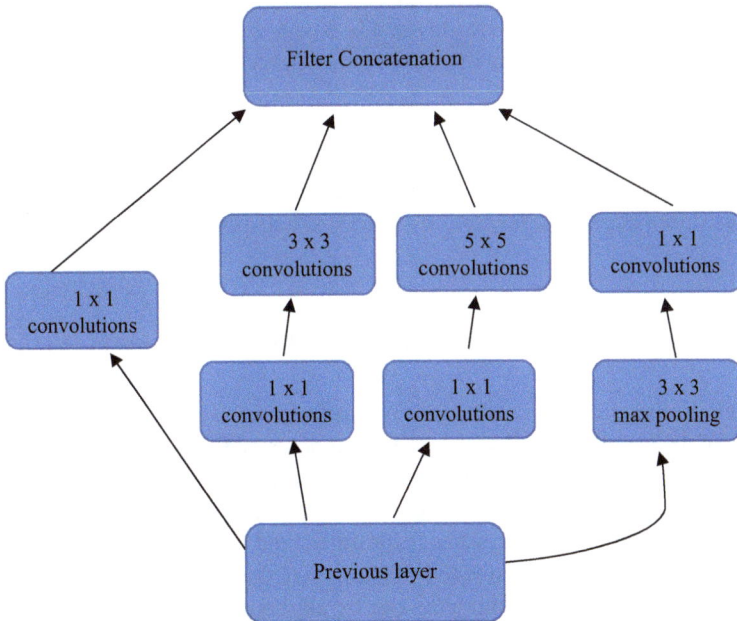

Fig. 4 Inception module with dimensionality reduction

Table 8 Performance of single model for detection

Team	mAP (%)
Trimps-Soushen	31.6
Berkeley vision	34.5
UvA-Euvision	35.4
CUHK DeepID-Net2	37.7
GoogLeNet	**38.02**
Deep insight	40.2

OXFORD-17 and 94 on the flower database OXFORD-102. Table 8 compares the proposed model with a variety of other standard models. It is concluded that the proposed model attains a state-of-the-art result.

The ILSVRC 2014 classification challenge used for testing the model involves the work of dividing the image into one of the 1000 nodes in the ImageNet list. There are approximately one million self-portraits, fifty thousand validation and hundred thousand test images. Seven models of the same GoogLeNet model were independently trained (including one comprehensive version) and worked to test with them. In the testing phase, a more vigorous framing proposal than the Krizhevsky et al. was used [24]. Specifically, there was enlargement of the image into 4 scales in which the minimum dimensions (height or width) are 256, 288, 320, and 352 respectively, and corners are taken with 224×224 center and 224×224 squares, and versions their entries. This results in $4 \times 3 \times 6 \times 2 = 144$ cultivars per image. The opportunities for softmax are found over most image crops and above all students who obtain the final test scores.

The above shown results provide strong evidence that achieving the right size structure for the available blocks is an effective way to optimize the computer's neural networks. In both differentiation and adoption, it is expected that the same results are more accessible for less cost-effective networks with the same width.

2.9 Residual Learning in Deep Networks

This section discusses a superfluous learning framework introduced to reduce the learning of deep networks [9]. A direct descent down is made with two-dimensional linear layers. The network concludes with a mid-layer waterproof layer with a 1000-layer fully integrated softmax. The aggregate number of heavy layers is 34. It is important to note that the model has fewer filters and lower complexity than VGG nets. A shortcut connection is added that converts the network into its equivalent remnant version. Ownership shortcuts can be used straight when input and output are of the exact size. While size increases there are two methods to look at:

- Shortcut continues to map ownership, with zero entries attached to size. This method did not introduce additional parameter

- The shortening is used to compute the magnitude (made by 1×1 convolutions).

In both options, while the shortcuts cross the two-dimensional feature maps, they are made with a 2-string. The image is reproduced on its short side and sampled in (256, 480) to enlarge the scale. The 224×224 crop is projected horizontally from the image or its horizontal flap, with each pixel mean extracted. Batch normalization (BN) is used shortly after each sentence and before it is activated. Weights were introduced as is and trained all exposed/remaining nets from scratch. SGD is used with a mini-batch size of 256. The models are trained up to 60×104 iterations. A weight loss of 0.0001 and a pressure of 0.9 was used. The approach was tested in the classification data of ImageNet 2012 with 1000 classes. The models were trained on one million training input images and were analyzed on fifty thousand validation images and got the result in hundred thousand test images. Examining the 18 main nets and 34 pillars is the beginning step. A net that is 34 ft wide is available. The 34-wide net has a high training error throughout the training process, or the 18-channel network solution space is the basis for that 34-layer one ensures forward propagation signal for a nonzero variance. It is also confirmed that the back-propagation gradients reflect healthy trends with BN. So, either the front or back sign does not disappear. In fact, a 34-foot plain network can still achieve competitive accuracy indicating that the solver is operating to some extent. The base structure is like the above-mentioned nets, except that the connectivity network is added to each of the 3×3 filters. There are three major observations:

- The thirty-four layer ResNet is superior than the eighteen-layer ResNet (by 2.8%).
- The thirty-four layer ResNet diminishes the top-1 error by 3.5% hence successfully reduced training error
- The eighteen-layer residual nets are comparably accurate but the eighteen-layer ResNet converges faster

On the COCO dataset, the proposed model obtains a 6.0% increase in standard metric ((mAP) [5.95]), which was a 28% relative improvement.

2.10 Inception-V4, Inception-ResNet

Most deep convolutional networks have been principal to tremendous improvements in image recognition [10]. An instance is the inception architecture which can be used to achieve the best results at relatively low computational cost [50]. There is some evidence that residual inception networks without a thin edge are more expensive than similarly expensive inception networks. The initial configuration of the inception model is very consistent, and hence, a lot of changes in the filters can be made through which one can affect the quality of a fully trained network. To improve the training speed, we carefully modify the layer size to balance the computation between different model sub-networks. In contrast, with the inclusion of TensorFlow, models can be practiced without duplication. This is partly enabled by the fresh developments

Table 9 Experimental outcome on single model using ILSVRC 2012 dataset

Network	Crops	Top-1 error (%)	Top-5 error (%)
ResNet-151	Dense	19.4	4.5
Inception-v3	144	18.9	4.3
Inception-ResNet-v1	144	18.8	4.3
Inception-v4	144	17.7	3.8
Inception-ResNet-v2	144	17.8	3.7

in the memory used by back-propagation, with careful consideration of what tensors are required for gradient calculation, and by configuring the calculation to turn down the number of such tensors. We decided to shed this unnecessary luggage for the Inception-V4 and make uniform choices for the inception modules for each phase. For other versions, we use low-cost inception modules than the original inception. Each inception block is accompanied by a filter-expansion layer, which is used to measure the dimension of the filter bank in addition to matching the depth of the input. The small technical dissimilarity within our variants is that in the case of inception response, we used volume normalization only on top of traditional layers but not on top of contractions. It is appropriate to look forward to the full use of volume normalization to be dominant, but we wanted to keep each sample replica in the same GPU. The memory footprint of layers with a large operational size turned out to consume an excessive amount of GPU-memory. Table 9 compares different versions of the proposed model with other high-end expensive models. A mix of different approaches results in a model with better precision and less error.

By avoiding volume normalization on top of those layers, the overall number of initial blocks was significantly increased. We believe that by making better use of computing resources, disabling this business will become unnecessary. When the number of filters exceeds 1000, the remaining variants begin to become unstable, and at the beginning of the exercise the network is "dead"; that is, the last layer before the average pool starts to form zeros after many tens of iterations. Measuring the residuals prior to adding them to the previous layer process appear to confirm the exercise. If the number of filters is so high, then even the lowest (0.00001) learning rate is insufficient to deal with instability, and training with a high learning rate has a chance to ruin its effects. We found it very reliable to measure the rest. Introducing the rest of the connections leads to a dramatically improved training speed for the inception models.

Figure 5 is a flowchart describing the working and schema of the Inception v-4 model.

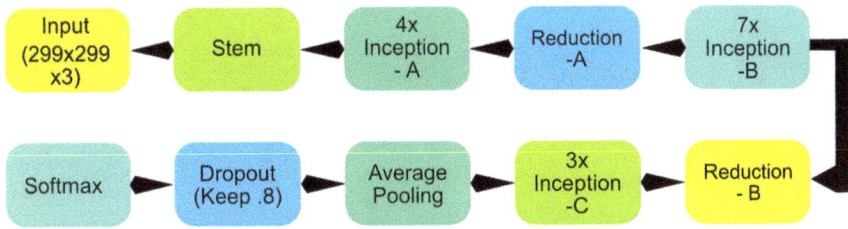

Fig. 5 Overall schema of the Inception-v4 network

2.11 Deep Learning in Medical Field

Image detection and recognition deal with the problem of detecting a certain element in a medical image [59, 60]. In many cases, the images are volumetric [61]. In this section, we shed some light on how deep learning is used in medical field continues from theory substructure for Applications [62]. We discuss the common grounds for the recognition of deep learning, in addition to many important advances in machine leaning followed by the fundamentals of perceptron networks [63]. Medical image processing among the one which is affected by rapid improvement in deep learning, especially image segmentation, recording, and computer-assisted detection [64].

Nonetheless, various approaches ignore preceding information and therefore carry the risk of creating unreliable results. Deep learning is well-suited for medical imaging. The new activation functions are bulging and covers large area of "nonzero derivatives" [65]. Many derivatives are required to calculate the slope of the deeper layers. The more the depth of the net, the more multiples needed. If multiple elements with this sequence are less than 1, the whole gradient deteriorate aggressively with the numeral layers [55]. Therefore, saturated spin-off is crucial for solving several problems, such that the diminishing gradients do not authorize for the training of deeper networks than three layers. Note that 0 is the classification boundary, so that each neuron does not lose its description as a classifier. Moreover, the "global approximation theorem still holds for a single-layer network with ReLUs" is a disadvantage, the ReLU does not differ across the complete domain of x. An unauthorized kink determining a peculiar slope at $x = 0$ is found. For optimization, the slope of a function is an important property that is oriented toward the direction of the steep climb. Therefore, adopting a negative direction will allow the activity to slow down [66]. This authorizes us to use "gradient fall algorithms" to improve these issues, if a sub-slope can be determined, that is, at the minimum, one occurrence of the direction of the optimal direction. For ReLU, all value among 0 and -1 will be accepted at $x = 0$ for the descending function. Figure 6 demonstrates the working of a traditional pattern recognizer. The flowchart flawlessly illustrators how the training phase combings with the testing phase of a model.

To be able to obtain such a direction, it is ensured that the optimization of the convergence to the convex problems using optimization programs. This allows us to be the backbone of optimization when using indistinguishable activation functions.

Test Phase

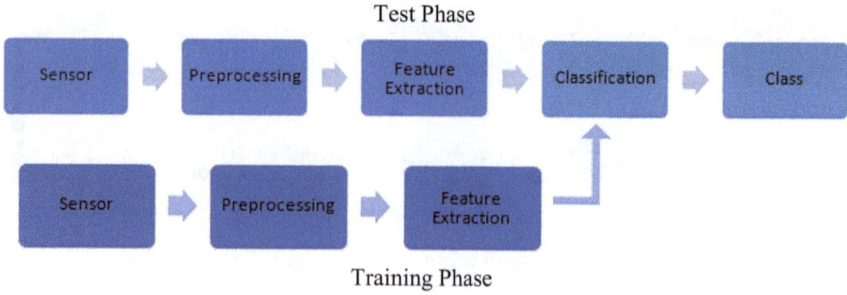

Training Phase

Fig. 6 Illustration of the traditional pattern recognition. Sensor data is finetuned, and "handcrafted" attributes are drawn out

If the loss rises rapidly, a common complication arises that the step size is adjusted too elevated. Setting the training rate too low can result in the loss of repetitions. In this case, we notice the slopes that disappear again. Another frequent miscalculation is done in the case of bias in training or test data. In principle, data used for testing should be viewed only when the structure, parameters and all other factors of determination are adjusted; otherwise, recurrent testing will lead to the system being overestimated. That is prohibitive as the inclusion of evaluation data in the training package. However, several other variables may affect the outcome of the classification. In this short instigation to deep learning, the aim was at two objectives at the same time. Firstly, to establish a field of deep learning and secondly, to provide a rundown on the sector and potential future applications.

2.12 Deconvolutional Network

Convolutional network models have lately shown good phase result in the ImageNet benchmark [12]. There is, however, no clear apprehension of why they work so well. This section describes a novel approach that provides understanding into the function of the layers of the convolutional networks and the functionality of the classifier. When using the diagnostic role, this visualization allows us to obtain high-quality architectural drawings of models that surpassed Krizhevsky et al. on the ImageNet data. Following this is an intensive study to determine the contribution of performance from different components of the models. Convolutional networks showed magnificent outcome in endeavor such as handwritten digit separations and face detection and other complex types, but we have no idea about the insights. Hence, a new visualization process is introduced that creates input that generate feature maps for each element in any model layer. The optical path suggests using the multilayered deconvolutional network (deconvnet). We use fully monitored standard models for the whole process. These map types show the color2D image input with x_i, in a series of layers, in probes that have different C levels. Each layer contains:

- Modification to extract the preceding layer by a collection of learned filters.
- Transferring responses with corrected line function [letter $(x) = \max(x, 0)$];
- [Optional] Max hangs over the neighborhood.
- [Optional] Local comparison functionality that makes the answers understandable on the maps of the whole feature. The cross-loss function, which is suitable for image segmentation, is used to compare $\hat{y}i$ and yi.

To test the functionality, each function of a layer is set to zero and feature maps are passed as input. After this, we do the following operations:

- Unpooling: In convnet, the performance of the max pooling does not affect; however, we can find exceptions by noting the locations of the maxima within each pool region. In deconvnet, the unrestricted operation uses these objects to rotate the rebuild from the top layer into the appropriate locations, preserving the stage architecture.
- Rectify: convnet uses line incompatibility, which corrects feature maps by ensuring that feature maps remain reliable. To obtain a valid feature reconstruction for each layer (which is also appropriate), we pass the reconstructed wrong signal to the next line.
- Filter: convnet uses the read-through filters to convince feature maps in the previous layer. Almost to this point, deconvnet uses modified versions of the same filters (as in other autoencoder models, such as RBMs), but it is used for optimized maps, not for output.

2.13 Architecture of Deconvnet

The first filtering is a mixture of very high and low frequency data, a small overlay of intermediate frequencies. In addition, the observation of layer 2 shows the computational complexity originated by the main line 4 operated in the 1st layer of convolutions. This new structure stores more information on the features of layers 1 and 2. Most importantly it improves the performance of the phases. As the model was trained to discriminate, they clearly showed all the parts that were discriminated for the input image.

The model was trained in the ImageNet 2012 training set. Each RGB image was enhanced by subtracting the smaller size to 256, positioning the center 256×256 in the region, subtracting each pixel size (in all images) and using less than 10 plants of 224×224 (corners + four center (out) horizontal blocks). From the stochastic expertise domain with a mini-batch size of 128 for reusable frameworks, it starts with a reading rate of 10–2, in conjunction with a quick term of 0.9. The observation of the first layer filters all over the training disclose that few dominates. To counteract this, we rearrange individual filter into the permissible layers of the RMS value exceeding the default 10–1 in the radius. Projections from each layer indicate the state of the optical properties of the network. Layer 2 acknowledges to corners and other combinations/colors. Layer 3 has a more compounded entry and arresting the

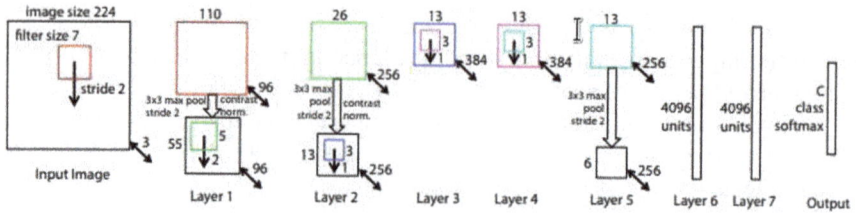

Fig. 7 Architecture of the proposed convnet model

Table 10 Error with many architectural differences on ImageNet 2012 dataset	Model	Top-1 error (%)	Top-5 error (%)
	Replica of Krizhevsky et al.	40.5	18.1
	Removed layer 7	40.0	18.4
	Removed layers 6, 7	44.8	22.4
	Our model	38.4	16.5
	Adjust layers 3, 4, 5	**37.5**	**16.0**
	Adjust layers 6, 7	38.3	16.9

same construction. Figure 7 illustrates the architecture of the proposed model. Each layer of the model is shown and explained separately.

Table 10 compares a wide variety of proposed model alterations and standard pre-built models. The proposed model achieved a state-of-the-art result when used on ImageNet data.

All of this introduced a novel way of visualizing the work done in a model. This disclose features that are far from arbitrary, less useful to see. A view that can be used to detect complications with a model and therefore get superior outcomes.

3   Analysis

It can be analyzed by the different models that are discussed in this paper like the deep CNN model which is enormous and can attain a new level of precision on complex and highly challenging datasets. CNN advances may be classified in several forms, including activation, lack of function, enhancement, regularization, learning algorithms, and design developments. CNN's processing ability has been greatly increased over the years by leveraging scope as well as other systemic improvements. Recent literature shows that the main boost in CNN performance has been achieved by replacing the conventional layer structure with blocks. As unsupervised pre-training was not used during this experiment, but it is expected that it will increase in much more precision and the model is very efficient for video sequencing.

In the next network, a very fresh kind of activation function named "maxout" is introduced which is specifically good for training alongside dropout. It was also proved that optimization reacts in a different way in case of dropout than in "pure SGD" (speech generating device). The approach performs well with inexpensive models for better precision.

There is another model which is a state-of-the-art advanced network called "network in network" which is introduced in this paper. The architecture contains mlpconv layers that model the local patches with greater precision and a global average pool layer which prevents overfitting. NIN works well in the field of object detection.

Overfeat is a new approach toward classification which includes multi-scaling and sliding windows. The approach involves maintainable changes to the network specifically designed for classification. Also, the approach can be further improved by using localization, $l2$ loss and alternate parameterizations. The approach exceeded expectations in various visual object detection.

In the R-CNN model, the performance of detection and classification came to absolute standstill before this paper. The model proposed here gave a significant boost in these operations by applying a high-capacity network to regions in depth and by introducing a paradigm for training when data is scarce. The model can be tuned to give accurate results on scarce data.

Deeper convNets performed an evaluation over a network for big scale image classification for up to nineteen weight layers. As it has shown the depth of a network is crucial for its precision and a model can improve significantly just after carefully increasing depth. The model generalizes well with a variety of image data to give consistent results.

Furthermore, there are other models like deep convNets with hashing for more robust and instant mobile visual search. We use this method of hashing. Based on hashing methods, we can showcase the remarkable power of automatic learning with compact binary code representation for binary visual search. The method of using such hashing improves models by as much as 30%.

As inception is an advanced model which was generated, and it is proved that averaging the predictable favorable architecture so far provided dense building units which is a feasible technique for enhancing computers' level of recognizing images. The model achieves top results for both classification and detection tasks.

Residual learning in deep networks is a state-of-the-art residual architecture of machine learning to speed up the time taking training session of the model by learning from reference of other layers and enhancing the depth of the model. The model works good for objects detection on both PASCAL and CIFAR data.

In inception-v4 and Inception-ResNet, there are total of 3 advanced inception models which were formulated with one being a hybrid Inception, one being an expensive hybrid but with much better outcome and last but none the least an inception model without the residual connections. The three new architecture provides similar accuracy in object detection even after a big difference in cost.

Deep learning in medical field, which was targeted toward beginners, a very short but knowledgeable introduction was provided in the field of medical image classification and some light was shed at the possible future applications of the same. The studies provided gave a clear vision about deep learning and it is working.

Deconvolutional network is a very enormous sized model which has a very precise path to visualize the activities of the model. It has been also adapted that the outcome may change and improve for the better if a different loss function would have been used. The model works well for object detection on datasets like ImageNet and Caltech.

After all this discussion, it can be analyzed that all the different models which are discussed above have their own pros and cons in different fields and at a certain level. Despite fast advances and encouraging results in the identification of artifacts, several problems remain available for potential research. Finally, it suggests a range of interesting possible paths to achieve a detailed understanding of the object detection environment.

The aim of this analysis was to explore and summarize the new techniques and to understand which model is best suited for what kind of research.

4   Conclusion

After carefully examining many models working in the field on computer vision and deep learning, the following decision can be made:

- A deep convolutional neural network can accomplish record shattering outcomes on a highly burdensome dataset using exclusively supervised learning.
- Dropout accomplishes a satisfactory estimate for averaging in deep models and optimization acts very contrastingly in the case of dropout.
- Use of an amalgamation of legacy tools from deep learning can guide to noteworthy achievements.
- Hash-like binary codes with deep learning framework can work for fast image retrieval.

References

1. LeCun Y, Bengio Y, Hinton G (2015) Deep learning. Nature 521(7553):436
2. Sze V, Chen Y-H, Yang T-J, Emer JS (2017) Efficient processing of deep neural networks: a tutorial and survey. Proc IEEE 105:2295–2329
3. Dahl GE, Yu D, Deng L, Acero A (2012) Context-dependent pre-trained deep neural networks for large-vocabulary speech recognition. IEEE Trans Audio Speech Lang Process 20:30–42
4. Silver D, Huang A, Maddison CJ, Guez A, Sifre L, Van Den Driessche G et al (2016) Mastering the game of go with deep neural networks and tree search. Nature 529:484
5. Wang XJ, Zhao LL, Wang S (2012) A novel SVM video object extraction technology. In: 2012 8th international conference on natural computation. IEEE, pp 44–48
6. Rish I (2001) An empirical study of the naive Bayes classifier. In: IJCAI 2001 workshop on empirical methods in artificial intelligence, vol 3, no 22, pp 41–46

7. Li J, Li J (2015) Fast image search with deep convolutional neural networks and efficient hashing codes. In: 2015 12th international conference on fuzzy systems and knowledge discovery (FSKD), Zhangjiajie, pp 1285–1290
8. Park SU, Park JH, Al-masni MA, Al-antari MA, Uddin Z, Kim T (2016) A depth camera-based human activity recognition via deep learning recurrent neural network for health and social care services. Proc Comput Sci 100:78–84
9. Baccouche M, Mamalet F, Wolf C, Garcia C, Baskurt A (2011) Sequential deep learning for human action recognition. In: International workshop on human behavior understanding. Springer, Berlin, Heidelberg pp 29–39
10. Zhao X, Shi X, Zhang S (2015) Facial expression recognition via deep learning. IETE Tech Rev 32(5):347–355
11. Xie S, Yang T, Wang X, Lin Y (2015) Hyper-class augmented and regularized deep learning for fine-grained image classification. In: Proceedings of the IEEE conference on computer vision and pattern recognition, pp 2645–2654
12. He K, Zhang X, Ren S, Sun J (2016) Deep residual learning for image recognition. In: CVPR, pp 770–778
13. Floyd MW, Turner JT, Aha DW (2017) Using deep learning to automate feature modeling in learning by observation: a preliminary study. In: 2017 AAAI spring symposium series
14. Tang C, Feng Y, Yang X, Zheng C, Zhou Y (2017) The object detection based on deep learning. In: 2017 4th international conference on information science and control engineering (ICISCE), pp 723–728
15. Bengio Y, Courville A, Vincent P (2013) Representation learning: a review and new perspectives. IEEE Trans Pattern Anal Mach Intell 35:1798–1828
16. Huang FJ, Boureau Y-L, LeCun Y, Huang Fu Jie, Boureau Y-Lan, LeCun Yann et al (2007) Unsupervised learning of invariant feature hierarchies with applications to object recognition. In: IEEE conference on computer vision and pattern recognition. CVPR'07. IEEE, pp 1–8
17. Fukushima K (1980) Neocognitron: a self-organizing neural network model for a mechanism of pattern recognition unaffected by shift in position. Biol Cybern 36:193–202
18. Kim Y (2011) Convolutional neural networks for sentence classification. arXiv:1408.5882
19. Zhou X, Gong W, Fu W, Du F (2017) Application of deep learning in object detection. In: 2017 IEEE/ACIS 16th international conference on computer and information science (ICIS). IEEE, pp 631–634
20. Ranjan R, Sankaranarayanan S, Bansal A, Bodla N, Chen J-C, Patel VM, Castillo CD, Chellappa R (2018) Deep learning for understanding faces: machines may be just as good, or better, than humans. IEEE Signal Process Mag 35(1):66–83
21. Milyaev S, Laptev I (2017) Towards reliable object detection in noisy images. Pattern Recognit Image Anal 27(4):713–722
22. Zhou X, Gong W, Fu W, Du F (2017) Application of deep learning in object detection, pp 631–634
23. Druzhkov PN, Kustikova VD (2016) A survey of deep learning methods and software tools for image classification and object detection. Pattern Recognit Image Anal 26(1):9–15
24. Krizhevsky A, Sutskever I, Hinton G (2012) Imagenet classification with deep convolutional neural networks, pp 1097–1105
25. Goodfellow I, Warde-Farley D, Mirza M, Courville A, Bengio Y (2013) Maxout networks. In: 30th international conference on machine learning, ICML, 1302
26. Bastien F, Lamblin P, Pascanu R, Bergstra J, Goodfellow I, Bergeron A, Bouchard N, Bengio Y (2012) Theano: new features and speed improvements. In: Deep learning and unsupervised feature learning NIPS 2012 workshop
27. Rifai S, Dauphin Y, Vincent P, Bengio Y, Muller X (2011) The manifold tangent classifier. In: NIPS'2011, student paper award
28. Zeiler MD, Fergus R (2013) Stochastic pooling for regularization of deep convolutional neural networks. In: International conference on learning representations
29. Glorot X, Biordes A, Bengio Y (2011) Deep sparse rectifier neural networks. In: JMLR W&CP: proceedings of the fourteenth international conference on artificial intelligence and statistics (AISTATS 2011), Apr 2011

30. Goodfellow IJ, Courville A, Bengio Y (2013) Joint training of deep Boltzmann machines for classification. In: International conference on learning representations: workshops track
31. Maier A, Syben C, Lasser T, Riess C (2019) A gentle introduction to deep learning in medical image processing. Zeitschrift für Medizinische Physik 29.https://doi.org/10.1016/j.zemedi.2018.12.003
32. Hong Z (2011) A preliminary study on artificial neural network. In: 2011 6th IEEE joint international information technology and artificial intelligence conference, vol 2, pp 336–338
33. Xu H, Han Z, Feng S, Zhou H, Fang Y (2018) Foreign object debris material recognition based on convolutional neural networks. EURASIP J Image Video Process 2018:21
34. Simonyan K, Zisserman A (2014) Very deep convolutional networks for large-scale image recognition. arXiv:1409.1556
35. Girshick R, Donahue J, Darrell T, Malik J (2013) Rich feature hierarchies for accurate object detection and semantic segmentation. Proceedings of the IEEE computer society conference on computer vision and pattern recognition. https://doi.org/10.1109/CVPR.2014.81
36. Yudistira N, Kurita T (2017) Gated spatio and temporal convolutional neural network for activity recognition: towards gated multimodal deep learning. EURASIP J Image Video Process 2017:85
37. Szegedy C, Liu W, Jia Y, Sermanet P, Reed S, Anguelov D et al (2015) Going deeper with convolutions. In: Proceedings of the IEEE conference on computer vision and pattern recognition, pp 1–9
38. Papakostas M, Giannakopoulos T, Makedon F, Karkaletsis V (2016) Short-term recognition of human activities using convolutional neural networks. In: 2016 12th international conference on signal-image technology and internet-based systems (SITIS). IEEE, pp 302–307
39. Sermanet P, Eigen D, Zhang X, Mathieu M, Fergus R, Lecun Y (2013) OverFeat: integrated recognition, localization and detection using convolutional networks. In: International conference on learning representations (ICLR) (Banff)
40. Chatfield K, Lempitsky V, Vedaldi A, Zisserman A (2011) The devil is in the details: an evaluation of recent feature encoding methods. In: BMVC
41. Sermanet P, Kavukcuoglu K, Chintala S, LeCun Y (2013) Pedestrian detection with unsupervised multistage feature learning. In: Proceedings of international conference on computer vision and pattern recognition (CVPR'13). IEEE, June 2013
42. Farabet C, Couprie C, Najman L, LeCun Y (2013) Learning hierarchical features for scene labeling. In: IEEE transactions on pattern analysis and machine intelligence (in Press)
43. He K, Zhang X, Ren S, Sun J (2016) Deep residual learning for image recognition. In: 2016 IEEE conference on computer vision and pattern recognition (CVPR), Las Vegas, NV, pp 770–778
44. Szegedy C, Ioffe S, Vanhoucke V, Alemi A (2016) Inception-v4, Inception-ResNet and the impact of residual connections on learning. In: AAAI conference on artificial intelligence
45. Xia X, Xu C, Nan B (2017) Inception-v3 for flower classification. In: 2017 2nd international conference on image, vision and computing (ICIVC). IEEE, pp 783–787
46. Szegedy C, Ioffe S, Vanhoucke V, Alemi AA (2017) Inception-v4, inception-resnet and the impact of residual connections on learning. In: Thirty-first AAAI conference on artificial intelligence
47. Jastrzebski S, Arpit D, Ballas N, Verma V, Che T, Bengio Y (2018) Residual connections encourage iterative inference. In: ICLR
48. Ioffe S, Szegedy C (2015) Batch normalization: accelerating deep network training by reducing internal covariate shift. In: ICML, pp 448–456
49. Zeiler M, Fergus R (2013) Visualizing and understanding convolutional neural networks. ECCV 2014, Part I, LNCS 8689. https://doi.org/10.1007/978-3-319-10590-1_53
50. Huang G, Sun Y, Liu Z, Sedra D, Weinberger KQ (2016) Deep networks with stochastic depth. In: ECCV. Springer, pp 646–661
51. Szegedy C et al (2015) Going deeper with convolutions. In: 2015 IEEE conference on computer vision and pattern recognition (CVPR), Boston, MA, pp 1–9

52. Pawlowski N, Ktena SI, Lee MC, Kainz B, Rueckert D, Glocker B et al (2017) DLTK: state of the art reference implementations for deep learning on medical images. arXiv:1711.06853

53. Everingham M, Van Gool L, Williams CK, Winn J, Zisserman A (2010) The pascal visual object classes (VOC) challenge. In: IJCV, pp 303–338

54. Girshick R, Donahue J, Darrell T, Malik J (2014) Rich feature hierarchies for accurate object detection and semantic segmentation. In: CVPR

55. Aman D, Payal P (2018) Image retrieval techniques: a survey. Int J Eng Technol 7(1.2):215–219

56. Lin T-Y, Maire M, Belongie S, Hays J, Perona P, Ramanan D, Dollar P, Zitnick CL (2014) Microsoft COCO: common objects in context. In: ECCV

57. He K, Zhang X, Ren S, Sun J (2014) Spatial pyramid pooling in deep convolutional networks for visual recognition. In: ECCV

58. He K, Zhang X, Ren S, Sun J (2015) Delving deep into rectifiers: surpassing human-level performance on imagenet classification. In: ICCV

59. Zhou SK, Greenspan H, Shen D (2017) Deep learning for medical image analysis. Academic Press

60. Lu L, Zheng Y, Carneiro G, Yang L (2017) Deep learning and convolutional neural networks for medical image computing. Springer

61. Zheng Y, Comaniciu D (2014) Marginal space learning. In: Marginal space learning for medical image analysis. Springer, pp 25–65

62. Gauthier J (2014) Conditional generative adversarial nets for convolutional face generation. In: Class project for Stanford CS231N: convolutional neural networks for visual recognition, Winter semester 2014

63. Ghesu FC, Krubasik E, Georgescu B, Singh V, Zheng Y, Hornegger J et al (2016) Marginal space deep learning: efficient architecture for volumetric image parsing. IEEE Trans Med Imaging 35:1217–1228

64. Ker J, Wang L, Rao J, Lim T (2018) Deep learning applications in medical image analysis. IEEE Access 6:9375–9389

65. Lin M, Chen Q, Yan S (2013) Network in network

66. Aman D, Payal P (2019) Analysis of non-linear activation functions for classification tasks using convolutional neural networks. Recent Patents Comput Sci 12:156. https://doi.org/10.2174/2213275911666181025143029

Chapter 7
Kidney Lesion Segmentation in MRI Using Clustering with Salp Swarm Algorithm

Tapas Si

1 Introduction

According to the Bulletin of the World Health Organization (WHO) [1], the Global Burden of Disease (GBD) 2015 study estimated that 1.2 million people died from kidney failure in 2015, an increase of 32% since 2005. Around 1.7 million people are thought to die from acute kidney injury each year. Overall, therefore, an estimated 5–10 million people die annually from kidney disease. Therefore, the present situation demands the fast diagnosis and treatment of the kidney disease to save human life. Radiological imaging of kidney plays a crucial role in kidney disease diagnosis [2]. The ultrasound (US) and computed tomography (CT) imaging of kidney are the widely used imaging techniques used for diagnosis as well as treatment planning of kidney cancer and other related disorders. Martĺin-Fernĺandez and Alberola-Lĺopez [3] proposed a method for boundary detection of kidney in 3D US images using Markov random field (MRF) and active contour models. Li et al. [4] made comparative studies of region growing, texture segment, active contour model, and MRF model for segmentation of kidney in US images. Kim and Park [5] developed a computer-aided detection (CADe) system of kidney tumor on abdominal CT image using region growing method. Skalski et al. [6] proposed a segmentation method based on hybrid level set method with elliptical shape constraints. Yang et al. [7] proposed an automatic method to segment the kidney and renal tumor in CT angiography images using 3D fully convolutional network (FCN). Choudhari et al. [8] proposed a supervised segmentation method of kidney tumor in CT image using U-Net deep learning model. Pang et al. [9] proposed a supervised segmentation method for kidney tumor in CT images using generative adversarial networks

T. Si (✉)

Department of Computer Science & Engineering, Bankura Unnayani Institute of Engineering, Bankura, West Bengal 722146, India

e-mail: c2.tapas@gmail.com

© The Author(s), under exclusive license to Springer Nature Singapore Pte Ltd. 2021   93
X.-Z. Gao et al. (eds.), *Applications of Artificial Intelligence in Engineering*, Algorithms for Intelligent Systems, https://doi.org/10.1007/978-981-33-4604-8_7

(GANs). A systematic review on computer-aided diagnosis (CADx) of renal lesions in CT images can be obtained from the study [10].

US and CT modalities are the first choices for renal imaging, and MRI is mainly used as problem-solving tool so far [11]. MRI has the advantage of having superior soft tissue contrast and thus provides a powerful tool in the detection and characterization of renal lesions. Li et al. [12] proposed segmentation of kidney in dynamic contrast-enhanced (DCE) MRI using wavelet features and K-means clustering algorithm. Rudra et al. [13] used graph cut and pixel connectivity to segment the kidney from low contrast MRI data. Yang et al. [14] proposed a segmentation method in which maximally stable temporal volume (MSTV) was used to detect the anatomical structures and to segment the kidney as whole. Then, voxels in the segmented kidney are described as principal components (PCs) which was used by K-means clustering to separate the voxels belong to internal renal structures (i.e., cortex, medulla, and renal pelvis). Will et al. [16] proposed an automatic segmentation method for entire kidney and its renal structures in MRI. First, non-contrast-enhanced T1-W and T2-W MR images were co-registered using rigid registration algorithm. Second, the thresholding algorithm was used to segment the entire kidney in T2-W MR images. Finally, the T1-W and T2-W MR images were used to separate the renal structures, i.e., cortex, medulla, and pelvis.

As per the best knowledge of the author and after making the literature survey, it is found that there is no work on kidney lesion segmentation in MRI. Therefore, the work in this paper is the first attempt to segment the kidney lesions in MRI. The objective of this paper is to develop a segmentation method for kidney lesions in MRI. Therefore, a hard-clustering technique with SSA is developed in this paper to segment the kidney MRI for lesion detection in kidney. In the same framework of the proposed segmentation method, K-means algorithm is used instead of SSA-based clustering technique in the second experiment. The results of two methods are compared both quantitatively and qualitatively. The proposed clustering technique performs better than K-means algorithm.

2 Materials and Methods

2.1 DCE-MRI Dataset

Total 4 T2-Weighted MRI slices have been taken from the Cancer Genome Atlas Kidney Renal Clear Cell Carcinoma (TCGA-KIRC) [17, 18]. All MRI slices have a size 256×256.

2.2 Proposed Method

The proposed methods have three steps as the following:

1. preprocessing
2. Segmentation
3. Postprocessing.

The outline of the proposed SSA-based segmentation method is given in Fig. 1.

2.3 Preprocessing

The segmentation process faces difficulties due to the presence of noise and IIHs in the MR images. Therefore, noise is removed using median filter [19]. Then IIHs are corrected using max filter-based method [20].

2.4 Segmentation Using Clustering with SSA

After preprocessing the MR images, region of interests (ROIs) are selected from the MR images. Then clustering technique is applied to ROIs to segment the lesions. Clustering is an unsupervised learning or exploratory data analysis technique widely used in biomedical applications like MRI image analysis [25]. Let $\mathbb{X} = \{\vec{X}_1, \vec{X}_2, \ldots, \vec{X}_{\mathcal{N}}\}$ is a set of input patterns where $\vec{X}_i = (x_{i1}, x_{i2}, \ldots, x_{id}) \in \mathcal{R}^d$ and

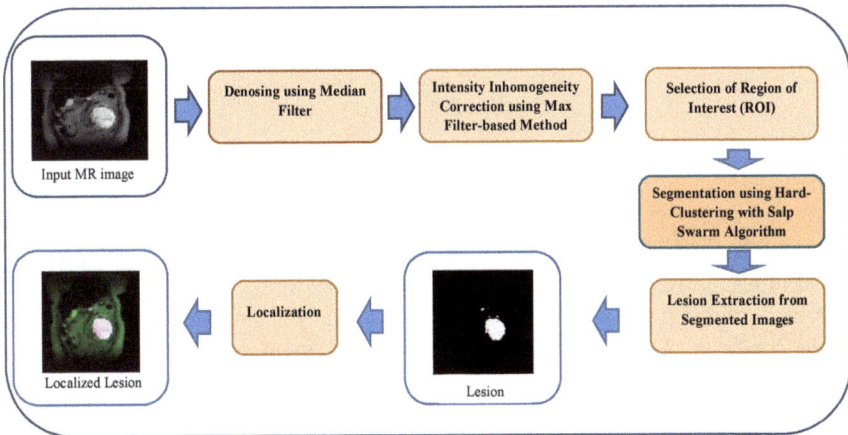

Fig. 1 Outline of the proposed method

each x_{ij} is the attribute of input patterns. Hard or partitional clustering attempts to seek a $\mathcal{K}$ partition of $\mathbb{X}$, $\mathbb{C} = \{C_1, C_2, \ldots, C_\mathcal{K}\}(\mathcal{K} \leq \mathcal{N})$ such that

1. $C_k \neq \phi, \quad k = 1, 2, \ldots, \mathcal{K}$
2. $\cup_{k=1}^\mathcal{K} C_k = \mathbb{X}$
3. $C_k \cap C_l = \phi, \quad k, l = 1, 2, \ldots, \mathcal{K}$ and $k \neq l$.

In this paper, a hard-clustering technique with SSA (CSSA) is developed to segment the kidney lesion in MRI. A clustering solution is represented by ith candidate solution in SSA as $\vec{y}_i = \{\vec{m}_1, \vec{m}_2, \ldots, \vec{m}_\mathcal{K}\}$ where $\vec{m}_j$ is the jth cluster center. The two-dimensional MR images are converted into one-dimensional data points. The size of the data points is $\mathcal{N} = \mathcal{P} \times \mathcal{Q}$ where $\mathcal{P}$ and $\mathcal{Q}$ are the numbers of rows and columns of the images, respectively. For $\mathcal{N}$ data points of d dimension and for a user-specified cluster number $\mathcal{K}$, candidate solutions in SSA are the real vector of dimension $\mathcal{K} \times d$. In this work, clustering is performed on gray values of MR images and therefore, $d = 1$. Hence, the resultant dimension of the candidate solutions is $\mathcal{K}$.

In a good clustering solution, a good *trade-off* between compactness (i.e., intra-cluster distance) and separation (i.e., inter-cluster distance) is needed. *DB-index* is the ratio of intra-cluster compactness and inter-cluster separation. It is discussed in Sect. 2.5.

2.4.1 SSA Algorithm

The salp swarm algorithm (SSA) is modeling the swarm behavior of salp species in nature [21], and it is designed to solve global optimization problem. In SSA, the population of salps are divided into two groups: leader and followers. The leader salp is at the front and the rest of the salps in chain follow the leader. Let assume the ith salp is represented by a vector $X_i = (x_{i1}, x_{i2}, \ldots, x_{id})$ where d is the dimension of the problem to be solved. The position x_{1j} of the leader, i.e., first salp in jth dimension, is updated by the following equation:

$$x_{1j} = \begin{cases} F_j + c_1((x_j^{max} - x_j^{min})c_2 + x_j^{min}) & c_3 \geq 0.5 \\ F_j - c_1((x_j^{max} - x_j^{min})c_2 + x_j^{min}) & c_3 < 0.5 \end{cases} \quad (1)$$

where x_j^{max} and x_j^{min} are the upper bound and lower bound in jth dimension. c_1 is the coefficient which balances the exploration and exploitation in the search process, c_2 and c_3 are uniformly distributed random numbers in the range $(0, 1)$. F_j is the position of the food source, i.e., swarm's target in the jth dimension.

The coefficient c_1 is updated using the following equation:

$$c_1 = 2e^{-(\frac{4t}{t_{max}})^2} \quad (2)$$

where t is the current iteration and t_{max} is the maximum number of iterations.

The position of the follower is updated by:

$$x_j^i = \frac{1}{2}(x_j^i + x_j^{i-1})$$ (3)

where $i \geq 2$. The flowchart of SSA is given in Fig. 2.

2.5 Objective Function

In this work, *DB-index* is used as an objective function in the CSSA. Previously, *DB-index* was used to validate the clustering performance in the segmentation of brain MRI [23–25]. *DB-index* is the ratio of intra-cluster compactness and inter-cluster separation. ith intra-cluster compactness is defined as:

$$S_{i,t} = \left[\frac{1}{N_i} \sum_{\vec{X} \in C_i} \| \vec{X} - \vec{m}_i \|^t \right]^{\frac{1}{t}}$$ (4)

Inter-cluster distance, i.e., separation between ith and jth cluster is defined as:

$$d_{ij,p} = \left[\sum_{i=1}^{d} |m_{i,k} - m_{j,k}|^p \right]^{\frac{1}{p}} = \| \vec{m}_i - \vec{m}_j \|^p$$ (5)

where $\vec{m}_i$ is the centroid of ith cluster, t, $p \geq 1$, t, p are integer, N_i is the number of elements in the ith cluster C_i. $R_{i,qt}$ is expressed as

$$R_{i,tp} = \max_{j \in K, j \neq i} \left\{ \frac{S_{i,t} + S_{j,t}}{d_{ij,p}} \right\}$$ (6)

Finally, *DB-index* is measured by the following equation:

$$DB(K) = \frac{1}{K} \sum_{i=1}^{K} R_{i,qt}$$ (7)

The smallest *DB-index* indicates a valid optimal clustering. In the current work, *DB-index* is used as fitness for salps in SSA.

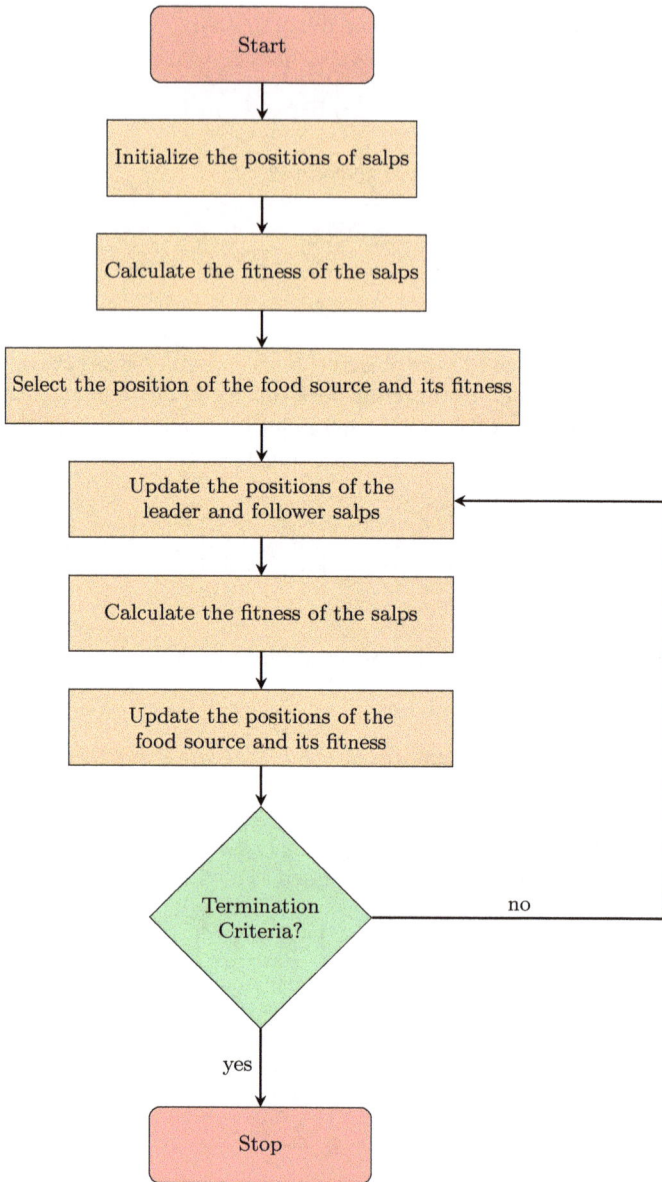

Fig. 2 Flowchart of SSA

2.6 Postprocessing

After obtaining segmented MR images using CSSA, the lesions are extracted from the segmented images. The pixels belong to the lesions have the hyperintensities in the MR images, and they are separated from other pixels based on that criteria. The resultant images having only lesions are overlaid with the original MR images to localize the lesions.

3 Experimental Setup

3.1 Parameters Settings

In SSA, number of salps is set to 30. The maximum number of function evaluations is set to 3000 for both CSSA and K-means algorithms.

3.2 PC Configuration

1. CPU: Intel(R) Core(TM) i7-9700K @3.60 GHz
2. RAM: 64 GB
3. Operating System: Windows 10 Professional 64-bit
4. Software: MATLAB 2018b.

4 Results and Discussion

The proposed method is applied to segment the kidney lesion in four MR images given in Fig. 3. First, the MR images are denoised using the median filter. Then the intensity inhomogeneities are corrected using Max filter-based correction method. Finally, the proposed hard-clustering technique CSSA is applied to the selected ROIs of the preprocessed MR images with cluster number 3. After clustering the images, the lesions are extracted from the segmented images. The lesions are overlaid with the original MR images to localize the lesions. The CSSA is run for 31 independent runs for each image. The results of CSSA are compared with that of K-means algorithm. The K-means algorithm is used in the proposed framework, and the same experiments are carried out with K-means algorithm. The performance is quantitatively measured using DB-index for the validation of the clustering method. The mean and standard deviation of DB-index values over 31 runs are given in Table 1. To statistically validate the results, nonparametric statistical test, Wilcoxon Signed Rank Test [26] has been conducted with confidence interval level $(\alpha) = 0.01$ using the results of 31 independent

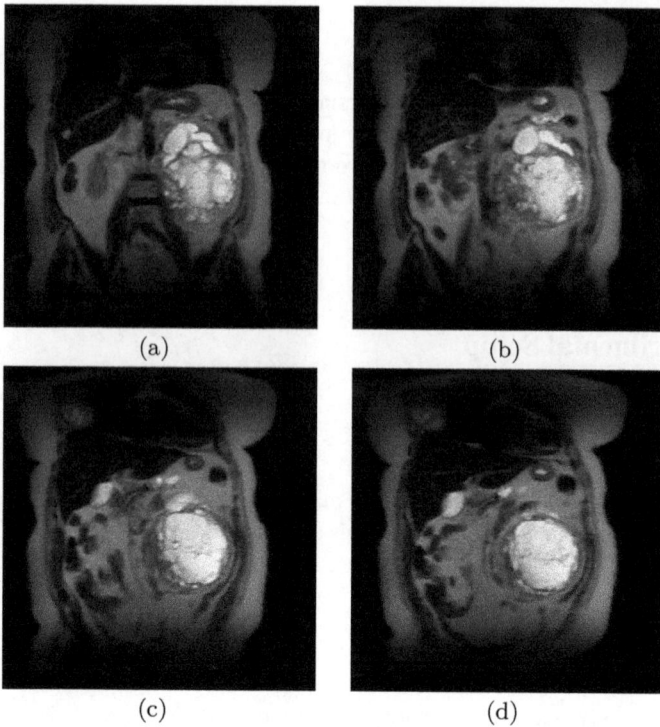

(a)   (b)

(c)   (d)

Fig. 3 MR images of kidney

Table 1 Mean and standard deviation of DB-index values over 31 independent runs

MRI#	K-means		CSSA		Statistics	
	Mean	Std.	Mean	Std.	p-value	h
1	0.08247656	**1.41E−17**	**0.071368028**	1.44E−05	1.17E−06 ≪ 0.01	1
2	0.07541229	**1.44E−14**	**0.065370794**	2.32E−06	1.12E−06 ≪ 0.01	1
3	0.074365711	**0**	**0.065246261**	2.76E−06	1.17E−06 ≪ 0.01	1
4	0.072564915	**5.64E−17**	**0.063653688**	4.76E−06	1.06E−06 ≪ 0.01	1

runs for each image. The test results are given in the last two columns of Table 1. The bold-faced fonts in this table indicate better results. The $h = 1$ in Table 1 indicates the rejection of the null hypothesis (i.e., there is no significant difference in the performance of two methods) and the acceptance of the alternative hypothesis. From Table 1, it is observed that the mean DB-index values of the proposed CSSA are better than K-means and it indicates that CSSA performs better clustering of MR images than K-means. But, the standard deviations of DB-index are lower than that of CSSA. From the statistical results, it is observed that the p-values are very less than 0.01 for all the images and it indicates that CSSA statistically outperforms K-means algorithm with the significance level 0.01.

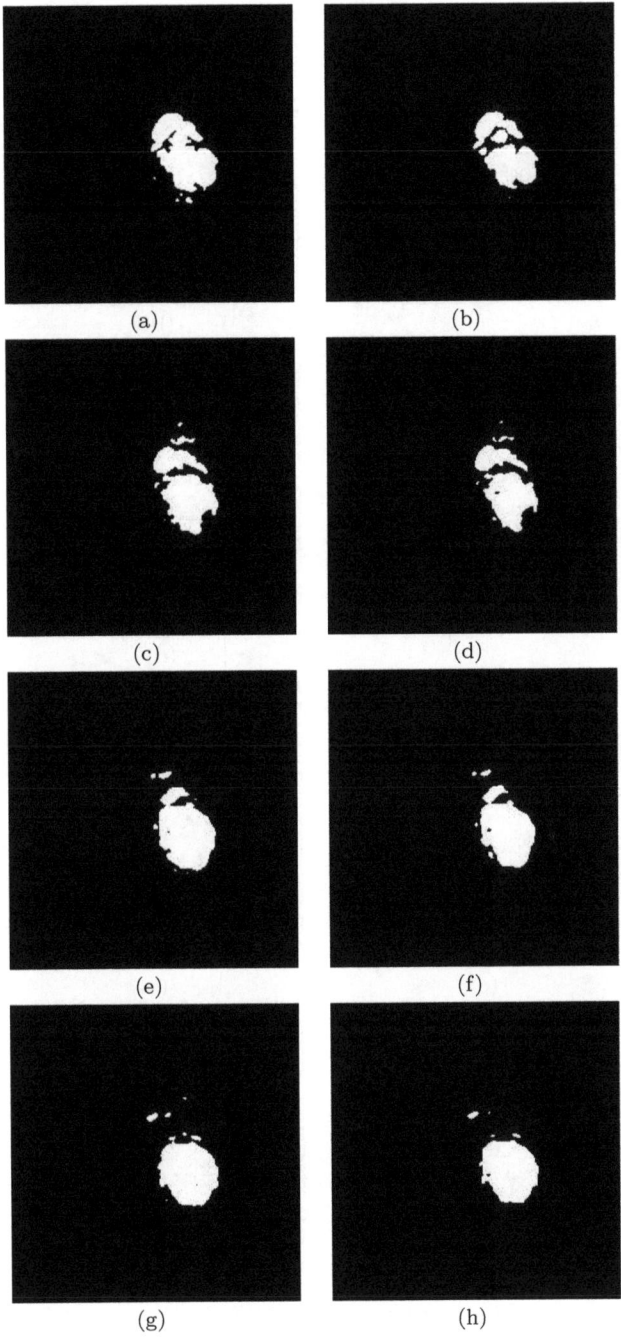

Fig. 4 kidney lesions. 1st column: K-means, 2nd column: CSSA

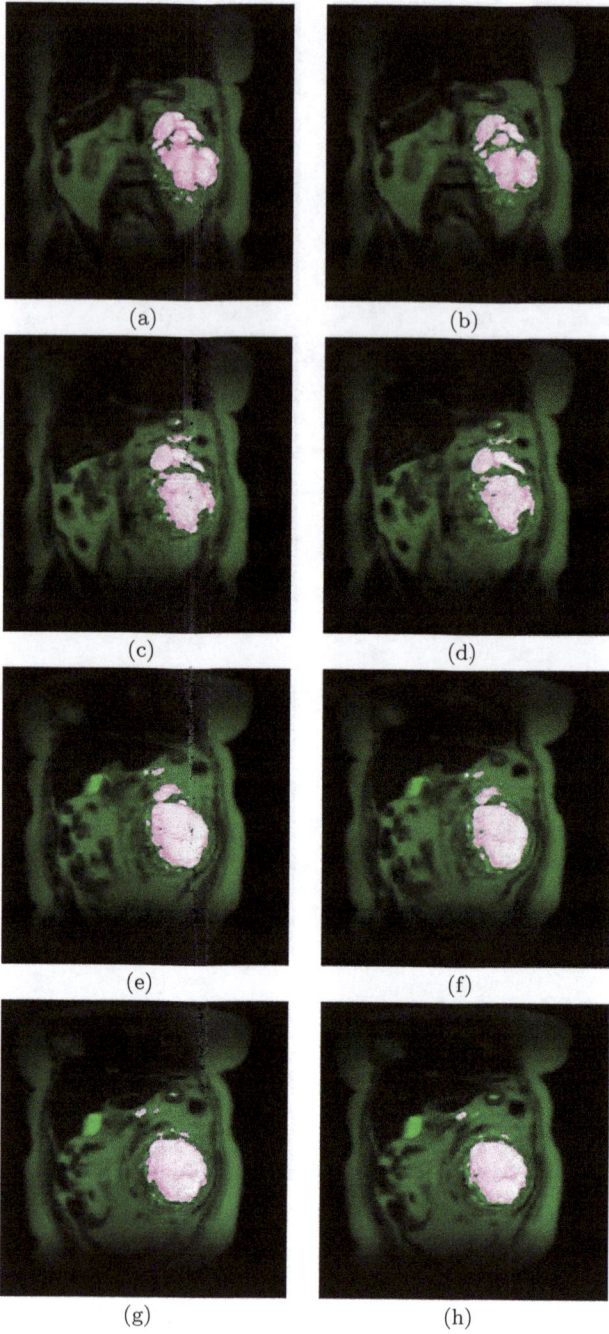

Fig. 5 Localized kidney lesions in the MRI. 1st column: K-means, 2nd column: CSSA

The qualitative (i.e., visual) results are given in Figs. 4 and 5. The images having extracted lesions from MR images are given in Fig. 4. The original images with localized lesions are given in Fig. 5. From Fig. 4, it can be observed that the proposed CSSA performs well in lesion segmentation. It can also be noticed that some healthy tissues are also segmented with the lesions by both the methods. But, compare to K-means algorithm, CSSA methods segment less number of healthy tissues as lesions.

The proposed method takes average 133.0820 s computational time, whereas K-means algorithm takes average 0.1415 s. Though the computational time cost of the proposed CSSA is higher than K-means, CSSA outperforms K-means algorithm in lesion segmentation.

After analysis both quantitative and qualitative results, it is found that the proposed CSSA performs better than K-means algorithm. K-means algorithm has worse performance than CSSA because K-means algorithm suffers from premature convergence in local optimum due to the selection of initial cluster center from the image data. On the other hand, the cluster centers are initialized randomly in the range [0, 255] and it is updated using the different steps of SSA. Finally, it may be concluded that the proposed method is robust as the standard deviations of DB-index values are very low and it is very efficient and effective in kidney lesion segmentation in MRI.

5 Conclusion

In this paper, a segmentation method is proposed to segment the lesions in kidney MRI. A hard-clustering technique with SSA is proposed to segment the MR images. The experimental results of the proposed method are compared with that of K-means algorithm. The proposed method statistically outperforms K-means algorithm based on quantitative results. It also qualitatively performs better than K-means algorithm. In this paper, only gray-level features of MR images are used in clustering. Wavelet features can be used in the future work. In this paper, single objective clustering technique is used. Multi-objective clustering can be used to segment the lesions in kidney MRI.

References

1. Luyckx VA, Tonelli M, Stanifer JW (2018) The global burden of kidney disease and the sustainable development goals. Bull World Health Organ 96:414–422D. https://doi.org/10.2471/BLT.17.206441
2. Cova MA, Cavallaro M, Martingano P, Ukmar M (2011) Magnetic resonance imaging of the kidney. In: Quaia E (ed) Radiological imaging of the kidney. Medical radiology. Springer, Berlin, Heidelberg. https://doi.org/10.1007/978-3-540-87597-0_6

3. Martlin-Fernlandez M, Alberola-Llopez C (2005) An approach for contour detection of human kidneys from ultrasound images using Markov random fields and active contours. Med Image Anal 9:1–23

4. Li L, Ross P, Kruusmaa M, Zheng X (2011) A comparative study of ultrasound image segmentation algorithms for segmenting kidney tumors. In: Proceedings of the 4th international symposium on applied sciences in biomedical and communication technologies. Article no 126, pp 1–5. https://doi.org/10.1145/2093698.2093824

5. Kim D-Y, Park J-W (2004) Computer-aided detection of kidney tumor on abdominal computed tomography scans. Acta Radiol 45(7):791–795

6. Skalski A, Jakubowski J, Drewniak T (2016) Kidney tumor segmentation and detection on computed tomography data. In: IEEE international conference on imaging systems and techniques (IST), Chania, Greece

7. Yang G, Li G, Pan T, Kong Y, Wu J, Shu H, Luo L, Dillenseger J-L, Coatrieux J-L, Tang L, Zhu X (2018) Automatic segmentation of kidney and renal tumor in CT images based on pyramid pooling and gradually enhanced feature modules. In: 24th international conference on pattern recognition (ICPR), Beijing, China. IEEE, pp 3790–3795. https://doi.org/10.1109/ICPR.2018.8545143

8. Choudhari K, Sharma R, Halarnkar P (2020) Kidney and tumor segmentation using U-Net deep learning model. In: 5th international conference on next generation computing technologies (NGCT-2019). https://doi.org/10.2139/ssrn.3527410

9. Pang S, Du A, Orgun MA, Yu Z, Wang Y, Wang Y, Liu G (2020) CTumorGAN: a unified framework for automatic computed tomography tumor segmentation. Eur J Nucl Med Mol Imaging. https://doi.org/10.1007/s00259-020-04781-3

10. Kaur R, Junejaa M, Mandal AK (2019) Computer-aided diagnosis of renal lesions in CT images: a comprehensive survey and future prospects. Comput Electr Eng 77:423–434

11. Nikken JJ, Krestin GP (2007) MRI of the kidney—state of the art. Eur Radiol 17:2780–2793. https://doi.org/10.1007/s00330-007-0701-3

12. Li S, Zöllnera FG, Merrema AD, Peng Y, Roervikc J, Lundervoldd A, Schada LR (2012) Wavelet-based segmentation of renal compartments in DCE-MRI of human kidney: initial results in patients and healthy volunteers. Comput Med Imaging Graph 36:108–118

13. Rudra AK, Chowdhury AS, Elnakib A, Khalifa F, Soliman A, Beache G, El-Baz A (2013) Kidney segmentation using graph cuts and pixel connectivity. Pattern Recogn Lett 34:1470–1475

14. Yang X, Minh HL, (Tim) Cheng K-T, Sung KH, Liu W (2016) Renal compartment segmentation in DCE-MRI images. Med Image Anal 32:269–280

15. Rusinek H, Lim JC, Wake N, Seah J, Botterill E, Farquharson S, Mikheev A, Lim RP (2015) A semi-automated "blanket" method for renal segmentation from non-contrast T1-weighted MR images. Magn Reson Mater Phys Biol Med 29:197–206

16. Will S, Martirosian P, Würslin C, Schick F (2014) Automated segmentation and volumetric analysis of renal cortex, medulla, and pelvis based on non-contrast enhanced T1- and T2-weighted MR images. Magn Reson Mater Phys Biol Med 27:445–454

17. Akin O, Elnajjar P, Heller M, Jarosz R, Erickson BJ, Kirk S, Filippini J (2016) Radiology data from The Cancer Genome Atlas Kidney Renal Clear Cell Carcinoma [TCGA-KIRC] collection. The Cancer Imaging Archive. https://doi.org/10.7937/K9/TCIA.2016.V6PBVTDR

18. Clark K, Vendt B, Smith K, Freymann J, Kirby J, Koppel P, Moore S, Phillips S, Maffitt D, Pringle M, Tarbox L, Prior F (2013) The Cancer Imaging Archive (TCIA): maintaining and operating a public information repository. J Digit Imaging 26(6):1045–1057

19. Mohana J, Krishnavenib V, Guo Y (2014) A survey on the magnetic resonance image denoising methods. Biomed Signal Process Control 9:56–69

20. Balafar MA, Ramli AR, Mashohor S (2010) A new method for MR grayscale inhomogeneity correction. Artif Intell Rev 34:195–204

21. Mirjalili S, Gandomi AH, Mirjalili SZ, Saremi S, Faris H, Mirjalili SM (2017) Salp Swarm Algorithm: a bio-inspired optimizer for engineering design problems. Adv Eng Softw 114:163–191

22. Davies DL, Bouldin DW (1979) A cluster separation measure. IEEE Trans Pattern Anal Mach Intell 1(2):224–227
23. Si T, De A, Bhattacharjee AK (2014) Brain MRI segmentation for tumor detection using Grammatical Swarm based clustering algorithm. In: Proceedings of IEEE international conference on circuits, power and computing technologies (ICCPCT-2014), Nagercoil, India, 2014
24. Si T, De A, Bhattacharjee AK (2015) Grammatical swarm based segmentation methodology for lesion segmentation in brain MRI. Int J Comput Appl 121(4):1–8
25. Si T, De A, Bhattacharjee AK (2016) MRI brain lesion segmentation using generalized opposition-based glowworm swarm optimization. Int J Wavelets Multiresolution Inf Process 14(5):1650041 (29 pp)
26. Derrac J, Garcla S, Molina D, Herrera F (2011) A practical tutorial on the use of nonparametric statistical tests as a methodology for comparing evolutionary and swarm intelligence algorithms. Swarm Evol Comput 1:3–18

Chapter 8
Smart Approach to Optical Character Recognition and Ubiquitous Speech Synthesis Using Real-Time Deep Learning Algorithms

Bhargav Goradiya, Yagnik Mehta, Nisarg Patel, Neel Macwan, and Vatsal Shah

1 Introduction

The genuine world contains an excessive number of critical messages and helpful data, however sadly most of them are written in diverse authority language relies upon the host nation. Here and there a billboard or some other notification could convey a significant message or even peril [1]. On the off chance that the message is inaccessible to humankind with various language foundations, it might make significant data be passed up a major opportunity. Other than that, it is badly designed for a voyager to carry on their errands in a remote nation on the off chance that they do not comprehend the language utilized in that nation. They must convey a pocket word reference or utilize online interpretation administration to comprehend the message [2].

Be that as it may, a pocket word reference probably will not be useful if the clients need to decipher a language that does not gather by letter sets [3]. It additionally implied a similar path in another examination that clients cannot compose the content of what they see. This issue may cause a correspondence breakdown for humanity from an alternate language foundation as they cannot comprehend the language even though the pocket word reference and online interpretation administration are provided.

Moreover, more than a quarter of the world's population—some 2.2 billion people—suffer from vision impairment out of which, one billion cases could have been prevented or have been left unaddressed, according to the first World Vision Report released by the World Health Organization (WHO) on October 8, 2019 [4]. It

B. Goradiya (✉) · Y. Mehta · N. Patel · N. Macwan · V. Shah
Birla Vishvakarma Mahavidyalaya Engineering College, V.V. Nagar, Anand, Gujarat 388120, India
e-mail: bhargav.goradiya@bvmengineering.ac.in

© The Author(s), under exclusive license to Springer Nature Singapore Pte Ltd. 2021   107
X.-Z. Gao et al. (eds.), *Applications of Artificial Intelligence in Engineering*, Algorithms for Intelligent Systems, https://doi.org/10.1007/978-981-33-4604-8_8

is likewise revealed that 90% of the world's outwardly disabled individuals are from low pay gathering.

One of the major issues looked by outwardly impeded individuals is that they are most certainly not fit in getting to printed content. Despite the fact that there are numbers of assistive innovation implied for outwardly debilitated, the greater part of these exceptional gadgets is not advantageous on the grounds that it requires custom alterations and some are excessively costly [5]. Practically, 70% of outwardly disabled individuals are jobless, and a large portion of them cannot utilize assistive innovation because of its expense.

This results for some users to botch the chance to get to significant content that is available on the planet to do day-to-day tasks productively. So as to beat these issues, this paper proposes to build up an Android application that catches a content-based picture that conveys significant messages from a genuine world and interprets them into the intended language and lastly articulates it [6].

2   Literature Review

Optical character recognition (OCR) programs provide users with the ability to capture written text and give the system an ability to convert it into audio [7]. OCR technology features scan, recognize, and read the text. Current OCR technical systems provide excellent accuracy and formatting power [8].

Nearly everyone you meet uses a variety of tools to increase their independence, efficiency, and their daily strength in life. We use gadgets, apps, devices, and all kinds of useful tools to help us share and find information and fulfill tasks. Many of us have reached a point where we feel we cannot return to life without them. People with special needs rely on any tool they can find to strengthen their independence.

Screen readers are a type of software that allows people to vocalize the text displayed on a computer screen. These days, even mobile devices and tablets can have screen readers on them. With some of them, the user sends commands by pressing a different combination of buttons on the keyboard that teaches the speech conjunction what part of the text to say aloud [9]. The command may instruct the compiler to spell or read a single word, line, or full-text screen.

Instructions can also order the synthesizer to announce the location of the computer cursor or object. Users can also use screen readers to find text in different colors, to read the highlighted text, to point to a menu, to use a spell check, or to read spreadsheet cells [10]. The limit to these types of tools is that you can only read text from a screen with software installed on it [9].

Similar in the sense to the screen readers mentioned above, these available devices offer a variety of new capabilities for people to use. Technology-based tools for visually impaired people come in many sizes. In addition, some are mobile and portable, while others are stationary. Some do one or two things, and only a few of them do many things at once.

The purpose of this paper is to provide a brief but comprehensive overview of text-to-speech synthesis by highlighting its computer vision for recognition text and natural language processing (NLP) with background work of WaveNet API in action.

3 Proposed Model

3.1 Implementation of Deep Learning Model for Text Recognition

Text recognition can computerize dreary information passage from newspaper articles, blogs, long PDFs, receipts, and novels. With the reinforced learning and firebase ML kit [11], we can likewise extricate content from pictures of reports, which you can use to expand availability or decipher archives. Applications can even monitor true items, for example, by perusing the numerical characters in the intended content. Text and optical character recognition assumes an undeniably significant job in numerous cutting-edge applications in the field of medication, money, transport, and security [9].

While numerous frameworks have been intended to recognize messages inside a picture, those strategies were intended to manage extraordinary kinds of pictures, for example, those from checked reports, site pages, and papers. With changeability in picture quality, content arrangement, and fluctuating degrees of goals, precisely separating information implanted in a picture is an open research issue that propelled analysts to concentrate on content recovery procedures. What is more, the acknowledgment of corrupted content imprinted on a reasonable plastic surface has not been tended to. The process of capturing the image and extracting the intended information is mentioned as the following texts.

3.1.1 Image Capturing and Preprocessing

Firebase ML kit is a portable SDK that brings Google's AI mastery to Android and iOS applications in an incredible yet simple to utilize bundle. ML kit makes it simple to apply ML strategies in our application by bringing Google's ML innovations, for example, the Google Cloud Vision API, Mobile Vision, and TensorFlow Lite, together in a solitary SDK [12]. Initially, pictures are caught in this stage by a camera or a few other gadgets. Extraction of examined pictures with a white foundation and dark character forefronts is anything but difficult to be recognized yet the camera caught pictures may contain commotion because of the natural reasons and low brilliance of the pictures. In this manner, there are a few methods like picture improvement, binarization, and commotion decrease to be done in the preprocessing stage to expand the presentation also, the precision of a character acknowledgment framework.

3.1.2 Text Recognition and Extraction

In this stage, after the character is sectioned, they are standardized by expelling commotion. In conclusion, optical character recognition extricates the character and remembers it. In spite of the fact that OCR is helpful, it is not impeccable with no issues. Scientists have found issues, for example, light condition, content slant, recognition contortion, misalignment of content, obscure, and trouble in perceiving written by hand record [9]. So these are the difficulties for the technologist to lead further investigate improving this innovation. With the developing and astounding innovation in the versatile gadgets, cell phones are presently proficient to catch high goals pictures with at any rate 1440 × 2960 pixels, which are increasingly good and have higher opportunities to be identified with OCR [11]. Executing OCR application on a cell phone could be practical as there are numerous continuous looks into this field. Nonetheless, cell phones have their downsides as well [13]. Some of the available models including the Tesseract OCR [14] are unable to interpret the content where there is artificial noise in the background, whereas our model here overcomes the problems which are currently faced by Tesseract OCR engine. As shown in the figure below, the model developed can recognize the text from the noisy image with 100% accuracy. The execution can be better understood from the image in Fig. 1.

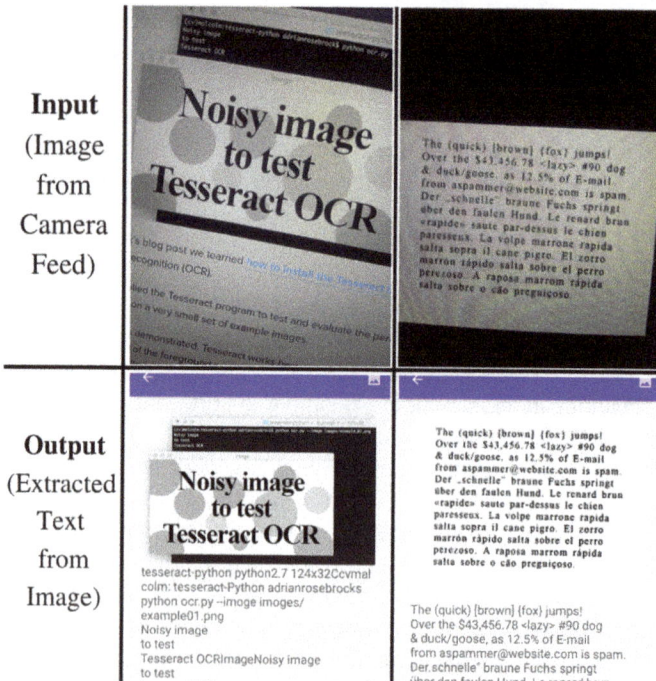

Fig. 1 Input and output screens for the text recognition and extraction process

3.2 Speech Synthesis with WaveNet API

The large portion of the language interpreters is accessible on the electronic application contrasted with versatile-based. Anyway, there are interpretation methods
that gave in the versatile condition a set number of bolstered dialects. As research
directed, there are three unique calculations of how interpreters can be performed.
They depend on publicly supporting, where the OCR came about the character will
be sent to a gathering of human laborers to do interpretation task, an online interpretation administration, where character separated will be conveyed to Google Translator
[9] to do the real content to content interpretation, and finally, constant interpreter,
which incorporates an interpreters and word reference on a cell phone to do the
interpretation.

In order to eliminate the barriers faced during the transformation, we have developed the app based on WaveNet API. This API has a unique selling point that nullifies
the process of sending the data over the cloud which also ensures the security and
is much convenient when in someplace the mobile device is not connected to the
Internet. Upon starting the app for the first time, it automatically downloads the
voice packs of 108 languages each differing with individual female and male tones
of various pitches. The app automatically recognizes the language in which the text
is extracted.

3.3 Architectural Design

Figure 2 presents the overall structure of the proposed application. After the user
captures a picture of the text Android phone's built-in camera, sent for image
processing. The image is processed to reduce noise and delivered to the computer
vision engine to identify the text [13, 15]. Once the text extraction is accomplished, now they are sent to the vocalizer. At this stage, the WaveNet APIs result in
interpreting the language and voice out through the phone's built-in speaker.

3.4 Flow of the System

Initially, when the user opens the app for any particular usage, the apps on create
behavior allow the system to directly open the camera to capture the intended article
or any page which the user wants to get information on. The system is designed with
a novelty that it will ask the user to try clicking the picture again if because of a blurry
image, no text is interpreted by the model. As soon as the application interprets the
text; it is extracted. The system flow for text recognition and extraction is designed
with the following flow as shown in Fig. 3.

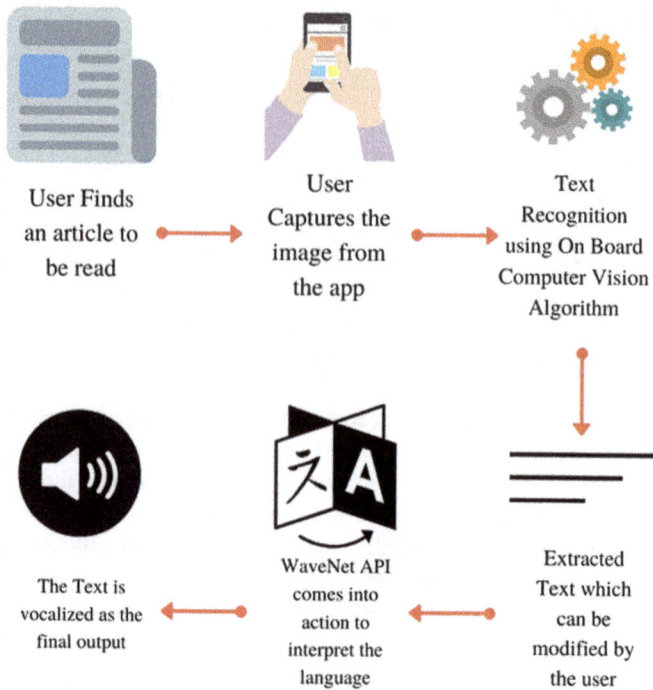

User Finds an article to be read

User Captures the image from the app

Text Recognition using On Board Computer Vision Algorithm

The Text is vocalized as the final output

WaveNet API comes into action to interpret the language

Extracted Text which can be modified by the user

Fig. 2 Graphical representation of the architecture and flow of the system

With the result of the extracted text as the output of the system then automatically passes the acquired information to the text view feed whereupon user's satisfaction, they can edit the text. The text is later analyzed by WaveNet API which interprets the language and propagates the text for speech synthesis. The model of which is deliberated with the flowchart shown in Fig. 4.

4 Implementation and Testing

Here to demonstrate the working and to give an epitome of the implementation on WaveNet API, we took the image of a verse from Shri Bhagwad Geeta; for the foremost, the image is captured from the camera. The captured image is then processed with the on-board machine learning algorithm to extract the text and pass it to a text view to the next feed the user there can manipulate the text according to his need and then he can ask deep to vocalize in the particular language. The working is shown as the graphical representation in Fig. 5.

Fig. 3 Flowchart of the text
recognition and extraction
process

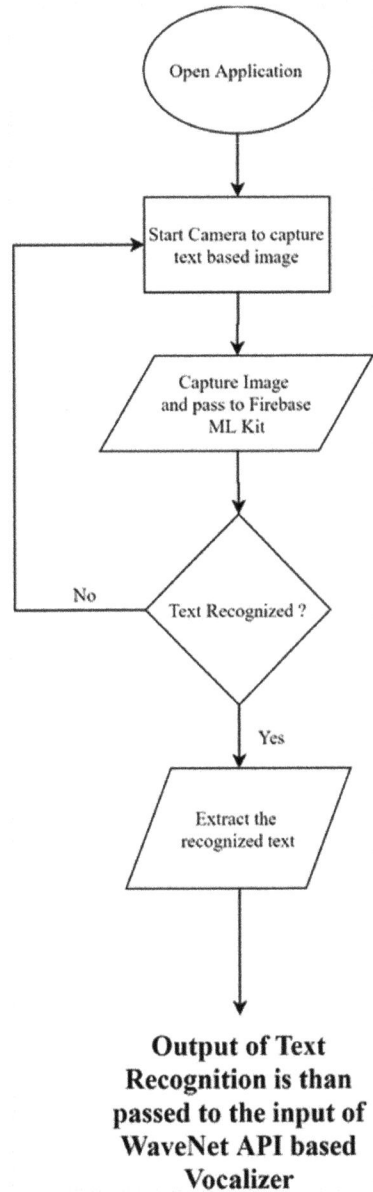

Open Application

Start Camera to capture
text based image

Capture Image
and pass to Firebase
ML Kit

Text Recognized ?

No

Yes

Extract the
recognized text

**Output of Text
Recognition is than
passed to the input of
WaveNet API based
Vocalizer**

Fig. 4 Flowchart of the
speech synthesis of extracted
text with WaveNet API

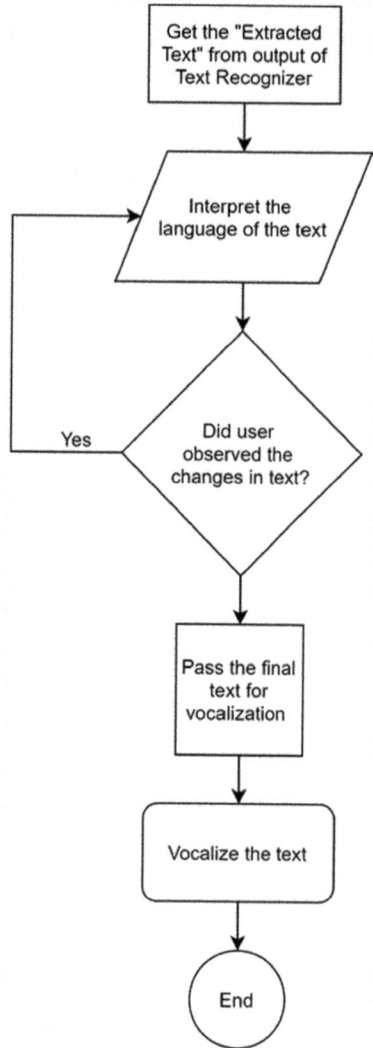

5 Experimental Results

5.1 Results

Here to verify the integrity and authenticity of the results obtained, we evaluated the
final vocalized audio by studying the spectrogram and analog signal of the signal,
which clearly augments the change in frequency upon a time to verify the results.
To augment the final vocalized output, we constructed the time–versus–frequency

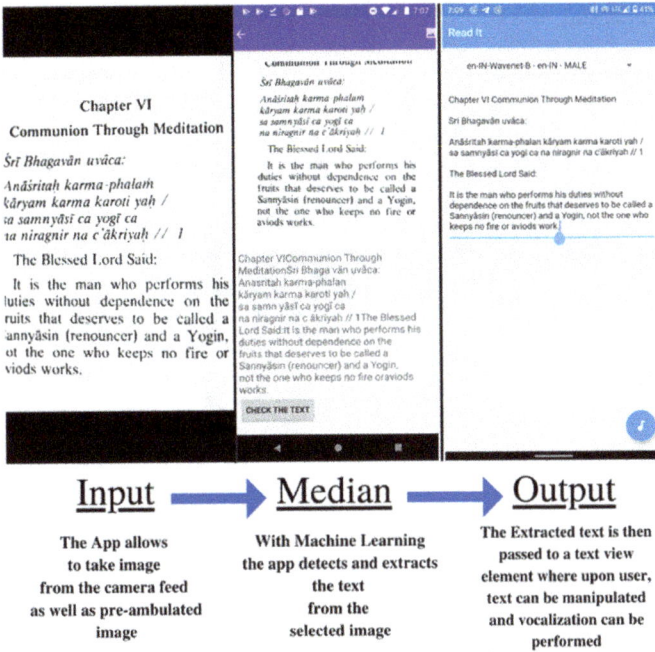

Fig. 5 Augmented process of the intended article to be vocally synthesized

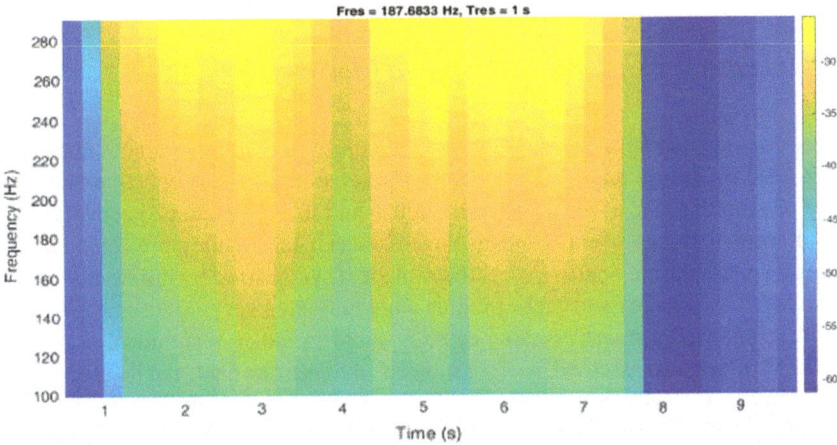

Fig. 6 Spectrogram of the mp3 output

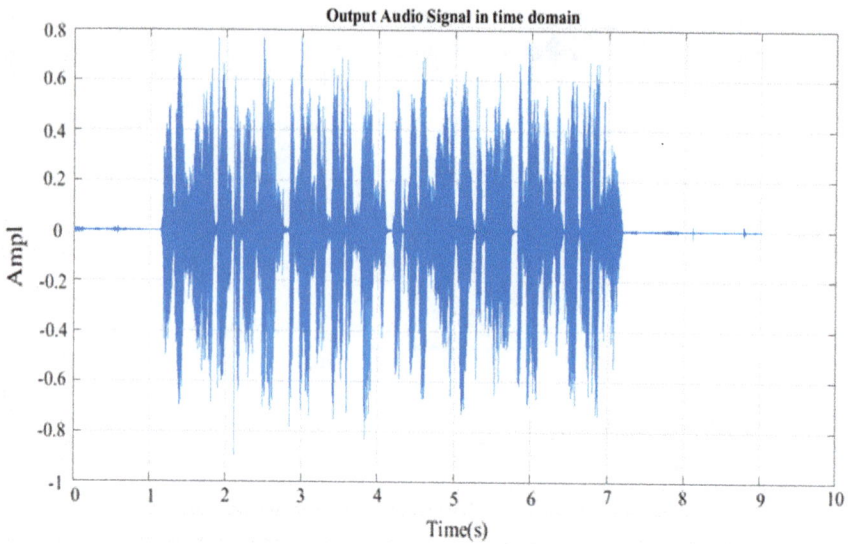

Fig. 7 Time domain analysis of the mp3 output

spectrogram and time domain analysis representations as shown in Figs. 6 and 7. The final output is also saved with a respective mp3 file for the user's further application.

5.2 Accuracy and Loss of Proposed Model

Subjective listening tests are blind and crowdsourced. Hundred sentences that were not included in the training data were used for evaluation. Each subject can predict the 8 and 63 stimuli to the north Australian English and Indian Hindi, respectively. Test stimuli were randomly selected and displayed for each item. In the paired comparison test, each pair of speech patterns was the same text synthesized by different models. In the MOS test, subjects in each stimulus were presented keeping it separate. Each paired comparison test was assessed by eight subjects and each stimulus.

The MOS test was assessed by eight subjects. Subjects are paid, and local speakers are doing the work. Those ratings (approximately 55%) that do not use headphones are excluded. Calculating preference and average feedback scores, Table 1 shows the complete details attached.

Preference scores for speech samples are between LSTM-RNN-based parametric statistics (LSTM), HMM-driven concatenative, and proposed WaveNet speech producers. Each row of the table specifies the two-dimensional experimental measurements compilers. Many filter scores were better than those competing in them as $p < 0.01$ level was indicated by liver type. Note that WaveNet (L) and WaveNet

Table 1 Subjective preference (%) in naturalness

Language	LSTM	Concat	WaveNet (L)	WaveNet (L + F)	No preference	p-value
Australian English	23.3	**63.6**			13.1	$\ll 10^{-9}$
	18.7		**69.3**		12.0	$\ll 10^{-9}$
	7.6			**82.0**	10.4	$\ll 10^{-9}$
		32.4	**41.2**		26.4	0.003
		20.1		**49.3**	30.6	$\ll 10^{-9}$
			17.8	**37.9**	44.3	$\ll 10^{-9}$
Indian Hindi	**50.6**	15.6			33.8	$\ll 10^{-9}$
	25.0		23.3		51.8	0.476
	12.5			**29.3**	58.2	$\ll 10^{-9}$
		17.6	**43.1**		39.3	$\ll 10^{-9}$
		7.6		**55.9**	36.5	$\ll 10^{-9}$
			10.0	**25.5**	64.5	$\ll 10^{-9}$

$(L + F)$ are compatible. WaveNet is restricted to language features and can be found in both language features and F_0 values.

6 Conclusion and Future Scopes

In this paper, an Android application for interpreting and extracting the textual content from the image captured of any article and its articulation with speech synthesis using WaveNet API has been designed and developed. Finally, deliveries have been evaluated, and the results obtained are promising to indicate that this study successfully dealt with problems discussed in the first section. This study proves that the app works easily for anyone who wants to grab the information on the go by just clicking a photo of an article and allowing the app to vocalize it, a traveler who is in the remote country can easily understand the signs, as well as partial independence, is provided to visually impaired by having someone click a photo of textual content for them and by which they can understand and gain knowledge of the particular topic, which earlier was not available to them as inefficient format.

The pinnacle of the app will be when visually impaired people can hold the information by just pointing the phone's camera toward the article and the app can understand and synthesize the content in their present language, in real time neither by capturing the photo nor touching any UI component within the app.

References

1. Smith R (2007) An overview of the tesseract OCR engine. In: Ninth international conference on document analysis and recognition (ICDAR 2007), vol 2. IEEE, pp 629–633
2. Fragoso V, Gauglitz S, Zamora S, Kleban J, Turk M (2011) TranslatAR: a mobile augmented reality translator. In: 2011 IEEE workshop on applications of computer vision (WACV)
3. Nakajima H, Matsuo Y, Nagata M, Saito K (2005) Portable translator capable of recognizing characters on signboard and menu captured by built-in camera. In: Proceedings of the ACL interactive poster and demonstration sessions
4. Shaik A, Hossain G, Yeasin M (2010) Design, development and performance evaluation of reconfigured mobile Android phone for people who are blind or visually impaired. In: Proceedings of the 28th ACM international conference on design of communication—SIGDOC'10
5. Arnab A, Torr PH (2017) Pixelwise instance segmentation with a dynamically in-stantiated network. In: Proceedings of the IEEE conference on computer vision and pattern recognition, pp 441–450
6. Who.Int (2020) WHO|visual impairment and blindness [online]. Available https://www.who.int/mediacentre/factsheets/fs282/en/. Accessed 27 May 2020
7. Canedo-Rodriguez A, Kim S, Kim J, Blanco-Fernandez Y (2009) English to Spanish translation of signboard images from mobile phone camera. In: IEEE Southeastcon 2009
8. Hymes K, Lewin J (2008) OCR for mobile phones
9. Erichsen (2020) Which online translator is best? About.com Education, 2015 [Online]. Available https://spanish.about.com/od/onlinetranslation/a/online-translation.htm. Accessed: 29 Apr 2020
10. Manikandan AVM, Choudhury S, Majumder S (2017) Text reader for visually impaired people: any reader. In: 2017 IEEE international conference on power, control, signals and instrumentation engineering (ICPCSI)
11. Firebase ML Kit [Online]. Available https://firebase.google.com/docs/ml-kit
12. Rashad M, El-Bakry H, Ismail IS, Mastorakis N (2010) An overview of text-to-speech synthesis techniques. In: International conference on communications and information technology—proceedings
13. Ramiah S, Liong TY, Jayabalan M (2015) Detecting text based image with optical character recognition for English translation and speech using Android. In: 2015 IEEE student conference on research and development (SCOReD)
14. Tejaswi C, Goradiya B, Patel R (2018) A novel approach of tesseract-OCR usage for newspaper article images. J Comput Technol Appl 9(3):24–29p
15. Dalwadi D, Mehta Y, Macwan N (2021) Face recognition-based attendance system using real-time computer vision algorithms. In: Hassanien A, Bhatnagar R, Darwish A (eds) Advanced machine learning technologies and applications. AMLTA 2020. Advances in intelligent systems and computing, vol 1141. Springer, Singapore

Chapter 9
Sine Cosine Algorithm with Centroid Opposition-Based Computation

Tapas Si and Debolina Bhattacharya

1 Introduction

The SCA is a population-based global optimization algorithm to solve the optimization problems, and it was developed by S. Mirjalili in the year 2016 [1]. The central theme of SCA is to utilize the sine and cosine functions to fluctuate the candidate solutions towards or outwards the best solution. The efficiently use of sine and cosine functions ensures the *trade-off* or *balance* between the exploration and exploitation in the search space. The SCA becomes one of the popular optimization algorithms because it is easy to implement and it has less parameter setting. But, like other metaheuristic algorithms, SCA also gets stuck in local optima due to that the operators used for exploration do not work efficiently [2]. Recently, several improved versions of SCA are developed in order to improve its performance. Elaziz et al. [2] proposed an opposition-based SCA (OBSCA) in which opposition-based learning (OBL) was used to make better exploration of the search space to achieve more accurate solutions. Meshkat and Parhizgar [4] proposed a weighted update rule for position update in place of classical position update rule in basic SCA to improve the exploration and exploitation abilities of SCA. In this modified algorithm, first the individual solutions are sorted based on their fitness and the better individuals are assigned higher weights. The average position is computed for a specific number of weighted individuals, and the average position is used to update the position of individuals by weighted position update rule. Qu et al. [3] developed a modified SCA in

T. Si (✉) · D. Bhattacharya
Department of Computer Science & Engineering, Bankura Unnayani Institute of Engineering, Bankura, West Bengal 722146, India
e-mail: shritapassi.ai@gmail.com

D. Bhattacharya
e-mail: debolinab987@gmail.com

which neighborhood search operator and greedy Lévy mutation strategy are incorporated. In this algorithm, first, exponential decreasing transmission control parameter and linear decreasing inertia weight were employed to balance the exploration and exploitation of the search process. Secondly, it used random individuals in place of best individual in the basic algorithm to avoid local optima and random individuals were selected from the neighborhood of the best individual. Thirdly, it used greedy Lévy mutation strategy for the best individual to enhance the local search capability of the algorithm. Gupta et al. [5] developed a modified SCA called MSCA in which a nonlinear time-varying transmission control parameter was employed firstly to make balance between the exploration and exploitation. Secondly, the classical equation for position updating was modified. Finally, a mutation operator was used to generate a new position to avoid getting trap in the local optima. Gupta et al. [6] developed a memory guided SCA (MG-SCA) in which each individual saved its personal best found so far in its memory. The number of guidance is decreased over time during the search. In the beginning of the search process, it facilitates the exploration around the personal best solution, whereas it helps the exploitation later on with less number of guidance. Gupta et al. [7] developed a hybrid SCA called HSCA in which SCA is hybridized with simulated quenching algorithm. This algorithm is used to solve CEC2014 and CEC2017 optimization problems, and it is also applied to train multi-layer perceptron (MLP). Huang et al. [8] developed an improve SCA in which chaotic local search mechanism and Lévy flight operator are used. Bairathi and Gopalani [9] applied OBSCA [2] to MLP training for classification and regression tasks, and it performed better than evolutionary strategy (ES), genetic algorithm (GA), particle swarm optimization (PSO), ant colony optimization (ACO), differential evolution (DE).

In this paper, an OBL scheme known as COBC [15] is incorporated in SCA to create more exploration in the search space to achieve more accurate solutions. The proposed algorithm is applied to solve the CEC2013 benchmark problems. The performance of proposed algorithm COBSCA is compared with classical SCA and OBSCA [2]. The experiment results demonstrate that the COBSCA statistically outperforms other algorithms for most of the problems in the CEC2013 benchmark suite.

The remaining of this paper is organized as follows: The SCA is discussed in Sect. 2. The proposed COBSCA is given in Sect. 3. The experimental setup is given in Sect. 4. The results and discussion are given in Sect. 4.4. The conclusion with future works is finally given in Sect. 5.

2 Sine Cosine Algorithm

The most conventional technique in the expansion of stochastic population-based optimization is the fission of optimization method to two phases, exploration versus exploitation. Combining the random solution in the set of solution abruptly with a high rate of randomness, an optimization problem finds the region of search space.

In these two phases, there are some changes. Some gradual changes occur in the random solutions of exploitation phase however the random variation comparatively less than those in the exploration phase. This algorithm illustrates two equations considering both phases as follows:

$$x_{ij}(t + 1) = x_{ij}(t) + r_1 \times \sin(r_2) \times |r_3 \times P_j(t) - x_{ij}(t)| \tag{1}$$

$$x_{ij}(t + 1) = x_{ij}(t) + r_1 \times \cos(r_2) \times |r_3 \times P_j(t) - x_{ij}(t)| \tag{2}$$

where $x_{ij}(t)$ is the position of the ith solution at jth dimension and tth iteration. P_j is the destination position at jth dimension. r_1, r_2, r_3 are random numbers, and $| \cdot |$ represents the absolute value. For illustrating, as a mathematical model these two equations are combined as follows:

$$x_{ij}(t + 1) = \begin{cases} x_{ij}(t) + r_1 \times \sin(r_2) \times |r_3 \times P_j(t) - x_{ij}(t)|, & r_4 < 0.5 \\ x_{ij}(t) + r_1 \times \cos(r_2) \times |r_3 \times P_j(t) - x_{ij}(t)|, & r_4 \geq 0.5 \end{cases} \tag{3}$$

r_4 is a random number in the range [0, 1], and it switches between sine and cosine components in the above equation. r_1 derives the updated position and find if it is in the space between the solution X_i and destination position P or outside of it. r_2 decides how far the movement will be towards or outwards the destination position. r_3 brings a random weight for the destination position to emphasize or deemphasize the effect of destination position in defining the destination. This formula defines a space between two solutions in the search space. And the most efficient factor of this formula is that it can have the expansion of the higher dimension. The cyclic pattern of sine and cosine function allows a solution to be repositioned around another solution. This can gage exploitation for sure. By changing the range of sine and cosine function, the solution should be able to search outside the space and its limited destination also for exploration. To reach the global optimum, any method should have the balancing efficiency to balance between exploration and exploitation to achieve the destination position in the search space. The range of sine and cosine function has to be changed accordingly for balancing. Using the following equation, the range is changed with need:

$$r_1 = a - a \times \frac{t}{t_{max}} \tag{4}$$

Here t represents the current iteration and t_{max} represents the maximum number of iterations and a is a constant. The SCA algorithm is given in Algorithm 1.

Algorithm 1: SCA

---

1 Initialize the population (X) of size N ;
2 **while** *(termination criteria)* **do**
3 $\quad$| $\quad$ Evaluate the objective function for each individual;
4 $\quad$| $\quad$ Update the destination position, i.e., best solution $(P = X^*)$ found so far;
5 $\quad$| $\quad$ Update the r_1, r_2, r_3 and r_4;
6 $\quad$| $\quad$ Update the position of the individuals using Eq.(3);
7 **return** *the best solution found so far.*

---

3 Centroid Opposition-Based SCA

3.1 Concept of OBL

The primary opposition concept has come to the era of machine learning in 2005 [10]. The different OBL schemes are successfully incorporated in metaheuristic algorithms for performance improvement [12–14]. The definitions of *type-I* opposite point [11] are given below:

Definition 1 Let x be a real number having ascertained in the interval $[a, b]$. The opposite number $\check{x}$ is defined as:

$$\check{x} = a + b - x \tag{5}$$

If $x = 2$ is point in the interval $[-5, 5]$, then the opposite of X is -2.

Definition 2 For each point $X(x_1, \ldots, x_D)$ in the D dimensional space, x_i is defined in the interval $[a_i, b_i]$. There is a unique opposite point $\check{x}_i$ defined as follows:

$$\check{x}_i = a_i + b_i - x_i \tag{6}$$

If $X = (2, 1)$ is a 2D point in the interval $[-5, 5]^2$, the opposite point of X is $\check{X} = (3, 4)$.

3.2 Concept of COBC

The COBC, a OBL scheme, is proposed by Rahnamayan et al. [15], and it is successfully incorporated in DE algorithm which performs better than its competitive algorithms [15]. While computing the centroid opposite points in metaheuristic algorithm, the entire population is considered. Let $(X_1, \ldots, X_N)$ be N points in D dimensional search space which are carrying unit mass in the search space Then the centroid of the body can be defined as follows:

$$M = \frac{X_1 + X_2 + X_3 + \cdots + X_N}{N} \tag{7}$$

The centroid point in jth dimension can be calculated as follows:

$$M_j = \frac{1}{N} \sum_{i=1}^{N} x_{i,j} \tag{8}$$

Having defined the centroid of a discrete uniform body as M, the opposite-point $\check{X}_i$ of a point X_i of the body is calculated as follows:

$$\check{X}_i = 2 \times M - X_i \tag{9}$$

The centroid approach can be employed with a better performance than min-max method. The estimated boundary which is based on the generated sample points is calculated as $[X_{\min}, X_{\max}]$.

3.3 Proposed COBSCA

The proposed COBSCA has centroid opposition-based population initialization and centroid opposition-based generation jumping. In the initialization phase, first the initial population X is randomly generated in the search space range. Then COBC-based opposite solutions $\check{X}(\check{x}_1, \check{x}_2, \ldots, \check{x}_D)$ are computed using Eq. (9) in the dynamic search space range $[a_j, b_j]$ where a_j and b_j are computed using the following equations:

$$a_j = \min_{\forall i}(x_{ij}) \tag{10}$$

$$b_j = \max_{\forall i}(x_{ij}) \tag{11}$$

where i and j are the index of the individual solution dimension, respectively. After computation of centroid opposite point $\check{x}_{ij}$, it may exceed the search space range. If it exceeds b_j, then it is repositioned to the search space by the following:

$$\check{x}_{ij} = M_j + (b_j - M_j) \times \text{rand}(0, 1) \tag{12}$$

If $\check{x}_{ij}$ exceeds a_j, then it is repositioned to the search space by the following:

$$\check{x}_{ij} = a_j + (M_j - a_j) \times \text{rand}(0, 1) \tag{13}$$

The fitness of original solutions and opposite solutions is calculated, and the N best solutions are selected from the set $\{X, \check{X}\}$ based on sorted fitness. After opposition-based initialization, centroid opposition-based generation jumping with a generation jumping probability P_{gj} is incorporated in SCA. In this phase, the COBC is applied to the original solutions to calculate the opposite solutions and the best N solutions are selected in the same way as in the opposition-based initialization phase. The positions of individuals are updated using basic position update rule as in basic SCA with a probability $(1 - P_{gj})$. The complete proposed algorithm COBSCA is given in Algorithm 2.

Algorithm 2: COBSCA

---

1 Initialize the population (X) of size N;
2 Calculate the opposites of individual position;
3 Calculate the fitness of the opposite solutions;
4 Select N best solutions from $\{X, \check{X}\}$;
5 **while** *(termination criteria)* **do**
6 | **if** $rand(0, 1) < P_{gj}$ **then**
7 | | Calculate the opposites of search agent's position;
8 | | Calculate the fitness of the opposite solutions;
9 | | Select N best solutions from $\{X, \check{X}\}$;
10 | **else**
11 | | Evaluate the objective function for each individual;
12 | | Update the destination position, i.e., best solution $(P = X^*)$ found so far;
13 | | Update the r_1, r_2, r_3 and r_4;
14 | | Update the position of the individuals using Eq.(3);

15 **return** *the best solution found so far.*

---

4 Experimental Setup

4.1 Benchmark Functions

The COBSCA is applied to solve the optimization problems from well-known CEC2013 benchmark suite [16]. There are total 28 functions having different level of complexities in the CEC2013 suite. There are 5 unimodal functions, 15 basic multimodal functions, and 8 composition functions.

4.2 Parameter Settings

Parameter settings of SCA, OBSCA, and COBSCA are given in Table 1.

Table 1 Parameter settings of SCA, OBSCA, and COBSCA

Algorithm	Population size (N)	a	P_j
SCA	30	2	–
OBSCA	30	2	–
COSCA	30	2	0.3

4.3 Termination Criteria

When any one of the following criteria is met, the executions of algorithms are terminated:

1. Maximum number of function evaluations (FEs) = 300,000, or
2. Best-error-runs $E = |f^*(x) - f(x)| \leq e$ where $f^*(x)$ is the global optimum, $f(x)$ is the best objective function values, e is the threshold error and it is set to 1e−8.

4.4 PC Configuration

1. CPU: Intel(R) Core(TM) i7-9700K @3.60 GHz
2. RAM: 64 GB
3. Operating System: Windows 10 Professional 64-bit
4. Software: MATLAB 2018b.

Results and Discussion The proposed COBCSA is applied to solve 28 CEC2013 benchmark functions. The results of COBSCA are compared with that of SCA and OBSCA. Each algorithm is executed for 51 independent runs. For making a fair comparison among the algorithms, the same initial population is used for the same run. The mean and standard deviation of the best-error-runs over 51 independent runs are given in Table 2. From this table, it is observed that the COBSCA performs better than others for unimodal functions f_1, f_2, f_5, basic multimodal functions f_6, $f_7, f_9 - f_{19}$, and composite functions $f_{21} - f_{28}$. The OBSCA performs better than others only for unimodal function f_3, whereas SCA performs better than others for unimodal function f_4, basic multimodal function f_8, composite function f_{20}.

To test the statistical significance of the results, a nonparametric test known as Wilcoxon Signed Rank Test [17] has been conducted pairwise with the significance level (α) = 0.5. The statistical results in terms of p-value, h, and significance (Sign.) are given in Table 3. In this table, $h = 1$ and $h = -1$ indicate the rejection of the null hypothesis, whereas $h = 0$ indicates the acceptance of the null hypothesis. $h = 1$ indicates COBSCA statistically performs better than other in the pair, whereas $h = -1$ indicates that other algorithm in the pair statistically performs better than COBSCA. In the same table, "++" symbol indicates that the performs of COBSCA

Table 2 Mean and standard deviation (in parenthesis) of the best-error-runs over 51 independent runs

F	SCA	OBSCA	COBSCA
1	11198.24652 (2424.58365)	29361.47256 (5208.615456)	**262.5418215 (238.7737216)**
2	139843050.7 (40300431.11)	820722486.6 (263096692.7)	**38018327.31 (24304687)**
3	31232642409 (11807399727)	**2.15052E+17 (7.33936E+17)**	6201005714 (4182891294)
4	**34861.01589 (5854.3273)**	71639.39358 (8017.4303)	41948.08959 (5846.0471)
5	2185.380909 (653.0384)	17596.13075 (6245.9470	**1201.349377 (504.8384)**
6	657.9055881 (204.4592)	4158.52503 (1029.8793)	**123.9566411 (39.3257)**
7	171.1760991 (36.5773)	174797.8054 (321234.5812)	**68.62181289 (17.0815)**
8	**20.93013125 (0.0499)**	21.37763794 (0.0760)	20.95379451 (0.0465)
9	39.1161094 (1.4029)	35.78950086 (1.6608)	**19.95105229 (3.1298)**
10	1527.294815 (307.8228)	4773.870774 (864.9714)	**119.47961 (81.1645)**
11	356.2387384 (34.9244)	704.6229585 (80.2031)	**94.27963981 (22.6794)**
12	380.8177762 (33.0978)	654.7526204 (80.0879)	**137.1278625 (29.2612)**
13	377.4257892 (33.0501)	683.8032663 (74.2906)	**155.2534239 (28.9159)**
14	7120.181257 (370.2006)	6917.052574 (359.8580)	**4614.71736 (1088.5494)**
15	7356.48517 (354.6053)	6668.995514 (442.9273)	**6272.449043 (954.3193)**
16	2.558968089 (0.2660)	3.655032532 (0.6998)	**2.550784349 (0.2998)**
17	485.5876162 (37.2305)	834.8849447 (68.0424)	**175.4917083 (18.9569)**
18	474.8779693 (43.8687)	832.5276337 (68.4394)	**212.8752027 (15.5913)**
19	2909.150501 (2107.8254)	501959.2978 (195714.1437)	**574.1575449 (1219.6852)**
20	**13.95093977 (0.4237)**	15 (0)	14.13122076 (0.3303)
21	1843.229908 (179.0569)	2420.172576 (69.6777)	**596.467004 (181.6159)**
22	7522.509925 (483.5919)	7071.354946 (470.4435)	**4630.314435 (1668.0191)**
23	7810.220263 (314.9683)	7571.412266 (536.3565)	**7274.922451 (907.6651)**
24	316.9016623 (6.4922)	362.6472609 (12.6735)	**240.4592579 (7.5375)**
25	328.2664964 (4.6563)	390.6408619 (13.6445)	**285.3624345 (7.2478)**
26	211.8836 (5.3840)	243.5038 (15.0460)	**202.1841026 (3.8760)**
27	1359.850344 (43.5924)	1569.407559 (143.1145)	**707.2639228 (111.2301)**
28	2591.279897 (193.6934)	6018.538158 (605.9221)	**934.2615568 (209.8668)**

have extremely higher significance over others when p-value is much less than 0.5. "+" indicates a higher significance when p-value is less than 0.5, whereas "$\approx$" indicates that there is no significant difference in the performance of the algorithms in the pair. From Table 3, it is observed that COBSCA has extremely higher significant performance over SCA for functions $f_1 - f_3$, $f_5 - f7$, $f_9 - f_{15}$, $f_{17} - f_{19}$, f_{21}, f_{22}, $f_{24} - f_{28}$. COBSCA has a higher significant performance over SCA for function f_{23}. On the other hand, SCA has extremely higher significant performance over COBSCA for functions f_4, and f_{20}. SCA has a higher significant performance

Table 3 Wilcoxon signed rank test results

F	COBSCA versus SCA			COBSCA versus OBSCA		
	p-value	h	Sign.	p-value	h	Sign.
1	7.56e−10	1	++	7.56e−10	1	++
2	9.07e−10	1	++	7.56e−10	1	++
3	8.53e−10	1	++	7.56e−10	1	++
4	1.81e−06	−1	++	7.56e−10	1	++
5	9.64e−08	1	++	7.56e−10	1	++
6	7.56e−10	1	++	7.56e−10	1	++
7	8.03e−10	1	++	7.56e−10	1	++
8	0.010093815	−1	+	7.56e−10	1	++
9	7.56e−10	1	++	7.56e−10	1	++
10	7.56e−10	1	++	7.56e−10	1	++
11	7.56e−10	1	++	7.56e−10	1	++
12	7.56e−10	1	++	7.56e−10	1	++
13	7.56e−10	1	++	7.56e−10	1	++
14	7.56e−10	1	++	7.56e−10	1	++
15	1.91e−07	1	++	0.050607988	0	≈
16	0.881059837	0	≈	6.02e−09	1	++
17	7.56e−10	1	++	7.56e−10	1	++
18	7.56e−10	1	++	7.56e−10	1	++
19	1.02e−07	1	++	7.56e−10	1	++
20	0.012582495	−1	++	7.56e−10	1	++
21	7.56e−10	1	++	7.56e−10	1	++
22	1.98e−09	1	++	3.17e−09	1	++
23	0.000151309	1	+	0.286113067	0	≈
24	7.56e−10	1	++	7.56e−10	1	++
25	7.56e−10	1	++	7.56e−10	1	++
26	9.64e−08	1	++	7.56e−10	1	++
27	7.56e−10	1	++	7.56e−10	1	++
28	7.56e−10	1	++	7.56e−10	1	++
Win		24			26	
Loss		3			0	
Tie		1			2	

over COBSCA for function f_8. There is no statistical significant difference in the performance of SCA and COBSCA for function f_{16}. COBSCA has extremely higher significant performance over OBSCA for functions $f_1 - f_{14}$, $f_{16} - f_{22}$, $f_{24} - f_{28}$. There is no statistical significant difference in the performance of COBSCA and OBSCA for functions f_{15}, and f_{23}. In the bottom of Table 3, numbers of win, loss, and tie are calculated. It is observed that COBSCA statistically outperforms SCA

Table 4 Computation complexities (in seconds) of SCA, OBSCA, and COBSCA for function f_{14} with dimension $= 30$

Algorithm	T_0	T_1	$\bar{T}_2$	$(\bar{T}_2 - T_1)/T_0$
SCA	0.087399	0.856036	1.8994	11.9379
OBSCA	–	–	1.4666	6.9859
COBSCA	–	–	1.6889	9.5294

for 24 functions and SCA statistically outperforms COBSCA for three functions, whereas there is a single tie. It is also observed that COBSCA statistically outperforms OBSCA for 26 functions. There are two ties between COBSCA and OBSCA.

The computational complexities are calculated as per the criteria given in [16]. The computational complexities (in seconds) of SCA, OBSCA, and COBSCA for function f_{14} with dimension $= 30$ are given in Table 4. From this table, it is observed that the computational complexity of OBSCA is lower than both SCA and COBSCA. The computational complexity of COBSCA is lower than SCA.

From the above analysis of the results, it is found that the proposed COBSCA statistically outperforms SCA and OBCSA for most of the problems in CEC2013 benchmark suite and it is computationally more cost effective than SCA. Therefore, it may be concluded that the proposed COBSCA is efficient and effective in numerical function optimization.

5   Conclusion and Future Works

In this paper, an improve version of SCA by incorporating COBC, COBSCA is proposed. The COBSCA is applied to solve CEC2013 benchmark optimization problems having different level of complexities. The performance of COBSCA is compared with SCA and OBSCA. The experimental results with statistical analysis demonstrate that the COBSCA statistically outperforms SCA and OBSCA for most of the problems. The future research is directed towards the improvement of SCA by incorporating other OBL schemes. The proposed COBSCA can be further improved by employing the adaptive strategies for computing the probability of generation jumping.

References

1. Mirjalili S (2016) SCA: a Sine Cosine Algorithm for solving optimization problems. Knowl-Based Syst. https://doi.org/10.1016/j.knosys.2015.12.022
2. Elaziz MA, Oliva D, Xiong S (2017) An improved Opposition-Based Sine Cosine Algorithm for global optimization. Expert Syst Appl 90:484–500

3. Qu C, Zeng Z, Dai J, Yi Z, He W (2018) A modified Sine-Cosine Algorithm based on neighborhood search and greedy levy mutation. Comput Intell Neurosci 2018. Article ID 4231647, 19 pp. https://doi.org/10.1155/2018/4231647
4. Meshkat M, Parhizgar M (2017) A novel weighted update position mechanism to improve the performance of sine cosine algorithm. In: 2017 5th Iranian joint congress on fuzzy and intelligent systems (CFIS). IEEE, pp 166–171
5. Gupta S, Deep K, Mirjalili S, Kim JH (2020) A modified Sine Cosine Algorithm with novel transition parameter and mutation operator for global optimization. Expert Syst Appl. https://doi.org/10.1016/j.eswa.2020.113395
6. Gupta S, Deep K, Engelbrecht AP (2020) A memory guided sine cosine algorithm for global optimization. Eng Appl Artif Intell 93:103718
7. Gupta S, Deep K (2020) A novel hybrid sine cosine algorithm for global optimization and its application to train multilayer perceptrons. Appl Intell 50:993–1026. https://doi.org/10.1007/s10489-019-01570-w
8. Huang H, Feng X, Heidari AA, Xu Y, Wang M, Liang G, Chen H, Cai X (2020) Rationalized Sine Cosine Optimization with efficient searching patterns. IEEE Access. https://doi.org/10.1109/ACCESS.2020.2983451
9. Bairathi D, Gopalani D (2017) Opposition-Based Sine Cosine Algorithm (OSCA) for training feed-forward neural networks. In: Proceedings of 13th international conference on signal-image technology and internet-based systems (SITIS). IEEE, pp 438–444. https://doi.org/10.1109/SITIS.2017.78
10. Tizhoosh HR (2005) Opposition-based learning: a new scheme for machine intelligence. In: Proceedings of the international conference on computational intelligence for modelling, control and automation, and international conference on intelligent agents, web technologies and internet commerce (CIMCA-IAWTIC'05), vol 1. IEEE, pp 695–701
11. Al-Qunaieer FS, Tizhoosh HR, Rahnamayan S (2010) Opposition based computing—a survey. In: Proceedings of international joint conference on neural networks (IJCNN), pp 1–7
12. Xu Q, Wang L, Wang N, Hei X, Zhao L (2014) A review of opposition-based learning from 2005 to 2012. Eng Appl Artif Intell 29:1–12
13. Rojas-Morales N, Rojas M-CR, Ureta EM (2017) A survey and classification of Opposition-Based Metaheuristics. Comput Ind Eng 110:424–435
14. Mahdavi S, Rahnamayana S, Deb K (2018) Opposition based learning: a literature review. Swarm Evol Comput 39:1–23
15. Rahnamayan S, Jesuthasan J, Bourennani F, Salehinejad H, Naterer GF (2014) Computing opposition by involving entire population. In: Proceedings of IEEE congress on evolutionary computation (CEC), pp 1800–1807
16. Liang JJ, Qu BY, Suganthan PN, Hernández-Díaz AG (2013) Problem definitions and evaluation criteria for the CEC 2013 special session on real-parameter optimization. http://www.ntu.edu.sg/home/EPNSugan/indexfiles/CEC2013/CEC2013.htm
17. Derrac J, Garcia S, Molina D, Herrera F (2011) A practical tutorial on the use of nonparametric statistical tests as a methodology for comparing evolutionary and swarm intelligence algorithms. Swarm Evol Comput 1:3–18

Chapter 10
Data Mining Techniques for Fraud Detection—Credit Card Frauds

Krishna Kushal, Greeshma Kurup, and Siddhaling Urolagin

1 Introduction

Fraud is an act of deliberately deceiving a victim to gain information that wouldn't be readily available without deception. It usually includes coverups, misleading data, false accusations, and other illegitimate activities that intern lead to disruption of customer relations, loss of reputation, and self-sacrifice.[2] There can be various types of fraud including credit card fraud, insurance fraud, tax evasion, investment fraud, medical-related fraud, and many more. These criminal deceptions may not even see the light of the law, and hence, organizations require fraud prevention and detection mechanisms [3]. Fraud preventions hinder the occurrences of the illicit activity in the first place, while detection systems deal with the impostors who have struck down the prevention systems. We are going to investigate specifically on credit card and related frauds, as it is a prominent mode of transaction. Being extremely convenient for users, credit card holders are increasing every day. But with all its benefits, it poses some predicaments as imposters can take over and deliberately rip you off. The state when an individual uses another individual's credit card without authorization is called a credit card fraud. Detection of this kind of fraud in terms of data mining is an information exhausting procedure that uses various techniques like clustering, classification procedures, and segmentation to find links, guidelines, and patterns in the data that would ultimately aid in detecting fraudulent activity [4]. The progression of technology as vastly increased the bulk of data that is required to be processed. Enormous amounts of data need to be manipulated, processed, and extrapolated

K. Kushal (✉) · G. Kurup (✉) · S. Urolagin (✉)
Birla Institute of Technology and Science, Pilani, Academic City, Dubai, UAE
e-mail: f20170056@dubai.bits-pilani.ac.in

G. Kurup
e-mail: f20170289@dubai.bits-pilani.ac.in

S. Urolagin
e-mail: siddhaling@dubai.bits-pilani.ac.in

© The Author(s), under exclusive license to Springer Nature Singapore Pte Ltd. 2021
X.-Z. Gao et al. (eds.), *Applications of Artificial Intelligence in Engineering*, Algorithms for Intelligent Systems, https://doi.org/10.1007/978-981-33-4604-8_10

efficiently by the detection system to make it sustainable in the real world. Using data mining techniques, we can train models to detect former illegitimate activities, and this can be further applied to predict any suspicious activities on a new dataset [5]. These techniques elevate the precision during detection and also drastically increases performance. And if the model is trained well, then it can predict with ease, even if it is not previously known data [1]. This mined information proves to be highly essential for organizations to make vital decisions that in due course help them thrive in this financial world. In this paper, various data mining methods used for credit card fraud detection are described in the following section. The methodology for the prototype tested in this project is elaborated in Sect. 3. The experimental observations and statistical results are documented in Sect. 4, followed by the conclusion of this paper.

1.1 Literature Survey

Credit card fraud is a common and frequent threat faced. With the advent of digital banking, credit cards are of two types, virtual and physical. A credit card fraud occurs when credit amount or account issued to a customer which is managed or interacted with, without the knowledge of the card holder. Out of the numerous data mining methods, few of them which are quite effective in this sort of fraud detection are summarized as decision trees, genetic algorithm (AI), hidden Markov model (HMM), K-means clustering method, and neural networks [1, 6].

Apart from the most widely used conventional data mining techniques, some effective ones used to minimize false positives are methods like K-Nearest Neighbors (KNN) and outlier detection. KNN is a part of supervised learning that handles problems based on regression and classification. Outliers appear in anomaly detection of data mining that are particularly removed for refined results with reduced false values. These can also be used for fraud prevention. The next method of concern is adversarial machine learning technique. This field of ML and data mining is used for training and fooling models based on harmful or erroneous inputs. Adversarial ML is used to strategize the best way of transaction, learn the nature of fraudulent activity, and use the information to predict future frauds and thereby take measures in advance to prevent further complications. Furthermore, streaming analytics is based on live or real-time analysis, detection and protection technique. Pattern recognition and even signal processing have been combined to compare the existing fraud evidences with the statuses of the credit card accounts and businesses to determine the authenticity of the transaction. Several clustering approaches are also incorporated to achieve the same. Apart from these unconventional methods, hybrids of conventional techniques such as ANN with perceptron and decision trees, etc. have proven to be innovative and effective in detecting similar credit card and business-related frauds. Parameter optimization is used to compare among different conventional methods based on cost, profit, performance, etc., [3–5].

As mentioned earlier, hybrids and combining different individual techniques can result in highly effective and optimized method for prevention and detection of frauds. Trends of ten years based on researches by various scholars and professors have given rise to different and creative ways. Some use subclasses of classification models, some use other models, and some use both. Some latest techniques among the described ones are a combination of

- HMM, K-mean and clustering without classification models (2019)
- Classification approaches consisting of decision trees, random forest, and naive Bayes (2019).
- Support vector machines, NN, regression, decision trees are used from classification along with genetic algorithm and rule induction (2018) [7] (Fig. 1).

Knowledge extraction in data mining is the crucial part of any model developed. They are executed in seven stages—Data cleaning, integration, selection, transformation, mining, pattern evaluation, and knowledge evaluation. These processes sequentially remove unwanted data initially, then combine and refine relevant data; patterns are created from the final data and analyzed and is finally translated into user's result. The working of conventional DM techniques is pivotal for their implementation into applications such as credit card frauds:

- Decision trees—recursive traversals among the roots and child nodes, in the case of credit card frauds, IP addresses, and transaction locations are tracked for detecting fraudulent activities.
- Neural networks—input layers go through optional hidden layer(s) to give an output (layer). The inputs would generally consist of the necessary information like card number or pin to successfully execute a transaction. Threshold values and activation functions would be chosen accordingly to compare the nature of transaction which is mostly unique according to the bank's algorithm.
- K-Mean clustering—involves finding alike patterns and making groups according to them, say normal and abnormal transactions, with other intermediate clustering

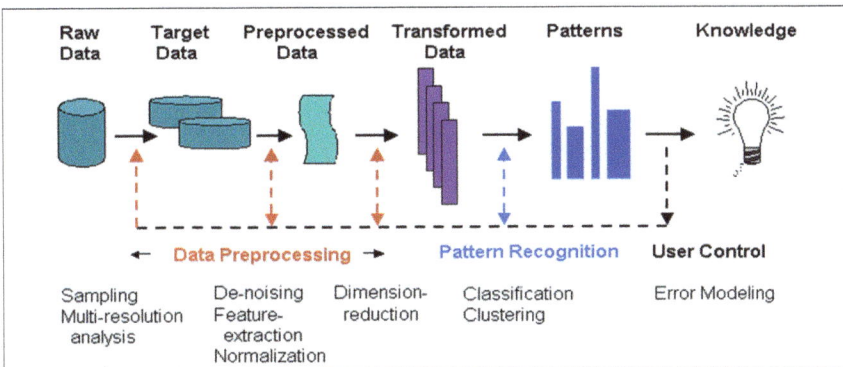

Fig. 1 Knowledge extraction steps in data mining [11]

based on some confidential information like the pin that is only supposed to be with the card holder.

- Hidden Markov model—a probability-based model which is only known to client (hidden from anyone else and hence the name) and tracks the expenditure rates (based on spending from the card) and classifies into low, normal, and high. These past information aid in the validation of future transactions and hence in fraud detection.
- Genetic algorithm—AI-based algorithm that finds critical values and then identifies whether card activity is authorized or not [2, 8].

Different kinds of fraud, however, related to financial frauds are businesses or insurance based. Automobile frauds are mostly concerning insurance forgery or illegal exploitation or even misuse of the benefits of vehicle insurance provided to the owners. Manual analysis and fraud detection are really time-consuming and in most cases, expensive and inefficient. Several data mining models have been researched ever since the need has come. A hybrid model was projected in 2004 based on supervised learning techniques such as meta classifiers and fuzzy logic to choose best models among base ones and test it on existing datasets. Five years later IAA model was proposed, based on unsupervised learning. Later, statistical models and related functions were researched to bring out a new model [9]. One of the business anomaly detection processes researched was based on UDMACTFDQ along with clustering was used; and corresponding mean and standard deviations were obtained to validate results [10].

2 Methodology

It is highly essential that credit card companies have the capability to identify fraudulent transactions. This ability will allow them to have a clear picture regarding the transactions of their customers and also aid in better customer service. Here, we will explain the methodology for detecting credit card fraud (Fig. 2).

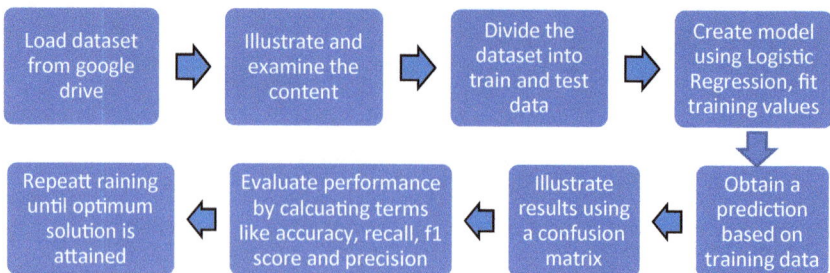

Fig. 2 Block diagram of methodology

Initially, we are required to load the dataset of transactions into our python code. This dataset contains 284,807 transactions, and amidst those, 492 cases are a scam. Which means that only 0.172 percentile of cases is fraudulent. All the entries except amount and time have undergone principal component analysis transformation to become numerical data as data in that form is more apt for data analysis. Principal Component Analysis (PCA) is a technique that uses an orthogonal conversion which converts a group of associated variables into a group containing unassociated variables [11]. In order to maintain discretion, the actual features aren't disclosed. The amount column is the actual quantity obtained during transactions, and the class column describes whether or not it's a fraud case.

After loading the data, we examine and evaluate out dataset contents using python methods like head(), info(), describe(), and so on. Further, we illustrate our transformed data in the form of graphs for better understanding. Subsequently, the dataset is split into two parts. One part contains all the features which are used for testing while the other has the target class values which are used for training. Now, we create a fraud detection model using logistic regression.

2.1   Logistic Regression

Logistic regression model is a probability-based model with only two possible results. There are numerous predictors that can be used and they aren't required to be normally distributed. Logistic regression based on odds ratio:

$$T/(1 - T) = e(a_0 + a_1 x) \qquad (1)$$

where a_0 is a constant, and a_1 is the slope. a_1 is also the ratio of the log odd changes with respect to x. After applying log to both sides, we can show that it can take care of any number of numerical elements. Odds ratio expanded form:

$$T = 1/(1 + e^{-(a_0 + a_1 x_1 + a_2 x_2 + a_3 x_3 + \cdots + a_T x_T)}) \qquad (2)$$

Since they have only two possible results, this model is best suited for our problem.

After creating this model, we fit this data with respect to the x and y training values and simulate our data with various testing and training sizes to obtain the best performance. We illustrate our evaluation with the help of a confusion matrix (as displayed in Fig. 6). Our model will make predictions, and the performance the various simulations are evaluated by analyzing the terms like precision, recall, $f1$ score, and accuracy. Precision is the skill to find pertinent data points while recall is the capability of discovering the applicable cases in a dataset.

- Precision and recall can be calculated using the following formula

$$\text{Precision} = \text{true positives}/(\text{true positives} + \text{false positives}) \qquad (3)$$

$$\text{Recall} = \text{true positives}/(\text{true positives} + \text{false negatives}) \qquad (4)$$

- $F1$ score of the other hand is the harmonic mean of recall and precision.

$$F1 = 2 * (\text{Precision} * \text{Recall})/(\text{Precision} + \text{Recall}) \qquad (5)$$

- Accuracy is the ratio of the accurate predictions to the summation of all predictions.

$$\text{Accuracy} = \text{Number of correct predictions}/\text{Total number of predictions} \qquad (6)$$

3   Statistical and Experimental Results

The dataset used here has 492 frauds out of all the cases, as mentioned in Sect. 3 (Table 1).

These plots are plotted against the unspecified credit card transaction features $V1$, $V2$ … $V28$ with respect to the number of seconds have gone between the current transaction and the initial transaction in the dataset (Figs. 3, 4, 5 and 6).

This confusion matrix takes a testing and training size of 50% and classifies them into four categories. 142143 are cases that are not sfraudulent and are correctly predicted as true cases. 96 cases are actually fraud cases but predicted incorrectly, also called true negatives. 34 are false positives, and finally, 131 are accurately detected as fraudulent cases (Figs. 7 and 8; Table 2).

A graphical representation of the targeted data to illustrate the true vs. fraudulent cases in our dataset. This dataset contains 284,807 transactions; out of which, 284,315 are true cases while the rest are fraudulent. Two descriptive tables are obtained as we try to explore the dataset. First step done is to acquire more information about the data used. This results in the number of rows, data type, and the data space of the set. The table below shows this information. The second table provides statistical data (provided in the statistical results section).

After training the model with different test and train sizes, we obtain the confusion matrices to find out the true classes and predicted classes of fraudulent and non-fraudulent credit card activities. To check the performance of the logistic regression model—the accuracy, recall, precision, and $F1$ scores are also obtained. To look into the trends, each of the result is plotted against the changing test size used while training. The figures are shown below.

In Figs. 9 and 10, the recall and $F1$ scores are plotted for different test sizes, where recall is the ratio of true positives to actual number of positives (true positives and false negatives), and for the case with training size of 80% out of the dataset, recall and $F1$ score are highest, showing that the results are better for that case.

In Figs. 11 and 12, precision and accuracy are plotted. As observed in the plots, the accuracies differ very minutely and is therefore not a useful measure to compare

Table 1 Statistical report of dataset (partial)

	Time	V1	V2	V3	V4	V5	V6	V7	V8	V9	V10
Count	284807	2.85E+05	2.85E+05	2.85E+05	2.85E+05	2.85E+05	2.85E+05	2.85E+05	2.85E+05	2.85E+05	2.85E+05
Mean	94813.85958	3.92E−15	5.69E−16	−8.77E−15	2.78E−15	−1.55E−15	2.01E−15	−1.69E−15	−1.93E−16	−3.14E−15	1.77E−15
std	47488.14596	1.96E+00	1.65E+00	1.52E+00	1.42E+00	1.38E+00	1.33E+00	1.24E+00	1.19E+00	1.10E+00	1.09E+00
min	0	−5.64E+01	−7.27E+01	−4.83E+01	−5.68E+00	−1.14E+02	−2.62E+01	−4.36E+01	−7.32E+01	−1.34E+01	−2.46E+01
25%	54201.5	−9.20E−01	−5.99E−01	−8.90E−01	−8.49E−01	−6.92E−01	−7.68E−01	−5.54E−01	−2.09E−01	−6.43E−01	−5.35E−01
50%	84692	1.81E−02	6.55E−02	1.80E−01	−1.98E−02	−5.43E−02	−2.74E−01	4.01E−02	2.24E−02	−5.14E−02	−9.29E−02
75%	139320.5	1.32E+00	8.04E−01	1.03E+00	7.43E−01	6.12E−01	3.99E−01	5.70E−01	3.27E−01	5.97E−01	4.54E−01
max	172792	2.45E+01	2.21E+01	9.38E+00	1.69E+01	3.48E+01	7.33E+01	1.21E+02	2.00E+01	1.56E+01	2.37E+01

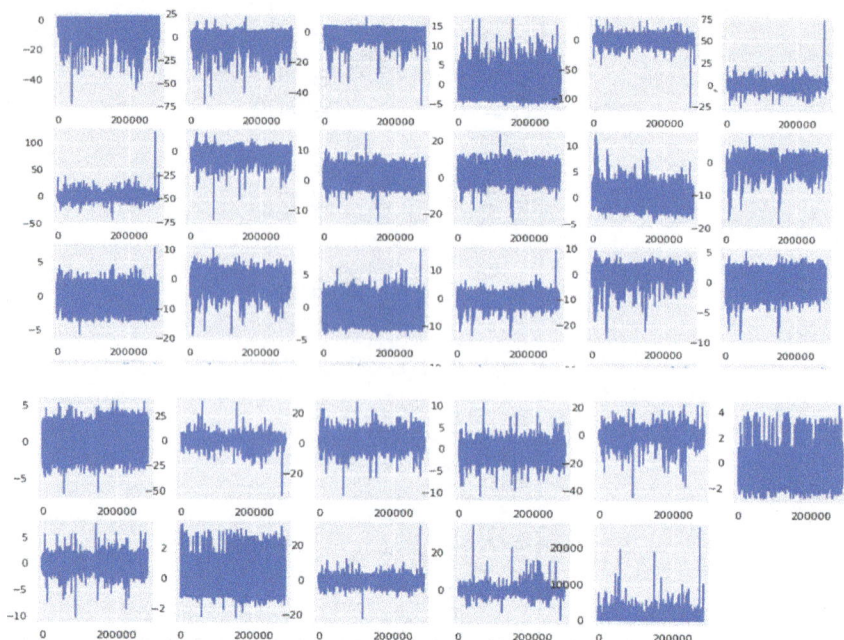

Fig. 3 Data plotted with respect to each of its column entries

```
Length of trainX is: 142403
Length of testX is: 142404
Length of trainY is: 142403
Length of testY is: 142404
```

Fig. 4 Length of train and test data (test and train size)) = 50%)

```
LogisticRegression(C=1.0, class_weight=None, dual=False, fit_intercept=True,
                   intercept_scaling=1, l1_ratio=None, max_iter=100,
                   multi_class='auto', n_jobs=None, penalty='l2',
                   random_state=None, solver='lbfgs', tol=0.0001, verbose=0,
                   warm_start=False)
```

Fig. 5 Logistic regression model used in code

the performance. Precision can be considered a good discriminator; however, recall is also a reliable one; therefore, both recall and precision are taken together as $F1$ score.

It is known that accuracy refers to the probability of true values or classes (true positive and true negatives) over the sum of all the classes. In case of this experimental model, from the plot, it can be observed that accuracy value does not convey much

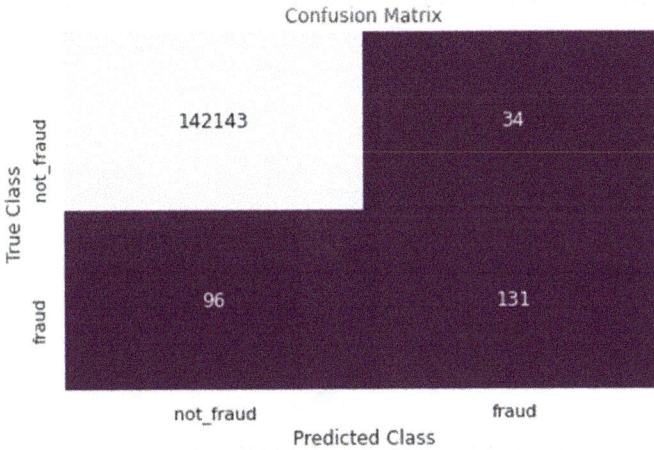

Fig. 6 Confusion matrix based on test-*y* values and predictions

```
Precision for Logistic Regression Model 1 : 0.79
Sensitivity/Recall for Logistic Regression Model 1 : 0.58
F1 Score for Logistic Regression Model 1 : 0.67               Accuracy
0.9990871042948232  ←
```

Fig. 7 Precision, recall, *f*1 score, and accuracy

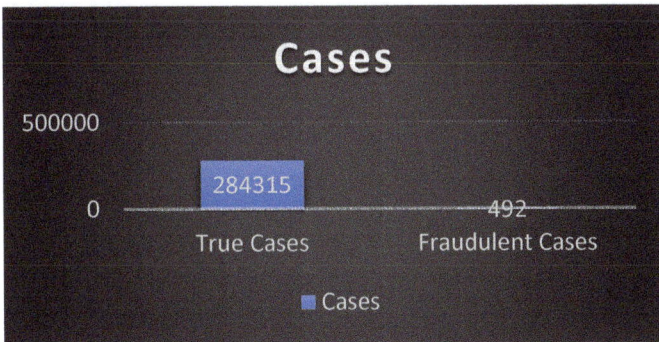

Fig. 8 Analysis of dataset—for true versus fraudulent cases

Table 2 Tabulation for precision and recall based on training and testing size

Training size (%)	Testing size(%)	Precision	Recall
50	50	0.79	0.58
60	40	0.80	0.57
80	20	0.85	0.6
70	30	0.86	0.56

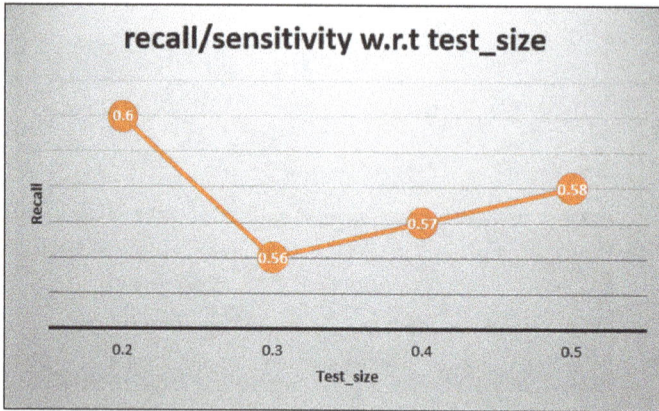

Fig. 9 Plot of recall versus test_size

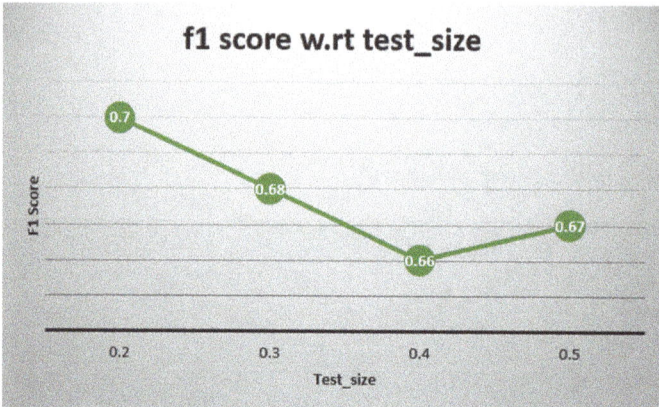

Fig. 10 Plot of $F1$ score versus test_size

about how well the model works. The change in accuracy is minute, and the values are really high; about 99% for all the cases, which give the impression that this model is highly accurate and shows the best performance. However, the model predictions are much lesser than the accuracy values. The recall, (which is the ratio of true positive cases over all the true cases) and $f1$ score, is observed to give a more realistic information on the performance of the model.

In the confusion matrix for each test case, we can find a few fraudulent cases being wrongly classified (upto 30%). This shows us that a **pure accuracy-based** evaluation of the performance is misleading. This is sometimes referred to as accuracy fallacy. The number of true or non-fraudulent cases in our dataset is really huge compared to the actual frauds. As a result, accuracy will be high, and thus, it is not a good measure to evaluate the performance. Based on research, $F1$ score has been found

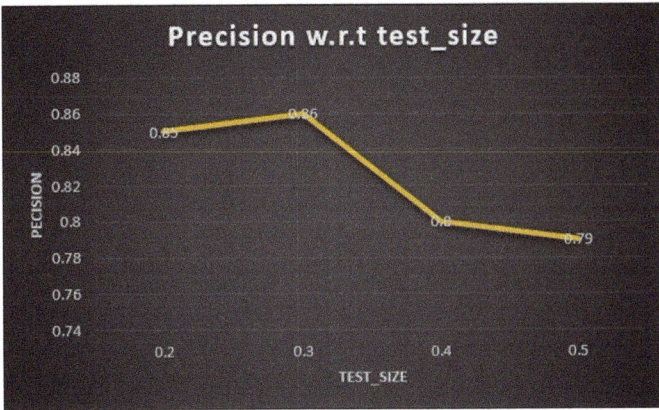

Fig. 11 Plot of precision versus test_size

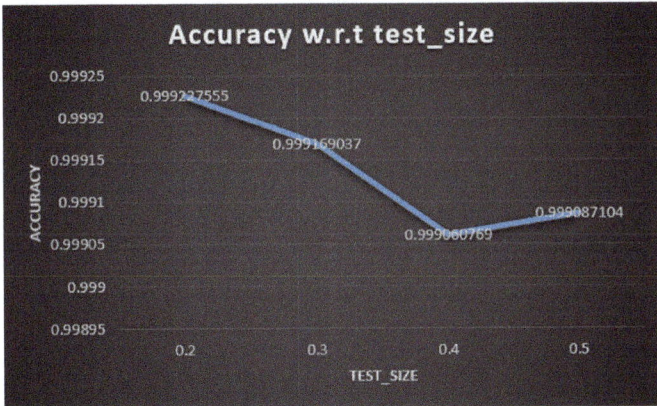

Fig. 12 Plot of accuracy versus test_size

to be a better measure to judge the performance of a model when the false positives, and false negatives are of more concern, as in the case of credit card frauds; since it is a harmonic average of the recall and precision values that give a better idea about the true classifications and misclassifications. Thus, **higher the $F1$ score, better the performance**. Figure 13 shows how $f1$ score can be more informative and reliable than the accuracy:

The accuracy is almost constant when compared to the $f1$ score and therefore is not as helpful as $f1$ score. Comparing both $F1$ score and accuracy, the former is a reliable and significant in measuring the performance.

Therefore, out of the cases tested in this assignment, the training case with $F1$-score of 0.7 is the best. The confusion matrices for each of the cases mentioned (in Table 3) are shown.

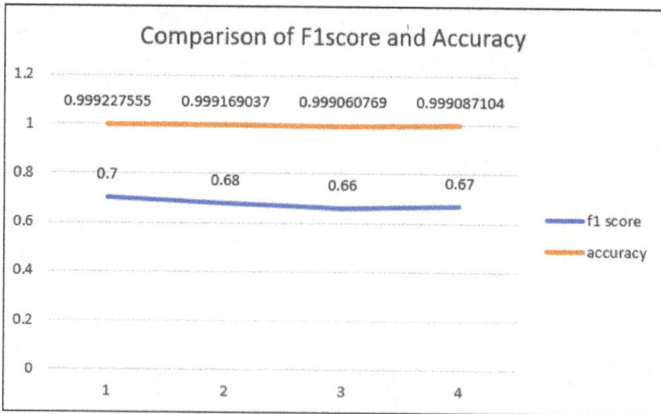

Fig. 13 Plot of accuracy versus $F1$ score

Table 3 Trends of $F1$-score and accuracy with train-test split

S. no.	Percentage of training data	Percentage of testing data	$F1$ score	Accuracy (in %)
1	0.8	0.2	0.7	99.92
2	0.7	0.3	0.68	99.91
3	0.6	0.4	0.66	99.90
4	0.5	0.5	0.67	99.91

It is observed that the confusion matrix (in Fig. 14) for the maximum $F1$ score obtained (0.7) has least false positives; i.e., the mismatch between true class and

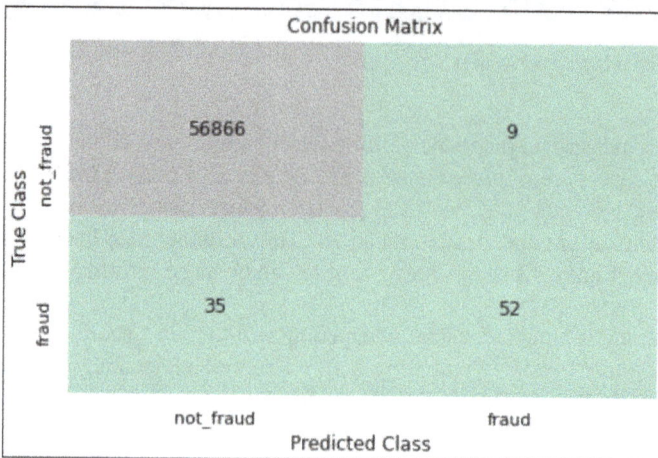

Fig. 14 $F1$ score $= 0.7$

predicted class compared to other cases. Thus, this shows that the $F1$ score is reliable to estimate how good the prediction is. Based on this, further intensive training can give better $F1$ scores and thus lesser false positives than the ones experimented here. This way we can achieve a more professional model to predict fraudulent activities effectively (Figs. 15, 16, and 17).

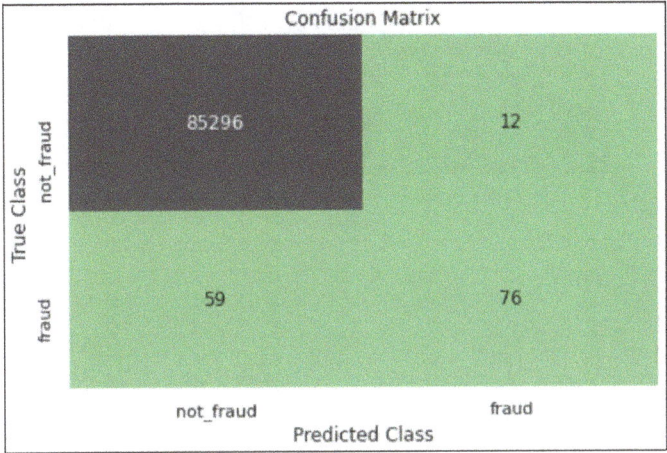

Fig. 15 $F1$ score $= 0.68$

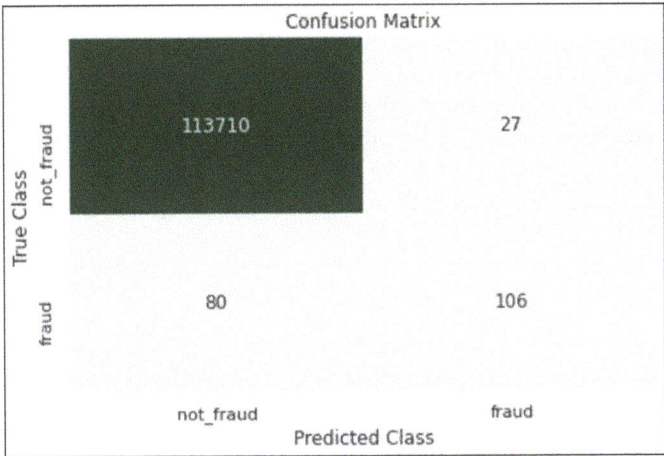

Fig. 16 $F1$ score $= 0.66$

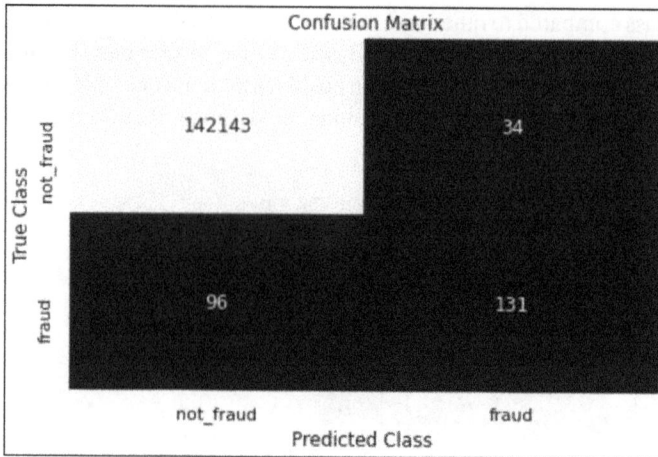

Fig. 17 $F1$ score $= 0.67$

4 Conclusion

Data mining is a powerful tool in various types of fraud detection models. It has proven to improve the efficiency in the process, over the years and continues to be. Although many new unconventional methods have been projected, conventional models like HMM, KNN, ANN, genetic algorithms, etc., continue to be in practice, but in more sophisticated and hybrid combinations. These data mining models can be extended across several other detection models such as intrusion, malware, and even healthcare. Other concepts of IT and computer science can be successfully combined with data mining to include the advantages of all these variant concepts into one model. These models have assuredly led to the utilization of the leverage of IT and CS concepts and have in turn led to the transition from manual audit practices of fraud detection that is highly inefficient and costly in this era of high-scale tech-savvy and progressed population.

References

1. Amanze BC, Onukwugha CG (2018) Data mining application in credit card fraud detection system. Int J Trend Res Dev 5(4):2394–9333
2. Arpita M, Chelsi S, Sunil K (2019) An overview of credit card fraud detection using data mining techniques. Int J Adv Res Comput Sci Manage 5(4):2395–1052
3. Aliza Z, Mehreen S (2018) A survey on application of data mining techniques. it's proficiency in fraud detection of credit card, research and reviews. J Eng Technol:2319–9873
4. Galina B, Helmut K (2018) Reducing false positives in fraud detection: combining the red flag approach with process mining. Int J Account Inf Syst 31
5. Kaiwen M, Sandip S (2018) Use of data mining methods to detect test fraud, research and reviews. J Eng Technol

6. Suresh G, Justin RJ (2018) A study on credit card fraud detection using data mining techniques. Int J Data Mining Techn Appl 7(1):2278–2419, 21–24
7. Mehak M, Sandeep S (2019) Detect frauds in credit card using data mining techniques. Int J Innov Technol Explor Eng 9(2):2278–3075
8. Randula K (2018) Data mining techniques for credit card fraud detection. Int J Adv Res Comput Sci Manage:337085824
9. Sharmila S, Suvasini P (2018) Detection of automobile insurance fraud using feature selection and data mining techniques. Int J Rough Sets Data Anal 5(3):326102521
10. Ezenwafor JIU, Frank O (2019) Utilization of data mining and anonymous communication techniques for fraud detection in large scale business organizations in delta state. Br J Educ 7(11):2054–6351
11. Credit Card Fraud Detection, Kaggle Data Page. https://www.kaggle.com/mlg-ulb/creditcardfraud. Last accessed 2020/06/26

Chapter 11
Performance Evaluation of β Chaotic Map Enabled Grey Wolf Optimizer on Protein Structure Prediction

Akash Saxena, Shalini Shekhawat, Ajay Sharma, Harish Sharma, and Rajesh Kumar

1 Introduction

Proteins are basic building blocks of cells. Proteins are responsible for all cell activities. Proteins are complex macro-molecules that are formed with amino acids and combined with peptide bonds. During synthesis, amino acid sequences can form a three-dimensional structure. The knowledge of this structure helps doctors to disease prediction. Moreover, the information of this structure is inevitable for drug design, pharmaceuticals and for other clinical investigations. Accurate knowledge of this structure is inevitable and evolved methodologies for finding these, have paramount importance in computational biology stream. The final 3-D structure obtained by protein represented as conformation of peptide chains considering freedom of rotations. This final structure is called native structure of protein and possess minimal free energy value [1].

A. Saxena (✉) · S. Shekhawat
Swami Keshvanand Institute of Technology Management and Gramothan, Jaipur, India
e-mail: aakash.saxena@hotmail.com

S. Shekhawat
e-mail: shekhawatshalini17@gmail.com

A. Sharma
Government Engineering College, Jhalawar, India
e-mail: ajay_2406@yahoo.com

H. Sharma
Rajasthan Technical University, Kota, India
e-mail: hsharma@rtu.ac.in

R. Kumar
Malaviya National Institute of Technology Jaipur, Jaipur, India
e-mail: rkumar.ee@mnit.ac.in

© The Author(s), under exclusive license to Springer Nature Singapore Pte Ltd. 2021
X.-Z. Gao et al. (eds.), *Applications of Artificial Intelligence in Engineering*, Algorithms for Intelligent Systems, https://doi.org/10.1007/978-981-33-4604-8_11

Other than metaheuristic algorithms, methods mostly used in prediction of native structure of protein are divided into two categories [2]. The first one is X-ray crystallography, based on crystallization, usually time consuming and do not guarantee on obtaining optimal conditions for a crystal of protein. Second one is Nuclear Magnetic Response based on treatment of protein in solution. These two experimental methods are mostly used for PSP, yet these are time consuming, not cost effective and require meticulous efforts.

In previous years, PSP problem has been solved with various metaheuristic algorithms. Recently, Enhanced Chaotic Grasshopper Optimisation Algorithms have been applied to PSP in Ref. [3]. A Sinusoidal chaotic bridging mechanism has been proposed and application of those on Protein problem was investigated in Ref. [4]. Variants of Artificial Bee Colony Algorithm have also been developed to solve PSP [5, 6]. An important study of fitness landscape analysis and choice of different algorithms to solve PSP has been given in Ref. [7]. Like wise an Immune algorithm has been proposed and developed to solve PSP in Ref. [8]. Some other examples of the application of metaheuristics in PSP problems are in Ref. [9–11]. In those references Wrapping Search, Grid based genetic algorithm and hybrid harmony search algorithm has been applied. A gradient search based gravitational search algorithm has been proposed by [12]. Recently, a powerful variant of Grey Wolf Optimizer named as β-GWO has been presented and evaluated on many real and challenging problems. The prime objective of writing this paper is to evaluate the performance of the variant on PSP. On the basis of the review, following investigation objectives are framed for this manuscript

- To solve PSP problem for a bench of artificial and real protein sequences of different lengths.
- To apply β-GWO on PSP problem and analyse the results.
- To derive a meaningful comparison between β-GWO and other GWO variants on the basis of calculated statistical attributes.

The paper is having following sections: Sect. 1 presents introductory details of PSP, Sect. 2 presents problem formulation and how PSP can be converted into an optimization problem. Section 3 includes details of native GWO and development steps for β-GWO. In Sect. 4 simulation results and analysis is presented. Lastly, Sect. 5 showcase major findings obtained from the work.

2 Protein Structure Prediction: An Optimization Approach

Due to complexity in protein native structure a lot of models have been developed and studied, out of which HP model is mostly used. HP model is based on two residues, non-polar end and polar end or hydrophobic (H) and hydrophilic (P) [13]. Although sometimes this model is not suitable for complex protein model as local interactions are neglected in this method. Hence AB-off lattice model has introduced which consider the local interaction also. In this model hydrophobic and hydrophilic ends are represented by A and B respectively.

In this model, r particles attached with each other through chemical bonding and formed a chain of $(r - 2)$ bend angles given as $[\phi_2, \phi_3 \ldots \phi_{r-1}]$. Here $\phi_p \in [180\cdot, -180\cdot]$, and can be rotate in clock wise and anti-clock wise direction. The intermolecular energy of a protein sequence can be given as

$$\epsilon = \sum_{p=2}^{r-1} \frac{(1 - \cos\phi_p)}{4} + 4\sum_{p=1}^{r-2}\sum_{q=p+2}^{r} [d_{pq}^{-12} - I(\alpha_p, \alpha_q)d_{pq}^{-6}] \tag{1}$$

Here first term represents bending energy of function and second term represents specific Leonard-Jones type of interaction. $I(\alpha_p, \alpha_q)$ represents characteristic of particular pth particle. d_{pq} is the length of space between pth and qth particle given as

$$d_{pq} = \sqrt{\left[1 + \sum_{k=p+1}^{q-1} \cos\left(\sum_{l=p+1}^{k} \theta_l\right)\right]^2 + \left[\sum_{k=p+1}^{q-1} \sin\left(\sum_{l=p+1}^{k} \theta_l\right)\right]^2} \tag{2}$$

$$I(\alpha_p, \alpha_q) = \frac{1}{8}(1 + \alpha_p + \alpha_q + 5\alpha_p\alpha_q) \tag{3}$$

The interaction between two particles is given by Eq. 3. The value of this coefficient $I(\alpha_p, \alpha_q)$ for different pairs of residues may be different. The values for AA pair is (1), BB pair is (0.5), for AB and BA are (−0.5) [14]. It can be observed that the correlation between AA particles is strong, while BB particles are weakly encouraged. In this way, the protein secondary structure prediction is transformed into a numerical optimization problem via AB OFF lattice model.

3 Grey Wolf Optimizer and Its Variants

Grey Wolf Optimizer (GWO) was proposed by Mirjalali et al. [15]. The algorithm is based on the hunting behavior of grey wolf and mimics the process through mathematical translation. Recently, this algorithm has been applied on many problems including harmonic estimation, strategic bidding and Automatic Generation control, Transmission network expansion planning and many more [16–19]. The important procedure is elaborated in the algorithm are as follows:

3.1 Encircling the Prey

The prey encircled by the grey wolf can be represented mathematically as:

$$S = |R.X_p(i) - X(i)| \tag{4}$$

$$X(i + 1) = X_p(i) - P.S \tag{5}$$

where current iteration is represented by i, X is position vector of grey wolf, coefficient vectors are P and R, and X_p is the position vector of the prey. We can calculate the vectors P and R as follows:

$$P = 2a.r_1 - a \tag{6}$$

$$R = 2.r_2 \tag{7}$$

where control vector a decreases linearly from 2 to 0 and r_1 and r_2 are the random numbers.

3.2 Hunting the Prey

In the search space, it is not possible to find proper location of prey. Hence, to require the social hierarchy, three best solutions (α, β & δ) are kept. This fact can be visualized by following mathematical representation:

$$S_\alpha = |R_1 X_\alpha - X|, \quad S_\beta = |R_2 X_\beta - X|, \quad S_\delta = |R_3 X_\delta - X| \tag{8}$$

$$X_x = X_\alpha - A_x.(S_\alpha), \quad X_y = X_\beta - A_y.(S_\beta), \quad X_z = X_\delta - A_z.(S_\delta) \tag{9}$$

$$X_{(i+1)} = \frac{(X_x + X_y + X_z)}{3} \tag{10}$$

The position updation represented by the equations according to alpha, where the distance of the prey are S_α, S_β and S_δ from α, β and δ wolves respectively and positions of α, β and δ wolves are X_x, X_y and X_z respectively.

3.3 Attacking Prey

This stage is accountable for exploitation and by linear decrement in a it can be controlled. Grey wolves are empowered by using linear decrement in a to assault the prey, while it stops moving. By changing a we can control the fluctuations in P i.e. there are more fluctuations in P if the value of a is higher.

3.4   Development Steps of β-GWO

Saxena et al. proposed a modified version of GWO in year 2019, in this version of GWO the bridging mechanism of GWO has been changed with a controlled chaotic series. This chaotic series has one property that it kept exploration and exploitation alive till last iteration. The procedure and different stages of development of this variant is illustrated below.

In the first stage, search agents find a route to explore the search space in a feasible way and in second stage, control vector **a** shrinks the search boundary with the help of linearly decreasing bridging mechanism. This work revolves around this linear diversity and uses β-chaotic course of action through a normalized function.

To develop β-GWO following steps are considered:

1.

$$\beta(y; \lambda, \eta, y_1, y_2) = \begin{cases} (\frac{y-y_1}{y_c-y_1})^\lambda (\frac{y_2-y}{y_2-y_c})^\eta, & \text{if } y \in [y_1, y_2] \\ 0, & \text{otherwise.} \end{cases} \tag{11}$$

where $(\lambda, \eta, y_1, y_2) \in R$ and $y_1 < y_2$. At any i iteration β-Chaotic sequence can be shown as:

$$y_{i+1} = k\beta(y_i; \lambda, \eta, y_1, y_2) \tag{12}$$

For an iterative process, Fig. 1b shows the chaotic sequence generated. Data for formulation of this chaotic sequence are taken from [19].

2. The Normalization function is used to disperse this arrangement among the Maximum and minimum values before the control vector **a** gets biased by it. This procedure is presented in Fig. 1a. At any i iteration the mathematical expression for this function can be given as:

$$S_m(i) = S_m^{max} - \left(\frac{S_m^{max} - S_m^{min}}{I} \right) i \tag{13}$$

where the maximum number of iterations are I, S_m^{max} is the highest value and S_m^{min} is the least value of standardization function. Mathematical form of normalize β-Chaotic sequence can be given as:

$$Z(i) = S_m(i) y_i \tag{14}$$

3. The chaotic sequence is added with the existing **c** of GWO which can be given as:

$$a^{\beta-GWO}(i) = a^{GWO}(i) + Z(i) \tag{15}$$

Figure 1c represents this mathematical procedure in a clear way. .

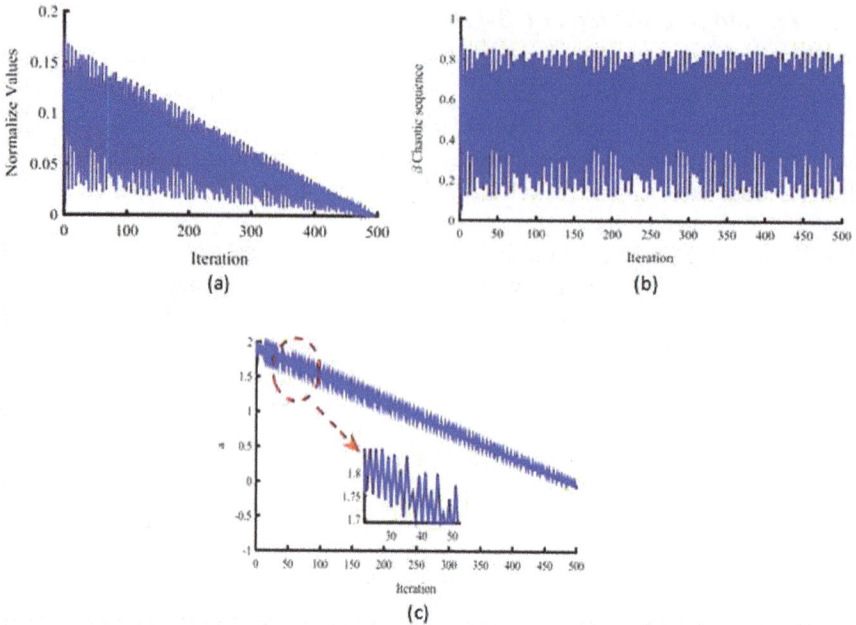

Fig. 1 **a** Normalized β-chaotic sequence, **b** β-chaotic sequence function, **c** Changes in control vector of β-GWO [19]

3.5 Participating Variants of GWO

In this work, comparison has been carried out for solving PSP by using following variants of GWO

- β-Chaotic Map enabled Grey Wolf Optimizer (β-GWO) [19]: this variant employs a critical beta chaotic sequence for enhancing the bridging between exploration and exploitation of the GWO. The variant has tested on strategic bidding problem, CEC functions and other real applications.
- Intelligent Grey Wolf Optimizer (IGWO)[20]: this variant employs a sinusoidal bridging mechanism between exploration and exploitation phase and for enhancing the exploration. Authors have applied opposition theory concept. The variant has been tested on standard bench and strategic bidding problem.
- Modified Grey Wolf Optimizer (MGWO) [18]: this variant employs a new position update equation that enables submissive wolves for hunting. The variant has been tested over conventional bench and applied over Transmission Network Expansion Planning (TEP) problem.
- Fast Convergent Grey Wolf Optimizer (FCGWO) [21]: this variant employs some modifications in position updating equation. Further, the variant has been applied on transmission expansion problem and on a conventional bench.

- Grouped Grey Wolf Optimizer (GGWO) [22]: developed a variant where they assumed that grey wolves can be divided into two groups one is random scout group and other one is cooperative hunting group. The categorized cooperative hunting group wolves perform all operations including, searching, encircling and harassment. Other group perform global search.
- Oppositional based Grey Wolf Optimization [23]: this variant employs opposition based learning concept for improving the GWO and further the variant is applied on economic load dispatch problem.
- Improved Grey Wolf Optimizer (ImGWO) [24]: this variant employed modified augmented Lagrangian (MAL) multiplier method to handle constraints and thus the algorithm was capable to discover capability the global optimum for constrained problems.

4 Simulation Results and Analysis

To evaluate the performance of β-GWO, a bench of 7 protein sequence is considered. These proteins are represented in Table 1. In this bench, protein of small length such as Asm1 and Asm2 (4) , a bit larger As1 and As2 (5) and medium length artificial proteins Am1 and Am2 (13 and 17) and one Real protein sequence Rs1 (1BXP) is considered for evaluation of the β-GWO. A comparison of β-GWO is conducted with different participating algorithms. The details of participating algorithms are given in previous section.

To maintain a fair environment for comparison, search agents numbers are kept 30 for β-GWO and all participants are run for 20 runs and having maximum number, or iteration 500. The fitness values are taken as results and these are obtained from 20 independent runs of optimization. Mean, maximum, minimum and standard deviation of these values are calculated and reported against protein sequences. Results of sequence Asm1 and Asm2 are shown in Table 2. From this Table, one can easily see that for these small length proteins mean free energy values obtained by β-GWO are optimal and it exhibits competitive performance with others. For the sake of clarity

Table 1 Protein bench for evaluation

S. No.	Name	Length	Sequence
1	Asm1	4	ABAB
2	Asm2	4	AAAA
3	As1	5	AAAAB
4	As2	5	AAAAA
5	Am1	13	ABBABBABABBAB
6	Am2	17	ABABBAABBBAAABABA
7	Rs1 (1BXP)	13	ABBBBBBABBBAB

Table 2 Optimization routine results for Asm1 and Asm2

Protein	Algorithm	Mean	Maximum	Minimum	St. Dev.
Asm1	GWO	−0.648997255	−0.641814863	−0.649375351	0.00169056
	mGWO	−0.649375051	−0.649372374	−0.64937535	6.50E−07
	β-GWO	**−0.649375236**	−0.649374743	−0.64937535	1.64E−07
	Im GWO	−0.647667812	−0.641715998	−0.649375308	0.003124217
	IGWO	−0.649090342	−0.643994098	−0.649375216	0.001201242
	FCGWO	−0.649142249	−0.64720112	−0.649373193	0.000644517
	OGWO	−0.644262743	−0.641815284	−0.649127201	0.002459829
	GGWO	−0.447829952	−0.069876349	−0.645607802	0.194479323
Asm2	GWO	−1.676278679	−1.675428788	−1.676326606	0.00020006
	mGWO	−1.6626414	−1.404811925	−1.676325795	0.060688431
	β-GWO	**−1.676419306**	−1.672574427	−1.676426869	0.000836964
	Im GWO	−1.66651311	−1.565986194	−1.676325407	0.027539084
	IGWO	−1.663077593	−1.412511305	−1.676325322	0.058977216
	FcGWO	−1.675393105	−1.672820901	−1.676280823	0.000900802
	OGWO	−1.557413587	−0.808684031	−1.676095078	0.189741616
	GGWO	−1.269775858	−0.679666859	−1.621852987	0.301690747

Table 3 Optimization routine results for As1 and As2

Protein	Algorithm	Mean	Maximum	Minimum	St.Dev.
As1	GWO	−1.561274597	−1.32763124	−1.589441903	0.064
	mGWO	−1.572548026	−1.322790354	−1.589434952	0.059
	β-GWO	**−1.581245194**	−1.32762324	−1.589439349	0.080
	Im GWO	−1.475868478	−1.213110198	−1.589288219	0.130
	IGWO	−1.518413425	−0.734948781	−1.589431663	0.195
	FCGWO	−1.5409409	−1.317798293	−1.588150234	0.095
	OGWO	−1.139485545	−0.713827259	−1.569977199	0.372
	GGWO	−0.599287538	0.026912479	−1.465512012	0.413
As2	GWO	−2.805014767	−2.438653555	−2.848281651	0.106
	mGWO	−2.815207749	−2.451008711	−2.84828084	0.094
	β-GWO	**−2.823202718**	−2.45108868	−2.848280698	0.123
	Im GWO	−2.692186898	−2.311489897	−2.84823202	0.167
	IGWO	−2.634047193	−2.181753657	−2.848196936	0.199
	FCGWO	−2.810720154	−2.683690091	−2.846926753	0.056
	OGWO	−2.429187782	−1.444993498	−2.786338729	0.321
	GGWO	−1.349356418	−0.833652719	−2.349357862	0.411

these values are written in bold faces in this Table. These values for β-GWO are for ASm1 **−0.649375236** and for Asm2 **−1.676419306**.

Like wise, results of As1 and As2 are shown in Table 3. Inspecting the values from these Tables, it can be concluded that mean free energy values are optimal. For As1 the mean of optimization routine is **−1.581245194** and for As2 it is **−2.823202718**. As it is clearly visible that these results are optimal as compared with native GWO

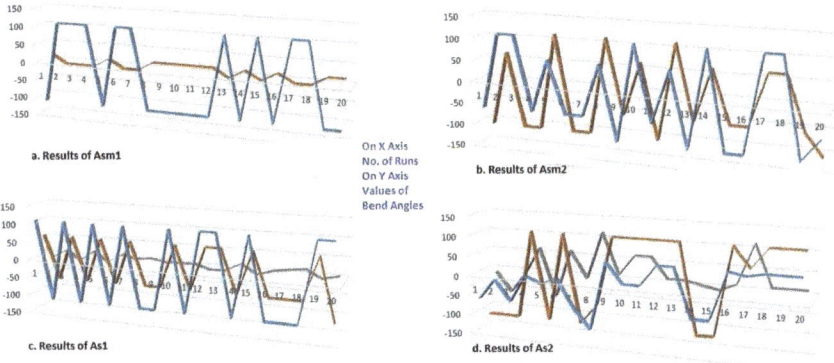

a. Results of Asm1

On X Axis No. of Runs
On Y Axis Values of Bend Angles

b. Results of Asm2

c. Results of As1

d. Results of As2

Fig. 2 Results of small protein sequences

Table 4 Optimization routine results for Am1 and Am2

Protein	Algorithm	Mean	Maximum	Minimum	St.Dev.
Am1	GWO	−3.414877599	−1.15092366	−6.444108436	1.22137168
	mGWO	−3.201986244	−1.428247587	−4.624119995	0.934881862
	β-GWO	**−3.892754995**	−1.252873471	−4.155012032	0.829761287
	Im GWO	−1.88764644	−0.313373877	−3.977050607	1.032557819
	IGWO	−2.69120342	−0.721035305	−5.506067717	1.396511697
	FCGWO	−1.587435548	−0.598203968	−3.858701016	1.000630274
	OGWO	−0.936759847	−0.071583936	−2.719205379	0.607336415
	GGWO	−2.33124427	−1.205050560	−4.18868609	0.62372369
Am2	GWO	−0.47405216	−0.012210802	−2.19430244	0.506352483
	mGWO	−0.554147647	−0.012210802	−1.575681627	0.368656049
	β-GWO	−0.677019835	−0.012210802	−1.624254607	0.480427375
	Im GWO	−0.600223761	−0.012210802	−1.556695772	0.333064434
	IGWO	−0.69948351	−0.012210802	−1.547932238	0.38876727
	FCGWO	−0.503746666	−0.012210802	−1.49276346	0.402300547
	OGWO	−0.323307704	−0.012210802	−0.540758443	0.261106635
	GGWO	−0.150810.104	−0.041619274	−1.051775955	0.108768922

and its variants. We also observe that mean values for GGWO are poor for As1 and AS2. Further, results of 20 runs of β-GWO are shown in Fig. 2. These are the obtained values of bend angles.

Further, the results of sequences in cases Am1 and Am2 are depicted in Table 4. It is observed from the Table that for Am1 β-GWO outperforms others and have the mean value **−3.892754995**. In case of Am2 it stands second and IGWO has optimal value **-0.69948351** and for β-GWO it is **−0.677019835**. Hence, we can conclude that for medium length proteins, performance of this algorithm is quite competitive.

Table 5 exhibits the optimization routine results for real protein sequence Rs1. This protein sequence has been taken from the Protein Data Bank (PDB, http://www.rcsb.org/pdb/home/home.do).

Table 5 Optimization routine results for Rs1

Protein	Algorithm	Mean	Maximum	Minimum	St.Dev.
Rs1	GWO	−0.723951444	−0.091483122	−1.627630924	0.599
	mGWO	−0.503209837	−0.091483122	−1.53647397	0.584
	β-GWO	**−0.887493886**	−0.091483122	−1.658946792	0.542
	Im GWO	−0.526306944	−0.091483122	−1.257772784	0.469
	IGWO	−0.38680568	−0.091483122	−1.547056664	0.498
	FCGWO	−0.453545973	−0.091483122	−1.231051198	0.474
	OGWO	−0.119191644	−0.091483122	−0.533597826	0.101
	GGWO	62691.48939	496574.3019	1.305146587	146166.715

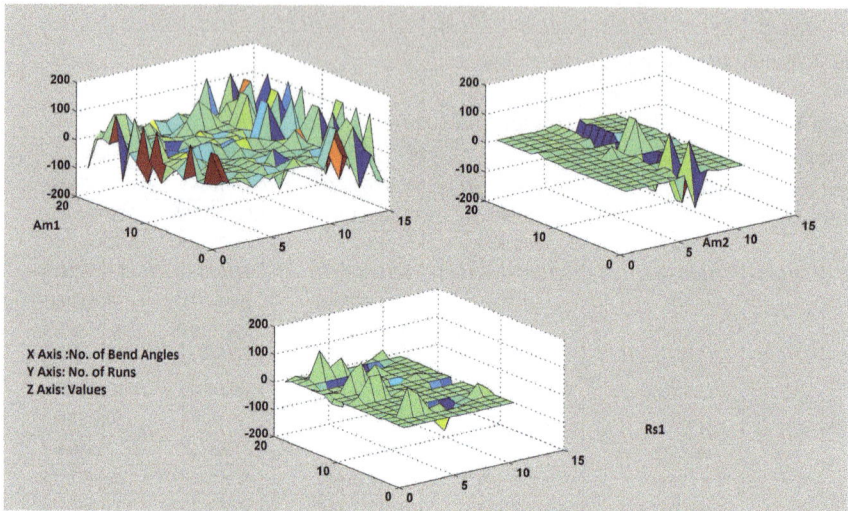

Fig. 3 Results of medium protein sequences

Inspecting the results of the Table it is concluded that the mean values are optimal for β-GWO and the value is shown in bold face **-0.887493886**. Further, the optimization routine results and bend angles obtained from optimization process are exhibited in Fig. 3. Sub figures of this Figure shows results of Am1, Am2 and Rs1. From the results obtained from these experiments, it is quite evident to state that the performance of β-GWO is satisfactory as compared with other contemporary powerful variants of GWO. It has also been observed that apart from power engineering problems, this variant provides fruitful results in other challenging problems.

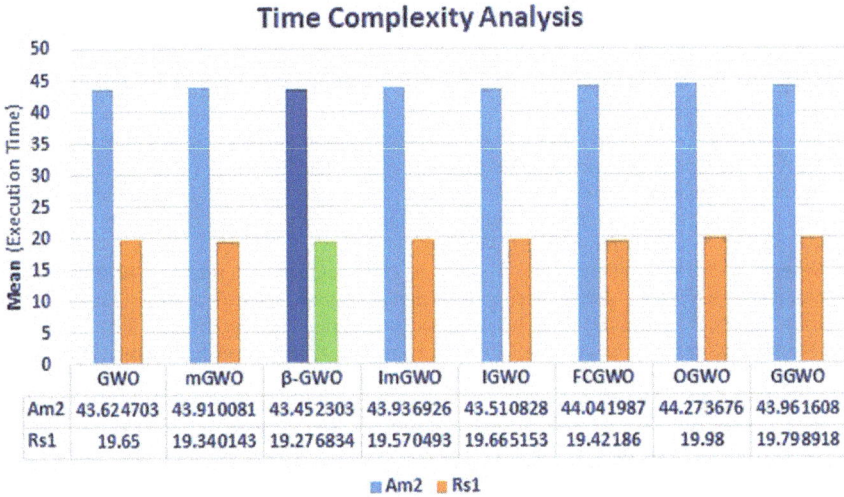

Fig. 4 Time complexity analysis of algorithms. Edited by Foxit Reader Copyright(C) by Foxit Corporation, 2005–2009 For Evaluation Only

It is empirical to evaluate time complexity of the competitive algorithms as PSP problem, is time consuming for large proteins. For showcasing the applicability of β-GWO, execution times for Rs1 and Am1 sequences are evaluated and mean values of time taken to complete runs are reported. We observe that execution time for β-GWO is competitive and same is depicted in Fig. 4. These simulations are carried out on Intel(R) Core(TM) i5-9400F CPU @ 2.90GHz processor.

5   Conclusion

An application of a powerful variant of GWO: β-GWO to protein structure prediction problem has been presented. A bench of artificial and real sequence of proteins has been considered in an evaluation bench. Results revealed that for almost all sequences the performance of β-GWO is quite competitive when compared with recently published variants of GWO and GWO itself. The performance evaluation of β-GWO has been done with the help of statistical attributes calculated from the optimization process. We observe that the performance is competitive as it obtains optimal free energy values for almost all protein sequences.

Acknowledgements The Authors are thankful for the fully financial support from the CRS, RTU (ATU), TEQIP-III of Rajasthan Technical University, Kota, Rajasthan, India. (Project Sanction No. TEQIP-III/RTU (ATU)/CRS/2019-20/33).

References

1. Anfinsen CB (1973) Principles that govern the folding of protein chains. Science 181(4096):223–230
2. Jana ND, Das S, Sil J (2018) Backgrounds on protein structure prediction and metaheuristics. In: A metaheuristic approach to protein structure prediction. Springer, pp 1–28
3. Saxena A, Kumar R (2020) Chaotic variants of grasshopper optimization algorithm and their application to protein structure prediction. In: Applied nature-inspired computing: algorithms and case studies. Springer, pp 151–175
4. Saxena A (2019) A comprehensive study of chaos embedded bridging mechanisms and crossover operators for grasshopper optimisation algorithm. Expert Syst Appl 132:166–188
5. Li B, Chiong R, Lin M (2015) A balance-evolution artificial bee colony algorithm for protein structure optimization based on a three-dimensional ab off-lattice model. Comput Biol Chem 54:1–12
6. Li B, Lin M, Liu Q, Li Y, Zhou C (2015) Protein folding optimization based on 3D off-lattice model via an improved artificial bee colony algorithm. J Mol Model 21(10):261
7. Jana ND, Sil J, Das S (2017) Selection of appropriate metaheuristic algorithms for protein structure prediction in ab off-lattice model: a perspective from fitness landscape analysis. Inf Sci 391:28–64
8. Cutello V, Nicosia G, Pavone M, Timmis J (2007) An immune algorithm for protein structure prediction on lattice models. IEEE Trans Evol Comput 11(1):101–117
9. Gonçalves R, Goldbarg MC, Goldbarg EF, Delgado MR (2008) Warping search: a new metaheuristic applied to the protein structure prediction. In: Proceedings of the 10th annual conference on Genetic and evolutionary computation, pp 349–350
10. Tantar A-A, Melab N, Talbi E-G (2008) A grid-based genetic algorithm combined with an adaptive simulated annealing for protein structure prediction. Soft Comput 12(12):1185
11. Abual-Rub MS, Al-Betar MA, Abdullah R, Khader AT (2012) A hybrid harmony search algorithm for ab initio protein tertiary structure prediction. Netw Model Anal Health Inf Bioinf 1(3):69–85
12. Dash T, Sahu PK (2015) Gradient gravitational search: an efficient metaheuristic algorithm for global optimization. J Comput Chem 36(14):1060–1068
13. Stillinger FH, Head-Gordon T, Hirshfeld CL (1993) Toy model for protein folding. Phys Rev E 48(2):1469
14. Li B, Li Y, Gong L (2014) Protein secondary structure optimization using an improved artificial bee colony algorithm based on ab off-lattice model. Eng Appl Artif Intell 27:70–79
15. Mirjalili S, Mirjalili SM, Lewis A (2014) Grey wolf optimizer. Adv Eng Softw 69:46–61
16. Gupta E, Saxena A (2015) Robust generation control strategy based on grey wolf optimizer. J Electr Syst 11(2):174–188
17. Saxena A, Kumar R, Mirjalili S (2020) A harmonic estimator design with evolutionary operators equipped grey wolf optimizer. Expert Syst Appl 145:113125
18. Khandelwal A, Bhargava A, Sharma A, Sharma H (2018) Modified grey wolf optimization algorithm for transmission network expansion planning problem. Arab J Sci Eng 43(6):2899–2908
19. Saxena A, Kumar R, Das S (2019) β-chaotic map enabled grey wolf optimizer. Appl Soft Comput 75:84–105
20. Saxena A, Soni BP, Kumar R, Gupta V (2018) Intelligent grey wolf optimizer-development and application for strategic bidding in uniform price spot energy market. Appl Soft Comput 69:1–13
21. Khandelwal A, Bhargava A, Sharma A (2019) Voltage stability constrained transmission network expansion planning using fast convergent grey wolf optimization algorithm. Evol Intell pp 1–10
22. Yang B, Zhang X, Tao Y, Shu H, Fang Z (2017) Grouped grey wolf optimizer for maximum power point tracking of doubly-fed induction generator based wind turbine. Energy Conv Manage 133:427–443

23. Pradhan M, Roy PK, Pal T (2018) Oppositional based grey wolf optimization algorithm for economic dispatch problem of power system. Ain Shams Eng J 9(4):2015–2025
24. Long W, Liang X, Cai S, Jiao J, Zhang W (2017) A modified augmented lagrangian with improved grey wolf optimization to constrained optimization problems. Neural Comput Appl 28(1):421–438

Chapter 12
Transmission Expansion Planning Using Teaching and Learning Based Optimization Approach

Jitesh Jangid, Aishwarya Mehta, Akash Saxena, Shalini Shekhawat, Rajesh Kumar, and Ajay Sharma

1 Introduction

Transmission Expansion Planning (TEP) is essential to accomplish the potential growth in load. The goal of the TEP problem is to determine the optimal circuit configuration that should be applied to the network in order to fulfill the load demand and operational constraints. Different mathematical models such as the static model, hybrid model, DC model are built for the solution of the TEP problem. The TEP is a dynamic problem of different forms of random and non-random variables, such as demand, generation price, capacity of the generator, availability of lines, transmission facilities for substitution, and business laws. Such complexities should be treated independently through the various expert structures such as artificial neural networks and fuzzy logic, thanks to their capacity to solve unknown problems.

The problems of Electrical power system are divided into two parts one is generation expansion planning (GEP) [1] and other is Transmission expansion planning

J. Jangid (✉) · A. Mehta · A. Saxena · S. Shekhawat
SKIT, Jaipur, India
e-mail: jj90946@gmail.com

A. Mehta
e-mail: aishu.sharma.0786@gmail.com

A. Saxena
e-mail: aakash.saxena@hotmail.com

S. Shekhawat
e-mail: shekhawatshalini17@gmail.com

R. Kumar
MNIT, Jaipur, India
e-mail: rkumar.ee@mnit.ac.in

A. Sharma
Government Engineering College, Jhalawar, India
e-mail: ajay_2406@yahoo.com

© The Author(s), under exclusive license to Springer Nature Singapore Pte Ltd. 2021   161
X.-Z. Gao et al. (eds.), *Applications of Artificial Intelligence in Engineering*, Algorithms for Intelligent Systems, https://doi.org/10.1007/978-981-33-4604-8_12

(TEP) [2]. Transmission in the power system is an important term, which not only provides a reliable environment but also establish link between distribution and generation. Nowadays TEP is a very complex problem in the power system. In the twenty first century, one and the major challenge is to provide clean and sustainable energy to meet the demand load. The Aim [3] of TEP is to improve the reliability and security of the power system to supply the demand load, excluding investment cost.

Capacity of the transmission system [4] could be increased by large integration of renewable energy sources (RES) [5] into the system. It is a difficult task to decide when, what and where reinforcement is required in power system. The new arrangement of transmission lines not only relieves traffic of the existing network, but also improves the efficiency of system. It is less expensive to connect a new generation station at nearby node of transmission line. Recently a blackout [6] has occurred, for example, a 56GW power generation station shut down due to the limited capacity of transmission. It is a reason for large economic loss not only to government but also to consumers.

This problem of system makes the market deregulated [7]. Due to this situation government sells their plants to private organizations and hence there is no control of government on the cost of distribution. For the last two decades, the term deregulation is introduced in many countries. Competition in power generation is initiated while the transmission remains a monopoly. Recently solar and wind concepts are used on a large scale of energy generation. But difficulties occur in the integration of the generated energy in the power system. After the Renewable energy sources integrated into the modern power system, the problem of TEP becomes more challenging. To overcome this we need to install power generation stations near node of transmission. This is a big challenge in front of planners.

During last decades such models are developed. Among these models Mathematical programming, Heuristic and Meta-heuristic are types of solution methods. For optimal solution of TEP, some parameters are required: Latest Designs, Investment constraints, Demand of electricity and generation forecast. Day by day solutions of TEP are becoming more complex. As above discussion this paper aims to present more information on TEP. The TEP problem seeks the attention of researchers for last five decades. In this duration, a lot of work has been reported. The first problem has been solved through Dynamic Programming (DP) [8] and Linear Programming (LP) [9] techniques to minimize the investment cost and balancing of power at each and every bus. For investment decision the latest TEP models used Mixed-Integer Linear Programming (MILP) [10] method with binary variables for the capitalization of the plant. Due to these variables, solution of the problem becomes more complex. To overcome this, we use technique of Benders Decomposition for large level of LP. According to planning, TEP is classified into two types: Dynamic transmission expansion planning (DTEP) [11] and Static transmission network expansion planning (STEP) [12].

In DTEP, solution process is divided into several intervals, for each interval the peak demand is calculated and the input of the generation unit is depends on the intervals to meet the demand load. In STEP planning, scope is initially defined, where the generation unit input and maximum demand load is calculated at the beginning of the planning.

On the basis of critical review of literature following research objectives are proposed.

- To address the issue of transmission expansion planning (TEP) by teaching learning based optimization algorithm (TLBO).
- To design an expansion model of transmission network and test the robustness of TLBO.
- To derive a meaningful comparison between contemporary optimizers for optimizing standard test networks.

The paper is having following sections: Sect. 1 presents introduction of TEP, Sect. 2 presents historical details and chronological development of previous work on TEP. Section 3 shows mathematical representation of TEP. In Sects. 4 and 5, we have given the details of TLBO algorithm and done assessment of this algorithm and in Sect. 6 the major findings obtained from the work are concluded.

2 TEP Solution Methods

Several techniques have been developed in regulated power system to solve TEP. Further, these techniques can be classified into two folds mathematical solution techniques and metaheuristic techniques.

2.1 Mathematical Solution Techniques

For the optimum expansion plan, mathematical solution techniques were used. During the calculation, it was hard to consider all constraints, so some selected constraints were used to obtain results. The solution process is first started by simple models and further switched to perfect models at the final step. These methods mainly are Dynamic Programming (DP), Quadratic Programming (QP) [13], Linear Programming (LP) [14], Non-Linear Programming (NLP) [15] etc. For the efficient solution of NLP and LP, interior point method (IPM) is used which completely depends on the Benders Decomposition method. For large-scale problems, user faced slow computational speed that was the main drawback of mathematical optimization techniques.

Formulation of problem by linear programming was firstly introduced by Soviet Economist Leonid Kantorovich during Second World War (1939). Garver proposed this technique to solve TEP problem [16]. LP method is basically designed for long term planning of power system. Solution steps of this method are divided into two parts first one is new circuit selection and other one is load flow estimation.

2.2 Meta-Heuristic Solution Technique

It is new and advanced prototype of the heuristic algorithm. It was introduced in the eighth decade of nineteenth century [17]. Meta-heuristic is a collection of searching algorithms. This is imposed to solve the complex problems. Meta-heuristics is a high level of heuristic to guide other heuristics in solving of the problem.

TEP has been addressed with the application of Harmony Search Algorithm [2], Fibonacci series based Spider Monkey algorithm [18], Modified Grey Wolf optimizer [19]. These all are fine examples of application of metaheuristics for solving TEP.

3 Mathematical Representation of TEP Problem

A DC model is utilized to demonstrate TEP problem. Security of system plays an important role in TEP. N-1 contingency analysis views system status after a single line failure. Security limitations mathematical representation of TEP of Eq. 1 in [20] should be possible and to be expressed as follows,

$$\min \quad v = \sum_{l \in \omega} c_l n_l \tag{1}$$

s.t.

$$Jt^s + u = p \tag{2}$$

$$q_v^s - \gamma\left(r_v^0 + r_v\right)\left(\delta\phi_v^s\right) = 0, \ q \text{ or } v \in 1, 2, \ldots, rv \ \& \ v \neq s \tag{3}$$

$$q_v^s - \gamma\left(r_v^0 + r_v - 1\right)\left(\delta\phi_v^s\right) = 0, \ f \text{ or } v = s, \tag{4}$$

$$\mid t_1^s \mid \leq \left(r_v^0 + r_v\right)\bar{q}_1, f \text{ or } v \in 1, 2, \ldots, rv \ \& \ v \neq s \tag{5}$$

$$\mid q_1^s \mid \leq \left(r_v^0 + r_v - 1\right)\bar{q}_1, \ q \text{ or } l = k, \tag{6}$$

$$0 \leq r_1 \leq \bar{r}_1 \tag{7}$$

$$r_1 \geq 0, \tag{8}$$

and integer, for $v \in 1, 2, \ldots, rv \ \& \ v = s$

$$\left(r_1 + r_1^0 - 1\right) \geq 0 \quad and \ integer, f \text{ or } v = s \tag{9}$$

where,

v = Overall Cost.

c = Cost of new line to be added.

n = New line to be added in network.

l = line between node i and j.

$s = 0$, Refers to the base case without any line interruption

J = Matrix displacement of branch-node events force Structure

q^s = elements of vector q_v^s

γ_v = The vulnerability of the circuit to be added in vth.

r_v^0 = base case circuits.

$\delta\phi_v^s$ = v th& sth line phase difference.

q_v^s = vth & sth total real power flown by circuit.

ω = collection of the right way.

rv = Circuits total branches.

In this paper the basic model of TEP is same as shown in [18, 21], different techniques to solve planning problem with security constrains. For this reason, the mathematical model was modified to include these below given techniques at each load bus to make the mathematical model more tractable.

4 Teaching Learning Based Optimization Algorithm (TLBO)

Firstly TLBO introduced by Rao et al. (2011) [22] which is inspired from teaching learning process of class room. Reason behind introducing this algorithm is inaccurate adjustment of different techniques specific parameters either to improve the computing effort or to deliver the optimum local solution, that's why author gives TLBO where it only requires standard input parameters such as population size and number of iterations for it's operation.

TLBO basically apportioned it in two parts where one shows teaching phase and other one shows learners phase. Both are described below in separate sections.

4.1 Teaching Phase

In this part learner gain some knowledge from teacher. Through out the course duration, teacher always tried to increase the average result of the class in the subject taught by him. Let's take for lth iteration, we have b subjects (e.g. design variables) and p number of learners. Population size is $h = 1, 2,\ldots,$p and $B_{a,l}$ is average result of the learners in a subject a $(a = 1, 2, \ldots, b)$. Overall result is $Z_{\text{sum}-hbest,l}$ considering all the subjects gained together in the whole population of the students can be mentioned as the result of good learner or student $hbest$.

In education community, a teacher is always considered as highly knowledgeable person and has the ability to give training to learners to improve their results. Among these learners, the best learner is chosen by the algorithm as a teacher.

The existing result of every subject and current obtained result difference is given by this formula.

$$\text{Diff\_mean}_{a,h,l} = w_l \left(Z_{a,hbest,l} - T_c B_{a,l} \right) \tag{10}$$

where,

$Z_{a,hbest,l}$ = Best learner result in subject a. T_c = Factor of teaching which determines the mean value to adjust.

w_i = Random number between [0, 1]

value of T_c can be either 1 or 2, this value is chosen randomly as per the equation demand.

$$T_c = \text{round} \{1 + \text{rand}(0, 1)(2 - 1)\} \tag{11}$$

T_c is not a input parameter of this technique, hence value of T_c is calculated using Eq. 11. After some experiments, it is observed that this technique gives better results when $1 \leq T_c \leq 2$.

On the basis of difference between (a, h, l), solution is updated in teaching part as

$$Z'_{a,h,l} = Z_{a,h,l} + \text{Diff\_mean}_{a,h,l} \tag{12}$$

where $Z'_{a,h,l}$ is updated solution. This solution only accepted when it gives better result.

4.2 Learning Phase

This is the second part of TLBO technique in which the learners meet each other and boost up their knowledge. During this process a learner meet with other learner randomly and obtain new knowledge from each other if the other one has more knowledge than him.

Let us assume a population size of p and select any two learners from the population randomly, i. e. R and S.

$$Z'_{sum-R,l} \neq Z'_{sum-S,l} \tag{13}$$

where $Z'_{sum-R,l}$ and $Z'_{sum-S,l}$ updated function values of $Z_{sum-R,l}$ for R and $Z_{sum-S,l}$ for S of teacher phase.

$$Z''_{a,R,l} = Z'_{a,R,l} + w_l \left(Z'_{a,R,l} - Z'_{g,S,l} \right) \tag{14}$$

if $Z'_{\text{sum}-R,l} < Z'_{\text{sum}-S,l}$ and

$$Z''_{a,R,l} = Z'_{a,R,l} + w_l \left(Z'_{a,S,l} - Z'_{a,R,l} \right) \tag{15}$$

if $Z'_{\text{sum}-S,l} < Z'_{\text{sum}-R,l}$

Above Eqs. 14 and 15 are used for minimization problems. These two equation are further used in this paper on transmission expansion planning.

5 Simulation Results and Analysis

In this section, application of TLBO is reported in two standard test networks. For carrying out this analysis, Some recently published approaches has been considered, those are namely Modified Grey Wolf Optimizer (MGWO) [19] ,Grey Wolf Optimizer (GWO) [19], Sine Cosine Algorithm SCA, Whale Optimization Algorithm WOA and Artificial Bee Colony (ABC) [23]. To make a fair competition parameters of algorithms such as swarm size, termination criterion and number of runs are kept constant for all participants. These are mentioned as:

- 30 = Number of Search Agents (SA), 100 = Number of runs,
- 200,000 = Number of Maximum Function Evaluation,
- The whole evaluation process is done on MAT-LAB 2015, on intel core i3 system with 8 GB ram and 500 GB HDD.

5.1 Graver's 6-Bus System

In order to solve TEP problem, evaluation has been done on Graver's 6-Bus system. This system contains 6-buses and 15 different candidate paths with a load of 760MW. For evaluation of this system all compulsory data is mentioned and taken in [24]. To solve TEP proposed methodology has been tested for 100 times and later the obtained data is sorted. From sorted data, particular values are stored as final result that appears most consistently in the array. Then the obtained result is compared with the previously reported results of MGWO and GWO [19] and artificial bee colony(ABC) [23] as shown in Table 1. In this table the results are shown in terms of some parameters where (AFE) used for average function evaluation, (OC) for operational cost in $ $\times 10^3$ and (SR) shows success rate of evaluation. After observing results reported in Table 1. It is observed that TLBO performs better in terms of parameters SR and AFE. It has been observed that for this system TLBO achieves 98% SR as compared to other competitors. It is also observed that WOA and SCA are not suitable choice for conducting research on TEP. Like MGWO, some modifications are required in WOA and SCA. However, it is worth to mention here that TLBO is a clear winner as

168

J. Jangid et al.

Table 1 Graver 6 bus system

Graver six bus system

Technique	Bus topology			O.C. ×10³ US $	A.F.E.	S.R.
	From	To	No. of Lines			
TLBO	**6**	**2**	**3**	**190.0**	**405.94**	**98**
	3	**5**	**2**			
	4	**6**	**2**			
MGWO [19]	2	3	1	190.0	49237	92
	3	5	1			
	4	6	2			
	6	2	3			
GWO [19]	2	3	1	190.0	91490	60
	3	5	1			
	4	6	2			
	6	2	3			
SCA	1	5	1	190.0	128150.7	40
	3	5	1			
	4	6	2			
	6	2	3			
ABC [23]	3	5	1	200.0	N/A	N/A
	4	6	2			
	6	2	4			
WOA	1	5	1	262.0	121441.5	23
	3	5	1			
	4	6	2			
	6	2	3			

the OC is optimal with highest success rate and optimal AFE. As per observations, reported in table, 3 lines can be added between bus 6 and 2, 2 lines between bus 1 and 5 and 2 lines between bus 4 and 6.

5.2 Brazilian 46-Bus System

Brazilian system is selected as second evaluation system that contains 79 lines and 46 number of bused with 6880MW load. To evaluate this, all compulsory data is taken from Ref. [24]. To solve TEP proposed methodology has been tested for 100 times and later the obtained data is sorted. From sorted data, particular values are stored as final result that appears most consistently in the array. Table 2 shows the results

Table 2 Brazilian-46 bus system

Brazilian-46 Bus System						
Technique	Bus topology			O.C. $\times 10^3$ US $	A.F.E.	S.R.
	From	To	No. of lines			
TLBO	2	3	1	**118.1**	**249.26**	**100**
	2	4	2			
	5	11	1			
	9	10	1			
	14	15	2			
	15	16	1			
	24	25	1			
	26	29	1			
	27	29	1			
	28	31	1			
	40	41	1			
MGWO [19]	2	3	1	118.8062	49237	88
	2	4	3			
	9	10	1			
	14	15	2			
	15	16	1			
	26	29	2			
	27	29	2			
	28	31	1			
GWO [19]	2	4	3	120.0531	89899	75
	4	11	1			
	5	6	1			
	5	11	1			
	9	10	1			
	14	15	3			
	26	29	1			
	28	31	2			
SCA	2	4	3	121.26	583.2	100
	5	11	1			
	9	10	3			
	14	15	2			
	24	25	1			
	26	29	1			
	28	31	1			
	46	10	1			

(continued)

Table 2 (continued)

Brazilian-46 Bus System

WOA	1	2	3	123.1	3268.2	99
	5	8	1			
	13	20	1			
	16	17	1			
	17	19	1			
	19	21	1			
	24	34	1			
	27	38	1			
	33	34	1			
	35	38	1			
	36	37	1			

Operational Cost Analysis of Test Cases

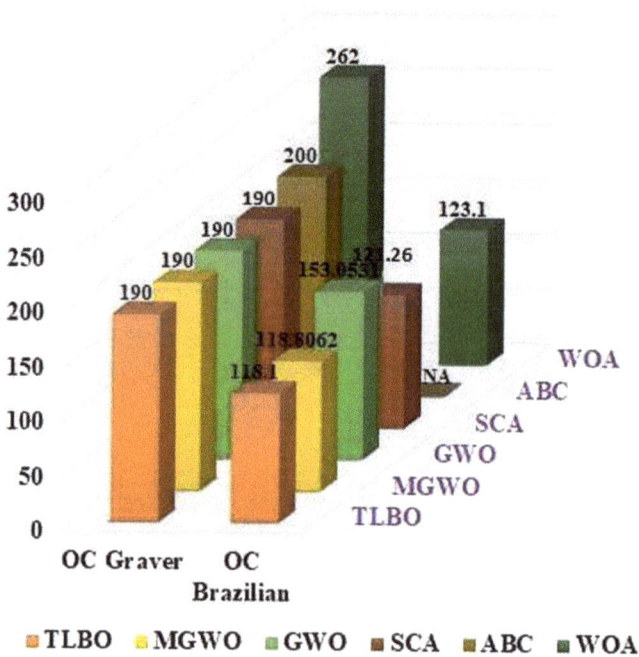

Fig. 1 Operational cost analysis

of optimization process. The results are shown in terms of AFE, SR and operational cost $ $\times 10^6$. It has been observed that TLBO secure 100 % SR as compared to other participants. However, the performance of SCA and WOA is comparatively better in this case as per SR but the OCs are not optimal as compared with TLBO. 13 new transmission line has been proposed by different algorithms. As per results reported in Tables 1 and 2 TLBO based solution propose 2 lines between buses 2 and 4, 1 line between buses 5 and 11, 1 line between buses 28 and 31, 1 line between buses 15 and 16, 1–1 lines between (24–25), (40–41), (2–3) and buses (9–10). The results of Operational Cost for both cases are shown in Fig. 1.

In this table, TLBO shows better AFE and SR then the other state-of-the-art algorithms.

6   Conclusion

TLBO is based on comprehensive search strategy that applies to a number of problems earlier. In this paper authors have investigated TLBO for TEP. The performance of algorithm has been investigated on standard test bus systems namely graver 6 bus and Brazilian bus system. It has been observed that TLBO finds better solutions than other participating algorithms in terms of operational cost and success rate. Average Function evaluations are also optimal for TLBO. A fine comparison has been derived with MGWO, GWO, SCA, and ABC. This comparison gives an idea that TLBO is fast in giving the best solutions in the less computational time. Based on simulation results, it is concluded that the TLBO algorithm has less computational time and ideal convergence properties. Investigation of more complex problems with bigger networks will be investigated in future.

References

1. Conejo AJ, Baringo L, Kazempour SJ, Siddiqui AS (2016) Generation expansion planning. In: Investment in electricity generation and transmission. Springer, pp 61–114
2. Verma A, Panigrahi BK, Bijwe PR (2010) Harmony search algorithm for transmission network expansion planning. IET Gener Transm Distrib 4(6):663–673
3. Rouhani A, Hosseini SH, Raoofat M (2014) Composite generation and transmission expansion planning considering distributed generation. Int J Electr Power Energy Syst 1(62):792–805
4. Hansen PB, Eskildsen L, Grubb SG, Stentz AJ, Strasser TA, Judkins J, DeMarco JJ, Pedrazzani R, DiGiovanni DJ (1997) Capacity upgrades of transmission systems by raman amplification. IEEE Photonics Technol Lett 9(2):262–264
5. Panwar NL, Kaushik SC, Kothari S (2011) Role of renewable energy sources in environmental protection: a review. Renew Sustain Energy Rev 15(3):1513–1524
6. Vassell GS (1990) The northeast blackout of 1965. Public Utilities Fortnightly; (United States) 126(8)
7. Zhang D, Wang Y, Luh PB (1999) Optimization based bidding strategies in the deregulated market. In: Proceedings of the 21st international conference on power industry computer appli-

cations. Connecting utilities. PICA 99. To the millennium and beyond (Cat. No. 99CH36351). IEEE, pp 63–68

8. Bertsekas DP, Bertsekas DP, Bertsekas DP, Bertsekas DP (1995) Dynamic programming and optimal control, vol 1. Athena scientific Belmont, MA

9. Dantzig GB (1998) Linear programming and extensions. Princeton University Press, Princeton

10. Zhang H, Vittal V, Heydt GT, Quintero J (2011) A mixed-integer linear programming approach for multi-stage security-constrained transmission expansion planning. IEEE Trans Power Syst 27(2):1125–1133

11. Hemmati R, Hooshmand R-A, Khodabakhshian A (2013) Comprehensive review of generation and transmission expansion planning. IET Gener Transm Distrib 7(9):955–964

12. Yu H, Chung CY, Wong KP (2010) Robust transmission network expansion planning method with Taguchi's orthogonal array testing. IEEE Trans Power Syst 26(3):1573–1580

13. El-Metwally MM, Al-Hamouz ZM (1990) Transmission networks planning using quadratic programming. Electric Mach Power Syst 18(2):137–148

14. Dorfman R, Samuelson PA, Solow RM (1987) Linear programming and economic analysis. Courier Corporation

15. Al-Hamouz ZM, Al-Faraj AS (2002) Transmission expansion planning using nonlinear programming. In: IEEE/PES transmission and distribution conference and exhibition, vol 1. IEEE, pp 50–55

16. Garver LL (1970) Transmission network estimation using linear programming. IEEE Trans Power Apparatus Syst 7:1688–1697

17. Sorensen K, Sevaux M, Glover F (2017) A history of metaheuristics. arXiv preprint arXiv:1704.00853

18. Sharma A, Sharma H, Bhargava A, Sharma N (2017) Fibonacci series-based local search in spider monkey optimisation for transmission expansion planning. Int J Swarm Intell 3(2–3):215–237

19. Khandelwal A, Bhargava A, Sharma A, Sharma H (2018) Modified grey wolf optimization algorithm for transmission network expansion planning problem. Arab J Sci Eng 43(6):2899–2908

20. Das S, Suganthan PN (2010) Problem definitions and evaluation criteria for cec 2011 competition on testing evolutionary algorithms on real world optimization problems. Jadavpur University, Nanyang Technological University, Kolkata, pp 341–359

21. Verma A, Panigrahi BK, Bijwe PR (2009) Transmission network expansion planning with adaptive particle swarm optimization. In: 2009 World congress on nature and biologically inspired computing (NaBIC). IEEE, pp 1099–1104

22. Rao RV, Savsani VJ, Vakharia DP (2011) Teaching-learning-based optimization: a novel method for constrained mechanical design optimization problems. Comput Aided Des 43(3):303–315

23. Rathore C, Roy R, Sharma U, Patel J (2013) Artificial bee colony algorithm based static transmission expansion planning. In: 2013 International conference on energy efficient technologies for sustainability. IEEE, pp 1126–1131

24. Romero R, Monticelli A, Garcia A, Haffner S (2002) Test systems and mathematical models for transmission network expansion planning. IEE Proc Gener Transm Distrib 149(1):27–36

Chapter 13
Harmonic Estimator Design Using Teaching Learning Based Optimization

Aishwarya Mehta, Jitesh Jangid, Akash Saxena, Shalini Shekhawat, and Rajesh kumar

1 Introduction

Harmonic estimation is the first step towards the most popular power quality issue in the power system. Harmonics are distorted electrical signals, which possess frequencies of multiples of the fundamental frequency (f) like $3f$, $4f$, $5f$, etc. Harmonic generation sources are non-linear devices, power converter, transformer and their sources. Harmonics have an adverse effect on power systems like losses in the electrical signal, reduce the life of electrical devices and functioning of electronic devices and I^2R losses in power system networks, etc.

In the Current Scenario, modern power system helps to fulfill the demand of consumers by reducing losses. This demand is increasing day by day which can be fulfill by non-linear devices. Non-linear devices help to transmit and distribute the bulk power from the power system. These non-linear devices do not follow Ohm's law and draw a large amount of power, which are cause of harmonics. Due to load variations, harmonics in the power system network vary continuously. Harmonics estimation is used to provide an accurate estimation of the phase and amplitude of the electrical signal. This accurate estimation helps to mitigate the losses of the

A. Mehta (✉) · J. Jangid · A. Saxena · S. Shekhawat
SKIT, Jaipur, India
e-mail: aishu.sharma.0786@gmail.com

J. Jangid
e-mail: jj90946@gmail.com

A. Saxena
e-mail: aakash.saxena@hotmail.com

S. Shekhawat
e-mail: shekhawatshalini17@gmail.com

R. kumar
MNIT, Jaipur, India
e-mail: rkumar.ee@mnit.ac.in

© The Author(s), under exclusive license to Springer Nature Singapore Pte Ltd. 2021   173
X.-Z. Gao et al. (eds.), *Applications of Artificial Intelligence in Engineering*, Algorithms for Intelligent Systems, https://doi.org/10.1007/978-981-33-4604-8_13

power system and also helps to design the mitigation filters. These mitigation filters help to remove unwanted frequency signals and improve the power quality of the power system. Artificial Neural Network (ANN) [1] method is used for harmonic estimation. In ANN neurons are connected to each other by directional link. When the number of neurons increases then directional links became complex which is a disadvantage of ANN.

In present time, various types of harmonic estimation techniques are available to solve power harmonic estimation problems. Many researchers use Fast Fourier Transform (FFT) [2] method. To estimate harmonics during picket fence effect, FFT fails to provide accurate estimation of harmonics. Wavelet Transform (WT) [3] techniques also used to handle the harmonic estimation design problem. In this method, mother wavelets (mother wavelets are formed by translation and dilation of function) help in accurate estimation but mother wavelets are randomly chosen, hence it is not beneficial for complex problems.

Traditional search methods can used to solve these types of problems. These methods are time consuming so these are not efficient for complex research problems. To solve complex research problem, we move on newly developed techniques, these new techniques are known as meta-heuristic techniques. Meta-heuristic methods are classified into different categories that are Evolutionary based methods, Swarm intelligence-based methods and Hybrid methods. Evolutionary based methods depend on the Charles Darwin evolution theory. Some evolutionary-based methods are Differential Evolution (DE) [4], Genetic Algorithm (GA) [5], and differential search methods, etc. Swarm intelligence based methods are nature-inspired algorithms which are based on population. Swarm based methods are also known as population-based methods. Some of them are Crow Search Algorithm (CSA) [6], Intelligent Crow Search Method (ICSA) [7], Grey Wolf Optimization (GWO) [8], Ant-Lion Optimization (ALO) [9], Evolutionary operators equipped Grey Wolf Optimizer (E-GWO) [10], Whale Optimization Algorithm (WOA) [11] and Artificial Bee Colony method (ABC) [12]. Hybrid algorithms are used to improve the result and made by hybridization of two or more techniques. These methods are Genetic Algorithm (GA) with Particle Swarm Optimisation (PSO) [13], Grey Wolf Optimisation (GWO) with PSO [14] and ACO with PSO [15].

Teaching learning-based optimization algorithm (TLBO) provides the optimal result in less number of iterations and also give accurate results in comparison to Genetic Algorithm (GA) [5] and Simulated Annealing (SA) [16]. TLBO is proposed by Rao et al. [17] inspired by the teaching and learning behaviour of class. TLBO algorithm is used to solve many real world problems like optimal coordination of directional over-current relay [18], unconstrained and constrained optimization problem [19] and others, to locate automatic voltage regulator in distribution system [20], optimal design of heat pipe [21] etc. Motivating through these successful utilization of TLBO in different types of fields, we use here TLBO algorithm for accurate estimation of phase and amplitude components of power harmonics which helps to design a powerful harmonic estimator.

After conducting literature review on applications of modern metaheuristic in harmonic estimation problem, following research objectives are formed for this work.

- To solve the harmonic estimation problem by teaching learning-based optimization algorithm.
- To develop a mathematical model of harmonic estimator design and test the robustness of TLBO for estimation design.
- To provide a comparative analysis of various contemporary methods based on different optimization parameters with TLBO based harmonic estimator model.

This paper is divided in different sections. Section 1 presents the literature of harmonic estimation problems and introductory details of the problem, Sect. 2 describes formulation of problem, Sect. 3 presents TLBO algorithm and brief literature review of TLBO, Sect. 4 contains the case studies of harmonic estimation problem and discussion on case 1 shown in Sect. 4.1 or on case 2 in Sect. 4.2. In last conclusion is given.

2 Problem Formulation of Harmonic Estimation in Power System

The Voltage or current waveform of power harmonics generally represented as

$$x(t) = \sum_{y=1}^{Y} Z_y \sin(w_y t + \theta_y) + Z_{dc} \exp(-\alpha_{dc} t) + \gamma(s) \tag{1}$$

where Y represents order of harmonics, $w_y = y 2\pi f^0$ in this w_y is angular velocity of y component, f^0 shows fundamental frequency, $\gamma(s)$ indicates AWGN (additive white Gaussian noise) [22] and $Z_{dc} \exp(-\alpha_{dc} t)$ denotes decaying term of the power system. θ_y and Z_y are unknown phase and magnitude values of yth harmonics. Now the signal $x(t)$ is sampled with sampling period T_i, this discrete time signal can be written as

$$x(s) = \sum_{y=1}^{Y} Z_y \sin(w_y s T_i + \theta_y) + Z_{dc} \exp(-\alpha_{dc} s T_i) + \gamma(s) \tag{2}$$

Now, Taylor series expansion theorem [10] is applied in Eq. 2 and simplified form of this equation is given as

$$x(s) = \sum_{y=1}^{Y} Z_y \sin(w_y s T_i + \theta_y) + Z_{dc} - Z_{dc} \alpha_{dc} s T_i + \gamma(s) \tag{3}$$

At last, a simplified and generalized formulation is obtained which helps to estimate the amplitude and phase of power harmonics, can be represented as follow

$$z(s) = \sum_{y=1}^{Y} [Z_y \sin (w_y sT_i) \cos \theta_y + Z_y \cos (w_y sT_i) \sin \theta_y] + Z_{dc} - Z_{dc} \alpha dc s T_i + \gamma (s)$$

(4)

Now, electrical signal is converted into parametric form as

$$x(s) = J(s)\theta(s)$$

(5)

where $J(s)$ is denoted as

$$J(s) = [\sin (w_1 sT_i) \cos (w_1 sT_i) \cdots \sin (w_y sT_i) \cos (w_y sT_i)1 - PT_i]^T$$

(6)

Unknown parameters are expressed as

$$\theta(s) = [\theta_{1y}\theta_{2y} \cdots \theta_{(2y-1)s}\theta_{2ys}\theta_{(2y+1)s}\theta_{(2y+2)s}]^T$$

(7)

$$\theta = [Z_1 \cos (\theta_1) Z_1 \sin (\theta_1) \cdots Z_y \cos (\theta_y) Z_y \sin (\theta_y) \cdots Z_{dc} Z_{dc} \alpha_{dc}]^T$$

(8)

Objective function H helps to find the optimal unknown parameters which is represented as

$$H = \min \left(\sum_{y=1}^{Y} f_y^2(y) \right) = \min \left(\sum_{y=1}^{Y} (x_y - \bar{x}_{y_{est}}^2) \right) = MSE(x_y - \bar{x}_{y_{est}})$$

(9)

Following section presents overview of TLBO algorithm.

3   Overview of TLBO Algorithm

TLBO is a population-based approach that helps to find an optimal solution to the problem. Many real-world problems of power systems are solved by TLBO in which some problems are like multi-objective problems of power system, controller tuning of voltage regulator [23], placement of DG in the distribution system [24], co-ordination of directional over-current relays [18], unit commitment problem [25], Optimal power flow [26] etc. In this method, the problem can be solved by two steps, first one is the teaching phase and second one is learning phase. These steps are explained below.

3.1 Teaching Phase

In teaching phase, student try to obtain more knowledge from the educator and educator try to give her/his best. Educator wants to increase the average knowledge (result) of class. Assume that j is iteration, i is number of subjects (problem variable attribute) and the number of learners are denoted by 'k'. Here, $m = 1, 2, \ldots, k$ are population size of the classroom and in a particular subject 'x' ($x = 1, 2, \ldots i$), average result is indicated as $A_{x,j}$.

In the teaching phase, this algorithm assumes that educators are highly educated and more knowledgeable in comparison to learners and helps to improve the knowledge of learners. This algorithm basically depends on the knowledge of educator and grasping power of learner. The efforts are to improve the result of learners. With respect to educator, learner who gives the best result in comparison to other learners, is called 'mbest'. Learner's best result is expressed as $Y_{\text{Total-mbest},j}$. In this result all the students with their all subjects, help to find the best result.

The existing average result of every subject and the respective results of the subject with the teacher differ from each other. This difference is written as

$$\text{Differ}_{\text{Avg}_{x,m,j}} = d_j(Y_{x,\text{mbest},j} - T_e Ax, j) \tag{10}$$

Here $Y_{x,\text{mbest},j}$ represents subject 'x' best result, T_e is teaching factor which is either 1 or 2. This teaching factor is randomly selected according the problem, d_j are random numbers which lies between 0 and 1.

$$T_e = \text{round}\{1 + \text{rand}(0, 1)(2 - 1)\} \tag{11}$$

Here T_e is neither an input argument nor an attribute of the TLBO algorithm. T_e can be randomly selected but after many hit and trial experiments on the benchmark functions, best results can be obtained at 1 and 2. So T_e can be 1 or 2 according to problem for best result.

In teaching phase, $\text{Differ}_{\text{Avg}_{x,m,j}}$ solution is updated by the following expression:

$$Y'_{x,m,j} = Y_{x,m,j} + \text{Differ}_{\text{Avg}_{x,m,j}} \tag{12}$$

Here, $Y'_{x,m,j}$ is updated value of $Y'_{x,m,j}$ and this updated value is acceptable when it gives better results.

3.2 Learning Phase

Learning phase depends on the teaching phase. In this phase, the learner tries to update the knowledge with the help of educators and other learners. By the interaction of learners with each other, they increased their knowledge.

Let consider "K" is the population size of class, from this class, two learners are randomly selected, they are R and S.

$$Y'_{total-R,j} \neq Y'_{total-S,j} \tag{13}$$

where $Y'_{total-R,j}$ and $Y'_{total-S,j}$ are updated values of $Y_{total-R,j}$ and $Y_{total-S,j}$ respectively R and S learners in teaching phase.

$$Y''_{x,R,j} = Y'_{x,R,j} + d_j(Y'_{x,R,j} - Y'_{x,S,j}) \tag{14}$$

when $Y'_{total-R,j} < Y'_{total-S,j}$

$$Y''_{x,R,j} = Y'_{x,R,j} + d_j(Y'_{x,S,j} - Y'_{x,R,j}) \tag{15}$$

when $Y'_{total-S,J} < Y'_{total-R,j}$. $Y''_{x,R,J}$ is useful when it gives better results for function. Equations 14 and 15 are used for the minimization of result for any problem.

$$Y''_{x,R,j} = Y'_{x,R,j} + d_i(Y'_{x,R,j} - Y'_{x,S,j}) \tag{16}$$

when $Y'_{total-S,j} < Y'_{total-R,i}$

$$Y''_{x,R,j} = Y'_{x,R,j} + d_i(Y'_{x,S,j} - Y'_{x,R,j}) \tag{17}$$

when $Y'_{total-R,j} < Y'_{total-S,i}$, Eqs. 16 and 17 also help to maximize the result of any problem.

Following section presents case studies and the discussion on the performance of TLBO.

4 Case Studies for Harmonic Estimator Design

This Compilation work is done on MAT-LAB 2015a on Intel Core i5 8th generation processor with 8 GB of ram. Parameters that are used for case studies are

- Search Agents: 30
- Maximum Iteration: 500.

We have considered exhaustion of maximum iteration as stopping criterion. Further, two case studies are presented to validate effectiveness of harmonic estimator with the help of TLBO by comparing the proposed design with GWO, ALO, and WOA based harmonic estimators. To develop a power harmonic estimator, we take into account two problems of harmonic estimator, these problems are $C_1(p)$ and $C_2(p)$, which are taken from the literature [10].

4.1 Case 1 Design Problem

The waveform considered for this case consists of fundamental, 3rd, 5th, 7th and 11th harmonics. Data of this waveform helps to measure amplitude and phase values respective to harmonic. Mathematical representation of case 1 problem is given as $C_1(p)$:

$$C_1(p) = \frac{3}{2} \times \sin(2\pi f^{(1)}p + 80) + \frac{1}{2} \times \sin(2\pi f^{(3)}p + 60) + \frac{1}{5} \times \sin(2\pi f^{(5)}p$$
$$+ 45) + \frac{3}{20} \times \sin(2\pi f^{(7)}p + 36) + \frac{1}{10} \times \sin(2\pi f^{(11)}p + 30) + \frac{1}{2} \times \exp(-5p)$$
$$+ \sigma_m$$

After running optimization routine, statistical attributes of obtained fitness values of 20 independent runs such as Mean, Maximum, Minimum, and Standard Deviation values are calculated. These values are represented in Table 1. These results prove that the Mean value of fitness for TLBO are optimal as compared to other algorithms (GWO, ALO, and WOA). Ranks obtained by the algorithms are mentioned in last column of this table.

For accurate estimation of phase and amplitude values, the error in estimation should be optimized. With this aim, comparison of performance of different algorithms is carried out. Comparative analysis of phase and amplitude values are reported in Table 2. Phase and amplitude values help to analyze the % error of phase and % error of amplitude components. Mathematical values of these errors are represented in Table 3. Figure 1 represents estimated wave, which is plotted with the help of original case 1 design problem and TLBO estimated wave. This estimated wave represents that estimated and original waves are almost similar and aligned with each other.

By the above observation and estimation, it can also be concluded that TLBO provides optimal values of phase and amplitude for case 1 design problem. These optimal values provide an accurate estimation of harmonic that is a current need of power system to reduce losses.

Table 1 Estimated results of case-1

Case-1					
Values	Mean	Max	Min	SD	Rank
TLBO	2.9044E−29	4.0155E−28	0.0000E+00	9.4548E−29	1st
GWO	1.7440E−06	5.3310E−06	4.4068E−07	1.1099E−06	3rd
ALO	2.4789E−07	1.0392E−06	2.7004E−09	2.9005E−07	2nd
WOA	7.3554E−06	1.2655E−05	6.0609E−07	3.6201E−06	4th

Table 2 Case-1 phase and amplitude

Case-1						
Algorithms	Parameters	Fun	3rd	5th	7th	11th
Actual	F (Hz)	50	150	250	350	550
	A (V)	1.5	0.5	0.2	0.15	0.1
	P (deg.)	80	60	45	36	30
BFO-RLS [27, 28]	A (V)	1.49	0.4956	0.2018	0.1526	0.0986
	P (deg.)	80.3468	58.5461	45.6977	34.8079	29.9361
PSOPC-LS [27, 28]	A (V)	1.482	0.488	0.182	0.1561	0.0948
	P (deg.)	80.54	62.2	46.6	34.621	27.31
GA-LS [27, 28]	A (V)	1.48	0.485	0.18	0.158	0.0937
	P (deg.)	80.61	62.4	47.03	34.354	26.7
MABC [29]	A (V)	1.5006	0.4997	0.1995	0.1498	0.1003
	P (deg.)	80.0187	60.0098	45.0905	36.0894	29.5511
TLBO	A (V)	**1.5000**	**0.5000**	**0.2000**	**0.1500**	**0.1000**
	P (deg.)	**80.0000**	**60.0000**	**45.0000**	**36.0000**	**30.0000**

Table 3 Case 1 % error result

Amplitude % error case-1					
Algorithms	Fundamental	3rd	5th	7th	11th
BFO-RLS [27, 28]	0.3840	0.2860	0.9020	1.7600	1.7500
PSOPC-LS [27, 28]	1.2000	2.4000	9.0000	4.0600	5.2000
GA-LS [27, 28]	1.3300	3.0000	10.0000	5.3300	6.3000
MABC [29]	0.0412	0.0553	0.2420	0.1550	0.3370
TLBO	**0.0000**	**0.0000**	**0.0000**	**0.0000**	**0.0000**

Phase % error case-1					
Algorithms	Fundamental	3rd	5th	7th	11th
BFO-RLS [27, 28]	0.4340	0.5660	1.5500	3.3100	0.2130
PSOPC-LS [27, 28]	0.6750	3.6700	3.5600	3.8200	8.9700
GA-LS [27, 28]	0.7630	4.0000	4.5100	4.5700	11.0000
MABC [29]	0.0234	0.0164	0.2010	0.2490	1.5000
TLBO	**0.0000**	**0.0000**	**0.0000**	**0.0000**	**0.0000**

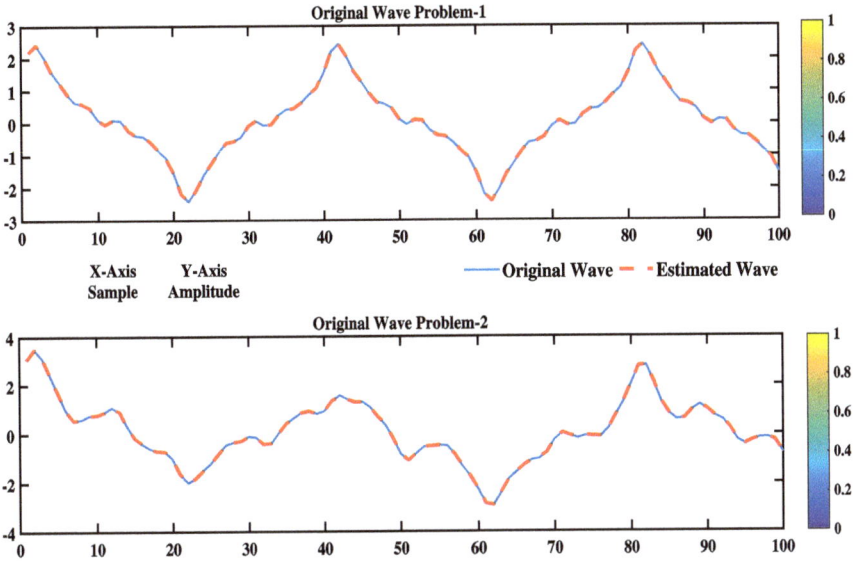

Fig. 1 Estimated waves

4.2   Case 2 Design Problem

The waveform of case 2 design problem consists of sub, fundamental, 3rd, inter-1, inter-2, 5th, 7th and 11th harmonics. When harmonic frequencies are less than fundamental frequency, then these contaminations are known as sub-harmonics. While inter harmonics is greater than fundamental harmonics. This design problem is used to measure amplitude and phase of power harmonics respectively. Case 2 design problem is mathematically written as

$$C_2(p) = \frac{101}{200} \times \sin(2\pi f^{(sub)}p + 75) + \frac{3}{2} \times \sin(2\pi f^{(1)}p + 80)$$
$$+ \frac{1}{2} \times \sin(2\pi f^{(3)}p + 60) + \frac{1}{4} \times \sin(2\pi f^{(inter1)}p + 80)$$
$$+ \frac{7}{20} \times \sin(2\pi f^{(inter2)}p + 20) + \frac{1}{5} \times \sin(2\pi f^{(5)}p + 45)$$
$$+ \frac{3}{20} \times \sin(2\pi f^{(7)}p + 36) + \frac{1}{10} \times \sin(2\pi f^{(11)}p + 30)$$
$$+ \frac{1}{2} \times \exp(-5p) + \sigma_m$$

Table 4 presents the Mean, Max, Min and SD values of case 2 design problem. Results of Table 4 show that the Mean value of TLBO is optimal as compared to GWO, ALO, and WOA. These values are used to rank the algorithms. We observe that TLBO secured first rank for case 2 design problem.

Table 4 Estimated results of case-2

Case-2					
Values	Mean	Max	Min	SD	Rank
TLBO	3.6241E−20	5.1778E−19	1.0932E−24	1.1811E−19	1st
GWO	2.9346E−05	5.7458E−05	1.0446E−05	1.5363E−05	3rd
ALO	1.3340E−05	4.9642E−05	2.0256E−06	1.2895E−05	2nd
WOA	2.5572E−04	4.0781E−04	1.2567E−04	6.6236E−05	4th

Table 5 Case-2 phase and amplitude

Case-2									
Algorithms	Parameters	Sub	Fun	3rd	Inter-1	Inter-2	5th	7th	11th
Actual	F (Hz)	20	50	150	180	230	250	350	550
	A (V)	0.505	1.5	0.5	0.25	0.35	0.2	0.15	0.1
	P (deg.)	75	80	60	65	20	45	36	30
BFO-RLS [27, 28]	A (V)	0.511	1.5029	0.4921	0.2581	0.3639	0.2009	0.1479	0.1015
	P (deg.)	74.81	79.9148	59.076	65.3445	19.8677	46.278	36.4473	30.0643
PSOPC-LS [27, 28]	A(V)	0.53	1.5049	0.281	0.24	0.377	0.211	0.165	0.11
	P(deg.)	73.51	79.45	58.12	63.28	18.23	48.1	37.109	31.87
GA-LS [27, 28]	A (V)	0.532	1.5083	0.472	0.238	0.381	0.251	0.172	0.117
	P (deg.)	73.02	79.23	57.55	62.41	17.64	48.33	38.78	32.56
MABC [29]	A (V)	0.5052	1.5008	0.4998	0.25	0.35	0.1997	0.15	0.1
	P (deg.)	74.9539	79.98	60.1254	64.9374	20.04	45.1636	35.9987	29.958
TLBO	A (V)	**0.5050**	**1.5000**	**0.5000**	**0.2500**	**0.3500**	**0.2000**	**0.1500**	**0.1000**
	P (deg.)	**75.0000**	**80.0000**	**60.0000**	**65.0000**	**20.0000**	**45.0000**	**36.0000**	**30.0000**

Estimated phase and amplitude values are shown in Table 5. A comparative analysis is used to find the optimal phase and amplitude values. Phase and amplitude values of original waves are used to get % error of phase and % error of amplitude. These error values are shown in Table 6. It has been observed that TLBO shows satisfactory performance. Further, Fig. 1 represents estimated wave which is plotted with the help of original case 2 design problem and TLBO estimated wave. This estimated wave represents that estimated and original waves are similar.

From the above discussion of case 1 design problem and case 2 design problem, it is evident that TLBO algorithm provides better results which are essential for accurate estimation of harmonics. In the modern power system, accurate estimation plays an important role to improve power quality by reducing losses. Harmonic estimator design with TLBO helps to develop mitigation filters to reduce harmonics and increase the overall efficiency of the power system. Following section presents conclusion of the work.

Table 6 Case 2 % error result

Amplitude % error case 2								
Algorithms	Sub	Fun	3rd	Inter-1	Inter-2	5th	7th	11th
BFO-RLS [27, 28]	1.1900	0.1950	1.5900	3.2400	3.9700	0.4540	1.4100	1.4800
PSOPC-LS [27, 28]	4.9500	0.3260	3.8000	4.0000	7.7000	5.5000	10.0000	11.0000
GA-LS [27, 28]	5.3500	0.5530	5.6000	4.8000	8.8500	7.5000	14.7000	17.0000
MABC [29]	0.0468	0.0530	0.0441	0.0133	0.0011	0.1390	0.0319	0.0184
TLBO	**0.0000**	**0.0000**	**0.0000**	**0.0000**	**0.0000**	**0.0000**	**0.0000**	**0.0000**
Phase % error case 2								
BFO-RLS [27, 28]	0.2530	0.1060	1.5400	0.5300	0.6610	2.8400	1.2400	0.2140
PSOPC-LS [27, 28]	1.9900	0.6870	3.1300	2.6500	8.8500	6.8800	3.0800	6.2300
GA-LS [27, 28]	2.6400	0.9620	4.0800	5.5200	11.8000	7.4000	7.7200	8.5300
MABC [29]	0.0615	0.0250	0.2090	0.0964	0.2000	0.3640	0.0036	0.1400
TLBO	**0.0000**	**0.0000**	**0.0000**	**0.0000**	**0.0000**	**0.0000**	**0.0000**	**0.0000**

5 Conclusion

To design harmonic filters, accurate estimation of phase and amplitude components are essential and hence harmonic estimation process is inevitable. Harmonic filters are useful to reduce the losses and make power supply clean and free from harmonic contamination to the consumers. This work reports an application of TLBO for estimation of phase and amplitude to design an effective harmonic estimator. Authors have validated TLBO based estimator on two case studies of harmonic estimator design that are taken from literature.

We observe that TLBO based harmonic estimator exhibits satisfactory performance. In these case studies, we have estimated the phase and amplitude of waves with their respective estimation errors. In this validation we concluded that accuracy of TLBO algorithm is better than other algorithms (GWO, ALO and WOA). From this application, it can be concluded that TLBO is a robust algorithm and can be a suitable choice for designing harmonic estimators. Investigations on more challenging problems such as regulator design, Transmission network expansion planning problem and other problems of bio informatics will be addressed in future by authors.

References

1. Hassoun MH et al (1995) Fundamentals of artificial neural networks. MIT Press, Cambridge, MA
2. Nussbaumer HJ (1981) The fast Fourier transform. Fast Fourier transform and convolution algorithms. Springer, New York, pp 80–111
3. Newland DE (1993) Harmonic wavelet analysis. Proc Roy Soc Lond Ser A: Math Phys Sci 443:203–225
4. Price KV (2013) Differential evolution. In: Handbook of optimization. Springer, pp 187–214
5. Darrell W (1994) A genetic algorithm tutorial. Stat Comput 4(2):65–85
6. Alireza A (2016) A novel metaheuristic method for solving constrained engineering optimization problems: crow search algorithm. Comput Struct 169:1–12
7. Shekhawat S, Saxena A (2019) Development and applications of an intelligent crow search algorithm based on opposition based learning. ISA Trans
8. Mirjalili S, Mirjalili SM, Lewis A (2014) Grey wolf optimizer. Adv Eng Softw 69:46–61
9. Mirjalili S, Lewis A (2016) The whale optimization algorithm. Adv Eng Softw 95:51–67
10. Saxena A, Kumar R, Mirjalili S (2020) A harmonic estimator design with evolutionary operators equipped grey wolf optimizer. Expert Syst Appl 145:113125
11. Mirjalili S (2015) The ant lion optimizer. Adv Eng Softw 83:80–98
12. Karaboga D, Akay B (2009) A comparative study of artificial bee colony algorithm. Appl Math Comput 214(1):108–132
13. Settles M, Soule T (2005) Breeding swarms: a GA/PSO hybrid. In: Proceedings of the 7th annual conference on genetic and evolutionary computation, pp 161–168
14. Singh N, Singh SB (2017) Hybrid algorithm of particle swarm optimization and grey wolf optimizer for improving convergence performance. J Appl Math 2017
15. Xiao G, Li S, Wang X, Xiao R (2006) A solution to unit commitment problem by ACO and PSO hybrid algorithm. In: 2006 6th world congress on intelligent control and automation, vol 2. IEEE, pp 7475–7479
16. Van Laarhoven PJM, Aarts EHL (1987) Simulated annealing. Simulated annealing: theory and applications. Springer, New York, pp 7–15
17. Venkata Rao R, Savsani VJ, Vakharia DP (2011) Teaching-learning-based optimization: a novel method for constrained mechanical design optimization problems. Comput-Aided Des 43(3):303–315
18. Singh M, Panigrahi BK, Abhyankar AR (2013) Optimal coordination of directional over-current relays using teaching learning-based optimization (TLBO) algorithm. Int J Electr Power Energy Syst 50:33–41
19. Venkata Rao R, Savsani VJ, Balic J (2012) Teaching-learning-based optimization algorithm for unconstrained and constrained real-parameter optimization problems. Eng Optim 44(12):1447–1462
20. Niknam T, Azizipanah-Abarghooee R, Narimani MR (2012) A new multi objective optimization approach based on TLBO for location of automatic voltage regulators in distribution systems. Eng Appl Artif Intell 25(8):1577–1588
21. Rao RV, More KC (2015) Optimal design of the heat pipe using TLBO (teaching-learning-based optimization) algorithm. Energy 80:535–544
22. Galdi V, Pierro V, Pinto IM (1998) Evaluation of stochastic-resonance-based detectors of weak harmonic signals in additive white Gaussian noise. Phys Rev E 57(6):6470
23. Chatterjee S, Mukherjee V (2016) PID controller for automatic voltage regulator using teaching-learning based optimization technique. Int J Electr Power Energy Syst 77:418–429
24. Sahu S, Barisal AK, Kaudi A (2017) Multi-objective optimal power flow with DG placement using TLBO and MIPSO: a comparative study. Energy Procedia 117:236–243
25. Khazaei P, Dabbaghjamanesh M, Kalantarzadeh A, Mousavi H (2016) Applying the modified TLBO algorithm to solve the unit commitment problem. In: 2016 world automation congress (WAC). IEEE, pp 1–6

26. Bouchekara HREH, Abido MA, Boucherma M (2014) Optimal power flow using teaching-learning-based optimization technique. Electr Power Syst Res 114:49–59
27. Singh SK, Sinha N, Goswami AK, Sinha N (2016) Robust estimation of power system harmonics using a hybrid firefly based recursive least square algorithm. Int J Electr Power Energy Syst 80:287–296
28. Ray PK, Subudhi B (2012) BFO optimized RLS algorithm for power system harmonics estimation. Appl Soft Comput 12(8):1965–1977
29. Kabalci Y, Kockanat S, Kabalci E (2018) A modified ABC algorithm approach for power system harmonic estimation problems. Electr Power Syst Res 154:160–173

Chapter 14
Real-Time Frame-to-Frame Jitter Removing Video Stabilization Technique

Madhura R. Shankarpure and Deepa Abin

1 Introduction

Image stabilization refers to algorithms used to stabilize and improve video quality by deleting undesirable camera shakes and jitters due to handheld devices. Image stabilization technology is used to eliminate visual effect loss of consistency by reduction of excessive unwanted noise. A computer capturing frame sequence with no impact of motion object [1]. It is especially important for handheld imagery appliances that are more influenced by shakes because of its smaller size.

The common source of distorted photos is by unnecessary jiggling of hand and deliberate camera panning, and unwanted fluctuations in position camera result in unstable image sequences. Applications video stabilization techniques guarantee high visual performance and secure video also non-optimal terms and conditions. Ideally, it is packed with imaging equipment mechanical means physically stopping camera shakes, or they use sophisticated electronic or optical means object movement first, and then compensate it by taking action the image sensor. For reducing the cost of imaging devices Effects of stabilization can be obtained via digital Image stabilization, using data drawn from footage images, without the need for additional Knowledge of physical motion on camera [1].

Digital evidence for all instances video stabilization should be applied the applications are real time and offline. The stabilization in video capture allows to lower the data bit rate compression because of projected motion output typically boost dramatically after stabilization subsequent pictures. The recording of video from handheld devices going from various unwanted and slow motions which significantly affect the performance of the newly processed video. To minimize shakes or frame-to-frame

M. R. Shankarpure (✉) · D. Abin
Department of Computer Engineering, PCCOE, Pune, India
e-mail: rshankarpure1968@gmail.com

D. Abin
e-mail: deepaabin@gmail.com

© The Author(s), under exclusive license to Springer Nature Singapore Pte Ltd. 2021   187
X.-Z. Gao et al. (eds.), *Applications of Artificial Intelligence in Engineering*, Algorithms for Intelligent Systems, https://doi.org/10.1007/978-981-33-4604-8_14

jitter from handheld mobile cameras, some external equipment such as a tripod stand and other hardware needs to be used. Consumer and professional videography are extremely important. Therefore, there are several different mechanicals, optical, and algorithmic solutions available. Stabilization can help take handheld pictures with long exposure times, including in still image photography.

Videos need to be synchronized in medical diagnostic applications such as endoscopy and colonoscopy, to determine the exact position and width of the problem. Similarly, in the military application and robotic application.

There are mainly three approaches to stabilize the video which are following. The mechanical approach for video stabilization, optical approach, and digital approach for video stabilization [2]. In mechanical approach for video stabilization systems, external special sensors such as gyroscope and accelerometers are used. In the optical video stabilization method, stabilization is achieved by moving parts of the lens instead of moving the entire camera. In the digital video stabilization system, no special sensors are needed to estimate camera movement. The motion estimation, smoothing, image composition are three key steps in digital video stabilization. The proposed algorithm inspired by a digital video stabilization system in which each frame from the input RGB colored video is converted into grayscale and the specified corresponding frame key point. Following this process, the main point transformation between the consecutive frame was followed and filtered out of unwanted motion by applying the smoothing technique plus reconstructed the sequence of frames. The use of digital videos such as telehealth, marketing, television, mechatronics, learning, self-driving cars, and safety is a large number of applications. Because of the vast amount of video shot, stored and distributed, the investigation of and develop effective multimedia treatment and techniques of research for indexing, searching, and downloading the video material [3].

The paper's main goal is to stabilize video using inbuilt mobile sensor and strictly prohibited the use of external to obtain a video sequence where jitter has been effectively frame-to-frame removed.

This paper is structured according to the following. Many basic definitions and the relevant work are illustrated briefly in Sect. 2. Details of the proposed method for video stabilization in Sect. 3. It presents and addresses experimental findings in Sect. 4. Finally, a few concluding comments and guidelines for future research are set out in Sect. 5.

2 Litrature Survey

Currently, there are many video stabilization techniques available in the market which consider various different parameters. Research has been carried out in this field and the following papers have been referred for the purpose of research and study. For 3D video stabilization of handheld camera Liu, jin, and Gleicher [4] proposed a content preserving warps algorithm. The key insight of the work is that for video stabilization purposes a carefully designed material that preserves warp can falsify pity shift in feature point but the output less accurate is not physically accurate by comparing the 2D video stabilization the major drawback of this method is that it requires first real-time structure from transformation as well as method is also very difficult heavier weight. The proposed algorithm allows effective usage for available information in image alignment as well as produces very reliable results in alignment. That is because every pixel in the frame is used to estimate global motion. The computational load is however heavy and the range of convergence is also small. In paper [5], Weina Ge, Tang, and Shum proposed the full-frame video stabilization system based on direct pixels with motion in painting. After reflecting the interframe error for adjacent frame, the Laplacian image pyramid level is established and error between frames is calculated. The estimation process involves SSD minimization with Gauss-Newton minimization, which uses the first-order expansion before squaring of the individual error quantities. The main limitation of this approach is that it relies heavily on the outcome of estimating global motion which can unreliable when a huge space covered by object with motion. Neighboring frames will not be distorted correctly for fast-moving objects, so there will be noticeable items at the borders. The Gauss-Newton minimization's ability to converge is also limited.

Pang et al. [6] suggested the dual-tree complex wavelet transform (DT-CWT) video stabilization. It performs motion estimation studying the relation between spatial domain movement and DT-CWT phase changes. Tang et al. [7] Chen suggested a simple video stabilization algorithm based on block-matching and edge finishing. A statistical approach helps achieving the global vibrant moving features after estimating the global motion vectors. This proposed algorithm's strength point works on real-time images. And limitation is that the proposed algorithm is complex, and high storage space is needed.

A digital video stabilization method through the validation and filtering of adaptive motion vectors [8] uses estimation of block motion and filters the unwanted motions by adaptive IIR filter. The strength of this paper is that tests for the proposed algorithm will show high efficiency. And the proposed limitation algorithm is that certain motion vectors' accuracy is not accurate, and the algorithm is complex. By filtering the accumulated global motion vectors, Rawat and Singhai [9] proposed an adaptive motion smoothing method which eliminates high-frequency jitters. An updated approach for the validation of motion vectors implemented with adaptive threshold on global motion vectors. The suggested solution not only eliminates the jitters but preserves details about the scene as well. The proposed method significantly reduces the areas missing from the image.

Javaria Maqsood, Asma Katiar, Lubna Ali proposed object tracking technique by using global motion estimation and Kalman filter; paper implements robust algorithm that find out and records the Moving Video artifacts. The proposed approach results present 94.73% accuracy [10]. Gustav Hanning proposed a system which removes the rolling shutter effect and stabilized the video. Using measurements from the sensors like accelerometer and gyroscope, the device corrects for roll-shutter distortions. Experimentally, the precision of the orientation projections was tested using ground-truth data from a motion capture device. Adaptive low pass filter algorithm is used by author to obtain a stable video while holding most of the material in view at the same time. In this paper, the precision of the orientation predictions was tested experimentally. The author performed a user analysis in which device performance, implemented in I phone operating system, was relate with incorrect video output [11].

Kornilova et al. [12] developed the general mathematical models needed to implement the MEMS sensor readings based on the video stabilization module. The main goal is to stabilize the video transmitted to the remotely controlled not only mobile robot operator but also to increase the accuracy of video-based navigation for autonomous sub-miniature models. The current stabilization methods using data from sensors were analyzed and viewed in a real-time environment from the application's point of view. The main limitation that arose during the experiments that the previous research papers did not resolve like, calibration and synchronization of camera as well as sensor, increasing the accuracy of sensor data determination of camera position. The authors propose potential solutions to these issues that would help improve existing algorithms quality, the main result of a system for applying video stabilization algorithms focused on the reading of MEMS sensors.

Jacobs, Baek [13] developed a new approach for stabilizing video by using the gyroscope. The author shows results for videos of large moving foreground objects, parallax, and low illumination using the proposed algorithm. It contrasts the paper approach with stabilization algorithms based on the commercial image. The proposed solution is more stable and cheaper in terms of computation. The proposed solution implemented on a smart phone and eliminate camera shake and roll-shutter artifacts in real-time by using the phone's built-in gyroscope and GPU. Neel Joshi, Sing Bing Kang [14] present a algorithm which deblur the image with use of attached external hardware. In an energy optimization framework, the proposed methodology uses a sensor like inexpensive gyroscopes and accelerometers to eliminate a blurriness, the author solves the camera movement and uses a joint optimization to infer the latent picture. The paper approach is automatic, and can deploy on every pixel. The author also provides a method for conducting camera motion blur "ground-truth" measurements used and the same in used to verify the deconvolution and hardware.

In the paper [15], an algorithm is suggested solution to remove jitters which occurring during video recording by smartphone. In this paper, the author founded the first silent feature points of the original video and then optimize and stabilize it by using video stabilization function. The proposed method presented in this paper decreases the jitters from many different types of situation and stabilize the video. The paper [16] to develop a stabilized video frame, a novel motion model, Steady Flow, was proposed. A Steady Flow shows particular optical flow by imposing strict spatial coherence for smoothing function. In this way, in a video stabilization device, the author can prevent fragile feature monitoring. The paper studies illustrate stabilization efficiency on images that are daunting in the real world.

3   Proposed Methodology

The accuracy of video stabilization mainly depends on the motion vectors in algorithms. For processing video stabilization good with good result, the algorithm used inbuilt mobile sensors which are accelerometer and gyroscope. The basic block diagram for the proposed algorithm is depicted in Fig. 1. First, read each input video frame and fetch inbuilt smartphone sensor data with respect to x, y, z axis. Now here might be chances of raw data so filtered out this data by applying Kalman filter. After this, for removing unwanted jitters and shakes, the camera coordinates and space coordinates are calculated.

Fig. 1 Basic block diagram of proposed algorithm

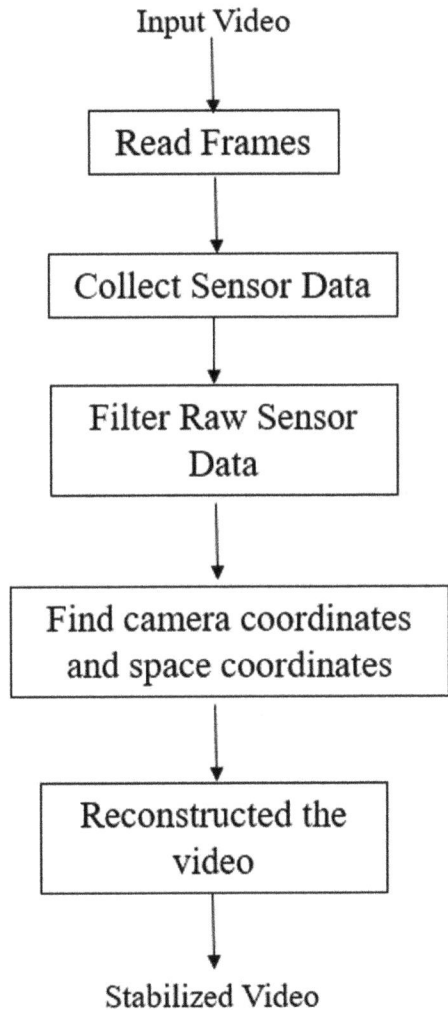

Input Video

Read Frames

Collect Sensor Data

Filter Raw Sensor Data

Find camera coordinates and space coordinates

Reconstructed the video

Stabilized Video

Finally, all new frames have been sequenced to reconstruct video output to achieve stabilized video. It is relatively common for a feature to switch from a frame to next because both cameras shake and camera shakes Intentional movements to panning. In this case, the Corresponding local motion vectors can lead to imprecise Estimation of motion, since stabilization of frames ought to only jitter on camera and not make up the needed moves. So as to stop the Problem, the proposed filtering of motion algorithms is used and the estimated movement would be assessed to recognize deliberate moves. The last step of a system for stabilizing the video is the process of composing frames. The process stabilizes the frame to frame sequence by compensate vector with motion estimation.

4 Experimental Setup and Results

Synchronization of sensor readings and camera is performed by an application, described in the corresponding section. The proposed system implemented the proto-type of the algorithm utilizing the synchronization of sensor readings and camera, performed by an application, on the android studio platform with hardware require-ment of i3 processor, with at least 5 GB hard disk with 1 GB ram and one android smartphone with at least 512 MB ram. The proposed algorithm has given synchro-nized sensor readings and frames, as well as intrinsic camera parameters, and has shown great significance as in Fig. 2. It shows the both original sensor graph of real-time data and stabilized sensor graph of proposed methodology in corrected data in which directly difference of shakes and jitters have been shown which seems in original data there are jerks present and in corrected graph, these all jerks are smoothened and stabilized very well.

Automatic evaluation of the performance of video stabilization is important for the tuning and analyzed by comparison of different techniques. Perfectly full stabilization of noisy video cannot be achieved in all cases. In addition, rather than removing the noise or jitters, the process itself frequently produce some blurriness in frames. Typically, automatic criteria for image quality are based on a simple pixel-by-pixel comparison between the ground truth image and the reference image. Such pixel-wise approaches are often extended to video quality assessment, simply by comparing still images on a frame-by-frame basis [18]. The most widely used quantitative video stabilization efficiency metric is the PSNR which means peak signal-to-noise ratio in the MSU Video quality management tool. In this paper, author used PSNR.

$$PSNR = 10. \log_{10} \frac{\text{MaxErr}^2 . w . h}{\sum_{i=1, j=1}^{w, h} \left(x_{i,j} - y_{i,j} \right)^2} \tag{1}$$

Equation 1 shows PSNR formula, video width, and video height are represented by w and h, respectively. Mixer stands for the maximum possible absolute value of color components difference [19]. The author recorded video for experimentation

Fig. 2 Graphical representation of real-time data [17]

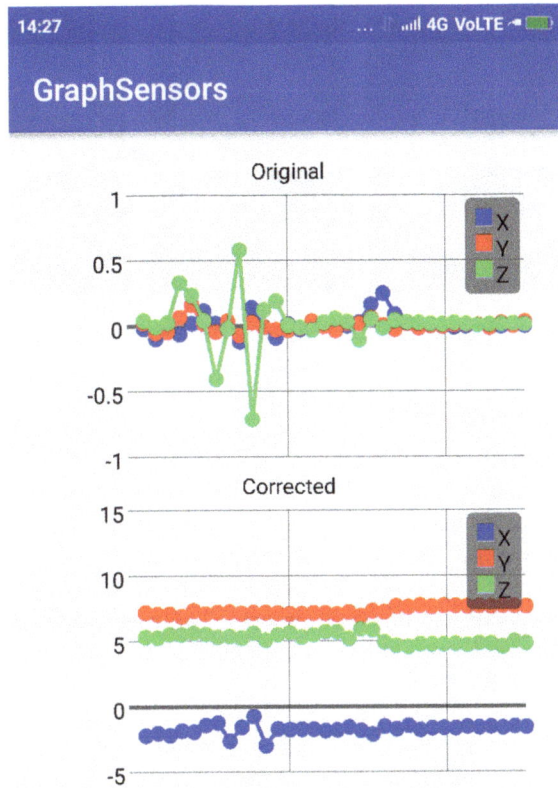

which has a resolution of 1280 × 720 and a framerate of 26.79 frames per second. The total duration of the video is about 7 s. The video is set indoors and is taken from a handheld device. The primary objects of interest in this video are the black fizz bottle. Figure 3 shows the selected snippets of stabilized video frame where top four frames are of normal jitter presenting video and bottom snippets are of stabilized jitter free frames.

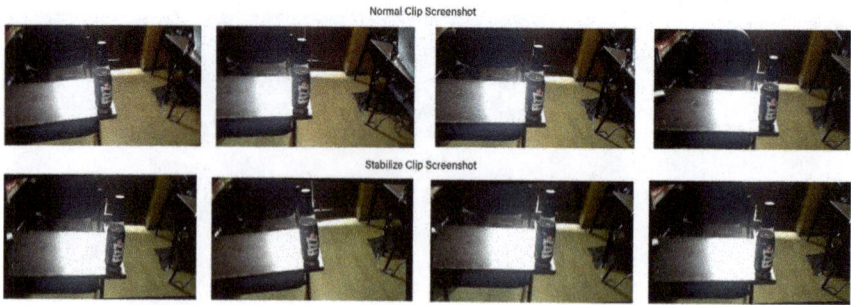

Fig. 3 Selected frames from normal video and stabilized video

5 Conclusion

The paper introduced the real-time video stabilization processing application and algorithm that is established by specifically banning the use of external equipment such as the tripod stand for general. Some critical issues were addressed throughout the article typically affecting all forms of video stabilization with different implications on the proposed process. The paper shows the original video values and stabilized video values in the graphical representation form to better visualization background work.

Paper studied scope, need, video stabilization approaches for short with their strength and limitations and a thorough study of the literature survey on various video stabilization techniques have been discussed. The proposed algorithm shows a little bit time consuming with good effects on normalization and removes noise from the recorded sequence of the output frames. There is a lot of scope for stabilizing the video in this area by observing the challenges and problems mentioned in the literature survey.

References

1. Vella F, Castorina A, Mancuso M, Messina G (2002) Digital image stabilization by adaptive block motion vectors filtering. IEEE Trans Consum Electron 48(3):796–801
2. https://www.learnopencv.com/video-stabilization-using-point-feature-matching-in-opencv/
3. Tico M, Alenius S, Vehvilainen M (2006) Method of motion estimation for image stabilization. In: Proceeding of the IEEE international conference on acoustics, speech and signal processing, pp 277–280
4. Liu F, Gleicher M, Jin H, Agarwala A (2009) Content preserving warps for 3D video stabilization. In: International conference proceedings of ACM SIGGRAPH 2009 papers. ACM, New York, NY, USA, pp 1–9
5. Matsushita Y, Ofek E, Ge W, Tang X, Shum HY (2006) Full frame video stabilization with motion in painting. IEEE Trans Pattern Anal Mach Intell 28(7):1163–1178
6. Pang D, Chen H, Halawa S (2010) Efficient video stabilization with dual-tree complex wavelet transform. EE368 Project Report, Spring

7. Tang C, Yang X, Chen L (2009) A fast video stabilization algorithm based on block matching and edge completion. In: IEEE 13th international workshop on Multimedia signal processing MMSP 2009
8. Tanakian MJ, Rezaei M, Mohanna F (2011) Digital video stabilization system by adaptive fuzzy filtering. In: 19th European signal processing conference, pp 318–322
9. Rawat P, Singhai J (2013) Adaptive motion smoothing for vide stabilization. Int J Comput Appl 72(20)
10. Maqsood J, Katiar A, Ali L (2018) Robust technique for object tracking by interference of global motion estimation and Kalman filter. Univ Sindh J Inform Commun Technol 2(3)
11. Hanning G (2011) Video stabilization and rolling shutter correction using inertial measurement sensors. Master Thesis at Linköping University, LiTH-ISY-EX–11/4464–SE
12. Kornilova AV, Kirilenko IA, Zabelina NI (2017) Real-time digital video stabilization using MEMS-sensors. Trudy ISP RAN/Proc ISP RAS 29(4):73–86
13. Karpenko A, Jacobs D, Baek J, Leyoy M (2011) Digital video stabilization and rolling shutter correcting using gyroscope. In: CSTR 2011, 171.67.77.70
14. Joshi N, Kang SB, Zitnick CL, Szeliski R (2010) Image deblurring using inertial measurement sensors. In: SIGGRAPH'10. ACM, New York, NY, USA, pp 30:1–30:9
15. Kulkarni S, Bormane DS, Nalbalwar SL (2017) Video stabilization using feature point matching. IOP Conf Ser J Phys Conf Ser 787:012017
16. Liu S, Yuan L, Tan P, Sun J (2014) SteadyFlow: spatially smooth optical flow for video stabilization. IEEE conference on vision and pattern recognition, September 2014
17. Shankarpure MR, Abin D (2020) Video stabilization by mobile sensor fusion. J Crit Rev 7(19):1012–1018
18. Gonzalez RC, Woods RE (2007) Digital image processing, 3rd edn. Prenctice Hall, NJ
19. Walha A, Alimi AM, Wali A (2013) Video stabilization for aerial video surveillance. In: AASRI conference on intelligent systems and control. Elsevier. https://doi.org/10.1016/j.aasri.2013.10.012

Chapter 15
A Review of Nature-Inspired Routing Algorithms for Flying Ad Hoc Networks

Amrita Yadav and Seema Verma

1 Introduction

Flying ad hoc network (FANET) is an ad hoc network between multiple unmanned aerial vehicles (UAVs). It can be defined as a new version of mobile ad hoc networks (MANET) in which the nodes are UAVs. FANETs provide a self-configured network for UAVs to link them with ground stations. These networks do not require pre-defined infrastructure since every UAV that are connected together are also connected to the ground station at the same time. There are certain characteristics of FANETs which makes it different from other ad hoc networks. FANETs have rapid change in its topology because of its high degree of mobility. The UAVs move at very high speed because of which they can lose connectivity. The node density in FANETs is also much lower than other ad hoc networks since the nodes are quite far from each other. FANETs also require sufficient energy so as to move the UAVs. FANET is useful in various fields including agriculture management, traffic monitoring, relay networks, etc. It is a multi-UAV system that can tell information regarding quality of crops so that which fertilizer has to be used may be identified in agriculture management system. Companies from various fields are using UAVs to collect information of ground regions and to minimize the cost and time of operation [1]. Multi-UAVs can be used by search and rescue teams to collect accurate knowledge base, which can result into improved efficiency of operations.

With such a rise in the applications of FANETs, there is a strong requirement for efficient and reliable routing algorithms to increase the communication between these

A. Yadav (✉)
Department of Computer Science, Banasthali Vidyapith, Banasthali, India
e-mail: amrita.yadav26@gmail.com

S. Verma
Department of Electronics, Banasthali Vidyapith, Banasthali, India
e-mail: seemaverma3@yahoo.com

© The Author(s), under exclusive license to Springer Nature Singapore Pte Ltd. 2021   197
X.-Z. Gao et al. (eds.), *Applications of Artificial Intelligence in Engineering*, Algorithms for Intelligent Systems, https://doi.org/10.1007/978-981-33-4604-8_15

UAVs. In FANET, the research is still going on to set up a proper communication network between the nodes.

Nature-inspired routing algorithms (NIA) appear promising to improve the issues of routing in FANETs. NIA is an intelligent algorithm which helps us to simulate through real biological systems. These algorithms are based on living organisms frameworks. Nature-inspired optimization algorithms represent an encouraging solution for explaining complex optimization problems. Some of NIA include ant colony optimization (ACO), artificial bee colony (ABC), particle swarm optimization (PSO), artificial fish swarm algorithm, firefly algorithm (FA), etc. These algorithms are used in various other domains as well. Some of them are used for routing in wireless ad hoc networks and have given much better results as compared to conventional algorithms.

2 Literature Review

To generate new networking family, few of the previously made wireless network protocols have been changed and few new protocols have been given for routing of FANETs. FANETs are used for highly sensitive applications [2]. Following are FANET protocols which are being cited in literature:

Sharma et al. [3] proposed distributed priority tree-based routing (DPTR) based on following factors: Packet delivery ratio, end-to-end delay, channel utilization, throughput, network connectivity time, which showed that the packet delivery ratio (PDR) was high for 11–15 inputs with negligible end-to-end delay. It also resulted in high throughput and network connectivity time.

Yassein and Damer [4], load carry and deliver routing (LCAD), used for ground-to-ground transfer which showed that it is secure and with no hops. Multi-level hierarchical routing (MLH) was also proposed by them which was used for ground station to UAVs, data transfer and based on clusters, for large network area. They also proposed data centric routing (DCR), one-to-many communication based on IDs of UAVs, in which UAVs does not need to know the ID, no need for UAVs to be online. Time slotted on-demand routing (TSOR) was also proposed by them which was based on following factors: Packet delivery ratio, communication time and the result concluded increase in PDR, decrease the communication time between UAVs.

Khan et al. [5], Zone routing protocol (ZRP) based on following factors: End-to-end delay(100 nodes with 25 m/s and 1 km × 1 km speed and area, network load area = 2 km × 2 km) which resulted into reduced end-to-end delay, worst throughput (speed of UAVs = 25 m/s, 60 m/s). Temporarily ordered routing algorithm (TORA) was also proposed by them. It is based on route discovery which minimizes the response to topological change and results into fast computation of paths.

Bujari and Ronzani [6] gave GPSR protocol based on following factors: Packet delivery ratio, response time, topology change and it showed lowest PDR at 150 nodes.

Palan and Sharma [7] presented geographic position mobility oriented routing (GPMOR) used for route discovery which shows that mobility relationship to select next hop is more accurate.

Qingwen et al. [8] invented an adaptive forwarding protocol (AFP), vector based forwarding protocol based on following factors: packet delivery fraction, end-to-end delay, energy consumption, simulation time $= 500$ s. The result shows that AFP achieves better performance.

Li and Huang [9], proposed a scheme known as adaptive beacon scheme (ABPP). This scheme is based on the geographic routing protocols. According to it, because of the rapid changing topology of FANET, the existing protocols keep transferring the packets to each other continuously for ease of selecting routes, which generates more packets and creates more overhead and collision of packets with delay in transmission. In order to avoid this situation, ABPP sends the packet dynamically and assures the prediction of UAV positions. It reduces the transmission delay and also lesser the data overhead.

Ghazzai et al. [10] categorized the routing problem in FANETs by mixed integer nonlinear programming problem. This algorithm is divided into two steps. Firstly, it tries to find the routing path of the UAVs by their initial position. The problem is converted into integer linear programming problem and then solved by a software. If this step does not succeed, then in its next step, it uses the Hooke-Jeeves algorithm which uses search patterns in repetitive manner. This algorithm helps to find paths for UAVs for different scenarios.

Rovira-Sugranes and Razi [11], proposed a modified Dijkstra's shortest path algorithm for optimal routing of UAVs. It uses prediction of distances between UAVs. The results show improvement of end-to-end delay in the network.

These are routing algorithms designed for FANETs keeping their characteristics in mind. The existing algorithms lack the intelligent solutions and coordination of multiple components. These algorithms are based on demand or table-driven approaches. The results of the above protocols show that some of the protocols reduces the communication time between the nodes but at the same time increases the packet delivery ratio or vice versa. Therefore, current results are not as efficient as required. With the increasing technology and world becoming more artificial intelligence based, there is a need of bringing AI into wireless communications as well. There exists huge demand of intelligent routing solutions which have a property of reasoning as well as decision making. Thus, Nature Inspired algorithms can be a new solution for finding optimal path for UAVs.

In literature, there are many works cited by scholars. Saleem et al. [12] proposed Formal Modeling of BeeAdHoc: a bio-inspired mobile ad hoc network routing protocol. The mathematical expressions of the performance metrics provide valuable insight about the behavior of BeeAdHoc in particular, and a typical ad hoc routing protocol.

Maistrenko et al. [13] compared AODV, DSDV, DSR, AntHocNet with each other and it was found that AntHocNet exceeds AODV and DSDV in terms of PDR, end-to-end delay.

Fiala [14], proposed multi-agent ant-based routing algorithm (MAARA) and compared it with conventional routing algorithms AODV, DSR, and AntHocNet. The results showed that MAARA outperformed all algorithms in terms of end-to-end delay and packet delivery ratio.

Leonov [15], they performed analysis for proving that bio-inspired algorithms can be effectively applied to solve the routing problem in FANET networks using BeeAdHoc and AntHocNet protocols. The results indicate that the performance of AntHocNet and BeeAdHoc is comparable to that of the protocols DSR, DSRV, and AODV. BeeAdHoc is as good as DSR and DSDV in throughput and routing overheads. AntHocNet possesses fine scalability.

Darwish [16], analyzed Bio-inspired computing—BIOA i.e. observation of behavior and reaction of animals, setting assumptions, develop pseudo code, test algorithm. The proposed analysis of these algorithms leads to some key observations such as—one of the mathematical methods that has been applied to improve the search process is chaos theory. The bio-inspired optimization algorithm could hybridize together with the methods like chaotic theory in future to enhance the performance of algorithm.

Khan et al. [17], proposed a bio-inspired clustering scheme for FANETs (BICSF), which uses the hybrid mechanism of glowworm swarm optimization (GSO) and krill herd (KH). The proposed scheme uses energy aware cluster formation and cluster head election on the basis of the GSO algorithm. The results show BICSF have better results as compared with the other bio-inspired clustering algorithms in terms of cluster building time, energy consumption, cluster lifetime, and probability of delivery success.

Tropea et al. [18], proposed protocol based on a bio-inspired approach, showing the performance of the proposed mechanism varying different algorithm parameters. This approach has good performance in terms of killed parasites and to be scalable in terms of bytes sent on the network.

Zhao and Ding [19], introduced the variable of link expiration time predicted by the extended Gaussian Markov mobility model into the fitness function of the bee colony algorithm, and then uses bee colony algorithm for the route discovery of FANETs in the three-dimensional environment. Through experimental simulation, the method improves the packet delivery rate and reduces the end-to-end delay.

Firefly algorithm has also been used for routing in wireless sensor networks. It is compared to various conventional routing algorithms like EADC (Energy Aware Distributed Clustering) etc. and showed better results in energy efficiency.

3   Nature Inspired Algorithms for FANET

Nature-inspired optimization algorithms represent a promising solution for solving complex optimization problems. The capability of NIA to work in real-time systems has made it to gain a wide popularity. NIA is basically divided into two parts: Genetic algorithms and swarm intelligence.

Genetic algorithms (GA) are a part of evolutionary computing [20]. "Genetic Algorithm is based on the Darwin's Theory about evolution" [21]. It is an efficient search method which can be used to select paths in a network. A GA is made with a set of solutions, which represents the chromosomes. This set is known as population. Following steps show the process of GA:

Initialization in which population is generated. Selection, where the population is calculated with the help of fitness function. Mutation in which the population from the previous step is added to the mating pool and termination, where once the stopping condition is reached, the algorithm is made to exit.

The other type of algorithm is swarm intelligence (SI). Swarm intelligence approach also indicates the results of better performance than using other routing protocols. "Swarm Intelligence is a subfield of Artificial Intelligence which is concerned with the intelligent behavior of swarms by the interaction of individuals in such environments to solve real world problems by simulating biological behaviors" [22]. It can also be defined as designing of algorithms which are intelligent in nature and are simulated by the behavior of various animal communities. There are certain algorithms which are upgraded by its improved version and performed better. The Improved ABC (Ant Bee Colony) algorithm is one such example which works better in real time [23]. Algorithms inspired by swarm intelligence are "flock-based congestion control, inspired on the behavior of bird flocks" [24], chicken swarm algorithm etc.

4   Discussion

According to previous research, it has already been proved that the performance of nature-inspired algorithms is comparable to that of protocols used in FANET. BeeSensor algorithm, which is swarm intelligence algorithm, delivered approximately 85% packets as compared to 60% of ad hoc on-demand distance vector (AODV) [25]. BeeAdhoc swarm intelligence algorithm is as good as dynamic source routing (DSR) and destination-sequenced distance vector (DSDV) in throughput and routing overheads. Swarm intelligence algorithms are used for developing secure routes in p2p networks with dynamic self-organization permits increasing the level of safety and dynamic decision making in routes. **Therefore, the solution based on nature-inspired algorithms can be considered to be the most promising solution and an advanced approach to solve the major routing issue in FANETs.**

4.1   Applications of NIA

Swarm intelligence algorithms are used for developing secure routes in p2p networks with dynamic self-organization permits increasing the level of safety and dynamic decision making in routes [26]. NIA will help us to learn that how intelligent routing techniques can help in the communication process between Flying Ad hoc Networks. Many domains may be benefitted by the usage of multiple UAVs like search and rescue operations, accidents tracking, etc. Intelligent routing techniques are analyzed and the optimal and effective solution of the routing problem is discovered.

5 Conclusion

This paper shows how conventional routing algorithms can be replaced by nature-inspired algorithms in FANETs. It represents the importance of nature-inspired algorithms for FANETs in current times. With the rising demand of artificial intelligence, there is a need to improve the routing technologies in FANETs by incorporating nature-inspired algorithms. Routing being a major issue in FANETs can be therefore resolved by using NIAs. These algorithms can be taken into consideration because of its various characteristics like easily adaptable, recognizing complex behaviors, survivability and moreover their real-time ability to solve complex problems. These algorithms have also showed better results when tested in various scenarios for wireless communication in ad hoc networks. Hence, NIAs can be served as a new and promising solution for routing in FANETs.

References

1. Avellar G, Pereira G, Pimenta L, Iscold P (2015) Multi-UAV routing for area coverage and remote sensing with minimum time. Sensors 15:27783–27803. https://doi.org/10.3390/s15112 7783
2. Umre SA, Mehta K, Malik L (2014) Performance improvement of communication in zone-based routing that uses cluster formation and bio-inspired computing in VANET. In: 2014 IEEE international conference on vehicular electronics and safety, Hyderabad, pp 147–151. https://doi.org/10.1109/icves.2014.7063739
3. Sharma V, Kumar R, Kumar N (2018) DPTR: distributed priority tree-based routing protocol for FANETs. Comput Commun 122. https://doi.org/10.1016/j.comcom.2018.03.002
4. Bani Yassein M, Damer NA (2016) Flying ad-hoc networks: routing protocols, mobility models, issues. Int J Adv Comput Sci Appl 162–168. https://doi.org/10.14569/ijacsa.2016.070621
5. Khan MA, Safi A, Qureshi IM, Khan IU (2017) Flying ad-hoc networks (FANETs): a review of communication architectures, and routing protocols. In: 2017 first international conference on latest trends in electrical engineering and computing technologies (INTELLECT), Karachi, pp 1–9. https://doi.org/10.1109/intellect.2017.8277614
6. Bujari A, Palazzi CE, Ronzani D (2018) A comparison of stateless position-based packet routing algorithms for FANETs. IEEE Trans Mob Comput 17(11):2468–2482. https://doi.org/10.1109/tmc.2018.2811490
7. Palan K, Sharma P (2015) FANET communication protocols: a survey. Int J Comput Sci Commun 7(1) 219–223. https://doi.org/10.090592/ijcsc.2016.034
8. Qingwen W, Gang L, Zhi L, Qian Qi (2015) An adaptive forwarding protocol for three dimensional flying ad hoc networks. 142–145. https://doi.org/10.1109/iceiec.2015.7284506

9. Li X, Huang J (2017) ABPP: an adaptive beacon scheme for geographic routing in FANET. In: 2017 18th international conference on parallel and distributed computing, applications and technologies (PDCAT), Taipei, pp 293–299
10. Ghazzai H, Feidi A, Menouar H, Ammari ML (2017) An exploratory search strategy for data routing in flying ad hoc networks. 1–7. https://doi.org/10.1109/pimrc.2017.8292474
11. Rovira-Sugranes A, Razi A (2017) Predictive routing for dynamic UAV networks. In: 2017 IEEE international conference on wireless for space and extreme environments (WiSEE), Montreal, QC, pp 43–47. https://doi.org/10.1109/wisee.2017.8124890
12. Saleem M, Khayam SA, Farooq M (2008) Formal modeling of BeeAdHoc: a bio-inspired mobile ad hoc network routing protocol. In: ANTS conference, pp 315–322
13. Maistrenko VA, Alexey LV, Danil VA (2016) Experimental estimate of using the ant colony optimization algorithm to solve the routing problem in FANET. In: 2016 international siberian conference on control and communications (SIBCON), pp 1–10
14. Fiala J (2016) A survey of bio-inspired wireless communication. https://www.cse.wustl.edu/~jain/cse574-16/ftp/biocomm/index.html
15. Leonov A (2017) Applying bio-inspired algorithms to routing problem solution in FANET. Bull South Ural State Univ Ser Comput Technol Automatic Control Radio Electron 17:5–23. https://doi.org/10.14529/ctcr170201
16. Darwish A (2018) Bio-inspired computing: algorithms review, deep analysis, and the scope of applications. Future Comput Inform J. https://doi.org/10.1016/j.fcij.2018.06.001
17. Khan AN, Aftab F, Zhang Z (2019) BICSF: bio-inspired clustering scheme for FANETs. IEEE Access 7:31446–31456
18. Tropea M, Santamaria AF, Potrino G, Rango FD (2019) Bio-inspired recruiting protocol for FANET in precision agriculture domains: pheromone parameters tuning. In: 2019 wireless days (WD), pp 1–6
19. Zhao B, Ding Q (2019) Route discovery in flying ad-hoc network based on bee colony algorithm. In: 2019 IEEE international conference on artificial intelligence and computer applications (ICAICA), pp 364–368
20. Lin Q, Song H, Gui X, Wang X, Su S (2017) A shortest path routing algorithm for unmanned aerial systems based on grid position. J Netw Comput Appl 103. https://doi.org/10.1016/j.jnca.2017.08.008
21. Seetaram J, Kumar PS (2016) An energy aware Genetic algorithm multipath distance vector protocol for efficient routing. In: International conference on wireless communications, signal processing and networking (WiSPNET), Chennai, pp 1975–1980. https://doi.org/10.1109/wispnet.2016.7566488
22. Sahingoz O (2014) Networking models in flying ad-hoc networks (FANETs): concepts and challenges. J Intell Rob Syst 74. https://doi.org/10.1007/s10846-013-9959-7
23. Tian G, Zhang L, Bai X, Wang B (2018) Real-time dynamic track planning of Multi-UAV formation based on improved artificial bee colony algorithm. 10055–10060. https://doi.org/10.23919/chicc.2018.8482622
24. Escalante LDS (2013) Swarm intelligence-based energy saving greedy routing algorithm for wireless sensor networks. In: CONIELECOMP 2013, 23rd International Conference on Electronics, Communications and Computing, Cholula, pp 36–39. https://doi.org/10.1109/conielecomp.2013.6525754
25. Saleem M, Farooq M (2007) BeeSensor: a bee-inspired power aware routing protocol for wireless sensor networks. In: Giacobini M (ed) Applications of evolutionary computing. EvoWorkshops 2007. Lecture notes in computer science, vol 4448. Springer, Berlin, Heidelberg
26. Krundyshev V, Kalinin M, Zegzhda P (2018) Artificial swarm algorithm for VANET protection against routing attacks. In: 2018 IEEE industrial cyber-physical systems (ICPS), St. Petersburg, pp 795–800. https://doi.org/10.1109/icphys.2018.8390808

Chapter 16
Performance Analysis of Different Machine Learning Classifiers in Detection of Parkinson's Disease from Hand-Drawn Images Using Histogram of Oriented Gradients

Akalpita Das, Himanish Shekhar Das, Anupal Neog, B. Bharat Reddy, and Mrinoy Swargiary

1 Introduction

Parkinson's disease (PD) is a neurodegenerative disorder that generally visible in old-age people. It affects their nervous system which leads to shaking, stiffness, difficulty in walking, balancing, as well as coordinating moments [1]. It also affects the voice and other cognitive difficulties. Motor as well as non-motor symptoms are the characteristic symptoms of Parkinson's disease. Motor symptoms include tremor, slowness of movement due to rigidity in muscle, gait disturbances and speech difficulties. On other hand, non-motor symptoms comprise of disturbances in sleep, mood as well as cognitive disturbances which includes loss of memory, sleep problems,

A. Das (✉)
Department of Computer Science and Engineering, GIMT Guwahati, Guwahati 781017, Assam, India
e-mail: dasakalpita@gmail.com

H. S. Das
Department of Computer Science and Information Technology, Cotton University, Guwahati 781001, Assam, India
e-mail: hsdas0815044@gmail.com

A. Neog · B. Bharat Reddy · M. Swargiary
Department of Computer Science and Engineering, Jorhat Engineering College, Jorhat 785007, Assam, India
e-mail: neog.anupal101@gmail.com

B. Bharat Reddy
e-mail: reddy.bharat.11@gmail.com

M. Swargiary
e-mail: mrinoy02@gmail.com

© The Author(s), under exclusive license to Springer Nature Singapore Pte Ltd. 2021   205
X.-Z. Gao et al. (eds.), *Applications of Artificial Intelligence in Engineering*, Algorithms for Intelligent Systems, https://doi.org/10.1007/978-981-33-4604-8_16

abstract thinking, problem solving capability, language and visual-emotional capacities. The disease occurs due to disintegration of a region called "substantia nigra" in the thalamic region of human brain. Dopamine hormone which acts as the neurotransmitter to transmit the neural signals is responsible for making the coordination between the brain and the body. In Parkinson's disease (PD)-affected patients, the generation of dopamine hormone gets reduced and it affects the coordination between the brain and body of the affected person.

For past few years, many researchers are carrying out their research in this field to find out an effective way for early detection of PD as there is no cure for PD until and unless it gets detected in an early stage. M. Hariharan et al. used the PD dataset from University of California-Irvine (UCI) machine learning database of voice signals to build a hybrid system to detect PD with the help of three supervised classifiers such as least-squares support vector machines (LS-SVM), probabilistic neural network (PNN) and general regression neural network (GRNN) [2]. The combination of feature pre-processing like Gaussian mixture modeling (GMM) and efficient feature reduction/selection methods like principal component analysis, linear discriminant analysis resulted to a maximum classification accuracy of 100% for the Parkinson's dataset. To differentiate the PD group from healthy group based on the voice data, Satyabrata et al. [3] used principal component analysis (PCA)-based feature sets along with nonlinear-based classification approach and obtained an accuracy of 97.57%. Musa Peker et al. [4] obtained an accuracy of 98.25% by designing a hybrid model where minimum redundancy maximum relevance (mRMR) is used for feature selection, and those features are fed to a complex-valued artificial neural network (CVANN).

Along with all other symptoms of PD, hand tremor is the mostly used symptom to detect PD from the hand-drawn images/sketches/handwriting by the PD patients. In past few years, lot of research works have been performed to detect PD using the handwriting and hand-drawn images. Drotár et al. [5] have used PaHaW Parkinson's disease handwriting database, consisting of 37 PD patients and 38 healthy persons performing eight different handwriting tasks include drawing an Archimedean spiral, repetitively writing orthographically simple syllables, words and sentence. In this study, three classifiers such as k-nearest neighbor (KNN), support vector machine (SVM) and AdaBoost were used to predict PD based on the conventional kinematic and handwriting pressure features, and they obtained an accuracy of 81.3%. C. Loconsole et al. [6] investigated a different method of using electromyography (EMG) signals and computer vision techniques such as morphology operators, image segmentation process, etc. In this study, ANN with two different cases (dataset 1 with 2 dynamic and 2 static features, dataset 2 with only 2 dynamic features) were carried out and obtained an accuracy of 95.81% and 95.52%, respectively. Pereira et al. [7] used convolution neural network (CNN) for the discrimination of PD group and healthy patients based on the data acquired by a pen with several sensors attached with it. The data were comprised of spiral and meanders drawn by both the PD and healthy individuals. Folador et al. [8] worked on hand-written drawings to classify the PD and healthy individual by using histogram of oriented gradients (HOG) descriptor and random forest classifier and reached a good accuracy. In this paper,

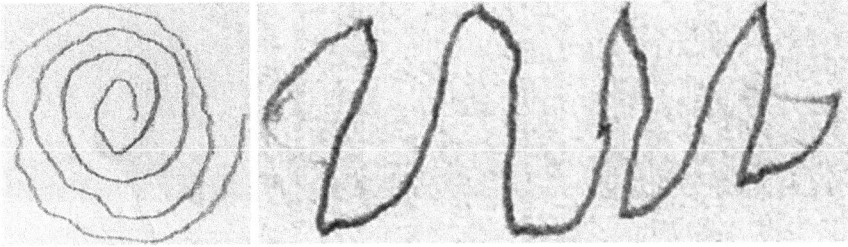

Fig. 1 Spiral and wave images of PD patients from dataset 1

we propose the method to discriminate the PD and healthy individual from the hand-drawn images by using the HOG descriptor and different machine learning classifiers. This work also includes the performance analysis of different classifiers based on various performance metrics like accuracy, specificity and sensitivity.

The organization of the paper is as follows: Sect. 2 presents the dataset description. Section 3 elaborates feature extraction methods and different classifiers being used. Experimental results and discussions are presented in Sects. 4 and 5 which conclude the paper.

2   Database Description

In this paper, for experimental work two different datasets have been used.

Dataset 1: It consists of a total of 204 images out of which 102 images are spiral in shape and the remaining 102 are wave images [9]. Among 102 spiral images, 51 spiral images were drawn by PD patients and remaining 51 were drawn by healthy person. The wave images were also taken in the same way. Sample images from Dataset 1 are shown in Fig. 1.

Dataset 2: It consists of a total of 332 images including cube, spiral and triangle shape. Among 332 images, a total of 166 images are drawn by PD patients and remaining 166 images are drawn by healthy person. Dataset 2 comprises of 54 cube, 54 spiral and 58 triangle images [10]. It consists of both dilated and skeletal images. Sample images from Dataset 2 are shown in Fig. 2.

3   Proposed Methodology

In this work, histogram of oriented gradients features has been extracted from the different images such as spiral, wave, cube and triangle from both the datasets. To extract HOG features, the configurations of image size, image orientations, cell size, pixels per cell and block normalization are taken as $200 \times 200, 9, 2 \times 2, 10 \times 10$ and

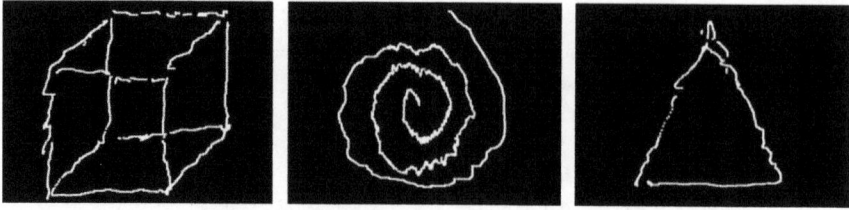

Fig. 2 Cube, spiral and triangle images of PD patients from dataset 2

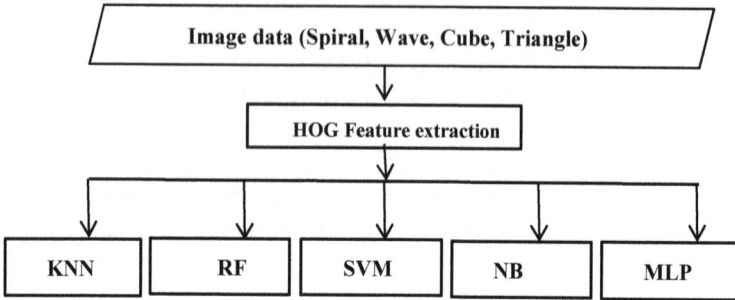

Fig. 3 Flowchart of the proposed methodology

Manhattan distance or "L1 Norm," respectively. Using these configurations, 12996-dimensional feature vector has been extracted from the images. These extracted features are then feed into different classification algorithms such as KNN, random forest (RF), SVM, Naïve Bayes (NB) and multi-layer perceptron (MLP), respectively. For each of the datasets, at first, the similar kind of images are trained and tested by the proposed model. In later phase, the model is trained and tested with the combination of images from the respective datasets to improve the influence of the proposed model. Figure 3 represents the proposed model.

3.1 Histogram of Oriented Gradients Features

The Histogram of oriented gradients is widely used as image feature descriptor for object detection. This feature descriptor counts the occurrences of gradient orientation in localized portions of an image. For this, the specific input image is decomposed into small squared regions, and gradient orientation of each of the regions is computed. The results, i.e., the computed gradients, are normalized using a block-wise pattern which returns a descriptor for each of the cells. In this paper, the HOG features of the images from both the datasets are computed, and these computed features are feed into the training model. HOG descriptor of the sample images from Dataset 1 is shown in Fig. 4.

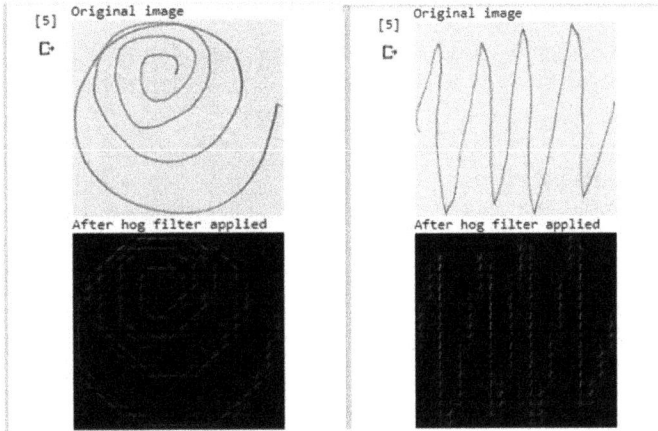

Fig. 4 HOG descriptor of spiral and wave image

3.2   Various Machine Learning Classifiers

After histogram of oriented gradients descriptor estimation from various images present in both the datasets, the data are ready to classify using some machine learning classifiers. In this paper, we have used five different types of classifiers, and their performance metrics are analyzed. A brief description of the various machine learning classifiers and their respective parameter values used in this work is as follows:

K-Nearest Neighbor
K-nearest neighbor classifier is efficiently used in different kinds of classification problems because of its simplicity and high convergence speed [11]. The main disadvantage with this algorithm is the selection of k value. In this paper, value of k has been taken as 2 and Minkowski distance has been chosen as the distance metric.

Random Forest
Breiman [12] introduced RF classifier, which generates large collection of binary decision trees on randomly chosen datasets. In order to build the RF, a random feature is generally applied in each node of the tree. In this work, for discrimination of PD and healthy person, the number of trees in the forest has been taken as 100.

Support Vector Machine
It is one of the most popular and efficient classification algorithms in solving various pattern recognitions and nonlinear function estimations. The main motivation to introduce SVM is to map training data nonlinearly into high dimensional feature space as well as to generate a hyperplane having maximum separation among classes. Different types of kernel function can be used to compute the separating hyperplane. In this paper, regularization parameter has been taken as 0.025, and linear kernel function is used.

Naïve Bayes
Naïve Bayes classifier is based on Naïve Bayes algorithm [13]. It is one of the classification methods which does not depend on any parameter and is based only on the statistics.

Multi-Layer Perceptron
MLP is one of the highly used techniques in artificial intelligence which imitates the concept of brain functions. It is characterized by its architecture, learning algorithm and various activation functions. It comprises of input layer, hidden layer and output layer [14]. In this work, we have considered the learning rate as 0.001, and maximum number of iteration has been set as 400.

4 Experimental Work and Result Analysis

To assess the performance of the proposed model, two different approaches have been applied on each of the datasets. In the first approach, for each dataset, uniform images are used to train and test the model. On the other hand, in the second approach, a mixture of different shapes of images of respective dataset is taken into consideration for training and then the trained model is tested with a random image. In this paper, the second approach has been opted to improve the accuracy despite of the less availability of image data.

For performance analysis of the proposed model, different performance metrics used in this paper are accuracy, sensitivity and specificity which are defined in Eqs. 1, 2 and 3.

$$\text{Accuracy} = \frac{TP + TN}{TP + FN + FP + TN} \tag{1}$$

$$\text{Sensitivity} = \frac{TP}{TP + FN} \tag{2}$$

$$\text{Specificity} = \frac{TN}{FP + TN} \tag{3}$$

The values of true positive (TP), false positive (FP), true negative (TN) and false negative (FN) values were calculated from respective confusion matrix.

4.1 Performance Analysis Using Dataset 1

Tables 1, 2 and 3 contain the result set for spiral, wave and mixture of spiral–wave images. Confusion matrices are shown in Fig. 5. Among spiral and wave images,

Table 1 Result set for spiral images (in %)

Parameters	KNN	SVM	RF	NB	MLP
Accuracy	70.0	79.7	84.0	53.3	76.7
Sensitivity	53.3	80.0	76.0	93.3	80.0
Specificity	86.7	73.3	92.0	13.3	73.3

Table 2 Result set for wave images (in %)

Parameters	KNN	SVM	RF	NB	MLP
Accuracy	73.3	73.3	74.7	53.3	73.3
Sensitivity	60.0	66.7	74.7	53.3	66.7
Specificity	86.7	80.0	74.7	53.3	80.0

Table 3 Result set for the mixture of spiral–wave images (in %)

Parameters	KNN	SVM	RF	NB	MLP
Accuracy	68.3	74.7	74.5	70.0	73.0
Sensitivity	73.3	66.7	71.3	73.3	72.7
Specificity	63.3	76.7	78.0	74.7	73.3

spiral image showed maximum accuracy rate of 84% and SVM classifier gives the most accurate result.

4.2 Performance Analysis Using Dataset 2

Tables 4, 5, 6 and 7 contain the result set for cube, spiral, triangle and mixture of cube–spiral–triangle images, respectively. Confusion matrices are shown in Fig. 6. From the following tables, it can be concluded that out triangle and spiral images give maximum accuracy, whereas overall SVM provide convincing accuracy rate.

The state-of-the-art classifiers are initially trained with all the training data available and later tested with that particular pattern of images. Tables 1, 2, 3, 4, 5, 6 and 7 reflect various approaches applied to find the suitable pattern, each pattern of images are trained and tested separately and in another approach, to verify the efficiency of the models, all patterns of images are trained together and tested. Since the database is well classified between two classes, i.e., Parkinson and healthy, for both training and testing, the models can easily differentiate the two classes after training with those particular patterns. From Tables 1, 2, 3, 4, 5, 6 and 7, it can be concluded that overall SVM performs better as compared to other models for both dataset 1 and dataset 2. The result set also signifies that SVM is better for two-class classification problem when number of features is less.

(a) KNN

(b) SVM

(c) RF

(d) NB

(e) MLP

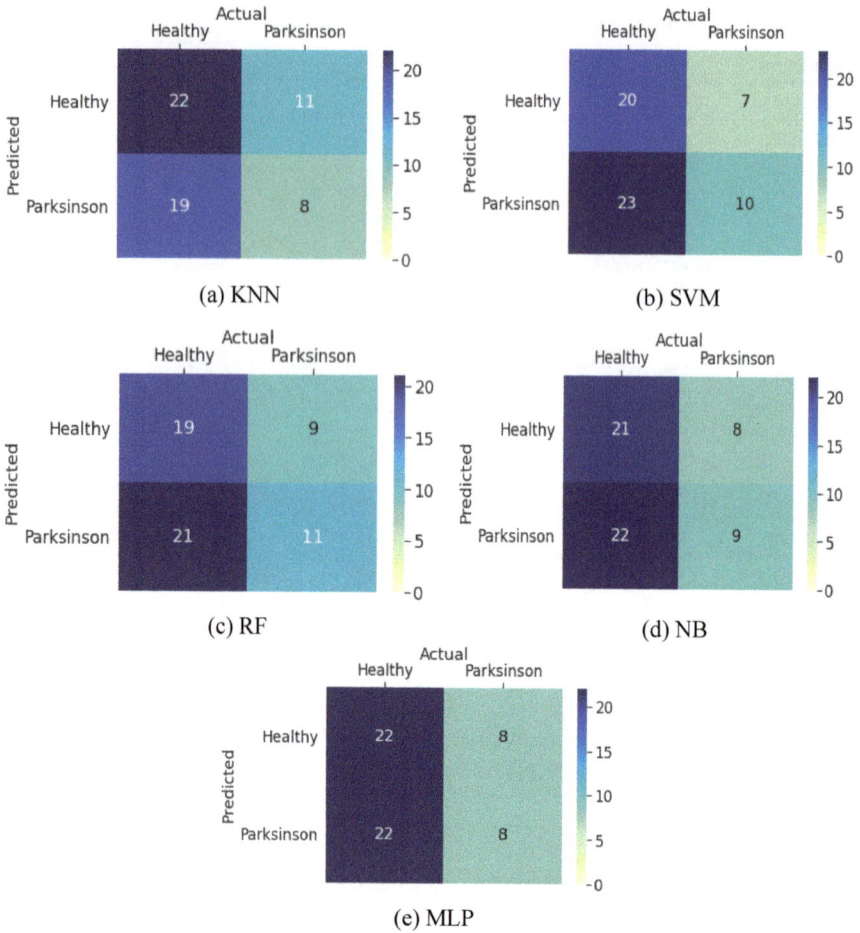

Fig. 5 Confusion matrices for mixture of spiral–wave images from dataset 1

Table 4 Result set for cube images (in %)

Parameters	KNN	SVM	RF	NB	MLP
Accuracy	81.2	87.5	85.4	82.8	86.9
Sensitivity	61.5	53.3	80.0	76.0	66.7
Specificity	75.0	50.0	75.0	33.3	60.0

Table 5 Result set for spiral images (in %)

Parameters	KNN	SVM	RF	NB	MLP
Accuracy	90.9	95.5	87.3	85.6	95.5
Sensitivity	76.0	83.3	71.9	67.0	53.8
Specificity	50.0	75.0	30.0	75.0	60.0

Table 6 Result set for triangle images (in %)

Parameters	KNN	SVM	RF	NB	MLP
Accuracy	95.5	95.5	93.6	90.9	95.5
Sensitivity	56.0	62.3	77.1	66.1	79.0
Specificity	75.0	75.0	65.0	50.0	75.0

Table 7 Result set for the mixture of cube–spiral–triangle images (in %)

Parameters	KNN	SVM	RF	NB	MLP
Accuracy	87.3	96.8	87.9	92.4	96.4
Sensitivity	100	100	96.8	100	100
Specificity	33.3	83.3	50.0	60.0	91.7

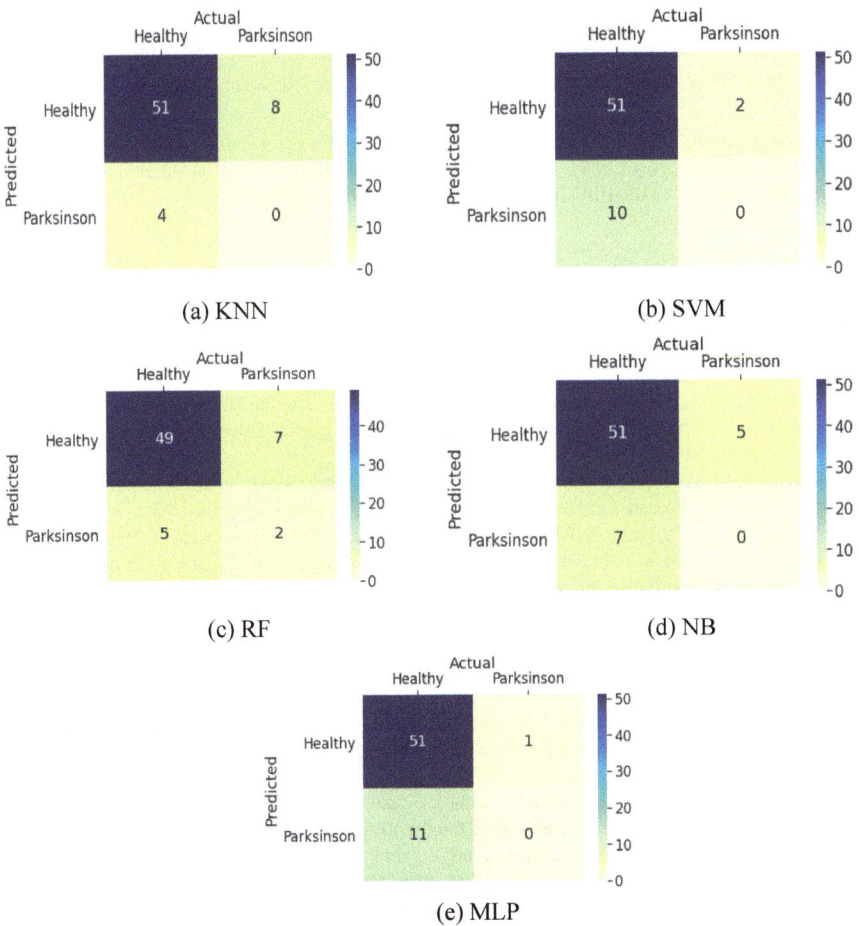

(a) KNN

(b) SVM

(c) RF

(d) NB

(e) MLP

Fig. 6 Confusion matrices for cube–spiral–triangle images from dataset 2

5 Conclusion

This paper proposes a model for early detection of Parkinson's disease from the images drawn by Parkinson's disease-affected people and healthy person. In this paper, different types of image shapes such as spiral, cube, wave and triangle from two different datasets have been used. Features extracted from HOG are feed into different classification algorithms such as KNN, SVM, RF, NB and MLP, respectively. The performance in terms of accuracy, sensitivity and specificity for each classifier is analyzed. It is found that for Dataset 1 and Dataset 2, for the proposed approach out of all the classifiers, SVM achieved an accuracy of 74.7% and 96.8%, respectively, for the overall mixture image set as well as for individual sets. This is the first study in which HOG descriptor of the input images is used along with five different types of machine learning classifiers for discrimination of PD patient and healthy person from hand-drawn images. In future, more hand-drawn images of few more variety of shapes can be collected to increase the dataset and along with that, the performance analysis can be done after fine-tuning of the parameters of the classification algorithms.

Acknowledgements This work is a part of the project that has been sponsored by Assam Science and Technology University (ASTU), Guwahati, under Collaborative Research Scheme of TEQIP-III via grant no. ASTU/TEQIP-III/Collaborative Research/2019/2479.

References

1. Mohamed GS (2016) Parkinson's disease diagnosis: detecting the effect of attributes selection and discretization of Parkinson's disease dataset on the performance of classifier algorithms. Open Access Libr J 3(11):1–11
2. Hariharan M, Polat K, Sindhu R (2014) A new hybrid intelligent system for accurate detection of Parkinson's disease. Comput Methods Programs Biomed 113(3):904–913
3. Aich S, Younga K, Hui KL, Al-Absi AA, Sain M (2018) A nonlinear decision tree based classification approach to predict the Parkinson's disease using different feature sets of voice data. In: 20th international conference on advanced communication technology (ICACT). IEEE, pp 638–642
4. Peker M, Sen B, Delen D (2015) Computer-aided diagnosis of Parkinson's disease using complex-valued neural networks and mRMR feature selection algorithm. J Healthcare Eng 6(3):281–302
5. Drotár P, Mekyska J, Rektorová I, Masarová L, Smékal Z, Faundez-Zanuy M (2016) Evaluation of handwriting kinematics and pressure for differential diagnosis of Parkinson's disease. Artif Intell Med 67:39–46
6. Loconsole C, Trotta GF, Brunetti A, Trotta J, Schiavone A, Tatò SI, Losavio G, Bevilacqua V (2017) Computer vision and EMG-based handwriting analysis for classification in Parkinson's disease. In: International conference on intelligent computing. Springer, pp 493–503
7. Pereira CR, Weber SA, Hook C, Rosa GH, Papa JP (2017) Deep learning-aided Parkinson's disease diagnosis from handwritten dynamics. In: 29th SIBGRAPI conference on graphics, patterns and images (SIBGRAPI). IEEE, pp 340–346

8. Folador JP, Rosebrock A, Pereira AA, Vieira MF, de Oliveira Andrade A (2019) Classification of handwritten drawings of people with Parkinson's disease by using histograms of oriented gradients and the random forest classifier. In: Latin American conference on biomedical engineering. Springer, Cham, pp 334–343
9. Zham P, Kumar DK, Dabnichki P, Poosapadi Arjunan S, Raghav S (2017) Distinguishing different stages of Parkinson's disease using composite index of speed and pen-pressure of sketching a spiral. Frontiers Neurol 8:435
10. Bernardo LS, Quezada A, Munoz R, Maia FM, Pereira CR, Wu W, de Albuquerque VHC (2019) Handwritten pattern recognition for early Parkinson's disease diagnosis. Pattern Recogn Lett 125:78–84
11. Athitsos V, Sclaroff S (2005) Boosting nearest neighbor classifiers for multiclass recognition. In: IEEE computer society conference on computer vision and pattern recognition (CVPR'05)-Workshops. IEEE, pp 45–45
12. Breiman L (2001) Random forests. Mach Learn 45(1):5–32
13. Bustomi MA, Faricha A, Ramdhan A, Faridawati (2018) Integrated image processing analysis and naive Bayes classifier method for lungs X-ray image classification. ARPN J Eng Appl Sci 13(2):718–724
14. Kanafiah SNAM, Ali H, Firdaus AA, Azalan MZ, Jusman Y, Khairi AA, Ahmad MR, Sara T, Amran T, Mansor I, Shukor SAA (2019) Metal shape classification of buried object using multilayer perceptron neural network in GPR data. IOP Conf Ser Mater Sci Eng 705(1):012028. IOP Publishing (2019)

Chapter 17
Coal Mine Monitoring Smart Rover Using LabVIEW and Wireless Sensor Network

Das Udeshna and Das Nipan Kumar

1 Introduction

Coal is one of the major sources of energy as well as a large industry worldwide. In coal mining, coal is extracted from the ground. The underground coal mine environment is very complex and because of the factors like high temperature, toxic gases, chances of fire, suffocation, it can create major risk to the safety and health of the workers. During mine operation methane is released which in presence of high temperature and certain range of oxygen, can create violent explosion which lead to big loss of property and human life. Therefore, it is very necessary to continuously monitor the mine working environment so that the workers can work under safe condition. For this, an accurate and effective sensing system is required. In recent decades, many techniques are adopted; among them mounting of semi conductor type gas sensors in coal mine is effective [1] but, sometimes, accidental damage of sensors can create problems in coal mines. Use of robot is another technique. A mine security detecting crawling robot is developed which based on wireless communication with multi-sensor in article [2]. In [3], a robot vehicle with a MQ4 sensor and a wireless camera is developed. The MQ4 gas sensor detects the level of methane and wireless camera monitors the operation in coal mines. The movement of robot is controlled by a remote through Bluetooth sending messages to the safety division. These robots are also quite effective, but, sometimes, their complexity, use in productive domains and expensive components, makes them a little unreliable [4].

Another effective technique for coal mine monitoring is smart safety helmet having an array of sensors which generates an alarm when the gas concentration exceeds critical level [5]. But, in this technique, sometimes, the worker wearing it may become the victim of the disaster when critical atmospheric condition occurs. Technologies that use microcontroller and software like LabVIEW are also operative

D. Udeshna (✉) · D. Nipan Kumar
Department of Electrical Engineering, Jorhat Engineering College, Jorhat, India
e-mail: udeshna80@gmail.com

© The Author(s), under exclusive license to Springer Nature Singapore Pte Ltd. 2021   217
X.-Z. Gao et al. (eds.), *Applications of Artificial Intelligence in Engineering*, Algorithms for Intelligent Systems, https://doi.org/10.1007/978-981-33-4604-8_17

in coal mine monitoring. A LabVIEW and microcontroller-based coal mine monitoring system is developed in article [6] which uses wireless technology like Zigbee and GSM. LabVIEW is used to continuously compare the values given by the sensors with the predetermined critical safety values and give alert indication signal to the GSM system for calling and sending messages to the safety division.

In this presented paper, an unmanned vehicle (Rover) with a sensor of array is developed which accomplishes rover's movement control, temperature and humidity detection, gas detection, exchange of data and provides real-time monitoring information. The rover is controlled wirelessly with the help of LabVIEW and an android platform through Bluetooth technology. Here, a wireless sensor network (WSN) is established using client and server communication between the Rover and LabVIEW for monitoring the coal mine parameter. Rover using WSN network in this paper makes it very effective and advantageous over the traditional methods and shows significant prospect from environment to discovery of disaster.

2 System Description

The proposed system consists of sensor unit, microcontroller unit, motor driver unit, power supply unit, and a wireless communication medium (HC-05) bluetooth module (Fig. 1).

2.1 Rover Structure

The unmanned vehicle (Rover) has two wheels which are run by a pair of DC motors of 100 rpm. L298N motor driver is used to control the speed and direction of the two motors. The device has a very lightweight frame and it can move in four direction, namely forward, backward, left, and right.

2.2 Sensor Unit

The system contains three sensors. The two gas sensors MQ7 and MQ4 are used to detect the concentration of carbon monoxide (CO) and methane (CH_4) gas, respectively, in coal mines. These sensors operate at 5 V DC supply and they can detect CO and CH_4 gas concentration from 200 to 10,000 ppm.

For temperature and humidity measurement, digital temperature and humidity (DHT11) are used. This low cost sensor also operates at 5 V DC supply and can measure temperature ranging from 0 to 50 °C and relative humidity from 20 to 90%.

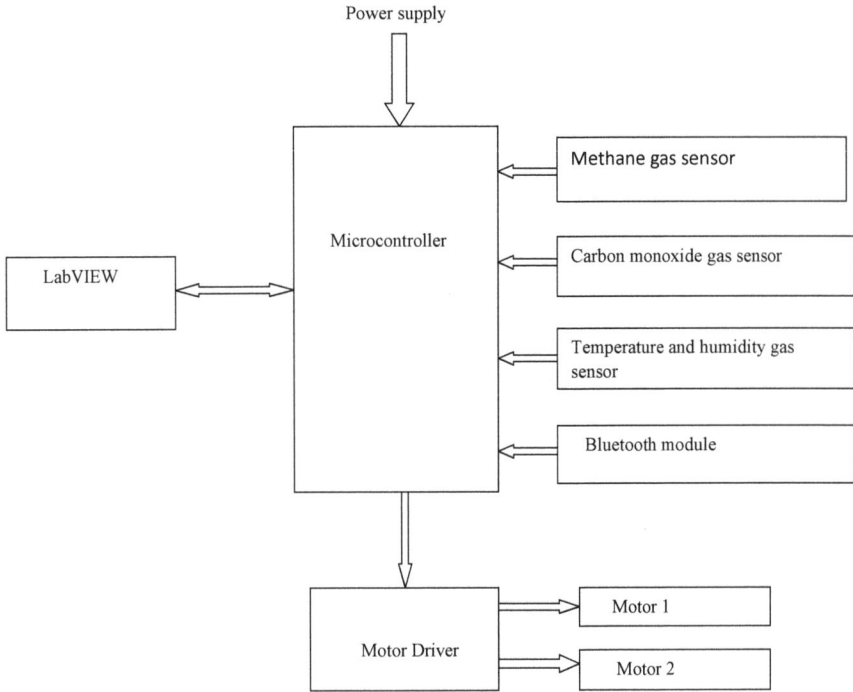

Fig. 1 Block diagram of coal mine monitoring rover

2.3   *Microcontroller Unit*

In this project, Arduino Pro Mini is used as microcontroller which is based on ATmega328. The board comes with 14 digital I/O pins out of which six pins are used to provide PWM outputs and it has six analog input pins. Arduino Integrated Development Environment (IDE) software is used to program the board and it contains serial communication interface that burns the program into Arduino Pro Mini. Operating voltage of the board is 3.3–5 V. Arduino Pro Mini is chosen for the prototype due to its smaller size, low cost, lightweight and also for its simplicity in interfacing with LabVIEW software in integration with Bluetooth module. The microcontroller basically collects the parameter values from the sensor unit and transfer the corresponding data to the monitoring station through the wireless communication module. It also received command from the monitoring station and transfer it to the rover for its movement.

2.4 Bluetooth Module

Here, HC-05 bluetooth SPP (Serial Port Protocol) module is used as the wireless communication medium. It transfers data from the microcontroller to PC or android platform and vise-versa. It is designed for transparent wireless serial communication set-up.

3 Working Principle

The aim of the proposed system is monitoring the variations in temperature, humidity, hazardous gas in coal mine environment. As shown in the circuit diagram in Fig. 2, the system consist of a 12 V DC power supply. The terminal of the power supply is connected to a H-bridge motor driver L298N. The driver has an inbuilt voltage regulator 7805 and it reduces the 12 V power supply to a 5 V dc output which drives the controller unit Arduino Pro Mini, Bluetooth module as well as all the sensors. The outputs of all the three sensors are connected to microcontroller. The HC-05 bluetooth module has two transmitter and receiver port. The transmitter port is connected to the receiver port of the arduino and vise-versa.

All the three sensors sense their respective parameter values and send them to microcontroller. The microcontroller basically converts the analog sensors value into digital ones through its inbuilt analog-to-digital converter (ADC). Thus, the signals obtained from the sensors are digitally converted sampled quantized value and are

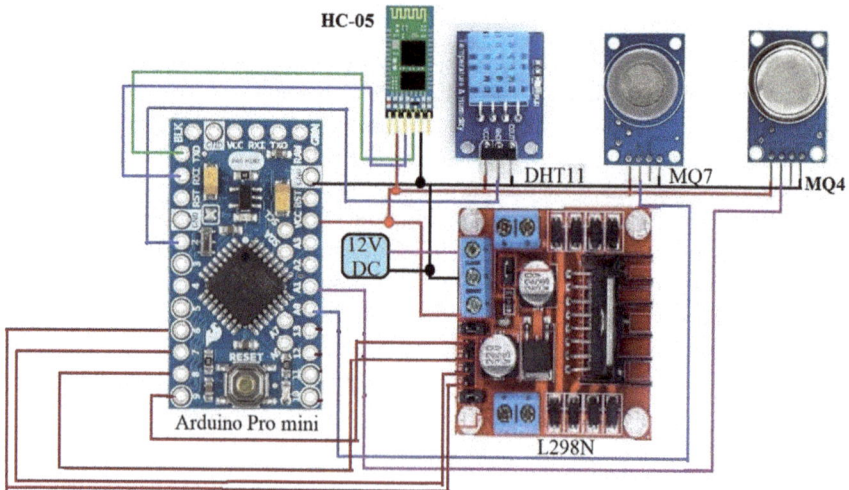

Fig. 2 Circuit diagram of the coal mine monitoring rover

transmitted over the wireless sensor network to a PC or android platform. Data can be continuously monitored by using LabVIEW software or by an android platform.

4 Experimental Analysis

4.1 Monitoring of Data Using LabVIEW

LabVIEW is a graphical programming software, a system design platform especially designed for Instrumentation engineer by NI (National Instruments). It can create human–machine interface (HMI) or graphical user interface (GUI). LabVIEW plays an important role in this project by monitoring and analyzing the coal mine parameters sensed by the sensors. It also control the movements of the Rover (Fig. 3).

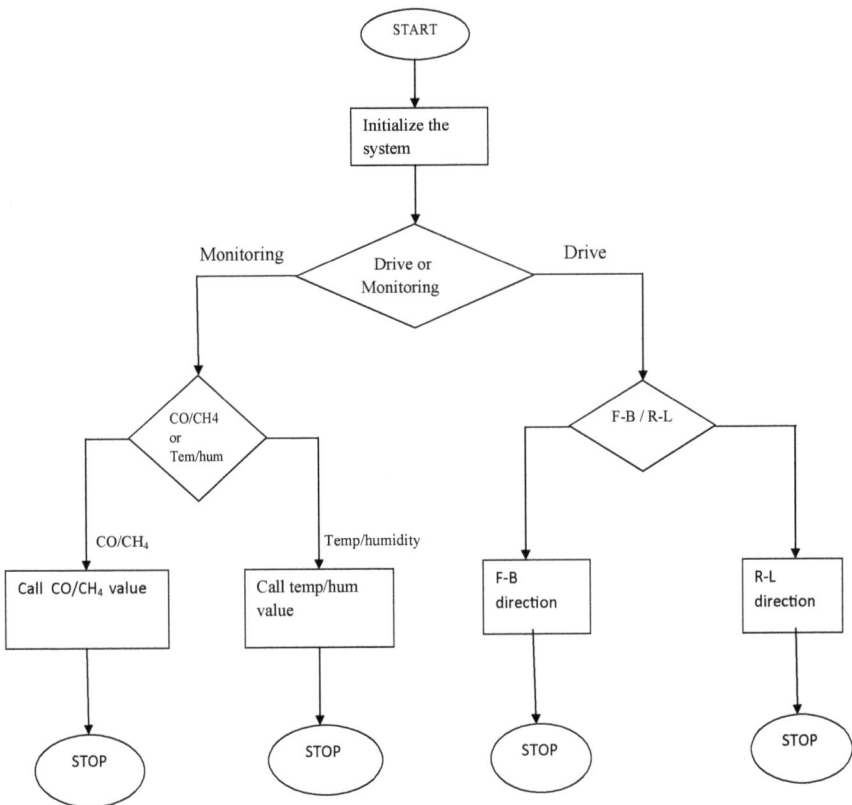

Fig. 3 Flowchart for monitoring and driving unit

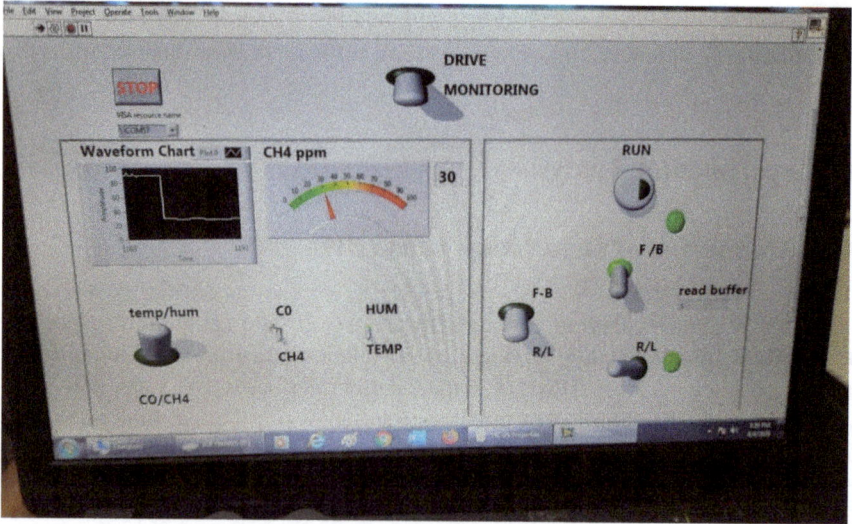

Fig. 4 Human–machine interface of the system using LabVIEW

At the monitoring station, the operator operates the system using LabVIEW. When it needs to drive the system, the switch is put on "Drive" mode and when it is monitoring, the switch is put on "Monitoring" mode. A single bit command is generated by the operator and it transmitted over the wireless sensor network to the rover. As soon as the rover receives the command, it immediately respond to the command and give an acknowledgement to the sender (server) and also value associated with the command. Thus, LabVIEW creates an human–machine interface (HMI) for the monitoring system which displays the different parameter value on the PC as well as controls the rover's movement (Fig. 4).

4.2 Android Platform

An android platform can also be used in the system to communicate directly by the operator using a HMI, in the form of tablet or android phones. The android platform displays values sensed by the sensors in the Rover and can also control the rover's movement by giving it command to move forward, backward, left, and right (Fig. 5).

5 Hardware Implementation

Figure 6 shows the hardware set-up of the proposed system implemented using this paper concept.

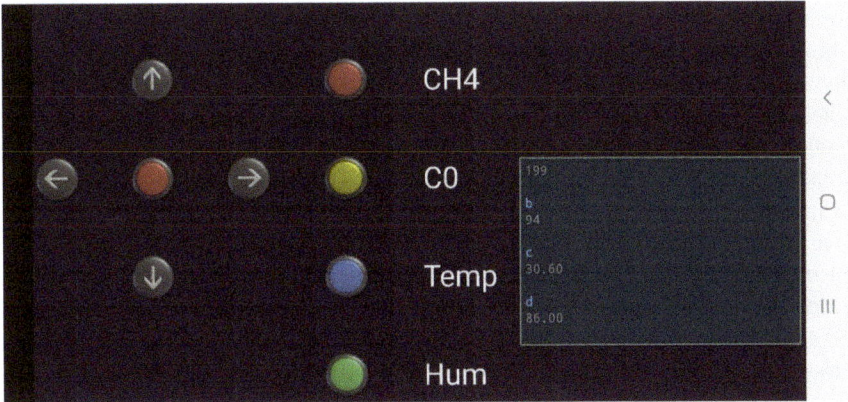

Fig. 5 Monitoring of data using android platform

Fig. 6 Coal mine monitoring rover prototype

6 Conclusion

The developed system prototype can detect concentration of humidity, temperature, various toxic gases concentration in underground coal mines. It can be sent to the hazardous environment of underground coal mine as a substitute of human to collect data of the environmental parameter where human arrival can be very dangerous. This can help the workers to get rid of any kind of disaster occurrence in coal mines. LabVIEW monitoring is more accurate but the software is little expensive. Using of wireless communication like Bluetooth improves the practical ability of the monitoring system. Also use of android platform makes it more user friendly and functional.

In the future work, the prototype system can be developed by using advanced smart sensors for monitoring the underground coal mine condition. It can be built as a robus t rover by implementing wide range of Bluetooth or new developing communication technologies for high speed data transmission incorporated with advanced smart sensors for sensing coal mine parameter. This will make it more proficient in coal mine safety and monitoring operation.

Acknowledgements The author gratefully acknowledge mentor and guide Mr. Nipan Kumar Das, Assistant professor, Department of Electrical engineering, Jorhat Engineering College, Jorhat, Assam for his precious guidance, encouragement and support for witting this paper and in completing the project.

References

1. Zhou G, Zhu Z, Chen G, Hu N (2009) Energy-efficient chain-type wireless sensor network for gas monitoring. In: Second international conference on information and computing science
2. He Z, Zhang J, Xu P, Qin J, Zhu Y (2013) Mine detecting robot based on wireless communication with multi-sensor. IEEE 978-1-4673-4933/13
3. Deepan A, Jayakrishna SV, Nagha Abirami S (2018) Detection of hazardous gas using Land rover in mines first. IRJET J 5(4):2735–2736
4. Vasanthi D, Logeshwari M, Priyadharshini S, Karthiga L (2019) Mine detecting robot using wireless technology and IOT. IOSR J Eng 22–26
5. Hazarika P (2016) Implementation of smart safety helmet for coal mine workers. In: IEEE international conference on power electronics, intelligent control and energy systems (ICPEICES), DTU, Delhi, 978-1-4673-8587-9/16
6. Putke AJ, Bhagat SN, Dr Nalbalwar SL (2017) LabVIEW based coal mine monitoring and alert system with data acquisition. IEEE 978-1-5386-2745-7/17

Chapter 18
A Survey on Spatiotemporal Co-occurrence Pattern Mining Techniques

S. Sharmiladevi, S. Siva Sathya, and Nangi Ramesh

1 Introduction

Space and time related data are recorded in various domains like earth science, public safety, epidemiology, neuroscience, social science, climate science, transportation, ecology, etc. The data obtained from these domains are more complicated compared to classical data which are discrete, because the spatiotemporal data are continuous and ST data instances are related to each other in context of space and time. Therefore, obtaining useful information from spatiotemporal dataset is difficult compared to obtaining knowledge from traditional transaction data.

In spatiotemporal data mining, the objects and features can be defined in different ways unlike the classical dataset with distinct data instance or objects. One way is to define spatial locations as objects and the measurement collected from them over times as feature. For example, grouping of forest areas based on the exploitation in flora and fauna measured over time [1]. Another way is to treat time points as objects and the measurements recorded in the spatial locations as features. For example, identifying time points in which human brain exhibit similar brain activity pattern from neuroimages [2]. In some cases, the events are treated as objects, and features are the space and time associated with that event. For example, discovering co-occurring social events and natural disaster events inorder to create a better traffic prediction system in smart cities [3]. Hence based on the kind of analysis to be performed on the ST data, the object and feature can be modelled accordingly. Therefore, spatiotemporal data mining presents a wide scope of opportunities and challenges across several domains.

Spatiotemporal knowledge discovery process can be categorized into

- Spatiotemporal outlier detection,
- Spatiotemporal associations,

S. Sharmiladevi (✉) · S. Siva Sathya · N. Ramesh
Pondicherry University, Puducherry, India
e-mail: sharmiladevi94@gmail.com

© The Author(s), under exclusive license to Springer Nature Singapore Pte Ltd. 2021
X.-Z. Gao et al. (eds.), *Applications of Artificial Intelligence in Engineering*, Algorithms for Intelligent Systems, https://doi.org/10.1007/978-981-33-4604-8_18

- Spatiotemporal prediction,
- Spatiotemporal clustering,
- Spatiotemporal hotspot detection,
- Spatiotemporal change.

A ST outlier detection or ST anomaly detection is the process of finding a ST object whose non-spatiotemporal value varies from the value found in its neighbourhood [4]. Anomalies are instances that are highly dissimilar from a set of ST instances. ST association mining or STCOP mining finds subsets of Boolean features whose instances frequently co-occur in both space and time [5]. For an ST dataset containing explanatory variables (input features) and dependent variables (output features), a ST predictive learning predicts the dependent variables from the explanatory variables by learning the mapping between them [6]. ST clustering is the process of grouping data instances that share similar properties [7]. ST hotspots are special cases of ST clustering, where the possibility of an event or activity occurring is high compared to outside the cluster. ST change detection is the process of identifying the particular time point or location where a system starts to behave differently compared to the past.

The existing surveys gives an overview about the various types of analysis carried on spatial data [8, 9] and on spatiotemporal data [10–12]. But none of these surveys provide dedicated attention to the spatiotemporal co-occurrence pattern mining problem. This survey will give a brief summary about the different STCOP techniques available with respect to the nature of the space and time data.

The rest of the paper is organized as follows. Section 2 explains the different ST data types available. Section 3 gives an idea about the STCOP mining techniques available. Section 4 concludes the paper.

2   Spatiotemporal Data

A spatiotemporal data has both the space and time factor in it, and there are few ways to define the object or instance and feature present in them depending upon the problem at hand. But based on the way in which space and time values are measured, they can be categorized into different ST data types. Hence, in a ST data, the underlying space value can be either a vector or raster or spatial network, and time factor can be a temporal snapshot model, temporal change model, temporal event, or process model [10]. A spatial vector consists of points, line, and polygon. A raster data consists of regularly spaced or irregularly spaced grids (Table 1). Spatial network is an extension of graph in which the vertices and nodes are spatial elements.

Table 1 ST data types with underlying space and time model

ST data type	Space model	Time model
Event data	Vector model	Temporal event model
Point reference data	Vector model/Raster model	Temporal snapshot
Trajectory	Vector model	Temporal snapshot
Raster data	Raster model	Temporal snapshot
Evolving data	Vector model	Temporal change

2.1   Event Data

A ST event data has a spatial location and a time associated with the event. The spatial location can be point [13] line or polygon [14]. It gives information about when and where the event has occurred. For example, in a crime dataset, the crime activity is an event which has information about the location where the crime has happened and the time when it took place. The event can have additional attributes associated with it, which gives more information about the event. Some of the examples of event data in real world applications are crime analysis, accident data, epidemiology data (disease outbreak events), climate science(hurricane), social media (trending tweets). ST event data can be used to find ordered, unordered, and totally ordered patterns. If event-instances have disjoint occurrence times then the ordering is total [15]. Otherwise, ordering is partial [16]. If the event evolves as time changes then they are called as evolving data [17]. Figure 1a is an example for event data, where

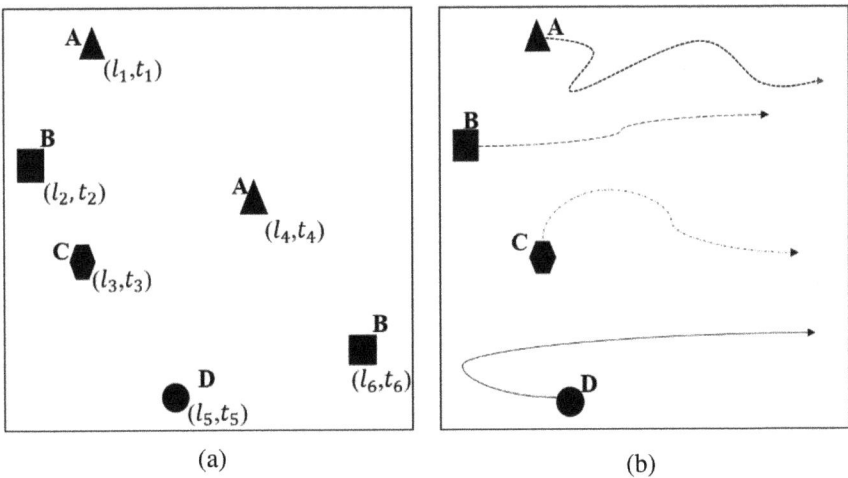

(a)   (b)

Fig. 1 Event data representation. **a** Events belonging to four types A(triangle), B(square), C(Hexagon), D(circle). **b** Trajectories of four moving bodies A, B, C, and D

Fig. 2 Evolving region
trajectories

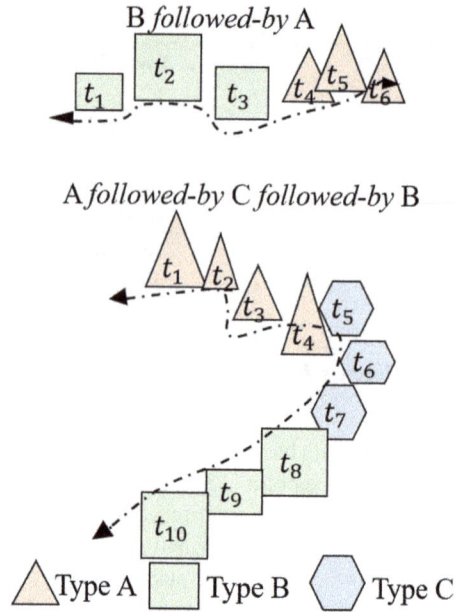

(A, B, C) are three different events. Each event has many instances where an event could have happened.

2.2 Trajectory

Trajectory is the path traced by the moving object in space over time [18]. Figure 1b shows the path traced by three different moving bodies. A trajectory data can be taxi pickup/drop data, a predator-prey movement, game tactics, crowd movement (Mobile GPS), and evolving region trajectory [18–20]. This kind of data is usually collected by mounting a sensor in the moving object. Totally ordered [20, 21] and unordered patterns [19, 18] are mined using trajectory data. An evolving event data can be treated as a trajectory to get totally ordered STCOP [20]. Figure 2 shows the trajectory of three different events that keeps evolving.

2.3 Point Reference Data

The point reference data comprises of continuous measurements of ST attributes like temperature, pressure, humidity, or vegetation (crop harvesting) recorded on a set of moving reference points. Hence, the spatial data (a vector or raster data) that is recorded on different snapshots is called point reference data. The spatial location keeps changing over time; buoy sensor measuring ocean variables is an example for this [22]. Generally, geostatistical data in spatial statistics is referred to as a point reference data.

2.4 Raster Data

A raster data is continuous or discrete measurements of ST field recorded from a fixed space at a fixed time point. The raster data consists of fixed location either distributed regularly or irregularly in space S, and the measurement is recorded at a fixed time stamp (T). The interval between timestamps can be regular or irregular. Thus, formally, a ST raster data is a spatiotemporal grid $S \times T$ and every vertex in the grid (s_i, t_i) has a distinct measurement. Ground-based sensors on ST field recording air quality data or weather information are examples of raster data.

2.5 Evolving Data

ST evolving data consists of a spatial vector, usually a line or polygon that experiences deformation or changes as time varies continuously. Figure 3 shows an example of evolving data having three different even types. Initially, the evolving data where considered as distinct events with an associated time stamp. Later, this data was seen as a trajectory with evolving polygon-based geometric regions where the size and location of the region varies in each time snapshot [23, 20]. Figure 2 is an illustration of evolving region trajectories involving three event. An example of this data type is solar events where the shape, size, and location of the solar events continuously evolve over time. Another example for evolving data is the time lapse image of embryo growth recorded in in vitro fertilization (IVF).

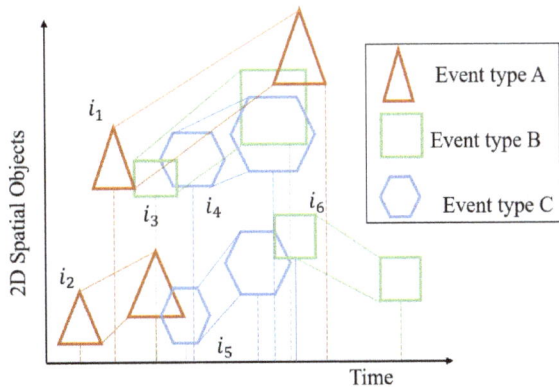

Fig. 3 Example of spatiotemporal evolving data

2.6 Data Type Conversion

ST data can be found in any of the above-mentioned types. ST data recorded in a particular data type can be converted to another data type based on the problem requirement. For example, an event data can be converted to raster by counting the number of events in a ST grid cell. A raster data can be converted to an event data by extracting the events occurring inside a ST grid. Similarly, a ST point reference data can be converted to ST raster data by interpolating the measurements over a ST grid. A ST raster data can be converted to ST point reference data by viewing every vertex of the ST grid as a ST reference point.

3 Spatiotemporal Co-occurrence Pattern Mining (STCOP Mining)

Spatiotemporal co-occurrence pattern mining or spatiotemporal frequent pattern mining or spatiotemporal association pattern mining corresponds to mining spatiotemporal features or objects whose instances are often found in close geographic and temporal proximity. STCOP mining techniques can be applied in traffic control, criminal detection, location-based services, and other areas. For example, predicting the traffic flow in an area at a time based on the congestion in nearby traffic signals.

STCOP techniques are classified based on the temporal ordering of the objects as

- Unordered STCOP
- Partially ordered STCOP
- Totally ordered STCOP (Fig. 4).

STCOP mining is an extension of the spatial co-location pattern (SCOP) mining techniques. SCOP were initially found using a transaction-based approach, but it resulted in a loss of the dependency relationship among different features. Inorder

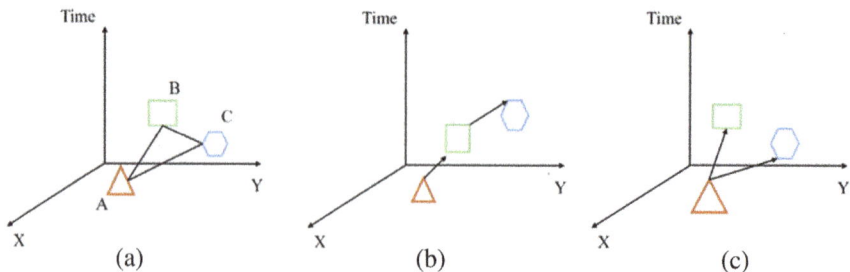

Fig. 4 Example of types of pattern. **a** Unordered pattern. **b** Totally ordered pattern. **c** Partially ordered pattern

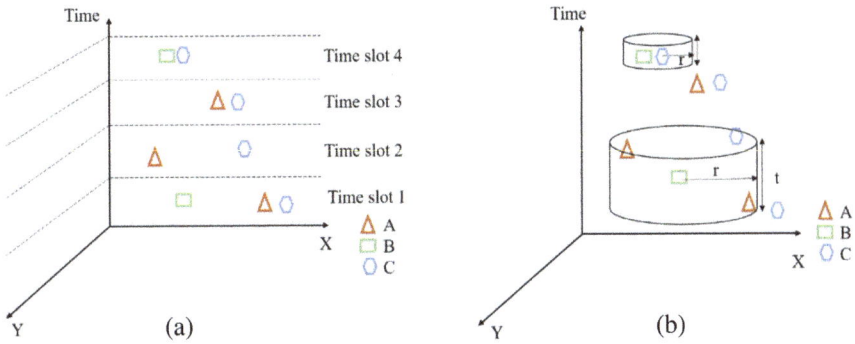

Fig. 5 STCOP mining approaches. **a** Divide and conquer. **b** Coupling

to overcome this problem, a transaction free model was developed for discovering SCOP [24, 25]. STCOP methods have been developed by adding a time factor to these transactions free model. Two types of STCOP detection methods are available

1. Divide and conquer [19, 26–28]
2. Coupling method [16, 29, 30].

3.1 Divide and Conquer

In divide and conquer, the temporal data is treated as a set of divided time slots. Figure 5a gives an illustration of divide and conquer technique. For example, in [31], the el-nino data set (1980–1998) was divided in 18 time slots, where each time slot contains the data recorded on a daily basis for a year. Frequently, co-occurring spatial patterns are found in each time slot, then the prevalence of these co-occurring patterns across all the time slots is checked to get the STCOP which are prevalent in space and time. In the divide and conquer methods, the forcible partitioning of the time dimension is likely to cause the loss of the spatiotemporal neighborhood relationship among features across adjacent time slots [31]. The existing divide and conquer strategies are useful in finding ordered and unordered STCOPs.

3.2 Coupling

Coupling methods use time as an alternate dimension. An ST neighborhood relation is used in coupling method where, a space and time constraints are used in this relation [16]. From Fig. 5b it can be seen that events which are close to each other with respect to space (*r*) and time (*t*) threshold are mined as STCOPs. Later, prevalence index value is calculated (based on the ordering it may be a Sequence index, Cascading index, etc.) for each pattern, and the prevalence threshold is used to find the frequent

STCOPs. Coupling methods can be used to mine all three types of patterns namely unordered, ordered, and partially ordered STCOPs.

4 Unordered STCOP

STCOPs whose temporal ordering is not taken into consideration is called unordered STCOP. They are also called as ST mixed drove patterns [19, 28]. Figure 4a gives an illustration of unordered STCOP involving three events. Unordered STCOP can be mined using both divide and conquer and coupling method. Finding unordered STCOP is useful in many applications like finding strategy in battlefield and games and tracking predator-prey relation [19].

4.1 Divide and Conquer

For mining unordered STCOP using the divide and conquer method, the entire data set is sliced into time slots. Spatially prevalent pattern in each time slot is got by comparing the spatial prevalence index value calculated for each pattern with user defined spatial prevalence threshold. To get STCOPs, the spatially prevalent patterns time prevalence is calculated and compared with time prevalence threshold. STCOPs can be mined using the above-mentioned strategy in three ways [19]

1. Naïve approach
2. MDCOP Miner
3. FastMDCOP Miner.

In Naïve approach, spatial prevalent patterns of all sizes are mined for each time slot and after that in post-processing step, time non-prevalent patterns are pruned out. In MDCOP miner, size k spatially prevalent patterns are mined in all time slot and then time non-prevalent patterns are pruned out. This size k MDCOP patterns are used to generate size $k + 1$ pattern in the next iteration. A FastMDCOP miner is a computationally efficient MDCOP miner, that prunes time non-prevalent MDCOP in between time slots. For example, if there is a total of 10 time slots and the time prevalence threshold is 0.5. It can be said that a size k pattern is prevalent if it is present in at least 5 time slots. If the time prevalence index of a pattern is 0 for the first six time slots, then there is no need to generate it and check its prevalence for the remaining time slots. Because, even if it is time persistent for the remaining four time slots, it will not be able to satisfy the given threshold. This MDCOP mining strategy is used to find partial spatiotemporal co-occurrence patterns (PACOP's) [26, 32]. PACOP's are used to find co-occurrence patterns that are partially present in the dataset. For example, mining interesting relationship between migrating and non-migrating birds while they are passing over a region. In PACOP mining method, a time participation ratio (TPR) and time participation index (TPI) measure are used instead of the time prevalence index of MDCOP. Determining a threshold value for finding MDCOPs is a difficult task for user. Therefore, a TopMDCOP miner was introduced which can mine the top $K\%$ STCOP without the need for any user defined threshold [28]. The MDCOP miner can be used for mining STCOPs in extended spatial objects like lines, polygons, etc. [33, 34].

4.2   Coupling

Mining unordered STCOP via a coupling approach overcomes the drawback of time slot approach (i.e.) the loss of spatiotemporal neighbourhood relation caused by forcible partitioning of time dimension. Initially, the coupling method was used to mine unordered topological patterns, using the spatiotemporal neighbourhood relationship which is defined based on a spatial radius and an unordered time window [29]. A summary like structure is used to find topological patterns from ST dataset, which does not require a costlier candidate generation step. The Topology Miner works by using the pattern growth logic. Later [31] proposed an algorithm to mine both unordered and ordered patterns, by using apriori approach. In this paper, an objective spatiotemporal participation ration and objective spatiotemporal participation index are introduced to mine prevalent ST patterns. It makes use of the weighted sliding model, where a weighted window moves across the time axis calculating the spatiotemporal participation index.

Coupling method is also used to find unordered STCOP from evolving data. First an apriori algorithm was proposed to find unordered STCOPs from this data, by treating the evolving data as distinct events with time stamps. Here, an overlap relation is used to determine whether two or more evolving events are co-occurring or not. Figure 3 illustrates how an evolving data is treated to mine unordered STCOP. Significance measure called co-occurrence co-efficient (CCE) is used to determine prevalence of a pattern. The calculation of this measure is similar to that of Jaccard co-efficient [35]. A Fast STCOP miner was introduced in [36] to mine computationally efficient STCOPs. A filter and refine strategy are used in this Fast STOCP miner to prune irrelevant patterns. The filter and refine strategy use OMAX to filter STCOP patterns based on the CCE threshold values given, later jaccard (J) values are calculated for this refined pattern and further filtered using the CCE threshold. Since the evolving data are complex in nature efficient indexing techniques like Scalable efficient trajectory index (SETI) and Chebyshev polynomial index (CPI) are used [37]. A combination of these indexing technique along with filter and refine strategy is used to mine unordered STCOPs from solar dataset in [17]. Finding a suitable CCE value for mining STCOP is a tedious job, [38] proposes the Top-$K\%$ STCOP approach to mine the top $k\%$ patterns without requiring the user to give a threshold value. An apriori algorithm require the candidate generation process to mine patterns, [39] proposed a pattern growth approach to mine STCOPs. In this approach, the costlier candidate generation step is replaced by the efficient proto-pattern generation step. Proto-patterns are subsets of event types, where the participating event types have at least one pattern instance. Proto-patterns reduce the search space. FP growth algorithm is used for generating the proto-patterns. J value is calculated for proto-patterns to find the significant STCOPs. Apart from J and OMAX various other significance measure like J^* [40] and J^+ [41] have been introduced to find significant STCOPs in evolving data.

5 Ordered STCOP

Ordered STCOP is also called as spatiotemporal sequential patterns. An ordered STCOP is a sequence of event types, with one leading to another. A ST sequential pattern in the form $E_1 \rightarrow E_2 \rightarrow \ldots E_k$ represents a chain reaction from event type E_1 to event type E_2 and then to event type E_3 until it reaches event type E_k. This "chain reaction" happens in a specific spatial and temporal manner. For example in epidemiology, the disease transmission follows a path connecting different events. The spread of Nipah virus could be traced to a sequence of event like Bats $\rightarrow$ fruits $\rightarrow$ human being, i.e., when a fruit eating bat consumes a fruit and the left over is consumed by human. ST sequence mining is useful in solar physics, bio-medical science, Web log sequence, DNA sequence, gene structure, targeted advertising, location prediction for taxi services, urban planning, etc.

In [30], a framework is proposed for mining ordered STCOP from event data. A prevalence measure named sequence index (density ratio based) is introduced to find ordered STCOPs. A follow predicate and tail events are defined to find STCOPs. Two algorithms namely STS-miner and Slicing-STS-miner is proposed. The STS-miner constructs a sequence tree to mine patterns, whereas in slicing-STS-miner, the database is sliced according to time and events are hashed into the slice. The sequential patterns are mined from each slice and, at the same time, merged with the patterns found in the previous slice through updating a common pattern tree. In some cases, events get priority based on where it occurs and the when it occurs. Sunitha and Reddy [42] proposed an algorithm for mining Weighted Regional Sequential Patterns (WRSPs) from spatiotemporal event databases. Two interest measures sequence weight and significance index are used for efficient mining of WRSPs. Weights are assigned to the spatiotemporal cells and the event types to get prioritised ordered STCOPs. In [43] a Spatiotemporal Sequence Miner (STSM) is proposed to mine sequence that are frequent in constrained space and time and may not be frequent in the entire dataset.

Sequence patterns can be mined from evolving trajectory data. In an evolving trajectory data, the spatiotemporal event instances are the chronologically ordered lists of time stamped geometry pairs. The geometries are region-based, and represented with polygons. Figure 2 is an example of evolving region trajectories where instances evolve with time [20]. A ST follow relation is used to find sequence in evolving trajectories, a tail and head window are constructed to find a follow relation. If two instances ins_i and ins_j are given, then a spatiotemporal follow relationship exists between them only when (1) the start time of ins_i is less than the start time of ins_j, and (2) there exists a spatiotemporal co-occurrence between the tail window of ins_i and the head of ins_j. A chain index (ci) is used for determining the significance of instance sequences. The ci value is calculated by using the J, J^*, J^+ and OMAX. A pattern growth-based approach in used in [39, 20], it adopts a graph-based mining algorithm, which transforms the sequences of spatiotemporal trajectories into a directed acyclic graph. An apriori-based algorithm is proposed to find ST event sequence (STES) in [20]. In [15] an algorithm called Top-($R\%$, K) is proposed for mining the most prevalent K sequences discovered from most significant $R\%$ of spatiotemporal follow relationships. This algorithm overcomes the threshold determining problem present in the previous work.

6   Partially Ordered STCOP

Cascading spatiotemporal patterns are referred to as partially ordered patterns. According to [16] cascading spatiotemporal patterns (CSTPs) are partially ordered subsets of event types whose instances are spatial neighbours and occur in a series of stages. For example, analysing a crime data may reveal that after bar closing assaults and drunk and drive incidents are recorded in nearby areas. Figure 6 shows the cascading events that are connected with a Cyclone. Hence, finding cascading patterns could be useful in various domains like public safety, disaster planning, climate change, and public health.

To mine CSTP from a ST data, a directed neighbourhood relation measure is defined, where the space and time are constrained by a threshold. For example, an assault and bar closing event can be called as a directed neighbour if assault happens after bar closing in a location that is at a reasonable distance from bar. Using this neighbourhood relation, cascading patterns of all possible size or length are mined. Next, Cascading participation index (CPI) value is calculated for these patterns and the prevalence threshold value got from user is used to obtain the frequent CSTPs. Two filter techniques namely upper bound filter and multiresolution ST filter are used to eliminate non-prevalent patterns before the CPI value is calculated. CSTPs are represented using a directed acyclic graph (DAG) [16]. Mining CSTPs requires multiple scan on the database depending upon the number of items involved. Hence,

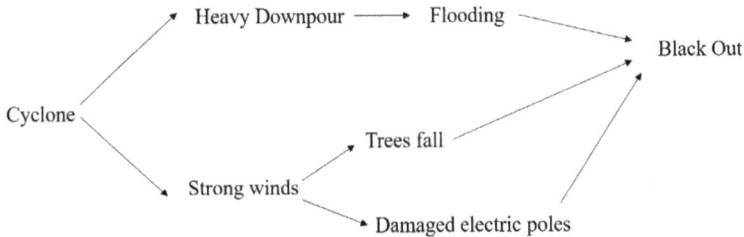

Fig. 6 Example of cascading patterns

a normalized-mutual-information-based mining technique for cascading patterns (M^3Cap) was proposed to address this challenge [44].

7 Conclusion

In this article, an overview of the STCOP mining strategies is given. The STCOPs can be mined from a variety of ST data like event data, point reference data, ST trajectories, ST raster data, ST evolving data. Based on the type of data at hand and ordering of the patterns to be mined an algorithm can be chosen by using this article as a guide. The existing STCOP mining algorithms generate numerous patterns, and hence this problem has to be addressed in the future. In spatial co-location pattern, mining many algorithms has been proposed to mine non-redundant patterns [45–47]. These works can be used to extend it to STCOP mining. ST data is dynamic in nature and grows exponentially in this era of big data by frequent updates. Therefore, an incremental maintenance of ST patterns is necessary; otherwise, the patterns can become obsolete as time goes by. Since STDM is an emerging field, there are a lot more problems yet be addressed and has scope for promising research directions.

Acknowledgements This work was supported by Council of Scientific and Industrial Research, India under the scheme Direct-SRF with grant File No: 09/559(0141)/19-EMR-I.

References

1. Kawale J, Steinbach M, Kumar V (2011) Discovering dynamic dipoles in climate data. In: Proceedings of the 11th SIAM international conference on data mining, SDM 2011, January 2014, pp 107–118
2. Liu X, Chang C, Duyn JH (2013) Decomposition of spontaneous brain activity into distinct fMRI co-activation patterns. Front Syst Neurosci 7(Dec):1–11
3. Tran-The H, Zettsu K (2017) Finding spatiotemporal co-occurrence patterns of heterogeneous events for prediction. In: Proceedings of the 3rd ACM SIGSPATIAL international workshop on the use of GIS in emergency management, EM-GIS 2017
4. Kratz L, Nishino K (2009) Anomaly detection in extremely crowded scenes using spatio-temporal motion pattern models. In: IEEE computer society conference on computer vision and pattern recognition workshops, CVPR workshops 2009, vol 2009. IEEE, pp 1446–1453
5. Xia D, Lu X, Li H, Wang W, Li Y, Zhang Z (2018) A MapReduce-based parallel frequent pattern growth algorithm for spatiotemporal association analysis of mobile trajectory big data. Complexity 2018
6. Cheng T, Wang J (2008) Integrated spatio-temporal data mining for forest fire prediction. Trans GIS 12(5):591–611
7. Birant D, Kut A (2007) ST-DBSCAN: an algorithm for clustering spatial-temporal data. Data Knowl Eng 60(1):208–221
8. Shekhar S, Evans MR, Kang JM, Mohan P (2011) Identifying patterns in spatial information: a survey of methods. Wiley Interdisciplinary Rev Data Min Knowl Disc 1(3):193–214
9. Zala RL, Mehta BB, Zala MR (2014) A survey on spatial co-location patterns discovery from spatial datasets. Int J Comput Trends Technol 7(3):137–142

10. Shekhar S et al (2015) Spatiotemporal data mining: a computational perspective. ISPRS Int J Geo-Inf 4(4):2306–2338
11. Aydin B, Angryk RA (2018) A gentle introduction to spatiotemporal data mining. In: SpringerBriefs in computer science, pp 1–7
12. Shekhar S, Varsavai RR, Celik M (2008) Spatial and spatiotemporal data mining: recent advances. In: Data mining: next generation challenges and future directions, AAAI Press
13. Maciag PS (2018) Efficient discovery of top-K sequential patterns in event-based spatia-temporal data. In: Proceedings of the 2018 federated conference on computer science and information systems, FedCSIS 2018, pp 47–56
14. Aydin B, Angryk RA (2018) Spatiotemporal co-occurrence pattern (STCOP) mining. In: SpringerBriefs in computer science, pp 55–69
15. Aydin B, Angryk RA (2018) Spatiotemporal event sequence (STES) mining. In: SpringerBriefs in computer science, pp 71–96
16. Mohan P, Shekhar S, Shine JA, Rogers JP (2010) Cascading spatio-temporal pattern discovery: a summary of results. In: Proceedings of the 10th SIAM international conference on data mining, SDM 2010, pp 327–338
17. Aydin B, Kempton D, Akkineni V, Angryk R, Pillai KG (2015) Mining spatiotemporal co-occurrence patterns in solar datasets. Astron Comput 13:136–144
18. Celik M, Azginoglu N, Terzi R (2012) Mining periodic spatio-temporal co-occurrence patterns: a summary of results. In: INISTA 2012—international symposium on innovations in intelligent systems and applications
19. Celik M, Member S, Shekhar S, Rogers JP, Shine JA (2008) Mixed-drove spatiotemporal co-occurrence pattern mining. IEEE Trans Knowl Data Eng 20(10):1322–1335
20. Aydin B, Angryk R (2016) Discovering spatiotemporal event sequences. In: Proceedings of the 5th ACM SIGSPATIAL international workshop on mobile geographic information systems, MobiGIS 2016, no 2, pp 46–55
21. Aydin B, Angryk RA (2017) A graph-based approach to spatiotemporal event sequence mining. In: IEEE international conference on data mining workshops, ICDMW, no 1, pp 1090–1097
22. Saha S et al (2010) Supplement to the NCEP climate forecast system reanalysis. Bull Am Meteorol Soc 9–25
23. Pillai KG, Angryk RA, Aydin B (2013) A filter-and-refine approach to mine spatiotemporal co-occurrences. In: GIS: proceedings of the ACM international symposium on advances in geographic information systems, pp 104–113
24. Yoo JS, Shekhar S (2004) A partial join approach for mining co-location patterns. In: GIS: proceedings of the ACM international symposium on advances in geographic information systems, pp 241–249
25. Huang Y, Shekhar S, Xiong H (2002) Discovering co-location patterns from spatial datasets: a general approach. IEEE Trans Knowl Data 612–626
26. Celik M (2015) Partial spatio-temporal co-occurrence pattern mining. Knowl Inf Syst 27–49
27. Celik M, Kang JM, Shekhar S (2007) Zonal co-location pattern discovery with dynamic parameters. In: Proceedings—IEEE international conference on data mining, ICDM, pp 433–438
28. Wang Z, Peng X, Gu C, Huang B (2013) Mining at most top-K% mixed-drove spatiotemporal co-occurrence patterns. In: 9th Asian control conference, ASCC 2013
29. Wang J, Hsu W, Lee ML (2005) A framework for mining topological patterns in spatio-temporal databases. In: International conference on information and knowledge management, proceedings, pp 429–436
30. Huang Y, Zhang L, Zhang P (2008) A framework for mining sequential patterns from spatio-temporal event data sets. IEEE Trans Knowl Data Eng 20(4):433–448
31. Qian F, Yin L, He Q, He J (2009) Mining spatio-temporal co-location patterns with weighted sliding window. In: Proceedings—2009 IEEE international conference on intelligent computing and intelligent systems, ICIS 2009, vol 3, pp 181–185
32. Celtic M (2011) Discovering partial spatio-temporal co-occurrence patterns. In: ICSDM 2011—proceedings 2011 IEEE international conference on spatial data mining and geograph-ical knowledge services, pp 116–120

33. Akbari M, Samadzadegan F, Weibel R (2015) A generic regional spatio-temporal co-occurrence pattern mining model: a case study for air pollution. J Geogr Syst 17(3):249–274
34. Akbari M, Samadzadegan F (2015) Identification of air pollution patterns using a modified fuzzy co-occurrence pattern mining method. Int J Environ Sci Technol 12(11):3551–3562
35. Pillai KG, Angryk RA, Banda JM, Schuh MA, Wylie T (2012) Spatio-temporal co-occurrence pattern mining in data sets with evolving regions. In: Proceedings—12th IEEE international conference on data mining workshops, ICDMW 2012, pp 805–812
36. Pillai KG, Angryk RA, Aydin B (2013) A filter-and-refine approach to mine spatiotemporal co-occurrences. In: Proceedings of the 21st ACM SIGSPATIAL international conference on advances in geographic information systems—SIGSPATIAL'13, pp 104–113
37. Aydin B, Kempton D, Akkineni V, Gopavaram SR, Pillai KG, Angryk R (2015) Spatiotemporal indexing techniques for efficiently mining spatiotemporal co-occurrence patterns. In: Proceedings—2014 IEEE international conference on big data, IEEE big data 2014, pp 1–10
38. Pillai KG, Angryk RA, Banda JM, Kempton D, Aydin B, Martens PC (2016) Mining at most top-k% spatiotemporal co-occurrence patterns in datasets with extended spatial representations. ACM Trans Spat Algorithms Syst 2(3)
39. Hamdi SM, Aydin B, Angryk RA (2017) A pattern growth-based approach for mining spatiotemporal co-occurrence patterns. In: IEEE international conference on data mining workshops, ICDMW, pp 1125–1132
40. Aydin B, Akkineni V, Angryk R (2015) Time-efficient significance measure for discovering spatiotemporal co-occurrences from data with unbalanced characteristics. In: GIS: Proceedings of the ACM international symposium on advances in geographic information systems, vol 03-06-Nove
41. Aydin B, Kucuk A, Angryk RA, Martens PC (2017) Measuring the significance of spatiotemporal co-occurrences. ACM Trans Spat Algorithms Syst 3(3)
42. Sunitha G, Reddy ARM (2016) WRSP-miner algorithm for mining weighted sequential patterns from spatio-temporal databases. Adv Intell Syst Comput 379:309–317
43. Campisano R et al (2018) Discovering tight space-time sequences. In: Lecture notes in computer science (including subseries lecture notes in artificial intelligence and lecture notes in bioinformatics), vol 11031. LNCS, pp 247–257
44. Xue C, Liu J, Li X, Dong Q (2016) Normalized-mutual-information-based mining method for cascading patterns. ISPRS Int J Geo-Inf 5(10)
45. Yoo JS, Bow M (2011) Mining top-k closed co-location patterns. In: ICSDM 2011—proceedings 2011 IEEE international conference on spatial data mining and geographical knowledge services, pp 100–105
46. Wang L, Bao X, Chen H, Cao L (2018) Effective lossless condensed representation and discovery of spatial co-location patterns. Inf Sci 436–437:197–213
47. Wang L, Bao X, Zhou L (2018) Redundancy reduction for prevalent co-location patterns. In: Proceedings—IEEE 34th international conference on data engineering, ICDE 2018, vol 30, no 1, pp 1777–1778

Chapter 19
The Classification and Comparative Study of Real-Time Task Scheduling Algorithms Based on Various Parameters

Jayna Donga, M. S. Holia, and N. M. Patel

1 Introduction

For any real-time system, to generate correct output, the task must be executed within its timing constraints, which is predefined [1, 2]. In other words, to meet the deadline that is the most critical criteria for every real-time system and real-time operating system provides help to real-time systems to meet their deadlines by providing a technique of the scheduling. The scheduler is responsible for making a decision about which task will be scheduled next in suitable order so that each task into the real-time system can be completed within its deadline. The real-time systems may be divided into three different types: (a) hard real-time system, (b) soft real-time system, (c) firm real-time tasks. The task is called hard real-time task; if it is having a hard deadline, that means if deadline miss occurs, the whole system will be destroyed. The major class of the hard real-time systems may be the safety-critical, so the deadline miss creates a loss of property, resources, and sometimes life also. The task can be called soft real-time tasks only when the task is having a soft deadline means that if deadline miss occurs, then it will not destroy the whole system, but it may reduce the usefulness of the result over the time period elapsed. If many deadlines miss occurs in the soft real-time system, then it degrades system performance. The deadline miss of a firm real-time task will not create whole system failure, but the utility of the result becomes zero [3]. Nowadays, in many application, domains like space systems, robotics, automotive electronics, medical systems, avionics, home automation, industrial automation, etc. are using the systems which are having hard

J. Donga (✉) · M. S. Holia · N. M. Patel
Gujarat Technological University, Ahmedabad, Gujarat, India
e-mail: jdonga@mbict.ac.in

M. S. Holia
e-mail: msholia@bvmengineering.ac.in

N. M. Patel
e-mail: nmpatel@bvmengineering.ac.in

© The Author(s), under exclusive license to Springer Nature Singapore Pte Ltd. 2021   239
X.-Z. Gao et al. (eds.), *Applications of Artificial Intelligence in Engineering*, Algorithms
for Intelligent Systems, https://doi.org/10.1007/978-981-33-4604-8_19

deadline bound extensively [4]. For any real-time systems, the scheduling approach is having an important role to play for achieving desired and predictable system behavior.

2 Literature Survey

The next section is dedicated to the research and researchers who have played an important role for the domain of real-time operating systems and especially in the area of real-time scheduling.

Ismail and Jawawi [5] have presented a hybrid scheduling algorithm for weakly hard real-time tasks that can be used in a multiprocessor environment in 2017. For scheduling the real-time jobs in a multiprocessor environment, there are two main approaches: (1) partitioning scheduling and (2) global scheduling. Though partitioned method having tolerable overhead, it does not provide any kind of guarantee of being optimal. While in the case of the global method, such kind of guarantee can be provided, but it has substantial overhead. In this paper, the researcher had proposed an unconventional scheduling tactic to take advantage of both the partitioning scheduling strategy and the global scheduling strategy. Hybrid algorithm having less overhead as it limits the number of context switches of a task and to increase the schedulability by defining an efficient allocation of jobs to the CPU. They observed that proposed algorithm gives optimal schedulability with minimum overhead.

Sanati and Cheng [6] have presented a multiprocessor scheduling algorithm for periodic soft real-time tasks for QoS optimization with the online semi-partitioned approach in 2016. In soft real-time operating systems, tasks which are meeting their deadline will gain rewords, that is, why the researcher emphasis on increasing the rewards to enhance quality of service. Parameters which are affecting to QoS: rewards, overall response time, deadline miss ratio. Semi-partitioning real-time scheduling approach extended partitioning method by permitting some of the tasks from the task set for migration, so less switching overhead. This algorithm uses two dynamic approximation algorithms like greedy algorithm and load balancing algorithm with an online semi-partitioned method for the soft real-time task, which is periodic in nature. Goal of this algorithm to improve the quality of service through decreasing deadline miss ratio and increasing total rewords.

Alhussian et al. [7] have published an optimal real-time scheduling algorithm for multiprocessor system in 2014. Utmost all presently existing real-time schedulers, which is providing optimal schedulability in a multiprocessor environment, obey the fairness rule, as a result of following the fairness rule results in a large number of task pre-emptions and migrations. It highly affects the predictability of the algorithm. In this algorithm, the fairness rule is completely relaxed, and the semi-greedy algorithm is introduced. The aim is to use of global approach, relaxation in fairness rule to reduce the number of migrations and pre-emptions and miss a very few deadlines.

Li and Ba [8] have presented the scheduling algorithm, which considers group priority earliest deadline first in 2012. In most of the priority scheduling algorithms, the assumption is taken that priority levels are unrestricted, but in some of the cases in which task set wants priority levels more than the system's capacity, so the same priority level will be assigned to the several jobs. To address this issue, the new scheduling algorithm GP-EDF has been designed in that group-wise priority is assigned, and the group which is having the earliest deadline will contain the highest priority and executed first. The proposed approach provides schedulability test to form a task group and the jobs within one group may change their execution order arbitrarily without dropping the Schedulability. In this paper, new proposed algorithm GPEDF was compared with the traditional algorithm EDF and the gEDF. The results show that the proposed algorithm is having less switching overhead, the shortest average response time, and it requires very few priority levels.

3 Scheduling Algorithms

Nowadays, we are using computers are multitasking and multiprocessing systems. We mean by multitasking is the system can run multiple tasks simultaneously. A computer system is full of resources, which may be either software or hardware. These resources will be shared among all the tasks into the systems. To take a decision about which resource will be given to which task at which instance of time is an important task in any computer system. The module of OS called scheduler is accountable for taking judgment about resource allocation and de-allocation during the task's execution so to design an efficient scheduling algorithm is an important operating system design issue. The goals of any scheduling algorithm are to distribute equal load among all the processors, maximize resource utilization, minimize response time, and maximize throughput [9]. As especially whenever we are talking about the real-time operating system, then important scheduling criteria are meeting the deadlines for all the tasks into the system.

3.1 Classification of Scheduling Algorithms Based on Various Parameters

There are many real-time scheduling algorithms designed by a number of researchers. They can be classified into different categories based on various parameters. The classification will be done according to the type of application on which the scheduling approach will be deployed. A consideration may be like, whether the job scheduling is to be done at the time of compilation or at the time of execution, job scheduling will be on a central node or on distributed nodes, whether the jobs are going to be executed on uniprocessor or multiprocessor environment.

3.1.1 Classification in User's Context

As per the user's aspect, scheduling approaches may be divided into three different categories.

Interactive Scheduling:

The task scheduling will be done with iterative technique, and the scheduling algorithms used for this approach are called as iterative scheduling algorithms [10]. Shortest job first, Round-Robin scheduling, SJF, lottery scheduling, etc. can be considered as examples of iterative scheduling algorithms.

Batch Scheduling:

Jobs are composed of the queue, which will form a batch. The scheduling will be performed batch-wise. The scheduling approaches deployed for this approach are called the batch scheduling algorithms. FCFS, SRTN, HRN, etc. can be the examples of batch scheduling algorithms.

Real-Time Scheduling:

For real-time scheduling, the rightness of the output would not rely upon the logical correctness of results only, but it must be created within the predefined time limit also.

The real-time scheduling approaches used to schedule real-time tasks [11]. Rate monotonic (RM), earliest deadline first (EDF), deadline monotonic (DM), Least Laxity First (LLF), etc. are the well-known real-time scheduling approaches.

3.1.2 Categorization Based on Scheduling Location

The categorization may be done based on the process scheduling location, whether it is from a central location or from distributed locations.

Centralized Scheduling:

The central site is responsible for all the scheduling decisions [12]. In any distributed system, the critical system failure may happen if the central scheduler goes down. In this case, the communication bottleneck may occur as the latest load information will be required on every node.

Distributed Scheduling:

In this policy, the chaining of various jobs into a synchronized workflow that covers several systems. Scheduling decisions are taken by various nodes but not by one centralized system. This approach of scheduling is used for non-interactive processes or jobs in distributed systems [10].

3.1.3   Classification Based on Pre-emption

Pre-emption is the method which permits to take resource back forcefully after assigning it to the process when a higher priority process comes. This classification may be based on whether the pre-emption is allowed or not by a particular scheduling algorithm.

Pre-emptive Scheduling:

In this scheduling, approach the scheduler is allowed to pre-empt a processor from the running task when a higher priority process comes. Such scheduling algorithms are called pre-emptive scheduling algorithms [9, 10]. The real-time kernel is pre-emptive, and all the real-time scheduling approaches are pre-emptive.

Non-Pre-emptive Scheduling:

In a non-pre-emptive approach, the scheduler cannot take CPU back from running tasks forcefully until the process releases it either voluntarily or completes its execution. The FCFS and SJF are examples of the non-pre-emptive scheduling algorithm.

3.2   Real-Time Scheduling Approaches

The various real-time schedulers are there; among them, we have presented very few well-known algorithms here.

3.2.1   Dynamic Real-Time Scheduling Algorithms with Fixed Priorities

1. **Rate Monotonic (RM)** [13, 14]
   RM is the optimal static priority real-time scheduler for uniprocessor systems. It handles all the tasks into the system, which are periodic in nature, and their deadline equals to its period. In this approach, every task will be allotted the static priority which related to its rate. If the rate is higher, then the priority is higher, and if the rate is lower, then priority is lower. The task's rate is inversely proportional to its period. The task set of n number of tasks and to meet their deadline, the necessary and required condition using this approach is:

$$U = \sum (c_i/p_i) \leq n(2^{1/n} - 1)$$

   Where C_i = Worst-Case Run Time, p_i = Period
2. **Deadline Monotonic (DM)** [15, 16]
   RM may not be an optimal algorithm if the deadline does not equal to the task's period for any periodic real-time tasks. The DM is an optimal algorithm for a

uniprocessor system with a fixed priority. DM is the extension of RM, where DM can also schedule the periodic tasks whose deadline not equals to its period. In this approach, the process will be allotted a static priority, which proportionates to its relative deadline. The RM is the special case of DM if the deadline is equal to the period.

3.2.2 Dynamic Real-Time Scheduling Algorithms with Dynamic Priorities

1. **Earliest Deadline First (EDF)** [13, 17]

   The earliest deadline first is an optimal algorithm for a uniprocessor system with dynamic priorities. It schedules all the periodic tasks into the system. In this approach, the absolute deadline will be taken into consideration. The task which is having the shortest deadline will be scheduled first. The task set of n tasks can meet their deadlines if

$$\text{Utilization } (U) = \sum_{}^{n} C_i / P_i \leq 1$$

   where C_i = worst-case computation time, p_i = period

2. **Least Laxity First (LLF)**

   One of the efficient algorithm for uniprocessor with dynamic priority is the LLF. The task's Laxity is the difference between deadline and worst-case run time. The Laxity can also say the slack time. Laxity for task i is:

$$L_i = D_i - C_i$$

   where

   D_i **is the deadline interval for Task i,** which is the duration between the task's deadline and the task's request instant.

   C_i **is computation time of task i.**

   In this approach, priorities are assigned based on slack time. The task of having large laxity will get low priority, and the task of having small laxity will get higher priority. This will cause an increase in context switching.

4 Conclusion

The survey about the various real-time scheduling approaches has been covered in this paper. The classification of scheduling algorithms is done based on various parameters. The paper mainly categories real-time scheduling algorithms into two categories, like dynamic real-time scheduling algorithms having static priority and dynamic

Table 1 Comparative study of different scheduling approaches for real-time system

Performance matric	Techniques			
	Rate monotonic (RM)	Deadline monotonic (DM)	Earliest deadline first (EDF)	Least Laxity First (LLF)
Implementation	Easiest	Easy	Complex	Complex
Priority	Static priority	Static priority	Dynamic priority	Dynamic priority
CPU utilization	Low	High as compared to RM	Full utilization	Full utilization
Scheduling parameters	Period of task	Relative deadline	Deadline	Laxity (slack time)
Effectiveness	Good transient overload handling capacity only for high priority tasks	Good transient overload handling capacity only for high priority tasks	Poor performance under transient overload	Poor performance under transient overload
Limitations	(a) Only supports periodic tasks (b) Not optimal when task periods and deadlines are different	Optimal algorithm for uniprocessor only with static priority	Cannot handle transient overload condition optimally	Cannot handle transient overload condition optimally
Optimal	Yes	Yes	Yes	Yes
Deadline miss chances	Average	Average	Average	Average
Number of context switching	Very less	Less	Average	Average

real-time scheduling algorithms having dynamic priority. It includes the study of very renowned real-time schedulers like RM, DM, EDF, LLF, and the comparison between all these algorithms we have analyzed the advantages and disadvantages of each based on various performance measure parameters. Basically, the existing algorithms give better performance for the uniprocessor system, but nowadays, we are working on a multiprocessor environment so one can focus on multiprocessor real-time systems and can design an efficient multiprocessor scheduling algorithm. Table 1 shows the comparative study of various real-time scheduling algorithms.

References

1. Buttazzo GC (2011) Hard real-time computing systems predictable scheduling algorithms and applications, 3rd edn. Springer, Berlin
2. Robert I, Davis R, Burns A (2013) A review of fixed priority and EDF scheduling for hard real-time uniprocessor systems. In: EWiLi'13, 26–27 Aug 2013, Toulouse, France

3. Lindh F, Otnes T, Wennerström J Scheduling algorithms for real-time systems
4. Davis RI, Burns A (2011) A survey of hard real-time scheduling for multiprocessor systems. ACM Comput Surv 43(4), 44 (Article 35)
5. Ismail H, Jawawi DNA (2017) A hybrid multiprocessor scheduling approach for weakly hard real-time tasks. Published in AsiaSim, Part II, CCIS 752. Springer, pp 666–678
6. Sanati B, Cheng AMK (2016) Online semi-partitioned multiprocessor scheduling of soft real-time periodic tasks for QoS optimization. In: Real-time and embedded technology and applications symposium (RTAS), IEEE
7. Alhussian H, Zakaria N, Hussin FA (2014) An efficient real-time multiprocessor scheduling algorithm. J Convergence Inf Technol
8. Li Q, Ba W (2012) A group priority earliest deadline first scheduling algorithm. Front Comput Sci 6(5):560–567
9. Buttazzo GC Hard real-time computing systems: predictable scheduling algorithms and applications, 3rd edn. Springer, Berlin
10. Kopetz H Real-time systems: design principles for distributed embedded applications, 2nd edn. Springer, Berlin
11. Brucker P Scheduling algorithms, 5th edn. Springer, Berlin
12. Bovet DP, Cesati M (2000) Understanding the Linux Kernel. O'Reilly online catalogue
13. Harkut DG, Agrawal AM (2014) Comparison of different task scheduling algorithms in RTOS: a survey. Int J Adv Res Comput Sci Softw Eng 4(7):1236–1240
14. Singh V, Vidhani K, Ganeshpurkar P, Khatri S (2014) Analytical study on scheduling algorithms for real time static system. In: Proceedings of the international conference on recent advances in communication, VLSI and embedded systems, pp 179–182
15. Audsley NC, Burns A, Richardson MF, Wellings AJ (1991) Hard real-time scheduling: the deadline monotonic approach. In: Proceedings of the 8th IEEE workshop on real-time operating systems and software, pp 133–137
16. Audsley NC (1990) Deadline monotonic scheduling. In: Technical report, department of computer science, University of York
17. Silberschatz A, Galvin PB, Gagne G (2012) Operating system concepts, 8th edn. Wiley, Hoboken

Chapter 20
Behavioral Analysis Using Gamification and Data Visualization

Lavina Kunder and Siddhaling Urolagin

1 Introduction

Machine learning or ML can be defined as a study of algorithms and different statistical models that can be used by multiple computer systems to perform tasks without explicit instructions being given by a user. It allows the system to draw information from data and patterns created while the system was in use earlier. This allows the computer system to "learn" and effectively take decisions on their own.

ML is divided into three main subtypes: mainly supervised, unsupervised, and reinforcement Learning. These techniques are applied onto various real-life situations, for example, Traffic alerts, product recommendations, etc. Although there are many such applications, there are a plethora of shortcomings as well, like the lack of data and resources, privacy problems, inconsistent tools and algorithms, etc.

Gamification may be defined as the application of game design elements and principles into non-game concepts. It is essentially a set of processes that are used to solve problems in different scenarios by applying characteristics of game elements. Gamification improves user engagement, productivity, learning, ease of use, evaluation techniques, etc. It also helps in identifying and profiling the behavior and the types of decisions they take in different circumstances. Gamification can be applied to various fields in daily life such as in product distribution, recruiting, marketing, lifestyle, and health. Some of the hindrances to these applications are the lack of planning and strategy, bad processes, poor design, and unrealistic expectations.

Machine learning being integrated into gamification creates a plethora of new opportunities and eliminates their individual shortcomings. Algorithms that form an

L. Kunder (✉) · S. Urolagin
Department of Computer Science, Birla Institute of Technology and Science—Dubai Campus,
Dubai International Academic City, Dubai, UAE
e-mail: lavina.kunder@gmail.com

S. Urolagin
e-mail: siddhaling@dubai.bits-pilani.ac.in

inherent part of every game code can be designed so as to optimize their "learning behavior" with different and repeated interactions with the user. This allows datasets to be formed that provides information on the behavior of the user in different scenarios and how he/she tackles them. This in turn also allows the game to develop itself and become more complex; therefore, making it a harder to solve the same problem in the same way, encouraging users to take other different approaches. While this occurs repeatedly, it helps us in profiling the user and his behavior while also allowing us to develop better games with better strategies and ideologies. This is what we aim to achieve through our research. In Sect. 2, we have the literature survey that explores the previous developments made in this field and conclusions drawn on a different research papers written to understand these topics better. Section 3 discusses the methodology of the game, how it was developed and what steps were taken so as to be able to understand the game as well as the dataset it generates better. In Sect. 4, we use graphs to better understand the dataset generated by the games. This helps in identifying patterns in which different people play. Section 5 deals with the experimental results of the games and how the players were profiled. It also explains the machine learning techniques and how they could be further improved. Section 6 draws a final conclusion to the whole research paper.

2 Literature Survey

Machine learning being integrated into gamification creates a plethora of new opportunities. Algorithms that form an inherent part of every game code can be designed so as to optimize their "learning behavior" with different and repeated interactions with the user.

This allows datasets to be formed that provides information on the behavior of the user in different scenarios and how he/she tackles them. This in turn also allows the game to develop itself and become more complex; therefore, making it a harder to solve the same problem in the same way, encouraging users to take other different approaches. While this occurs repeatedly, it helps us in profiling the user and his behavior while also allowing us to develop better games with better strategies and ideologies. This is what we aim to achieve through our research.

In [1], we find that gamification has shown pattern of supporting user engagement, increasing user activity, encouraging social interactions, etc. It enables methods for generating leads in marketing and improving customer engagement. Startups have also come up with the strategy of providing services focused on adding gamified layer to core activities of their client companies (e.g., CodeAcademy: Uses game like elements to helps users learn how to code).

Delving into the process of designing an algorithm for personalized gamification, [2] focuses on automating multiple personalization tasks and strategies that help in compiling a profile of the player of the game. Such algorithms are widely used in the education system and to motivate students and encourage more interaction between fellow students. It also helps in assessing student personalities and their strengths and weaknesses in certain subjects.

Through [3], we understand the importance of machine learning and gamification in the recruiting industry. The continued interest in gaming techniques is determined by the wide research done in order to find the solutions of specific special and economic problems of an organization. It also helps in increasing the level of competitiveness, training and development of staff, and most importantly, in hiring new recruits that fit the ideas and working of the company. It also specifies how gaming may be used to understand buyers as well as attract new ones. But the most important aspect of gamification, as per the paper, is stated to be is ability to increase labor productivity which directly affects labor costs and speed of production.

The need for gamification in the industrial sector is explored through [4]. Since most modern companies focus not only on the qualifications of the employee but as well as their skills and qualities like collaboration and teamwork. These skills and qualities cannot be easily judged through an interview. This is where gamification comes in. How the possible recruit plays the game with respect to his team members or quizzes based on logical reasoning can easily help in identifying these skills and qualities.

In [5], the paper focuses on highlighting the response of people entering the labor market and their attitude toward being recruited using gamification. Most people have replied with positive responses as they declare it as an attractive and interactive form of recruitment. It focuses not only on their qualifications but also on their skills, knowledge, and decision-making capabilities. It also proves to be a single step platform through which many judgements of the recruit can be made through one iteration of the game as opposed to standard interviews that have multiple rounds before deciding on the final candidate.

Providing a deeper look into gamification and machine learning [6] explores how they work together in the recruiting field (HR). These tools not only help in identifying the candidates with the best skill set but they also enable the recruits understand themselves better and find which skill set they would have to improve on. It also enables them to business reality and organizational dynamics amongst their own colleagues. The paper provides examples on how people hired through gamification techniques are found to be more prepared and able to handle tough decision-making scenarios than those employees recruited in traditional ways as they have been exposed to such scenarios before and therefore understand the dynamics.

Through [7], we explore how machine learning and gamification are linked through a game that based on basic classification principles. Machine learning algorithms require examples that are labelled to be trained and tested, but labelling is costly and also requires a large amount of time. To counteract these disadvantages, gamification is applied to the classification problem to create labelled datasets at affordable costs. This explains how both these ideas go hand in hand to reduce costs, amount of work, and automate the system.

The world of adaptive gamification is explored through [8]. Machine learning methods are used to collect task information. The player's facial expression data is also collected to predict his/her performance on a gamified task. The new data collected is used to update the model in every iteration thus improving the analytic abilities of the model. Data is gathered in discrete time intervals through quizzes and questionnaires as current gamification techniques do not have the ability to dynamically capture data of a player's interaction with an application.

In [9], we deal with how machine learning techniques provide a way to improve the behavioral and interaction abilities of a computer-controlled game by automating certain processes and behavior, thus enhancing its capabilities and allowing the developer to create more engaging and challenging games. While most of the focus lies on statistical machine learning techniques, application of neural networks withing digital games has shown varying amounts of success. The paper also explores multiple machine learning techniques applied to gaming environments and the results that they have given. This enables a deeper understanding how these algorithms function within a game and how different they are from one another.

Motivational outcomes of gamification are addressed through [10]. The concept of need satisfaction was applied to the gamification successfully to show that it is a two-dimensional construct since the game design elements affected only certain parts of the motivational autonomy. It also states that it is important to investigate and constantly improve game design elements on the surface level so that they may be manipulated and functional independently of each other.

While designing games, algorithms can be designed to as to allow the machine player to adapt to the opponent player's moves and exploit his/her weaknesses. This concept is explored further in [11], where models of the behavior of the human player are created and algorithms are used to enhance its performance. Furthermore, this profiling method can be plugged easily into multiple platforms which allows easier data collection and player profiling. This concept is also employed on a small scale in the Tic-Tac-Toe game developed in tandem with this research paper.

With data from 32 studies, [12] focuses on providing a gamification model for developing gamified applications. Gaming elements were also identified which may be adopted and applied to present methods of recruitment and job training processes. This will help developers choose specific elements based on the environment and the context of the job. The paper also focuses on understanding the importance of gaming elements that focus on motivating users, so that recruits as well as present employees focus on doing better on their day-to-day work life. In [13], machine learning algorithm is applied to first person shooter games to create the best possible experience for a player playing against a bot. The algorithm learns combat behavior from an expert and then this is applied to the bot playing against the player. Multiple learners were used to see which on the them would fare the best. It helps in providing varying levels of difficulties to the player and based on new game iterations, the learning could be improved by adapting new techniques from each new player.

Gamification and machine learning have come a long way even in the education sector explored further in [12]. It has been observed to influence efficiency amongst students and also helps professors and teachers understand their students and their approaches to situations better. Learning mechanisms are employed within the game to allow students to understand them easily and at a much faster pace. Although gamification and machine learning initiatives have been carried out at different education levels, a strong prevalence exists mostly at the universities.

In [14], machine learning in gaming is explored. Multiplayer games provided a wide range of network traffic with varying skills which generate large sets of data for machine learning algorithms to learn. This allows the machine to eventually imitate a human player in real time and create a more challenging gaming environment which may then be used in a variety of sectors. The algorithm learns how to adapt to fast reactions and develops planning and counter reactive abilities. This concept is sought to be applied to the games developed in tandem with this research paper as well.

3   Methodology

This section explores the method of development of the three games and the datasets that were generated by them. It helps in understanding the basic rules and nooks of the game to get a better grasp on the ideas given by data set visualization in the next section.

3.1   Hangman

It follows the basic approach wherein you have to guess the word provided by the computer in a limited number of turns.

There are a number of advantages to playing games like Hangman like improved vocabulary, increased thinking power, and faster response time. These notions can be exploited while creating the player profile from the data.

Difficulty levels have been set into ensure there is variation in the data collected and to see how the players fare individually within those difficulty levels. First; Easy with Word Length < 8 and Turns Allowed = 12. Second; Difficult with Word Length > 8 and Turns Allowed = 10.

The dataset records four main aspects. First: Difficulty which is to analyze which level is most preferred by the players. Second: Result which depending on the difficulty tells us whether the player lost or won. Third: Total attempts which gives the number of total attempts to guess the word correctly or till the turns ran out and lastly: Failed attempts which gives the number of incorrect guesses to reach the final word or terminating the game before reaching the final word.

3.2 Tic-Tac-Toe

Tic-tac-toe is a paper-and-pencil game for two players, X and O, who take turns marking their respective characters onto a 3 × 3 grid (9 spaces). The player who succeeds in placing three of their marks in a horizontal, vertical, or diagonal row is the winner.

The game requires only basic understanding in order to find an optimal strategy. The optimal strategy being that if both players play their best game, then the game ends with a TIE.

The game has been looped to allow more collection of data. The player can choose his/her own character [x or o]. To ensure an even distribution of data to analyze the game better, the player that goes first is chosen at random by the computer.

The dataset has been divided into two separate blocks for better data mining.

Records the winner and the first player for each game played. If the game is played more than once by a player, multiple entries are made. In the FIRST dataset, there are 57 different entries which records the total number of times played by the player and whether the result of those games was a win, loss, or a tie. The SECOND dataset has the results for 57 games played by 23 players.

3.3 Logical Reasoning Quiz

This game is constructed in a quiz like format that contain APTITUDE questions which require a certain level of skill, logic, and sense of analysis to arrive at the correct answer.

Most of the questions are constructed on basic concepts and the others are out of the box thinking ones.

These games are mainly used by companies while RECRUITING as a part of the interviewing process to check if the applicants have basic cognitive and analytical skills along with their education to solve daily problems and tasks.

It helps in profiling the personality of the applicant and understand them better as people so as to provide a base for the hiring process.

The quiz has ten questions which are split into two main types which are verbal reasoning, which is the ability to extract useful information and implications in a problem, analyze and use them to find the solutions. It helps in understanding the ANALYTICAL skills of the applicant. Then, verbal reasoning which is the ability to evaluate figures, shapes, series, etc., and find hidden information that can be used to find the solution. It helps in understanding the COGNITIVE skills of the applicant. Each of these two types has five questions each from different areas within them so as to help in collecting data with a wider reach into the personality of the applicants.

The dataset has entries from 57 players. It collects the total score based on how many correct or wrong options were chosen. With the correct options chosen, it also collects information on how many of verbal and non-verbal reasoning questions each were answered correctly. This allows us to understand whether the applicant has better cognitive skills or analytical skills. The capability column examines exactly that and has an extra provision if the applicant displays a balanced aptitude for both the skills.

4 Data Set Visualization

Through data mining techniques, several graphs were obtained from the dataset, which will help us better understand the players and also find methods on improving the game.

4.1 Hangman

This graph is to understand the distribution of the length of the words available. While playing the game, a word is picked randomly from the data bank of words based on the difficulty level chosen. There is a higher concentration of words of the length 5–12. This is because this range falls in both the EASY and DIFFICULT categories and shall therefore ensure that the words are not repeated in multiple executions of the game (Figs. 1 and 2).

Total attempts and failed attempts take a dip when a game is won and vice versa when a game is lost. With this, we are able to understand that the computer progressively gives harder words to guess which leads to the increase in the failed and total

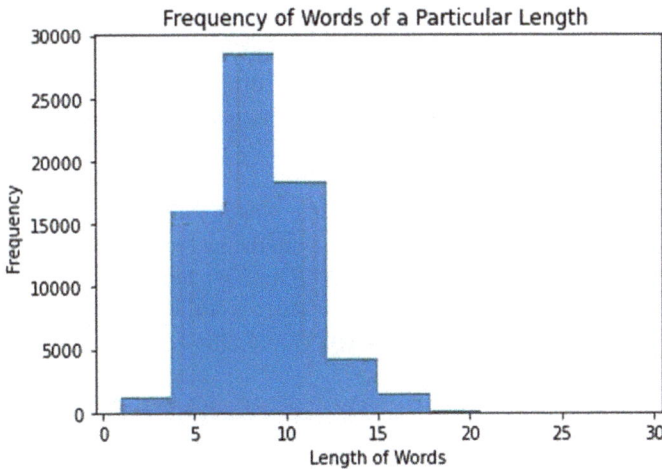

Fig. 1 Displays frequency of words of different lengths

RESULT as a function of TOTAL and FAILED ATTEMPTS

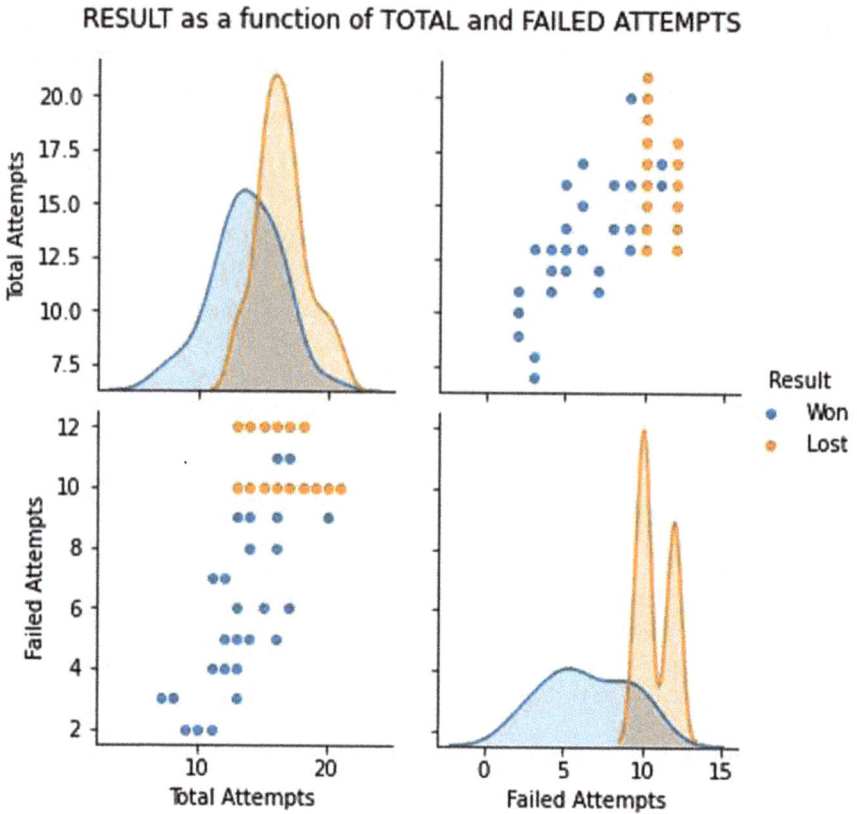

Fig. 2 Results shown as a function of total and failed attempts

attempts. While some of these games are won, most of them are lost which provides the computer with the insight of what length and difficulty of the word would make the game challenging as well as enjoyable. In subsequent iterations of the game, the words chosen are challenging to guess in both difficulty levels offered in the game (Fig. 3).

This boxplot shows the range of each of the bins in failed attempts. The bins are as follows. First, low spanning from 8 to 16 with the median at 12 followed by Med spanning from 11 to 17 with the median at 14. Third, we have high spanning from 13 to 20 with the median at 16 and lastly, there are two outliers, one located at a value less than 8 and the other at 21. Total attempts and failed attempts are shown to be directly proportional.

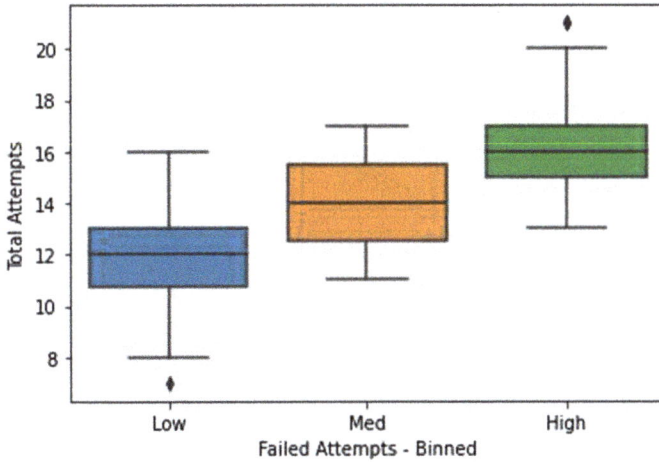

Fig. 3 Failed attempts binned into three different categories

4.2 Tic-Tac-Toe

These graphs show the percentage frequency distributions for the number of times played, wins, losses, and ties for different players as a whole. While the graph for ties is equally distributed, the graphs for wins and losses lean more toward the lower side. This means that most players have an equal number of ties with the computer out of the total times they have played. Wins and losses, on the other hand, seem to differ from player to player with some players winning more games than losing and the others, vice-versa. Since ties are the ideal outcome, this shows that most games are balanced between the computer and the player (Figs. 4 and 5; Table 1).

This graph shows that most games played end up in a tie. Thus, most games are ensured to be ideal as discussed earlier. This attributes to the fact that both the player and the computer mostly prove to be of equal capabilities showing that algorithm can be capable of learning from its opponents with each iteration of the game eventually proving to be an extremely hard opponent to defeat (Fig. 6; Table 2).

The computer also ensures that all games are fair to allow a better dataset to be collected for analyzing. Both the player and the computer have gotten equal chances to start first in all of the games. This shows that the game is impartial and assigns equal opportunities. This also allows the dataset to be unbiased which in turn leads to easier analysis and better grasp on the final results (Table 3).

The columns in the dataset were grouped using the played first column to give a better understanding of the data. The counts for each possible outcome are given in the table giving a clear-cut idea on the scenarios that resulted in a win, loss, or a tie between the computer and the player (Fig. 7).

Derived from the table discussed earlier, this graph provides a pictorial representation of the number of games each player wins, loses, or ties depending on the factor

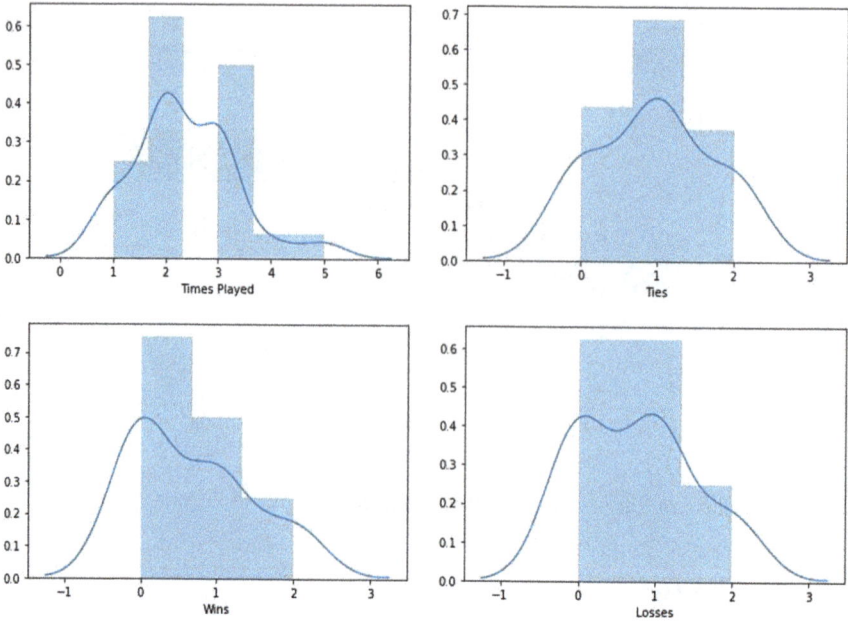

Fig. 4 Frequency distributions for number of times played, wins, losses, and ties

Fig. 5 Distribution of winners

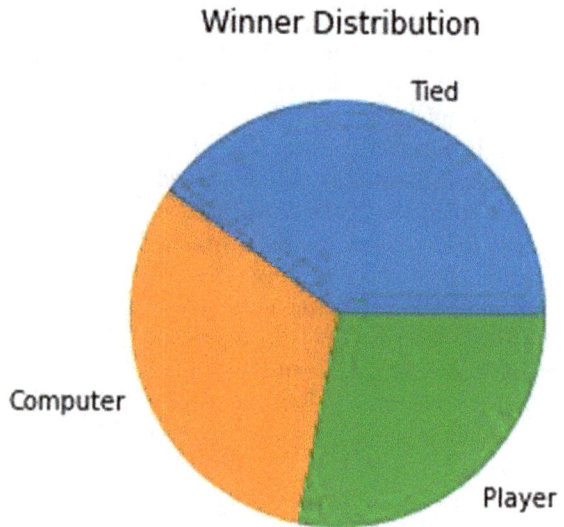

Winner Distribution

Table 1 Winner

Tied	23
Computer	18
Player	16

Fig. 6 Distribution showing
who played first

Distribution of who Played First

Table 2 Played first

Player	29
Computer	28

Table 3 Grouped data

Played first	Winner	Name: count
Computer	Computer	14
	Player	4
	Tied	10
Player	Computer	4
	Player	12
	Tied	13

of who played first. Both the computer and the player take the lead in winning the games when they are the ones to start. It is interesting to note that when the player plays first, there are a lot of games that also end up being a tie. This shows that the computer also seems to learn from its past mistakes and work toward a better play.

4.3   Logical Reasoning Quiz

This graph shows the distribution of questions answered correctly by all players. This includes both the verbal and non-verbal reasoning questions. It is observed that

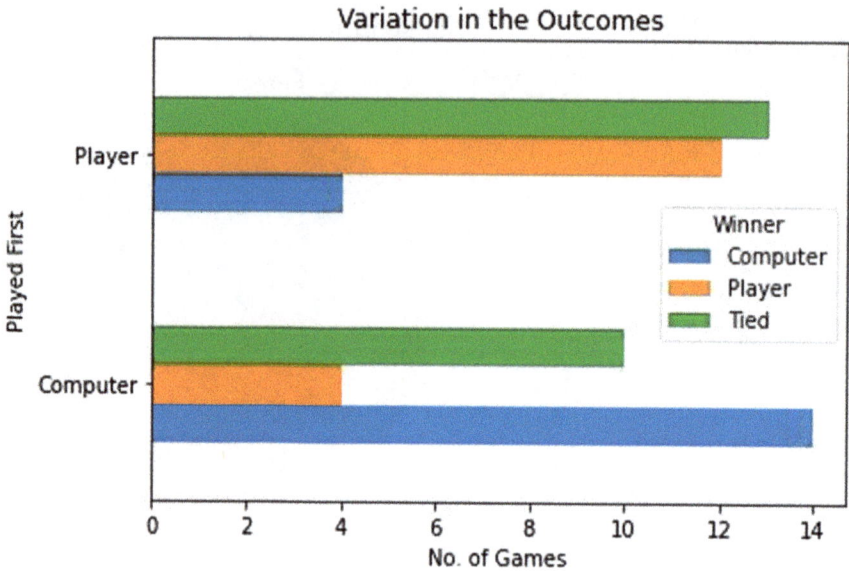

Fig. 7 Result variations depending on who played first

most players have answered up to six questions correctly closely followed by eight questions answered correctly out of the ten total questions (Figs. 8 and 9).

This graph shows the distribution of questions answered incorrectly by all players. This includes both the verbal and non-verbal reasoning questions. It is observed that

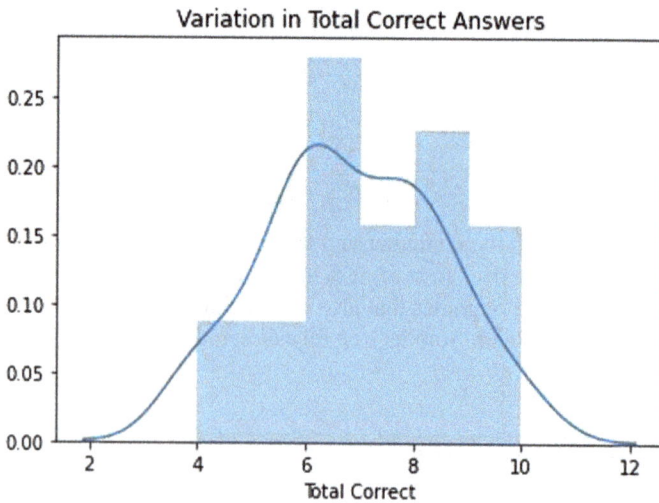

Fig. 8 Correct answers variation

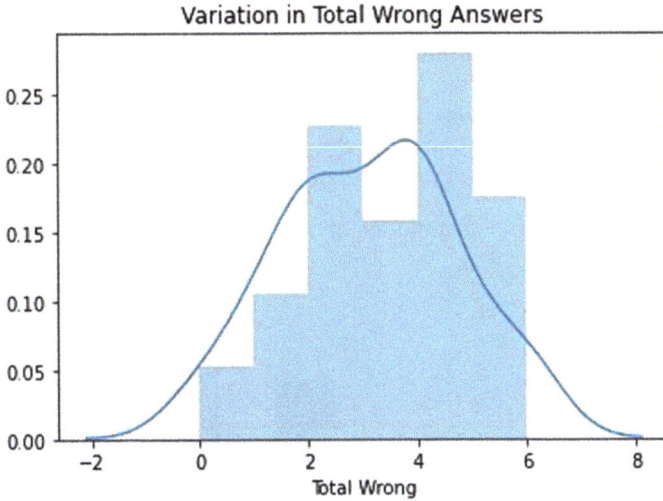

Fig. 9 Wrong answers variation

most players have answered up to four questions incorrectly closely followed by two questions answered incorrectly out of the total ten questions (Tables 4 and 5).

This graph shows the distribution in the non-verbal scores which is used to analyze cognitive skills of the players. Thirteen people have answered all of them correctly which averages out to 22.80% of all the 24 players. This graph shows that most players have a mediocre to excellent cognitive skills with very few players faring relatively bad in this section of the quiz.

This graph shows the distribution in the verbal scores which is used to analyze analytical skills of the players. Only seven people have answered all of them correctly

Table 4 Non-verbal score distribution

Score	Frequency
3.0	19
4.0	16
5.0	13
2.0	6
1.0	3

Table 5 Verbal score distribution

Score	Frequency
3.0	19
4.0	16
2.0	13
5.0	6

which averages out to 12.28% of all players. None of the players have answered all five correctly and most of the demographic falls below the average score. This shows that the analytical section of the quiz is much harder to answer than the cognitive section. This section requires thorough understanding of the question to be answered correctly (Figs. 10 and 11).

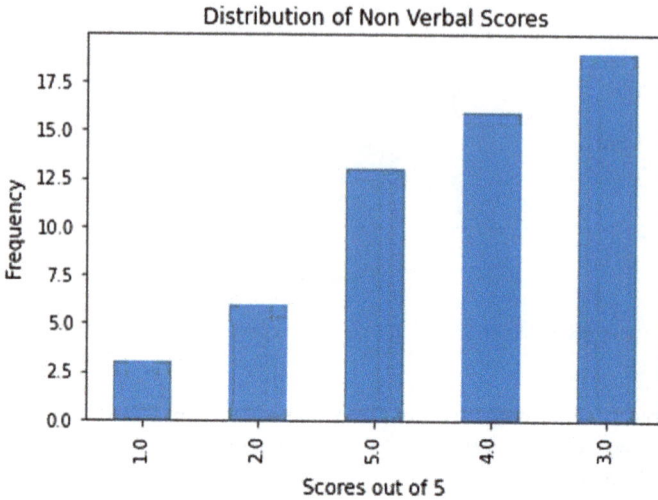

Fig. 10 Non-verbal individual score frequency

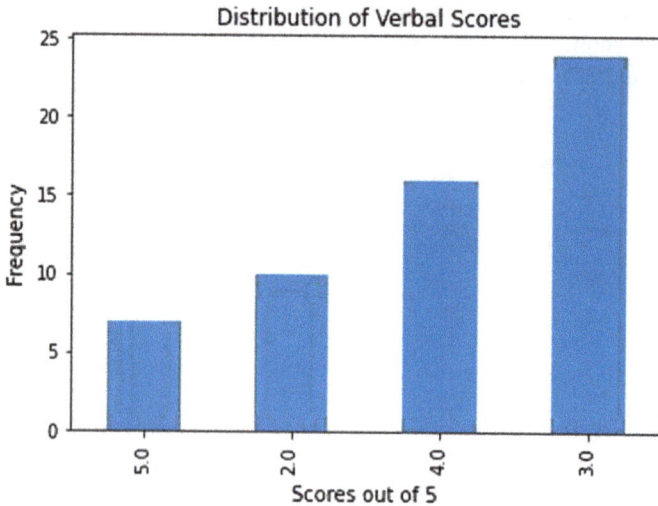

Fig. 11 Verbal individual score frequency

5   Experimental Results

Based on the analysis of the dataset, an interpretation is drawn with respect to the profile of the players and for further development of the games. They are explained in detail for each of the games with two important divisions: Player profiling and machine learning and further improvements.

5.1   Hangman

Player Profiling

Players prefer HARD mode over EASY mode. There could be various reasons for this. Mainly being better vocabulary amongst people now due to increased social interactions, better education systems, etc. Although HARD mode is chosen often, they are usually unable to correctly guess the final word. So, a two different profile can be created, them being (a) Choose Hard and can correctly guess the word AND (b) Choose Hard but are unable to arrive at the correct word (Fig. 12).

As a result of more players choosing the harder level, the number of games lost is also comparatively larger than the number of games won. This again helps in differentiating the vocabulary level amongst those who did choose the harder level, but could not overcome it.

Machine learning and further improvements

With increase in the density of total and failed attempts, it is clear that the computer is focusing on giving comparatively harder and longer words in both the difficulty levels. Over time, a more refreshing balance between a challenging and an enjoyable game can be established through collection of more datasets and identifying the trends

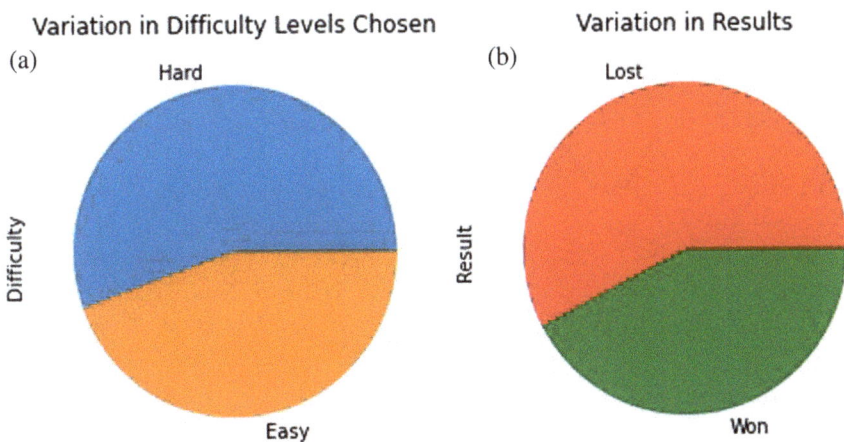

Fig. 12   **a** Distribution in the difficulty levels. **b** Distribution in the final results

in those datasets. As an improvement, by analyzing player attempts and adding an age factor, we would be able to obtain a more established dataset. This would enable us to improve the game by providing more relevant words based on the age group that chooses to play.

5.2 Tic-Tac-Toe

Player Profiling

This graph aims at showing the performance of each player based on the total number of games they played and out of those the number of times they won, lost or tied with the computer. Both the computer and the player are equally good at the game by analyzing the distribution of wins, losses, and ties. Players usually take the lead when they start to play first. This establishes the fact that even though they know the game well, they take a hit when the computer starts. A better understanding of the game could help them achieve the ideal strategy. Players mostly choose to play the game only a few times before they stop. They have been noticed to continue playing multiple iterations only if they are winning (Fig. 13).

Machine learning and further improvements

When the player usually starts the game, the computer loses more frequently. More training based on datasets and machine learning can help the computer prove to be a worthy opponent. With the high number of wins in the computer's bank when it starts playing proves that it has used machine learning to better the playing experience.

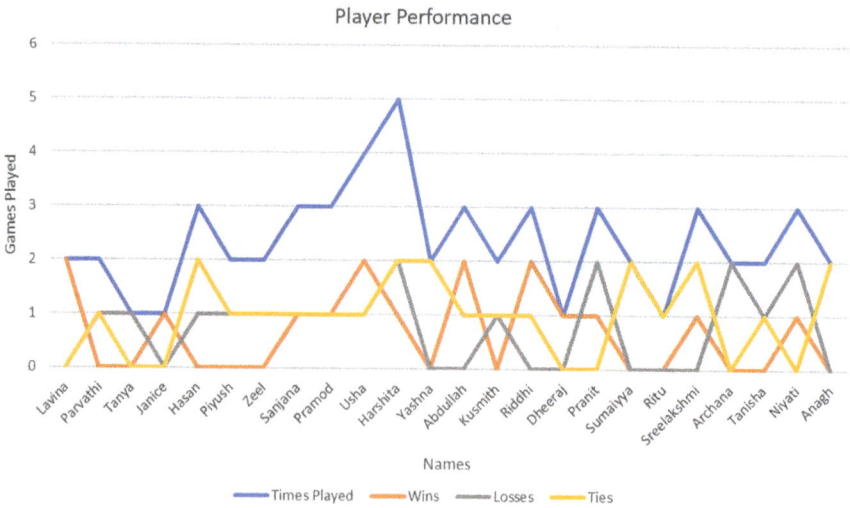

Fig. 13 Plot showing the individual performances

Fig. 14 Capability variation

Variation in the Capabiity of the Players

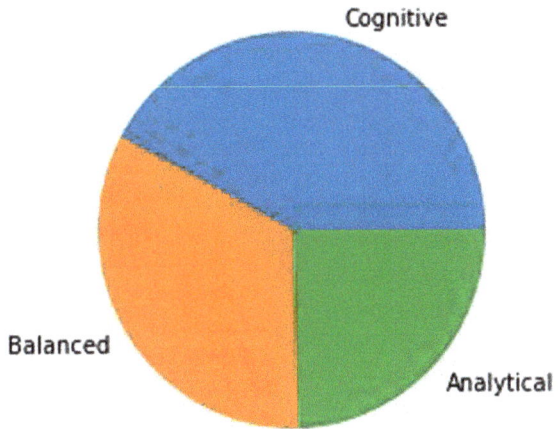

5.3   *Logical Reasoning Quiz*

Player Profiling

This graph shows that variation in capabilities amongst the 57 players of the game. Through the pie chart, it is easily noticeable that most players fare well in the non-verbal reasoning questions and get better scores than they do for verbal reasoning questions. This shows that most players have better COGNITIVE skills followed by a BALANCED skillset and then ANALYTICAL skills. Verbal questions are found to be harder to find accurate solutions for as often the options given are similar to each other and sometimes even misleading. This dataset therefore shows consistency with known facts.

There are also players who show a balance for both analytical and cognitive Skills. Helps in understanding the player's thinking capacity beyond their education. This helps companies in the recruiting process. They are able to understand if the applicant will be a good fit with the existing employees, their knack for solving simple day-to-day problems and their thinking processes (Fig. 14).

6   Conclusion

Machine learning and gamification have come a long way together. But they still have more potential and new heights to reach.

Together they will be able to create more immersive experience not only in the video game world but also, in those sections of society where gamification has proved to be of abundant help. Especially in fields like recruiting, healthcare, and education.

In all the games, more levels and questions can be included. Depending on previous answers, and how the player is faring in general in the game; the questions can be designed to be harder or easier to help profile the player more easily. Interactive questions that include puzzle solving, maze solutions, etc., can be set and timed so as to analyze the time taken to complete these tasks and then be compared with the other players.

Taking the games a step further, machine learning can be automated using better techniques and algorithms in comparison to the one used right now so as to give a better and easier approach to profiling the player. This will also help in developing models which will work more efficiently to give the desired results faster. All of the three games developed allow a peek at the decision-making skills, knowledge, and capability of the player in one way or the other. This information can be used in various fields such as recruitment, education, etc. This hints to the fact that machine learning and gamification sync together to provide a plethora of uses and while they have come far in the last decade, they still have the potential to be developed further to make daily human operations and work much quicker and easier.

References

1. Hamari J, Koivisto J, Sarsa H (2014) Does gamification work?—a literature review of empirical studies on gamification. In: 2014 47th Hawaii international conference on system sciences. https://doi.org/10.1109/hicss.2014.377
2. Knutas A, Roy RV, Hynninen T, Granato M, Kasurinen J, Ikonen J (2018) A process for designing algorithm-based personalized gamification. Multimedia Tools Appl 78(10):13593–13612. https://doi.org/10.1007/s11042-018-6913-5
3. Joseph MM (2017) An investigation into gamification as a tool for enhancing recruitment process. Ideal Res 3(1)
4. Abdullah H, Younis S, Hossan C (2015) Gamifying recruitment process: a qualitative study aimed at bridging the skills gap in UAE jobs market. J Bus Manag 4(2):7–27
5. Krasulak M (2015) Use of gamification in the process of selection of candidates for the position in the opinion of young adults in Poland. Jagiellonian J Manag 1. https://doi.org/10.4467/245 0114xjjm.15.015.4472
6. Perinot C (2015) Gamification in the field of human resource management—gamified solutions for recruitment. In: Final Thesis for Master's Degree programme in economics and management of companies, Ca' Foscari University of Venice, Venice, Italy
7. Di Nunzio GM, Maistro M, Zilio D (2016) Gamification for machine learning: the classification game. In: Proceedings of the third international workshop on gamification for information retrieval co-located with 39th international ACM SIGIR conference on research and development in information retrieval (SIGIR 2016), Pisa, Italy, pp. 45–52, 21 July 2016. http://ceur-ws.org/Vol-1642/paper7.pdf
8. Lopez C, Tucker C (2018) Towards personalized adaptive gamification: a machine learning model for predicting performance. IEEE Trans Games 1–1. https://doi.org/10.1109/tg.2018.2883661
9. Galway L, Charles D, Black M (2008) Machine learning in digital games: a survey. Artif Intell Rev 29(2):123–161
10. Sailer M, Hense JU, Mayr SK, Mandl H (2017) How gamification motivates: an experimental study of the effects of specific game design elements on psychological need satisfaction. Comput Hum Behav 69:371–380. https://doi.org/10.1016/j.chb.2016.12.033

11. Aiolli F, Palazzi CE (2008) Enhancing artificial intelligence in games by learning the opponent's playing style. In: IFIP International Federation for Information Processing, vol 279, New Frontiers for Entertainment Computing; Paolo Ciancarini, Ryohei Nakatsu, Matthias Rauterberg, Marco Roccetti. Springer, Boston, pp 1–10
12. Araya R, Ortiz EA, Bottan NL, Cristia JP (2019) Does gamification in education work?: Experimental evidence from chile. https://doi.org/10.18235/0001721
13. Geisler B (2004) Integrated machine learning for behavior modeling in video games. In: Fu D, Henke S, Orkin J (eds) Challenges in game artificial intelligence: papers from the 2004 AAAI workshop. AAAI Press, Menlo Park, pp 54–62
14. Thurau C, Bauckhage C (n.d.) Exploiting the fascination: video games in machine learning research and education. In: Proceedings of the 2nd international workshop in computer game design and technology
15. Obaid I, Farooq MS, Abid A (2020) Gamification for recruitment and job training: model, taxonomy, and challenges. IEEE Access 8:65164–65178. https://doi.org/10.1109/access.2020.2984178

Chapter 21
Smart Agricultural Monitoring and Decision Support System

Shahnaz Akhtar Hussain and Papu Moni Saikia

1 Introduction

Agriculture is the most important part of the Indian Economy. Indian agriculture sector explains 18% of India's Gross Domestic Product (GDP) and provides employment to 50% of the countries population. India is the world's largest producer of pulse, rice, wheat, spices and also advanced in many organic products [1]. But, continuous growing population in India brings challenges to produce more and with large number of fields being compromised to industries and housing needs. Deep researches have been going on which helps the researchers to produce new varieties in order to increase the yield. Use of modern techniques can help the farmers to not only remotely monitor their crop but also take timely action. Therefore, the future of farming in agriculture sector in India should be made fewer; educated farmers should employ highly efficient farms to produce high-value goods with advanced technology. Amidst the pandemic season in the whole nation, the country has brought the economic activity to a near halt. But the government is expecting that the only the agriculture sector could be a silver lining for the Indian Economy at this difficult time as it is estimated to grow at a rate of 3% for the year 2020–21 according to NITI Aayog. Therefore, agriculture sector must adapt the concept of smart farming or precision agriculture with a proper monitoring of the real-time data intelligently for speed in productivity with less chance of loss and damage. Smart agriculture provides more strength to the agricultural industry by introducing the advanced technology that includes big data, the cloud, and Internet of Things for tracking, monitoring, automating, and analyzing operations. Also, smart farming is sensor monitored and software managed [2]. To fulfill the demand of high gain in agriculture sector, the

S. A. Hussain (✉) · P. M. Saikia
Jorhat Engineering College, Jorhat, Assam, India
e-mail: shahnazahussain@gmail.com

P. M. Saikia
e-mail: papums123@gmail.com

© The Author(s), under exclusive license to Springer Nature Singapore Pte Ltd. 2021   267
X.-Z. Gao et al. (eds.), *Applications of Artificial Intelligence in Engineering*, Algorithms for Intelligent Systems, https://doi.org/10.1007/978-981-33-4604-8_21

agricultural companies should move toward the modern technologies that provide more accuracy, less time consumption, low cost, and less human interactions [3]. On the other hand, Decision Support System (DSS) plays a vital role in smart agriculture. DSS is completely a software-based system that allows the user to analyze and collect a variety of data sources to help the complexity of crop production. The future of DSS in smart agriculture will become more robust with increase connectivity of technologies and addition of smart devices on the field that triggers the storage and collection of data [4].

2 Literature Survey

Experts and researchers have made an analysis from the collected data to find a mutual relation between the environmental work and yield to produce a quality work. They give more intense on collecting the information of temperature and crop monitoring data from the environment and further made a proper analysis to reduce the crop losses and to improve the crop production. Some researchers have suggested in the year 2016 to collect the real-time data from the agriculture production field and give an easy access to the user which also adds facilities like Short Messaging Service (SMS) alerts and suggestions on weather pattern, crops, etc., [5] Rayan Nurbadi proposed A Smart Agriculture System Based On Service-Oriented Architecture in the year 2018 with the expectation to answer the problems that exist in the agricultural sector in Indonesia, especially in rural area far from urban areas. And with the use of this application, they expected to improve the standard of living of rural farmers in various regions in Indonesia [6]. In the mean while, some other researchers have made a model in 2017 to help the farmers in monitoring the fields effectively and help them to take corrective measure for the protection and better yield of the crop not only from distant locations but also improves the quality of crops [7]. To erase the traditional methods of farming, an evolving technology combining the smart agriculture and automation, a MMS alerts-based system developed in 2017 interfacing a camera to capture the on filed pictures and sends it to the farmers [8]. Sushanth has described a system that is useful in water limited isolated geographical areas that help the farmers to monitor the real-time data and sensing the movement of any animals toward the agricultural field resulting to a SMS alerts to mobiles as well as sends a notification to the farmers smartphone, where a mobile application is been develop [9]. Patel et al. [10] develop a system that circulates the real-time data wirelessly to the receiver placed at a remote location. Though the agriculture is expanding in worldwide and leading its role, it is still immature in ecosystem control technology with lack in exact information and communication. To overcome those problems, another system describes with no restrictions of day and night to supply automatic water through irrigation process to the fields by avoiding the water wastage problem without human interaction [11]. In [12], the experts have made a survey on the changes of traditional technology into modern IoT technologies for smart agriculture and suggested IoT as a backbone of smart agriculture. Monitoring the real-time data is not enough to bring a rapid variation growth in agriculture filed. Some researchers utilize solar power for functioning the combination of IoT, data mining, and mobile application system for low-income farmers to use it in smart climate farming practice [13]. To increase the productivity, a decision support system is also required. Decision Support System (DSS) acts as a computer-based interactive system that helps the user/farmer in making firm decision for increase in productivity if the field and decrease in production cost with minimum environmental impacts [14]. So keeping the knowledge of rapid growth in crop production, a model in the year 2017 developed with an objective to assist the farmers for selection of a crop for cultivation mapping using different ground parameters like soil type, average weather required, required water consumption, temperature range, etc., [15].

After a detailed study of the existing works, this paper proposes a model for smart agriculture with less energy, less time consumption, and more accurate output results. This project work uses NodeMCU as a controller for controlling all the activities that take place on the board and keeps the system further use for IoT purpose with the knowledge of a few more sensors. The monitoring and decision together make the system more profitable and helpful to the farmers. This research work highly focuses on the high yield in agriculture field and helps in lowering the problem of poverty. This system in our project work is designed with more security to avoid unethical issues as a local server in the system designer is introduced. The database can be accessible only by the user and has an option to add any new plants, crops, etc., to the database that helps the farmers to monitor and take appropriate decisions immediately from a huge aggregate of flora or fauna.

3 Proposed Model Statement

This paper presents a proposed model for smart agriculture with a local server (instead of using a third party server) to develop the real-time monitoring system for the properties like temperature, humidity, carbon dioxide, and soil moisture content. The data obtained from the real-time monitoring will be compared to an existing database which is already pre-installed in the local server. With the implementation of microcontrollers and SMS-based alerts in the designing of the system, it helps the user to improve the quality and quantity of their farm yield with less chances of loss or damage. The system will also be possible to control various operations of the field remotely from anywhere, anytime by mobile as well as web application.

4 Proposed Model Architecture

The system architecture mainly consists of four main blocks—A power supply, a controller (NodeMCU), an array of sensors viz. temperature and humidity sensor

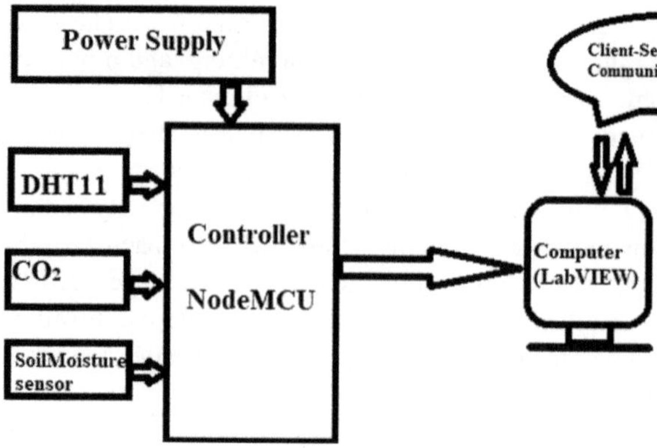

Fig. 1 Block diagram of the proposed model

(DHT11), capacitive type soil moisture sensor, carbon dioxide gas sensor (MQ-135), and a serial communication network as shown in Fig. 1. The software consists of LabVIEW acting as a server which includes setting up of the profile for the pre-defined database based on the seasons or on daily or weekly or monthly or on a yearly mode. The software has also been programmed to give a notification to the user whenever the physical parameters sensed are below the threshold value, and based on the users input, a control signal will be sent to the controller to take an immediate action. NodeMCU which acts as a controller controls all the activities that takes place on board. Sensors sense all the physical parameters and convert the analog value to digital value. Temperature and humidity sensor sense and measure the environmental temperature and humidity, respectively, on field. The soil moisture sensor is of capacitive type, measures the moisture content of the soil. The controller used here is NodeMCU which works on a 5 V dc supply. It has a built-in wi-fi chip which allows the system to connect the controller directly to the Internet or local client–server system when needed. A digital gas sensor is used to measure the carbon dioxide on filed. The analog and digital sensors are directly connected to the controller which gives a response to the environmental parameters in the form of analog or digital impulses. These voltage signals are further sampled and quantized in the NodeMCU itself and later transmitted over serial communication network in the form of digital data.

5 System Description

5.1 Circuit Diagram of the Proposed Model

As shown in Fig. 2, the system consists of a microcontroller, NodeMCU, DHT11 temperature and humidity sensor, a digital MQ-135 carbon dioxide sensor, and a

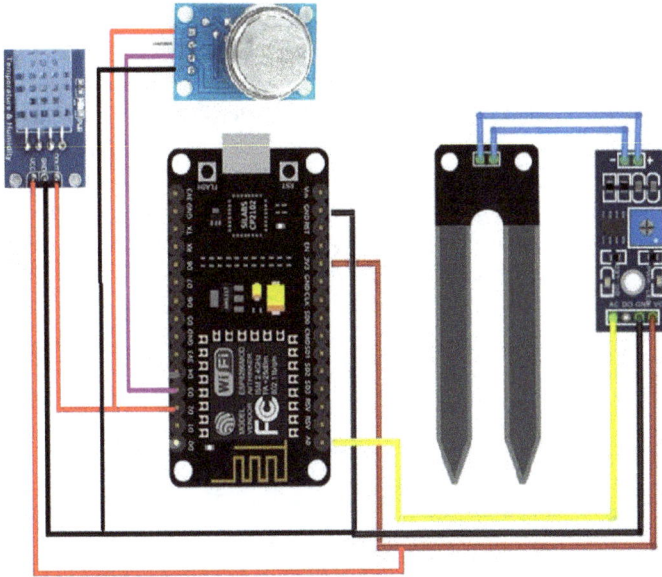

Fig. 2 Circuit diagram of the proposed model

capacitive soil moisture sensor. The supply to the controller is fed from an external 5 V 1 A dc source. The source also energized the array of sensors. The output of the DHT11 sensor is connected to pin number D_2 of NodeMCU. Output from the carbon dioxide sensor is fed to an analog pin A_0 of the controller. An output of the moisture sensor is fed to another analog pin of NodeMCU. Data from all these sensors are collected in an array and continuously fed to the controller. The controller saves and transmits the data over a serial communication when there is a pull up request from the server.

5.2   *Working Principle of the Proposed Model*

The model "Smart Agricultural Monitoring And Decision Support System" is based on the principle of client–server communication system. Here, LabVIEW is acting as a server. The hardware part consists of an array of sensors which act as a client in the overall system model. The number of client can be one or more depending on the user. But, in this model, a single-node client–server is being used. The environmental parameters are collected through the sensors and fed to the controller in a continuous manner. The server then pull up the request in the form of some characters like "a," "b," "c," and "d." These characters are then transmitted from the server to the client. The client receives the character and return the data of the associated sensors along with an acknowledgment request to the server. When there is a mismatch in the acknowledgment or if there is no data received from the sensor/client unit, then the server again sends the pull up request, and thus, this process continues (as shown in the flowchart Fig. 3). The data received from the signal is graphically represented in a Graphical User Interface (GUI) in the server unit called LabVIEW. The system model not only monitors but also records the data in an interval of 1 min (i.e., user-defined). These data in the form of an excel sheet can be recorded, saved, and analyzed when needed for further research and development. A website is also designed to monitor and control the data anywhere from the world using Internet explorer and makes the system more smarter and beneficial.

Fig. 3 Flowchart of the proposed model

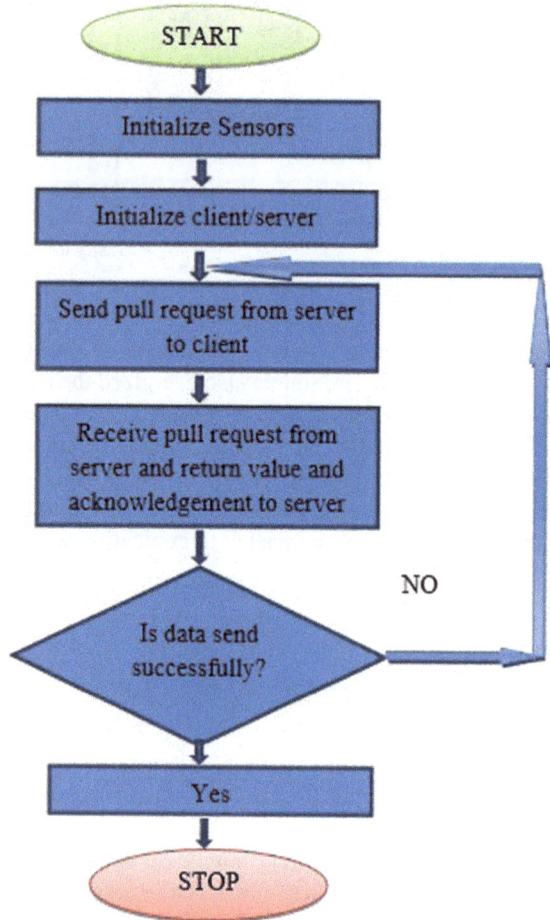

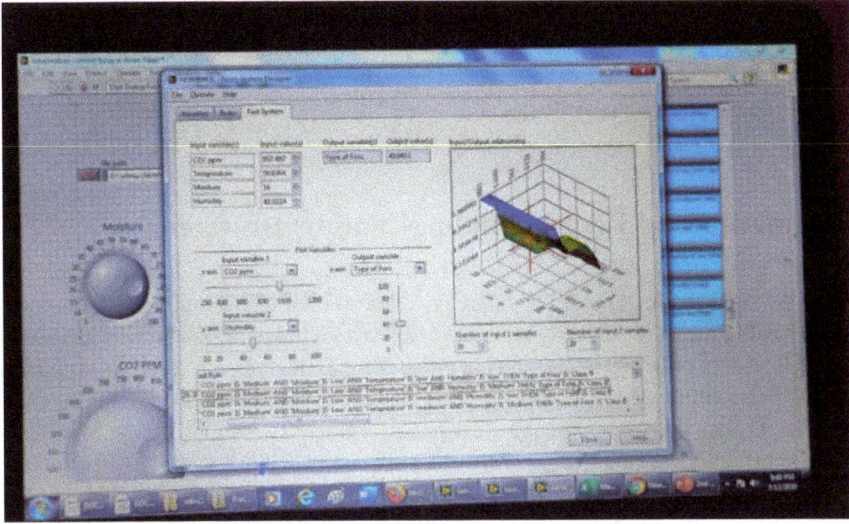

Fig. 4 A picture of the system designer in *XY* plotter using two input variables

6 Brief Study on the Parameters Used

The model uses four parameters viz. temperature, humidity, CO_2 level, and soil moisture content for the monitoring and control process. Monitoring always gives us a better understanding of the occurrence, distribution, and status of the crops, plants, trees, and shrubs. A good amount of all the parameters with proper plant monitoring helps to preserve the common species more common, and that rare, threatened and endangered species receive a continuous protection and assistance. If the necessary temperature, relative humidity, carbon dioxide content, and soil moisture requirements are not received from the environment, then the growth in plants may get affected or even it may die.

Temperature monitoring is one of the most important which is responsible for the growth of plants since temperature required for appropriate growth differs from plant to plant as some requires high, whereas some need low. Therefore, it becomes necessary to maintain a requisite temperature for all plants.

Humidity plays an another important factor to maintain the growth of plants. As the plants grow, they give out oxygen and process CO_2 which makes the greenhouse environment naturally moist and warm. Therefore, some plants require a high, some need less moisture level. Humidity level should be well monitored, controlled, and maintained since excess moisture in plants leads to its inappropriate growth and diseases which can be sometimes extremely harmful for the crops, plants, trees, and shrubs.

Soil moisture is a primary factor in field or farm productivity. So, providing the right and proper amount of water to the plants is quite necessary as low moisture may results in poor production of the yield.

In addition to temperature, humidity and soil moisture content, CO_2, are quite very necessary as it helps the plants for the photosynthesis process. Monitoring and adjusting the CO_2 level to a required level helps the plants to mature earlier and helps in harvesting more annually [10].

7   Experimental Results, Findings, and Its Discussion

Figure 5 shows the final model of smart agricultural monitoring and decision support system and is working perfectly and able to extract all the parameters from a jasmine flower. The data obtained in monitoring process are extracted and stored safely which will further compare with the database (pre-installed) created in the system designer. In order to test the system in *XY* test panel, we use two input variables; one is CO_2 (ppm), and the other is humidity (%). These two input variables can further be replaced by other input variables for test purpose. When we manually changes the input level of CO_2 and humidity, the system formulates its output according to the fuzzy rules developed by the fuzzy system designer. In the system, we have four different parameters; each parameter has three membership function as low, medium, and high. Therefore, total number of rules for formulating will be 3^4 and that equals to 81 number of rules. These rules are designed in the system designer in such a way that the output will be a single parameter with the type of flora, i.e., we are developing Multiple Input Single Output (MISO) system. The output obtained is the

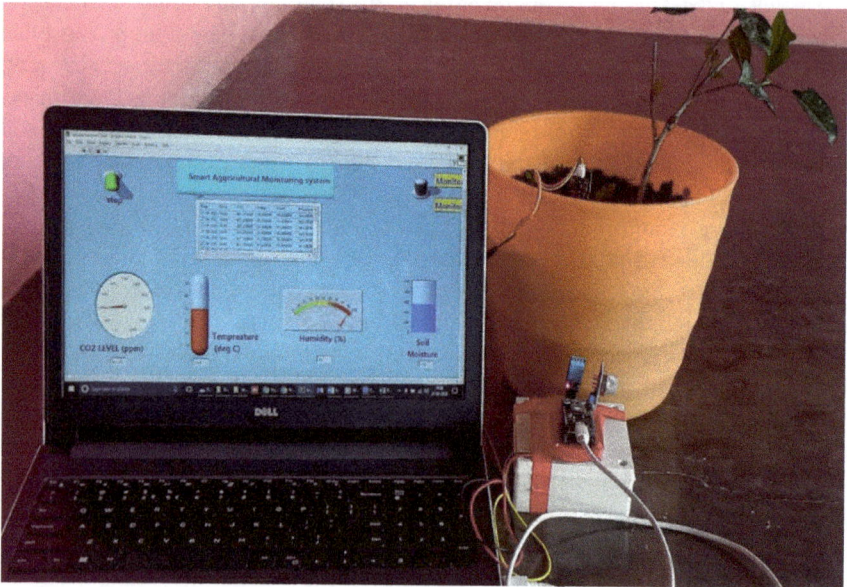

Fig. 5   Final working of the model

Table 1 Outputs observed from different samples

Observations	Input variable 1, temperature (°F)	Input variable 2, humidity (%)	Input variable 3, CO_2 (ppm)	Input variable 4, soil moisture (%)	Experimental result	Expected result	Remarks
1	30	50	NA	NA	Class A	Class A	Accurate
2	NA	NA	330	60	Class C	Class C	Accurate
3	60	NA	NA	40	Class E	Class B	Error
4	NA	NA	357	60	Class B	Class B	Accurate
5	75	NA	NA	80	Class A	Class A	Accurate
6	55	NA	1200	NA	Class E	Class E	Accurate
7	37	65	NA	NA	Class A	Class B	Error
8	NA	95	NA	50	Class C	Class C	Accurate
9	NA	90	350	NA	Class E	Class E	Accurate
10	35	NA	200	NA	Class B	Class A	Error

type of flora and is sub-divided into five membership function as Class A, Class B, Class C, Class D, and Class E. The table for every classes are pre-installed in the system designer. For example, the picture shown in Fig. 4 is indicating the condition of CO_2 and humidity level that points out the type of flora which is under class B. Similarly, by changing the inputs and level of the parameter, we can achieve another output, and each outputs are tabulated Table 1:

It has been observed that the system has been working perfectly with less amount of error. The system takes two parameter inputs at one time and gives out a single output which indicates the type of flora. For the observation, we had taken ten samples for our experiment, and the resulting output is coincides the theoretical output.

8   Conclusion and Future Work

Smart farming in agricultural sector helps to minimize the human efforts, simplifies techniques of farming since live knowledge of environmental parameters which are well experimented. Along with these features, smart farming can also help to grow the market for farmers with minimum efforts. This system model "Smart Agricultural Monitoring and Decision Support System" is fabricated and designed successfully for a more secure and safe use by the farmers since a local server is designed in the system itself without involving a third-party server which makes the system more reliable and suitable in research and development sector. As the system is also cheaper in cost and consumes less power, it is, therefore, more efficient and beneficial for the farmers to use in practical field.

In future, this system can be further enlarged and made IoT based with the implementation of a few more sensors. The database creation can be linked with a third-party server that enables to easily access the system anywhere from the world with a mobile application. It will also help the user with a more desired and productive result as well as less time consumption.

References

1. Agriculture Wikipedia homepage. https://en.wikipedia.org/wiki/Agriculture_in_India
2. Uva Dharini P, Monisha S, Narrmadha K, Saranya K (2018) IOT based decision support system for agriculture yield enhancements. Int J Recent Technol Eng (IJRTE)
3. Yadav Y, Mahjabin, Kumar L, Chaudhary S, Malik S (2018) An overview on internet of things (IoT) as a technological boon for sustained agriculture farming. Int J Agric Innovations Res
4. https://www.iof2020.eu/latest/news/2018/03/dss-for-agriculture#:~:text=DSS%20are%20soft ware%2Dbased%20systems,issues%20related%20to%20crop%20production
5. Patil KA, Kale NR (2016) A model for smart agriculture using IoT. In: International conference on global trends in signal processing, information computing and communication
6. Nurbadi R, Amyus A, Fajar AN (2018) A smart agriculture systems based on service oriented architecture. In: 3rd International conference on information technology, information system and electrical engineering (ICITISEE)
7. Jha RK, Kumar S (2017) Field monitoring using IoT in agriculture. In: 2017 International conference on intelligent computing, instrumentation and control technologies (ICICICT)
8. Prathibha SR, Hongal A, Jyothi MP (2017) IoT based monitoring system in smart agriculture. In: International conference on recent advances in electronics and communication technology
9. Sushanth G, Sujatha S (2018) IOT based smart agriculture system. IEEE
10. Patel NR, Behrani A, Kale PD (2014) Smart design of microcontroller based monitoring system for agriculture. In: International conference on circuit, power and computing technologies (ICCPCT)
11. Vanaja KJ, Suresh A, Srilatha S, Vijay Kumar K, Bharath M (2018) IOT based agriculture system using NodeMCU. Int Res J Eng Technol (IRJET)
12. Mekala MS, Dr. Viswanathan P A survey: smart agriculture IoT with cloud computing. 978-1-5386-1716-8/17/$31.00 ©2017 IEEE Xplore
13. Subash TD, Kumar S (2019) Smart agro farm solar powered soil and weather monitoring system for farmers. In: 2019 International multi-conference on computing, communication, electrical and nanotechnology, I2CN
14. Venkatalakshmi B, Devi P (2014) Decision support system for precision agriculture. IJRET
15. Khadke N, More D, Patil P, Patil H (2017) Agriculture decision support system using data mining. In: 2017 International conference on intelligent computing and control (I2C2)
16. Cambra Baseca C, Sendra S, Lloret J, Tomas J (2019) A smart decision system for digital farming

Chapter 22
Characterizing Shoulder Implants in X-Ray Images

Sudarshan S. Chawathe

1 Introduction

Shoulder implants are prostheses that are used to treat severely damaged shoulder joints in shoulder arthroplasty procedures [18]. This treatment requires several pre- and post-operative X-ray images of the affected area. Such images are also used in case of later complications or recurring problems. Automated analysis of such images to gather information such as the manufacturer and model of an implant has the potential to reduce the time and effort spent by physicians on these tasks.

This paper studies such automated analysis of X-ray images of shoulder implants. This analysis task is composed of several sub-tasks that form an approximate pipeline (with some stages benefiting from iteration), ranging from detecting an implant, isolating it from the rest of the image, and determining its orientation, through extracting features and identification of specific models. At the first consideration, it may seem that these tasks should be amenable to well-established image processing and classification techniques to enable, for instance, automated determination of the implant model from the X-ray image. In practice, however, several of the sub-tasks are very challenging due to the complexity of the imaged environment, composed of both natural and artificial elements, as well as imaging issues related to contrast, noise, and other artifacts.

S. S. Chawathe (✉)
School of Computing and Information Science & Climate Change Institute,
University of Maine, Orono, ME 04605-5711, USA
e-mail: chaw@eip10.org

© The Author(s), under exclusive license to Springer Nature Singapore Pte Ltd. 2021   277
X.-Z. Gao et al. (eds.), *Applications of Artificial Intelligence in Engineering*, Algorithms for Intelligent Systems, https://doi.org/10.1007/978-981-33-4604-8_22

1.1 Contributions

This paper's *contributions* may be summarized as follows:

- It motivates the task of automated analysis of X-ray images of the region surrounding a shoulder implant.
- It outlines a pipeline of image processing and classification sub-tasks for this purpose.
- Using concrete examples from a recently published dataset, it describes the characteristics of the data that pose challenges to several sub-tasks.
- It presents solutions to sub-tasks such as segmenting and orienting of implants within images.
- It describes methods for extracting features to enable classification of implants.
- It presents an experimental evaluation of this method applied to the task of determining the manufacturers of implants from their X-ray images.

1.2 Results

The key *results* may be summarized as follows:

- Careful composition of well-known image processing algorithms and similar careful tuning of their parameters yield methods that are effective for sub-tasks such as proper detection and orienting implants within images, despite the challenging aspects of the problem.
- Extracting simple grid-based features from processed images, followed by an application of well-established classification algorithms, yields accuracies comparable to those of similar methods in prior work. This result suggests that higher accuracies may be obtainable from the use of more sophisticated feature extraction following the earlier stages of processing described here.

1.3 Outline

The dataset of X-ray images that is the concrete focus of this study is described in Sect. 2 using a few illustrative samples that exhibit some of the challenging aspects. The early stages of the image analysis pipeline outlined above are addressed by Sect. 3. In preparation for classification and related tasks, feature extraction is outlined in Sect. 4. Classification of images by manufacturer of the shoulder implants within them is addressed by Sect. 5, which summarizes some results of an experimental evaluation. Related work is described in Sect. 6, followed by the a summary of the work in Sect. 7.

2 X-Ray Images of Shoulder Implants

This section describes the publicly available X-ray image dataset from prior work [17] used by this study. The dataset is composed of 597 X-ray images digitized as 250 × 250 grayscale JPEG images. It is plausible that higher resolution images would be available in some applications and that such resolution may ameliorate some of the problems described later. Nevertheless, methods that are suitable for lower-resolution images are likely to be valuable in cases where higher resolutions are unavailable (e.g., due to age of image) or beneficial to avoid for performance reasons. The images typically cover the shoulder joint, upper arm and proximal chest areas, but there are numerous exceptions. There is a mix of left and right shoulder joints in the images, although the proportion does not appear to be even. Most images appear to feature the arm in a natural, neutral position, but there are several other positions as well.

The images and captions in Figs. 1, 2, 3 and 4 illustrate some of the properties of the dataset, and problem domain in general, that pose challenges to automated analysis.

3 Detecting, Orienting, and Isolating Implants

This section addresses some early sub-tasks in the pipeline outlined in Sect. 1. The first is that of detecting the presence of an implant in an X-ray image. As illustrated by some samples in the previous section, artifacts such as low contrast, blooming, noise, and clutter complicate what may otherwise appear to be a simple task. This work detects implants by detecting the presence of the long rectilinear features they have in common, coupled with the curvilinear feature of size in certain proportion to the width of the rectilinear feature. The rectilinear and curvilinear features themselves are detected using a combination of well-established methods. Unfortunately, the dataset does not include any images without implants making it difficult to test how

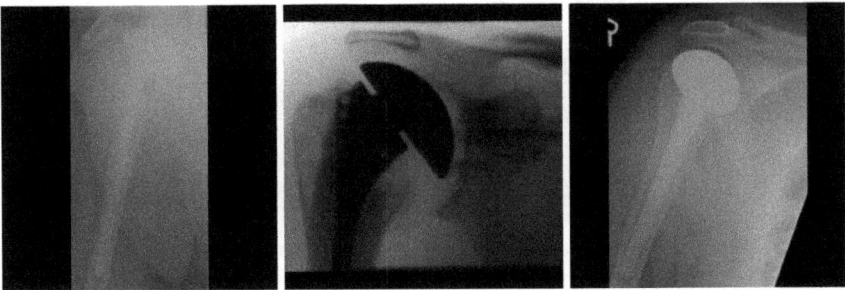

Fig. 1 Sample images from the dataset [17] illustrating some of the challenges: Left: very low contrast. Center: lightness inverted. Right: cropping at an angle, resulting in a strong false edge near the bottom right corner

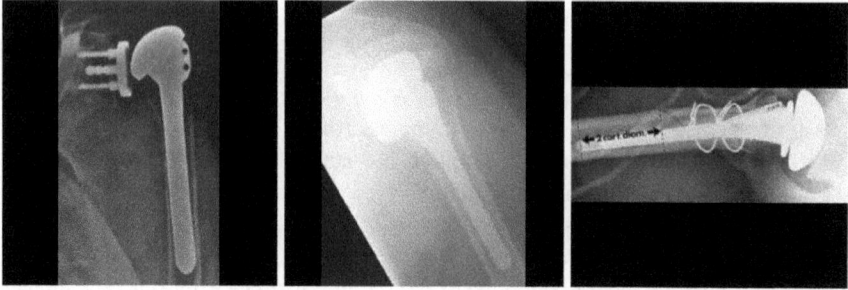

Fig. 2 Sample images from the dataset [17] illustrating further challenges. Left: inclusion of the glenoid component [18] of the implant (typically absent) with strong rectilinear features that compete with the features of the main implant. Center: strong rectilinear features resulting from cropping at an unusual angle. Right: additional objects with strong curvilinear features that compete with curvilinear features near the top (right in image) of the main implant

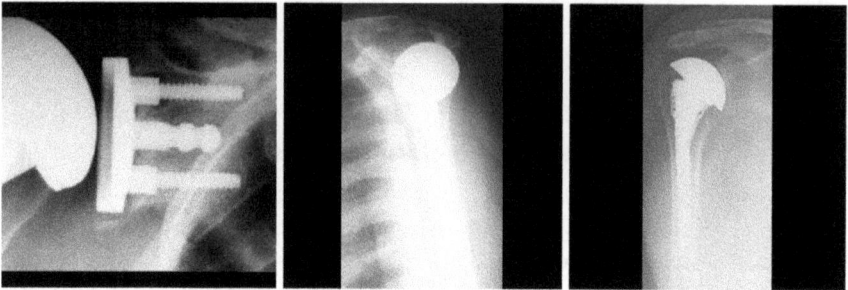

Fig. 3 Sample images from the dataset [17] illustrating some of the challenges: Left: Very close cropping to just the joint region and a predominance of the glenoid portion of the implant, with only a small part of the forearm portion visible. Center: Low image contrast and visual merging of the implant and the ribs. Right: Blooming artifact near the bottom of the image, resulting in indistinct boundaries near the lower part of the implant

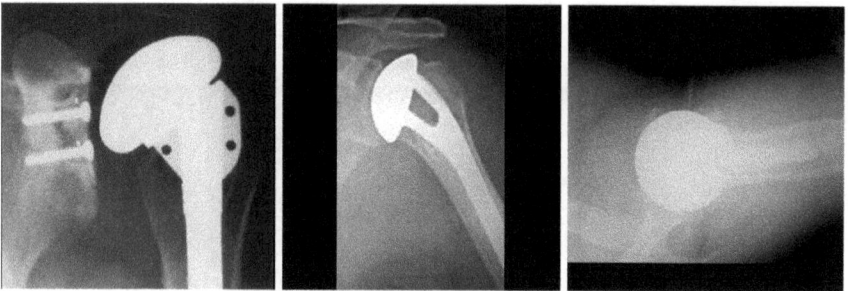

Fig. 4 Sample images from the dataset [17] illustrating some features. Left: Hole features not often discernible this clearly. Center: Non-solid regions. Right: Unusual imaging angle

well this method detects the presence of an implant. Instead, the method is evaluated by visually comparing the details it provides, such as orientation and boundary with those in the image. (Augmenting the dataset to include images of approximately the same body area without implants is a promising topic for future work.)

In more detail, the image is first cropped to remove any borders. It is then normalized for brightness and contrast. Next, the Canny edge detection algorithm [5] is used to detect edges. As is often the case for image processing applications, the tuning of parameters is a critical aspect. For the studied dataset, setting the threshold parameters for the Canny algorithm to 96 and 160 (on a scale of 0–255 as used by the OpenCV implementation), and using an aperture size of 3, produces good results. This edge detection is followed by contour detection using the Suzuki-Abe algorithm [16] with the "simple" contour approximation method. The resulting contours are further simplified using the Ramer–Douglas–Peucker algorithm [6, 10] with $\epsilon = 5$. This simplification step was found to be crucial for obtaining good results on this dataset. Without such simplification, the contours are either too numerous and short or too few, with inadequate coverage of the implant. Once a suitable set of contours has been found, it is easy to compute and measure the strongest rectilinear feature using the geometry of its points and likewise for curvilinear features.

This work has developed an interactive implementation of the above method that is invaluable for purposes of tuning and evaluation. This implementation uses the popular OpenCV library [4] with Java bindings coupled with the Kawa Scheme system [2, 3] for ease of iterative development and tuning.

4 Feature Extraction

The needs of some applications may be met using only the image processing tasks described earlier. A simple example is an application that visually enhances the images, reorients them to present either the implant or the torso in a preferred orientation, delineates the implant and any other notable features, etc. However, other applications require the extraction of additional features from the images. An example of the latter is the automated detection of the type of implant, its manufacturer, model, etc.

There are several popular feature extractors for images, such as SIFT, SURF, and ORB [11]. However, preliminary experiments with these feature detectors did not provide encouraging results. While it was possible to tune the process to yield descriptors focusing on the implant (in contrast to other distracting aspects of the image) for a small number of samples, extending such tuning to the entire dataset did not succeed. A deeper investigation into the tuning of these feature detectors is a topic of ongoing work. Nevertheless, these difficulties suggest considering other features that may be used either instead of or in addition to the above. The present work uses a simple method of gridding and sampling the processed images after the earlier steps of normalization, orientation, etc. It uses a 16×16 grid and selects the mean pixel value (integer in $[0, 255]$) for each grid cell.

5 Classification by Manufacturer

The images in the dataset of this study are labeled only with the manufacturers of the implants they feature and do not include any further details such as model numbers. Therefore, the experimental study of classification is limited to predicting the manufacturer from an analysis of the image. The implants featured in the dataset are from four manufacturers (counts in parentheses): Cofield (83), Depuy (294), Tornier (71), and Zimmer (149). The skewed distribution across the manufacturers poses the usual challenges for the classification task. However, more severe challenges are posed by the images themselves, due to the characteristics noted earlier and, in particular, due to the (visual, intuitively observed) high in-class variance and inter-class similarities. Figures 5, 6, 7 and 8 depict the first 10 images for each of the four manufacturers (using the order of the sequence number in the file names, which appears uncorrelated to any substantial features). Unlike what is often the case in visual recognition tasks, the mapping from images (implants) to manufacturers is not visually apparent.

The experimental study of classification (by manufacturer) of the images using the features described earlier uses a collection of well-established algorithms and, in particular, their implementations in the WEKA workbench [8]. The selection features

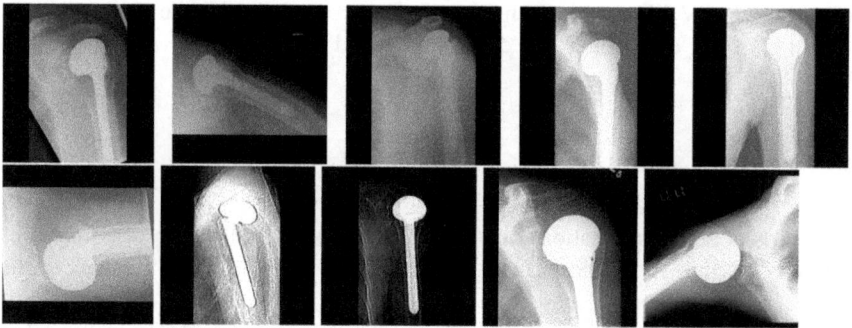

Fig. 5 A sample of X-ray images from the dataset [17] with implants from *Cofield*

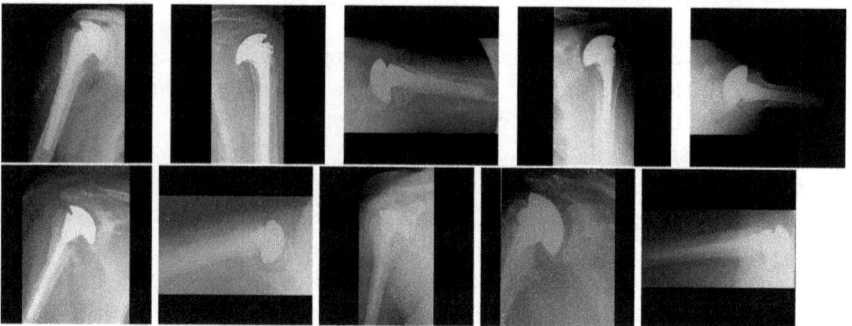

Fig. 6 A sample of X-ray images from the dataset [17] with implants from *Depuy*

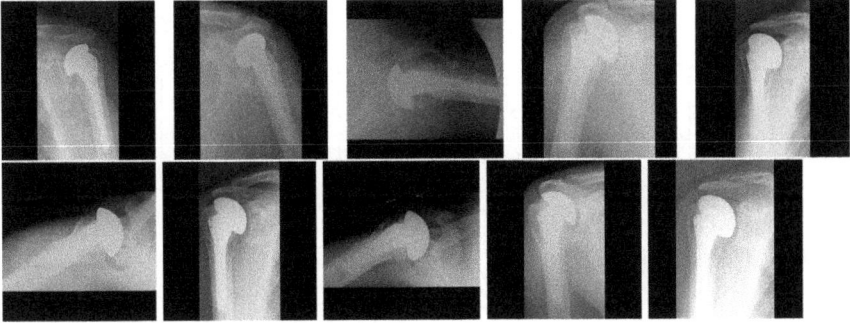

Fig. 7 A collection of X-ray images from the dataset [17] with implants from *Tornier*

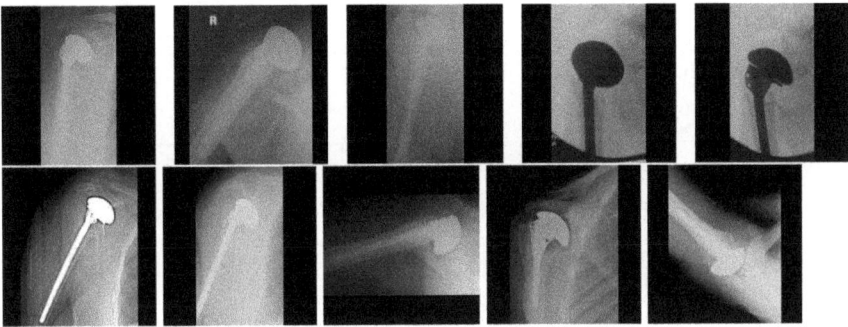

Fig. 8 A sample of X-ray images from the dataset [17] with implants from *Zimmer*

the following representatives from different families of classification algorithms. The abbreviations are used in to identify the classifiers in Figs. 9, 10 and 11. The methods ZeroR and OneR provide useful baselines. ZeroR ignores all attribute values and always predicts the most frequent class, which is Depuy for this dataset. OneR uses a single statically determined (at training time, not testing time) attribute for classification and ignores all others. Rule-based classifiers include PART, Decision Tables (DTab), and the RIPPER (JRip). Tree-based classifiers include J48, random tree (RTre), and random forest (RFor, also an ensemble method). Function-based methods include Simple Logistic (SLog) and SVM with sequential minimal optimization (SMO). Bayesian methods include Naive Bayes (NBay) and Bayes Net with single parents (BNt1) or up to five parents (BNt5). Finally, lazy (instance-based) methods are represented by the k-nearest neighbor (IBk) and locally weighted learning (LWtL). Further details of the methods and their specific implementations used in this study appear elsewhere [7] and so are omitted here.

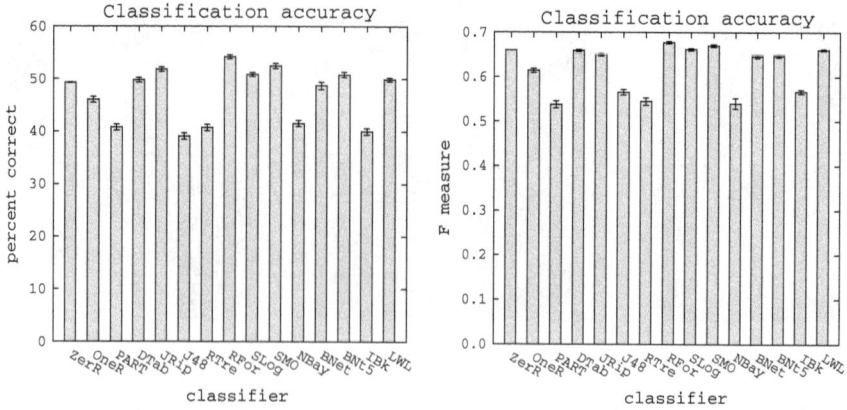

Fig. 9 Classifier accuracy using the base scheme and the metrics: percent correctly classified (left) and F-measure (right)

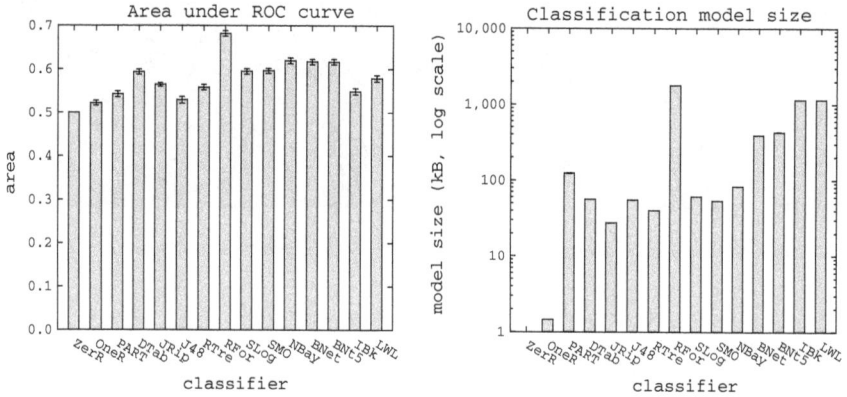

Fig. 10 Area under the receiver operating characteristic (ROC) curve (left) and the learned model size (right) for classifiers using the base scheme

All experimental results use conventional tenfold cross-validation. The error bars in the charts (barely discernible) represent the standard error of the mean (SEM). Figures 9 and 10 summarize classification accuracy using the metrics of percent correctly classified, F-measure, and area under the receiver operating characteristic (ROC) curve. A general observation here is that the accuracies of competitive methods such as random forest are only very modestly higher than those of the baseline methods. This result underscores the difficulty of the problem and agrees well numerically with results for similar methods reported by prior work [17]. Figures 10 and 11 summarize performance of the classifiers using the metrics of time required for training (model building), time required for testing (class prediction), and model size. (Note the logarithmic scales on the vertical axes.) The complementary relationship of training and testing times of eager and lazy methods is intuitive given that

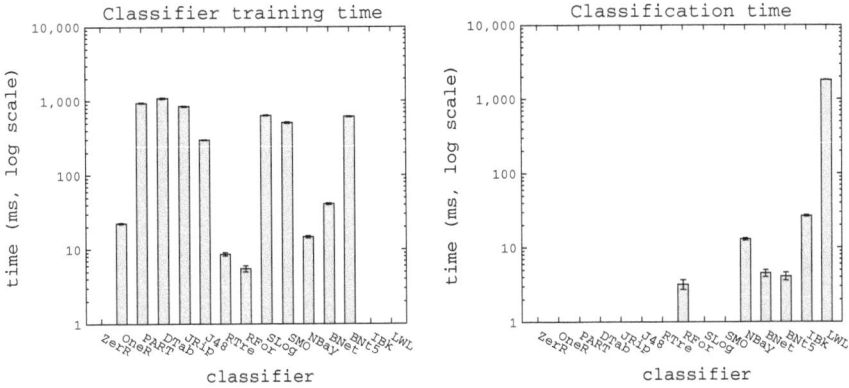

Fig. 11 Classifier training (model building) time (left) and classification (testing) time (right) using the base scheme

the lazy methods require no training and perform all the work at testing time instead. The main observation here is that the time and space performance characteristics are not significant problems for the prototype implementation on this dataset, although for some of the classifiers these issues may become significant when extended to much larger datasets.

6 Related Work

This work was motivated and enabled by the dataset provided by prior work on the topic [17]. That work focuses on the use of deep learning methods for the classification of the images by manufacturer, but also reports on classification accuracy of some conventional methods that are comparable to those in Sect. 5. Although the exact methods used for preprocessing the images and extracting features in that case are not reported, the numerical results are in general agreement with those in this work. Other work related to that dataset [15] is closer in spirit to the work reported here, although the details of the image processing methods used are different. It should be useful to compare the methods used there with those described in this work.

The image processing stages of this work build upon well-known algorithms and implementations for edge detection [5], contour finding [16], curve simplification (approximation), feature detection [11] and more and especially their implementations in the OpenCV [4] library. Similarly, the classification stages build upon well-known methods as implemented in the WEKA workbench [8]. Rapid development and interactive testing and tuning of parameters are facilitated by the use of the Kawa Scheme [2] system running in an OpenJDK Java environment with access to OpenCV via Java bindings that use JNI. In very performance-sensitive environments, it would be worth exploring native implementations of these methods. In a broader context, the use of automation to assist in health care has received significant attention in diverse forms, such as using data from the Internet of things (IoT) for healthcare monitoring [13] and using Bluetooth and WiFi for real-time biometric monitoring [9].

7 Conclusion

This paper has motivated the problem of automated analysis of X-ray images of shoulder implants and summarized the results of an experimental study that uses a recently published collection of such images. Using concrete illustrative examples, it has described the dataset and the features that pose challenges to image processing and classification tasks. It has described an experimentally tuned and validated sequence of image processing steps, based on well-established algorithms that are effective for this dataset. It has outlined difficulties in feature extraction and described a simple feature extraction method to enable classification of images by manufacturer of implant. Although the feature extraction is simple, the classification results are in agreement with those reported for similar methods in recent work. This observation suggests that higher accuracy may be achievable by using more sophisticated feature extraction, and such investigation is a topic for ongoing and future work.

Acknowledgements This work was supported in part by the US National Science Foundation and benefited from the reviewers' comments.

Conflict of Interest The authors declare that they have no conflict of interest.

References

1. Basith II, Kandalaft N, Rashidzadeh R, Ahmadi M (2013) Charge-controlled readout and BIST circuit for MEMS sensors. IEEE Trans Comput-Aided Des Integr Circuits Syst 32(3):433–441. https://doi.org/10.1109/TCAD.2012.2218602
2. Bothner P (1998) Kawa: compiling dynamic languages to the Java VM. In: Proceedings of the usenix annual technical conference, FREENIX Track. USENIX Association, Berkeley, CA, USA, pp 41–41. http://dl.acm.org/citation.cfm?id=1268256.1268297
3. Bothner P (2017) The Kawa Scheme language. Manual for version 3.0. https://www.gnu.org/software/kawa/
4. Bradski G (2000) The OpenCV library. Dr. Dobb's J Softw Tools
5. Canny J (1986) A computational approach to edge detection. IEEE Trans Pattern Anal Mach Intell PAMI-8(6):679–698. https://doi.org/10.1109/TPAMI.1986.4767851
6. Douglas DH, Peucker TK (1973) Algorithms for the reduction of the number of points required to represent a digitized line or its caricature. Cartogr Int J Geogr Inf Geovisualization 10(2):112–122. https://doi.org/10.3138/FM57-6770-U75U-7727
7. Frank E, Hall MA, Witten IH (2016) The WEKA workbench. Online appendix for data mining: practical machine learning tools and techniques, 4th edn. Morgan Kaufmann, p 2016

8. Hall M, Frank E, Holmes G, Pfahringer B, Reutemann P, Witten I (2008) The WEKA data mining software: an update. ACM SIGKDD Expl Newslett 11:10–18
9. Kandalaft N, Bajracharya A, Neupane S, Larson C, Saha HN (2018) Real time monitoring system for vital pediatric biometric data. In: IEEE 9th annual information technology, electronics and mobile communication conference (IEMCON), pp 1065–1069. https://doi.org/10.1109/IEMCON.2018.8615036
10. Ramer U (1972) An iterative procedure for the polygonal approximation of plane curves. Comput Graph Image Process 1(3):244–256. https://doi.org/10.1016/S0146-664X(72)80017-0
11. Rublee E, Rabaud V, Konolige K, Bradski G (2011) ORB: an efficient alternative to SIFT or SURF. In: 2011 international conference on computer vision, pp 2564–2571. https://doi.org/10.1109/ICCV.2011.6126544
12. Saha HN, Mandal A, Sinha A (2017) Recent trends in the internet of things. In: 2017 IEEE 7th annual computing and communication workshop and conference (CCWC), pp 1–4. https://doi.org/10.1109/CCWC.2017.7868439
13. Saha HN, Paul D, Chaudhury S, Haldar S, Mukherjee R (2017) Internet of thing based healthcare monitoring system. In: 2017 8th IEEE annual information technology, electronics and mobile communication conference (IEMCON), pp 531–535. https://doi.org/10.1109/IEMCON.2017.8117245
14. Saha J, Saha AK, Chatterjee A, Agrawal S, Saha A, Kar A, Saha HN (2018) Advanced IOT based combined remote health monitoring, home automation and alarm system. In: 2018 IEEE 8th annual computing and communication workshop and conference (CCWC), pp 602–606. https://doi.org/10.1109/CCWC.2018.8301659
15. Stark MBCG (2018) Automatic detection and segmentation of shoulder implants in X-ray images. Master's thesis, San Francisco State University, San Francisco, CA
16. Suzuki S, Abe K (1985) Topological structural analysis of digitized binary images by border following. Comput Vis Graph Image Process 30(1):32–46. https://doi.org/10.1016/0734-189X(85)90016-7
17. Urban G, Porhemmat S, Stark M, Feeley B, Okada K, Baldi P (2020) Classifying shoulder implants in X-ray images using deep learning. Comput Struct Biotechnol J 18:967–972. https://doi.org/10.1016/j.csbj.2020.04.005
18. Wand R, Dear K, Bigsby E, Wand J (2012) A review of shoulder replacement surgery. J Perioper Pract 22(11):354–359. https://doi.org/10.1177/175045891602201102 PMID: 23311021

Chapter 23
Human-Understandable Classifiers for COPD from Biosensor Data

Automated Analysis of Saliva Samples for Chronic Obstructive Pulmonary Disease

Sudarshan S. Chawathe

1 Introduction

Advances in biosensor technology continually increase the diversity and capabilities of sensors while reducing their costs. As a result, there is a steady stream of novel and low-cost sensors that are suitable for deployment not only in controlled environments such as hospitals and well-equipped homes but also in unpredictable and remote environments. An example of such a novel sensor is a recently developed portable dielectric biosensor that rapidly senses changes in viscosity and appears suitable for analysis of human saliva samples in vitro [16]. This low-cost sensor and others like it have the potential to greatly increase the chances of early diagnosis of chronic obstructive pulmonary disease (COPD), which often goes undiagnosed until later stages of the disease. Such early diagnoses are especially important in the context of pandemics such as the recent and ongoing (at the time of writing) COVID-19 pandemic.

However, effective use of such novel biosensors requires complementary development of methods for analyzing the data they generate in order to produce information that is suitable for clinical medical use. While there have been significant recent and ongoing advances in the development of such data analysis methods, many of them produce only a final result (such as a diagnosis of infected or not) without an accompanying explanation that is meaningful to a human expert. In particular, much recent emphasis has been on connectionist methods such as those based on various deep learning networks. While these methods are often impressively effective at classification and related tasks, their results are also completely opaque to human comprehension, which is a particularly a significant shortcoming in the medical domain.

S. S. Chawathe (✉)
School of Computing and Information Science and Climate Change Institute,
University of Maine, Orono, ME 04605-5711, USA
e-mail: chaw@eip10.org

Problem statement

This paper addresses this problem of, in general, devising classifiers for biosensor data that provide human-understandable results. That is, the results include not only the predicted class but also the reasoning leading to that prediction. It focuses in particular on a novel dielectric biosensor from recent work that has been used to generate a dataset from human saliva samples and auxiliary data.

Contributions

- This paper motivates the need for human-understandable classifiers that permit effective medical use of biosensors.
- It describes the characteristics of data from a recently developed dielectric biosensor (prior work) that rapidly senses viscosity variations and has been applied to human saliva samples.
- It reports on the results of exploratory experiments that use well-established classification algorithms to analyze this biosensor data for the purpose of classifying the underlying saliva samples into four categories relative to their COPD status: COPD, healthy, asthma, and infected.
- Guided by the emphasis on understandable classifiers and the results of the exploratory experiments, it summarizes detailed experiments on rule-based classifiers and, further, those that use rough sets and evolutionary computation. An important aspect of this work is a study of not only the numerical results using the usual metrics such as accuracy and running time, but also the concrete classification rules, trees, or other structures produced in the context of the studied dataset.

Results

- The concrete classifiers produced by rule-based methods seem well suited to human interpretation, as illustrated by the illustrative samples.
- Contrary to what may seem intuitive, the restriction of human-understandable classification does not appear to incur a large penalty in other criteria such as accuracy and running time.
- Rule-based methods that use rough sets and, further, evolutionary algorithms provide promising improvements in metrics such as accuracy without sacrificing human understandability.
- There is a need for larger and more diverse biosensor datasets that permit these classification methods to be assessed more widely.

Outline

Section 2 describes the biosensor from prior work that was used to generate the dataset studied here. It summarizes the salient features of the dataset and provides a few illustrative instances. Section 3 addresses the task of classification using a diverse array of well-established classification methods and summarizes the results of an experimental study. Section 4 focuses on rule-based classifiers. It illustrates the benefits of such classifiers in the specific context of this dataset by providing illustrative examples of rule-based classifiers generated by these methods. It also presents the results of a detailed experimental study of the effect of some key parameters on the accuracy and performance of these classifiers. Classifiers based on rough sets and evolutionary computation are studied in Sect. 4, which also summarizes the results of detailed experiments on the effect of parameters such as number of generations and population size. Related work is addressed in Sect. 6 followed by a summary in Sect. 7.

2   Dielectric Biosensor Data

Although it is likely that the methods described here are widely applicable, an important motivation of this work is providing numerical and nonnumerical results of their use on real data in a concrete setting. To that end, this section briefly describes a few aspects of the biosensor that are necessary to understand the dataset. A much fuller description appears in the paper introducing the sensor [15].

The main principle of operation of the sensor is as follows (Fig. 1): Changes in the permittivity of a saliva sample are sensed by capacitive elements that are part of a resonant LC tank circuit. Changes in permittivity lead to changes in capacitance and in turn to changes in the resonant frequency of the tank. This frequency change is sensed using a frequency discriminator and converted into a DC signal as the real component of the permittivity. In addition, the oscillator's output power is also detected and provided as the imaginary component of the permittivity.

The dataset [17] includes the average and minimum values of the above real and imaginary components for each sample (human saliva). In addition, it includes information on the gender (0 or 1 for female or male), age (integer), and smoking status (0, 1, or 2 indicating nonsmoker, past smoker, or active smoker) of the sample donor. Each instance also includes a categorization into one of the four classes that are

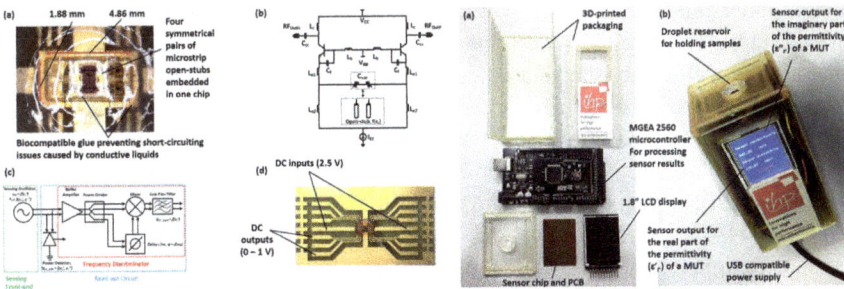

Fig. 1 Biosensor for testing saliva samples as reported in work [16] generating the dataset [17]. (Images credit CC-BY [16].) The image on the left depicts the core of the sensor platform, composed of capacitive elements (top left) that are part of an LC resonant tank (top right) and the readout circuit (bottom left). The image on the right depicts the sensor integrated into a microcontroller-based device that accepts samples in a receptacle and provides immediate readings on the attached display

Table 1 A few illustrative instances from the experimental dataset [17], with numerical precision curtailed for brevity

id	imin	iavg	rmin	ravg	Male	Age	Smoking	Diagnosis
301-4	−320.61	−300.56	−495.26	−464.17	1	77	2	COPD
302-3	−325.39	−314.75	−473.73	−469.26	0	72	2	COPD
14-5	−303.86	−288.04	−514.40	−453.22	0	50	2	HC
A218-1	−320.61	−311.91	−473.73	−469.12	1	57	1	Asthma
143-5	−320.61	−312.07	−476.12	−470.81	0	28	1	Infected

The sensor data is represented by the four attributes *imin*, *iavg*, *rmin*, and *ravg*, representing the minimum and average values of the imaginary and real components of the complex-valued sensor readings. There are 399 instances in total, and a significant number of them (299) have missing values. Key for *male*: 0 denotes male, 1 denotes female. Key for *smoking*: 0, 1, 2 denote non-smoker, past smoker, active smoker, respectively

the target of discovery: *COPD*, healthy control (*HC*), *Asthma*, or *Infected*. Table 1 provides a small excerpt of items in the dataset. The table and what follows use the abbreviations *imin*, *iavg*, *rmin*, and *ravg* for the minimum and average values of the imaginary and real components of permittivity described earlier. The dataset is of modest size ($N = 399$) and, although not apparent from Table 1, has a large proportion of records with missing values.

3 Classification

Although this work emphasizes human-understandable classification methods, it is useful to briefly study the application of a more diverse collection of classification methods to the data described earlier. The purpose of this initial exploration is to better understand the performance and other implications of the restriction to human-understandable methods as well as to identify the promising candidates for further study. This section reports on experiments conducted to that end, using a collection of well-established classification methods that includes representatives of several categories. In particular, it uses baseline methods (with abbreviations used in charts) 0R constant predictor (ZerR) and 1R predictor which uses a single attribute (OneR); rule-based methods Decision Table (DTab), FURIA fuzzy (rough-sets) rules (FURI), RIPPER rule induction (JRip), and PART lists (PART); tree-based methods J48/C4.5 (J48T), random tree (RanT), and random forest (RanF), function-based methods Simple Logistic (SLog), Logistic (Logi), and sequential minimal optimization for support vector machines (SMOp); Bayesian methods Naive Bayes (NBay), Bayes Net with single parent per node (BNe1), and Bayes Net multiple-parents per node (BNtP); and lazy methods K-NN (IBkN), K-Star (KStr), and locally weighted learning (LWtL).

The experiments use the primary dataset of Section 2 denoted as *Full* in the charts. They also use a *Pruned* version of the dataset that excludes all items with missing attribute values. This version was motivated by some preliminary experiments that suggested significant effects of missing values. The experiments use conventional tenfold cross-validation when evaluating classification accuracy. They use the implementations of classification methods from the WEKA workbench [5], driven from Kawa Scheme [2, 3] and running on the OpenJDK [10] Java 1.8 platform on a Debian GNU/Linux system. In the charts that follow, the error bars, which are often small and barely discernible, represent the standard error of the mean (SEM).

Figs. 2, 3 and 4 summarize the results on classifier accuracy using the conventional metrics of percentage correctly classified, F-measure, and area under the receiver operating characteristic (ROC) curve. Two observations are notable here: First, it is clear that missing values play a significant role, as evidenced by the large difference in accuracy of some methods for the full and pruned datasets. Particularly noteworthy in this regard is this difference for the random forest method, which is a popular and competitive choice. Second, the human-understandable methods (e.g., FURIA, JRip,

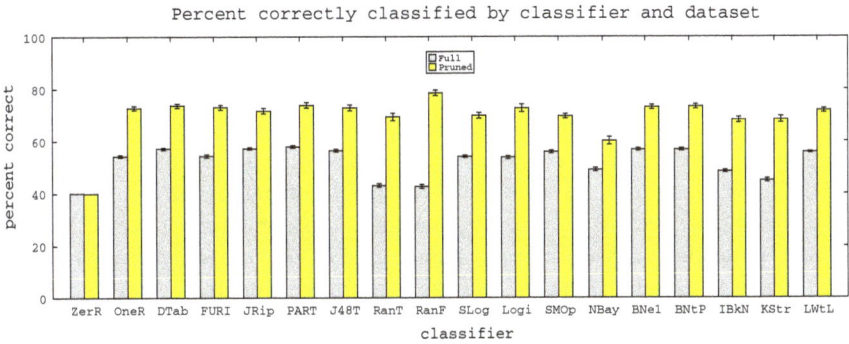

Fig. 2 Classification accuracy of diverse classifiers as measured by the percentage of correctly classified instances of the full and pruned datasets

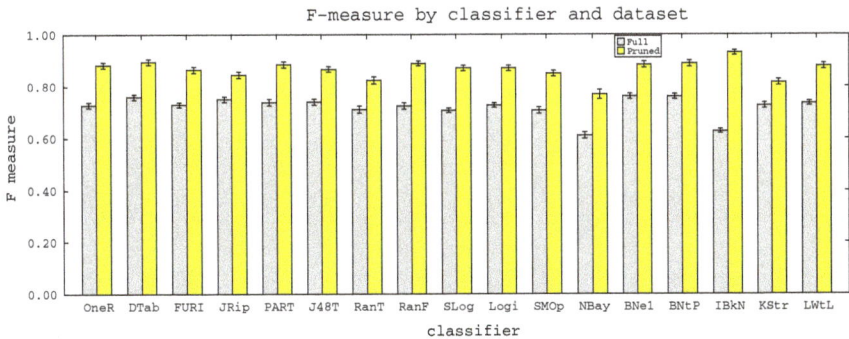

Fig. 3 Classification accuracy of diverse classifiers as measured by the F-measure for the full and pruned datasets

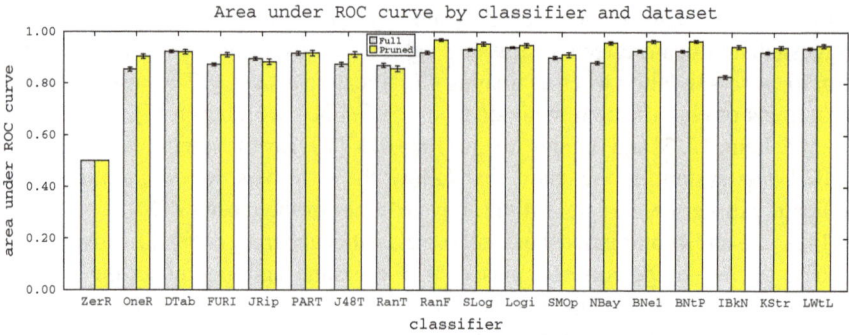

Fig. 4 Classification accuracy of diverse classifiers as measured by the area under the receiver operating characteristic (ROC) curve for the full and pruned datasets

PART) provide accuracy that is competitive in general and better than many other methods especially for the full dataset.

The time required for training (model building) for different classifiers is summarized by Fig. 5 with a logarithmic scale on the vertical axis. It is not surprising that this time is negligible for the lazy (instance-based) methods. The random forest implementation was used with its multithreaded option on a 4-hyperthread 2-core processor. In general, these results should be taken as indicative of the particular WEKA implementations and serve mainly to illustrate that the time requirements of these methods are modest. For the lazy methods, it is appropriate to study the time required for testing (classification) as summarized by the inset chart to the right in. (This time is too small to meaningfully measure for the other methods.)

For the lazy methods, it is appropriate to study the time required for testing (classification) as summarized Fig. 5. (This time is too small to meaningfully measure for the other methods.)

The final chart in this series, Fig. 6, summarizes the sizes of the models built by these methods. The significantly larger size (note logarithmic axis) for ensemble-based methods such as random forest is apparent here. However, for this dataset, none of the sizes is the cause for concern.

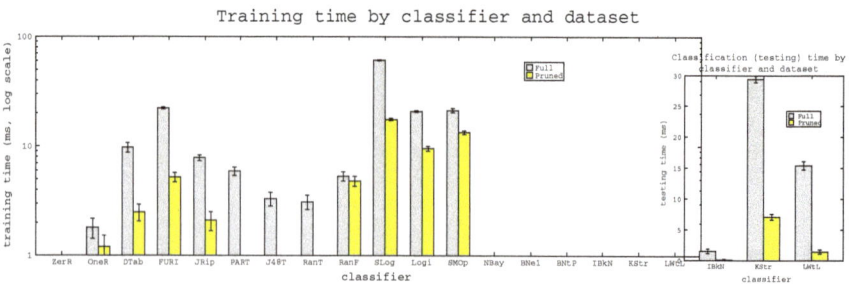

Fig. 5 Classifier training time of diverse classifiers for the full and pruned datasets, with a logarithmic scale on the vertical axis

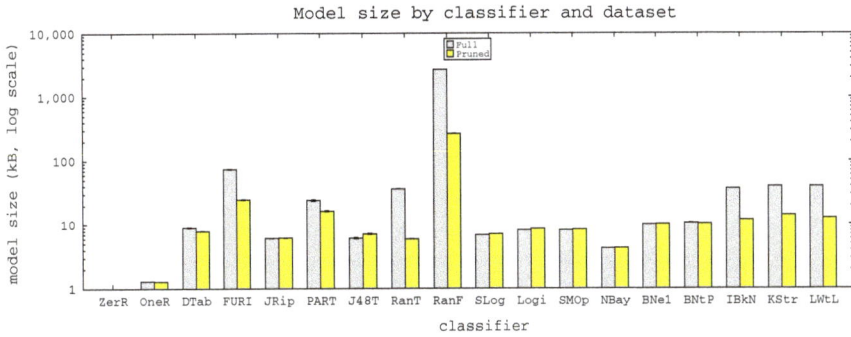

Fig. 6 Model sizes of diverse classifiers for the full and pruned datasets, with a logarithmic scale on the vertical axis

4 Rule-Based Classifiers

The benefits of rules-based and other human-understandable classifiers are illustrated Fig. 7, which depicts the classifier produced using the *J48* implementation of C4.5 decision trees [11]. The decision process is evident from the tree and a valuable addition to the prediction made by the classifier. The corresponding confusion matrix appears in Fig. 7. Similarly, the instances of the ordered rules produced by *JRip*, listed in Fig. 8, and unordered rules produced by FURIA, listed in Fig. 9, are easy to interpret. (In the case of FURIA, the four bracketed numbers denote a trapezoidal-shaped fuzzy interval with vertices at those coordinates.)

Assigned Cl.	Actual Class			
	COPD	HC	Asthma	Infected
COPD	37	3	0	0
HC	3	30	3	4
Asthma	2	6	1	1
Infected	0	2	1	7

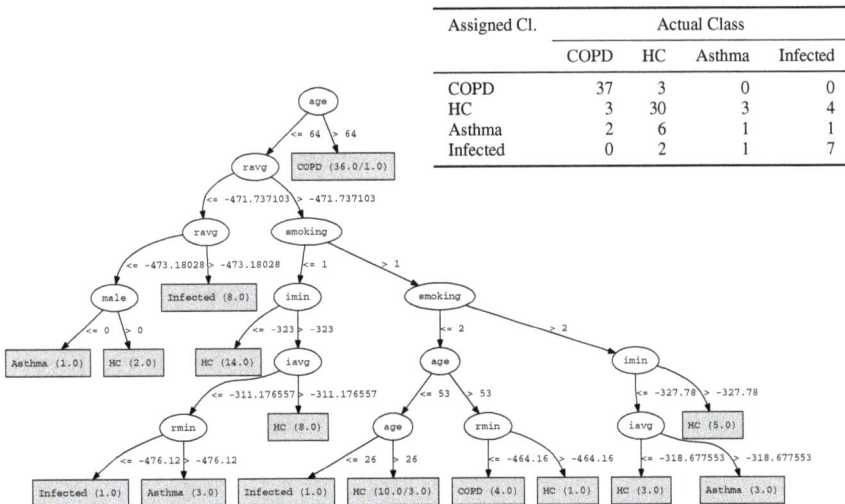

Fig. 7 A tree classifier produced using *J48* on the dataset of Table 1 and its confusion matrix

```
(age <= 17) => diagnosis=Asthma (1.0/0.0)
(imin >= -320.6) and (iavg <= -311.9) => diagnosis=Asthma (5.0/1.0)
(ravg <= -471.8) => diagnosis=Infected (14.0/6.0)
(age <= 55) => diagnosis=HC (39.0/5.0)
 => diagnosis=COPD (41.0/6.0)
```

Fig. 8 Ordered rule classifier produced by JRip

$$(age \in [63, 64, \infty, \infty]) \Rightarrow COPD(0.92)$$
$$(age \in [-\infty, -\infty, 55, 56]) and (ravg \in [-470.8, -470.4, \infty, \infty]) \Rightarrow HC(0.78)$$
$$(ravg \in [-\infty, -\infty, -467.1, -466.5]) and (imin \in [-323, -320.6, \infty, \infty]) and (rmin \in [-476.1, -473.7, \infty\ \infty]) \Rightarrow Asthma(0.69)$$
$$(ravg \in [-\infty, -\infty, -471.8, -471.6]) and (age \in [-\ \ , -\infty, 50, 65]) and (ravg \in [-473.2, -473.1, \infty, \infty]) \Rightarrow Infected(0.82)$$

Fig. 9 An unordered fuzzy rule classifier produced using FURIA

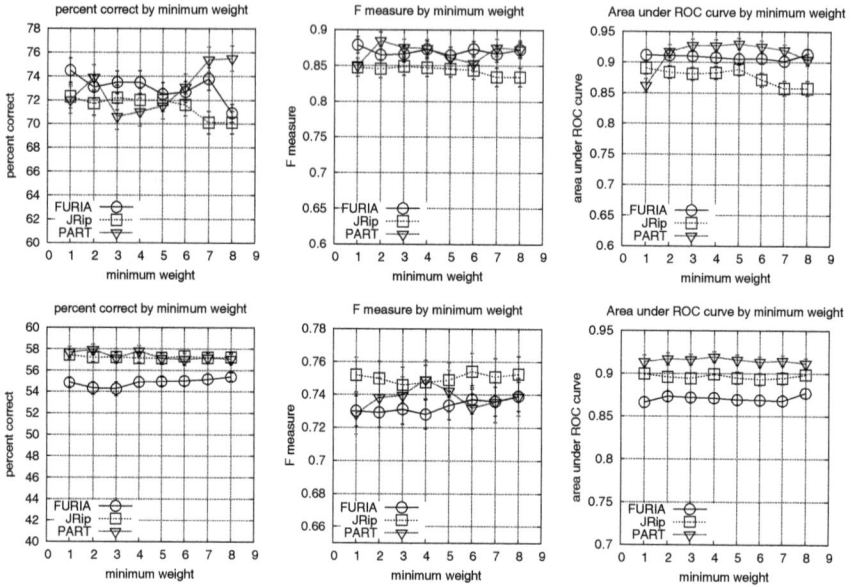

Fig. 10 Effect of the minimum weight parameter on the classifier accuracy (percent correctly classified, F-measure, and area under ROC curve) for the classifiers generated by FURIA, JRip, and PART

An important parameter in training rule-based classifiers is the minimum weight of a partition made by the classifier. Intuitively, it may be thought of as the minimum number of instances that must belong to each partition. The charts in Fig. 10 summarize the results of an experimental study of the effect of varying this parameter on the accuracy and performance metrics of three rules-based classifiers: FURIA, JRip, and PART.

5 Evolutionary Fuzzy Classification

The promising accuracy, performance, and understandability characteristics of the rule-based classifiers of Section 4, and in particular the fuzzy (rough sets) classifier FURIA, suggest further exploration in that direction. This section briefly reports on further experiments conducted to that end using a method based on the use of evolutionary algorithms to train a fuzzy rule-based classifier: ENORA [8]. Fig. 11 depicts a concrete instance of the (complete and unedited) rules generated by ENORA on the dataset of Table 1. The interpretation of the rules, even in this unedited format, is intuitive.

Evolutionary methods are typically characterized by a few parameters, and selecting suitable values for these parameters often poses difficulties in practice. For ENORA, the key parameters are: the number of evolutionary *generations*, the maximum number of *rules*, the *population size*, and the maximum number of *labels*. Fig. 12 is a representative result of an extensive set of experiments conducted to study the effect of these parameters on the accuracy and performance metrics of classifiers produced by ENORA. Additional results are omitted here for brevity.

6 Related Work

A recently published dataset [17], in turn based on earlier work on the underlying sensor [15, 16, 18], is the impetus for this work. That dataset is valuable for being both novel and real (generated by sampling human saliva). Nevertheless, it is very modest in size which limits the confidence of this and other work that relies on it. The problem is acerbated by a larger proportion of missing values. It should be fruitful to expand this study to other similar datasets. The implementation of the prototype system

```
RULE 1:
IF       imin    IS          Moderately Low  (Center: -293.5267, S.D.: 7.9315)
    AND iavg    IS          Very High       (Center: -237.9089, S.D.: 16.0278)
    AND rmin    IS          High            (Center: -178.4411, S.D.: 46.6161)
    AND ravg    IS          Very High       (Center: -70.3491, S.D.: 35.6996)
    AND male    IS          Moderately High (Center: 0.5899, S.D.: 0.0596)
    AND age     IS          Very High       (Center: 87.8695, S.D.: 4.8771)
    AND smoking IS          Moderately High (Center: 2.1702, S.D.: 0.0667)
THEN      diagnosis IS COPD
RULE 2:
IF       imin    IS          Moderately High (Center: -258.2567, S.D.: 15.9922)
    AND iavg    IS          Medium          (Center: -279.4895, S.D.: 8.0826)
    AND rmin    IS          Medium          (Center: -344.9725, S.D.: 42.0096)
    AND ravg    IS          Moderately High (Center: -215.9445, S.D.: 14.331)
    AND male    IS          Moderately Low  (Center: 0.4073, S.D.: 0.0961)
    AND age     IS          Low             (Center: 34.7277, S.D.: 4.0962)
    AND smoking IS          High            (Center: 2.6086, S.D.: 0.2483)
THEN      diagnosis IS HC
```

Fig. 11 First two rules of a classifier generated by ENORA. The two other rules are similar and are omitted for brevity

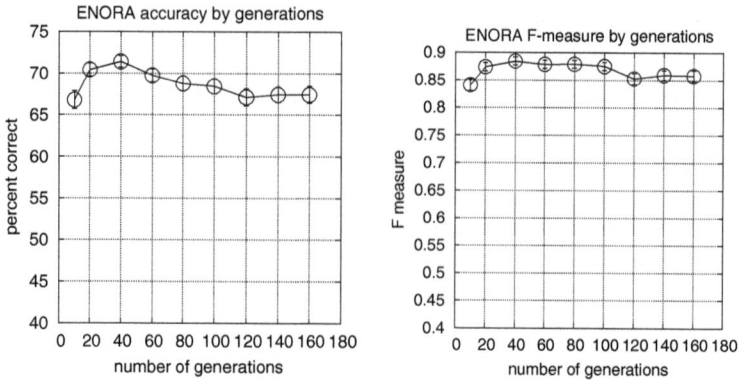

Fig. 12 Effect of the number of evolutionary generations on the accuracy of the ENORA classifier

used for the experimental evaluation relies heavily on well-established algorithms for machine learning [4, 6, 7] as well as more modern ones [8] and, in particular, on their implementations in the WEKA workbench [5].

Hardware aspects have not been the focus of this work; nevertheless, the connections between recent advances in sensors and the biometric sensor are used to generate the data in this study [1, 9]. In a broader context, there is a growing body of recent work that uses advances in sensors, communications, and microcontrollers to address diverse healthcare challenges [9, 12–14].

7 Conclusion

This paper has motivated the need for human-understandable classifiers that operate on data from novel biosensors to produce not only an accurate prediction but also accompanying information that is meaningful to a human expert. It has used a biosensor and accompanying dataset from recent work to study the characteristics of such data, and the interpretability, accuracy, and performance of the concrete classifiers produced by several rule-based methods. The results are encouraging in that they indicate that limiting attention to classification methods that produce such interpretable information need not result in lower accuracy or performance. While these results must be tempered by the limited size and scope of the studied dataset, they hold for at least that real dataset and likely for others similar to it. Extending the study to larger and more diverse datasets is a topic of ongoing work.

Acknowledgements This work was supported in part by the US National Science Foundation and benefited from the reviewers' comments.

Conflict of Interest The authors declare that they have no conflict of interest.

References

1. Basith II, Kandalaft N, Rashidzadeh R, Ahmadi M (2013) Charge-controlled readout and BIST circuit for MEMS sensors. IEEE Trans Comput Aided Des Integr Circ Syst 32(3):433–441. https://doi.org/10.1109/TCAD.2012.2218602
2. Bothner P (1998) Kawa: Compiling dynamic languages to the java VM. In: Proceedings of the USENIX annual technical conference, FREENIX track. USENIX Association, Berkeley, CA, USA, pp 41–41. URL http://dl.acm.org/citation.cfm?id=1268256.1268297
3. Bothner P (2017) The Kawa scheme language. Manual for version 3.0. https://www.gnu.org/software/kawa/
4. Cohen WW (1995) Fast effective rule induction. In: Proceedings of the twelfth international conference on machine learning. Morgan Kaufmann, pp 115–123
5. Hall M, Frank E, Holmes G, Pfahringer B, Reutemann P, Witten I (2008) The WEKA data mining software: an update. ACM SIGKDD Explor Newslett 11:10–18
6. Hühn J, Hüllermeier E (2009) FURIA: an algorithm for unordered fuzzy rule induction. Data Min Knowl Discov 19(3):293–319. https://doi.org/10.1007/s10618-009-0131-8
7. Hühn JC, Hüllermeier E (2010) An analysis of the FURIA algorithm for fuzzy rule induction. Springer Berlin Heidelberg, Berlin, Heidelberg, pp 321–344. https://doi.org/10.1007/978-3-642-05177-7_16
8. Jiménez F, Sánchez G, Juárez JM (2014) Multi-objective evolutionary algorithms for fuzzy classification in survival prediction. Artif Intell Med 60(3), 197–219. https://doi.org/10.1016/j.artmed.2013.12.006. URL http://www.sciencedirect.com/science/article/pii/S0933365713001668
9. Kandalaft N, Bajracharya A, Neupane S, Larson C, Saha HN (2018) Real time monitoring system for vital pediatric biometric data. In: IEEE 9th annual information technology, electronics and mobile communication conference (IEMCON), pp 1065–1069. https://doi.org/10.1109/IEMCON.2018.8615036
10. Oracle Corp., et al (2017) OpenJDK. http://openjdk.java.net/
11. Quinlan R (1993) C4.5: programs for machine learning. Morgan Kaufmann Publishers, San Mateo, CA
12. Saha HN, Mandal A, Sinha A (2017) Recent trends in the internet of things. In: 2017 IEEE 7th annual computing and communication workshop and conference (CCWC), pp 1–4. https://doi.org/10.1109/CCWC.2017.7868439
13. Saha HN, Paul D, Chaudhury S, Haldar S, Mukherjee R (2017) Internet of thing based healthcare monitoring system. In: 2017 8th IEEE annual information technology, electronics and mobile communication conference (IEMCON), pp 531–535. https://doi.org/10.1109/IEMCON.2017.8117245
14. Saha J, Saha AK, Chatterjee A, Agrawal S, Saha A, Kar A, Saha HN (2018) Advanced IOT based combined remote health monitoring, home automation and alarm system. In: 2018 IEEE 8th annual computing and communication workshop and conference (CCWC), pp 602–606. https://doi.org/10.1109/CCWC.2018.8301659
15. Soltani Zarrin P, Jamal FI, Guha S, Wessel J, Kissinger D, Wenger C (2018) Design and fabrication of a BiCMOS dielectric sensor for viscosity measurements: A possible solution for early detection of COPD. Biosensors 8(3):78 (2018). https://doi.org/10.3390/bios8030078. URL https://pubmed.ncbi.nlm.nih.gov/30134577. 30134577[pmid]
16. Soltani Zarrin P, Jamal FI, Roeckendorf N, Wenger C (2019) Development of a portable dielectric biosensor for rapid detection of viscosity variations and its in vitro evaluations using saliva samples of COPD patients and healthy control. Healthcare 7(1):11. https://doi.org/10.3390/healthcare7010011. URL https://doi.org/10.3390/healthcare7010011

17. Soltani Zarrin P, Roeckendorf N, Wenger C (2020) Exasens: a novel dataset for the classification of saliva samples of COPD patients. https://doi.org/10.21227/7t0z-pd65

18. Soltani Zarrin P, Wenger C (2019) Pattern recognition for COPD diagnostics using an artificial neural network and its potential integration on hardware-based neuromorphic platforms. In: Tetko IV, Kůrková V, Karpov P, Theis F (eds) Artificial neural networks and machine learning—ICANN 2019: workshop and special sessions. Springer International Publishing, Cham, pp 284–288

Chapter 24
Dimensionality Prediction for Word Embeddings

Korrapati Sindhu and Karthick Seshadri⊚

1 Introduction

Most of the machine learning algorithms are not capable of working directly with text data, so there is a need of converting text data into numerical data. Text document is a collection of words. Word embedding techniques maps the word in a text document to a real valued vector. Word embedding preserves the syntax and semantic information of words. Another way of defining word embedding is learning low-dimensional word representations from the input text corpus. Word embeddings are useful in many problems in NLP tasks like information retrieval [1–3], recommendation systems [4–7], sentiment classification [8] and measuring document similarity [9].

Dimensionality of word embeddings determines the representational ability of the embedding and exerts a heavy impact on the performance of the NLP tasks that use these embeddings. A word embedding (word vector) with a low dimensionality may not capture all possible word associations where as word embeddings with a high dimensionality may lead to over fitting and exploitation of memory. If word embeddings are learned using a neural network, then an increase in the dimensionality of the word embedding leads to increase in the number of parameters to be learned, number of computations, and the training time of the model.

Different word embedding algorithms use disparate dimensionalities. Popular choices for the dimensionality in the research literature [8–12] are 200, 300, and 500. Typically, the dimensionality for a word embedding is chosen either randomly or through grid search [13]. Grid search is used in hyper parameter optimization problems. In grid search, the ranges of values for hyper parameters are predefined and all possible combinations of hyper parameters forms a grid, and performance

K. Sindhu (✉) · K. Seshadri
NIT Andhra Pradesh, Tadepalligudem, India
e-mail: sindhukorrapati.sclr@nitandhra.ac.in

K. Seshadri
e-mail: karthick.seshadri@nitandhra.ac.in

© The Author(s), under exclusive license to Springer Nature Singapore Pte Ltd. 2021   301
X.-Z. Gao et al. (eds.), *Applications of Artificial Intelligence in Engineering*, Algorithms for Intelligent Systems, https://doi.org/10.1007/978-981-33-4604-8_24

of the algorithm is evaluated with each combination. Subsequently, the value which results in a better performance is selected.

The problems associated with grid search approach are the following: (i) grid search involves computational overhead when the search range for hyper-parameters is large, and for each potential combination of hyper-parameters, the NLP task has to be executed to select the optimal combination. (ii) Due to random sampling of potential values for hyper parameters, all possible dimensionalities are not verified. (iii) The inferred word embeddings may be suitable only for a specific NLP task.

In order to overcome the above-mentioned setbacks, in this paper, we propose a regression-based optimal dimension prediction (RODP) model for predicting the dimensionality of word embeddings for a given input text corpus. The dimensionality prediction problem addressed by the proposed model involves the following sub tasks: (i) feature engineering, (ii) word embedding dimensionality prediction, (iii) generating word embeddings, and (iv) Evaluation of word embeddings.

Evaluation of word vectors generated from a word embedding model can be done either by intrinsic or extrinsic methods. Intrinsic evaluation methods compare word embeddings with human judgment's regarding relationships among words. Intrinsic evaluation methods include word similarity, word analogy, and concept categorization. Extrinsic evaluation methods evaluate the word embeddings by considering word embeddings as an input to NLP downstream tasks. Downstream tasks are the supervised learning tasks that use pre-trained word vectors. For extrinsic evaluation, any downstream task can be used. Extrinsic evaluation methods include sentiment analysis, named entity recognition, part-of-speech tagging (PoS tagging), and paraphrase identification. If word embeddings are trained for a specific downstream task and if another NLP task is chosen for evaluation, then typically, the extrinsic evaluation method fails.

1.1 Contributions

The key contributions of this paper are the following:

i.   We identified the features from the corpus that significantly affect the dimensionality of word embeddings.
ii. We proposed an answer to the open research problem of predicting optimal dimensionality.
iii. By predicting the optimal dimensionality, we have minimized the computational and memory footprints of the NLP tasks.

Rest of the paper is organized as follows. Section 2 provides a survey on word embedding methods. Section 3 discusses the proposed model. Experimental setup and results are summarized in Sect. 4.

2   Related Work

This section outlines the different word embedding models that uses neural network to find the word vectors. Different word embedding methods include but are not limited to neural network language model (NNLM), word2vec (Skip-gram model and Continuous Bag-of-Words model), GloVe.

2.1   Neural Network Language Model

The first word embedding based on neural network is proposed by Wei Xu and Alex Rudnicky [14] and Yoshua Bengio et al. [15]. Basic idea is similarity between the words is high if they appeared in similar contexts. The approach used in NNLM follows the steps below.

1. Each word in the corpus is represented as a feature vector with a randomly selected dimension m.
2. A corpus is a sequence of words, the joint probability function of sequence of words is computed as the product of conditional probabilities of the next word given the previous words.
3. Neural network is trained with an objective of maximizing the log-likelihood of training data.

Typically, in the context of word embeddings, a network with one hidden layer is used. The training data consists of a sequence of words, let W be the sequence of words, and then, the probability of W can be represented using the Bayes rule as shown in Eq. (1)

$$p(W) = \prod_{i=1}^{V} p(w_i/w_1, \ldots, w_{i-1}) = \prod_{i=1}^{V} p(w_i|h_i) \tag{1}$$

where $p(w)$ is the joint probability distribution of the word sequence, V represents the vocabulary size and h_i represents the context words of ith word. If Bigram model is used to find the conditional probabilities for the next word, then the probability of W can be represented using the Bayes rule as shown in Eq. (2)

$$p(w_i|w_1^{i-1}) \approx p(w_i|w_{i-1}) \tag{2}$$

If trigram model is used to find the conditional probabilities for the next word, then the probability of W can be represented using the Bayes rule as shown in Eq. (3)

$$p(w_i|w_1^{i-1}) \approx p(w_i|w_{i-2}w_{i-1}) \tag{3}$$

If n-gram model is used to find the conditional probabilities for the next word, then the probability of W can be represented using the Bayes rule as shown in Eq. (4)

$$p(w_1^n) \approx \prod_{k=1}^{n} p(w_k | w_{k-n+1}^{k-1}) \tag{4}$$

w_i^j denotes the sub sequence $(w_i, w_{i+1}, ..., w_j)$

The problem with NNLM model is it's high computational complexity because it uses non-linear hidden layer.

2.2 Skip-Gram Model with Negative Sampling (SGNS)

Skip-gram model [11] predicts the context words given the target word, i.e., for each word, it predicts the surrounding words in a window of size m. The objective function of skip-gram model is maximizing the probability of any context word given the current target word, i.e.,

$$J(\theta) = \prod_{t=1}^{T} \prod_{-m \leq j \leq m \wedge j \neq 0} p(w_{t+j} | w_t; \theta) \tag{5}$$

where T represents the number of words in the training sequence, m is the context window size and w_t is the target word. The objective function can be achieved by maximizing the average log probability, i.e.,

$$J'(\theta) = \frac{1}{T} \sum_{t=1}^{T} \sum_{-m \leq j \leq m \wedge j \neq 0} \log p(w_{t+j} | w_t) \tag{6}$$

The probability $p(w_c | w_t)$ can be defined as specified in Eq. (7)

$$p(w_c | w_t) = \frac{\exp(u_c^T v_t)}{\sum_{i=1}^{V} \exp(u_i^T v_t)} \tag{7}$$

where w_c represents context word, w_t is the target word, u_c represents vector associated with context word, v_t represents the vector associated with the target word, and V is the number of words in the vocabulary. It is difficult to optimize the objective function in Eq. (6) because it uses softmax function to predict the likelihood of each context word. Computational complexity of softmax function depends on the vocabulary size. Updating or evaluating the objective function will take $O(|V|)$ time. This issue can be addressed just by sampling several negative examples [16].

The basic idea of negative sampling lies in modifying the objective function to treat the problem as a classification problem, where the pair (context word, word) are treated as training examples. Any given word and context word pair forms a positive sample, a word with non-context word forms a negative sample. The new objective function tries to maximize the probability of a word and context word in the corpus and maximize the probability of a word and context word not being in the corpus, i.e., replace log $p(w_{t+j}|w_t)$ in Eq. (6) with $\log \sigma(u_{c-m+j}^T v_c) + \sum_{k=1}^{K} \log \sigma(\tilde{u}_k^T v_c)$ where K indicates the number of negative samples. Parameters of the objective function are learnt using the stochastic gradient descent (SGD) method.

The problem with SGNS model is that it computes the noise distribution for negative sampling from the entire training data, so this model cannot perform incremental updates, i.e., whenever the new training data is added, the model has to process the entire training data. To overcome this problem, Nobuhiro Kaji and Hayato Kobayashi [17] proposed the incremental skip-gram model with negative sampling. Incremental SGNS model does not pre-compute the noise distribution. While performing SGD, incremental SGNS reads the training data word by word to for updating word frequency distribution and noise distribution.

2.3   Continuous Bag-of-Words Model (CBOW)

CBOW model predicts the target word based on context words. Let m denotes the context window size and to predict the word at position t, CBOW model maximizes the probability of a word given context, i.e., CBOW model maximizes $p(w_t|w_{t-m}, w_{t-m+1}, \ldots, w_{t-1}, w_{t+1}, \ldots, w_{t+m})$. The steps followed in the CBOW model are as specified below

1. Initially, for each word in the size of context window, one-hot vector is created. If context window size is m, then 2 m vectors are created $(h_{t-m}, h_{t-m+1}, \ldots, h_{t-1}, h_{t+1}, \ldots, h_{t+m})$
2. Subsequently, for each word in context window, a word embedding vector $(e_{t-m}, e_{t-m+1}, \ldots, e_{t-1}, e_{t+1}, \ldots, e_{t+m})$ is created by multiplying the input word matrix $V \in \Re^{N \times |V|}$ with one-hot vector of a word in the context window. N represents the embedding dimensionality of a word and each column in V represents the N-dimensional embedding vector for each input word in the vocabulary.
3. Average the embedding vectors of context words $e_{avg} = \frac{e_{t-m} + \cdots + e_{t+m}}{2m}$
4. Calculate $z = e_{avg}U$ where U represents the output word matrix and find the output $\hat{y} = \text{softmax}(z)$
5. Loss function can be calculated as $H(\hat{y}, y) = -\sum y_i \log(\hat{y}_i)$.

The objective function of CBOW model is to minimize log likelihood of conditional probability as shown in Eqs. (8) and (9)

$$J = -\log \ p(w_t|w_{t-m}, w_{t-m+1}, \ldots, w_{t-1}, w_{t+1}, \ldots, w_{t+m}) \qquad (8)$$

$$J = -\log \ p(u_c|e_{avg}) \qquad (9)$$

In this approach, dimensionality of the word embedding vector is chosen randomly.

2.4 GloVe

Window-based log linear models (CBOW or Skip-Gram) capture the co-occurrences of word in one window at a time, i.e., it uses local information only. However, a global co-occurrence matrix can be computed from the entire corpus. Singular value decomposition (SVD) can be applied on the co-occurrence matrix and the word vectors can be represented in a k-dimensional space. These k vectors are considered sufficient to learn the syntactic and semantic meaning of the words. The problems with this approach are the following: (i) it involves more computations if the vocabulary size is large (ii) words with a low frequency are not handled properly (iii) efficient representation of words depends on the value of k.

Global vectors for word representations (GloVe) [18] combines the advantages of both the SVD-based models and log-linear models (CBOW or skip-gram) for learning word representations. Glove model is trained on non-zero elements in a word-word global co-occurrence matrix. The objective function of GloVe is modeled to maximize Eq. (10)

$$J = \sum_{i=1}^{V} \sum_{j=1}^{V} f(X_{ij})(u_j^T v_i - \log \ X_{ij})^2 \qquad (10)$$

Word2Vec [11] uses a neural network with one hidden layer. Number of neurons in the input layer depends on the vocabulary size. Number of neurons in the hidden layer is equal to the dimensionality used to represent a word in the vector space model. The dimensionality is selected using grid search.

Wang [19] proposed a method to select the number of dimensions for word embeddings based on Principal Component Analysis (PCA). In his research, initially an upper bound for the embedding dimension is selected, i.e., 1000. The model is trained to obtain word embeddings of dimensionality 1000. Then, the embeddings are transformed using PCA and the dimensions that have little impact on performance are removed incrementally, and the performance of the embeddings is measured, to determine the optimal dimensionality. Skip-gram model is used as the embedding algorithm and two bench mark datasets Text8 and WikiText-103 are used for experimentation. In this article, the initial dimensionality is chosen randomly and also involves computational overhead since performance metrics are computed after pruning each insignificant dimension.

Das [20] proposed an empirical approach to choose the dimensionality of word embeddings for a specific corpus. Initially, words are represented in an n-dimensional space where n is a small positive integer. Subsequently, PCA is applied and the number of significant components is obtained. Significant components are the principal components with an eigen value greater than some threshold. If the number of significant components in an n-dimensional space, and the $n - 1$ dimensional sub-space are equal then the process is stopped and 'n' is taken as the word embedding dimension otherwise the value of n is incremented by one, and the process is repeated. The experiments are done on a movie reviews dataset and the author proved that the quality of word vectors generated using word2vec with 48 dimensions is similar to the quality of word vectors with 300 dimensions. The performance of this approach is impacted by the randomly chosen initial embedding size and the significance threshold on eigen values.

Yin and Shen [21] proposed a novel pairwise inner product (PIP) loss metric to find the dissimilarity between word embeddings. Optimal dimensionality is set to the value which minimizes the PIP loss. Selection of optimal dimensionality is based on an iterative process that checks embeddings of different dimensionality for minimizing the PIP loss and selects the dimensionality that yields the least loss.

The methods cited above, select the optimal dimensionality either by ad hoc techniques or through grid search. Instead of choosing a random initial value for the embedding dimension, it will be beneficial if we have a model that determines the optimal dimensionality of word embeddings to be used for a corpus, and in this paper, we propose such a model.

3 Proposed Method

In this section, we describe the methodology used to find the optimal dimensionality to represent the words of a given corpus in a vector space. The proposed method involves the following subtasks

   i.   Pre-processing
  ii.   Feature engineering
 iii.   Optimal dimensionality prediction
 iv.   Generating and evaluating word embeddings

The overall process of the proposed model is shown in Fig. 1.

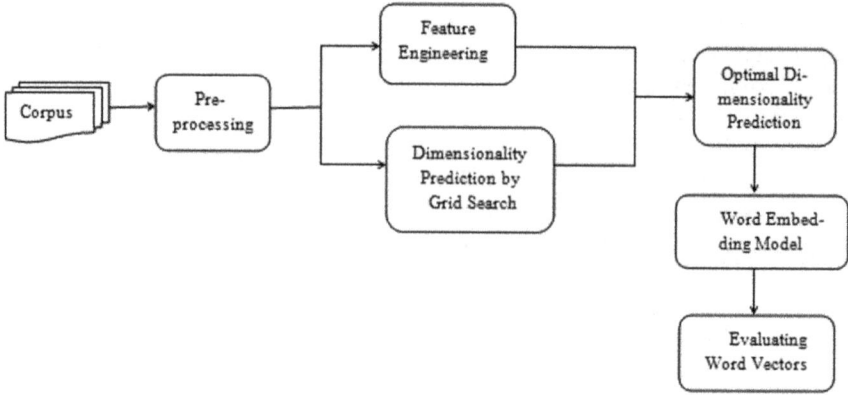

Fig. 1 Architecture of the proposed model

3.1 Pre-processing

Text data is collected from different resources like news articles, scientific articles, product reviews, movie reviews, and hotel reviews. Once the data is collected pre-processing is to be performed. Each document is split into a set of sentences and each sentence is tokenized into words. The pre-processing techniques used are: (i) stop word removal and (ii) lemmatization

Stop Word Removal. Stop words (a, an, the, in …) are the most common words in any natural language. In NLP tasks stop words will not have any influence. Hence, we removed the stop words using NLTK library in python. We have also removed special characters and numbers. Each word in the text corpus is converted into lower case.

Lemmatization. Lemmatization is the process of reducing the word to its basic form. Lemmatization is performed using morphological analysis by taking word structure and grammar relations into account. By reducing words into its root form, lemmatization reduces the vocabulary. For example, the words "playing," "plays," and "played" are converted into the word "play."

3.2 Feature Engineering

Once the corpus is pre-processed, the next step is the extraction of features from the corpus. In the proposed model, we used the following features corpus size (f_1), vocabulary size (f_2), frequency distribution (f_3), context distribution [22] (f_4), and PoS distribution (f_5).

(i) **Corpus Size.** It is defined as the total number of words in the text document.

(ii) **Vocabulary Size**. It is defined as the number of unique words in the input corpus.

(iii) **Frequency Distribution.** It is a triplet that comprises of the mean, standard deviation and the third moment of the frequencies of the words in the corpus. The second and third moments are denoted by M_2 and M_3, respectively.

(iv) **PoS Distribution**. For each document, we calculate the Jenson Shannon distance between POS distribution in the document and that in the corpus. If corpus is a single document, then each paragraph is treated as a document.

(v) **Context Distribution**. To estimate the distribution of context words, we use the entropy metric, the entropy of a word w is calculated using Eq. (8):

$$\text{Entropy}(w) = -\sum_{w_i \in c_w} \frac{p(w_i).\log p(w_i)}{\log |c_w|} \tag{11}$$

C_w denotes the set of context words which co-occur with a word w with-in a window size 10.

After calculating the context distribution of each word, we calculate the mean and the second moment of context distribution of words in the corpus. Algorithm 1 illustrates the steps adopted for training data set creation for the regression model.

Before building the regression model, we must prepare the dataset for the model. All the features that are extracted from the corpus are taken as independent variables and dimension for the corpus is taken as the dependent variable.

`Algorithm1 createDataset( Corpora C)`

Input: Set of text corpora collected from different domains

Output: Set of features extracted from each corpus $[F_1, F_2, F_3, F_4, F_5, F_6, F_7, F_8, F_9, Y]$

1.$n \leftarrow |C|$

2. for i = 1 to n do

3. for each $c_i \in C$ do   c

 3.1 F_1=corpus_size (c_i)

 3.2 F_2 = vocabulary_size(c_i)

 3.3 for each word w ϵ vocabulary(c_i)

   3.3.1 frequency[w] = count(w)

   3.3.2 entropy[w] $= - \sum\limits_{w_i \in c_w} \dfrac{p(w_i).\log p(w_i)}{\log |c_w|}$

 3.4 F_3=μ(frequency)

 3.5 F_4= σ(frequency)

 3.6 F_5= M_3(frequency)

 3.7 F_6=μ(entropy)

 3.8 F_7=M_2(entropy)

 3.9 (word,pos_tag)=pos_tagger(C_i)

 3.10 Find a vector V_t which represents the part of speech distribution over C_i

 3.11 P_{count}= no_of_paragraph(C_i)

 3.12 for k =1 to P_{count}

   3.12.1 V_k=POS_distribution(paragraph$_k$)

   3.12.2 d[k]=JensenShannon_dist(V_t,V_k)

 3.13 F_8=μ(d)

 3.14 F_9=M_2(d)

3.15 find the embedding dimensionality (y) for c_i.

To predict the dimensionality of the corpus, we perform the following steps:

(i) For each corpus in the training data, we generate word vectors using word2vec with randomly initialized parameters.

(ii) Depending on the corpus, we perform either classification or clustering by giving the word vectors generated in the step(i) as input. In our experiments,

for classification, we used gated recurrent unit (GRU) with ADAM optimizer [23]
(iii)   If the NLP task is classification, then we use accuracy as a performance measure, if the task is clustering then RandIndex is used.
(iv)   This process is repeated with different dimensionality values and the dimension which gives better performance is taken as the optimal dimension for the corpus.

3.3   Optimal Dimensionality Prediction

Once the features are extracted from the corpus, the next step is to build a regression model which predicts the optimal dimensionality for an input corpus. We have used a random forest-based regression model. Random forest is an ensemble method that can be used for classification or regression. Random forest-based regression models are known to exhibit a lower root mean square error (RMSE) when compared to other regression models. After learning the parameters of the regression model's hypothesis, for any corpus one can determine the optimal dimensionality for word embeddings. For a new corpus, first we extract the features from the corpus and the optimal dimensionality is predicted using the regression model.

3.4   Generating and Evaluating Word Embedding

We have used Word2Vec [11] to generate word vectors with the dimensionality predicted by the proposed regression model.
    We have evaluated the word embeddings in different ways:

- For a test corpus, we find the optimal dimensionality for the word embeddings using the proposed method and also by grid search and subsequently we compare both the results.
- We generate word vectors with the dimensionality predicted by the RODP model and with the commonly used dimensionality 300 using word2vec, and then, we select a word (w) from a corpus, find the most similar words to the w (similar words are found using cosine similarity between word vectors) and compare the results.

4   Experimental Setup

In this section, we describe the experiments carried out to predict the optimal dimensionality for a given corpus.
    Experiments are done on an Intel Core i3 processor running at 1.9 GHz. A detailed specification of the machine used for the performance analysis experiments is as

follows: Machine name: HP Pavilion 15-cc1xx, Operating System (OS): Windows 10, cores:4, RAM: 8 GB, RAM speed:1800 MHz, processor Intel(R) Core (TM) i5-8250U CPU @ 1.60 GHz. The proposed model was implemented using python.

4.1 Dataset Description

We have collected data from different sources [24, 25]. Once the data is collected, we partitioned each corpus in the dataset into multiple segments. Each segment is preprocessed. The different corpora used in the experiments are listed in Table 1.

Initially, each corpus is divided into partitions of different sizes and each partition is pre-processed as explained in Sect. 3.1. From the pre-processed data, the features are extracted using the algorithm1 described in Sect. 3. After extracting the features, the dimensionality of the word embedding is selected for each partition using the method described in Sect. 3.2. Table 2 shows part of the dataset we have created.

Table 1 Details of the corpus

S. No.	Corpus name	Language	Vocabulary size
1	IMDB [24]	English	88,584
2	AG-News[a]	English	70,396
3	Hotel reviews[b]	English	20,310
4	Movie reviews[c]	English	46,319

[a]http://groups.di.unipi.it/~gulli/AG_corpus_of_news_articles.html
[b]https://github.com/kavgan/nlp-in-practice/blob/master/word2vec/reviews_data.txt.gz
[c]https://www.cs.cornell.edu/people/pabo/movie-review-data/

Table 2 Sample dataset created from corpora

S. No.	F1	F2	F3	F4	F5	F6	F7	F8	F9	Y
1	18,325	6915	2.65	5.95	35.52	0.18	0.05	−0.21	0.44	150
2	91,506	14,460	6.32	89.69	8045.84	0.08	0.13	−0.41	0.11	100
3	30,323	5222	5.80	22.47	505.04	0.18	0.08	−0.41	0.23	300
4	613,480	50,491	12.15	102.71	10,551.05	0.13	0.03	−0.34	0.12	275
5	252,748	27,213	9.28	115.64	13,373.05	0.08	0.13	−0.38	0.20	200
6	929,602	46,319	20.06	504.12	254,142.9	0.08	0.13	−0.31	0.08	400
7	1,462,622	80,709	18.12	192.69	37,131.43	0.13	0.03	−0.27	0.12	50
8	18,801	6876	2.73	6.30	39.80	0.18	0.05	−0.23	0.43	300
9	4,144,378	142,724	29.03	408.71	167,046.5	0.13	0.03	−0.19	0.10	250
10	5,234,709	162,126	32.28	484.28	234,536.5	0.13	0.03	−0.17	0.10	300

4.2 Optimal Dimensionality Prediction

Once the data set is prepared, we built a regression model to predict the dimensionality for any given corpus. We have tested with different regression models such as linear regression, random forest regression model, Lasso regression model, ridge regression, and gradient boosting regression. Linear regression is the basic form of regression models, in this model, dependent variable is continuous and independent variables can be binary, discrete or continuous, and independent variables are linearly dependent on the dependent variable [26]. Lasso and Ridge regression models are simple forms of linear regression technique, which are used to offset the problem of over fitting. Lasso regression uses L1 regularization and ridge regression uses L2 regularization [27]. We have trained the regression model using 80% of the data set created using Algorithm1 as the training data and 20% as the test data. Among these models, the random forest regression model performed better. Comparison of the RMSE value achieved by the different regression models is depicted in Fig. 2.

From Fig. 2, it is observed that random forest regression model gives the least root mean square error, and hence in our experiments, we used the random forest regression model. From Table 2, it is observed that the relation between features and target variable is non-linear. If the relationship is non-linear, then linear regression may not be employed to make a good prediction [26].

Execution times of different regression algorithms on the data set created using the Algorithm1 are compared. Time taken (in micro seconds) to build the different regression models namely linear regression, random forest regression, Lasso regression, ridge regression, and gradient boosting regression are 16,990, 17,843, 45,401, 18,226, 1,312,947, respectively. Eventhough linear regression takes less time when compared with random forest regression model but random forest model is good at handling non-linear relations between independent and dependent variables [27].

We have conducted a number of experiments to identify the features that significantly affects the embedding dimensionality by calculating the correlation between

Fig. 2 Comparison of RMSE of different regression models

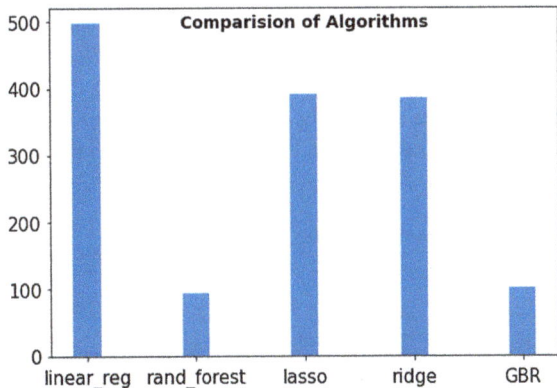

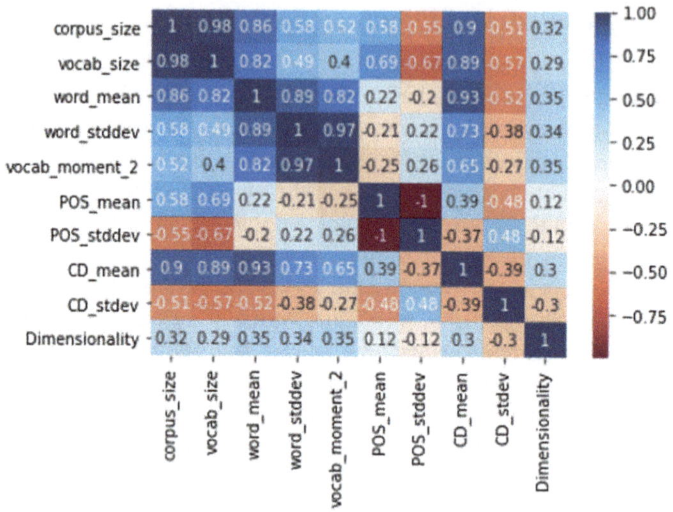

Fig. 3 Correlation between independent variables and dependent variable

the independent features extracted from the corpus and the dimensionality. Correlation is used to model the relationship between the two variables. If the correlation value between two variables is positive, it indicates increase in one variable leads to increase in another variable. This correlation is depicted in Fig. 3.

From Fig. 3 it is observed that the mean of frequency distribution, third moment of the frequency distribution, corpus size, and mean of the context distribution are highly correlated with the embedding dimensionality.

Table 3 shows the dimensionality predicted by the RODP model for different corpora, and the predicted dimensionality is compared with the dimensionality predicted by grid search. In grid search, for each corpus, we have tested with different dimensionalities ranging from 50 to 1000 and, with each dimensionality the accuracy of the classifier is measured. The dimensionality which yielded the maximal classification accuracy is selected as the optimal dimensionality.

Gated recurrent unit (GRU) is used for classification. Keras embedding layer is used with the vocabulary size as the input dimension and the word embedding dimension size as the output dimension. In our experiments, we have used softmax as the activation function and cross entropy as the loss function.

Table 3 Dimensionality predicted by RODP model

S. No.	Corpus	Dimensionality predicted (proposed model)	Dimensionality predicted (grid Search)
1	Twitter airline sentiment	123	123
2	BBC news	337	330
3	Movie reviews	373	375

Table 4 Performance of RODP for text classification task

Corpus name	Dimension and accuracy				
Twitter airline sentiment	Dimension	50	100	**123**	200
	Accuracy	59.78	60.69	**68.61**	68.41
Movie reviews	Dimension	100	200	**373**	400
	Accuracy	49.94	48.56	**51.23**	50.88

4.3 Performance Evaluation

Quality of the word embeddings generated with the optimal embedding size predicted by the RODP model is tested using text classification and word similarity tasks. We have used the Word2Vec model to find the word embeddings.

Text Classification. Text classification is the process of assigning pre-defined labels to text. We used Twitter Airline Sentiment dataset[1] for this task. The optimal dimensionality predicted by RODP model for the twitter airline sentiment corpus is 123. We generated word vectors with dimensionality 123 using word2vec model and for classification we used gated recurrent unit (GRU). Table 4 shows the accuracy for different values of embedding dimensionality.

From Table 4, it is observed that the dimensionality predicted by RODP model yields a better classification accuracy.

Word similarity. In the word similarity task, word embeddings are evaluated in two ways namely: (i) By computing the Spearman Rank Correlation [28] between the cosine similarity between the word vectors and the ground truth similarity. Spearman correlation coefficient measures the strength between two ranked variables. For word similarity evaluation, we have used MEN [29] and wordsim353 [30] test datasets which contains human judgment similarity values between pairs of words. We have trained the word2vec model with text8 corpus and the optimal dimensionality predicted for text8 corpus is 243.

(ii) Finding the most similar words for a word using the dimensionality predicted by RODP model and the generally used dimensionality 300.

To evaluate the word embeddings generated, we have taken a word plane from the twitter airline sentiment corpus and we compared the most similar words ranked based on the cosine similarity for the dimensionalities 300 and 123 (predicted by our model). Results are shown in Table 5.

5 Conclusion

In this paper, we propose a model to predict the optimal dimensionality of word embeddings for text corpora. The proposed model entitled regression-based optimal

---

[1]https://www.kaggle.com/crowdflower/twitter-airline-sentiment.

Table 5 Similar words for the word "plane" based on cosine similarity

Word chosen from corpus	Dimension: 300		Dimension: 123	
	Similar word	Cosine similarity	Similar word	Cosine similarity
Plane	Pilot	0.9688	Pilot	0.9627
	Board	0.9667	Board	0.9620
	Sit	0.9547	Sit	0.9543
	Landed	0.9523	Landed	0.9454
	runway	0.9460	Runway	0.9388

dimensionality prediction (RODP) uses a random forest regression model to predict the optimal dimensionality. Dimensionality prediction helps in reducing time and space requirements of the subsequent NLP tasks. We also identified the features that exert a significant influence on the dimensionality of the word embeddings. In future, we plan to investigate the impact of semantic factors on the embedding dimensionality.

Acknowledgements The authors would like to thank Ms. Chalamarla Pavana Sai Sree and Mr. Gaurav who are undergraduate students in NIT Andhra Pradesh, for their help with dataset collection and experimental setup.

References

1. Salton G (1971) The SMART retrieval system—experiments in automatic document processing. Prentice-Hall, Inc
2. Salton G, Buckley C (1988) Term-weighting approaches in automatic text retrieval. Inf Process Manage 24(5):513–523. https://doi.org/10.1016/0306-4573(88)90021-0
3. Jones KS (1972) A statistical interpretation of term specificity and its application in retrieval. J Documentation 28(1):11–21. https://doi.org/10.1108/eb026526
4. Breese JS, Heckerman D, Kadie C (1998) Empirical analysis of predictive algorithms for collaborative filtering. In: Proceedings of the 14th conference on Uncertainty in Artificial intelligence (UAI'98), Morgan Kaufmann Publishers Inc., San Francisco, CA, USA, pp 43–52
5. Ozsoy MG (2016) From word embeddings to item recommendation. arXiv: 1601.01356
6. Musto et al (2016) Learning word embeddings from wikipedia for content-based recommender systems. https://doi.org/10.1007/978-3-319-30671-1_60
7. Yin Z, Chang K, Zhang R (2017) DeepProbe: information directed sequence understanding and chatbot design via recurrent neural networks. In: Proceedings of the 23rd ACM SIGKDD international conference on knowledge discovery and data mining. ACM, pp 2131–2139
8. Lin Y, Lei H, Wu J, Li X (2015) An empirical study on sentiment classification of chinese review using word embedding. In: 29th Pacific Asia conference on language, information and computation, pp 258–266
9. Kusner MJ, Sun Y, Kolkin NI, Weinberger K (2015) From word embeddings to document distances. In: Proceedings of the 32nd international conference on machine learning, vol 37, pp 957–966
10. Pennington J, Socher R, Manning CD (2014) GloVe: global vectors for word representation. In: Empirical methods in natural language processing (EMNLP), pp 1532–1543

11. Mikolov T, Chen K, Corrado G, Dean J (2013) Efficient estimation of word representations in vector space. In: ICLR
12. Ling S, Song Y, Roth D (2016) Word embeddings with limited memory. In: Proceedings of the 54th annual meeting of the association for computational linguistics, pp 387–392
13. Aghaebrahimian A, Cieliebak M (2019) Hyper parameter tuning for deep learning in natural language processing. In: 4th Swiss text analytics conference (SwissText 2019), Winterthur
14. Xu W, Rudnicky A (2000) Can artificial neural networks learn language models? In: International conference on statistical language processing, pp 202–205
15. Bengio Y, Schwenk H, Senécal JS, Morin F, Gauvain JL (2003) A neural probabilistic language model. J Mach Learn Res 3(6):1137–1155
16. Mikolov T, Sutskever I, Chen K, Corrado GS, Dean J (2014) Distributed representations of sentences and documents. In: Proceedings of ICML
17. Kaji N, Kobayashi H (2017) Incremental skip-gram model with negative sampling. In: Proceedings of the conference on empirical methods in natural language processing, pp 363–371
18. Pennington J, Socher R, Manning C (2014) Glove: global vectors for word representation. In: Proceedings of the 2014 conference on empirical methods in natural language processing (EMNLP), pp 1532–1543
19. Wang Y (2019) Single training dimension selection for word embedding with PCA. In: Proceedings of the 2019 conference on empirical methods in natural language processing and the 9th international joint conference on natural language processing, pp 3597–3602
20. Das S, Ghosh S, Bhattacharya S, Varma R, Bhandari D (2019) Critical dimension of Word2Vec. In: 2nd International conference on innovations in electronics, signal processing and communication (IESC), Shillong, India, pp 202–206. https://doi.org/10.1109/iespc.2019.8902427
21. Yin Z, Shen Y (2018) On the dimensionality of word embedding. In: Advances in neural information processing systems, pp 895–906
22. Lee GE, Sun A (2020) Understanding the stability of medical concept embeddings. arXiv: Computation and Language
23. Kingma D, Ba J (2015) Adam: a method for stochastic optimization. In: International conference on learning representations (ICLR)
24. Maas A, Daly RE, Pham PT, Huang D, Ng AY, Potts C (2011) Learning word vectors for sentiment analysis. In: Proceedings of the 49th annual meeting of the association for computational linguistics, human language technologies, vol 1. Association for Computational Linguistics, pp 142–150
25. http://mattmahoney.net/dc/textdata.html. Last accessed 1 Sept 2011
26. Schneider A, Hommel G, Blettner M (2010) Linear regression analysis: part 14 of a series on evaluation of scientific publications. Deutsches Arzteblatt international 107(44):776–782. https://doi.org/10.3238/arztebl.2010.0776
27. Hastie T, Tibshirani R, Friedman JH (2001) The element of statistical learning. Springer, New York. https://doi.org/10.1007/978-0-387-21606-5
28. Dodge Y (2018) Spearman rank correlation coefficient. In: The concise encyclopedia of statistics. Springer, New York. https://doi.org/10.1007/978-0-387-32833-1
29. Bruni E, Boleda G, Baroni M, Tran NK (2012) Distributional semantics in technicolor. In: Proceedings of the 50th annual meeting of the association for computational linguistics (Volume 1: Long Papers), Jeju Island, Korea. Association for Computational Linguistics, pp 136–145 (2012)
30. Finkelstein L, Gabrilovich E, Matias Y, Rivlin E, Solan Z, Wolfman G, Ruppin E (2002) Placing search in context: the concept revisited. ACM Trans Inf Syst 20(1):116–131

Chapter 25
Classification of ECG Arrhythmia Using Different Machine Learning Approach

Asma Parveen, R. M. Vani, P. V. Hunagund, and Maisoun Ali Soher-wardy

1 Introduction

The World Health Organization (WHO), heart diseases are the number one cause of death. Over 17.7 million people died from CVDs which is an about 31% of all deaths, and over 75% of these deaths occur in low- and middle-income countries [1]. Arrhythmia is a type of cardiovascular disease that refers to any irregular change from the normal heart Beat.

An electrocardiogram (ECG) is a medical test which detects heart abnormality by measuring the electrical activity of the heart. A heart stimulate tiny electrical impulses which spread through the heart. These impulses can be detected by an ECG graph machine. An ECG machine is used to record the electrical activity of the heart and displays this data as a trace on a graph paper. ECG helps to find the cause of symptoms or chest pain and also helps to detect abnormal heart Beat.

The normal healthy heart have a characteristic shape. Any irregularity in the heart beat or damage to the heart muscle can change the electrical activity of the heart, so shape of the ECG gets changed.

ECG is the recording of the electrical activity of the heart beats and has become one of the most important tool in the diagnosis of heart diseases. The high mortality

A. Parveen (✉) · R. M. Vani · P. V. Hunagund · M. A. Soher-wardy
Gulbarga University Kalaburagi, Gulbarga, Karnataka, India
e-mail: asmamaamsfamilyg@gmail.com

Pri University College of Science Kalaburagi, Gulbarga, Karnataka, India

R. M. Vani
e-mail: vanirm123@rediffmail.com

P. V. Hunagund
e-mail: prabhakar_hunagund@yahoo.co.in

M. A. Soher-wardy
e-mail: maamsfamilyg@gmail.com

© The Author(s), under exclusive license to Springer Nature Singapore Pte Ltd. 2021   319
X.-Z. Gao et al. (eds.), *Applications of Artificial Intelligence in Engineering*, Algorithms for Intelligent Systems, https://doi.org/10.1007/978-981-33-4604-8_25

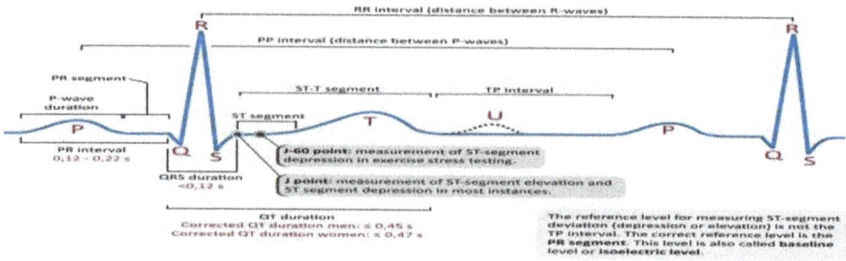

Fig. 1 Normal ECG waveform [5]

rate, the early detection and precise discrimination of ECG signal is essential for the treatment of cardiac patients. ECG signals can be classify using machine learning algorithms can provide the doctors to confirm the diagnosis. Classification and detection of arrhythmia types can help in identifying the abnormality in ECG signal of a patient. Major issues [2, 3]. Cardiac arrhythmias indicate most of the cardiovascular problems that may lead to chest pain, cardiac arrest or sudden cardiac death [4]. In ECG, categorizing are lack of order of ECG features, variability amongst the ECG features, individuality of the ECG patterns, non existence of optimal classification rules for ECG classification, and variability in ECG waveforms of patients.

Figure 1 shows the graphical representation of ECG signal.

In this paper, two algorithm are used like SVM and K-NN classifier are used to classify the arrhythmias into 5 classes. The UCI dataset is used for experiment purpose contain 187 attributes of ECG signal [6]. ECG categorizing includes steps namely preprocessing, feature extraction, feature normalization, and classification.

2 Related Work

Yu Hen Hu et al., Describe the "mixture-of-experts" (MOE) proceed to develop personalized electrocardiogram (ECG) beat classifier in an attempt to further improve the perfor mance of ECG processing and to offer individualized health care. A small customized classifier is developed supported brief, patient-specific ECG data. It is then combined with a worldwide classifier, which is tuned to an outsized ECG database of the many patients, to make a MOE classifier structure. Tested with MIT-BIH arrhythmia database [3].

Hu YH et al., This paper deals with the applications of artificial neutral networks for electrocardiographic QRS detection and beat classification. The authors used an adaptive multi-layer perception structure to model the non-linear background noise so on enhance the QRS complex. Preliminary results using the MIT-BIH (Massachusetts Institute of Technology/Beth Israel Hospital, Cambridge, MA) arrhythmia database are stimulated [7].

Mi Hye Song et al., Describes the algorithm of linear discriminate analysis (LDA) and a support vector machine (SVM)-based classifier. Seventeen original input features were bring out from pre-processed signals. The performance of the SVM classifier with reduced features by LDA showed above by principal component analysis (PCA) and even with original features. This classifier was compared with multilayer perceptrons and fuzzy inference, the accuracy of discrimination of normal sinus rhythm, arterial premature contraction, supraventricular tachycardia, premature ventricular contraction, ventricular tachycardia, and fibrillation were 99.307%, 99.274%, 99.854%, 98.344%, 99.441%, and 99.883%, respectively. The SVM classifier offered better performance than the MLP classifier [8].

Kemal Polat et al., Describe ECG arrhythmias using principal component analysis and least square support vector machine. The approach system has two stages. In the first stage, the 279 features is reduced to fifteen features using principal component analysis. In the second stage, diagnosis of ECG arrhythmias was conducted by using LS-SVM classifier, dataset is used UCI. Classifier system consists of three stages of test set and training set. 50 test—50% training-test dataset, 70 test—30% of training dataset and 80 of test—20% of training dataset, subsequently, the obtained classification accuracies 96.86%, 100%, 100%. The end benefit would be to assist the physician to make the final decision without hesitation [9].

Prajwal Shimpi et al., Describe the approach to classify the ECG data into one of the sixteen types of arrhythmia using machine learning. The proposed method uses the UCI machine learning respiratory dataset of cardiac arrhythmia to train the system on 279 different attributes with five classes. In order to increase the accuracy, the method uses principal component analysis for dimensionality reduction, bag of visual words approach for clustering and compares different classification algorithms like support vector machine, random forest, logistic regression, and K-nearest neighbor algorithms, thus choosing the most accurate algorithm, support vector machine [10].

Asma Parveen et al., Asma Parveen et al., Electrocardiogram (ECG) demonstrate the electrical activity of the heart. Feature extraction and segmentation in ECG plays a significant role in diagnosing most of the cardiac disease. From the literature survey, it is observed that many author studied and analysis an ECG signal, by studying the several algorithm have developed for analysis of ECG. Nowadays, artificial intelligence very much involved for analyzing of ECG. The several machine learning and deep learning approach are used for classification the normal and abnormal ECG signals. The classification performance of different algorithms is based on accuracy, sensitivity, selectivity, specificity. The research is carried out on the dataset taken from University of California at Irvine (UCI) and MIT-BHI arrhythmias database [6, 11].

2.1   Classification of Machine Learning Algorithms

The machine learning algorithms are used for classification of ECG signal. We investigate and compare the performance of two machine learning algorithms namely k-nearest neighbor (K-NN), support vector machines (SVM) machine learning algorithms. K-nearest neighbor (K-NN) algorithm, lack of training, easy to perform, analytically traceability, adaptable to local information, suitable for parallel operation, noisy resistant to educational data classification applications with many advantages particularly preferred.

Support vector machine (SVM) is an algorithm that can be used easily in difficult and complex data thanks to its high accuracy. SVM is generally known as non-parametric models, but this is considered because the parameters available in SVM are not previously defined and their numbers depend on the training data used. SVM faces optimization problems but this is solved by a binary formulation that makes it dependent on the number of support vectors. This eliminates the algorithmic complexity that can be used in a linearly separable state. With these advantages, SVM is one of the most widely used classification methods in mathematical optimization [12]. SVM can be used in a variety of domains such as face and speech recognition, face detection, and image recognition. It's simple algorithm, as outlined in [13], makes it an easy to use classification technique.

K-nearest neighbor (K-NN) algorithm, lack of training, easy to perform, analytically traceability, adaptable to local information, suitable for parallel operation, noisy resistant to educational data classification applications with many advantages particularly preferred. The K-nearest neighbor algorithm (K-NN) is one of the basic, functional, and popular classification algorithms. K-NN algorithm, the training set examples with n-dimensional numerical attributes identified. In this method, the similarity of the data to be classified to the normal datasets in the learning set is calculated; classification is made according to the threshold value obtained by averaging the k data considered to be the closest.

The biggest disadvantage is that each of the nearest k is equally important in the K-NN algorithm. The closer the neighbor is, the more likely it is that f vector is in this neighbor's class. Therefore, it is more accurate to assign neighbors with different voting weights by looking at their distance from the f vector [14].

2.2 Performance Matrix

In this section, we will examine the results of various classification techniques. Classification results of K-NN, SVM, are included. The results of the classifiers were evaluated according to the results of accuracy. Train set and test set values were also examined. Parameter configurations were tested for each classifier to determine the most successful performance values. In the performance metrics, the confusion matrix "Table 1" used for binary classification was calculated.

Table 1 Confusion matrix

Estimation class		
Actual class	Positive	Negative
Positive	TP (True Positive)	FN (False Negative)
Negative	FP (False Positive)	TN (True Negative)

Table 2 Comparison of various classification Techniques

Classifier	Classes	Accuracy
SVM	0, 1, 2, 3, 4	97.87
K-NN	0, 1, 2, 3, 4	96.85

2.3 Accuracy

The most commonly used classification accuracy to determine classification performances is used to measure the overall effectiveness of the classifier.

$$\text{Accuracy} = \frac{\text{TP} + \text{TN}}{\text{TP} + \text{FP} + \text{FN} + \text{TN}} * 100$$

2.4 Train and Test Set

The train set is used for training the classifier, and the test set is used for the evaluation of the classifier. The classifier with high accuracy for the train set is over fit if the test set has low accuracy. An algorithm must undergo training-test steps in order to make predictions and to make predictions in numerical or classification. In this study, the dataset was formed as 80% training and 20% test group. The test and train results of all classification algorithms are shown in Table 2.

2.5 Block Diagram of ECG Classification Using Machine Learning

The block diagram is consists of 5 Block, in first block is contain CSV file of ECG signal, second block is feature are extracted from the file, pre-processing the signals then applying the algorithms and then train the model and checking the accuracy (Fig. 2).

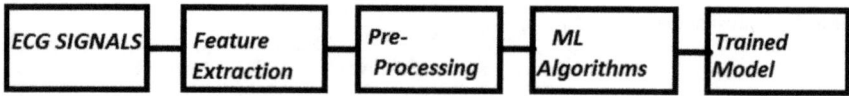

Fig. 2 Block diagram of ECG classification using ML algorithms

3   Results and Conclusion

In this section, we will examine the results of various classification techniques. Classification results of K-NN, SVM are included. The results of the classifiers were evaluated according to the results of accuracy. The accuracy of the SVM is 97.87%, and K-NN is 96.85 parameter configurations were tested for each classifier to determine the most successful performance values.

We have discussed the two machine learning approach in ECG analysis like SVM, K-NN, finally, identify that SVM classifier are giving better result.

References

1. World Health Organization (2017) Cardiovascular diseases (CVDs). http://www.who.int/med iacentre/factsheets/fs317/en/ Accessed 18 Apr 2018
2. Sharma LN, Tripathy RK, Dandapat S (2015) Multiscale energy and eigenspace approach to detection and localization of myocardial infarction. IEEE Trans Biomed Eng 62(7):1827–1837
3. Hu YH, Palreddy S, Tompkins WJ (1997) A patient-adaptable ECG beat classifier using a mixture of experts approach. IEEE Trans Biomed Eng 44(9):891–900
4. Afkhami RG, Azarnia G, Tinati MA (2016) Cardiac arrhythmia classification using statistical and mixture modeling features of ECG signals. Pattern Recogn Lett 70:45–51
5. Google.com. (2020) [Online]. Available: http://ECGgraphnormal
6. UCI machine learning repository: Arrhythmia data set (1998). [Online]. Available: https://arc hive.ics.uci.edu/ml/datasets/Arrhythmia. Accessed 10 Feb 2017
7. Hu YH, Tompkins WJ, Urresti JL, Afonso VX (1994) Applications of artificial neural networks for ECG signal detection and classification
8. Song MH, Lee J, Cho SP, Lee KJ, Yoo SK (2005) Support vector machine based arrhythmia classification using reduced features. 3(4):571–579
9. Uyar A, Gürgen F (2007) Arrhythmia classification using serial fusion of support vector machines and logistic regression. IEEE
10. Shimpi P et al (2017) A machine learning approach for the classification of cardiac arrhythmia. In: Proceedings of the IEEE 2017 international conference on computing methodologies and communication (ICCMC)
11. https://physionet.org/content/mitdb/1.0.0/MIT-BIHArrhythmiaDatabase
12. Srujan Raju K et al (2018) Support vector machine with K-fold cross validation model for software fault prediction. Int J Pure Appl Math 118(20):321–334
13. Karamizadeh S et al (2016) Advantage and drawback of support vector machine functionality. In: 2014 International conference on computer, communications, and control technology (I4CT) (2014): n. pag. Web. 5 Dec 2016

14. Guo G et al, KNN model-based approach in classification. School of Computing and Mathematics, University of Ulster Newtownabbey, BT37 0QB, Northern Ireland, UK
15. Parveen A et al (2019) Machine learning and deep learning approach for analysis of ECG signal: a review. In: International conference on signal processing and big data analysis (ICSPBD) held on 27 July 2019

Chapter 26
Role of Cloud Computing to Support Big Data and Big Data Analytics for Education Par Excellence

Anjum Zameer Bhat, Baldev Singh, and Teba Fadhil

1 Introduction

Big data and big data analytics in the recent past have dominated the IT world finding their place almost at every enterprise irrespective of scale and type. Big data is being implemented in lot of areas to achieve the insights of unknown facts. As the saying goes "If you can see the invisible, you can achieve the impossible." The big data and big data analytics has find its implementation in various areas like online shopping, retail business, marketing, advertisements, healthcare, e-commerce, and certainly the higher education. Cloud computing has been utilized in higher education for variety purposes [1] to support student learning and to support various information systems in a typical higher education setup. The implementation of cloud computing in higher education inevitably yields enormous benefits for higher education establishments and instead of various limitations and hurdles in the implementation of cloud computing in higher education [2] it has been adapted at various higher education establishments. Cloud computing has been implemented for providing the expensive software resources to students in higher education that enable students to work with expensive high end software applications/multimedia applications without necessitating a purchase for such high end and expensive applications [3]. Cloud computing can be utilized for big data and big data analytics in higher education establishments. Teaching and learning have seen an unprecedented transformation over the last several decades with enhanced tools and aids used for teaching and learning.

A. Z. Bhat (✉) · B. Singh
Vivekananda Global University, Jagatpura Jaipur, India
e-mail: azameer@mec.edu.om

B. Singh
e-mail: baldev.vit@gmail.com

T. Fadhil
Teba Fadhil Middle East College, Knowledge Oasis, Seeb, Oman
e-mail: teba@mec.edu.om

© The Author(s), under exclusive license to Springer Nature Singapore Pte Ltd. 2021   327
X.-Z. Gao et al. (eds.), *Applications of Artificial Intelligence in Engineering*, Algorithms for Intelligent Systems, https://doi.org/10.1007/978-981-33-4604-8_26

Innovative techniques implemented for teaching and learning to achieve excellence in education [4]. Big data analytics has been utilized in the past, and lot of research is currently going on for implementing big data in higher education. The implementation of big data in higher education can yield lot of benefits as there is always a need to enhance teaching and learning and apply appropriate teaching methods and pedagogies which are suitable to a particular cohort of students. An exhaustive study for knowing the appropriateness of teaching methods employed in a particular environment of student learning is utmost important, and more importantly, this exercise should be a continuous process that will systematically enhance one's teaching experience and would provide insights about the appropriateness of teaching methods in a particular environment or level of apprentices.

A typical higher education institution nowadays deploys many information systems and learning management systems for variety of purposes. Although these information systems provide some good capabilities to higher education stake holders to operate in an effective and efficient manner, however, there are still some limitations of these systems because of which higher education administrators and management face some challenges. With the latest technology available in twenty-first century, every organization including higher education establishments want to yield maximum through the use of technology. Planning in higher education establishment is not a cake walk, rightly so as modern higher education establishments are complex with respect to immediate and long term future requirements in terms of infrastructure, equipment, educational trends, industry requirements, employability, and so on. It becomes utmost difficult at times to plan certain things for future and any incorrect decision of management can have serious repercussions on the development of the institution, the career of the students, employability of the graduating students, and on other important aspects. There has to be a technology aid that can assist management in taking critical decisions and in policy making so that appropriate decisions could be taken in a timely manner which will have a positive impact on the development of the institution. Big data analytics can provide the management of the institution with certain facts and insights with a possible visualization and drill down capabilities making it easier for management to take decisions based on certain preconditions. This will provide management the preemptive and preventive capabilities that would be very beneficial for the development of the institution. However, implementing such a system is not a straight forward task, it involves a complex infrastructure which is a combination of data collection devices or sensing devices, type of data to be collected, big data storage mechanism, cloud services, big data analytics tools, and application for visualization and results.

Cloud computing has a significant role to play in higher education establishments and in future IT implementations cloud computing in its varied flavors and service models may find prevalent implementation in higher education establishments. Higher education establishments may subscribe to the services provided by various public clouds or may opt to go for the implementation of their own private cloud for variety of reasons. The complexity of information systems and support systems that are needed in a typical higher education institution is too much and increasing at a greater pace which poses a significant challenge for implementers, administrators, and other IT officials. The complexity of these systems not only causes dissatisfaction of services that are provided but tenders a substantial threat to the privacy, confidentiality, security, and availability of important resources. Cloud computing provides a substantial and optimum solution to most of these concerns and in modern era of technology assisted teaching and learning there are various new areas of research and development that are being tested and implemented in higher education. One of such areas is the implementation of big data and big data analytics in higher education. It is primarily important here to mention that the implementation of big data cannot be materialized with having proper mechanism and technology for data storage and retrieval. This data storage and retrieval should be efficient enough so that it can be easily assessed from anywhere. The fetching of data from the storage can be made by an application which processes a data for big data analytics, predictive analytics, and visualization.

In the modern era of information and communications technology, the role of cloud computing in higher education has significantly increased with the advent and inclusion of various new technologies into education sector. The implementation of IoT also is inevitable and has already seen some implementations and productive uses in education in general and higher education in particular. Cloud-based technologies are having capabilities to store huge amount of data which generally gets created in big data and IoT implementations. The technology assisted teaching and learning is going to pose a very significant challenge as far as maintaining and handling of data is concerned. This creation huge amount of data in higher education establishments may well result in a major justification for investing huge amount of money on private cloud implementation in higher education establishments. It may also increase the use and subscription of public cloud services for higher education establishments especially for such type of services where privacy, confidentiality, and other security related issues are not a major concern for higher education establishments. As a matter of fact both public and private cloud implementation are going to be widespread in higher education establishments all over the world bearing the fact that new and innovative technologies are being implemented to achieve excellence in the field of education. Keeping these aspects in mind cloud computing has a significant role to play in higher education not only in providing services that are directly based on cloud technology, however, supporting various innovative technologies which have a significant dependence on cloud-based systems.

2   Related Work

Big data and cloud computing have been a hot potato in the recent past as far as the IT research is concerned. A lot of research has been done on big data and big data analytics and the implementation of big data in higher education [5, 6]. Cloud computing in education has also attracted a lot of research work in which various researchers have discussed various aspects of cloud computing in education. The better use of public cloud computing for business, government, and education now seems predictable due to lower cost and greater ease of access [7]. Learning analytics can pierce the fog of indecision around how to allot resources, foster competitive benefits, and most significant, improve the quality and value of the learning experience [8]. While big data and analytics are not panaceas for resolving all of the matters, decisions and assessments confronted by higher education executives, these technologies can certainly become element of the solutions and resolutions which can be integrated into organizational and instructional purposes [9]. Cloud computing is a prevailing technology to accomplish massive-scale and complex computing. It excludes the necessity to conserve costly computing hardware, dedicated space, and software. Immense growth in the range of data or big data generated through cloud computing has been observed [10]. There is a significant evidence of big data implementation in higher education institutions for academic improvements and for decision support to the management for better planning and policy making. Big data analytics has clearly indicated that it is having capabilities to provide preemptive and predictive facilities to the higher education establishment administration for better decision making. The planning in a higher education establishment can be cumbersome and painful job, with big data and analytics the management of higher education institutions can yield a great benefit of a technology aid making them wiser to take critical policy and future decision which can have significant impact on the development of the institution. Upcoming edification is rather frequently associated with innovative technologies like ubiquitous computing devices and flexible class room design. However, the utmost impounding feature which is ignored and which would revolutionize the higher education arena is "data." Higher education is a domain where incredible volumes of data are obtainable. Analyzing this data would lead to accomplishment of advantages of greater bound. Many higher education establishments all around the world fail to efficiently utilize the power of data and yield something significant out of it by effective implementation of predictive and big data analytics [11]. Learning Analytics (LA) is an arena of inquiry that comprises the assembly, evaluating and reporting of data associated to learners and their environments with the determination of enhancing the learning experience. Stakeholders in learning analytics are students, educators, researchers, institutions, and government agencies. The means by which analytics data flows from students to other stakeholders comprises a hierarchy, where all participants are able to offer input and provide suggestions to improve the learning process for the benefit of the students. Challenges met by stakeholders include the movement of conventional analytics to learner-centered analytics, working with datasets across various settings, addressing issues with technology, and resolving ethical concerns [12]. The big data analytics and predictive analytics are going to be the norm for future higher education establishments when it comes to planning, infrastructure development, teaching and learning methods, innovative teaching strategies, and devising plans to enhance teaching experience. Big data along with few more innovative and imminent technologies like Internet of Things, cloud computing, and machine learning can contribute in a significant manner in higher education establishments in terms of enhancing teaching and learning as well as for decision support and overall research and development in higher education establishments.

3   Cloud-Based Architecture to Support Big Data and Big Data Analytics in Higher Education Establishments

The architecture of cloud which can support the big data implementation in higher education establishments should be required components and applications to provide analytics capabilities to the higher education establishment as per as specific requirement. Before analyzing the cloud architecture for big data in educational establishments let us first analyze and understand the exact role of the cloud component in big data analytics implementation. Refer to Fig. 1. Which shows a typical architecture of the big data implementation in higher education and how the data collected can be processed with different layers. The architecture shown in Fig. 1 has five different levels or layers in the architecture. Each layer of the architecture has a specific function to perform in an overall architecture. The layer 1 pertains to the mechanism

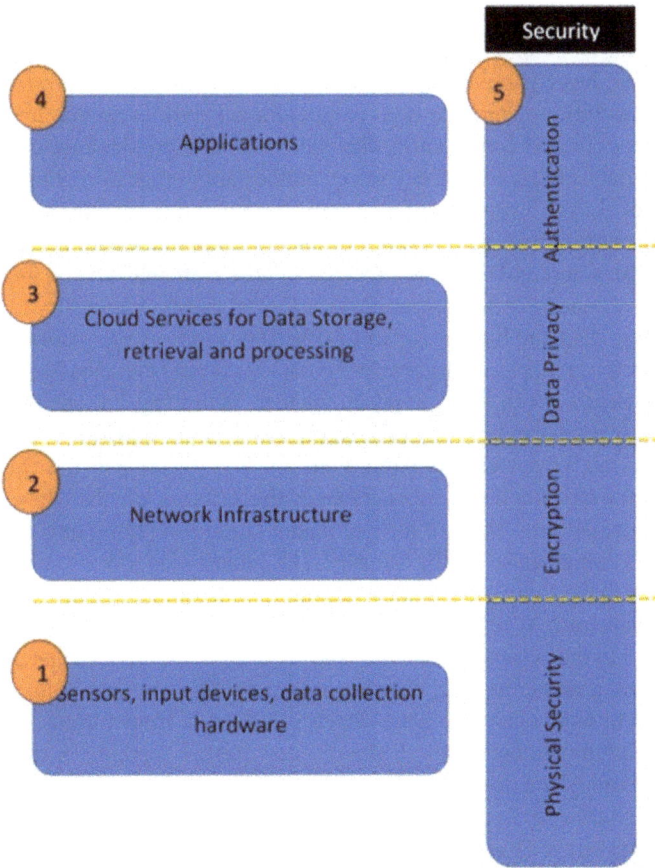

Fig. 1 A typical architecture of big data systems in higher education

and the devices that are used to attain and collect the data that is used for big data analytics, this is one of the difficult areas and in fact the most critical one as far as the success of big data analytics is concerned. The type of devices used for collection of data and the appropriate data that needs to be collected for big data analytics is very important for overall success of any big data implementation. There are numerous devices and ways by which data collection can take place in a higher education institution. The data can be collected by using sensing devices that are embedded at different places. The data collection can happen from more conventional devices or systems, e.g., feedback of students can be transformed into a particular format or can be taken as unstructured data. Sensors in educational establishments have been used in the past to collect the data that can be formulated into some sensible inferences which can benefit the higher education establishments [13].

Figure 1 shows layer 2 as the network infrastructure layer which contains the network devices and protocols to connect various devices in the layer 1 to the cloud services layer 3. The functionality of the network layer is to make communication possible between two layers, i.e., layer 1 and layer 3. It may not be out of place to mention here that cloud services may be hosted in campus or off campus. These services can be hosted in a public cloud or a private cloud. The network layer 2 has the responsibility of provide high bandwidth capabilities so that huge amount of data that is collected continuously through sensors, input devices, unstructured, and structure data, etc., is seamlessly stored on the data storage available through the cloud services. Layer 3 provides the system some unique capabilities not only for data storage and retrieval, however, for various types of operations and processing. A cloud infrastructure in layer 3 is expected to host high end servers, applications, storage devices, analytics tools, databases, etc. Databases specifically are capable of handling big data and big data analytics. Various types of applications are made available which can represent the processed data retrieved from cloud services layer 3 and can present facts in an appropriate manner. Many applications that are used for data visualization can be a part of Layer 4 which is the application layer.

Security in all IT systems is paramount and any lacking in the security can have significant repercussions. Irrespective of the services whether they pertain to networks, information systems or for that matter cloud computing or big data, there cannot be any compromise on the security. The security layer 5 provides optimum security levels to the complete system at various levels by implementing and adapting various mechanisms. At layer 1, it is utmost mandatory to ensure that data that is being inputted to the system is authentic and security of the devices that are responsible for fetching and collecting data is paramount. The physical security at layer 1 is provided to ensure that the data, which is collected from various devices and sensors, is a legitimate data, and all the devices that are working in this layer are physically secured. The devices and sensors utilized are working in an appropriate manner to fetch the correct data; i.e., devices are not having error which can result in collection of incorrect data. Physical security also ensures continuous monitoring of all the devices that are used for data collection purpose. The safety of these devices from fire, water, humidity, etc., is also the responsibility of the physical security layer. At layer 2, security layer ensures that encryption of data is done so that data which is transmitted using the network layer is secure from spoofing, man in the middle, and some other attacks. Security at this layer ensures that data integrity does not get compromised and the same time ensure confidentiality of the transferred data. Security mechanism that is deployed at the cloud services layer ensures that privacy of the data is maintained in all the transactions that take place during data access and fetching of processed and analyzed data. The application that are used by various people at different levels provide appropriate security mechanisms and checks and balances. The access is granted to the application for authenticated users only. These applications can provide different facilities to different stake holders in the higher education institution. The faculty members may be provided with relevant data that pertains to teaching and learning or conclusions and inferences that indicate a particular teaching method is more suitable for a specific type of a cohort or level of the students. Teachers may get an idea about which method of teaching a particular topic attracts maximum attention of students and which methods were most successful in making students understand some complex concepts related to a particular subject area. Similarly, an entirely different kind of facility may be provided to the people in administrative and management positions or policy makers. They will be getting insights about planning future course of action with respect to launching of courses, infrastructure development, admission trends, employability, forecasting, and enhancement of research and development at the institutional level. The applications provided different authorizations to different users assessing various applications.

4   Cloud Architecture for Higher Education Institutions

Cloud architecture or the layer 3 in Fig. 1 needs to be explained in quite a detail to understand the cloud implementation and its characteristics, the cloud implementation that is suitable for big data implementation in higher education should have specific components to provide expected services and support to big data implementation. Figure 2 shown below depicts various layers of the cloud with an additional layer $L_r$aaS (Learning Resources as a Service) [1]. There are certain services and facilities that are added to the learning resources layer. The services and applications provide capabilities to the cloud to be able to handle the big data and big data analytics. Various big data applications. Apache storm is an open source real-time distributed computing system, Apache spark is an open source general purpose distributed framework for cluster computing. Mahout is a linear algebra framework and Apache MAPR or Map Reduce is a powerful framework for processing large, distributed sets of structured or unstructured data on a Hadoop cluster [14]. Hadoop is an open source software collection that enables using network of many computer to solve a particular problem which usually involves huge amount of data and processing/computation.

As there are various service architecture which have been proposed in the past by the researchers in cloud computing which may suite the requirements of a specific cloud implementation based on the type of services that are required [15]. As three

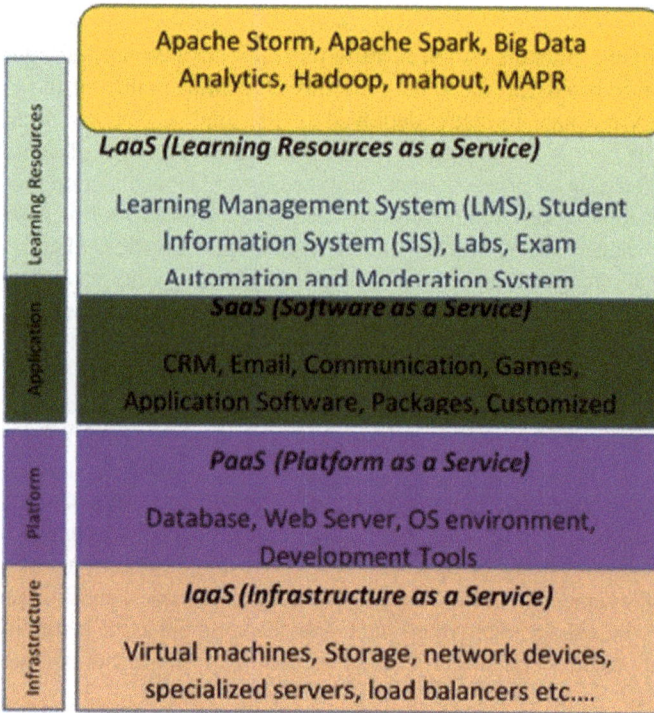

Fig. 2 Cloud architecture for higher education institutions

basic cloud service models are not able to provide use oriented services which concentrate on the specific requirements of a particular field like education. A discrete cloud architecture is required to appropriately deliver the services. Learning resources as a service ($L_r$aaS) provides apposite environment for higher education establishments to acquire relevant cloud services by subscribing to a public cloud service with learning resources as service ($L_r$aaS). The need for introduction of new services models has been proposed by many researchers in the past. Providing an appropriate service architecture not only facilitates with relevant services, however, it results in better management and delivery of cloud services as well [16]. The cloud model proposed in Fig. 2. is a comprehensive solution referred by Mohssen M. Alabbadi as complete cloud computing formation (C^3 F) [17]. Higher education has shown an unprecedented transformation in the recent past toward utilization of Information Technology resources and embedding of latest technology and concepts which are beyond the capabilities that can be expected from usual IT infrastructures in typical higher educational institutions. To meet the demands of Education 4.0, it is imperative to provide appropriate technology resources suitable and desired for teaching and learning in higher education. The above architecture will enable higher education institutions not only to avail the standard cloud services, however, would provide a pragmatic higher education setup having capabilities of utilizing advanced technological concepts in higher education without much of the administrative overhead and investment on infrastructure. The above architecture and the services are the set of services required for a typical higher education institution in the twenty-first century.

The above architecture is expected to provide facilities to higher education establishments to successfully yield enormous benefits of IoT, big data, and big data analytics. The architecture has all the feature and tools which have capabilities to store, handle, manage, and process the big data. The model has equal capabilities of doing big data analytics which can foster excellent capabilities to higher education establishments toward faster development. The above architecture takes into consideration the specific requirements of a higher education establishments and that is the reason a separate service architecture is shown which consists of one more service; i.e., learning resources as a service apart from other main service models that are available. However, this architecture puts in place in LraaS the capabilities so that big data analytics can be performed on the data that is taken from the higher education establishment. A typical higher education establishment can have learning resources like learning management system (LMS), Student Information System, and virtual labs that are based on the cloud infrastructure. The architecture provided in Fig. 2 shows apart from all these applications the critical hardware and applications that are required to manage big data and big data analytics.

Big data and big data analytics is finding lot of implementation in higher education and there seems to be lot scope for big data and data analytics in higher education. However, these technologies may not be implemented in isolation and need the underlying infrastructure for higher education establishments. The cloud architecture proposed will be suitable to provide these requirements for higher education establishments and may yield significant benefits in near future. Such type of implementations are very useful for academic institutions as indicated by recent research as well [18, 19].

5   Conclusion

Cloud Computing is being utilized by many organization irrespective of their scale and nature of business. The cloud computing has been efficiently utilized in higher education and many higher education establishments all over the world have implemented cloud models successfully and are yielding a lot of benefit from it. Over the past few years, the innovative technologies like big data and IoT are making footsteps in higher education and very soon it will mostly likely become inevitable for education establishments to implement big data and IoT for achieving academic excellence and for greater benefit of students and overall development of the institution. Cloud architecture can be provided to higher education institutions that can support the big data and big data analytics. Cloud computing has certainly a role to play in the development of higher education establishments, one of the roles of cloud computing models is to provide a supporting aid for new and innovative technologies that can significantly impact the education as it stand today. Cloud computing models can support by providing storage, analytics and some other unique capabilities which are utmost mandatory as far as big data, IoT, and big data analysis is concerned.

Acknowledgements Countless thanks to thank Almighty for bestowing us the knowledge to carry out this research work. We are extremely thankful to all our friends, colleagues and family members who supported and encouraged us throughout this research work. Appreciation is also due to Vivekananda Global University and Middle East College for providing support and research culture.

References

1. Bhat AZ, Singh B, Singh A (2017) Learning resources as a service (LraaS) for higher education institutions in sultanate of Oman, Noida
2. Bhat AZ, Singh B, Pandow B (2017) Economic hurdle for implementation of cloud computing in higher education in Sultanate of Oman, Dubai (2017)
3. Bhat AZ, Naidu VR, Singh B (2019) Multimedia cloud for higher education establishments: a reflection, vol 841. Springer, pp 691–698
4. Zameer A (2014) Inspiring creative minds. Springer
5. Chihi H, Chainbi W, Ghdira K (2015) Cloud computing architecture and migration strategy for universities and higher education. In: 12th International conference of computer systems and applications (AICCSA), IEEE
6. Bhat AZ, Ahmed I (2016) Big Data for institutional planning, decision support and academic excellence. In: IEEE international conference on big data and smart city, Muscat
7. Weber AS (2013) Cloud Computing in Education," Ubiquitous and mobile learning in the digital age, pp 19–36
8. Siemens G, Long P (2011) Penetrating the fog: analytics in learning and education. ERIC 46(5):30–32, 34, 36, 38, 40
9. Picciano AG (2012) The evolution of big data and learning analytics in american higher education. ERIC 16(3):9–20
10. TargioHashem IA, Yaqoob A, Anuara NB, Mokhtara S, Gania A, Khan SU (201) The rise of "big data" on cloud computing: Review and open research issues, vol 47. Elsevier, pp 98–115
11. Tulasi B (2013) Significance of big data and analytics in higher education. Int J Comput Appl 68(14)
12. Reyes A (2015) The skinny on big data in education: learning analytics simplified. TechTrends 59(2):75–80
13. He J, Lo DC-T, Xie Y, Lartigue J (2016) Integrating Internet of Things (IoT) into STEM undergraduate education: case study of a modern technology infused courseware for embedded system course. In 2016 IEEE frontiers in education conference (FIE), Eire, PA USA
14. MAPR (2018). [Online]. Available: https://mapr.com/products/product-overview/mapreduce/
15. Moussa AN, Ithnin N, Zainal A (2018) CFaaS: bilaterally agreed evidence collection. J Cloud Comput 7(1):1–9
16. Chen X, Zhang Y, Huang G, Zheng X, Guo W, Rong C (2014) Architecture-based integrated management of diverse cloud resources. J Cloud Comput Adv Syst Appl 3:11
17. Alabbadi MM (2011) Cloud computing for education and learning: Education and learning as a service (ELaaS). In: 2011 14th International conference on interactive collaborative learning, Piestany, pp 589–594. https://doi.org/10.1109/icl.2011.6059655

18. Muhsin TF, Bhat AZ, Ahmed I, Khan MS (2019) Systematic approach for development of knowledge base in higher education. J Student Res. Retrieved from https://23.235.216.135/index.php/path/article/view/999
19. Bhat AZ (2018) shodhgangotri.inflibnet.ac.in, Role of cloud computing in higher education and implementation challenges in higher education establishments in Oman and India

Chapter 27
Comprehensive Survey for Energy-Efficient Routing Protocols on WSNs Based IoT Applications Enhancing Fog Computing Paradigm

Loveleen Kaur and Rajbir Kaur

1 Introduction

WSNs play a significant role in today's modern era of Internet of things. The improvement of WSN technologies created the interest for the recent IoT applications that creating another unavoidable of smart IoT applications. Internet of things (IoT) has been emerging as the next immense thing in Internet. Initially, IoT was presented to community through supply chain management in 1999 [1]. Internet of things provide [2] numerous developments that make network reliable at every place and about everything. IoT is based on the standards that things or objects interface what's more, take an interest with one another by using wireless associations with ensure universal interchanges [3]. Moreover, IoT is being used in various applications such that healthcare services, military services, home, transports, smart cities, agriculture, etc. [4]. In addition of this, IoT is envisioned that billions of physical things or objects will be outfitted with different kinds of sensors and actuators and connected to the Internet via heterogeneous access networks enabled by technologies such as embedded sensing and actuating, radio frequency identification (RFID), wireless sensor networks, real-time and semantic Web services. [5]. The IoT will comprise of billions of advanced gadgets, individuals, administrations and other physical items having the ability to perfectly associate, interact and exchange data about themselves and their condition. This will make our lives less complex through a computerized domain that will be delicate, versatile and receptive to human needs. It will join the strength of all comprehensive network availability with sensors and actuators in the physical world. IoT [6] has a broad collection of fields and one of the overwhelming obstructions to completing such an affected arrangement is giving enough energy to work the system in a free manner without exchanging off the quality of service

L. Kaur (✉) · R. Kaur
Punjabi University, Patiala, Punjab, India
e-mail: loveleenchhina@gmail.com

(QoS). Thus, it is fundamental need to revamp the energy efficiency and effectiveness of different gadgets in IoT.

1.1 Energy Efficiency in IoT

In today's scenario, energy efficiency becomes the elemental factor in WSNs and IoT environment. WSNs can be deployed in human destructive and remote areas conditions to offer various promising types of assistance with the development of wireless communication and sensor advancements. Be that as it may, they need to duel with naturally remarkable difficulties, for example, energy constraints and extreme calculation [7, 8]. In addition, unresolved natural conditions are presented more challenge to energy preservation issues. Subsequently, energy effectiveness issues are as yet getting consideration from the researchers in these days. Expanding energy effectiveness and broadening network lifetime of WSNs has become a key research region for some IoT applications [9]. Energy efficiency is a significant reason for concentration during the network configuration because sensor nodes are fabricated with non-rechargeable batteries. In the routing process, the route selection is difficult or serious issue while transmitting the data from one node to another in the network. During the transmission process, a large amount of energy is used of sensor nodes. Various energy-efficient routing protocols are available for compelling in tackling the issue of maximum energy utilization. Due to high utilization of energy, some of the sensors nodes are fail to operate. Most of the attention, however, sensor nodes failure will create a situation of hotspot in the network. Indeed, the behavior of such type of network creates issue during the transmission process. To handle such type of critical conditions, there must be some mechanisms that can solve the problem of nodes failure. Here, fog computing paradigm is being introduced to handle such type of problems.

1.2 Fog Computing

Fog computing is the extension of cloud computing. Basically, fog word comes from the concept of fog computing. Actually, fog computing concept was introduced by Cisco [10]. Fog computing is used to compute, store and explain various IoT applications. As shown in Fig. 1, fog computing acts as the intermediary between the cloud and the edge devices which provides the cloud-like services nearby the edge devices [11, 12]. Fog computing is a collection of various fog nodes. Fog nodes are the basic components of fog computing that are installed in various techniques where the well preeminent requirement in IoT applications [13].

Fog nodes can be of any devices which have processing, storage and networking capabilities such as routers, switches, embedded servers and video surveillance cameras. In addition, various characteristics are well defined in fog computing that

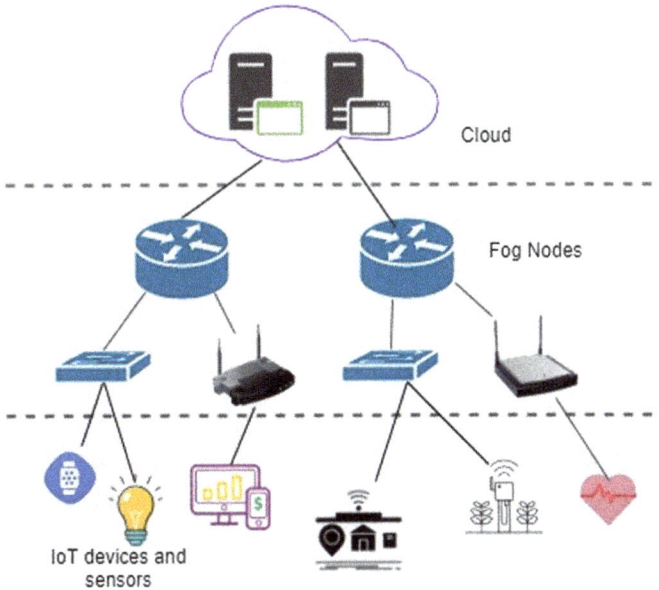

Fig. 1 Fog computing environment

are included as low latency, heterogeneity, mobility, security and privacy, low energy consumption, reduced network bandwidth and geographical distribution [14, 15].

The main contributions of this survey are explained as follows:

1. This paper describes energy-efficient routing protocols; mainly WSNs, IoT protocols and also describes fog computing paradigms.
2. A comprehensive study of protocols with their methods, performance metrics and future scope is presented. The performance of these protocols is compared on basis of various metrics.
3. This paper describes the various performance metrics related to this research domain.
4. Finally, to conclude the paper, the innovative research directions have been proposed for more improvement in this domain.

2 Literature Survey

Various techniques have been designed and developed for routing and evaluate the energy consumption and network performance, etc. The surveys of few of the energy-efficient routing protocols are as follows:

Qiu et al. [16] proposed a routing protocol for emergency response IoT based on global information decision (ERGID). ERGID is used to improve the exhibitions of reliable data transmission and efficient emergency response in IoT. Especially, authors introduced a method called delay iterative method (DIM), which is used for estimating delay and also to solve problem to take no notice of reliable paths. Furthermore, authors also introduced a scheme called residual energy probability choice (REPC), which is used to stable the network load by concentrating on node residual energy.

Zhou et al. [17] introduced a routing protocol named as E-CARP; E-CARP is developed to accomplish the location free and avaricious hop-by-hop packet sending system. Also, the PING-PONG methodology in CARP can be improved for choosing the most appropriate relay node at each time point, when the system topology is comparatively fixed. An experimental result comes about approve that our strategy can diminish the communication cost essentially and increment the network ability to a specific degree.

Wen et al. [18] proposed an adaptive distributed routing technique, which is successfully taken care of the issue that unmanned aerial vehicle (UAV) in FANETs, creates unconventionality and expanding the complexity for real-time routing in various applications. This scheme introduced a mathematical optimization problem. Also, it estimates the single-hop delay at relay node for every transmission, furthermore, utilize the double deterioration technique to change the centralized issue into disseminated issue which enables the relay nodes to utilize as it were the nearby or local channel data and evaluated delays to route the packets in this scheme. Experimental results show that the proposed scheme is very much suitable to enhance the network performance in case of energy efficiency, throughput and also end-to-end delay in the routing process.

Shah et al. [19] proposed a cluster-based aware energy routing that is used to expand the system throughput by diminishing the end-to-end delay, less packet drop proportion and enhancing system lifetime. It also introduces the proactive strategies of CLEARMaxT in network processing and also shows the performance efficiency of CLEARMaxT with the other three algorithms such as V-LEACH, TL-LEACH and E-LEACH in the proposed routing scheme.

Fouladlou and Khademzadeh [20] proposed an energy-efficient clustering algorithm for various sensor devices in Internet of things (IoT). This algorithm provides the platform to enhance the energy efficiency of various smart devices and also introduces the genetic algorithm for clustering. Simulation results show that the proposed routing algorithm has better outcomes on the basis of energy consumption, bit error rate, end-to-end delay, throughput, etc.

Anamalamudi et al. [21] proposed a cognitive AODV routing protocol that helps to find the channel route from the LBR to the goal that is associated inside the cognitive radio networks, i.e., the most encouraging network technology to transmit the obliged IoT information deftly in authorized PU free channels from LLN-LBR to the non-obliged systems. Simulated results show that the proposed cognitive AODV protocol provides the better results than the conventional-based networks.

Elappila et al. [22] introduced an energy-efficient routing technique for WSN for IoT applications in particular named as survivable path routing. The proposed protocol is assumed to work in the systems with high traffic on the grounds that different sources endeavor to send their packets to a goal at a similar time, which is a run of the mill situation in IoT applications for monitoring remote healthcare systems. Further, for choosing the next hop node, the calculation utilizes a basis which is an element of three variables: signal to interference and noise ratio of the link, the survivability factor the path from the next hop node to the destination, and the congestion level at the next hop node in the network. Simulation results recommend that the proposed works better concerning the network throughput, end-to-end delay, packet delivery ratio and the rest of the energy level of the nodes in the routing process.

Li et al. [23] proposed a novel routing algorithm in energy harvesting sensor networks to maintain the network lifetime and also handles the energy consumption. Also, authors introduce an energy harvesting sensor networks outline an optimal power control methodology to optimize the outage probability for bidirectional sub-channels. At that point, it also raises a rational goal function to address the worries of highly effective and adjustable energy utilization. Experimental outcomes are given to affirm the better performances of our proposed strategy from the points of view of sensors network lifetime.

Kaur and Sood [24] proposed an energy-efficient architecture for IoT, which comprises of three layers named as sensing and control layer, information layer and presentation layer. The design of architecture enables the framework to anticipate the sleep interval of sensors in view of their residual battery level, their past utilization history and nature of data required for a specific application. Simulated results show that a lot of energy saving on account of sensor nodes and also enhanced asset usage of cloud assets.

Ejaz et al. [25] proposed an energy-efficient optimization and scheduling framework for IoT-based smart cities that provide the extraordinary increment in urbanization over the recent years requires feasible, proficient and smart solutions for transportation, administration, environment, personal satisfaction, etc. Experimental results come out for the better optimization scheduling and wireless power exchange in the smart cities of IoT.

Mostafa et al. [26] introduced a new energy-efficient optimized scheduling model that provides the improved scheme is to minimize the energy consumption and communication overhead of observing every node in the IoT. Also, authors introduce the traveling salesman path problem that provides to minimize the state transitions of nodes that are sequenced across the time periods for optimizing the energy consumption. Simulation result shows that the proposed model is tried on randomly produced instances and provides the better optimizing scheduling in IoT systems.

Dhtore et al. [27] proposed routing protocol in the field of IoT-based WSNs applications. In this protocol, the energy efficiency is achieved as well as the QoS parameters using the clustering methods. Simulated results show the energy efficiency and QoS in terms of energy consumption and throughput.

Tabassum et al. [28] introduced efficient routing protocol that works with the scheme of greedy forwarding. In this scheme, the CHs use the greedy choice to select the next node in the sensor network. Furthermore, this proposed method reduces the void issue which occurs during the transmission process. The result outcomes show the better performance in terms of energy efficiency.

Safara et al. [29] focused on the energy-efficient routing scheme based on the RPL method that is used to determine the route in the network. Furthermore, this scheme focuses on the prevention of congestion. The outcomes demonstrate the reduction in energy consumption and end-to-end delay.

Jaiswal et al. [30] presented the routing scheme in WSNs that enhances the energy efficiency and quality of service (QoS) in the routing with huge traffic network with the consideration of network lifetime, reliability and traffic load to select the optimal route. Simulated outcomes show better energy efficiency, network lifetime, etc.

Nguyen et al. [31] introduced MAC-LEACH protocol that work on fuzzy logic system for the selection of cluster head based on number of transmissions and residual energy. Simulated outcomes show the proposed system enhances energy efficiency.

Thangarasu et al. [32] presented energy-efficient technique that uses the chaotic whale optimization method to enhance the energy efficiency by reducing the distance of sensor nodes that transfers the data between them.

Rani et al. [33] proposed clustering approach for IoT applications. This technique uses genetic algorithms for the election of best sensor node in cluster and improves the frame relay nodes. To do so, this method enhances the slot utilization, throughput, etc., in the transmission process.

Zhou et al. [34] proposed a routing protocol based on ant colony optimization. Authors consider the residual energy of the nodes for data forwarding and balance the energy consumption between the nodes.

Liu et al. [35] introduced an energy-efficient routing protocol, which provides the uniform energy consumption and avoids the "hot zone" problem and extends the network lifetime of the network.

Liu et al. [36] proposed two schemes based on LBE and cryptographic puzzle for intelligent traffic light control using fog computing. In two schemes, assuming traffic lights are as fog devices. In first scheme, traffic light needs to produce and confirm one puzzle for every vehicle in a schedule opening time. To diminish the calculation and communication overhead of the traffic light, authors proposed an enhanced plan, i.e., second scheme or improved scheme in which a traffic light just needs to broadcast a single puzzle and also the traffic light just needs to perform lightweight tasks in the system. The practicability of the enhanced plan is affirmed by experiments.

Bittencourt et al. [37] introduced the scheduling problems of resource allocation considering the hierarchical infrastructure of composing fog and cloud computing. After that, it also introduces various scheduling approaches that consider user mobility and edge computing capacity in terms of infrastructure in fog computing. Also, authors introduce the application model and two example applications which are used for evaluation and also present allocation policies and simulation results are produced by experiments.

Chen et al. [38] proposed a reliable routing protocol underwater acoustic sensor network against hotspot based on fog systems. For the reliable routing, the sensor nodes are transmitted messages to the surface sink which act as fog nodes. In this paper, an effective algorithm RRAHB is proposed to reduce the packet loss rate. Firstly, fuzzy decision algorithm (FDA) is used for selecting nodes. Based on FDA, the random selection and hotspots avoidance mechanism are presented. Simulation result shows the effectiveness of RRAHB in network with dynamic topology.

Sharma et al. [39] introduced the secure task allocation protocol for industrial IoT systems using fog computing. This protocol emphasizes on security using crypto-graphic methods to secure the data in the fog layer. The result outcomes show the improvement in latency, network lifetime and security aspects.

Shen et al. [40] proposed the data aggregation and decryption methods for dynamic groups in the fog computing environment. Furthermore, the fog device enhances the network bandwidth and filters the irrelevant data for processing in the network. More-over, this scheme results show the effective energy utilization in terms of dynamic groups.

Borujeni et al. [41] proposed routing protocol P-SEP based on fog computing to find the optimal cluster heads (CHs) depending upon the residual energy of sensor nodes. Furthermore, fog nodes compute and store data locally so there is no need to send all data to cloud. With this, the proposed method improves the energy efficiency and also deal with the latency-sensitive applications.

Kadhim et al. [42] introduced a new protocol based on SDN and fog computing. In this protocol, scheduling and classification algorithms are used. Furthermore, these two algorithms are used in SDN controller to deal with the multicast requests. To do so, this method improves the overall energy efficiency.

Oma et al. [43] presented the method that is based on fog computing which reduces the delay time and heavy traffic from various nodes in the network. However, this method enhances network lifetime and reduces the energy consumption of various nodes.

Yura et al. [44] proposed a routing protocol for street parked vehicles using fog computing. Authors consider the vehicle density, vehicle speed and parking duration for forwarding the packets with the use of various protocols and provides better longer parking duration.

Hou et al. [45] proposed a fog-based architecture, which not only provides the latency model but also emphasizes on the reliability of the network. Authors also explained the task allocation problem to minimize the energy consumption in the network and also enhance the network reliability.

Bonadio et al. [46] presented an integrated framework using fog communication and computing-based blockchain for vehicular networks. Authors concentrate on P2P network formation which allows efficient topology control with chord protocol.

AI Ahmad et al. [47] proposed a scheduling method to enhance the energy effi-ciency in the fog-assisted cloud environment using dynamic scheme. Simulation result shows the better energy saving by the fog systems.

Alharbi et al. [48] introduced an energy-efficient framework for improving the QoS of cloud system by extending the cloud to the edge of considering the various parameters by applying mathematical models.

Shahid et al. [49] proposed an energy-efficient approach that provides the better energy consumption and reducing delay by fog computing using caching techniques. The proposed approach shows better overall network efficiency in fog networks.

Mukherjee et al. [50] introduced the Internet of health system for indoor and outdoor environments using fog computing by reducing the delay and energy consumption. Simulation result shows the better performance as compared to existing cloud healthcare system.

Singh et al. [51] proposed an energy-efficient routing protocol for load balancing using fog computing. Authors proposed fuzzy load balancer for designing and managing the network traffic.

Various routing techniques and frameworks are compared and analyzed in Table 1. Table 2 introduces the fog computing paradigms with some techniques. To find the commonalities among the studies, classification techniques and performance metrics have been considered. Studies [16–35] present various different routing techniques and make energy-efficient routing protocols and also considered energy-efficient frameworks; [36–51] authors have focused on fog computing-based techniques. Apart from [24–26 and 48] who has worked on real-time system, all others have worked on various simulators like MATLAB, NS-2, OPNET, OMNET++ and ifogsim, etc. Various researchers have been already focused on different parameters for routing in sensor networks. In addition, according to the current works, research work has been done and analyzed through routing and taking care of the issues as far as system lifetime, bit rate ratio, throughput, average delay, latency and energy consumption for saving energy in sensor networks. Based on the observations, none of the researchers has been focus on the concept of energy-efficient sensor routing through decision-based fog computing paradigm. This paper will lead the new concept on behalf of previous studies that all discussed in Tables 1 and 2.

3   Performance Evaluation Metrics

In this paper, this section covered the different performance metrics are evaluated in routing protocols. Various parameters are taken as effectiveness of evaluation of performance is as follows:

Throughput: Number of bits are transferred as per second. It is measured in kilobits per second (Kbps).

End-to-end delay: defined as the time period between the sender node and the receiver node for transmission of packets.

Packet delivery ratio: defined as a number of packets received by destination node to the number of packets send by the sender node. It is measured in percentage (%).

Latency: defined as the delay in time between the sender and the receiver during the transmission of data. It is measured in seconds.

Energy consumption: defined as the total energy consumed by nodes during the transmission of packets in the network. It is measured in Joules (J).

Table 1 Various sensor routing techniques for WSN and IoT

Citation	Simulation	Real-time system	Performance metrics	Protocols used	Techniques	Future research directions
[16]	Yes	No	End-to-end delay, less packet loss, energy consumption, network lifetime	ERGID routing protocol	DIM and REPC method	Energy consumption consideration in large scale
[17]	Yes	No	Network reliability, energy consumption	E-CARP routing protocol	PING-PONG methodology	At some extent network performance for some applications in consideration
[18]	Yes	No	End-to-end delay, energy, network performance	Adaptive routing protocol	Double deterioration technique	Consideration on void region problem
[19]	Yes	No	End-to-end delay, less packet loss, network lifetime	CLEARMaxT protocol.	Proactive cluster-based technique	More improvement in cluster formation
[20]	Yes	No	Energy consumption, bit error rate, end-to-end delay, throughput	Energy efficiency clustering algorithm in IoT	Genetic algorithm for clustering	Improvement in energy saving
[21]	Yes	No	Network lifetime, energy efficiency	Cognitive AODV routing protocol	Hybrid-control channel	More achievable energy saving
[22]	Yes	No	Throughput, end-to-end delay, packet delivery ratio.	Survivable path routing.	Interference aware energy-efficient technique	More consideration on power schemes
[23]	Yes	No	Network lifetime, energy consumption	Energy efficiency routing algorithm in energy harvesting	Optimal power control strategy	Consideration on energy saving and improve network lifetime

(continued)

Table 1 (continued)

Citation	Simulation	Real-time system	Performance metrics	Protocols used	Techniques	Future research directions
[24]	No	Yes	Energy consumption, network lifetime	Energy-efficient architecture for IoT	Java-based Bootstrapping technique	More concentration on bandwidth utilization and latency
[25]	No	Yes	Energy consumption, network lifetime	Energy-efficient framework for IoT-based smart cities	Scheduling optimization techniques	More energy saving
[26]	No	Yes	Network lifetime, energy consumption	Energy-efficient scheduling model	Traveling salesman path problem	Different heuristics algorithm will be applied for better energy utilization
[27]	Yes	No	Energy efficiency, network lifetime, throughput	IoT-based protocol.	OSEAP and HY-IoT	Improvement in current method
[28]	Yes	No	Network lifetime, packet delivery ratio	IEEE802.15.4 protocol	Greedy forwarding technique	Improve more security
[29]	Yes	No	Energy consumption, end-to-end delay, routing overhead	PriNergy protocol	RPL method	Introduce metaheuristic algorithm
[30]	Yes	No	Energy consumption, end-to-end delay, packet delivery ratio, throughput, network lifetime	EOMR: MAC protocol	Data-centric technique	Emphasize on heterogeneous environment
[31]	Yes	No	Average energy consumption, network lifetime, throughput	MAC-LEACH protocol.	Fuzzy logic inference system	Emphasize on multihop WSNs

(continued)

Table 1 (continued)

Citation	Simulation	Real-time system	Performance metrics	Protocols used	Techniques	Future research directions
[32]	Yes	No	Network lifetime, energy efficiency	Integration of WSN-IoT protocol.	Chaotic whale optimization technique	Improve energy efficiency
[33]	Yes	No	Slot utilization, throughput, standard deviation.	DCRN-GA protocol	Genetic analysis approach	Emphasize on optimization of weighted sum
[34]	Yes	No	Energy consumption, network lifetime	LEACH protocol	Genetic algorithm and ant colony optimization	More emphasize on energy consumption
[35]	Yes	No	Energy consumption and network lifetime	LEACH protocol	MAC layer-IEEE802.11DCF	Work on real-time systems and more concentration on energy consumption

Table 2 Various fog computing techniques

Citation	Simulation	Real-time system	Performance metrics	Protocols used	Techniques	Future research directions
[36]	Yes	No	Latency, communication overhead, computation overhead	Secure intelligent traffic light control approach	Diffie–Hellman and hash collision	More security improvements
[37]	Yes	No	Delay, network usage, energy efficiency	FCFS and delay priority strategy	Edge word placement technique	More concentration on scheduling
[38]	Yes	No	Energy efficiency, network lifetime	RRAHB protocol based on FDA	Fuzzy decision algorithms	More improvement in underwater networks
[39]	Yes	No	Average latency, network lifetime, user satisfaction, energy consumption, security strength	FaCIIoT protocol	2FBO^2 and MOWO and SHA-3 techniques	Consideration on real-time applications, i.e., health care
[40]	Yes	No	Computational cost, communication cost	Data aggregation protocol	HMAC-SHA512 cryptographic function	Work on complex data types
[41]	Yes	No	Energy consumption, network lifetime	P-SEP uses FECR and FEAR algorithms	Pegasis-based and ACO-based techniques	Work on security issues and optimal resource allocation
[42]	Yes	No	End-to-end delay, transmitted ratio, overhead load, energy consumption, packet loss ratio	EEMSFV protocol	Mix integer programming (MIP)	Consideration on security issues
[43]	Yes	No	Energy consumption, execution time ratio	TBFC protocol	Height-balanced k-ary tree	More work on energy consumption
[44]	Yes	No	Throughput, end-to-end delay and packet data ratio	AODV, DSR, DSDV, FSR, DREAM and LAR	IEEE 802.11p	Reducing computational and storage burden for edge and cloud

(continued)

Table 2 (continued)

Citation	Simulation	Real-time system	Performance metrics	Protocols used	Techniques	Future research directions
[45]	Yes	No	Latency, energy consumption and reliability	Proximal Jacobi direction method (ADMM)	LP-based technique.	More emphasize on energy consumption and reliability
[46]	Yes	No	Latency, scalability	Flooding and chord protocols	Blockchain technique	Concentration on more energy consumption
[47]	Yes	No	Energy consumption	Greedy algorithm	Dynamic programming scheme	Improved scheme will be implemented on large networks
[48]	No	Yes	Energy consumption, minimize CPU utilization	MILP model	AT & T and BT networks as use case.	More concentration on reduce power consumption
[49]	Yes	No	Energy consumption and latency	Content filtration and load balancing algorithm	Caching technique	More improvement in energy consumption and performance by fog layer
[50]	Yes	No	Average delay, average jitter and energy consumption	MQTT protocol	Game theory	Proposed technique can be applied on nano-sensor-based health data collection and also in disease detection using deep learning
[51]	Yes	No	Energy consumption, traffic load, delay sensitivity	Traffic management algorithms	Fuzzy logics.	Work extension on handle routing failures

Energy efficiency: defined as a ratio of total packets delivered at the destination node to total energy consumed by sensor nodes. It is mentioned in kilobits per Joule (Kb/J).

Standard deviation: defined as the average variation among the energy levels of all sensor nodes in the entire network.

Network lifetime: defined as the time to the startup of the network to the time of the first node runs out of energy in the network.

4 Conclusion and Future Work

Inferable from their adaptability and effective features, routing protocols assume a significant job in the activity of an energy-efficient WSNs. Subsequently, they are promising research region. In view of energy utilization and network architecture, various routing protocols have been presented that guarantee energy efficiency to extend the network lifetime. This paper presents a brief review of energy-efficient routing protocols for WSNs based IoT applications. This paper covered key aspects, architecture and frameworks of various energy-efficient routing techniques for sensor routing. Moreover, this paper focuses on various fog computing techniques that are energy efficient. None of the researchers has been focus on the energy-efficient sensor routing through fog computing. In the future, a unique decision-based energy-efficient routing technique using fog computing paradigm for sensor routing will be designed with focus on residual energy of sensor nodes, thus, improving the consideration of characteristics of different routing services. This technique will be more energy-efficient routing technique as compared to existing routing techniques. Thus, it also improves the lifetime of the network.

References

1. Ashton K (2009) That 'internet of things' thing. RFID J 22(7):97–114
2. Atzori L (2010) The Internet of things: a survey. Comuput Netw 5415:2787–2805
3. Machado K, Rosário D, Cerqueira E, Loureiro A, Neto A, de Souza J (2013) A routing protocol based on energy and link quality for internet of things applications. Sensors 13(2):1942–1964
4. Li X, Lu R, Liang X, Shen X, Chen J, Lin X (2011) Smart community: an internet of things application. IEEE Commun Mag 49(11):68–75
5. Wang H, Jafari R, Zhou G, Venkatasubramanian KK, Sun JS, Bonato P, Wu D (2015) Guest editorial: special issue on internet of things for smart and connected health. IEEE Internet Things J 2(1):1–4
6. Kamalinejad P, Mahapatra C, Sheng Z, Mirabbasi S, Leung VC, Guan YL (2015) Wireless energy harvesting for the Internet of Things. IEEE Commun Mag 53(6):102–108
7. Rault T, Bouabdallah A, Challal Y (2014) Energy efficiency in wireless sensor networks: a top-down survey. Comput Netw 67:104–122

8. Mohamed Reem E, Saleh Ahmed I, Abdelrazzak Maher, Samra Ahmed S (2018) Survey on wireless sensor network applications and energy efficient routing protocols. Wirel Pers Commun 101(2):1019–1055
9. Fantacci R, Pecorella T, Viti R, Carlini C (2014) A network architecture solution for efficient IoT WSN backhauling: challenges and opportunities. IEEE Wirel Commun 21:113–119
10. Mahmud R, Kotagiri R, Buyya R (2018) Fog computing: a taxonomy, survey and future directions. In: Internet of everything, Springer, Singapore, pp 103–130
11. Ivanov S, Balasubramaniam S, Botvich D, Akan OB (2016) Gravity gradient routing for information delivery in fog wireless sensor networks. Ad Hoc Netw 46:61–74
12. Dastjerdi AV, Gupta H, Calheiros RN, Ghosh SK, Buyya R (2016) Fog computing: principles, architectures, and applications. In Internet of things, Morgan Kaufmann, pp 61–75
13. Byers CC (2017) Architectural imperatives for fog computing: use cases, requirements, and architectural techniques for fog-enabled iot networks. IEEE Commun Mag 55(8):14–20
14. Aazam M, St-Hilaire M, Lung CH, Lambadaris I, Huh EN (2018) IoT resource estimation challenges and modeling in fog. In: Fog computing in the internet of things, Springer, Cham, pp 17–31
15. Firdhous M, Ghazali O, Hassan S (2014) Fog computing: will it be the future of cloud computing? In: The third international conference on informatics and applications (ICIA2014)
16. Qiu T, Lv Y, Xia F, Chen N, Wan J, Tolba A (2016) ERGID: an efficient routing protocol for emergency response internet of things. J Netw Comput Appl 72:104–112
17. Zhou Z, Yao B, Xing R, Shu L, Bu S (2015) E-CARP: an energy efficient routing protocol for UWSNs in the internet of underwater things. IEEE Sens J 16(11):4072–4082
18. Wen S, Huang C, Chen X, Ma J, Xiong N, Li Z (2018) Energy-efficient and delay-aware distributed routing with cooperative transmission for internet of things. J Parallel Distrib Comput 118:46–56
19. Shah SB, Chen Z, Yin F, Khan IU, Ahmad N (2018) Energy and interoperable aware routing for throughput optimization in clustered IoT-wireless sensor networks. Future Gener Comput Syst 81:372–381
20. Fouladlou M, Khademzadeh A (2017) An energy efficient clustering algorithm for wireless sensor devices in internet of things. In: 2017 Artificial intelligence and robotics (IRANOPEN), IEEE, pp 39–44
21. Anamalamudi S, Sangi AR, Alkatheiri M, Ahmed AM (2018) AODV routing protocol for cognitive radio access based internet of things (IoT). Future Gener Comput Syst 83:228–238
22. Elappila M, Chinara S, Parhi DR (2018) Survivable path routing in WSN for IoT applications. Pervasive Mobile Comput 43:49–63
23. Li F, Xiong M, Wang L, Peng H, Hua J, Liu X (2018) A novel energy-balanced routing algorithm in energy harvesting sensor networks. Phys Commun 27:181–187
24. Kaur N, Sood SK (2015) An energy-efficient architecture for the Internet of Things (IoT). IEEE Syst J 11(2):796–805
25. Ejaz W, Naeem M, Shahid A, Anpalagan A, Jo M (2017) Efficient energy management for the internet of things in smart cities. IEEE Commun Mag 55(1):84–91
26. Mostafa B, Benslimane A, Saleh M, Kassem S, Molnar M (2018) An energy-efficient multi-objective scheduling model for monitoring in internet of things. IEEE Internet Things J 5(3):1727–1738
27. Dhtore VA, Verma AR, Thigale SB (2020) Energy efficient routing protocol for IOT based application. In: Techno-Societal 2018, Springer, Cham, pp 197–204
28. Tabassum A, Sadaf S, Sinha D, Das AK (2020) Secure anti-void energy-efficient routing (SAVEER) protocol for WSN-Based IoT network. In: Advances in computational intelligence, Springer, Singapore, pp 129–142
29. Safara F, Souri A, Baker T, Al Ridhawi I, Aloqaily M (2020) PriNergy: a priority-based energy-efficient routing method for IoT systems. J Supercomputing 1–18
30. Jaiswal K, Anand V (2019) EOMR: an energy-efficient optimal multi-path routing protocol to improve QoS in wireless sensor network for IoT applications. Wireless Pers Commun 1–23

31. Nguyen T, Hoang TM, Pham VQ, Nguyen TT, Nguyen NS (2019). Enhancing energy efficiency of WSNs through a novel fuzzy logic based on LEACH protocol. In: 2019 19th International symposium on communications and information technologies (ISCIT), IEEE, pp 108–112

32. Thangarasu G, Dominic PDD, Bin Othman M, Sokkalingam R, Subramanian K (2019). An efficient energy consumption technique in integrated WSN-IoT environment operations. In: 2019 IEEE student conference on research and development (SCOReD), IEEE, pp 45–48

33. Rani S, Ahmed SH, Rastogi R (2019) Dynamic clustering approach based on wireless sensor networks genetic algorithm for IoT applications. Wireless Netw 1–10

34. Zhou Q, Zheng Y (2020) Long link wireless sensor routing optimization based on improved adaptive ant colony algorithm. Int J Wireless Inf Netw 27(2):241–252

35. Liu S (2019) Energy-Saving optimization and matlab simulation of wireless networks based on clustered multi-hop routing algorithm. Int J Wireless Inf Netw 1–9

36. Liu J, Li J, Zhang L, Dai F, Zhang Y, Meng X, Shen J (2018) Secure intelligent traffic light control using fog computing. Future Gener Comput Syst 78:817–824

37. Bittencourt LF, Diaz-Montes J, Buyya R, Rana OF, Parashar M (2017) Mobility-aware application scheduling in fog computing. IEEE Cloud Comput 4(2):26–35

38. Qiuli C, Wei X, Fei D, Ming H (2019) A reliable routing protocol against hotspots and burst for UASN-based fog systems. J Ambient Intell Humaniz Comput 10(8):3109–3121

39. Sharma S, Saini H (2020) Fog assisted task allocation and secure deduplication using 2FBO2 and MoWo in cluster-based industrial IoT (IIoT). Comput Commun 152:187–199

40. Shen X, Zhu L, Xu C, Sharif K, Lu R (2020) A privacy-preserving data aggregation scheme for dynamic groups in fog computing. Inf Sci 514:118–130

41. Borujeni EM, Rahbari D, Nickray M (2018) Fog-based energy-efficient routing protocol for wireless sensor networks. J Supercomputing 74(12):6831–6858

42. Kadhim AJ, Seno SAH (2019) Energy-efficient multicast routing protocol based on SDN and fog computing for vehicular networks. Ad Hoc Netw 84:68–81

43. Oma R, Nakamura S, Duolikun D, Enokido T, Takizawa M (2018) An energy-efficient model for fog computing in the internet of things (IoT). Internet Things 1:14–26

44. Yura AMI, Newaz SS, Rahman FH, Au TW, Lee GM, Um TW (2020) Evaluating TCP performance of routing protocols for traffic exchange in street-parked vehicles based fog computing infrastructure. J Cloud Comput 9:1–20

45. Hou X, Ren Z, Wang J, Zheng S, Cheng W, Zhang H (2020) Distributed fog computing for latency and reliability guaranteed swarm of drones. IEEE Access 8:7117–7130

46. Bonadio A, Chiti F, Fantacci R, Vespri V (2020) An integrated framework for blockchain inspired fog communications and computing in internet of vehicles. J Ambient Intell Humaniz Comput 11(2):755–762

47. Al Ahmad M, Patra SS, Barik RK (2020) Energy-efficient resource scheduling in fog computing using SDN framework. In: Progress in computing, analytics and networking, Springer, Singapore, pp 567–578

48. Alharbi HA, Elgorashi TE, Elmirghani JM (2020) Energy efficient virtual machines placement over cloud-fog network architecture. IEEE Access 8:94697–94718

49. Shahid MH, Hameed AR, ul Islam S, Khattak HA, Din IU, Rodrigues JJ (2020) Energy and delay efficient fog computing using caching mechanism. Comput Commun

50. Mukherjee A, De D, Ghosh SK (2020) Fogioht: a weighted majority game theory based energy-efficient delay-sensitive fog network for internet of health things. Internet Things 100181

51. Singh SP, Sharma A, Kumar R (2020) Design and exploration of load balancers for fog computing using fuzzy logic. Simul Model Pract Theory 101:102017

Chapter 28
A Review on Applications of Machine Learning in Health Care

Aikendrajit Ningthoujam and R. K. Sharma

1 Introduction

Machine Learning (ML) can be considered as a sort of skill that a machine possesses to learn without being programmed explicitly. It can be referred to as a wide range of algorithms that predicts a result intelligently based on some given data(s) [1]. ML algorithms are tools that help generalize set of data and have proven to solve problems without having much knowledge of a particular domain [2]. ML is a subset of artificial intelligence, and its existence in the field of computer science and AI dates back to 1950s. In this century, healthcare domain is facing issues of improper distribution of medical resources. This could be solved using machine learning algorithms.

2 Types of Machine Learning

2.1 Supervised Learning

This learning in ML is a study for the machine to learn a function that maps an input $x^{(i)}$ variable, also known as feature to an output $y^{(i)}$ variable known as target based on a labelled data or a training set $\left\{ \left(x^{(i)}, y^{(i)} \right) \right\}$ where i belongs to a natural number ranging between 1 and n which is the number of training examples [3]. The ultimate goal is to predict the output. So, a training set is provided initially to learn from this labelled input–output data set. Later, a test set is provided where the machine

A. Ningthoujam (✉)
ESD, NIT Kurukshetra, Kurukshetra 136119, India
e-mail: akens206@gmail.com

R. K. Sharma
ECE Department, NIT Kurukshetra, Kurukshetra 136119, India
e-mail: mail2drrks@gmail.com

© The Author(s), under exclusive license to Springer Nature Singapore Pte Ltd. 2021
X.-Z. Gao et al. (eds.), *Applications of Artificial Intelligence in Engineering*, Algorithms for Intelligent Systems, https://doi.org/10.1007/978-981-33-4604-8_28

Fig. 1 Pictorial
representation of predicting
an outcome y [3]

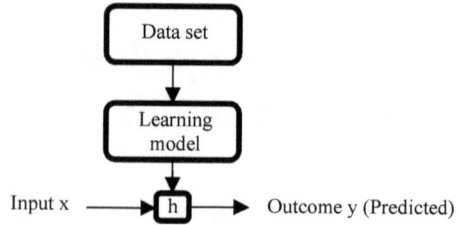

especially a computer program tries to predict the output with the highest possible
accuracy. Hence, the learner analyses the given example thoroughly and develops an
ability to classify an unlabelled set of data.

Let X and Y denote the space values of input and output where $X, Y \in R$. For a
given training set, the goal is to learn a function $h: X \rightarrow Y$ so that $h(x)$ is a good
predictor for the corresponding value of y. Here, h is known as the hypothesis [3].
The process is pictorially described (Fig. 1).

Like trained professional who is good in certain field say hand writing recognition
or document classification, supervised learning is trying to take that role and approximate the performance of that trained person. Let us consider an example where
there are a set of images of animals (say, tiger, lion, leopard). And let us assume
that the identity of these set of images is known to the learner. The above consideration is valid for a training set. Again, let us consider some more unlabelled images
of animals which are of the same class with respect to the trained set, this comes
under the test set. The objective of this programme is to build a rule to predict those
unlabelled animals correctly. There are many machine learning algorithms to predict
the output. There are two learning problems which are well suited for supervised
learning, they are regression problem and classification problem [3, 4].

2.2 Unsupervised Learning

In unsupervised learning, the training set is unlabelled. That is, it consists of unlabelled inputs and outputs. There is no reference provided for an input about the
corresponding output. Hence, in this learning, the machine tries to find patterns
which are similar and group them together. This concept of grouping similar data
together is known as clustering. Clustering is an algorithm which is used quite effectively in unsupervised learning. One of the applications of this algorithm is grouping
documents that fall under the same topic.

2.3   Semi-supervised Learning

This learning in ML is a study for the machine to learn from a labelled and an unlabelled data. As the name suggests, this learning stands between supervised learning and unsupervised learning. This learning model consists more of unlabelled data set rather than those labelled ones. It has a lot of attention in the medical research since this learning requires less labelled data [5]. The training data consists of m labelled instances $\{(x^{(i)}, y^{(i)})\}_{i=1}^{m}$ and n unlabelled instances $\{x^{(j)}\}_{j=m+1}^{m+n}$, where we assume that $m \gg 1$ [6]. Considering an example, it took nearly a week with reference to a neuroradiologist's time to manual segmentation of each brain MRI scan in the isointense infant brain [7]. In contrast, it is relatively easier to obtain large unlabelled data in many domains including medical imaging.

2.4   Reinforcement Learning

In reinforcement learning, machine is kept in a dynamic environment where it should discover which action would result better reward or appreciation. It is a learning based on the mapping of things to actions say a in this case, so as to get the maximum credit. The more the credit obtained, the better the learning is. Features like trial and error search and delayed reward are considered to be important in reinforcement learning [8, 9].

Figure 2 pictorially describes a reinforcement learning model. From Fig. 2, we can observe that the agent receives an input in the form of a perceived state, s from the environment. Then, it puts an effort a to obtain an outcome. This effort a changes s, which is the perceived state. This results in the generation of a reinforcement signal, r to the agent. The primary objective of a is to realize a policy π in order to map the perceived state, s through action, a which will maximize the measure of reinforcement, r [10].

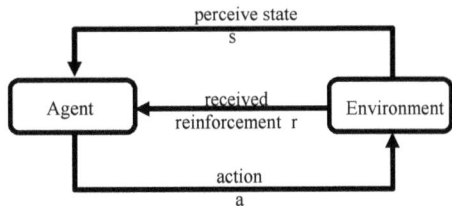

Fig. 2 Pictorial representation of reinforcement learning model

Fig. 3 Representation of a
decision tree model

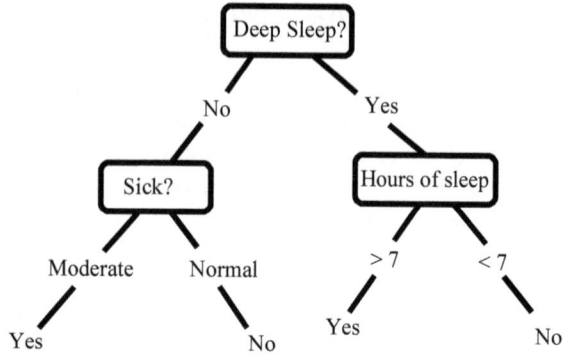

3   Different Types of Algorithms Used in ML

3.1   *Decision Tree*

This algorithm is one of the most commonly used algorithms for supervised learning.
It is generally used to solve classification problems. Like a tree that has leaves and
branches, decision tree consists of node, leaf and branch. It is similar to a flowchart
having top-down approach in recursive way [11]. Figure 3 pictorially describes an
example of a decision tree. For better diagnosis of patients, decision tree algorithm
can be used in hospitals [12]. We can use this algorithm in market analysis too.

3.2   *Support Vector Machine (SVM)*

In ML, SVM technique is used especially in recognition problems. It is a type of
supervised learning model that learns rules like classification, regression, etc., from
some given data [13]. This algorithm is based on statistical learning theory for solving
problems but not the sophisticated ones [14].

3.3   *Artificial Neural Network (ANN)*

History of artificial neural network dates back to 1940s when trying to describe human
neurons using electrical circuits [15]. The pattern of ANN algorithm is contrived on
the structure of human brain [16]. In biology, dendrites receive electrical signals from
axons while in the artificial model, these signals are represented by numerical values.
An artificial neuron is also known as a perceptron. It is calculated as the weighted
sum of inputs. Let us consider two inputs p and q for a certain perceptron P. The
weighted sum is $Mp + Nq$ where M, N are the weights for p and q, respectively. The

output of P will be non-zero for a condition where the weighted sum is greater than a threshold O.

$$\text{Output of } P = \begin{cases} 1, & Mp + Nq > O \\ 0, & Mp + Nq <= O \end{cases} \tag{1}$$

3.4   Naïve Bayes Classifier

Naïve Bayes classifier is one of the most effectual algorithm used in ML classification [17]. This algorithm is based on probability theory. This classifier is based on the application of Bayes theorem [18]. This classifier is characterized by certain features like computational efficiency, low variance, incremental learning, etc. [19].

3.5   K-Means

K-means algorithm is used for clustering data. It falls under the category of unsupervised learning. One of the most popular clustering algorithms used in machine learning is k-means algorithm [20].

3.6   K-Nearest Neighbour

KNN algorithm is commonly used in unsupervised learning. Application of this algorithm includes pattern recognition. This algorithm gives good performance characteristics and less time to get trained [21].

4   Importance of Machine Learning in Health Care

Machine learning has driven progress in medical health care. It shows excellent results in management, patient monitoring, decision making and care of patient [22]. With better hardware and ML algorithm, it is very effective at prediction and classification in the field of medical health. ML model could speculate the genesis of sepsis by using as many as millions of data sets from health records [23].

4.1 Detection of Alzheimer's Disease (AD)

Alzheimer is a type of central nervous system disorder that affects the thinking ability and impairs memory. It is the most common disease that causes dementia in man [24]. In 2017, there were around 50 million people across the world that has dementia. Its figure is doubling in every 20 years and is expected to cross 131.5 million by 2050 [25]. Diagnosis of this disease requires careful observation of physical and mental health of a person.

For the diagnosis of AD, several traditional analyses like survival analysis, polynomial regression analysis and multinomial logistic regression analysis were used [26–28]. The performance of these traditional methods could be increased using machine learning like supervised learning. Algorithms like random forest, extra trees classifier are used to classify whether a person is a control subject or an Alzheimer patient using medical data sets (MRI scan, etc.) [29].

4.2 For Sleep Analysis

There are many types of sleep disorders of which Obstructive Sleep Apnoea (OSA) is the most common type. It is caused due to the blockage of airway path while asleep. Accurate data for sleep of a person is required to classify the subject's health condition [30]. Supervised and unsupervised learning of ML are used to access the sleep parameters like total duration of sleep, actual sleep duration, etc. Even sleep efficiency of a person is being estimated using this learning. It is found that, the result of unsupervised classification via clustering is the best when k-means model is used on normalized activity count summary representation [31].

The k-means model under unsupervised classification via clustering method gives 85% accuracy for the prediction of sleep/wake state [31]. This prediction is exactly the same that is given by supervised learning model using labelled data set. Using ML, it is found that the result of sleep/wake state analysis is very close in comparison with polysomnography study [31].

4.3 Diagnosis of Cancer

We all are familiar with the disease cancer. It can be elucidated as an unwonted growth of cell in a certain part of a body and destroys tissues. It has many types, which includes lung cancer, breast tumour, skin tumour, leukaemia, etc. ML is not new for the detection of tumour. It is found that deep learning system achieved an accuracy of 0.7 based on Gleason scoring as compared to 0.61 accuracy obtained by a group of twenty-nine experts [32]. The application of machine learning algorithms in the diagnosis of cancer has improved by 15–20% in the past few years [33, 34].

Table 1 Accuracy of diagnosis with different ML techniques

Reference and date	Health problem	Technique of ML	Data used	Accuracy of diagnosis (%)
[40] 2020	Alzheimer's disease	Random forest	NHIS-NSC	82.3
[41, 42] 2015, 2019	Alzheimer's disease	SVM	841 patients	96.5
[43] 2019	Alzheimer's disease	SVM	Neuroimaging data of patients	95.9
[44] 2018	Sleep apnoea hypopnea severity	Multiple hidden layers neural network (apnoea hypopnea Index 30)	MrOS sleep data	92.69
[45] 2017	Sleep apnoea detection	ANN	Sleep disorders clinic St. VUH, Dublin	92.18
[46, 47] 2020, 2012	OSA	SVM	Apnoea ECG database, Physionet	92.9
[48] 2019	Breast cancer	KNN	RCBC Omid hospital, Iran	84.1
[49, 50] 2016, 2018	Breast cancer	SVM	PubMed, ProQuest, EBSCO	97.13

4.4   In the Field of Medical Imaging

Machine learning enables machines to extract useful patterns from a given image which is quite similar to human intelligence. In few cases, machines could observe beyond human perception which is applicable for its use in the field of medical imaging. ML algorithms are used for diagnosis of breast cancer with mammography, segmentation of brain tumour with medical resonance imaging (MRI), in the detection of Alzheimer's disease with MRI [35–39] (Table 1).

5   Research Challenges

ML is quite popular these days with its rapid growth. From providing better results to effective utilization of resources, machine learning models have done a good task in the medical field. Though the applications of machine learning have been effectively used in health care system, there are limitations of it too. Some of these are discussed here. As we know that ML models learn from data of patient's past records, its algorithms just imitate the past protocols [51]. So, on the basis of the prediction, there is no ideal time for the next intervention. In the field of tumour diagnosis, several machine learning models have already been used in many researches. However, there is need of more financial investment for further research and development (R&D). We have seen the applications of ML methods in medical imaging, where accuracy of the outcome is quite appreciable. Still there are errors generated by these models which could be improved through further development.

6 Conclusion

In this review, we have summarized different types of learning models and algorithms of machine learning. An outline on some applications of ML models used in sleep analysis, tumour diagnosis, Alzheimer's disease and in the field of medical imaging with several references provided. Besides that, some research issues faced with the use of machine learning in medicine is briefly explained. The improvement in accuracy of ML in terms of prediction is included with references from several research works. Though there are many developments in the field of machine learning, there are limitations too. There should be balance between manpower and machines towards R&D.

References

1. Nichols JA, Chan HWH, Baker MAB (2018) Machine learning: applications of artificial intelligence to imaging and diagnosis. In: International union for pure and applied biophysics (IUPAB) and Springer, GmbH Germany, part of springer nature
2. Sendak M, Gao M, Nichols M, Lin A, Balu S (2019) Machine learning in health care: a critical appraisal of challenges and opportunities. J Electron Health Data Methods 7(1):1, 1–4
3. Ng A (2018) Computer scientist, Standford Univeristy CS229 Fall 2018 lecture notes
4. Blum A, Hopcroft J, Kannan R (2020) Foundations of data science. Cambridge University Press
5. Enguehard J, O'Halloran P, Gholipour A (2019) Semi supervised learning with deep embedded clustering for image classification and segmentation. IEEE Access, pp 1–1
6. Zhu X, Goldberg AB (2009) Introduction to semi-supervised learning. Synth Lect Artif Intell Mach Learning 3(1):1–130. Morgan & Claypool Publishers
7. Wang L, Li G, Lin W, Shen D (2017) 6-months Infant brain MRI segmentation. (Online), Available: http://iseg2017.web.unc.edu/
8. Maia TV (2009) Reinforcement learning, conditioning and the brain: Successes and challenges. Cogn Affect Behav Neurosci 4:343–364
9. Hussein A, Elyan E (2017) Mohamed medhat gaber. Deep reward shaping from demonstrations, IJCNN IEEE, Chrisina Jayne
10. Fürnkranz J, Hüllermeier E, Cheng W, Park S-H (2012) Preference-based reinforcement learning: a formal framework and a policy iteration algorithm. Mach Learn 89:123–156
11. Zhong Y (2016) The analysis of cases based on decision tree. In: 7th IEEE international conference on software engineering and service science (ICSESS) INSPEC Accession no. 16760605
12. Navada A, Ansari AN, Patil S, Sonkamble BA (2011) Overview of use of decision tree algorithms in machine learning. In: IEEE control and system graduate research colloquium INSPEC Accession no. 12182936

13. Mohammadi V, Minaei S (2019) Artificial intelligence in the production process. In: Engineering tools in the beverage industry, pp 27–63, Elsevier Inc
14. Wilson MD (2008) Support vector machines. Encycl Ecol 3431–3437
15. Huang Tie-Jun (2017) Imitating the brain with neurocomputer a "New" way towards artificial general intelligence. Int J Autom Comput 14(5):520–531
16. Cox DD, Dean T (2014) Neural networks and neuroscience-inspired computer vision. Current Biol 24:R921-R929
17. Bužić D, Dobša J (2018) Lyrics classification using naive bayes. MIPRO IEEE opatija croatia 1011
18. Peng F (2003) Augmenting naive bayes classifiers with statistical language models. Comput Sci Dept Fac Publ Ser, Paper 91
19. Sammut C, Webb GI (2017) Encyclopedia of machine learning and data mining. Springer
20. Lei Y, Yu D, Bin Z, Yang Y (2017) Interactive K-Means clustering method based on user behavior for different analysis target in medicine. Comput Math Methods Med 1–9
21. Wang L, Khan L, Thuraisingham B (2008) An effective evidence theory based k-nearest neighbor (KNN) classification. In: IEEE/WIC/ACM international conference on web intelligence and intelligent agent technology
22. Rowe M (2019) An introduction to machine learning for clinicians. Acad Med 94(10)
23. Uddin S, Khan A, Hossain ME, Moni MA (2019) Comparing different supervised machine learning algorithms for disease prediction. BMC Med Inf Decis Mak 19(281)
24. Valladares-Rodríguez S, Anido-Rifón L, Fernández-Iglesias MJ, Facal-Mayo D (2019) A Machine learning approach to the early diagnosis of alzheimer's disease based on an ensemble of classifiers. ICCSA, Springer, pp 383–396
25. Prince MJ (2015) World alzheimer report 2015: the global impact of dementia: an analysis of prevalence, incidence, cost and trends. Alzheimer's Disease Int
26. Ding D, Zhao Q, Guo Q et al (2016) Progression and predictors of mild cognitive impairment in Chinese elderly: a prospective follow-up in the Shanghai aging study. Alzheimer's Dement Diagn Assess Dis Monit 4:28–36
27. Cloutier S, Chertkow H, Kergoat M-J et al (2015) Patterns of cognitive decline prior to dementia in persons with mild cognitive impairment. J Alzheimer's Dis 47:901–913
28. Han JW, Kim TH, Lee SB et al (2012) Predictive validity and diagnostic stability of mild cognitive impairment subtypes. Alzheimer's Dement 8:553–559
29. Williams JA, Weakley A, Cook DJ, Schmitter-Edgecombe M (2013) Machine learning techniques for diagnostic differentiation of mild cognitive impairment and dementia. In: Workshops at the Twenty-Seventh AAAI Conference on AI
30. Jung DW, Lee YJ, Jeong DU, Park KS (2017) New predictors of sleep efficiency. Chronobiol Int 34:93–104
31. El-Manzalawy Y, Buxton O, Honavar V (2017) Sleep/wake state prediction and sleep parameter estimation using unsupervised classification via clustering. IEEE BIBM
32. Machine learning in cancer diagnostics. EBioMedicine 45:1–2. Elsevier (2019)
33. Cruz JA, Wishart DS (2006) Applications of machine learning in cancer prediction and prognosis. Cancer Inf2(59)
34. Kourou K, Exarchos TP, Exarchos KP, Karamouzis MV, Fotiadis DI (2015) Machine learning applications in cancer prognosis and prediction. Comput Struct Biotechnol J 13:8–17. Elsevier
35. Dheeba S, Tamil Selvi S (2011) A CAD system for breast cancer diagnosis using modified genetic algorithm optimized artificial neural network. SEMCCO, Part I, LNCS 7076:349–357
36. Bauer S, Wiest R, Nolte LP, Reyes M (2013) A survey of MRI-based medical image analysis for brain tumor studies. Phys Med Biol 58(13):R97–R129
37. Mitchell TM, Shinkareva SV, Carlson A et al (2008) Predicting human brain activity associated with the meanings of nouns. Science 320(5880):1191–1195
38. Davatzikos C, Fan Y, Wu X, Shen D, Resnick SM (2008) Detection of prodromal Alzheimer's disease via pattern classification of magnetic resonance imaging. Neurobiol Aging 29(4):514–523

39. Kim D, Burge J, Lane T, Pearlson GD, Kiehl KA, Calhoun VD (2008) Hybrid ICA-Bayesian network approach reveals distinct effective connectivity differences in schizophrenia. Neuroimage 42(4):1560–1568
40. Park JH, Cho HE, Kim JH, Wall MM, Stern Y, Lim H, Yoo S, Kim HS, Cha J (2020) Machine learning prediction of incidence of alzheimer's disease using large-scale administrative health data. NPJ Dig Med Article No. 46
41. Lee J, Lee JS, Park S-H, Shin SA, Kim KW (2015) Cohort profile: the national health insurance service–national sample cohort (NHIS-NSC), South Korea. Int J Epidemiol 1–8
42. Albright Jack (2019) Forecasting the progression of Alzheimer's disease using neural networks and a novel preprocessing algorithm. Alzheimer's Dement Trans Res Clinical Interv 5:483–491
43. Jo T, Nho K, Saykin AJ (2019) Deep learning in alzheimer's disease: diagnostic classification and prognostic prediction using neuroimaging data. Front Aging Neurosci 11(220)
44. Lakhan P, Ditthapron A, Banluesombatkul N, Wilaiprasitporn T (2018) Deep neural networks with weighted averaged overnight airflow features for sleep apnea-hypopnea severity classification. In: TENCON 2018 IEEE conference
45. Shewinvanakitkul P, Buchner M (2017) Automated detection and prediction of sleep apnea events. Doctoral Dissertation, Case Western Reserve Univ
46. Apnea ECG-Database Physionet. https://physionet.org/content/apnea-ecg/1.0.0/. Accessed 17 July 2020
47. Almazaydeh L, Elleithy K, Faezipour M (2012) Obstructive sleep apnea detection using SVM-based classification of ECG signal features. In: 34th Annual international conference of the IEEE EMBS
48. Nekouie A, Moattar MH (2019) Missing value imputation for breast cancer diagnosis data using tensor factorization improved by enhanced reduced adaptive particle swarm optimization. J King Saud Univ.-Comput Inf Sci 31(3):287–294
49. Asri H, Mousannif H, Al Moatassime H, Noel T (2016) Using machine learning algorithms for breast cancer risk prediction and diagnosis. Procedia Comput Sci 83:1064–1069 (2016)
50. Nindreal RD, Aryandono T, Lazuardi L, Dwiprahasto I (2018) Diagnostic accuracy of different machine learning algorithms for breast cancer risk calculation: a meta-analysis. Asian Pac J Cancer Prev 19(7):1747-1752
51. Jarrett D, Stride E, Vallis K, Gooding MJ (2019) Applications and limitations of machine learning in radiation oncology. Br J Radiol

Chapter 29
Face Mask Detection Using VGG-16 Net for Low Computation Power Devices

V. Narmadha and R. Bharath Raam

1 Introduction

The aim of artificial intelligence is to make machine to work, think, and behave like human being. It is a field of science which models human brain. Therefore, the main factors of AI could be understanding the language that human speak, interacting with the surroundings and making decision and creative thinking which can lead to a solution for a current problem.

Machine learning is a branch of computer science where the principles of artificial intelligence are applied to a problem without writing elaborate and explicit code. It uses data to understand, predict and gives solution to the complex problems.

The automatic analysis and understanding of images and videos, a field called computer vision, occupies significant role in applications including health care, entertainment, mobility, security, etc. The deep learning methods have revolutionized the field of computer vision, making new developments increasingly closer to deployment that benefits human life by applying these methods to real-world applications.

The WHO situation report [1] has evidently shown that Corona Virus/Covid-19 is a growing global pandemic, which requires constructive technological intervention for prevention and cure. Face mask has become the first line of defense in corona virus prevention, and every store and facility out there has brought in precious manpower to do a repetitive job of checking if the customer is wearing a mask when entering. In this ever evolving world, that is stumped because of this pandemic, there is way for

V. Narmadha (✉)
Department of Computer Science, Women's Christian College, Chennai 600006, India
e-mail: vnarmadha@yahoo.co.in

R. Bharath Raam
Department of Electronics and Communication, Loyola-ICAM College of Engineering and Technology, Chennai 600034, India
e-mail: bharathram.22ec@licet.ac.in

© The Author(s), under exclusive license to Springer Nature Singapore Pte Ltd. 2021   365
X.-Z. Gao et al. (eds.), *Applications of Artificial Intelligence in Engineering*, Algorithms for Intelligent Systems, https://doi.org/10.1007/978-981-33-4604-8_29

everyone to contribute towards this combat. With artificial intelligence and machine learning making huge waves in protein structure research to self-driving cars, it can also help us in this pandemic.

The aim is to find if a person is wearing a mask or not. It is not sufficient that a person wears a mask we need to ensure that its worn properly, this model is tailored to ensure that too. We have deployed transfer learning in VGG-16 net as we are working with dataset having 20 classes, where most are interconnected. This has yielded desirable results, and the model is easily deployable. With certain improvement, we can move forward to geometric feature as it would be more usable in the field of face recognition security which is currently hit a road block due to wearing masks as they cover most of the face.

2 Literature Survey

The proposed method by Jiang et al. [2] called retina face mask method gives achieve very high accuracy. But this model still brute forces onto find the mask, it does not look for context of the mask. This model agrees with wearing mask improperly, i.e., not covering the nose and mouth with the mask, and gives a result that the person is wearing the mask. Meenpal et al. [3] validated the use of VGG-16 net for mask detection, as its capable of image segmentation which will help in further development. The model is capable of significant transfer learning and since it is a lightweight network, it makes it easily deployable. This paper states the need of research in face mask detection, as we can evidently see face-id failing to recognize when wearing mask.

The pros and cons of the holistic approach are discussed by Grassi and Faundez-Zanuy [4]. They considered the image as a high-dimensional vector and geometry feature-based methods that try to identify the position and relationship between face parts. With the mask covering the predominant part of the face, according to me, the geometric feature-based detection will soon seem a boom, as the feature of most pixels in face will be just for mask and not the face as it would defeat the purpose face feature detection.

Simonyan and Zisserman [5] have made two best-performing ConvNet models to exercise use of deep visual representations in computer vision. Furthermore, the representation depth is beneficial for the classification accuracy.

Viola and Jones proposed a framework in their paper which can detect the face rapidly with high accuracy. They used image representation called 'Internal Image' [6] which uses less amount of memory.

Segmentation concept [7] is used by Kaihan Lin to detect face and remove the background using R-CNN and the framework called G-Mask. It was possible for them to retain eigen-spatial position by RoIAlign and generate binary masks using the fully convolutional network (FCN).

The R-CNN developed in [8] has recently demonstrated impressive results on various object detection benchmarks. By training a faster R-CNN model on the large-scale WIDER face dataset, we report state-of-the-art results on two widely used face detection benchmarks,

A system based on iris acquisition devices solves the main problems encountered in face recognition due to variability between different acquisitions which was proposed by Faundez-Zanuy [9]. The system developed was based on statistical approach in which he has used direct cosine transformation.

Li et al. [10] presented a CNN cascade for fast face detection. The detector evaluates the input image with minimum resolution and swiftly rejects regions which did not contain parts of the face and carefully process the challenging regions at higher resolution. They have used Calibration nets in the cascade to accelerate detection and improve bounding box quality.

3 Methodology

3.1 Transfer Learning

Transfer learning is a type of supervised learning that uses previously trained model; for a similar task, it was trained for. It is implemented when there is insufficient data for a new domain, for which you require a neural network but there is a preexisting data pool that can be transferred to your problem. It drastically reduces the time required for training and feature engineering. To implement transfer learning, we have to delete 'loss output' layer, which is the final layer used to make predictions and replace with a new 'loss output' layer for the required prediction.

3.2 Model Description

Transfer learning was implemented on VGG net. A preprocessing layer was added to the VGG output layer and a dense layer with output size, 20, i.e. number of classes, as shown in Fig. 1 is added. The model was compiled with 'categorical crossentropy' as loss function, Adam optimizer was used as the optimizer, and the accuracy is as the metric for the model. ImageNet weights were used for VGG layers, and they were fixed. The model has 14,806,868 parameters, of which 14,714,688 parameters are trainable and 92,180 are non-trainable parameters.

3.3 Model Architecture

VGG-16 net has a simple model architecture as it does not have a much of hyperparameters. The input to cov1 layer is of fixed size 100 × 100 RGB image. The

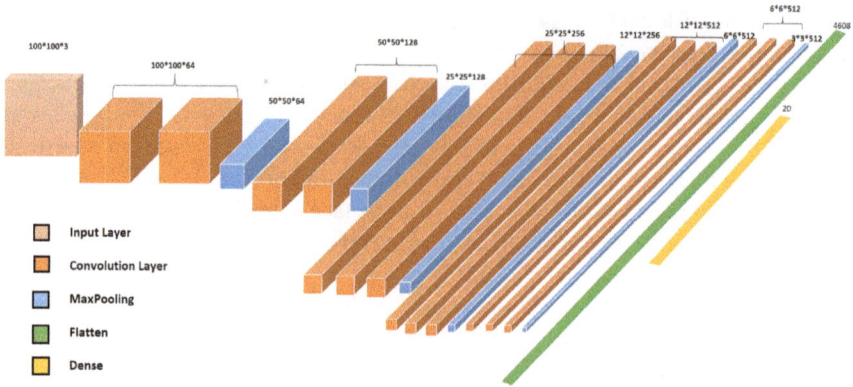

Fig. 1 Model architecture

image is passed through a stack of convolutional layers, where the filters were used with a very small receptive field: 3 × 3 which is the smallest size to capture the notion of left/right, up/down, center. In one of the configurations, it also utilizes 1 × 1 convolution filters, which can be seen as a linear transformation of the input channels followed by nonlinearity. The convolution stride is fixed to 1 pixel; the spatial padding of convolutional layer input is such that the spatial resolution is preserved after convolution, i.e., the padding is 1-pixel for 3 × 3 convolutional layers. Spatial pooling is carried out by five max-pooling layers, which follow some of the convolutional layers (not all the convolutional layers are followed by max-pooling). Max-pooling is performed over a 2 × 2 pixel window, with stride 2 as shown Fig. 2.

Three fully connected (FC) layers follow a stack of convolutional layers which has a different depth in different architectures: the first two has 4096 channels each, the third performs 1000-way ILSVRC classification and thus contains 1000 channels (one for each class). The final layer is the softmax layer. The configuration of the fully connected layers is the same in all networks.

All hidden layers are equipped with the rectification (ReLU) nonlinearity. It is also noted that none of the networks except for one contain local response normalization (LRN), such normalization does not improve the performance on the ILSVRC dataset, but leads to increased memory consumption and computation time. A 100 × 100 × 3 input layer is added to the standard VGG-16 net and a dense layer of dimension 20 which is the number of classes of classification.

3.4 Novelty of the Model

The model has a special class dedicated for masks worn incorrect, and this makes sure that the model detects that the mask is worn properly instead of just detecting mask without any context. Before using VGG Net, a two layer and six layer neural

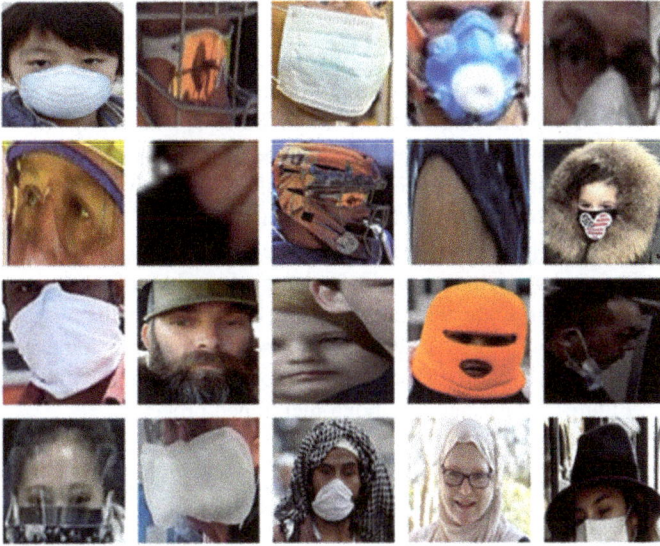

Fig. 2 Sample data of each class

networks were trained for this task, but they yielded very poor results compared to VGG Net. With custom neural net, the training time was very high. The model has a very good accuracy even though most of the classes are interconnected or subclasses of each other. VGG-16 net is simple model that can easily be run without a GPU, even in a Raspberry Pi.

4 Experiment Results

4.1 Data Description

A dataset of 20 classes with 4326 photographs and 15,411 regions of interest are used to construct the face mask detection model, which was taken from kaggle [11]. All the 4326 photographs were annotated with all the regions of interests and their class. The dataset was split into train and test datasets. The model was trained on 11095 samples and validated on 2774 samples, and the rest 1542 samples were used as test data. The pictures are read in RBG, and it was reshaped into 100×100 and normalized and appended into a numpy array. A sample from each class is shown in Fig. 2. The classes are further split into mask and no mask categories. Mask surgical, mask colorful, face shield, and face with mask constitute the 'mask' category and other classes constitute 'no mask' category.

4.2 Data Preprocessing

The dataset has 4326 images which belong to one of the 20 classes. The 20 classes and the number of samples are represented in Table 1. The region of interest for each is extracted from the images, which are specified in the annotations. Numpy arrays are created to hold the extracted region of interest and the target. A total of 15411 region of interest is extracted. The extracted region of interests is resized to 100×100 while still preserving it in RGB. The resized region of interests is then normalized. The goal of normalization is to change the values of numeric columns in the dataset to a common scale, without distorting differences in the ranges of values. Then the data and target numpy arrays are split into test and train datasets. The train dataset is then funnelled into the model and put to prove its worth by testing with the train dataset. Figure 3 illustrates the above-mentioned process.

Table 1 Summary of data classification

Class name	Number of samples
mask surgical	1381
Mask colorful	1481
Face no mask	974
Face with mask incorrect	130
Face other covering	1112
Other	36
Face shield	129
Balaclava ski mask	123
Gas mask	43
Turban	80
Helmet	157
Sunglasses	327
Eyeglasses	748
Hat	679
Goggles	150
Hijab niqab	125
Hood	137
Face with mask	2622
Hair net	146
Scarf bandana	235

Train data $\Rightarrow$ Read Image as RGB values in Numpy Array $\Rightarrow$ Extract ROI $\Rightarrow$ Resize the Image $\Rightarrow$ Normalize the data $\Rightarrow$ Store Normalized data in an array

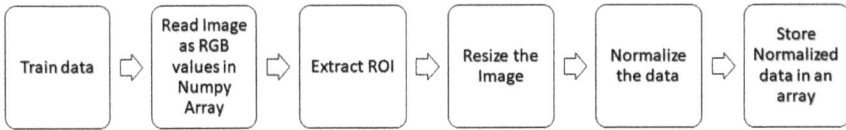

Fig. 3 Data preprocessing workflow

4.3   Experimental Setup

The model was designed using Keras and TensorFlow in Python 3.7. The dataset was imported onto Google-colab using kaggle API for training. With the GPU enabled in colab, the model was trained with the pre-trained weights of VGG-16 net. After training the model was exported as a '.h5' weight file. For validation and practical purposes, the model is loaded with the '.h5' weight file and the input image is taken from the desired source. The region of interest is extracted from the input image, which is resized to 100×100, normalized and fed into the model. The prediction is made and which is shown on top of the bounding box of the region of interest in the image taken.

4.4   Practical Workflow

The model is designed keeping in mind ease of use in practical situations and low computation power. The usage of VGG-16 net is user friendly and easily malleable. It is highly adaptable. The input image read by using OpenCV, in case of implementation through python. The input images should be in RGB as it is requisite for the model. Haar Cascade is used to detect one or multiple faces in the input image, and these are the region of interest. Each region of interest is extracted from the image, and the region of interests is resized to $100 \times 100x3$ and normalized. The normalized data is fed into the model one by one, and the predictions are made. This process is shown in Fig. 4.

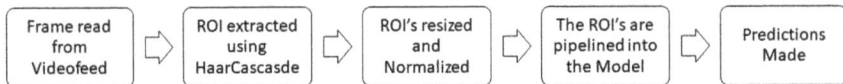

Frame read from Videofeed $\Rightarrow$ ROI extracted using HaarCascasde $\Rightarrow$ ROI's resized and Normalized $\Rightarrow$ The ROI's are pipelined into the Model $\Rightarrow$ Predictions Made

Fig. 4 Practical workflow

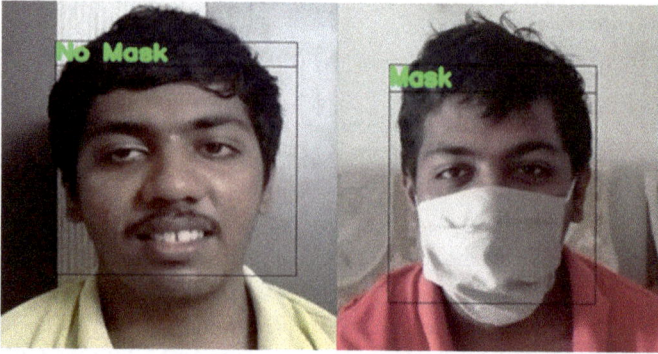

Fig. 5 Real-time result

Fig. 6 Accuracy versus epoch graph

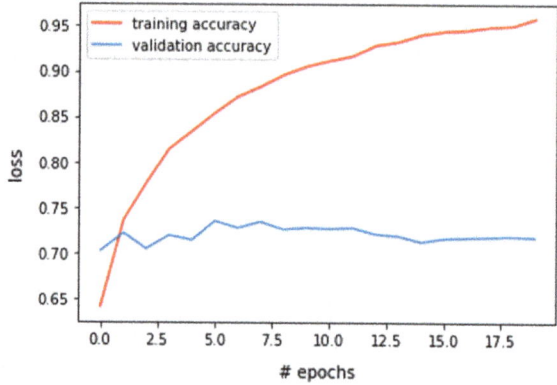

4.5 Results

It has a validation accuracy of 71% at the end of 20 epochs, as observed from Fig. 6, since the dataset is imbalanced to an extent, that is bearable, and has 20 classes achieving an accuracy of 71% is remarkable. The training time is around 1330 s on an average per epoch. It is capable of handling real-time video feed and photos. To check the practical capabilities of the model in real-time applications, live photographs taken from a webcam were processed and fed into the model. Sample of the results are shown in Fig. 5.

5 Conclusion

In the current state of the model, it can be deployed for face mask detection for customers entering store, basically to check if people are wearing masks. Face-id and other face recognition-based securities have hit a road block due to mask wearing, and this is small step forward to tackle that. The model is capable of handling real-time situations which is evident form the real-time result, as shown in Fig. 5. If we can detect a person wearing a mask, we can move on to design a system that can work of the rest of the facial features that are visible in future.

References

1. WHO Covid-19 situation report: https://www.who.int/docs/default-source/coronaviruse/situat ion-reports/20200725-covid-19-sitrep-187.pdf?sfvrsn=1ede1410_2
2. Jiang M, Fan X, Yan H (2020) RetinaMask: a face mask detector. https://arxiv.org/pdf/2005. 03950.pdf
3. Meenpal T, Balakrishnan A, Verma A (2019) Facial mask detection using semantic segmentation. In: 2019 4th International conference on computing, communications and security (ICCCS), Rome, Italy, pp 1–5. https://doi.org/10.1109/cccs.2019.8888092
4. Grassi M, Faundez-Zanuy M (2007) Face recognition with facial mask application and neural networks. In: Sandoval F, Prieto A, Cabestany J, Graña M (eds) Computational and ambient intelligence. IWANN 2007. Lecture Notes in Computer Science, vol 4507. Springer, Berlin, Heidelberg
5. Simonyan K, Zisserman A (2014) Very deep convolutional networks for large-scale image recognition. https://arxiv.org/pdf/1409.1556.pdf
6. Viola P, Jones MJ (2004) Robust real-time face detection. Int J Comput Vision 57:137–154. https://doi.org/10.1023/B:VISI.0000013087.49260.fb
7. Lin K, Zhao H, LV, Li C, Liu X, Chen R, Zhao R (2020) Face detection and segmentation based on improved Mask R-CNN. In: Discrete dynamics in nature and society, vol 2020, Article ID 9242917. https://doi.org/10.1155/2020/9242917
8. Jiang H, Learned-Miller E (2017) Face detection with the faster R-CNN. In: 2017 12th IEEE international conference on automatic face and gesture recognition (FG 2017), Washington, DC, pp 650–657. https://doi.org/10.1109/fg.2017.82
9. Faundez-Zanuy M, Espinosa-Duro V, Ortega JA (2005) A low-cost Webcam and personal computer opens doors. IEEE Aerospace Electron Syst Mag 20(11):23–26. https://doi.org/10. 1109/MAES.2005.1576071
10. Li MH, Lin Z, Shen X, Brandt J, Hua G (2015) A convolutional neural network cascade for face detection. In: Proceedings of the IEEE conference on computer vision and pattern recognition (CVPR), pp 5325–5334
11. Dataset Link: https://www.kaggle.com/wobotintelligence/face-mask-detection-dataset

Chapter 30
Mining Morphological Similarities for Translation Lexicon Augmentation

Kavitha Mahesh Karimbi, Vaishnavi Naik, Sahana Angadi, Sandra Satish, Suman Nayak, and Evita Coelho

1 Introduction

An important characteristic of translation lexica is the coverage it provides. Fundamental to the discussions in this paper are the bilingual translation lexica for language pairs English-Hindi (EN-HI), English-Sanskrit (EN-SN), Konkani-Hindi (KO-HI) and Konkani-Sanskrit (KO-SN) acquired from Centre for Indian Language Technology (CFILT), IIT Bombay.[1] However, the available lexica are not complete as each of these do not contain all possible translation pairs. For relatively less-researched, morphologically rich languages such as KO and SN, prevalence of various inflected forms of word specifically poses challenges in cross-lingual information retrieval and machine translation tasks. Data sparseness is hence a concern and is severe for asymmetrical language pairs and relatively less-researched languages. Missing

---

[1] http://www.cfilt.iitb.ac.in/~sudha/bilingual_mapping.tar.gz.

K. M. Karimbi (✉) · V. Naik · S. Angadi · S. Satish · S. Nayak · E. Coelho
Department of Computer Science and Engineering, St. Joseph Engineering College,
Mangaluru 575028, India
e-mail: kavithakmahesh@gmail.com; kavithakmahesh@rediffmail.com

V. Naik
e-mail: vaishnavi.naik05@gmail.com

S. Angadi
e-mail: sahanamangadi@gmail.com

S. Satish
e-mail: sandra97andrade@gmail.com

S. Nayak
e-mail: nayaksuman23@gmail.com

E. Coelho
e-mail: evitac@sjec.ac.in

© The Author(s), under exclusive license to Springer Nature Singapore Pte Ltd. 2021 375
X.-Z. Gao et al. (eds.), *Applications of Artificial Intelligence in Engineering*, Algorithms
for Intelligent Systems, https://doi.org/10.1007/978-981-33-4604-8_30

translations in lexica could also be attributed to the deficiency in the techniques employed for translation extraction.

The current study draws motivation from the below stated problems in the context of less-researched languages:

1. Large coverage translation lexica are limited for Indian languages. To facilitate cross-lingual information sharing, developing parallel resources are essential.
2. Words in morphologically rich languages can have several different forms, and not all different forms are seen during training.

In this paper, we discuss a combination of approaches for completing the existing translation lexica for missing regular translation forms. The specific objective is to suggest new word and phrase forms along with their translations through a generative approach. The morphological similarities between translations from the initially available lexicon are used as clues in learning bilingual subword units (stem and suffix pairs). Further, clusters of bilingual stem and suffix are identified to enable generalization. Word translations are suggested by productive concatenation of bilingual stems and suffixes belonging to the same clusters. Phrase translations are suggested by first learning translation templates using bilingual stem pairs, followed by template generalization using bilingual stem–suffix clusters.

2 Related Work

Related works which exploit bilingual and multi-lingual subword correspondences in translating unknown terms are discussed below.

In one of the earliest discussed works [16], translations are predicted for unknown words based on inductive learning mechanism. On similar lines, the bilingual approach discussed in this paper relies on learning bilingual subword units by bilingual suffixation applied to pair of similar translation equivalents. Bilingual subword units thus learnt represent stem pairs and suffix pairs, with the units in a pair essentially the translations of one another.

In translating German compound words, Koehn and Knight [11] used translations of subword units, reporting an accuracy of 99.1% for German-English noun phrase translation task. In determining the translations of compound words from their parts, the authors constrained that parts should have been observed as whole word translations in the training corpus. We use common and different parts in translation pairs as clues for initial segmentation and base the final segmentation decision on the frequency distributions of candidate bilingual subword units initially identified.

Snyder and Barzilay [18] proposed simultaneous morphology learning from multiple languages for the discovery of abstract morphemes. A multi-lingual corpus of short parallel phrases serves as knowledge base for automatic segmentation and morpheme alignment. The bilingual subword units referred to in our study correspond to the abstract morphemes discussed in Snyder and Barzilay's [18] work. Bilingual morphemes in our work rely on pairing translation equivalents in the initially available translation lexicon. The segmentation model proposed by Snyder and Barzilay on the other hand relies on pairing words in two or multiple languages, where a pair represents a translation.

Habash et al. created a rule set of morphological inflection maps that allowed relating OOV word analyses to in-vocabulary words. The study focused on learning those morphological features in source language that are irrelevant to target language for the language pair under consideration [7, 8]. Toutanova et al. [20] showed improvements in phrasal and syntax-based machine translation by integrating independent model of morphology generation with an SMT system.

By augmenting the phrase table with all verb forms for EN-HI, Gandhe et al. [5] reported improvements of up to 1.5 BLEU points. The approach employed stemmers and monolingual data to generate new phrase table entries covering different variations of a verb.

Pushpananda et al. [17] report significant reduction in untranslated terms using unsupervised morphological analysis [3] as compared to the traditional word-based approach. Factored statistical machine translation models have been used to tackle the problem of data sparsity due to morphologically rich target languages [19]. Experiments in translating texts from English to morphologically rich Indian languages affirm the role of morphology injection on translation quality in terms of both adequacy and fluency [1].

In a recent work, authors report BLEU gains of up to 2.0 by adding subword translations of out-of-vocabulary words to statistical machine translation systems [13]. The experiments involved 14 languages in low-resource settings.

2.1 Template-Based Approaches

To learn structural correspondences, Cicekli et al. [2] use analogical reasoning between pair of translations. Hu's approach [9] relies on extracting semantic groups and phrase structure groups from the language pairs under consideration. The phrase structure groups upon alignment are post-processed to yield translation templates.

Gangadharaiah et al. [6] induce generalized templates combining different approaches. One of the approaches requires identifying similar or dissimilar portions of text in a groups of sentence pairs. Others are based on finding semantically similar words and identifying syntactic correspondences employing dictionaries and parsers. Semantically related phrase pairs are grouped based on the contexts, and templates are induced by replacing clustered phrase pairs by their class labels. The subword units representing bilingual stem pairs and suffix pairs induced by pairing similar translations in our study roughly align with the common and different parts referred to in the work by Gangadharaiah et al. [6] and are utilized in induction of translation templates. Bilingual stem pairs are clustered based on bilingual suffixes that associate with them.

3 Approach

To complete the available translation lexicon for missing forms, we use word-based and phrase-based translation suggestion strategies. The approach involved in learning missing word-word translations is discussed in Sect. 3.2 and that for generating phrase translations is thereafter discussed in Sect. 3.3. Each of these generation schemes fundamentally use bilingual subword units comprising of bilingual stems and suffixes, and the mechanism for learning bilingual subword units is discussed in Sect. 3.1.

3.1 Learning Bilingual Subwords

Bilingual subwords consisting of stem and suffix pairs are induced by applying the below mentioned steps over word-word translations taken from the initially available lexicon. The approach loosely follows the bilingual learning approach for identification of bilingual segments [10].

1. Given a pair of translations, determine the similarity between word pairs in Languages L1 and L2 separately using edit distance [12].
2. If the similarity score exceeds 0.5, identify the bilingual subword units representing stem pairs by applying the longest common prefix (LCP) [15] over word pairs in L1 and L2.
3. Validate the induced stems in consultation with the WordNet [4] for each of the languages[2] in the language pair under consideration.
4. For each bilingual stem identified in Step 3, determine the bilingual suffixes that associate with it.
5. Group all the bilingual suffixes that attach to a bilingual stem and determine bilingual suffix transformations rules applicable to a bilingual stem.
6. Filter redundant groups by retaining stem pairs sharing higher number of suffix transformations.
7. Determine bilingual clusters of stem pairs and associated suffix pairs by grouping those bilingual stems sharing identical suffix transformations.

CLUTO,[3] the clustering toolkit, was used to determine clusters of bilingual stems and suffixes, and the partition approach for clustering [21] was used in the experiments presented in this paper. Each bilingual stem is characterized by its associated bilingual suffixes. IndoWordNet, a linked WordNet for Indian languages, was used to check

---

[2]English-https://wordnet.princeton.edu.

Indian—IndoWordNet: http://www.cfilt.iitb.ac.in/indowordnet/.

[3]http://glaros.dtc.umn.edu/gkhome/views/cluto.

the validity of induced stems for Indian languages. In Python, we access the WordNet using the nltk[4] package and IndoWordNet using the pyiwn[5] package.

3.2 Word-Word Translation Suggestion—A Concatenative Approach

The following resources are produced following the approach presented in Sect. 3.1:
Bilingual subwords: These include the list of bilingual stems (columns 1, 2 in Tables 3 and 4) and bilingual suffixes (columns 3, 4 for EN-SN and 5, 6 for EN-HI).
Bilingual Stem and Suffix Associations: By grouping all bilingual suffix pairs that attach to the same stem pair, we identify the bilingual suffix replacement rules allowing one translation form to be obtained from the other.
Bilingual Stem and Suffix clusters: Clusters of bilingual stems and suffixes are formed by grouping all stem pairs that undergo same suffix transformations.

New translations are suggested using the clusters of bilingual stems and suffixes identified during the learning phase. Missing word-word translations are suggested by direct concatenation of stem pair and suffix pair belonging to the same cluster. The newly generated translations upon validation serve as training data for the subsequent iterations. Steps in generating word-word translations are summarized as follows:

1. Learn bilingual subword correspondences involving stem pairs and suffix pairs (as discussed in Sect. 3.1).
2. Evaluate the learnt bilingual subword units.
3. Generate new surface translations by productively combining the bilingual subword units using clusters of bilingual stems and suffixes [10].

3.3 Generating Phrase Translations

In this section, we present the translation template-based approach for automatic generation of translation equivalents using bilingual subword units consisting of stem pairs. Steps involved in generating phrase translations are as outlined below:

1. Induce phrase templates by replacing the bilingual stems with wild card character $[6] [14] in the phrase translations available in the lexicon.
2. For each bilingual stem utilized in template induction, identify the cluster to which the bilingual stem belongs.
3. Generate new phrasal forms by replacing '$' with all the remaining stem pairs belonging to the corresponding cluster (identified in Step 2).

---

[4]https://www.nltk.org/.

[5]https://github.com/riteshpanjwani/pyiwn.

[6]$ represents the place holder for stem.

Table 1 Statistics of input dataset

Language pair	Number of word pairs
EN-HI	19,024
EN-SN	11,995
KO-HI	22,623
KO-SN	13,026

Table 2 Statistics of the dataset used in learning bilingual subword units (stems–suffixes) and the associated bilingual subword units learnt

Language	Input dataset	No. of unique stem pairs induced		No. of Unique suffix pairs induced	
Pair	(No. of words)	Seen in the WordNet	Not seen in the WordNet	Seen in the WordNet	Not seen in the WordNet
EN-HI	26,247	12,900	1795	1171	842
EN-SN	15,242	1830	1883	260	571
KO-HI	22,623	7976	2037	4300	6875
KO-SN	13,026	1268	1753	1202	2931

4 Experimental Setup

Datasets used in the experiments were obtained from Centre for Indian Language Technology (CFILT), IIT Bombay.[7] The statistics of dataset for various language pairs used in our experiments are as shown in Table 1.

5 Results and Evaulation

Section 5.2 shows the experimental results obtained in the context of word-word translation generation via clustering. Phrase translations generated are presented in Sect. 5.3. The bilingual subword units learnt are presented in Sect. 5.1.

5.1 Bilingual Subword Units

Word-word translation generation relies on learning bilingual stems and suffixes. Statistics of bilingual subword units (unique stems and suffixes) learnt for language pairs EN-HI, EN-SN, KO-HI and KO-SN are presented in Table 2. The unique bilingual subword units induced were validated in consultation with WordNet.

---

[7] http://www.cfilt.iitb.ac.in/~sudha/bilingual_mapping.tar.gz.

Table 3 Bilingual Stems and Suffixes for EN-SN

Stem$_{EN}$	Stem$_{SN}$	Suffix1$_{EN}$	Suffix1$_{SN}$	Suffix2$_{EN}$	Suffix2$_{SN}$
assist	उपकृत	' '	'ि:'	ance	म्
clean	उज्वल	ness	ता	up	न
belated	विलम्ब	' '	ि॒त	ly	ॉन

Table 4 Bilingual stems and suffixes for EN-HI

Stem$_{EN}$	Stem$_{HN}$	Suffix1$_{EN}$	Suffix1$_{HN}$	Suffix2$_{EN}$	Suffix2$_{HN}$
accept	स्वीकार	' '	ना	ed	ॉय
assist	इमदाद	ance	' '	ed	ॉी
armor	अस्त्र	er	कार	y	शाला
belated	विलंब	' '	ि॒त	ly	तः

Table 5 Bilingual stems and suffixes for KO-HI

Stem$_{KO}$	Stem$_{HI}$	Suffix1$_{KO}$	Suffix1$_{HI}$	Suffix2$_{KO}$	Suffix2$_{HI}$
अक्षर	अक्षर	माळ	ॉती	वळख	ॉी
अज्ञान	अज्ञ	पण	ता	ॉी	ॉयः
अधिकार	अधिकार	ॉाची	ॢपूर्ण	ॉी	ॉी
प्रेम	प्रिय	ि॒का	ॉा	ॉी	तम

Tables 3 and 4 present the intermediate results (bilingual subword units) obtained after processing the initially available bilingual lexica for the language pairs EN-SN and EN-HI respectively. Table 3 shows randomly selected bilingual stems and suffixes for language pair EN-SN. Columns 1 and 2 in Table 3 show the bilingual stems learnt for language pair EN-SN. Similarly, the associated bilingual suffixes are shown in columns 3, 4 and 5, 6, respectively.

From Table 3, it follows that replacement of one suffix pair with another pair enables generation of other surface forms. Consider the first row in Table 3. In the surface form 'assistance' ↔ 'उपकृतम्', replacing 'ance' ↔ 'म्' with '' ↔'ि:' enables us to infer the form 'assist' ↔ 'उपकृति:'. We see that 'ance' ↔'म्' and '' ↔'ि:' represent bilingual suffixes that attach to the same bilingual stems and hence enable inferring one form from other. Similar phenomena are observed with respect to EN-HI, KO-HI and KO-SN and are shown in Tables 4, 5 and 6, respectively.

Table 6 Bilingual stems and suffixes for KO-SN

Stem$_{KO}$	Stem$_{SN}$	Suffix1$_{KO}$	Suffix1$_{SN}$	Suffix2$_{KO}$	Suffix2$_{SN}$
अक्षर	अक्षर	माळ	माला	शा	श:
अज्ञान	अज्ञ	पण	ता	◌ी	◌य:
अधिकार	अधिक	◌ाची	◌ृत	◌ी	◌ारी
प्रेम	प्रिय	ि◌का	तमा	◌ी	◌:

Table 7 Most frequent suffix pairs for the language pairs EN-HI and EN-SN

Suffix pair (EN-HI)	Count	Suffix Pair (EN-SN)	Count
' ', ' '	915	' ', ' '	231
' ', '◌ी'	156	' ', '◌:'	59
' ', '◌ा'	101	' ', 'म्'	26
' ', 'ना'	63	'er', ' '	20
' ', 'क'	52	' ', '◌'	19
' ', '◌ा'	47	'e', ' '	18
'er', ' '	41	' ', '◌ा'	17
'e', ' '	37	'ed', ' '	16
'ed', ' '	31	'ing', ' '	15
'ing', ' '	28	' ', 'ता'	11
' ', 'िया'	28	'al', ' '	8
'er', '◌ी'	27	' ', '◌ीकृ'	7

The precision of induced bilingual subword units is estimated as shown in Eq. 1. The precision is computed as the ratio of bilingual stems generated and seen in WordNet ($N_{WordNet}$) to the number of unique bilingual stems generated ($N_{Generated}$). Table 8 shows the precision for generated subword units across language pairs.

$$\text{Precision} = N_{\text{WordNet}}/N_{\text{Generated}} \tag{1}$$

Table 8 Evaluation of induced bilingual stems

Language pair	Precision
EN-HI	87.78
EN-SN	49.26
KO-HI	79.66
KO-SN	41.97

Table 9 Bilingual clusters for EN-HI

Paradigm	Cluster Number	Stem Pair	Suffix Pair
Noun	18	oppress:कठोर contradict:असंगत	ion:ता ion:चि
	48	citizen:नागरिक depress:उदास	ship:ता ion:ी
Adjective	36	friend:स्नेह	ly:शील
	49	immoral:बुरा cordial:सच्चा	ity:ई ity:ई
Verb	11	resign:संतोष	ed:ी
	34	open:खोल	ed:ी, ed:ी, ed:े
Adverb	20	equal:बराबर	ly:ी
	22	firm:मजबूत tender:कोमल	ly:ी ly:ता

5.2   Generated Word-Word Translations

Word-word translations are generated by productive concatenation of bilingual stems and suffixes belonging to the same cluster. The bilingual stem and suffix clusters for EN-HI are as shown in Table 9. The statistics of generated EN-HI word-word translations are as shown in Table 10.

5.3   Phrase Translation Generation

As an application of bilingual subword units learnt, we further experimented regarding their usage from the perspective of phrase translation generation. The statistics of

Table 10 Word-word translations generated via clustering

Language pair	Number of word-word translations	Number of validated word-word translations	Previously present in EN-HI lexicon	New word-word translations	Examples
EN-HI	173,601	1855	4	1851	Actually↔यथार्थता Oldness↔प्राचीनता Thriller↔रोमांचक

Table 11 Statistics of input dataset and induced phrase templates

Language pair	Number of input bilingual phrase pairs	Number of phrasal translation templates induced	Number of unique phrasal translation templates induced
EN-HI	28,732	2637	1020
EN-SN	10,025	243	137

the phrase translations used in induction of phrase translation templates are shown in Table 11. The last column in the table shows the number of unique phrase templates learnt for EN-HI and EN-SN.

Induced translation templates for EN-HI and EN-SN are shown in Table 12 along with the bilingual stems (last two columns) employed in the induction process. By identifying the bilingual stems used in template induction and further identifying the clusters to which they belong, it is possible to generate new phrase translation forms by replacing all the other bilingual stems of the associated cluster in the templates learnt. Newly generated phrase translations are as shown in Table 13.

6 Conclusion and Future Work

In this paper, we have discussed a tool for suggesting word and phrase translation forms using a combination of word-based and phrase-based approaches. In one of the approaches, clusters of bilingual subword units are employed in suggesting word-word translation variants to existing bilingual word-word translations by simple concatenation of stem pairs and suffix pairs. Translations at phrasal level are suggested by initially inducing phrase translation templates using bilingual stems and thereafter generating new phrasal forms employing bilingual stem and suffix clusters.

The approach involves identifying bilingual subword units having similar meaning using the initial available lexica. The system induces bilingual clusters of stem

Table 12 Phrase translation templates induced

Language Pair	Phrase Templates		Stem Pairs	
L1-L2	**L1**	**L2**	**L1**	**L2**
	$ out	$ওা	start, pick	चला, चुन
	$ in	$ওाना, $ना	colour	रंग
EN-HI	$ down	$ होना, $ करना	rain, shut	बाररश, बंद
	republic of $	$ गरॉज्य	singapore	संगापुर
	capital of $	$ सिटी	panama	पनामा
	ten $	दश$म्	thousand	सहस्र
	$ star	$तारा	north	ध्रुव
EN-SN	$ tense	$कालः	past	भूत
	$ man	$जनः	young	युव
	$ enough	$सु$	sure	निश्चितम्

Table 13 Phrase translations generated for EN-HI

EN	HI
wrap up	लपेटना
take harm	नुकसान होना
dry out	सुखाना
do feel	महसूस करना
join in	मिलाना, मिलना
calm down	शांत होना, शांत करना
hunt down	शिकार होना, शिकार करना

and suffix pairs and suggests morphological variations for word and phrase translations using the learnt bilingual subword units and clusters. Thus, the approach employs available knowledge bases to learn new translation forms that are missing in the lexicon via clustering and concatenation scheme. Dealing with non-contiguous phrasal forms would help in enhancing the lexicon further.

Acknowledgements Research supported by Vision Group on Science and Technology (VGST), Karnataka Science and Technology Promotion Society (KSTePS), Department of Information Technology, Bio Technology and Science & Technology, Karnataka, India, through Research Grant, ref. KSTePS/VGST/RGS/F/2017-18/GRD-693/54/2018-19.

References

1. Bhattacharyya P et al (2017) Role of morphology injection in smt: a case study from indian language perspective. ACM Trans Asian Low-Resour Lang Inf Process (TALLIP) 17(1):1
2. Cicekli I, Güvenir HA (2011) Learning translation templates from bilingual translation examples. Appl Intell 15(1):57–76
3. Creutz M, Lagus K (2005) Unsupervised morpheme segmentation and morphology induction from text corpora using Morfessor. Report A81, Helsinki University of Technology
4. Fellbaum C (1998) WordNet: an electronic lexical database. Bradford Books
5. Gandhe A, Gangadharaiah R, Visweswariah K, Ramanathan A (2011) Handling verb phrase morphology in highly inflected indian languages for machine translation. In: Proceedings of 5th international joint conference on natural language processing, pp 111–119
6. Gangadharaiah R, Brown RD, Carbonell J (2011) Phrasal equivalence classes for generalized corpus-based machine translation. Springer, Berlin, pp 13–28
7. Habash N (2008) Four techniques for online handling of out-of-vocabulary words in arabic-english statistical machine translation. In: Proceedings of the 46th annual meeting of the association for computational linguistics on human language technologies: short papers. Association for Computational Linguistics, pp 57–60
8. Habash N, Metsky H (2008) Automatic learning of morphological variations for handling out-of-vocabulary terms in urdu-english machine translation. In: Proceedings of the association for machine translation in the Americas (AMTA-08), Waikiki, HI
9. Hu R, Zong C, Xu B (2006) An approach to automatic acquisition of translation templates based on phrase structure extraction and alignment. IEEE Trans Audio Speech Lang Process 14(5):1656–1663 Sept
10. Kavitha KM, Gomes L, Lopes JGP 920140 Identification of bilingual segments for translation generation. In: Advances in intelligent data analysis XIII, volume 8819 of LNCS. Springer, pp 167–178
11. Koehn P, Knight K (2003) Empirical methods for compound splitting. In: Proceedings of the tenth conference on EACL-Volume 1. ACL, pp 187–193
12. Levenshtein VI (1966) Binary codes capable of correcting deletions, insertions, and reversals. Soviet Phys Doklady 10:707–710
13. Liu NF, May J, Pust M, Knight K (2018) Augmenting statistical machine translation with subword translation of out-of-vocabulary words. arXiv preprint arXiv:1808.05700
14. Mahesh KK, Gomes L, Lopes JPJ (2018) Exploring the relevance of bilingual morph-units in automatic induction of translation templates. In: Advances in artificial intelligence—IBERAMIA 2018. Springer, Cham, pp 417–429
15. Manber U, Myers G (1993) Suffix arrays: a new method for on-line string searches. SIAM J Comput 22(5):935–948
16. H.S.K.A.Y. Momouchi and K. Tochinai. Prediction method of word for translation of unknown word. In *Proceedings of the IASTED International Conference, Artificial Intelligence and Soft Computing, July 27 to August 1 1997, Banff, Canada*, page 228. Acta Pr, 1997
17. Pushpananda R, Weerasinghe R, Niranjan M (2015) Statistical machine translation from and into morphologically rich and low resourced languages. In: International conference on intelligent text processing and computational linguistics. Springer, pp 545–556
18. Snyder B, Barzilay R (2008) Unsupervised multilingual learning for morphological segmentation. ACL, pp 737–745
19. Sreelekha S, Dungarwal P, Bhattacharyya P, Malathi D (2015) Solving data sparsity by morphology injection in factored smt. In: Proceedings of the 12th international conference on natural language processing, pp 95–99
20. Toutanova K, Suzuki H, Ruopp A (2008) Applying morphology generation models to machine translation. In: Proceedings of ACL-08: HLT, pp 514–522
21. Zhao Y, Karypis G (2002) Evaluation of hierarchical clustering algorithms for document datasets. In: Proceedings of the eleventh CIKM. ACM, pp 515–524

Chapter 31
Spade to Spoon: An IoT-Based End to End Solution for Farmer Using Machine Learning in Precision Agriculture

Mahendra Swain, Rajesh Singh, and Md. Farukh Hashmi

1 Introduction

The history of agriculture dates back to several of thousands years ago when the early human race started to gather and regrow the wild seeds and cultivate them. The agriculture thus then became a means of living which was much more sustainable and a surplus option for survival, than the hunting and gathering lone did. Since then agriculture became an integral part of the lives of humans. From then to now, many advancements have taken place in the field of agriculture in terms of crops, fertilizers, pesticides, and much more. Agriculture has not only become a means of survival but over the years has also become the source of any country's economy. Hence, it is not wrong to say that the agricultural development of a nation speaks for the nation. The investments put into the betterment of the agricultural activities also benefit the employment of the country. But despite all the plus of healthy and better agriculture, some problems persist and many of the farmers have to go through all these each day passing. These issues include low or poor yield of crops, the inability of being able to maintain the proper storage solutions after harvesting, manually checking whenever to water the plants or crops and a proper check on thefts and intruders trespassing the fields may it be the animals as well. In a broader aspect, this research paper focuses on the project which can be used to assist the other smart automation taking place in the field such as smart irrigation, spraying of fertilizers and pesticides. Thus, the

M. Swain (✉) · R. Singh
Department of Electronics and Communication Engineering, Lovely Professional University,
Phagwara, Punjab, India
e-mail: er.mahendraswain@gmail.com

R. Singh
e-mail: srajsssece@gmail.com

Md. Farukh Hashmi
National Institute of Technology, Warangal, India
e-mail: mdfarukh@nitw.ac.in

© The Author(s), under exclusive license to Springer Nature Singapore Pte Ltd. 2021   387
X.-Z. Gao et al. (eds.), *Applications of Artificial Intelligence in Engineering*, Algorithms
for Intelligent Systems, https://doi.org/10.1007/978-981-33-4604-8_31

paper features various sensors to develop a device that would act as the decisive unit for the other system as above-stated ones in real time. These are a few sets of factors that affect the productivity to a great extent. It has been evident that how every single process has arrived at an improved yield through automation which is much higher than the manual work. This paper thenceforth proposes a system that is useful in monitoring as well as controlling the field data and operations which provides ultimate flexibility to agriculture.

2 Literature Survey

A literature survey has been conducted on the latest peer-review journal articles. In this article, the scalability of LoRa has been investigated concerning the number of device nodes against a single LoRa gateway. Simulation model is designed to determine the data rate among physical nodes [1]. The article shows the popularity of LPWAN for IoT applications. The scaling of LoRa is conducted through various simulation models [2]. LoRa flexible is mentioned in this article. Low-power and wide area technology enable us to connect multiple sensor nodes. A mathematical formula is formulated to enhance and to optimize LoRa networking. Average throughput is calculated by configuring LoRa end devices [3]. The challenges of the LoRa are discussed. When the number of sensor nodes increases to 1000 per gateway, it is very difficult to simulate the network with a high spreading factor. The common LoRa channel gets used by all sensor nodes randomly, which creates reliability issues [4]. In LoRa, received signal strength indicator (RSSI) plays a very crucial role to locate all sensor end nodes deployed at filed or outdoor. Due to obstacles and noisy environments, the RSSI changes dynamically [5]. This article highlights the adaptive technique to slice network dynamically to improvise quality of service (QoS) in LoRa-based network. In concerned with spreading factor and transmission power, parameter simulation mechanism is used [6]. Low-power wide area networks (LPWAN) enable a wide range of applications. In this article, a method to improve spreading factor and carrier frequency has been shown on the simulation platform using a mixed integer linear programming (MILP). Packet collision rate and data extraction rate are taken into consideration concerning heavy network trafficking [7]. Cell-edge nodes have a high chance of losing packets due to collisions, especially when high spreading factors (SFs) are used that increase time on air. Moreover, LoRa networks face the problem of scalability when they connect thousands of nodes that access the shared channels randomly [8]. Industrial application using LoRa is demonstrated in this article [9]. This review article explains the physical layer of LoRa, and SWOT analysis is presented along with the challenges that LoRa and LoRaWAN still face [10]. In this paper, ANS3 simulation is demonstrated to calculate the performance of LoRa over power consumption along with the LoRa testbed [11]. In the current scenario of drying up potable water and the water which is utilized in many other ways, there needs to be a system that can judge the moisture and temperature of soil which can be used for optimum watering techniques [12]. A logic is developed such that it can monitor and guide the irrigation system to water at regular intervals only after checking the field and necessity [13]. The system can automatically control temperature and humidity. Even a mechanism can be developed for the sunshine system at the greenhouse and plant factories. The system can be monitored through an app or PC, resulting in proper yield [14]. A security system can also be implemented by interfacing the ultrasonic sensor and can be used along with a camera to view the data in real time [15]. Thermal imaging can be implemented to keep a keen check on various activities happening on the field through a very convenient cloud platform [16]. Climate changes and rainfall have been disturbing factors over the decades. Thus, monitoring the rainfall through a rain detector and using further automation techniques can be a big savior [17]. A very different and less entertained issue in agriculture has been the fluctuating temperature and humidity of the warehouses where the post-harvest is stored. Thus, through temperature and humidity monitoring, the proper temperature and humidity can be maintained in the warehouse and the post-harvest can be protected [17]. Through intensive research and on and off the field, the researchers have concluded that the yield is becoming poor each passing day. Hence, the paper not only provides an optimal solution to the problem but also is effective in decreasing the manpower and workload of the farmers [18]. The research is done in hopes of improving agriculture.

3 LoRa Testbed

LoRa is gaining a lot of attention when it comes to the remote monitoring application. The customized-designed LoRa testbed is shown in Fig. 1. It includes field discrimi-

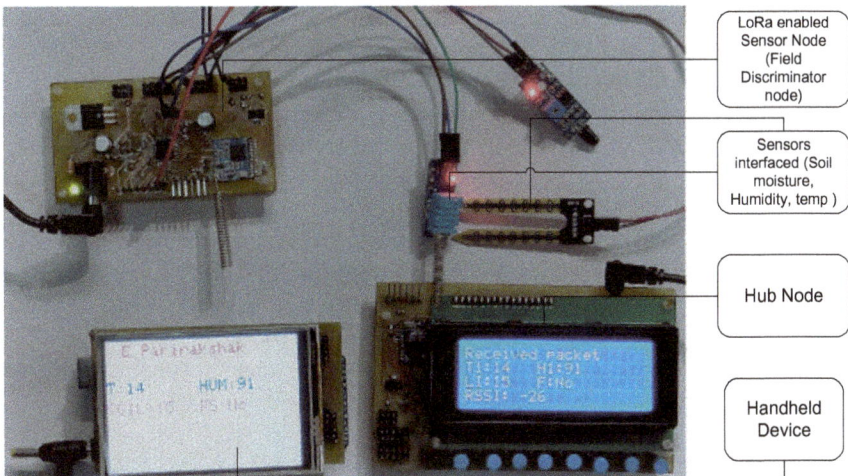

Fig. 1 LoRa testbed customized designed for agriculture field monitoring

nator node (end node), gateway (hub node) and handheld device. Field discriminator node is designed to deploy at farming land. It comprises a microcontroller, LoRa modem and GPIO for inputs and outputs. The operating voltage of the device is 3.3–5 V. These can be operated with batteries for the long life of the device. Hub node is the gateway among all the end nodes and handheld devices. Information gets collected at hub node for further data uploaded into a cloud platform where it records the data and store into the cloud. The intelligent device is the handheld device, and it comprises microcontroller, touch screen, and control buttons. Data collected from the cloud and end nodes locally get displayed at this device. The device is LoRa-enabled as well as WiFi-enabled, which adds an extra advantage to the classic LoRa testbed. The device has the feature to record the data and further process the data using machine learning algorithms.

4 Methodology

4.1 Simulation Platforms

Proteus 8 Simulator: Proteus 8 is one of the best simulation software for various circuit designs of microcontrollers. It has almost all microcontrollers and electronic components readily available in it, and hence it is a widely used simulator. It can be used to test programs and embedded designs for electronics before actual hardware testing. The simulation of the programming of the microcontroller can also be done in Proteus. Simulation avoids the risk of damaging hardware due to wrong design.

The block diagram of the LoRa transmitter has been shown in Fig. 2. It comprises power supply unit both 5 and 12-V variants. ATMEGA 328 controller has been used to design this transmitter and custom fabricated on PCB. There are multiple pins dedicated to sensor interfacing both analog and digital sensors like pH, ultrasonic, temperature, humidity, pressure sensor, rainfall sensor, gas sensors, etc.

The block diagram consists of power supply unit. On board power supply pins +5 V, +12 V ATMEGA 328 has been interfaced with liquid crystal display where all data can be visualized. LoRa module has been interfaced to transmit the data as shown in Fig. 3.

The end nodes data are collected and recorded at a cloud-based platform. An optimized technique of sensor localization is shown in Fig. 4. The steps involved are initialization of Lora end node localizations using various localization algorithms. Depending the node density of the sensor nodes, data transmission takes place among sensor end nodes and gateway. Using frequency hopping, the data reach the destination node without losing information.

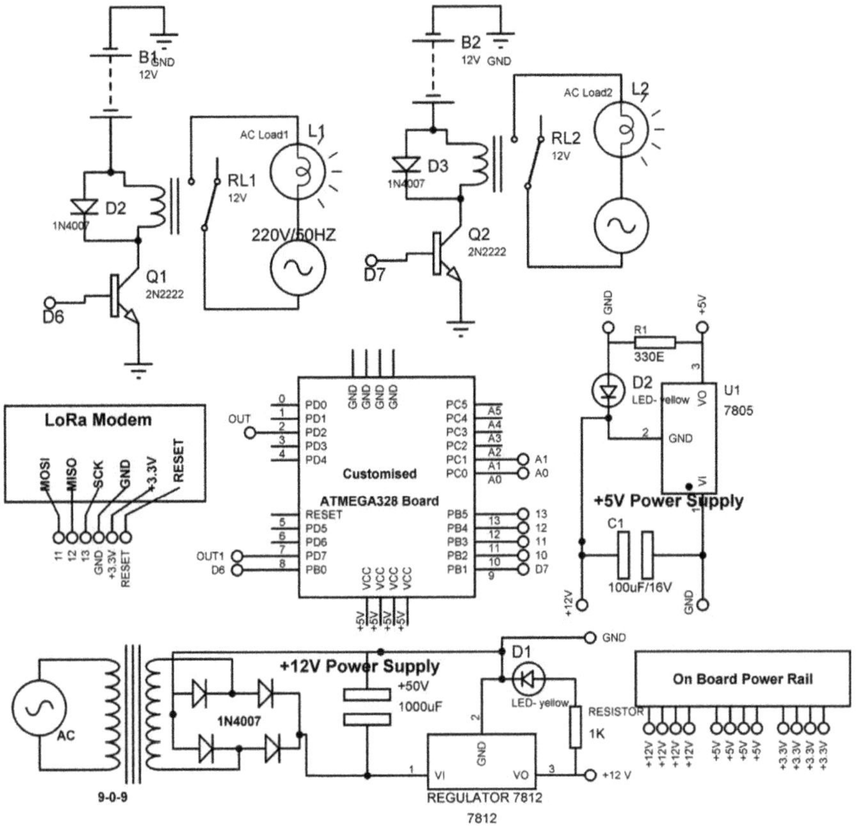

Fig. 2 Circuit layout of customized sensor node on the simulator

5 Result and Discussion

Data collected from sensor nodes have been displayed in Fig. 4. Further detailed data are shown in Fig. 5. The result obtained from the current study is displayed in Fig. 6. The data have been classified and form a cluster to recognize suitable crop species. The support vector machine (SVM) algorithm is a supervised machine learning model. It is suitable to train a model with independent variables. Here we have considered two parameters which are changing concerning various areas. Like if we will consider the southern part of India, the temperature and humidity change dynamically w.r.t soil salinity, whereas the same has been observed for the northern part of India.

Identifying the right hyperplane in Fig. 6, it is used as a regressive classifying technique. The hyperplane separates data into two clusters by eliminating data over-fitting and underfitting which has been shown in Fig. 7. If the data are nonlinear in nature, SVM uses a kernel trick where data get divided into a three-dimensional view.

Fig. 3 Block diagram of the
LoRa-based sensor node
deployed at the farm

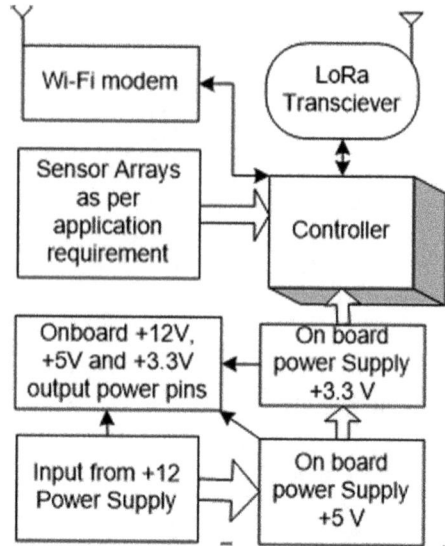

Kernel tricking helps the algorithm to identify the right hyperplane with minimal error. Further, a confusion matrix is calculated to calculate the mean square error concerning the training data and testing data.

6 Conclusion

Sensing technology and advanced machine learning are key technology to predict suitable crop species. SVM is one of the most popular algorithms to do the same. The data collected from sensor nodes have been validated before feeding into the trained model. SVM is the most reliable tool to implement in prediction. The mean square error has minimized and has a high rate of accuracy.

So, in agriculture context to design a predictor model or framework, machine learning model is useful. From the sensor data, we can identify the suitable crop species concerning the environmental parameters like temperature, humidity, soil pH, rain level, etc. Further advanced deep learning models can be used for optimization.

Fig. 4 Flow diagram of data
collection and analysis

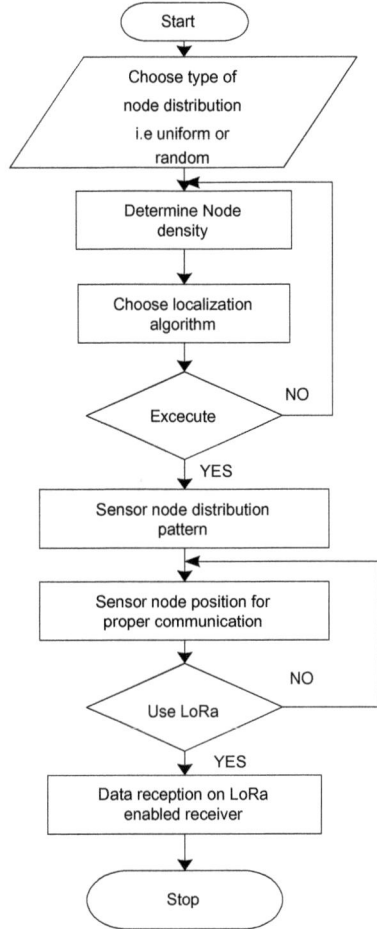

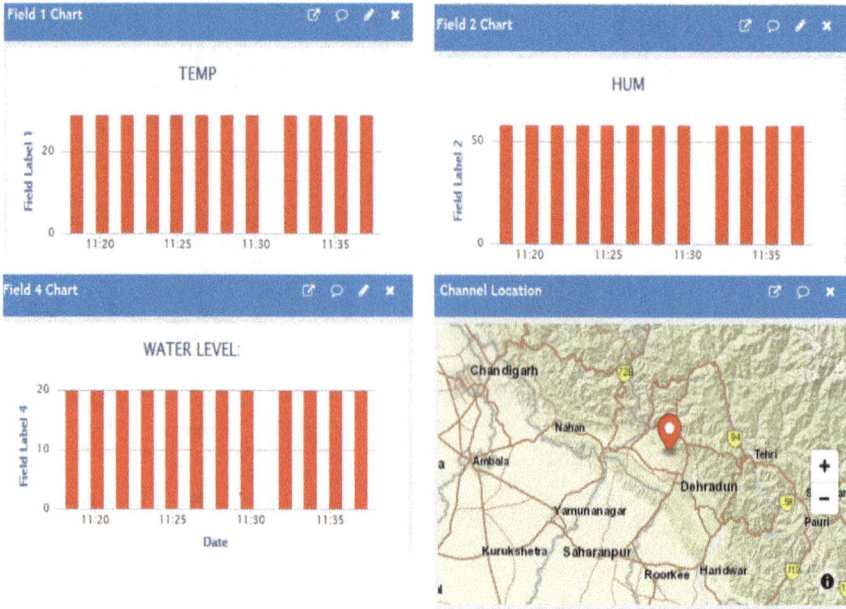

Fig. 5 Data showed on a cloud platform with location

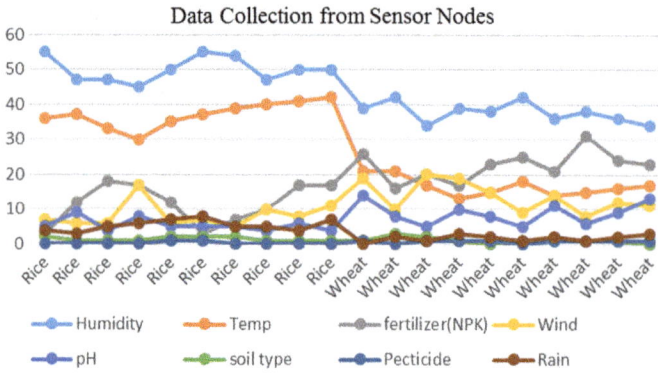

Fig. 6 Data collection from LoRa sensor nodes

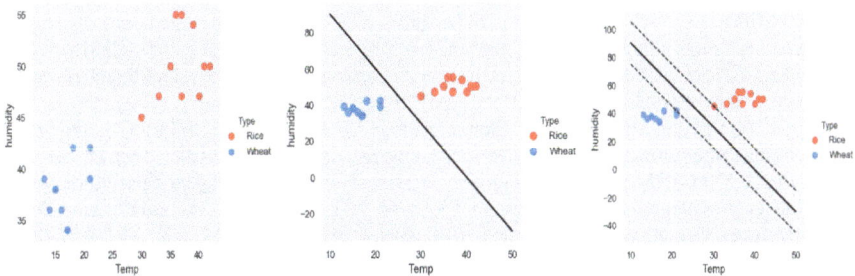

Fig. 7 Implementation of a machine learning algorithm on collected data from the sensor node for prediction of suitable crop species

References

1. Dawaliby S, Bradai A, Pousset Y (2019) Adaptive dynamic network slicing in LoRa networks. Future Gener Comput Syst 98:697–707
2. Zorbas D, Abdelfadeel K, Kotzanikolaou P, Pesch D (2020) TS-LoRa: time-slotted LoRaWAN for the Industrial Internet of Things. Comput Commun 153:1–10
3. Sandoval RM, Garcia-Sanchez AJ, Garcia-Haro J (2019) Optimizing and updating LoRa communication parameters: a machine learning approach. IEEE Trans Netw Serv Manage 16(3):884–895
4. Mahmood A, Sisinni E, Guntupalli L, Rondón R, Hassan SA, Gidlund M (2018) Scalability analysis of a LoRa network under imperfect orthogonality. IEEE Trans Industr Inf 15(3):1425–1436
5. Lam KH, Cheung CC, Lee WC (2017, October) LoRa-based localization systems for the noisy outdoor environment. In: 2017 IEEE 13th international conference on wireless and mobile computing, networking and communications (WiMob), pp 278–284. IEEE
6. Dawaliby S, Bradai A, Pousset Y (2019) Joint slice-based spreading factor and transmission power optimization in LoRa smart city networks. Internet Things 100121
7. Sallum E, Pereira N, Alves M, Santos M (2020) Improving quality-of-service in LoRa low-power wide-area networks through optimized radio resource management. J Sens Actuator Netw 9(1):10
8. Reynders B, Wang Q, Tuset-Peiro P, Vilajosana X, Pollin S (2018) Improving reliability and scalability of lorawans through lightweight scheduling. IEEE Internet Things J 5(3):1830–1842
9. Luvisotto M, Tramarin F, Vangelista L, Vitturi S (2018) On the use of LoRaWAN for indoor industrial IoT applications. Wirel Commun Mobile Comput (2018)
10. Haxhibeqiri J, De Poorter E, Moerman I, Hoebeke J (2018) A survey of lorawan for iot: from technology to application. Sensors 18(11):3995
11. Duda A, To TH (2018, May) Simulation of LoRa in NS-3: improving LoRa performance with CSMA. [1] TongKe F (2013) Smart agriculture based on cloud computing and IOT. J Convergence Inf Technol 8(2)
12. Channe H, Kothari S, Kadam D (2015) Multidisciplinary model for smart agriculture using internet-of-things (IoT), sensors, cloud-computing, mobile-computing & big-data analysis. Int J Comput Technol Appl 6(3):374–382
13. Gondchawar N, Kawitkar RS (2016) IoT based smart agriculture. Int J of Adv Res Comput Commun Eng (IJARCCE) 5(6):177–181
14. Kapoor A, Bhat SI, Shidnal S, Mehra A (2016, October). Implementation of IoT (Internet of Things) and image processing in smart agriculture. In: International conference on computation system and information technology for sustainable solutions (CSITSS), pp 21–26. IEEE

15. Lee J, Kim S, Lee S, Choi H, Jung J (2014) A study on the necessity and construction plan of the internet of things platform for smart agriculture. J Korea Multimedia Soc 17(11):1313–1324
16. Roopaei M, Rad P, Choo KKR (2017) Cloud of things in smart agriculture: intelligent irrigation monitoring by thermal imaging. IEEE Cloud Comput 4(1):10–15
17. Suma DN, Samson SR, Saranya S, Shanmugapriya G, Subhashri R (2017) IOT based smart agriculture monitoring system. Int J Recent Innov Trends Comput Commun 5(2):177–181
18. Dan LIU, Xin C, Chongwei H, Liangliang J (2015, December) Intelligent agriculture greenhouse environment monitoring system based on IOT technology. In: 2015 international conference on intelligent transportation, big data and smart city (ICITBS), pp. 487–490. IEEE

Chapter 32
Bacteria Foraging Optimization-Based Geographical Routing Scheme in IoT

J. Shreyas, Chethana S. Reddy, P. K. Udayaprasad, Dhramendra Chouhan, and S. M. Dilip Kumar

1 Introduction

The collaboration of physical objects attached with the embedded devices having the capacity of sensing, processing and transmitted data to Internet forms a highly distributed network called Internet of things. The distribution of node needs to be uniformly placed across the monitoring region. The typical application includes vehicle monitoring, weather monitoring, earthquake monitoring, etc. The improper design of deployment causes non-uniform power uses that lead to the challenges like error-prone link and thus improper routing. The challenge is to control the medium access to avoid collision. Therefore, in a proactive approach, the localization information of the nodes required making reliable routing protocols. A type of routing protocol, namely geographical routing protocol, uses the approach of maintaining one-hop, n-hop information that reduces the use of storage and forwarding the packet to the sink in the closest route. Whereas to maintain the proactive details of neighbor location flood of control message to be broadcasted that consume more energy and bandwidth, which leads to the problem of weakest link problem.

J. Shreyas (✉) · C. S. Reddy · P. K. Udayaprasad · D. Chouhan · S. M. Dilip Kumar
Department of Computer Science and Engineering, University Visvesvaraya College of Engineering, Bengaluru, India
e-mail: joseph.shreyas3@gmail.com

C. S. Reddy
e-mail: chethana9318reddy@gmail.com

P. K. Udayaprasad
e-mail: udayaprasad43@gmail.com

D. Chouhan
e-mail: dharmu2007@gmail.com

S. M. Dilip Kumar
e-mail: dilipkumarsm@gmail.com

Thereby, this reduces the network performance in terms of more packet loss, maximizing delays and most importantly the network life reduction due to excess use of energy. In IoT, the nodes decide the route deepening upon distance, number hops and the loss channelization.

2 Related Works

The usage of geographic hash table is also reported to offer better precision toward information dispersal process in IoT. The work carried out by [1] has presented an evolutionary approach in order to offer load balancing using geographic routing scheme. The existing system also reported that geographic routing significantly assists in heterogeneous routing scheme; however, they are also reported to be associated with message replication that induces overhead. This problem is addressed in the work of [2] where dynamic routing information is used over a sparse network emphasizing on the energy efficiency factor too. The usage of geographic protocol also assists in an effective traffic balancing process using a discrete design of load balancer in distributed system. The work of [3] has used optimization-based strategy using stochastic approach in order to develop an efficient traffic management over cloud environment. The study also introduces a unique network resource allocation process emphasizing over the energy factor too.

Precision is a demanding factor while working on geographic routing approach in IoT; however, it is quite challenging to ensure this with respect to domains within it, e.g., wireless sensor network with constraint nodes. This issue has been investigated by [4] who has used convex partitioning approach for the purpose of improving the precision factor of the location of the sensor nodes. Exclusive analysis of the energy factor over the geographic routing scheme is an essential criterion to enhance the network lifetime of the IoT devices, e.g., a sensor node. A unique geographic routing scheme has been presented by [5, 7] that makes use of the location details and energy information in order to ensure energy efficiency while performing routing. The study also offers solution toward resisting routing holes in networks. It was also found that truncation of the cross-links significantly assists in performance of geographic routing system. The work of [6] has addressed this problem in order to resist planarizing network as well as routing hole problem and thereby improve performance of geographic routing.

3 Problem Statement and Objectives

The problem considered in the proposed work is to find an communication efficient geographical routing scheme for IoT with the following objectives:

1. To reduce the delay.

2. To increase the throughput.
3. To increase the packet delivery ratio.

4 Proposed System

The complete routing protocol implementation will be carried out with respect to geographic and energy efficiency considering all the possible constraint of dynamic routing conditions over peak traffic condition in IoT. The implementation will also consider that there are fair chances of intermittent link breakage as well as uncertainty in traffic condition caused due to unknown and unpredicted events. This use case will be considered as the core for the model validation, and the complete implementation will be carried out in the following stages:

- **Stage-I: Developing an Analytical Routing Model** This is the first stage of implementation where an IoT ecosystem will be developed considering a two-dimensional topology for sensor node random deployment. The routing model will essentially consider design allocation of functions on the basis of the multiple layers over the protocol stack of IoT. The proposed model will initiate its design from selecting a sensor node to perform estimation of an attribute which associate distance with path loss between two motes.
- **Stage-II: Introducing Effective Traffic Management** The proposed system implements a standard queuing framework (Markov) in order to ensure an effective traffic management while performing the proposed routing. A stochastic process will be implemented over a discrete time in order to develop the queuing process with finite size of queue. An objective function along with constraint modeling will also be implemented toward ensuring the assurance of routing objectives in peak traffic condition and adherence to practical resource condition of sensor nodes in IoT ecosystem, respectively.
- **Stage-III: Designing Resource Evaluation System** This part of the implementation emphasizes on computing the effective energy being drained toward applying the proposed cross-layered routing protocol. For this purpose, various forms of energy-based attributes will be investigated and considered. Some of the tentative attributes will be associated with transition states of the sensor, size of packet, amplification energy, signal-to-noise ratio, rate of transmission, duty cycle, etc. The proposed system will apply the formulation of energy in such a way that there should be an inclusion of energy being depleted for all the sensors with the probability of steady states of respective sensors.
- **Stage-IV: Modeling of Bacterial Foraging Optimization Paradigm** The study realizes the scope of optimization to enhance the communication performance of IoT with energy-efficient communication and attempts to apply swarm intelligence and their collective behavior to solve distributed problem of futuristic IoT.
- **Stage-V: Model Performance Validation** The implementation of the proposed model will be carried out in MATLAB using simulation-based approach. Prior

to performing model validation, the proposed system will initiate simulate the scenario of sensor network and validate it. Apart from this, the proposed system will also construct a model for wireless communication links in IoT ecosystem considering the random deployment of the sensor nodes. The simulation will also consist of specific number of sensors with formulation of IoT ecosystem along with consideration of interference and channel condition.

5 Implementation Scenario

The core objective of this chapter is to brief the implementation scenario of the overall project module with adequate execution flow. The notion of the implementation in the context of this project performs numerical analysis of a cross-layer design approach to enhance the performance of quality-of-services (QoS) supported routing in the area of Green Internet of Things (IoT). The implementation structure aims to attain the planned design goal that includes energy-efficient optimal path formulation satisfying the QoS factors in IoT systems. The green computing model in IoT designs and conceptually develops the routing strategy for the optimal on-demand path allocation to different communication-exhaustive IoT devices without compromising the QoS aspects. The following are the implementation details associated with the project development.

a. *Analytical Formulation*

The numerical modeling of the formulated concept for QoS advancing routing applies a multi-hop communication-based traffic system which mimics the scenario with Markov discrete-time M/M/1 model. The system basically implements two different core modules: (i) It introduces an optimized swarm intelligence bio-inspired clustering paradigm based on bacteria foraging optimization algorithm (BFOA) and (ii) numerical modeling and implementation scenario of an energy-aware routing with green computation in the context of IoT systems.

Initially, the IoT systems are deployed with a fully connected network topological structure. Here, communication requirements of each IoT node differ from another. After the distribution of nodes over a specific region of interest, the density of nodes within that region is concerned. Therefore, the implementation scenario mimics the collaborative distributed networks of IoT that consists of a set of heterogeneous smart devices. The heterogeneous smart devices are scattered in an area of deployment for the purpose of monitoring application-specific activities in the real sense.

The proposed work here addresses the problems which arise in the antenna systems and their configuration as their design complexity triggers higher energy consumption from the sensor node attributes. This in long term affects the network lifetime of the system and negatively influences the wireless channel with intermittent link breakage and channel errors. Thereby, design proper routing mechanism with energy-efficient low-cost implementation can reduce the possibilities of error-prone links and incompetent routing.

The density factors within a smaller region basically forms a multi-hop network topological structure with a closed pattern of structure. Further, the number of intermediate relays is used to compute the networking performance. The prime intention of this computational design is into incorporate the energy-aware optimized delay-constrained routing strategy with optimized cross-layer design by involving (a) physical layer, (b) MAC layer, (c) routing layer and (d) application layer. The numerical modeling analyzes the fact that in the single-hop communication, the place where the event triggers, if a sensor node A is being deployed, then it usually has the longest transmission range and hence can waste extra amount of energy due to direct communication and retransmission.

Thereby, the prime reason of retransmission in this scenario occurs due to the unreliable links which is needed to be fixed. Thereby, this study numerically optimizes the energy allocation by estimating the design of multi-hop routing with end-to-end retransmission strategy and finally as bit error rate (BER). It attempts to fix the route structure with optimized cost-oriented adaptive switching policy which ensures minimal path loss component. The optimal forwarding of packets with respect to QoS factor optimization basically targets to minimize the power per bit allocation along with reliable cost-optimized link which ensures higher reliability of data transmission with lesser delay constraints.

b. *Discussion on BFO algorithm for energy-optimized clustering in IoT*

BFOA is nature-inspired computational technique which is influenced by forging behavior of *Escherichia coli* bacteria which live in human intestine. BFOA mimics the four fundamental processes that are chemotaxis step, swarming, reproduction and elimination-dispersal step as based on a real bacterial system present in the human intestine which helps to achieve the ideal solution for the complex problem.

- **Chemotaxis process**

This is a fundamental process generated by the swimming and tumbling of the bacterium in search of food. In this, flagellated cells have a sensory system that enables them to swim toward favorable surrounding that contains attractant such as nutrients and away from the noxious environment that contains repellent signals like as harmful product, and this type of sensory system is called as chemotaxis. Furthermore, when a bacterium meets with higher nutrient surface and noxious-free surface, they continue their swim process in the way of getting higher nutrient. But when they meet with a noxious surface, then it changes the direction and moves in the random direction with the fixed small steps.

$$\theta^i(j+1, k, l) = \theta^i(j, k, l) + C(i)\frac{\delta(i)}{\Delta t(i)\Delta(i)} \tag{1}$$

So let us assume in solution space that initial bacteria is an ith bacterium that is $i = 1$, and for each value of i, it considers as $i = j + 1$, where the jth parameter will consider as chemo-tactic steps, kth reproduction, lth is an elimination-dispersal process. Also, let us $J(i,j,k,l)$ be the fitness value with the position of the ith bacterium $\theta^i(j,k,l)$ here, (θ^i can be consider as the position of ith bacterium). So, the movement of bacterium at chemo-tactic process, reproductive and elimination-dispersal process can be represented in mathematical terms as:

- **Swarming Process**

In this process, each bacterium moves and releases a signal to attract the other bacterium to form swarm together. *E. coli* bacteria has a unique sensing and decision-making technique which they use to send a signal to other bacteria, and every bacteria release an offensive substance to suggest other bacteria to keep secure distance among each of them. The repelling effect and cell-to-cell attraction can be represented as:

$$
\begin{aligned}
Jcc(\theta, P(j,k,l)) &= \sum_{i=1}^{S} Jcc(\theta, \theta i(j,k,l)) \\
&= \sum_{i=1}^{S} \left[-dattract exp(-\omega attract) \sum_{m=1}^{P} (\theta m - \theta im)^2 \right] \\
&+ \sum_{i-1}^{S} \left[-hrepelent exp(-\omega repelent) \sum_{m=1}^{P} (\theta m - \theta im) \right] \quad (2)
\end{aligned}
$$

- **Reproduction process**

In this, bacteria arrange themselves according to their health value or cost value in which the half of bacteria die and remaining half of bacteria are ready to survive and for reproduction which means the weaker bacteria (which require more energy to sustain) die and the remaining healthier bacteria (which need the least energy or which having low-cost value) divide each of them into two bacteria. This process maintains the bacteria population size constant. The health of bacteria is calculated by the sum of fitness that is $Jihealth(h)$

$$
Jihealth(h) = \sum_{j=1}^{Nc+1} J(i,j,k,l) \quad (3)
$$

- **Elimination and dispersion process**

This process takes after the reproduction process. Due to the occurrence of an unpredicted change in the environment, all bacteria in that region are died, or they dispersed into a new environment. The newly born bacteria occupy the location of eliminated bacteria. The highest fittest bacteria or the bacteria with minimum cost value represent the optimal solution for an objective problem. The above pseudo-code shows the core computational design of the integrated BFO-based energy and delay-aware clustering in IoT. It also shows how efficient selection of CH can enhance the reliable link formation along with optimizes energy and delay performance.

Algorithm 1 BFO Clustering strategy in IoT

1: **Procedure**
2:     Initialize of BFO parameter
3:     Initialize global best and local best cost
4:     Start BFO main loop
5:     **For** (I = 1 to number of elimination dispersal)
6:         **For**(K = 1 to reproduction limit)
7:             Evaluate chemotactic loop
8:             **If** (chemotactic>eliminate dispersal) **then**
                 Jump to number 6
9:             **Else**
                 Compute the cost value for each bacteria
10:             Update value of bacteria and apply its limits
11:             Update position and bacterial mirror effect
12:             Apply bacterial position
13:             Evaluate personal best and global best cost
14:             Select the CH based on distance and energy factor
15:         Make global best solution as best solution
16:         Repeat for entire IoT network dies
17:         Evaluates the base parameters
18: **end procedure**

6 Performance Analysis

The section analyzes the performance using different parameters of QoS by examining the pragmatic scenarios of deployment of IoT in a wide-spread region, hence evincing the geographic routing of IoT. In most single-hop and multi-hop communication precis, single-hop transmission is considered to be more efficient with respect to energy within the radio range. Reducing the energy consumption can be assumed that it is the core issue, and it can reduce the power overhead through usage of innovative strategy for multipath transmissions. Here, considering the hop distances and energy utilized metrics of path loss, average packet delay, energy consumption in each node, throughput which ensures green efficiency. The following are the parameters that are examined with simulated results from mathematically calculated values and by plotting them.

A. *Analysis of Average delay*

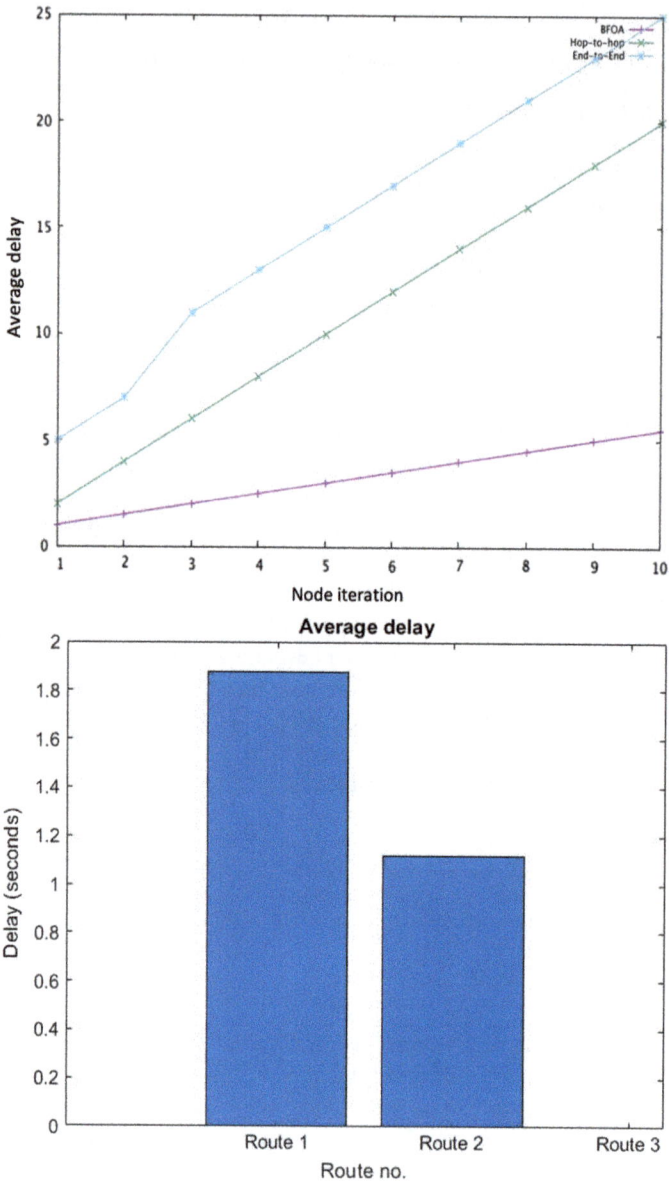

Fig. 1 Analysis of delay

Figure 1 shows that the average delay is very less in route-3 as compared to route 1 and 2, while the formulated hop-by-hop retransmission got applied. The delay could be happening in various stages such as when the node tends to transmit, it could have a delay depending on the packet size with respect to the distance it has to travel. It could occur as end-to-end transmission stage, where packets should deliver complete information from cluster node to the sink. Another instance of delay is during the queuing process. Where packets have route blocked hence to avoid packet drop, the queue is maintained. All the factors are considered and plotted with data checked from three algorithms.

B. *Analysis of Packet Delivery Ratio*

Here, Fig. 2 shows that the packet delivery ratio is very higher in the case of both transmission strategies. It directly relates to packet drop happening in the route. No of packets delivered are based on the energy left in the system as well as the information carried over to the sink. Hence considered this, the proposed algorithm has a benefit and has high packet delivery ratio with respect to both the strategies discussed.

C. *Analysis of Average throughput*

Figure 3 shows the amount of throughput obtained from different route establishments using the formulated retransmission strategies.

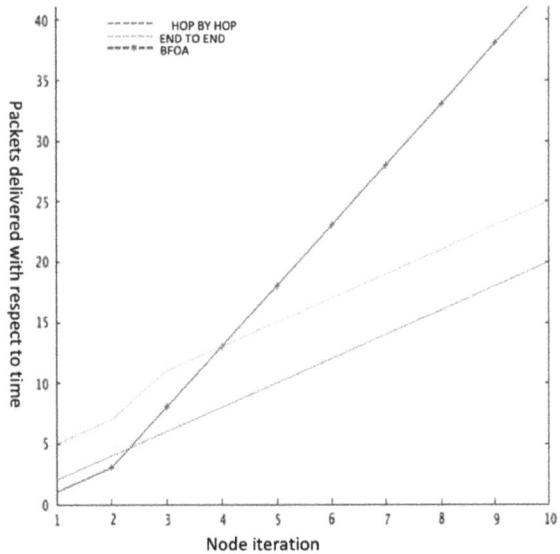

Fig. 2 Analysis of packet delivery ratio

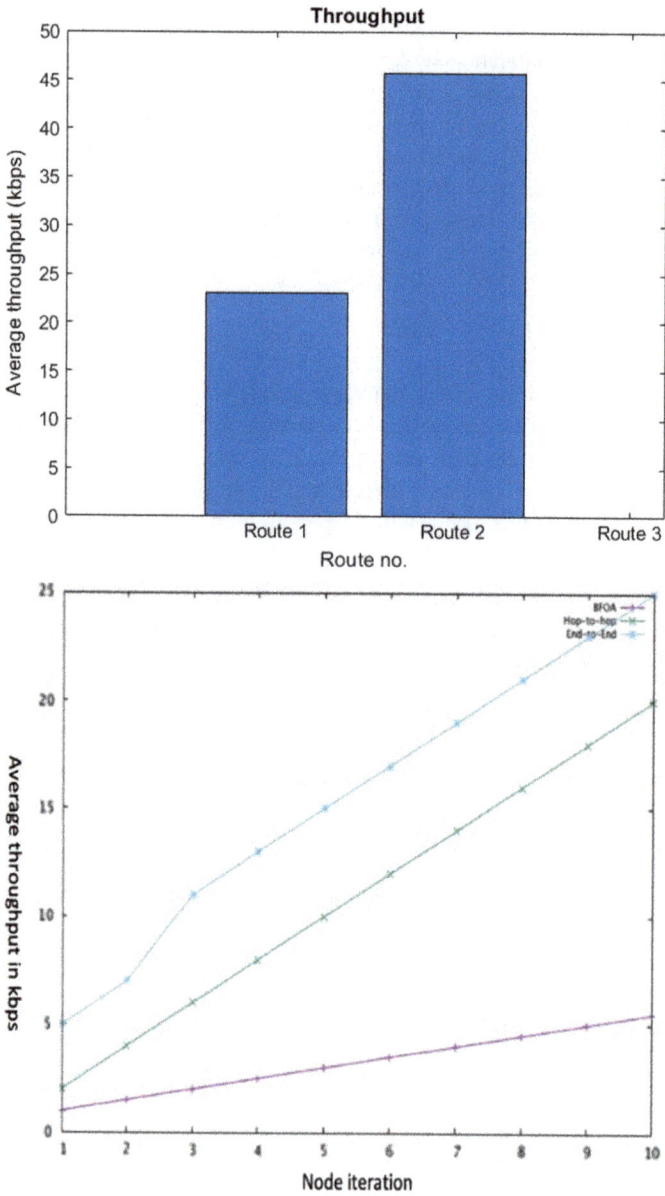

Fig. 3 Analysis of average throughput

7   Conclusion

The traditional geographical routing approach ensures efficient routing as it adopts proactive approach neighborhood discovery, whereas this process of neighborhood discovery requires control message flooding which exhausts the energy and severely affects the link quality; therefore, the geographical routing needs to be optimal to handle weak link problem. The weak link problem is aimed to be optimized by normalizing the objective function considering parameters like distance, number of hop counts, energy distribution across the zones of deployment geography, modulations, path loss, radio state from MAC layer, etc. In this stage of study, a suitable routing protocol modeling is simulated in a numerical computing platform to establish the route considering energy of zones for hop selection in the route accordingly to the event movement. In the next stage, the project aims to realize the objective functions for performance analysis of defined metrices. The study also attempts to numerically model a swarm intelligence-based clustering approach which targets to optimize the energy and delay constraints.

References

1. Shreyas J et al (2020) Route optimization with guaranteed fault-tolerance in IoT using ant inspired hierarchical LCA. IJAST 29(2):1669–1684
2. Fan J, Hu Y, Luan TH, Dong M (2017) DisLoc: a convex partitioning based approach for distributed 3-D localization in wireless sensor networks. IEEE Sens J 17(24):8412–8423
3. Shreyas J et al (2019) Congestion aware algorithm using fuzzy logic to find an optimal routing path for IoT networks. In: 2019 international conference on computational intelligence and knowledge economy (ICCIKE). IEEE
4. Huang H, Yin H, Min G, Zhang J, Wu Y, Zhang X (2018) Energy-aware dual-path geographic routing to bypass routing holes in wireless sensor networks. IEEE Trans Mob Comput 17(6):1339–1352
5. Shreyas J, Dilip Kumar SM (2019) A survey on computational intelligence techniques for internet of things. In: International conference on communication and intelligent systems. Springer, Singapore
6. Li S, Gao H, Wu D (2016) An energy-balanced routing protocol with greedy forwarding for WSNs in cropland. In: 2016 IEEE international conference on electronic information and communication technology (ICEICT), Harbin, 2016, pp 1–7
7. Shreyas J, Anand Jumnal, Dilip Kumar SM, Venugopal KR (2020) Application of computational intelligence techniques for internet of things: an extensive survey. Int J Comput Intell Stud 9(3):234–288

Chapter 33
Face Expression-Based Result Prediction in Talent Shows Using Deep Learning

Vakada Naveen⊙**, Nekkalapu Anvitha**⊙**, G. P. Sri Harsha**⊙**, and K. L. Sailaja**⊙

1 Introduction

We are living in an era where estimating one's true feelings truly are rather difficult. As almost everyone out there has mastered to conceal their inner turmoil and put on facade ahead of the world. But with today's technology anything is possible, so this project is our small attempt to reach a conclusion of one's decision regarding any case using expression classification and facial recognition.

Here, in this project, we are considering the case of a talent show Britain's Got Talent which is a famous US TV show. As everyone knows that the result declaration time makes people quite anxious. So, in this project, we collected several images of Judge Simon making it a dataset one dataset of images where he pressed the buzzer to yes and other datasets where he is not quite satisfied with the performances. Using this data as input to the neural network, we predicted the decision of Simon beforehand which was quite exciting.

V. Naveen (✉) · N. Anvitha · G. P. Sri Harsha · K. L. Sailaja
Velagapudi Ramakrishna Siddhartha Engineering College, (Autonomous) Kanuru, Vijayawada, Andhra Pradesh 52007, India
e-mail: vakadanaveen@gmail.com
URL: https://www.vrsiddhartha.ac.in

N. Anvitha
e-mail: anvithanekkalapu@gmail.com
URL: https://www.vrsiddhartha.ac.in

G. P. Sri Harsha
e-mail: gopisettisriharsha@gmail.com
URL: https://www.vrsiddhartha.ac.in

K. L. Sailaja
e-mail: sailaja0905@gmail.com
URL: https://www.vrsiddhartha.ac.in

Face recognition is one of the advanced technologies available in a conference, Lekdioui et al. gave a brief [1], intra-face algorithm is used to extract the main components of the face. On these components, a pre-processing state is performed in order to divide them into blocks and to adjust their size. Later on, feature extraction is done for building a face feature descriptor or else local descriptors are used for extracting the face features. In the final step of the face recognition, the features extracted are led to multiclass SVM to perform the recognition task.

Another method for the recognition was given by Raheel et al. [2] goes like, mostly, humans express their opinion or emotions through facial expressions. Electroencephalography is used to collect the brain signals which are in turn useful for the facial expression recognition. Initially, Emotiv headset records the EEG data whilst humans are watching a video. In the second step, noise should be removed from the EEG data in the pre-processing stage. In the third step, several features are extracted from the pre-processed, e.g., data. In this last step, the extracted features and classification algorithms such as SVM, K-nearest neighbour (KNN), Naive Bayes multilayer perceptron (MLP) are used to classify the EEG data. Of all the classification algorithms, KNN shows the highest classification accuracy of 81.06% for facial recognition.

A further discussion was given by Kim et al. [3] as, at each facial region, the features like histogram of oriented gradients and Haar are to be extracted. These extracted features are given to a classifier namely SVM to obtain the resulted output, i.e., expression recognition. The expression recognition is perfectly done by the system and had an F1 score of 0.8759.

Another implementation was given by Borkar et al. where [4] hybrid face recognition algorithm is used with the combination two other face recognition methods like linear discriminant analysis (LDA) and principle component analysis (PCA). Eigenvector, which is necessary for the PCA and LDA algorithms is calculated through Jacobi method. For face recognition techniques, Raspberry Pi 3 can be used. Combination of two or more techniques for face recognition gives more accuracy. The accuracy obtained by using the above techniques is 97%.

A much more connected approach was given by Lasri et al. [5]; here, CNN can be used to detect facial emotion recognition. It can be trained with a large number of images as input and the emotion as output. To detect the objects of the image or video that are given as input, Haar cascade algorithm is helpful. The faces that are detected in the first phase are fitted to size and normalised. These facial images are given to CNN and the resultant can be any of facial expressions like neutral, fear, happy, anger, disgust and surprise. This model has given an accuracy of 70% on the FER 2013 database.

Another explanation was given by Vyas et al. [6], static single images and image sequences are the two approaches for face expression recognition. For face expression recognition (FER), there are different classification algorithms like MLP, SVM and KNN. Different types of features regarding texture, eigenvectors and facial landmarks are extracted by classification techniques. CNN plays a very crucial role in face expression recognition among several classification techniques. FER mainly depends on the type of image that is given as an input. There can be frontal images, posed images and spontaneous images that are available online. After the gathering of data, pre-processing of images and data augmentation of images should be performed. Pre-processing of data means face detection and normalisation of the images. Data augmentation is technique for obtaining the synthesised signals from the original images. After all these, data is fed into the CNN model to get the resultant output.

Another approach given by Zhong et al. [7] gives temporal information preserving a model that is mainly based on facial and physiological expressions is proposed. Information from several periods of time can be combined in order to get good performance. Both physiological signals and facial expressions give the best performance to achieve emotion recognition.

A further correlated work was given by Tong et al. [8], broad learning system is a new model for recognising facial expressions. Initially, random features are to be extracted by using pre-set feature mappings from the feature nodes. In the next step, to strengthen the architecture enhancement, nodes were expanded. Feature nodes and enhancement nodes are connected with the output layer to show the union of normalisation methods. Extended Cohn-Kanade dataset was used in order to test the model.

Another related concept was given by Xie et al. [9], the proposed method is named as DCMA-CNN. The proposed framework involves the CNN's two branches. Local features are extracted from the image patches by the one branch and holistic features are extracted from the expressional image by another branch. By combining both the extracted features, we can get the output more precisely. Expressional transformation in variation (ETI) is proposed in the training phase to remove the noises. To study more about the expressional image patches of local features and to imagine the regions that are applicable for change of expression, a strategy is proposed. For this implementation datasets like Japanese female facial expression and extended Cohn-Kanade database are used.

One more approach was given by Gupta et al. [10], here extended Cohn-Kanade database and Cohn-Kanade database are used for detection of the facial emotion. These two datasets have a large number of static images. The resultant facial expression is completely facial action coding system (FACS) and the emotions are corrected and justified. Initially, faces are to be recognised in images or videos by using the Haar filter. If faces are identified then they need to be normalised, i.e. resized and processed for the detecting the facial landmarks. The detected face landmarks are used to train the datasets with the help of SVM classifier to classify the facial emotions. By this model, the accuracy obtained is 93.7%. For getting more accuracy, improvement of facial landmarks can also be done.

A new approach was given by Chan et al. [11] for expression classification. As several techniques provided deals mostly with small-scale datasets, image recognition for large data is still a struggle. Here, it offers a deep neural network architecture over several noted standard datasets. Model is a single component structure that captures facial images as input and then groups into six basic or the neutral expressions.

Nagpal et al. [12] proposed an approach for classification of expressions in children. An approach named mean supervised deep Boltzmann machine is used for this purpose. This classifies the image given to the model into various expression classes. They evaluated model on Radbaud faces and CAFE datasets.

Another approach given by Lv et al. [13], in their paper recognised facial expressions with components using face parsing. The elements that are mobile in expression disclosure are used for expression recognition. The training is done using a deep network and adapted by logistic regression. Also, a deep architecture pre-trained with an autoencoder is applied.

Another method for recognising facial expression was given by Yoshitomi et al. [14], where thermal image processing as well as neural networks were used. Infrared rays were used for detecting the facial temperature distribution. Input image is normalised based on location and size. To estimate the temperature difference, some of the facial squares are chosen by basing on the FACS-AU, particularly, whilst the squares including cheekbone area are chosen tentatively for improving acknowledgment exactness. The temperature difference brought by the adjustment of face muscle and internal temperature change is used as information for a neural network. Usage of backpropagation gives 90% accuracy for sad, surprised, happy and neutral facial expressions.

Another model was given by Thomas et al. [15] in which technologies used were MATLAB and neural networks. Backpropagation algorithm was used by the feed-forward network to identify the expressions of the face. Facial expressions like surprise, sad, neutral and happy are recognised in this model. This method obtained a performance rate of 0.0198 against a target of 0.0200.

2 Methodology

Here, in this project, we use deep neural networks which is a machine learning model [16] that uses several layers to steadily pull out salient characteristics from the raw input. Consider an instance of image processing, lower layers might detect edges, whilst the higher layers might detect the notion of human numbers, alphabets or faces. Majority of the deep learning prototypes are found on artificial neural networks, convolutional neural networks, though they might latent variables organised by layer-wise in deep generative models such as nodes in deep belief networks and deep, propositional formulas.

In the deep learning model, we use convolutional neural network model, CNN is an algorithm that accepts an image as input and then several features of the image are considered, and based upon the result, we can distinguish one image to the other. We use the same concept to differentiate the Simon's expressions and predict the end result.

Convolutional layers in stereotypes allow layers nearer to the input recognise features like lines and layers a bit away recognises the features like shapes or specific objects.

A major drawback is that they note accurate spot of features in the input. It causes the slightest of changes in position will arise in a completely new feature map.

Pooling layer is an additional layer appended to convolutional layer. The order of layers in the stereotype is input image, convolutional layer, nonlinearity and pooling layer (Fig. 1).

Here, in the project, we have taken several images (Fig. 2) of judge expressions and prepared a dataset using those images. Then, these images are used to train the neural network with the architecture as shown in Fig. 3. This neural network has three convolutional layers and three pooling layers. Pooling layers are used to reduce the dimensionality of the features, whereas convolution layers are used to find out the patterns in the input. As we go from the first layer to last layer, the complexity of the features recognised increases.

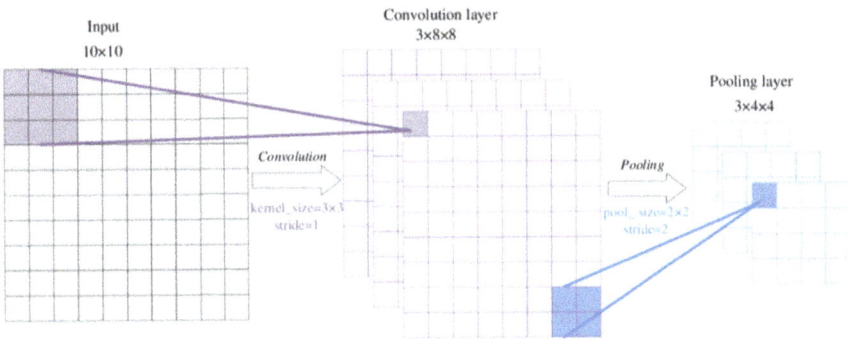

Fig. 1 Basic structure of convolutional neural networks [17]

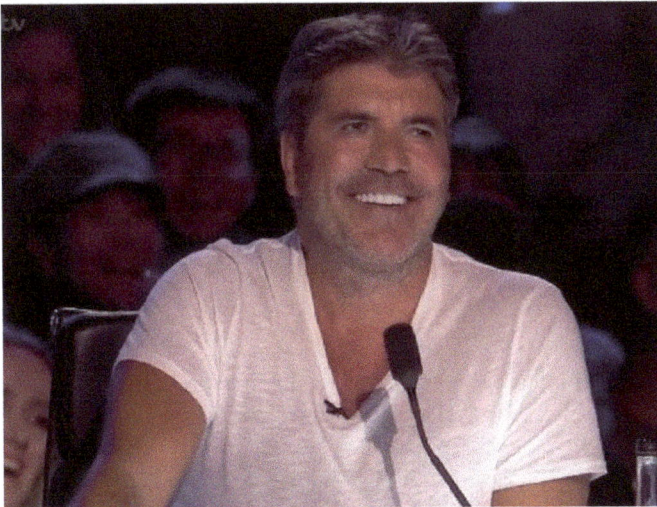

Fig. 2 Image from dataset [18]

```
Layer (type)                      Output Shape                    Param #
=========================================================================
conv2d (Conv2D)                   (None, 148, 148, 16)            448

max_pooling2d (MaxPooling2D)      (None, 74, 74, 16)              0

conv2d_1 (Conv2D)                 (None, 72, 72, 32)              4640

max_pooling2d_1 (MaxPooling2      (None, 36, 36, 32)              0

conv2d_2 (Conv2D)                 (None, 34, 34, 64)              18496

max_pooling2d_2 (MaxPooling2      (None, 17, 17, 64)              0

flatten (Flatten)                 (None, 18496)                   0

dense (Dense)                     (None, 512)                     9470464

dense_1 (Dense)                   (None, 1)                       513
=========================================================================
Total params: 9,494,561
Trainable params: 9,494,561
Non-trainable params: 0
```

Fig. 3 Architecture of the neural network (*Note* param = Parameters)

Here, in our model, initially, we consider the input image of standard size of 150 X 150 with 3 bytes colour. By changing the filters, we use the combination of convolution layer and max pooling layer after the fifth combination we use the flatten fun to flatten the results obtained so as to feed into the dense neural network.

Finally, we use two dense layers, firstly, 512 neuron hidden layer with activation of ReLu and the second layer of only 1 output neuron where it contains a value from 0 to 1 having 0 for 1 class ('yes') and 1 for the other ('no') having activation function of sigmoid. A total of 10 iterations have been done to reach the result.

Figure 2 is categorised into Simon yes category based upon the judgement he passed at the end of performance. In the similar fashion, we have populated data into two sets based on the end result.

3 Results and Analysis

The trained model is now used to make predictions on the test data. The test set used for the evaluation of model accuracy contains 50 images of judge expressions. The confusion matrix or otherwise called error matrix generated to determine the performance of test data is shown in Fig. 4.

From the matrix, the results obtained are,

(1) Accuracy = (TP + TN)/(TP + TN + FP + FN) = 0.88
(2) Recall = TP/(TP + FN) = 0.9

Fig. 4 Displaying the
confusion matrix

		PREDICTED	
		YES	NO
ACTUAL	YES	36	4
	NO	2	8

(3) Specificity = TN/(TN + FP) = 0.8
(4) Precision = TP/(TP + FP) = 0.947

We have taken a dataset of 50 images which is small dataset but when you move to larger datasets then the accuracy levels increase with the size. Accuracy tells us the level up to which we can trust the system's decisions regarding the classification or judgement. The model developed by us gives an accuracy of about 88% on test data.

4   Conclusion

This paper explains how deep learning technologies can be helpful in the real-world applications like analysis of image data to find interesting results and applications. A neural network model is developed which is used for expression classification. The model acts as a solution for several real-time situations of analysis and prediction. The best part is almost instantaneous solution delivery which can be quite helpful and bring solace to souls.

References

1. Lekdioui K, Ruichek Y, Messoussi R, Chaabi Y, Touahni R (2017) Facial expression recognition using face-regions. In: 2017 international conference on advanced technologies for signal and image processing (ATSIP), pp 1–6. IEEE
2. Raheel A, Majid M, Anwar SM (2019) Facial expression recognition based on electroencephalography. In: 2019 2nd international conference on computing, mathematics and engineering technologies (iCoMET), pp 1–5. IEEE
3. Kim S, An GH, Kang S-J (2017) Facial expression recognition system using machine learning. In: 2017 international SoC design conference (ISOCC), pp 266–267. IEEE
4. Borkar NR, Kuwelkar S (2017) Real-time implementation of face recognition system. In: 2017 international conference on computing methodologies and communication (ICCMC), pp 249–255. IEEE
5. Lasri I, Solh AR, Belkacemi ME (2019) Facial emotion recognition of students using convolutional neural network. In: 2019 third international conference on intelligent computing in data sciences (ICDS), pp 1–6. IEEE
6. Vyas AS, Prajapati HB, Dabhi VK (2019) Survey on face expression recognition using CNN. In: 2019 5th international conference on advanced computing & communication systems (ICACCS), pp 102–106. IEEE

7. Zhong B, Qin Z, Yang S, Chen J, Mudrick N, Taub M, Azevedo R, Lobaton E (2017) Emotion recognition with facial expressions and physiological signals. In: 2017 IEEE symposium series on computational intelligence (SSCI), pp 1–8. IEEE
8. Zhang T, Liu Z, Wang X-H, Xing X-F, Philip Chen CL, Chen E (2018) Facial expression recognition via broad learning system. In: 2018 IEEE international conference on systems, man, and cybernetics (SMC), pp 1898–1902. IEEE
9. Xie S, Hu H (2018) Facial expression recognition using hierarchical features with deep comprehensive multipatches aggregation convolutional neural networks. IEEE Trans Multimedia 21(1):211–220
10. Gupta S (2018) Facial emotion recognition in real-time and static images. In: 2018 2nd international conference on inventive systems and control (ICISC), pp 553–560. IEEE
11. Mollahosseini A, Chan D, Mahoor MH (2016) Going deeper in facial expression recognition using deep neural networks. In: 2016 IEEE winter conference on applications of computer vision (WACV), pp 1–10. IEEE
12. Nagpal S, Singh M, Vatsa M, Singh R, Noore A (2019) Expression classification in children using mean supervised deep Boltzmann Machine. In: Proceedings of the IEEE conference on computer vision and pattern recognition workshops
13. Lv Y, Feng Z, Xu C (2014) Facial expression recognition via deep learning. In: 2014 international conference on smart computing, pp 303–308. IEEE
14. Yoshitomi Y, Miyawaki N, Tomita S, Kimura S (1997) Facial expression recognition using thermal image processing and neural network. In: Proceedings 6th IEEE international workshop on robot and human communication. RO-MAN'97 SENDAI, pp 380–385. IEEE
15. Nisha T, Mathew M (2012) Facial expression recognition system using neural network and MATLAB. In: 2012 international conference on computing, communication and applications, pp 1–5. IEEE
16. Jurtz VI, Johansen AR, Nielsen M, Almagro Armenteros JJ, Nielsen H, Sønderby CK, Winther O, Sønderby SK (2017) An introduction to deep learning on biological sequence data: examples and solutions. Bioinformatics 33(22):3685–3690
17. https://www.researchgate.net/figure/The-sub-convolution-pooling-neuralnetwork_fig2_326 963855. Last accessed 2 July 2020
18. https://www.digitalspy.com/tv/reality-tv/a28776883/simon-cowell-olympic-britains-got-tal ent-spin-off/. Last accessed 2 July 2020

Chapter 34
Performance-Based Adaptive Learning Rate Scheduler Algorithm

Vakada Naveen[ORCID]**, Yaswanth Mareedu**[ORCID]**, Neeharika Sai Mandava**[ORCID]**, Sravya Kaveti**[ORCID]**, and G. Krishna Kishore**[ORCID]

1 Introduction

The neural network resembles the architecture of brain. The neural network contains input layer, dense layer and output layer. Each layer contains neurons. Neurons in each layer are connected to other layers with weights. The input is given to the neural network in the input layer. The dense layers learn the features of the given input data by adjusting the weights of each layer. The output will be obtained at output layer. The main aspect here is training a neural network to learn features. For learning the features by the neurons in the neural network in each layer, the hyperparameters play a vital role. The knowledge should be gained by the neuron from the input data.

Generally, training a neural network model is challenging. While coming to backpropagation algorithm due to fixed learning rate and two passes of information, training time is too long. Other disadvantages in backpropagation include slow convergence. In a backpropagation algorithm with adaptive learning and momentum coefficient [1], they had come up with an idea of adaptive learning where after every iteration, based on the error derivative the learning rate coefficient and momentum are adjusted. In learning process of the model, we can do this by adjusting the synaptic

V. Naveen (✉) · Y. Mareedu · N. S. Mandava · S. Kaveti · G. Krishna Kishore
VR Siddhartha Engineering College, Kanuru 520007, India
e-mail: vakadanaveen@gmail.com

Y. Mareedu
e-mail: yaswanthyashu333@gmail.com

N. S. Mandava
e-mail: neeharika.mandava@gmail.com

S. Kaveti
e-mail: kavetisravya87@gmail.com

G. Krishna Kishore
e-mail: gkk@vrsiddhartha.ac.in

© The Author(s), under exclusive license to Springer Nature Singapore Pte Ltd. 2021
X.-Z. Gao et al. (eds.), *Applications of Artificial Intelligence in Engineering*, Algorithms for Intelligent Systems, https://doi.org/10.1007/978-981-33-4604-8_34

weight. The algorithm designed in this paper [1] has been tested on three problems. The parity problem, optical character recognition and two-spiral problem. In the parity problem using this algorithm, it took around 174 epochs for $h = 0.05$ and $a = 0.75$ where as for Rumelhart backpropagation algorithm it took around 9259 epochs for the same weights and h, a values. When compared with Rumelhart algorithm, convergence speed is faster in this algorithm. In optical character recognition, the theme is the model should identify the character even with high noise. So they trained the model with two groups of ideal images and two groups of noisy images. This also showed a good outcome with a maximum error rate of 5.6%. In case of two-spiral problem, the training set consists of 200 X-Y values arranged in a spiral format. This algorithm has learned the spiral pattern in 1330 epochs, where as the traditional backpropagation algorithm did not do even in 5000 epochs. Hence, the proposed algorithm in this paper showed better performance with faster convergence rate than other traditional methods.

The error surface in backpropagation (BP) is either flat surfaces or extreme steeps. Fixed learning rate in the BP algorithm makes it less efficient. In this paper, learning rate is calculated dynamically based on derivative information. Using information from forward and backward propagation, the derivatives are calculated. Many approaches to optimize the learning rate are proposed in dynamic learning rate optimization of backpropagation algorithm [2] like linear expansion of outputs, linear search and Newton-like methods. Since the derivatives are obtained by simple calculations there is no extra storage burden on the networks and also there is an increase in the convergence rate. In backpropagation algorithm, if there are more than one hidden layers in the network the convergence becomes slow due to activation function's saturation property. So learning rate should be modified with changes in the gradient level. Learning rate optimization approaches via linear expansion, learning rate search, learning rate optimization approaches via polynomial approximation, learning rate optimization approaches using Newton-like method, determination of learning rate, and momentum factor are the methods used for learning rate optimization in this paper. These approaches are applied on the problems like parity check problem, classification of 2D mashed and disconnected regions and function approximation problem. Out of the proposed approaches, Newton-like method showed the best results with faster convergence rate and moderate increase in computational complexity.

BP becomes slow when there are more than one hidden layers in the network. So, the stochastic gradient is used to train the multilayer neural network. In conventional digital hardware, backpropagation (BP) algorithm becomes inefficient because of the physical distance between arithmetic and memory module. To overcome this, memristors are used because they can integrate processor and memory. In adjusting learning rate of memristor-based multilayer neural networks via fuzzy method [3] using memristors and two MOSFET transistors, a circuit is designed which reduces the area consumption by 92–98% when compared with CMOS-only circuit. A fuzzy method is also designed that showed an increase of 2–3% in learning accuracy.

Learning rate parameter has a significant impact on the accuracy of the model. Learning rate also effects the training speed. This paper shows how the learning rate effects the training speed and helps us to choose the learning rate that would maximize the accuracy. In gradient descent algorithm at each weight space, the error gradient is calculated. Further, weights will be moved based on this value in the opposite direction. But we cannot know how far to move in the same direction safely before the error gradient starts increasing again. If the learning rate is too large, it moves in correct direction for a long run but will overlook on the error surface. This results in low accuracy. Because of this, the model has to be trained again for increasing the accuracy, resulting in wastage of computation power. Hence, if the learning rate is small, step by step we can reduce the error gradient and improve the accuracy.

In the need of small learning rates on large problems [4], they have trained the neural network with different learning rates, for one epoch to find the learning rate with minimum error. Then, continue to train the model with this learning rate until the accuracy becomes smooth. Now, decrease the learning rate by a factor and repeat the same until you get a accuracy that is less that the previously calculated once. So, finally, at this learning rate, we have to train the model until we reach a maximum accuracy and stop training the model when the accuracy stops increasing or before it starts decreasing.

Learning rate is an important parameter for training a neural network. In cyclical learning rates for training neural networks [5], a cyclic method to find an ideal learning rate is introduced. Instead of decreasing the learning rate monotonically, it is varied cyclically between a range of values. Increasing the learning rate may have a negative impact but in a long run it helps. Here, the triangular learning rate policy is used to set the minimum and maximum boundaries for the cycle. We have a saddle point from where the learning process slows down. So the assumption is that the optimal learning rate will be between these bounds. For each of the architectures like Resnet, CIFAR-10, CIFAR-100 this method is applied and the results are efficient than the traditional methods.

An empirical study of learning rates in deep neural networks in speech recognization [6], we mainly see about AdaGrad which is one of the learning rate scheduling schemes and a new variant called AdaDec which helps in have results in high frame accuracies. Gaussian mixture model (GMM) systems have the same fundamental architecture. We consider minibatch stochastic gradient descent (SGD) on which random neural networks are initiated and trained. Predetermined piecewise constant learning rate, exponential scheduling, power scheduling, performance scheduling, AdaGrad and AdaDec are some of the methods of learning rate scheduling. The voice search task is examined with data. The network which is trained on a graphics processing unit (GPU) has an input of consecutive frames and the number of future frames is limited to 5. We use word error rate (WER) for measuring performances of the system. The WER must be low to have success of our system. The system can train the networks about 35,000 frames per second. With experimenting along AdaGrad, we have no practical improvements. Poor performance and having extra computations are its drawbacks. Hence, we came up with AdaDec. The self-regulating performance scheduling is found to be more effective. Thus, our system produced a 0.2% reduction of WER compared to its previous test.

The improvement of BP neural network learning algorithm [7] we mainly focus on backpropagation (BP) algorithm which is mostly used for training of multilayer feed forward artificial neural networks (FFANS). The gradient descent method is considered which has advantage of minimizing the sum of square errors and disadvantages as well. The learning rate parameter is constant which is directly proportional to weight. Small learning rate is recommended to avoid the problem of oscillation. The learning factor and initial factor are key factors in the convergence of backpropagation neural network. Hence, we present an adaptive BP algorithm which automatically updates them by dynamically trained error rate of change and are changed depending upon the signs of gradient direction during two consecutive iterations. By this, we can overcome the disadvantage by obtaining faster convergence rate. The fuzzy control method on the adaptive learning can also be used to control them. Thus, dynamically trained error rate of change brings advantages through adaptive BP algorithm.

The improved training algorithm of backpropagation neural network with self-adaptive learning rate [8] shows the functionality relation between the total quadratic testing error change and connection weights. The backpropagation (BP) network has some disadvantages which can be overcome by selecting various techniques to increase the learning rate. One of them is to start the training with high learning rate and later on gets decreased with the establishment of weights and biases. Here, learning rate is first raised with the advancement of training and falls down at final stage which is not self-adaptive. The single sample, average and total quadratic error are used to measure the error in the algorithm. To avoid single sample error, we use batch BP learning algorithm where the weight is updated after all samples being processed, differentiates it from the standard BP algorithm. The Taylor formula is followed to observe the change in the total quadratic error. The self-adaptive learning rate is therefore achieved from the original equation. An improved algorithm of the BP network for proving the effectiveness is considered to observe the change in the average quadratic error. In case of small error curve surface, there are a large measure of gradient and learning rate to decrease the iterations occur. We can obtain the fast convergence rate of the self-adaptive BP algorithm compared to traditional BP algorithm.

On adaptive learning rate that guarantees convergence in feed forward networks [9] represents the problem of training a multilayer feed forward neural network. The backpropagation (BP) method, a popular method for training a feed forward network has disadvantages regarding convergence rate. Though certain techniques such as quasi-newton and nonlinear optimization are proposed to overcome the disadvantages but are not as close to BP algorithm in terms of ease and simplicity. Lyapunov stability theory helps in guarantee of convergence and to have a constant learning law in case of multilayer dynamic neural network which is similar to BP algorithm but has an additional term about the identification error. By using Lyapunov stability theory, LF I and LF II algorithms are proposed. Since local minima are not avoided along the trajectory of convergence in LF I, to make this possible a second-order term, i.e., acceleration term to be added in the weight update algorithm which has computational difficulties. To overcome these difficulties to some extent, LF II came into existence where local minimum can be excluded by refusing the changes in updating of weight. These two algorithms are tested on the following and are compared with backpropagation and EKF, namely XOR, 3-bit parity, 8-3. Therefore, the algorithm to produce the desired result and approximates the 2D Gabor function is compared with its gradient descent counterpart. We can conclude that the convergence rate is speeded up of the BP algorithm by having adaptive learning rate and LF II helps to have this convergence in the all cases. By the results after testing, we can say that the convergence is better with LF I compared to BP and is more improved with LF II.

Optimal adaptive k-means algorithm with dynamic adjustment of learning rate [10] discusses about K-means clustering algorithm which is one of the commonly used competitive learning algorithm and also reduces the mean squared error (MSE) cost function and is simple to analyze and can be performed faster. It also used for dividing the domain of pattern vectors where a reference pattern is considered which represents these pattern vectors in each region. The algorithm only assures with the local optimality. Leaky learning and consistence learning law are used to yield the divisions depending on the optimal condition with respect to MSE cost function. The algorithm is mainly depended on the learning rate in case of online mode. Learning rate is directly proportional to the residual deviation. Trail-and-error methods are used to derive the optimal learning rate. Later, search-then-coverage schedule is proposed to overcome slow convergence rate. Learning rate is recommended to adapt dynamically to solve the problem with respective characteristics. With the help of delta-bar-delta rule, a fuzzy logic controller is proposed for controlling the vigilance parameter. The algorithm introduces two novel mechanisms which explain that the bad minima can be escaped without misrepresenting the cost function and to adjust learning rate dynamically which helps in obtaining accuracy in approximating an optimal solution. Here, we discuss about the following adaptive K-means clustering algorithms, namely Optm, Cons, Sqrt and Tard. The training patterns in these algorithms are generated from synthetic probability distributions which consist of stationary and non-stationary statistics. The non-stationary distributions consist of constantly rotating distribution, constantly translating distribution and abrupt change in distribution. By maintaining the quality of the image, vector quantization image coding is used which reduces transmission bit rate and data storage where the images are divided into small block called vectors. Since the two novel mechanisms are generic, special parameter controls are not required.

Learning rate plays a key role in deep neural network as mentioned in PACL: piecewise arc cotangent learning rate for deep neural network training [11]. Learning rate is an important hyperparameter for better performance of the model. The PACL has a cyclic mechanism where the loss is reduced and increase the performance of the model in training. This mechanism works in the following way, in the training process for every few iterations, the learning rate is re-initialized with cotangent function. For every cycle in this process, the learning rate decay is minimized. This process is more efficient for deep neural networks because it does not require any additional computational power for training the DNN.

Optimizing neural network learning rate by using a genetic algorithm with per-epoch mutations is also one of the methods for learning rate tuning [12]. The performance of convolutional neural networks has been increased with different learning methods and various new techniques like auto-coding. But there are some problems like decision about learning rate that is, if we have to fix the learning rate or vary it. And the other problem is controlling the use of bad local minimum. The learning method discussed in this paper "learning rate-optimizing genetic backpropagation (LOG-BP)" solves the above-mentioned problems by using a genetic algorithm.

Scheduling the learning rate via hypergradients: new insights and a new algorithm [13] discusses about the structure of the hypergradients and online gradient-based adaptive schedules along with a new algorithm named MARTHE. The aim of this algorithm is to have minimum validation error for the production of models and have an automatic system to figure out a learning rate scheduler for hypothetical optimization methods in online format against the learning task provided.

This paper presents a new algorithm to update the learning rate by based on the performance of the model.

2   Methodology

Some of the hyperparameters we have for a neural network are learning rate, batch size, number of epochs, activation function, bias, number of layers, number of neurons in a given layer, etc. There could be a large number of combinations available for hyperparameters, even for a small neural network. Hyperparameter tuning plays a major role in increasing the accuracy. The hyperparameters tuning means choosing the hyperparameters that are optimal for a neural network. Learning rate is one such hyperparameter which greatly effects the accuracy of a neural network.

The learning rate value chosen will be very small like $0.01, 0.005$, etc. It takes small steps to learn and minimize the loss function. Learning rate is the important hyperparameter in that. The learning rate determines size of each epoch while reducing the loss error. The main objective here is changing the learning rate according to the loss of the model in the previous iteration.

It is an optimization algorithm to fit the learning rate to its optimal value and change according to the loss in each iteration of neural network. All the hyperparameters are fit their optimal value, we get the best accuracy for the model which we have trained. Hyperparameter tuning is to be done for obtaining the optimal values for every hyperparameter in neural network.

The new algorithm that is proposed here takes the performance of the model until the current epoch as the factor in tuning the learning rate of the model. If the cost of the model increases than the previous epoch the algorithm takes a small step in learning rate of the model. In the next iteration, the same process is repeated to tune the learning rate of the model.

Algorithm for learning rate scheduler

Step 1:   Initialize Δlr to vary the learning rate (like $10^{-3}, 10^{-4}, 10^{-5}$...etc.)
Step 2:   If the current iteration loss is greater than previous iteration loss.
Step 3:   Change the sign of Δlr.

Fig. 1 Learning rate
scheduler

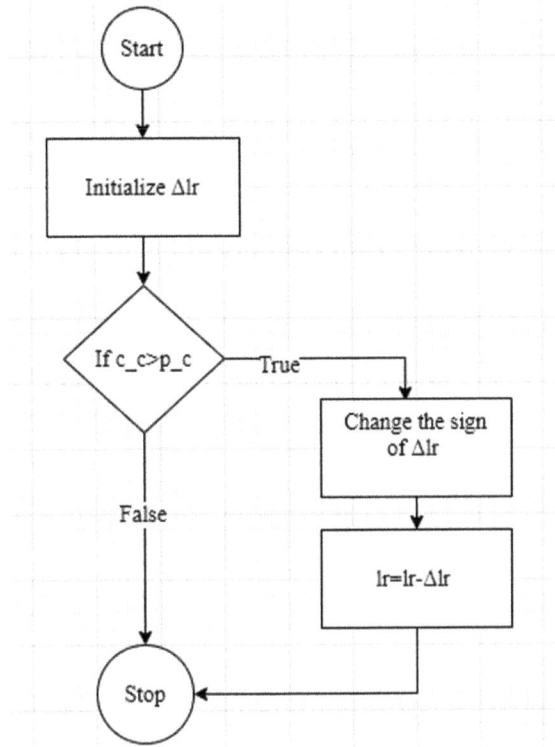

Step 4: $lr = lr - \Delta lr$.

The flowchart for the performance-based learning rate scheduler algorithm is given in Fig. 1.

The details of the symbols used in Fig. 1 is shown below:

c_c = current step loss

p_c = previous step loss

lr = learning rate

3 Experimental Results

The cats and dogs dataset is used to measure the performance of this new algorithm. A neural network with the proposed algorithm is created and tested with the dataset. The training accuracy and test accuracy of the model are collected and presented in the tables given (Table 1, 2, 3, 4, 5 and 6).

Table 1 Accuracy versus learning rate for Lr = 0.005

Δlr	10^{-7}	10^{-6}	10^{-5}	10^{-4}	10^{-3}	10^{-2}
Accuracy	87.5598086124402	87.5598086124402	87.5598086124402	87.5598086124402	87.5598086124402	87.5598086124402

Table 2 Accuracy versus learning rate for Lr = 0.004

Δlr	10^{-7}	10^{-6}	10^{-5}	10^{-4}	10^{-3}	10^{-2}
Accuracy	91.3875598080861244	91.3875598080861244	90.909090909090909	90.909090909090909	92.3444976076555	92.82966507717703

Table 3 Accuracy versus learning rate for Lr = 0.006

Δlr	10^{-7}	10^{-6}	10^{-5}	10^{-4}	10^{-3}	10^{-2}
Accuracy	84.68899521531101	84.68899521531101	84.68899521531101	84.68899521531101	84.68899521531101	84.68899521531101

Table 4 Accuracy versus learning rate for Lr = 0.007

Δlr	10^{-7}	10^{-6}	10^{-5}	10^{-4}	10^{-3}	10^{-2}
Accuracy	81.81818181818181	81.81818181818181	81.81818181818181	82.29665071770334	84.21052631578948	92.82296650717703

Table 5 Accuracy versus learning rate for Lr = 0.003

Δlr	10^{-7}	10^{-6}	10^{-5}	10^{-4}	10^{-3}	10^{-2}
Accuracy	84.6889952153110l	84.6889952153110l	84.6889952153110l	84.6889952153110l	84.6889952153110l	84.6889952153110l

Table 6 Accuracy versus learning rate for Lr = 0.0055

Δlr	10^{-7}	10^{-6}	10^{-5}	10^{-4}	10^{-3}	10^{-2}
Accuracy	91.3875598080861244	91.3875598080861244	91.3875598080861244	91.3875598080861244	92.8229665071770	93.3014354066986

These results have proved to be promising in helping the developers choose the performance-based adaptive learning rate algorithm as tool to improve the model performance by tuning the learning rate of the model (Figs. 2, 3, 4, 5, 6 and 7).

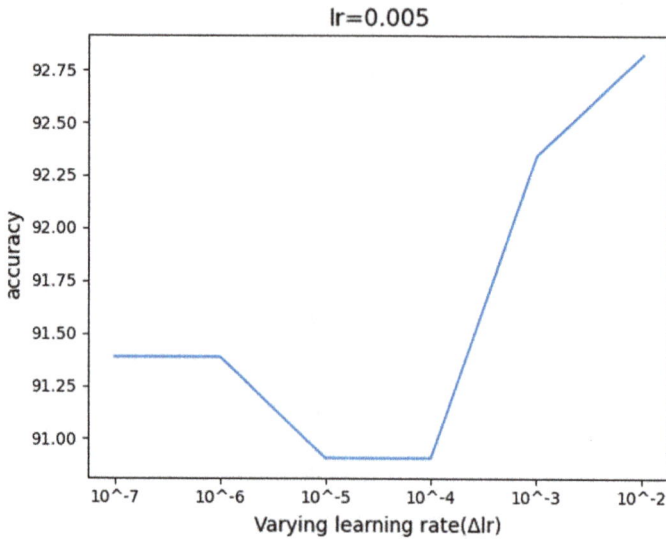

Fig. 2 Graph for accuracy versus learning rate for Lr $= 0.005$

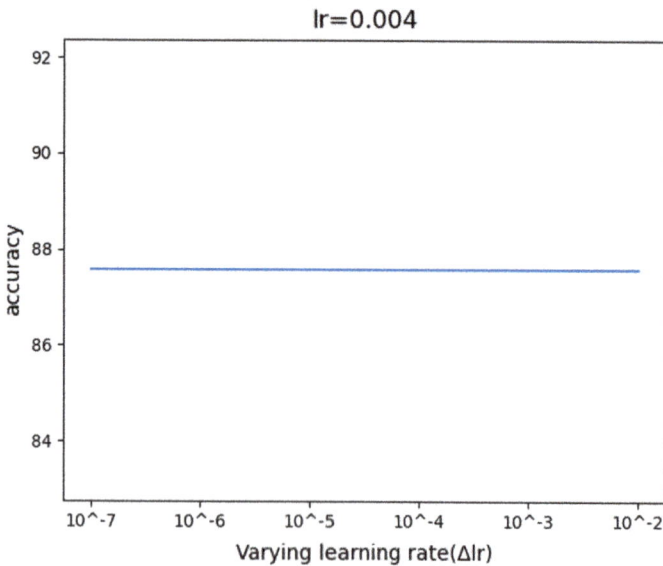

Fig. 3 Graph for accuracy versus learning rate for Lr $= 0.004$

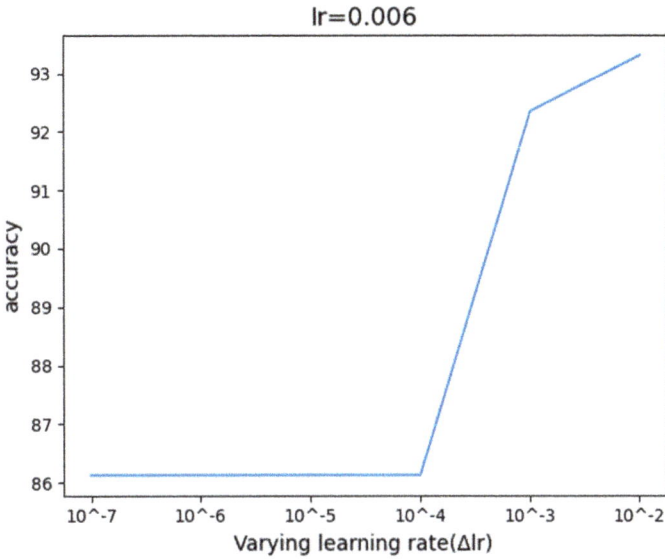

Fig. 4 Graph for accuracy versus learning rate for Lr $= 0.006$

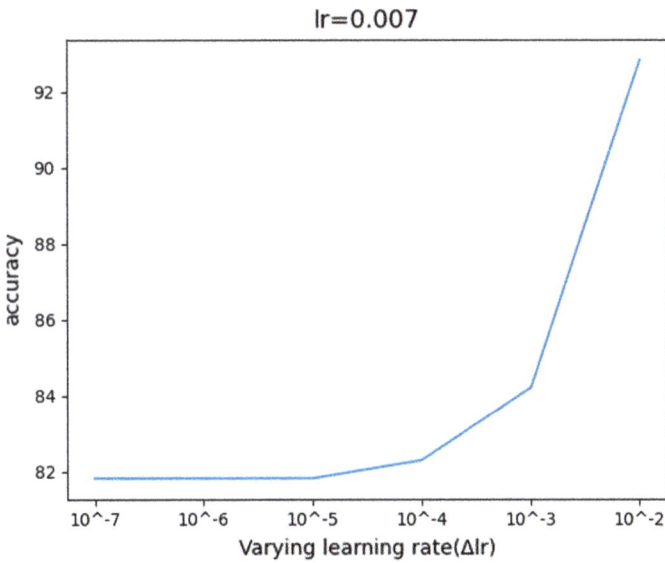

Fig. 5 Graph for accuracy versus learning rate for Lr $= 0.007$

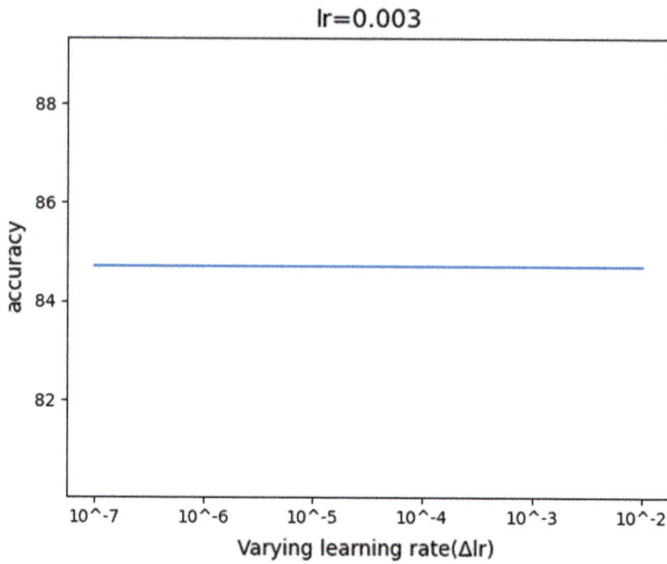

Fig. 6 Graph for accuracy versus learning rate for Lr = 0.003

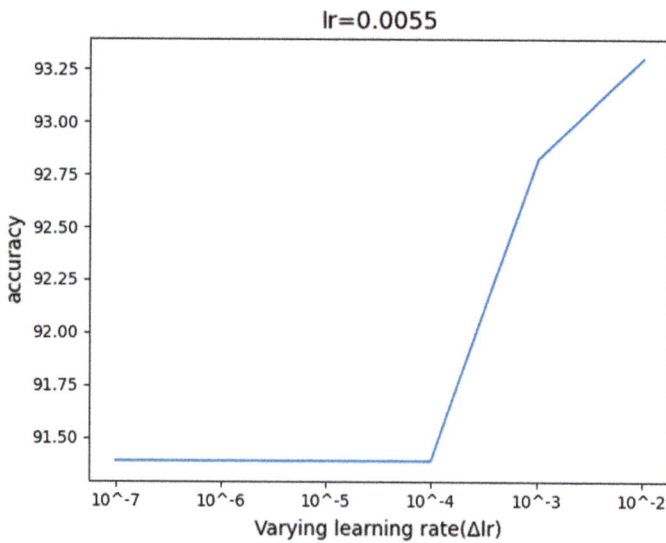

Fig. 7 Graph for accuracy versus learning rate for Lr = 0.0055

4 Conclusion

In this paper, a novel algorithm for improving the learning rate of a neural network is given. The algorithm is tested by using a neural network on a real-world dataset. The results of the new algorithms show that adaptive learning rate algorithms can help to increase the efficiency of a model and help to reach the minimal solution in less number of steps. This algorithm is very simple to implement and can be used as a learning scheduler for a neural network.

References

1. Yu C-C, Liu B-D (2002) A backpropagation algorithm with adaptive learning rate and momentum coefficient. In: Proceedings of the 2002 International Joint Conference on Neural Networks. IJCNN'02 (Cat. No. 02CH37290). vol 2. IEEE, 2002
2. Yu X-H, Chen G-A, Cheng S-X (1995) Dynamic learning rate optimization of the backpropagation algorithm. IEEE Trans Neural Netw 6(3):669–677
3. Wen S, Xiao S, Yang Y, Yan Z, Zeng Z, Huang T (2018) Adjusting learning rate of memristor-based multilayer neural networks via fuzzy method. IEEE Trans Comput Aided Design Integr Circuits Syst 38(6):1084–1094
4. Wilson DR, Martinez TR (2001) The need for small learning rates on large problems. In: IJCNN'01 International Joint Conference on Neural Networks. Proceedings (Cat. No. 01CH37222), vol 1, pp 115–119. IEEE, 2001
5. Smith LN (2017) Cyclical learning rates for training neural networks. In: 2017 IEEE Winter Conference on Applications of Computer Vision (WACV), pp 464– 472. IEEE, 2017
6. Senior A, Heigold G, Ranzato M, Yang K (2013) An empirical study of learning rates in deep neural networks for speech recognition. In: 2013 IEEE international conference on acoustics, speech and signal processing, pp 6724–6728. IEEE, 2013
7. Wen J, Zhao JL, Luo SW (2000) The improvement of B Pneural network learning algorithm signal processing proceedings. In: WCCC_ICSP International Conference, Germany
8. Li Y, Fu Y, Li H, Zhang S-W (2009) The improved training algorithm of back propagation neural network with self-adaptive learning rate. In: 2009 International Conference on Computational Intelligence and Natural Computing, vol 1, pp 73–76. IEEE, 2009
9. Behera L, Kumar S, Patnaik A (2006) On adaptive learning rate that guarantees convergence in feedforward networks. IEEE Trans Neural Netw 17(5):1116–1125
10. Chinrungrueng C, Sequin CH (1995) Optimal adaptive k-means algorithm with dynamic adjustment of learning rate. IEEE Trans Neural Netw 6(1):157–169
11. Yang H, Liu J, Sun H, Zhang H (2020) PACL: piecewise arc cotangent decay learning rate for deep neural network training. IEEE Access 8:112805–112813
12. Kanada Y (2016) Optimizing neural-network learning rate by using a genetic algorithm with per-epoch mutations. In 2016 International Joint Conference on Neural Networks (IJCNN), pp 1472–1479. IEEE, 2016
13. Donini M, Franceschi L, Pontil M, Majumder O, Frasconi P (2019) Scheduling the learning rate via hypergradients: new insights and a new algorithm. arXiv preprint arXiv:1910.08525

Chapter 35
Urdu QA: Question Answering System for Urdu Language

Mohammad Khalid Pandit and Azra Nazir

1 Introduction

After TREC-8, lot of research have been conducted for exploring many methods of answering the questions posed in natural language. This research have focused on how to retrieve a concise information from a set of usable documents for answering each TREC's question. The TREC's questions were usually focused on the factoid questions. Factoid questions are based on simple facts and can be usually answered in few words, e.g., which vitamin is present in milk? Or what is the name of chairman of FIFA? These questions inquire about person, location, quantity, etc. Many Question Answering (QA) systems have been built such as Shapaqa [1], AnswerBus [2], MultiText [3], AskMSR [4], or expanded such as START [5]. In 2006, the Ephyra [6] was proposed as a framework for answering the factoid questions in the open domain. In 2010, Watson [7] was introduced as an impressive Q and A system which had good experiment results in Jeopardy quiz show. The method of answering non-factoid questions such as why and how questions are very tricky and are still to be researched widely. The answers to these questions require a lot of reasoning and deep semantic analysis, and it also depends on the chain of events that occurred before that event about which question is asked. It also depends on the history of questions asked by the user. In order to improve the precision and the confidence of the answers for factoid as well as non-factoid questions, the semantic analysis and knowledge inference should be applied. Also machine learning techniques like classification are very useful for answering non-factoid questions [12].

Users often have specific questions in mind and they want precise and short answers, often users prefer to ask the questions in their own native language without

M. K. Pandit (✉)
AI & ML Group, IUST, Awantipora, India
e-mail: khalidpandit@gmail.com

M. K. Pandit · A. Nazir
Department of CSE, National Institute of Technology, Srinagar, Srinagar, India

being restricted to specific query language. This approach for developing question answering systems involves analysis of questions from the linguistic point of view so that we can understand what user actually means.

UrduQA is the result of combining information retrieval with Natural Language Processing (NLP) techniques. The IR system treats the question posed in Urdu language as a query in an attempt to identify the proper documents that contain the candidate answers. NLP techniques employed by the QA systems typically include the POS tagging, parsing, and name entity extraction. These NLP techniques are used to parse the questions and analyze the top ranked documents returned by the IR systems.

NLP techniques in Urdu language are still in the its initial stage compared to other languages like English and Chinese, which have already benefited a lot from the research in the field. Employing NLP techniques in Urdu language is difficult due to following reasons:

1. The writing direction is from right to left.
2. Urdu language follows the Nastaliq style [11] which makes it challenging among the languages that follow the Arabic script.
3. In Nastaliq style, most of the characters acquire different shapes depending upon the position in the ligature.
4. Capitalization is not used in Urdu, which makes it difficult to identify proper names and abbreviations.
5. Nouns in Urdu language have two types of gender [10] (masculine/feminine).

In addition to the above issues, also the lack of Urdu corpora that is essential in advanced research.

2   Question Answering Systems

A question answering system is a program written to answer questions asked by a user. It does so by querying a structured database of information which is also known as a knowledge base (corpus). To return the best possible answer, the system may formulate the result by combining information obtained from local databases and from documents from the web [1]. Question answering is becoming a fast developing groundwork area that combines results from information retrieval, information extraction, and Natural Language Processing (NLP) [2, 14]. Urdu QA consists of three important phases which are question processing, document processing, and answer extraction (Fig. 1).

Document retrieval is performed from the natural language question, and depending on the question type, the information retrieval engine searches for the documents containing the possible answers from the knowledge base. Answer extraction makes use of several NLP techniques to obtain the correct answer such as prepare a set of candidate answer, compute their weight, and rank the answers [5, 11].

Fig. 1 Question answer
systems

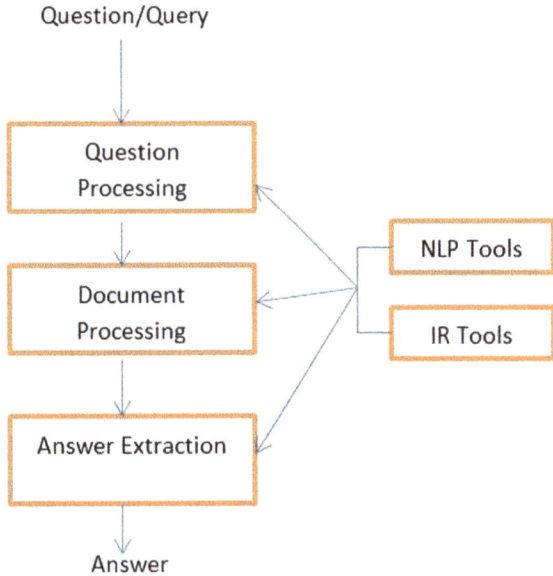

Question/Query

Question
Processing

Document
Processing

NLP Tools

IR Tools

Answer Extraction

Answer

3   Related Work

Many Question Answering (QA) systems have been developed only for English
or many western languages, such as Shapaqa [1], AnswerBus [2], MultiText [3],
AskMSR [4], START [5]. In 2006, the Ephyra [6] was proposed as a framework
for answering the factoid questions in the open domain. In 2010, Watson [7] was
introduced as an impressive Q and A system which had good experiment results in
Jeopardy quiz show. The initial step toward multilingual QA systems was AnswerBus
[2], it accepts a question in any language [2], and if the latter is not in English, then
it uses a tool called BabelFish in order to translate the question to English.

QARAB is the system which describes the approach for question answering in
Arabic language. Since Urdu being a Arabic script [] language mainly derived from
Arabic and Persian, the methodology remains almost the same but semantic analysis
and root word extraction differs.

Daud et al. [11] provide the detailed survey about the Urdu language processing
including the datasets, parsing tools, orthographical, and morphological analysis.
Challenges in writing stemmer for Urdu was first taken up by [13] saying that stemming in Urdu is challenging due to the following facts: (1) diverse nature of Urdu
(2) Farsi and Arabic stemmers can't be used for Urdu (3) correction methods based
on dictionary can't be used for Urdu due to lack of machine readable resources for
Urdu. Tokenization in Urdu is challenging task [13] but significant work has been
done in Arabic [14] which is similar to Urdu because of same script type. POS
tagging is very challenging in those languages which are rich in morphology like
Arabic, Urdu. There are three main tag sets designed for Urdu, CRULP tag set, U1

tag set, and sajjad tag set [15]. The U1 tag set is released as part of EMILLE corpus [8], based on EAGLES standard [16]. The POS tagger for Urdu was developed by [17] using EAGLES standard.

4   UrduQA

4.1   Overview

The amount of Urdu data in the text format is increasing on the web from past one decade. Urdu soft wares for web browsing is improving, and the volumes of text data in Urdu on web in form of newspapers, books, and other formats are increasing. Earlier most of the data available in major south Asian languages was in form of images including Urdu, which made it difficult for processing.

The main idea of Urdu QA is to identify candidate documents that answer the questions posed in Urdu language. The task of Urdu QA can be summarized as follows:

Given a set of questions posed in Urdu language, find answers to the question with following assumptions:

1. Answers exist in EMILLE/CILL [8] corpus (Enabling Minority Language Engineering) dataset released by Lancaster University.
2. The answers do not span through multiple documents (i.e., information about answer lies in single document).
3. The answer is short (to the point and concise).

The basic QA processing in UrduQA is composed of following:

1. Processing the question.
2. Retrieving the candidate documents.
3. Processing each of the candidate documents and returning sentences that may contain the answer.

5   Processing in the IR System

This step involves stop word removal, tokenizing, root extraction (stemming), and term weighting.

The initial task in the IR task is the word segmentation or tokenization. This task is non-trival for those languages which are scarce on machine readable resources like Urdu. Text tokenization and word boundary are difficult in Urdu as compared to languages like English and French as spaces, commas, and semi colons represent a word boundary. But in Asian languages like Urdu and Chinese, space is not used consistently.

Stemming is also main data preprocessing task. The aim of stemming is to reduce the word into origin of root (root word) stemming which consists of reducing given word into its stem base or root word, e.g., the stem of کتـاب ی (books) is کتـاب (book).

6 Question Processing in UrduQA

Question understanding requires deep semantic processing capabilities which is a non-trivial task in Urdu Language Processing (ULP). Urdu language processing does not have extensive research at semantic level; therefore, UrduQA does not understand the question content at a deep semantic level.

The incoming question in the UrduQA is treated as a *"bag of words"* [10]. The index file is searched to obtain the list of ranked documents that may possibly contain the answer. Processing begins by performing tokenization (cite) to extract individual terms. The stop words are then removed. The part of speech tagger is applied in an attempt to focus on the main words that should appear in the expected answer. Efforts are to be made to identify the proper names as they are the best way to identify the possible answer.

The question Q can be considered as a string $(q1, q2, q3,.......)$ and answer A as $(a1, a2, a3, ...)$; answer to any question depends upon two types of features [9]: $T = T(Q)$ and $I = I(Q)$ as follows:

$$P(A|Q) = P(A|T, I).$$

where $T = t1, t2.....$ are the set of $|T|$ features relating to "*question type*" as when کـب, why (کـیوں), how (کـیسـے), (Table 1 identifies the question types of Urdu QA), and $I = i1, i2, ...$ consists of information bearing part of the question asked, e.g., م دودھیں کـون وثـامن موجـود ہـے؟ and

کـی ا دودھ وثـامن ہـے؟. The information bearing part is seem inn both questions, whereas the question type is different.

Table 1 Feature selection for *Question type* query

Question starting with		Type
کـون	Who	Person
کـب	When	Date, Time
کـیا	What	Name, Event
کهـاں	Where	Location
کتنـا	How much	Quantity, Number

6.1 Question Expansion

For better retrieval results, the question is expanded to include all terms (verbs فعـل)
and (nouns اسـم) that occur in the index file having same roots. The result of this step
is passed to the IR system to retrieve the list of ranked documents that match the
terms in the question.

6.2 Question Type

The question $Q = q1, q2\ldots\ldots$ can be classified based on question types.
$T = T(Q)$ and $I = I(Q)$ as follows:

$$P(A|Q) = P(A|T, I).$$

where $T = t1, t2\ldots..$ are the set of $|T|$ features relating to "*question type*" as when
(کـب), why (کیوں), how (کیسـے). These question types help us in determining the
processing needed and the type of answer that is to be expected. UrduQA identifies
these types of questions:

7 Answer Processing in UrduQA

Input to the UrduQA answer generator module is the query in the natural language
and the ranked set of documents. The question is first processed by tagging all the
words; then, the ranked set of documents that may contain the answer is retrieved
by the IR system. The relevant documents that match the bag of words formed by
the question. Usually, the answer contains most of the terms containing in the query
including the proper nouns that appear in the query.

The following document extracted from the corpus was processed by the IR
system.

اکثر آپ کا ڈاکٹر آپ کا ٹیسٹ کرتا ہے ۔
اگر آپ مرد داکٹر سے معائنہ نہیں کرانا چاہتے تو بہت سی پریکٹس میں لیڈی ڈاکٹر
اور پریکٹس نرس ہوتی ہیں جنہیں سمیر ٹیسٹ لینے کی ترتیب دی جاتی ہے -
اور ویل وومن کلینک
بھی یہ ٹیسٹ لے سکتے ہیں اور ان میں زیادہ تر خواتین ڈاکٹر ہوتی ہیں
یہ ٹیسٹ مباشرت کرنے والی ۲۰ سال سے زائد عمر کی تمام خواتین کیلے ہے ۔
بچہ دانی نکالے جانے کے باوجود بھی خواتین کو سمیر ٹیسٹ کی ضرورت ہو سکتی ہے جس کے لئے
آپ اپنے ڈاکٹر سے معلومات حاصل کر سکتے ہیں

Practitioners in the practice of lady doctor and practice nurse that are set to take
a symmetric test.

They can also take this test, and most of them are doctors.

This test is for all women over 20 years old.

Assume the question posed by the user is as follows:

کون ٹیسٹ کرتا ہے؟

Who performs the tests?

Query processing:

The query is processed as follows:

Token	POS	Stop word
ٹیسٹ	فعلVerb	No
کون	اسم Noun	No
کرتا	فعلVerb	No
ہے		Yes

UrduQA generates the question as bag of words and passes it to the IR system.

ٹیسٹ
کون
کرتا

Assume that the QA system returns the following top ranked documents:

اکثر آپ کا ڈاکٹر آپ کا ٹیسٹ کرتا ہے ۔
اگر آپ مرد داکٹر سے معائنہ نہیں کرانا چاہتے تو بہت سی پریکٹس میں لیڈی ڈاکٹر
اور پریکٹس نرس ہوتی ہیں جنہیں سمیر ٹیسٹ لینے کی ترتیب دی جاتی ہے ۔
اور ویل وومن کلینک

Determine the expected type of answer

کون:: Who: Person (Name)

Generating the answer:

The answer generator looks for the keywords that help in identifying the person name or the person type.

اکثر آپ کا ڈاکٹر آپ کا ٹیسٹ کرتا ہے ۔

This paragraph has the most query terms and keywords that might identify the person type. Therefore, this is returned as the potential answer.

8 Conclusion

In this paper, the question answering system for Urdu language has been explored that provided short answers for factoid questions using both language processing elements and information retrieval techniques. The system developed is capable of producing accurate results for simple factoid questions with any closely related and relevant corpus. However, a larger training corpus can significantly improve accuracy of the model. The ultimate success of the UrduQA depends upon the actual available tools developed for Urdu language, which are still in their developing phase. Future work will explore the role played by weight assignment to different tokens and its effect on the overall time complexity of the system for role-based retrieval and integration of a much wider range of questions.

References

1. Buchholz S, Daelemans W (2001) Shapaqa: shallow parsing for question answering on the world wide web. In: Euro conference recent advances in natural language processing. Bulgaria, pp 47–51
2. Zheng Z (2002) AnswerBus question answering system. In: 2nd international conference on human language technology research. USA, pp 399–404
3. Clarke C, Cormack G, Kemkes G, Laszlo M, Lynam T, Terra E, Tilker P (2002) Statistical selection of exact answers (multitext experiments for TREC 2002). TREC. NIST, pp 823–831
4. Brill E, Dumais S, Banko M (2002) An analysis of the AskMSR question answering system. In: ACL-02 conference on empirical methods in natural language processing. USA, pp 257–264
5. Katz B, Marlon G, Borchardt G, Brownell A, Felshin S, Loreto D, Rosenberg J, Lu B, Mora F, Stiller S, Uzuner O, Wilcox A (2005) External knowledge sources for question answering. In: Proceedings of the 14th annual text REtrieval conference. Gaithersburg, MD
6. Schlaefer N, Gieselmann P, Schaaf T, Waibel A (2006) A pattern learning approach to question answering within the Ephyra framework. In: Proceedings of the ninth international conference on text, speech and dialog. Springer, Berlin, Heidelberg, pp 687–694
7. Ferrucci D, Brown E, Chu-Carroll J, Fan J, Gondek D, Kalyanpur A, Lally A, Murdock JW, Nyberg E, Prager J, Schlaefer N, Welty C (2010) Building Watson: an overview of the DeepQA project. AI Mag 31(3):59–79
8. Language Resource "The EMILLE/CIIL Corpus, ELRA catalogue". http://catalog.elra.info. ISLRN: 039-846- 040-604- 0, ELRA ID: ELRA-W0037
9. Heie MH, Whittaker EWD, Furui S (2012) Question answering using statistical language modelling. Comput Speech Lang 26(3):193–209
10. Hammo B, Abu-Salem H, Lytinen S (2002) QARAB: a question answering system to support the Arabic language. In: Proceedings of the ACL-02 workshop on Computational approaches to semitic languages. Association for Computational Linguistics
11. Daud A, Khan W, Che D (2017) Urdu language processing: a survey. Artif Intell Rev 47(3):279–311
12. Nguyen CT, Nguyen DT (2016, September) Towards an argument-based method for answering why-question in Vietnamese language. In: 2016 3rd National Foundation for Science and Technology Development Conference on Information and Computer Science (NICS). IEEE, pp 130–134
13. Riaz K (2010) Rule-based named entity recognition in Urdu. In: Proceedings of the 2010 named entities workshop, pp 12–35

14. Attia M (2007) Arabic tokenization system. In: Proceedings of the Urdu2007 workshop on computational approaches to semitic languages: common issues and resources, pp 65–72
15. Sajjad H (2007) Statistical part of speech tagger for Urdu. Master unpublished thesis: National University of Computer and Emerging Sciences. Lahore, Pakistan
16. Baker A, Hardie P et al (2003) Corpus data for south Asian language processing. In: Proceedings of the 10th annual workshop for South Asian language processing, pp 1–8
17. Hardie A (2003) Developing a tagset for automated part-of-speech tagging in Urdu. In: Proceedings of conference on corpus linguistics, Lancaster

Chapter 36
Review of Classifiers Used for Identification and Classification of Plant Leaf Diseases

G. Gangadevi and C. Jayakumar

1 Introduction

Plant malady is one of the significant factors which causes noteworthy decrease in the quality and amount of plant generation. Location and grouping of plant sicknesses are significant undertaking to expand plant profitability and financial development. Because of plant maladies the quality and amount of horticulture item are decreased [1]. The majority of these ailments made a decision according to a specialist around there based on their manifestations [2, 3]. The regular methods for illness the executives involve ranchers and the plant pathologists. The conclusion and utilization of the pesticide are all the more regularly done in the fields.

This procedure is tedious, testing, and more often than not brings about off base analysis with unacceptable exercise of the pesticides. With the coming of computer vision (CV), machine learning (ML), and artificial intelligence (AI) innovations, progress has been accomplished in creating computerized models engaging, exact and opportune recognizable proof of the plant leaves ailment. In the most recent decade, AI and ML innovations have accomplished a colossal enthusiasm with the accessibility of various elite registering processors and gadgets [4, 5]. Numerous endeavors have been applied to the snappy and exact finding of leaf infections [6]. The utilizations of efficient fungicides are generally lethal, non-degradable and contaminate the air by spread out noticeable all around and hoard in the dirt [7]. At long last, in view of the few classifier, the tainted piece of the leaf was ordered. The strategy created in this framework is utilized to accomplish picture handling and delicate

G. Gangadevi (✉)
Department of Information Technology, Meenakshi College of Engineering, Chennai, India
e-mail: ganga2007mtech@gmail.com

C. Jayakumar
Department of Computer Science and Engineering, Sri Venkateswara College of Engineering, Chennai, India
e-mail: cjayakumar2007@gmail.com

figuring for different sick rice plants. In any case, the acknowledgment exactness on the four informational index by and large, which implies that despite everything it needs further improvement. In [8], picture preparing and AI systems were utilized to screen seedlings with rickets in a non-damaging way so as to enhance include determination and model parameters. We took 40 papers with various plant leaf databases for our contemplate, based on the following pattern point from Sects. 2.1–5 our search review papers were discussed in extraction techniques and feature, as well as those that included classification techniques, those based on particular plant leaf, those that included a combination of shape, color, and texture also included a combination of texton, and those that worked to resolve the problem of accuracy. We list here the paradigms used for our study, and it shows the number of papers included for this detailed analysis so as to handle different problems in plant leaf disease detection research.

2 Survey on Plant Disease Detection

This segment depicts image acquisition, image preprocessing, image segmentation, and feature extraction systems utilized in different takes a shot at plant sickness recognition. This area introduces a study of 40 papers of plant malady location. Furthermore, arrangements will be portrayed in Sects. 3 and 4 (Fig. 1).

2.1 *Image Acquisition*

The photographs are delivered by blend of a lighting up source and the reflection or absorption of the essentialness by the parts of scene being imaged. Lighting up may be started by radar, infrared imperativeness source, PC created essentialness

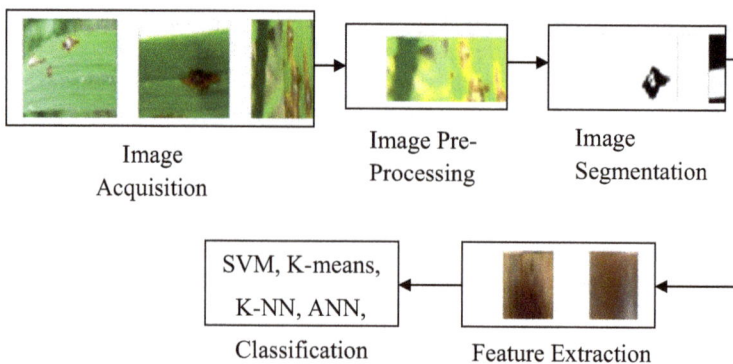

Fig. 1 Plant disease identification from leaf images

plan, ultrasound imperativeness source, X-pillar imperativeness source, etc. To identify the image, we use sensor as showed by the possibility of edification [9]. The system of picture sense is called picture verifying. By the sensor, basically edification imperativeness is changed into modernized picture. The idea is that moving toward edification essentialness is changed into voltage by the mix of information electrical imperativeness and sensor material that is open to the particular essentialness that is being recognized. The yield waveform is response of sensor and this response is digitalized to get mechanized picture.

A picture is described by two-dimensional limit $f(x, y)$. The value or sufficiency of f at spatial headings (x, y) is a positive scalar sum. Exactly when an image is delivered from a physical strategy, its value is relating to essentialness transmitted by physical source. As a result, $f(x, y)$ must be nonzero and restricted; that is,

$$0 < f(x, y) < \infty \tag{1}$$

It is in like manner discussed that for an image $f(x, y)$, we have two factors: The proportion of source edification scene on the scene being imaged and the proportion of lighting up reflected or devoured by the thing in the scene allow us to address it by:

$$i(x, y) \quad \text{and} \quad r(x, y) \tag{2}$$

The limit $f(x, y)$ may be portrayed by two sections: (1) the proportion of source lighting up scene on the scene being seen and (2) the proportion of illumination reflected by the things in the scene. These are called light and reflectance parts implied by $i(x, y)$ and $r(x, y)$ independently. The two limit join as thing to shape $f(x, y)$:

$$f(x, y) = i(x, y) \cdot r(x, y) \tag{3}$$

where $0 < i(x, y) < \infty$ and $0 < r(x, y) < 1$ $r(x, y) = 0$ means complete ingestion $r(x, y) = 1$ means all out reflectance. We call the force of a monochrome picture at any directions (x, y) the dull level (l) of the image by at that point. That is $l = f(x, y)$. The between time of l ranges from $[0, L - 1]$, where $l = 0$ exhibits dim and $l = 1$ shows white. All the widely appealing regards are hides of diminish moving from dull to white. It infers lighting up will be a nonzero and constrained sum and its sum depends upon illumination source.

$$0 < r(x, y) < 1 \tag{4}$$

Here 0 demonstrates no reflection or absolute ingestion and 1 implies no retention or complete reflection.

2.2 Image Preprocessing

For showing signs of improvement brings about further advances, and picture preprocessing is required in light of the fact that residue, dewdrops, creepy crawly's fertilizers might be available on the plant; these things are considered as picture clamor
[10]. Besides, caught pictures may have mutilation of some water drops and shadow
impact, which could make issues in the division and highlight extraction [10] stages.
Impact of such contortion can be debilitated or evacuated utilizing distinctive clamor
expulsion channels. There might be low complexity in caught pictures; for such
pictures differentiate upgrade calculations can be utilized. Now and again foundation expulsion strategies may likewise be required if there should arise an occurrence of district of intrigue should be removed. If there should arise an occurrence of
commotion, for example, salt and pepper, a middle channel can be utilized. Weiner
channel can be utilized to evacuate an obscuring impact. In the event of the pictures
caught utilizing top notch cameras, the size of the photos may be extremely huge,
for that decrease of picture size is required. Additionally, picture decrease helps in
diminishing the registering memory power [11].

2.3 Image Segmentation

Picture division can play an imperative and significant job in plant sickness location.
Picture division intends to isolate the picture into specific areas or articles [12, 13].
The essential point of division is to dissect the picture information so one can extricate
the valuable highlights from the information. There are two different ways to do the
picture division: (1) in light of discontinuities and (2) in view of similitudes. In
the principal way, a picture is parceled dependent on unexpected changes in power
esteems, e.g., done through edge discovery. While in the subsequent manner, pictures
are segment dependent on the particular predefined criteria, e.g., thresholding done
utilizing Otsu's technique.

2.4 Feature Extraction

The component extraction part of picture examination centers around recognizing
innate attributes or highlights of articles present inside a picture. These highlights
can be utilized to portray the article. By and large, considering the three classes
are removed: shading, shape, and edges and surface. The shading is a significant
component since it can separate one malady from another. Besides, every illness may
have diverse shape; in this manner, framework can separate sicknesses utilizing shape
highlights. Some shape highlights are territory, pivot, and edge. Surface methods how
shading examples are dispersed in the picture.

2.4.1 Color

Shading assumes a basic job for blossom examination than for leaf investigation. Shading is utilized each day to recognize among objects. For shading portrayal, we utilize different techniques: "shading histograms" which was powerful to revolution and interpretation, "shading minutes" where basic, "shading correlogram" and shading lucidness vector." Here, we utilized two models of histogram which are nearby and worldwide. In [14], shading histogram, edge histogram, and zone were utilized for restorative plant species recognizable proof; as we referenced, most investigations dependent on leaf in plants distinguishing proof, we found a particular leaf includes to be specific, and leaf venation which was utilized generally in plants ID draws near. Shading minutes speak to shading highlights to describe a shading picture. Highlights that can be included are mean (μ), standard deviation (σ), skewness (θ), and kurtosis (γ). For RGB shading space, the three highlights are removed from each plane R, G, and B. The recipes to catch those minutes:

$$u^m = \frac{1}{MN} \sum_{i=1}^{M} \sum_{j=1}^{N} P_{ij} \tag{5}$$

$$\sigma = \sqrt{\frac{1}{MN} \sum_{i=1}^{M} \sum_{j=1}^{N} (P_{ij} - \mu)^2} \tag{6}$$

$$\theta = \frac{\sum_{i=1}^{M} \sum_{j=1}^{N} (P_{ij} - \mu)^3}{MN\sigma^3} \tag{7}$$

$$\gamma = \frac{\sum_{i=1}^{M} \sum_{j=1}^{N} (P_{ij} - \mu)^4}{MN\sigma^4} \tag{8}$$

M and N are the dimension of image. P_{ij} is value of color on column ith and row jth.

2.4.2 Shape

Every single article has its own shape, and in some other content say as blueprint, a superior shape data must be relentless to scale, interpretation, revolution, and reflection. We isolated shape portrayal into two models: "Area based" and "Limit based" [15, 16]. Leaf location framework is dependent on leaf shape, and foods grown from the ground were utilized to distinguish plant leaf class (Fig. 2).

$$\text{leaf slight} = \frac{11}{12} \tag{9}$$

$$\text{Roundness} = \frac{4\pi A}{P^2} \tag{10}$$

Fig. 2 Irregular shape of
leaf

2.4.3 Edges and Texture

Edges are basic features of leaf pictures similarly as assessing sharp assortments in pictures. The sobel edge acknowledgment heads were used to isolate edge features from pictures in [17]. From the edges, incorporate centers were found which meet the edges and achieve 98% accuracy with 13 unmistakable plants. The model decides damages to veins. Corner features [18] are useful to find the resemblance of leaf pictures since corners are crossing purposes of two extraordinary edges or interest centers under various different headings and lighting conditions. The procedure used to evacuate the edge and surface information is using another image as reference. The reference picture is either entire white picture or a totally dim picture [19]. In our work, we have used entire dull picture as the reference picture. Allow the dim reference picture to be connoted as "B." How about we mean the image containing only the leaf picture in the wake of applying the division estimation. Consider another image with all the pixel regards to be zeros.

2.4.4 Feature Extraction Through Neural Network

The prediction and classification are done by an architecture which is capable of learning by supervised and semi-supervised data. Huge number of research has been done using neural network model for detection of diseases [20]. The initial stage of extraction of features is derived from the informative data which is not redundant, using that iterative learning that were helped for better understanding of human interpretation. Feature extraction directly encourages dimensionality reduction. As soon as the image is fed into the model and based on the filter size, the entire image is trained, it can be converted into a reduced set of features. Multiple CNN models were used, such as fine-tuned pre-trained AlexNet CNN model, AlexNet CNN model, and then proposed D-Leaf CNN model. Repetitive and new data images are introduced to learn the pattern of diseases using supervised algorithm. The morphological feature of the plant is observed nexus and extracted 62 vein features such as, number of branching points, number of ending points, number of branches, and number of areoles. Then the leaf portion is observed for density calculation of veins, branching points, ending points, and areoles [21]. The feature which is extracted might not be the required part of diseases always, but which might be a distinctive leaf characteristics, such as the venation. So the model which gets trained in multiple ways could be biased for crop-specific pattern especially when plant features might appear to be more discriminative [22].

3   Survey on Plant Disease Classifications

Extensive survey has been conducted to compare plant disease detection and classification techniques in machine learning. We studied support vector machine classification technique, K-means classifier, K-nearest neighbor classification technique, artificial neural network classification technique, fuzzy logic classifier, and neural network classification methods used in detection of plant diseases and its efficiency.

3.1   SVM Classifier

SVM classifier is directed learning strategy in machine learning; here investigated information is utilized for order. The accompanying creators utilized SVM classifier in infection discovery of various yields [23]. Detection of maladies on citrus trees incorporates grapefruit, lemons, lime, and oranges leaf assault by blister and anthracnose infections. The exploratory outcome acquired 95% of certified acknowledgment rate [24]. Grape plant sicknesses Downy Mildew and Powdery Mildew identified and give 88.89% normal exactness for both the maladies [25]. Oil palm leaf ailments Chimera and Anthracnose location accomplishes exactness of 97 and 95% individually [26]. Potato plant infections are late scourge and early curse identification more than 300 publically accessible pictures with exactness 95% [27]. Grape leaf illnesses black rot, esca, and leaf blight are characterized with exactness utilizing highlights from both LAB and HSI shading model [28]. A strategy is developed to distinguish illnesses in tea plants. Three distinct kinds of illnesses with less in highlights are recognized utilizing SVM classifiers. The created strategy arranged the ailments with precision of 90% [29]. Soybean culture is used to identify three distinct sicknesses Downy Mildew, Frog eye, and Septories leaf curse. They detailed with normal characterization exactness roughly 90% utilizing large dataset.

3.2 K-Means Classifier

K-implies calculation is an iterative calculation that attempts to segment the dataset into K pre-characterized particular non-covering subgroups (bunches) where every datum point has a place with just one gathering. It attempts to make the between group information focuses as comparative as could reasonably be expected while additionally keeping the bunches as various (far) as would be prudent. It allots information focuses to a group with the end goal that the entirety of the squared separation between the information focuses and the bunch's centroid (number juggling mean of the considerable number of information focuses that have a place with that group) is at the base. The less variety we have inside groups, the more homogeneous (comparative) the information focuses are inside a similar bunch [30].

The way k-means algorithm works is as follows:

1. Specify number of groups K.
2. Initialize centroids by first rearranging the dataset and afterward haphazardly choosing K information focuses for the centroids without substitution.
3. Keep emphasizing until there is no change to the centroids, i.e., task of information focuses to bunches is not evolving.
4. Compute the total of the squared separation between information focuses and all centroids.
5. Assign every datum point to the nearest group (centroid).
6. Compute the centroids for the groups by taking the normal of the all information focuses that have a place with each bunch.

The methodology k-implies pursues to take care of the issue called expectation-maximization. The E-step is relegating the information focuses to the nearest group. The M-step is processing the centroid of each bunch. The following is a breakdown of how we can explain it numerically.

The objective function is:

$$J = \sum_{i=1}^{m} \sum_{k=1}^{K} w_{ik} \left\| x^i - \mu_k \right\|^2 \tag{11}$$

where $w_{ik} = 1$ for data point x_i if it belongs to cluster k; otherwise, $w_{ik} = 0$. Also, μ_k is the centroid of x^i's cluster.

$$\frac{\partial J}{\partial w_{ik}} = \sum_{i=1}^{m} \sum_{k=1}^{K} \left\| x^i - \mu_k \right\|^2$$

$$\Rightarrow w_{ik} = \begin{cases} 1 \text{ if } k = \arg\min_j \left\| x^i - \mu_j \right\|^2 \\ 0 \text{ otherwise} \end{cases} \tag{12}$$

$$\frac{1}{mk} \sum_{i=1}^{mk} w_{ik} ||x^i - \mu_c k||^2 \qquad (13)$$

3.3 KNN Classifier

K-nearest neighbors have been utilized for design acknowledgment, measurable estimation, and arrangement in machine learning. We made study on plant infection location utilizing KNN classifier as pursues [31]. A calculation for ID of ailment in sugarcane culture is proposed. Picture handling calculations are utilized to include extraction. It verified a precision of 95% for leaf sear malady location in sugarcane leaf [32]. A strategy is developed to assess seriousness and identification of cotton plant sickness Gray Mildew malady accomplished with exactness of 82.5% utilizing 40 pictures. A calculation is proposed for plant ailment location utilizing GLCM highlight extraction strategy and KNN classifier. The KNN classifier is proposed instead of SVM classifier to order information in various classes. The exhibition is tried in the terms of exactness is in-wrinkled contrasted and SVM classifier (Fig. 3).

Fig. 3 Plant disease detection using KNN classifier

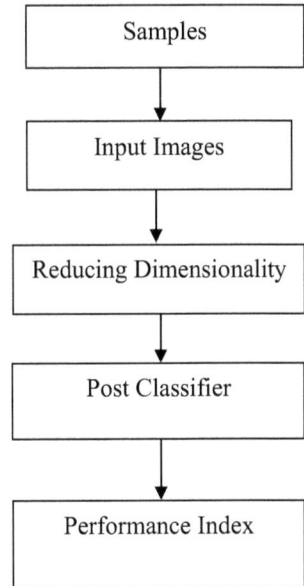

3.4 ANN Classifier

Artificial neural network is computational model in AI and example acknowledgment. Related work on plant infection identification utilizing ANN classifier as pursues [33]. It evaluated a proposed work for acknowledgment of plant ailments utilizing feedforward backspread calculation and it performed well with an accuracy of around 93%. They tried arrangement on early sear, cottony form, late burn, and minor whiteness sicknesses which impact on plants. They developed a model to build the exactness in ID for two kinds of infections brought about by parasite which are Downy Mildew and Powdery Mildew in cucumber plant. They introduced a framework to perceive and arrange ailments like leaf spot, bacterial scourge, natural product spot and organic product decay infections of pomegranate plant utilizing back-engendering calculation and the exploratory outcome appears around 90% exactness. And they proposed a work on recognizable proof of groundnut plant malady Cercospora (leaf spot) utilizing neural system back-engendering strategy. The test results and perception appear out of 100 example ailing leaf pictures and they characterized four kinds of illnesses and verified 97.41% of exactness, also proposed a technique to identify pomegranate.

3.5 Fuzzy Classifier

Related work on fuzzy classifier in plant malady recognition is a creator exhibited a technique to recognize the nearness of contamination in wheat crop pictures utilizing fuzzy classifier. This calculation is tried with the dataset of sound and unfortunate leaves. The arrangement of sound and unfortunate leaves found with exactness of 88% and acknowledgment of malady precision is 56%.

Let us accept that our example order issue is a n-dimensional issue with M classes and m given preparing designs $x_p = (x_{p1}, x_{p2}, ..., x_{pn})$, $p = 1, 2, ..., m$. Without loss of sweeping statement, we expect each quality of the given preparing examples to be standardized into the unit interim $[0, 1]$; that is, the example space is a n-dimensional unit hypercube $[0, 1]$ n. In this examination, we utilize fluffy in the event that rules of the accompanying sort as a base of our fluffy standard-based arrangement frameworks:

Rule R_j: If x_1 is A_{ji} and ... and x_n is A_{jn} at that point Class C_j with CF_j, $j = 1, 2, ..., N$

where R_j is the name of the jth rule, $A_{j1}, ..., A_{jn}$ are precursor fluffy sets on the unit interim $[0, 1]$, C_j is the subsequent class (e.g., one of the M given classes), and CF_j is the evaluation of conviction of the fluffy on the off chance that standard R_j. As forerunner fluffy sets we utilize triangular fluffy. We show a segment of the unit interim into three fluffy sets.

Fuzzy rule generation

Let us assume that m training patterns $x_p = (x_{p1}, ..., x_{pn})$, $p = 1, ..., m$, are given for an n-dimensional C-class pattern classification problem. The consequent class C_j and the grade of certainty CF_j of a fuzzy if–then rule are determined in the following two steps:

1. Calculate $\beta_{\text{class } h}(j)$ for class h as

$$\beta_{\text{class } h}(j) = \sum_{x_p \in \text{class } h} \mu_j(x_p) \tag{14}$$

where

$$\mu_j(x_p) = \mu_{j1}(x_{p1}) \ldots \mu_{jn}(x_{pn}) \tag{15}$$

and $j_n(\cdot)$ is the membership function of the fuzzy set A_{jn}. Find Class $\hat{h}$ that has the maximum value of $\beta_{\text{class } h}(j)$:

$$\beta_{\text{class } h}(j) = \max\{\beta_{\text{class } k_{1 \leq k \leq c}}(j)\} \tag{16}$$

3.6 Recurrent Neural Network

The recurrent neural network (RNN) is with a single unidirectional covered layer. In the specific utilitarian sort of how inert components and observations interface was to some degree emotional. This is unquestionably not a significant issue as long as we have enough flexibility to show different sorts of associations. With a singular layer, in any case, this can be very trying. By virtue of the acknowledgment we fixed this issue by including more layers. Inside RNNs this is increasingly questionable, since we first need to pick how and where to incorporate extra nonlinearity. Our trade underneath bases essentially on long short-term memory (LSTM) yet it applies to other course of action models, too. We could add extra nonlinearity to the gating parts. That is, as opposed to using a lone perceptron we could use various layers. This leaves the instrument of the LSTM unaltered. Or maybe it makes it progressively current. This would look good in case we were convinced that the LSTM instrument depicts some kind of by and large acknowledged actuality of how inactive variable autoregressive models work. We could stack various layers of LSTMs more than each other. This results in a framework that is dynamically versatile, due to the mix of a couple of clear layers. In particular, data might be huge at different degrees of the stack. For instance, we should keep raised level data about cash-related monetary circumstances (bear or emphatically slanting business division) available at a critical level, while at a lower level we simply record shorter-term transient components.

Similarly as with multilayer observations, the quantity of shrouded layers L and number of concealed units h are hyperparameters. Given a succession information $x = (x_1, x_2, ..., x_T)$, where x_i is the information at ith time step, a RNN refreshes its intermittent concealed state h_t by

$$h_t = \begin{cases} 0, & \text{if } t = 0 \\ \varphi(h_{t-1}, x_t), & \text{otherwise} \end{cases} \tag{17}$$

where ϕ is a nonlinear capacity, for example, a calculated sigmoid capacity or hyperbolic digression work.

$$h_t = o_t \tanh(c_t) \tag{18}$$

where $\tanh(\cdot)$ is the hyperbolic tangent function and o_t is the output gate that determines the part of the memory content that will be exposed. The output gate is updated by

$$O_t = \sigma(W_{oi} x_t + W_{oh} h_{t-1} + W_{oc} c_t) \tag{19}$$

where $\sigma(\cdot)$ is a logistic sigmoid function and W terms denote weight matrices; e.g., W_{oi} is the input–output weight matrix and W_{oc} represents the memory-output

$$c_t = i \Theta \tilde{c}_t + f \Theta c_{t-1} \tag{20}$$

Input gate modulates the extent to which the new memory information is added to the memory cell. The degree to which content of the existing memory cell is forgotten is decided by the forget gate f_t. The equations that calculate these two gates are as follows:

$$i_t = \sigma(W_i i x_t + W_{ih} h_{t-1} + W_{ic} c_{t-1}) \tag{21}$$

$$f_t = \sigma\left(W_f i x_t + W_{fh} h_{t-1} + W_{fc} c_{t-1}\right) \tag{22}$$

4 Reference Gap

See Table 1.

Table 1 Survey on various plant disease detection with several classifiers

Classifiers	Plant type	No. of diseases	Performance analysis
			Accuracy (%)
SVM	Citrus [23]	2	95
	Grape [24]	2	88.89
	Oil palm [25]	2	95
	Potato [26]	2	90
	Tea [27]	3	93
	Soybean [28]	3	90
K-means	Species [30]	6	88
K-NN	Sugarcane [31]	1	95
	Cotton [32]	1	82.5
ANN	Species [33]	5	93
	Cucumber	2	95
	Pomegranate	4	90
	Groundnut	4	97.41
Fuzzy	Wheat	1	56
RNN	Apple	–	91

5 Conclusion

This paper displays a study on various strategy for plant leaf infection identification utilizing picture handling system. There are numerous strategies in computerized or PC vision for illness location and grouping yet at the same time there this examination point is needed. All the infection cannot be distinguished utilizing single strategy. From investigation of above arrangement strategies we think of following end. The significant strategies for location of plant sicknesses are: SVM, K-means, K-NN, ANN, fuzzy and recurrent neural network. These procedures are utilized to investigate the sick plants leaves in the beginning time. A portion of the issue in this present system's, viz. impact of foundation in the subsequent picture, division of unhealthy spot in plant leaf, and mechanization of the procedure for consistent checking of plant leaf illnesses. The convolutional neural network (CNN) strategy is maybe the most straightforward of all calculations for foreseeing the class of a test model. A conspicuous weakness of the CNN strategy is the time multifaceted nature of making expectations. Furthermore, neural systems are tolerant to uproarious data sources. The survey recommended that this infection location strategy shows a decent outcome with a capacity to recognize plant leaf sicknesses.

References

1. Chouhan SS, Kaul A, Singh UP, Jain S (2018) Bacterial foraging optimization based radial basis function neural network (BRBFNN) for identification and classification of plant leaf diseases: an automatic approach towards plant pathology. IEEE Access 6:8852–8863
2. Hughes G, McRoberts NEIL, Madden LV, Nelson SC (1997) Validating mathematical models of plant–disease progress in space and time. Math Med Biol: J IMA 14(2):85–112
3. Singh UP, Chouhan SS, Jain S, Jain S (2019) Multilayer convolution neural network for the classification of mango leaves infected by anthracnose disease. IEEE Access 7:43721–43729
4. Zhang X, Qiao Y, Meng F, Fan C, Zhang M (2018) Identification of maize leaf diseases using improved deep convolutional neural networks. IEEE Access 6:30370–30377
5. Ponmurugan P (2016) Biosynthesis of silver and gold nanoparticles using *Trichoderma atroviride* for the biological control of Phomopsis canker disease in tea plants. IET Nanobiotechnol 11(3):261–267
6. Zhou G, Zhang W, Chen A, He M, Ma X (2019) Rapid detection of rice disease based on FCM-KM and faster R-CNN fusion. IEEE Access 7:143190–143206
7. Jiang P, Chen Y, Liu B, He D, Liang C (2019) Real-time detection of apple leaf diseases using deep learning approach based on improved convolutional neural networks. IEEE Access 7:59069–59080
8. Mishiba K (2020) Fast depth estimation for light field cameras. IEEE Trans Image Process 29:4232–4242
9. Gil Rodríguez R, Vazquez-Corral J, Bertalmío M (2020) Color matching images with unknown non-linear encodings. IEEE Trans Image Process 29:4435–4444
10. Hosu V, Lin H, Sziranyi T, Saupe D (2020) KonIQ-10 k: an ecologically valid database for deep learning of blind image quality assessment. IEEE Trans Image Process 29:4041–4056
11. Matysiak P, Grogan M, Le Pendu M, Alain M, Zerman E, Smolic A (2020) High quality light field extraction and post-processing for raw plenoptic data. IEEE Trans Image Process 29:4188–4203
12. Sandeep Kumar E (2012) Leaf, color, area and edge features based approach for identification of Indian plants. Indian J Comput Sci Eng 3(3). ISSN: 0976-5166
13. Kadir A, Nugroho LE, Susanto A, Santosa PI (2011) A comparative experiment of several shape methods in recognizing plants. Int J Comput Sci Inform Technol 3(3)
14. Zhang D, Lu G (2004) Review of shape representation and description techniques. Pattern Recogn 37(1):1–19
15. Hu R, Jia W, Ling H, Huang D (2012) Multiscale distance matrix for fast plant leaf recognition. IEEE Trans Image Process 21(11):4667–4672
16. Hu C, Liu X, Pan Z, Li P (2019) Automatic detection of single ripe tomato on plant combining faster R-CNN and intuitionistic fuzzy set. IEEE Access 7:154683–154696
17. Lee S, Liu Y (2012) Curved glide-reflection symmetry detection. IEEE Trans Pattern Anal Mach Intell 34(2):266–278
18. Tan Jw, Chang S, Abdul-Kareem S, Yap HJ, Yong K (2020) Deep learning for plant species classification using leaf vein morphometric. IEEE/ACM Trans Comput Biol Bioinform 17(1):82–90
19. Wang H, Wang J, Zhai J, Luo X (2019) Detection of triple JPEG compressed color images. IEEE Access 7:113094–113102
20. Masazhar ANI, Md Kamal M (2017) Digital image processing technique for palm oil leaf disease detection using multiclass SVM. In: IEEE 4th international conference on smart instrumentation, measurement and applications (ICSIMA), Malaysia, 2017, pp 1–6
21. Tan JW, Chang S-W, Abdul-Kareem S, Yap HJ, Yong K-T (2019) Deep learning for plant species classification using leaf vein morphometric. IEEE Trans Comput Biol Bioinform 17(1):82–90
22. Lee SH, Goëau H, Bonnet P, Joly A (2020) New perspectives on plant disease characterization based on deep learning. Comput Electron Agric 170:105220

23. Hossain MdS, Mou RM, Hasan MM, Chakraborty S, Abdur Razzak M (2018) Recognition and detection of tea leaf's diseases using support vector machine. In: IEEE 14th international colloquium on signal processing & its applications (CSPA), Malaysia, 2018, pp 150–154

24. Kumari CU, Prasad SJ, Mounika G (2019) Leaf disease detection: feature extraction with K-means clustering and classification with ANN. In: 2019 3rd international conference on computing methodologies and communication (ICCMC). IEEE, pp 1095–1098

25. Eaganathan U, Sophia J, Lackose V, Benjamin FJ (2014) Identification of sugarcane leaf scorch disease using K-means clustering segmentation and KNN based classification. Int J Adv Comput Sci Technol (IJACST) 3(12):11–16

26. Parikh A, Raval MS, Parmar C, Chaudhry S (2016) Disease detection and severity estimation in cotton plant from unconstrained images. IEEE In: International conference on data science and advanced analytic, Canada, 2016, pp 594–601

27. Kaushal Gautham, Bala Rajini (2017) GLCM and KNN based algorithm for plant disease detection. Int J Adv Res Electr Electron Instrum Eng 6(7):5845–5852

28. Al Bashish D, Braik M, Bani-Ahmad S (2010) A framework for detection and classification of plant leaf and stem diseases. In: IEEE international conference on signal and image processing (ICSIP), Chennai, 2010, pp 113–118

29. Vakilian KA, Massah J (2013) An artificial neural network approach to identify fungal diseases of cucumber (*Cucumis sativus* L.) Plants using digital image processing. Arch Phytopathol Plant Prot 46(13):1580–1588

30. Dhakate M, Ingole AB (2015) Diagnosis of pomegranate plant diseases using neural network. In: IEEE 5th national conference on computer vision, pattern recognition, image processing and graphics (NCVPRIPG), Patna, 2015

31. Ramakrishnan M, Sahaya Anselin Nisha A (2015) Groundnut leaf disease detection and classification by using back propagation algorithm. In: IEEE international conference on communications and signal processing (ICCSP), Melmaruvathur, 2015, pp 0964–0968

32. Majumdar D, Ghosh A, Kole DK, Chakraborty A, Majumder DD (2015) Application of fuzzy C-means clustering method to classify wheat leaf images based on the presence of rust disease. In: Proceedings of the 3rd international conference on frontiers of intelligent computing: theory and applications, vol 327, pp 277–284

33. Sladojevic S, Arsenovic M, Anderla A, Culibrk D, Stefanovic D (2016) Deep neural networks based recognition of plant diseases by leaf image classification. Comput Intell Neurosci 2016

34. Jeger MJ, Van Den Bosch F, Madden LV, Holt J (1998) A model for analysing plant–virus transmission characteristics and epidemic development. Math Med Biol: J IMA 15(1):1–18

35. Khan MA, Lali MIU, Sharif M, Javed K, Aurangzeb K, Haider SI, Altamrah AS, Akram T (2019) An optimized method for segmentation and classification of apple diseases based on strong correlation and genetic algorithm based feature selection. IEEE Access 7:46261–46277

Chapter 37
Artificial Intelligence-Based Job Applicant Profile Quality Assessment and Relevance Ranking Using Clusters in Talent Acquisition Process

G. M. Sridevi and S. Kamala Suganthi

1 Introduction

The rapid development of the electronic recruitment industry led to the enormous growth of the number of resumes and organization positions on the recruitment website. Scrutinizing suitable profiles according to job description from a huge database are a big challenging task for the talent acquisition specialist. Organizations always need to spend a lot of time to shortlist the resumes, and as a result, the recruitment process efficiency has been reduced. The biggest challenges for talent acquisition specialists are to hire the right candidate with minimal cost and less time to be in par with this current industry. Technology keeps changing and followed by the business requirement. The talent acquisition specialist should be able to hire the latest skilled resource as per the latest requirement of technology with minimal resources. As there are major challenges that are require to overcome, to bring efficiency to the complete process. It involves the process of scrutinizing relevant profiles from irrelevant profiles. Followed by choosing the best profile among better profiles (Ranking) and knowing the candidates can do the job before hiring them.

In Roy et al. [9], the researcher mainly focus on automated "Resume Classification and Matching" and recommendation regarding resumes. Researchers used ML-based technologies like NER, NLP, and text classification using n-grams. Application of Artificial Intelligence (AI) in the recruitment process increases the efficiency of the recruitment process, and also the productivity of employees is found. Samples are collected from HR, and Anova statistical tools are applied in Vedapradha et al. [13]. In Dixit et al. [2], a block diagram represented in this article regarding resume sorting using AI and explained advantages of application tracking system used in the recruitment process. AI-based chatbots are new revolutionary tools in the recruitment

G. M. Sridevi · S. Kamala Suganthi (✉)
PG Department of Management Studies and Research Centre, Atria Institute of Technology, Bangalore 560024, India
e-mail: kamalasuganthi@gmail.com

process. Chatbots are very useful tools in the hiring process and create an engaging environment for job applicants and increase employer branding which is discussed in Nawaz and Gomes [7]. This Ruby Merlin [10] is a comprehensive approach to AI application in the different functions of HR like recruitment, training, performance management, and retention. The automated HR process is a new concept in the workforce. This Wang et al. [14] publication focus on the vector representation of resumes including criteria like age, education, and vector representation of job description. Genetic Algorithm (GA) used to establish a relationship between two vectors for application of resume recommendation in online recruitment. The authors Faliagka et al. [3] proposed a system extracts a set of objective criteria from the different applicants' LinkedIn profiles and analyzes their personality traits using linguistic analysis on their blogs. Its more perform consistently compared to talent acquisition specialist and automation of applicant ranking and personality mining. The Nawaz [6] discusses AI swapping human resource in the recruitment process. AI intervention made the recruitment process more effective and competitive in the industry. The Sanyal et al. [11] NLP has been developed for e parser parses all the essential information from the candidate profiles and autofills a form for the recruiters to proofread. After user confirmation, the resume will be saved in the database and ready to show it to hiring managers. This saves time and cost. Predictive human resource candidate ranking system ranks the curriculum vitae according to different job requirement. This facilitate human resource department in hiring process. It considers candidates knowledge, education, and different vital aspect to perform the job. In [4], HR assessment tool system rank curriculum vitae supporting details in many area like student hobbies, strength, weakness, or system to conduct fifteen to sixteen queries temperament predication purpose. This CV ranking policy support human resource department in manpower hiring process in Sonar et al. [12]. The Pandita [8] digitalization of recruitment process has been discussing in this article. Evidence-based review was conducted to analyze impact of digitalization in hiring. In this research, AI model is developed using various techniques such as text-based preprocessing, NLP, clustering, and distance measure to assess the quality of candidate profile and rank them to facilitate hiring decisions. In Sect. 2, the data analysis is discussed; the Sect. 3 provides architecture of AI model; in Sect. 4, the experimental results are given followed by conclusion in Sect. 5.

2 Data Analysis

In this section, data analysis has been carried out on candidate profiles and job descriptions. The dataset has been obtained from Kaggle (Database, n.d.). There are 14,906 candidate profiles, and 8 job descriptions are present in our dataset. Table 1 illustration of the job descriptions is given. It consists of many job descriptions in which we selected a few latest technological job descriptions based out of Bengaluru, Karnataka. Job description like machine learning includes responsibilities like designing and developing machine learning and deep learning, etc. Specific

Table 1 Illustration of job description

Title	Company	City	State	Description	Responsibilities	Education	Preference _skills
Machine learning	Involvio	Bengaluru	Karnataka	NONE	Designing and developing machine learning and deep learning ….	B.Sc. in Computer Science, Mathematics, or similar	Data structures, data modeling, software, …
Data scientist	IBM	Bengaluru	Karnataka	NONE	Real-time anomaly detection solutions that proactively, …..	Degree in Statistics, Mathematics, Computer Science	R, Python, SQL, regression, simulation, scenario,…
Data analyst	Amazon	Bengaluru	Karnataka	Highly proficient in Microsoft Office and Windows, SQL …	Understanding of data warehousing, data modeling concept.	Bachelor's degree in Computer Science, Engineering, Operations Research, Mat	Data warehousing, data modeling, Advance SQL, R, Python,
Embedded developer	Winfoglobal tech	Bengaluru	Karnataka	Experience working within a product development	Design, develop embedded applications in Linux, C, C++	BE/B. Tech or ME/M. Tech in Computer science…	C, C++, Linux, Embedded Linux, RTOS,

educational criteria and preferred skills are also mentioned in the table. Similarly, few job descriptions are explained in the above Table 1.

In Table 2, the candidates with different job titles are illustrated in the dataset corresponding to job profiles consists of various fields such as resume title, city, state, description, work experiences, educations, skills, links, certificates, and additional information. A few examples of candidate profiles are shown in the above table. In Fig. 1, the frequency of candidates applies to various job titles such as data analyst, data scientist, embedded developer, full stack developer, Java developer, machine developer, Php developer, and python developer are shown in above Fig. 1.

In Fig. 2, there are 152 applicants from Pune, Maharastra for different job titles which are among the highest compared to other cities. There are 136 candidates from Bangalore, Karnataka. Local candidates are easy for processing for local job descriptions. There are only 5 candidates from Barakpur and Cuttack, where relocation interest to consider.

3 Candidate Profile Quality Measure and Selection

We developed an AI model to candidate profile selection and quality assessment in talent acquisition processes using the Natural Language Processing (NLP), clusters, distance measurement. This model scrutinizes relevant resumes from a huge database and organizes it in more suitable to less suitable to the job description, thus providing the ranking of resumes (Fig. 3).

The architecture for resume selection and quality measurement has been developed in this research as shown in Fig. 1. A resume database of candidates has been collected; initially, the tokenization process of resume has been carried out of obtaining a list of words from resume. Then, a resume subjected to preprocessing and filtering such as elimination of stop words and lemmatization. The Natural Language Tool Kit NLTK (Loper and Bird, n.d.) is utilized to determine part of the speech of sentences of the resume. The next four clusters of words are prepared by taking reference to part of speech. The collected words are segregated into adjectives, adverbs, skilled words, and non-preferred skilled words to create four clusters. The same procedure of cluster creation is also carried out for job descriptions. The text distance measuring techniques Jaccard, cosine, Elken, Longest Common subsequence (LCSEQ) similarity are used to measure the closeness between clusters of job descriptions and candidate profiles. Based on the distance measurement, candidate profile which is closer to the job descriptions can be selected. The quality of candidate profiles is computed as inversely proportional distance between clusters of job description to candidate profiles (Fig. 4).

Four clusters from profiles and three from job descriptions are created as shown in Fig. 2. The adjectives and adverbs are collected from fields of resumes such as description, work experience, education, skills, and additional information thus collected adjectives and adverbs form two clusters. One more cluster is created using the field of preferred skills; to obtain this cluster, these stop word elimination has

Table 2 Candidates profiles

Resume title	City	State	Description	Work experiences	Educations	Skills	Certificates
Java developer	Tirur	Kerala	To prove myself dedicated, worthy, and energetic in a progressive …	{"wtitle:": "Java Developer"}, {"wcompany:": "N"}, ….	[{"e_title:": "Bachelor's in Bachelor of Computer Application (BCA)"}, {"e_schoolname:": …	["Java (Less than 1 year)," " Jsp (1 year)," "JavascriC++"	[{"c_title:": "Java developer"}, {"c_duration:": "June 2019 to Present"}, {"c_descr:": as Junior and …
Java developer	Mohali	Punjab	Optimistic approach Con dent, flexibility. Adaptive to …	{0: [{"wtitle:": "Java developer"}, {"wcompany:":	{0: [{"e_title:": "B. Tech"},	["Spring MVC," "spring boot," " Hibernate," "MySQL"	None
Java developer	Hyderabad	Telangana	• Knowledgeable with core Java implementation of web	{0: [{"wtitle:": "Java developer"}, {"wcompany:": "N"}, {"wcity:": "NONE"}, {"wstate:":	{0: [{"e_title:": "MCA in Computers"}, {"e_schoolname:": "JNTUA Anantapuram"},	["Web services," "Html," "J2Ee, "Jquery," "Spring, "SpringBoot," "Java," "Css"]	None
Java developer	Patna	Bihar	My recent objective is to work with the company, where…	{0: {"wtitle:": "Java developer, angular developer"}, {:': and…	{0:[{"e_title:": "B.C.A in T.I.H.S Patna Khagaul:"	["Core Java," "C," "Jdbc," "servlet"]	[{"c_title:": " Angular8"}, {"c_duration:": "NONE"}, {"c_descr:":"NONE"}]]

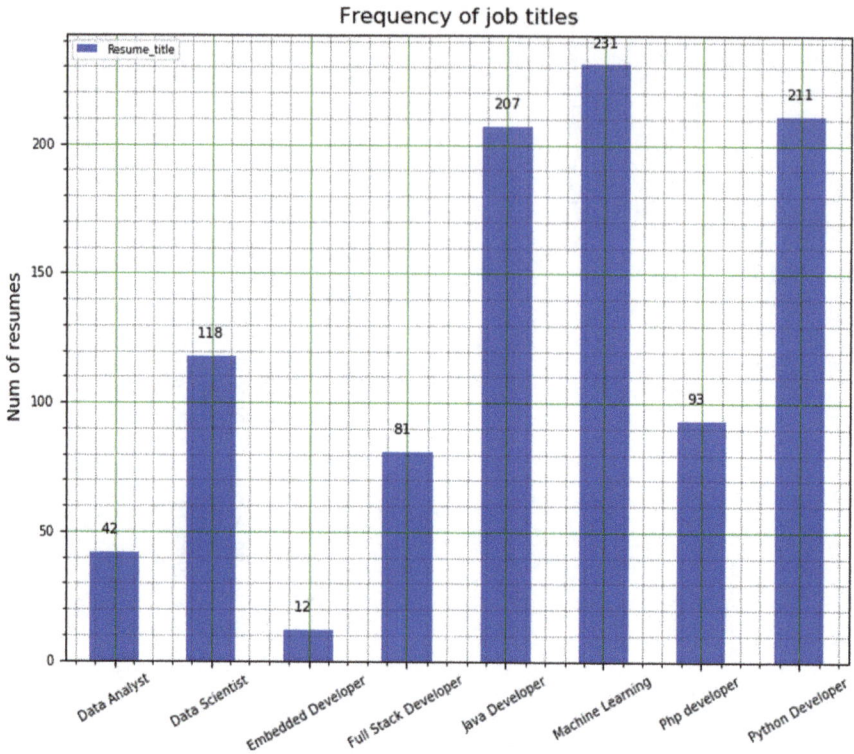

Fig. 1 Frequency of job applications to various job titles

been carried out on the preferred skills provided by the candidates, and from that filtered text, the nouns are collected to create a cluster of preferred skills. From the same field of the candidate profiles, the skills which are not mentioned in the job description are used to create one more cluster that is known as non-preferred skills. Similarly, three clusters of words are created from job descriptions; adjective and adverb are collected from the field of job descriptions such as responsibilities, educations, preference skills, description, and additional information. Then, using the field of preference skills, all skills are collected to form a preferred skill cluster. These clusters of words are used to measure the similarity between profiles and job descriptions. A list of profiles resumes is created according to the order of the similarity measurement between the cluster. Then, the quality of candidate profiles is measured using the cluster of words to signify the most suitable candidate profile concerning the job description as given in Eq. (1).

$$\text{Quality} = \left(\frac{1}{\text{Jaccard Dist}} \right) + (\text{Num of Adjective and Adverbs}) \times C \qquad (1)$$

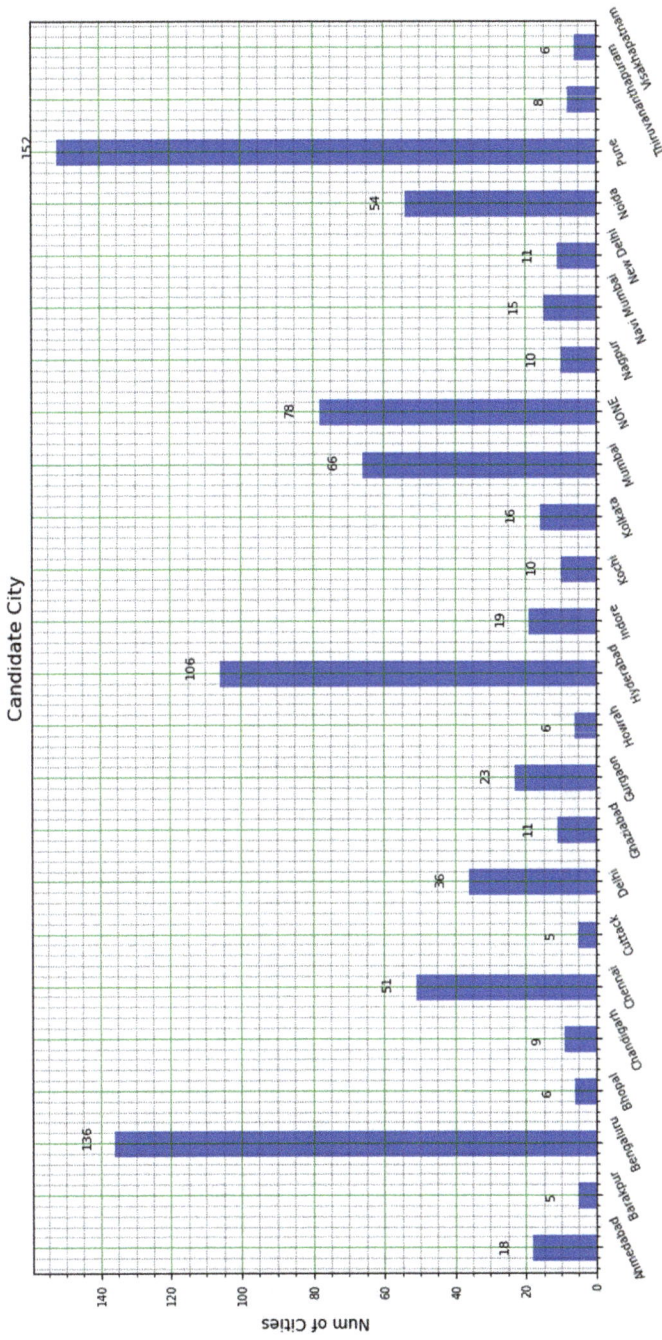

Fig. 2 Candidate location distribution

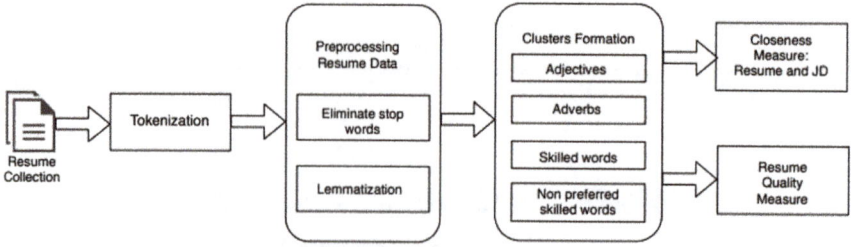

Fig. 3 Architecture of resume selections and quality measure

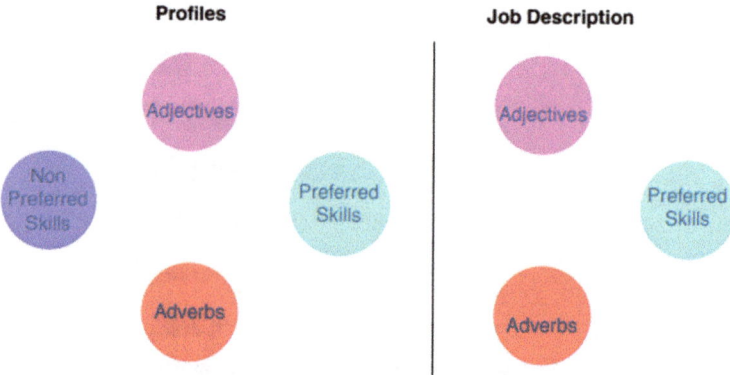

Fig. 4 Cluster of words from profiles and job descriptions

4 Experimental Results

In our experiment, we have used dataset corresponding Kaggle resume data (Database, n.d.). This data consists of 14,806 number of candidates profiles. Each profile has various fields such as resume title, city, state, description, work experiences, educations, skills, links, certificates, additional information. Few examples are shown in Table 2.

The job description database consists of eight job descriptions having various fields such as title, company, city, state, description, responsibilities, education, preference, skills, and additional information as shown in Table 1. The candidate profile data further processed to consider only title corresponding to eight job titles "machine learning," "data scientist," "data analyst," "embedded developer," "full stack developer," "Java developer," "Php developer," and "Python developer." Thus, there are 995 candidate profiles which are present in the above-mentioned titles. From each job description, cluster of words consisting of various skills is created.

Similar processing carried out to prepare a cluster of words from candidate profiles. The distance measurement between the cluster of the job description and candidate profiles has been performed various distance matrices used such as Jaccard,

Cosine, Elken, LCSEQ similarity which are computed. For each job description, the resumes with similar job titles are searched in the candidate profile dataset and computed cluster distances. The distances are sorted ascending order according to Jaccard distance to choose the closure candidate profile for the job description. In Table 3, the computed Jaccard distance for four job descriptions are shown. In this table, five closest candidate profiles and corresponding distances are tabulated.

The clusters of adjectives and adverbs are created as discussed in Sect. 3. The quality for each candidate is now determined as the closest candidate profile with attributes mentioned or specified in a cluster of adjectives and adverbs. The quality now computed using Eq. (1). In Table 4, the quality measure is shown along with a cluster of adjectives and adverbs. The quality computed as inversely proportional to distance plus number of positive adjectives and adverbs multiplied by a constant. The quality measure for top-five candidate profiles for four job descriptions is given in Table 4.

In Fig. 5, the cosine, Jaccard, and quality measures are depicted for five candidate profiles. In Fig: 5a, the quality measure for job description 1 is shown. Here, the candidate profile 1604 is the most suitable with the quality measure as 3.71; then,

Table 3 Cluster of words-based distance measurement between job description and candidate profile

JD Num.	CV Num.	Jaccard distance	Cosine distance	Elkan distance	LCSEQ
1	1604	0.37	0.22	0.99	0.58
1	1721	0.83	0.62	1.00	0.89
1	2048	1.00	0.99	1.00	1.00
1	1811	0.99	0.92	1.00	1.00
1	1972	0.99	0.93	1.00	1.00
2	1603	0.52	0.33	0.99	0.70
2	1820	0.94	0.78	1.00	0.95
2	1900	0.97	0.82	1.00	0.98
2	1973	1.00	0.93	1.00	1.00
2	2016	1.00	0.97	1.00	1.00
3	1829	1.00	0.95	1.00	1.00
3	914	0.53	0.33	0.99	0.69
3	1642	0.80	0.57	1.00	0.88
3	2244	1.00	1.00	1.00	1.00
3	2406	1.00	0.97	1.00	1.00
4	2270	0.73	0.55	0.99	0.81
4	2173	0.59	0.41	0.98	0.75
4	2493	0.82	0.66	0.99	0.85
4	2893	0.93	0.77	1.00	0.95
4	7225	1.00	0.97	1.00	1.00

Table 4 Cluster of words-based Jaccard distance and quality measurement

JD Num.	Profile Num.	Jaccard distance	Positive adjectives	Positive adverbs	Quality
1	1604	0.37	"efficient," "good," "intellectual," "responsible"	"well"	3.71
1	1721	0.83	"effective," "innovative," "useful"	"well"	2.01
1	2048	1.00	"best," "certain," "effective," "good," "responsible"	None	2.00
1	1811	0.99	"effective," "natural," "useful"	None	1.61
1	1972	0.99	"optimize," "protect"	"well"	1.61
2	1603	0.52	"dynamic," "engage"	None	2.32
2	1820	0.94	"responsible," "rich"	"dynamically"	1.66
2	1900	0.97	"good," "intelligence"	None	1.43
2	1973	1.00	None	"actively," "clearly"	1.40
2	2016	1.00	"good"	"well"	1.40
3	1829	1.00	"advanced," "easy," "favorite," "important," "solid," "sure"	"easily"	2.40
3	914	0.53	None	"well"	2.08
3	1642	0.80	"challenge," "responsible," "sentimental"	"actively"	2.04
3	2244	1.00	"certain," "challenge," "huge"	"well"	1.80
3	2406	1.00	"good," "intellectual," "proactive"	None	1.60
4	2270	0.73	"desire," "good," "smart," "special," "splendor"	"easily," "well"	2.77
4	2173	0.59	None	"well"	1.89
4	2493	0.82	"effective," "strong"	None	1.62
4	2893	0.93	"efficient," "strong"	None	1.48
4	7225	1.00	"effective," "strong"	None	1.40

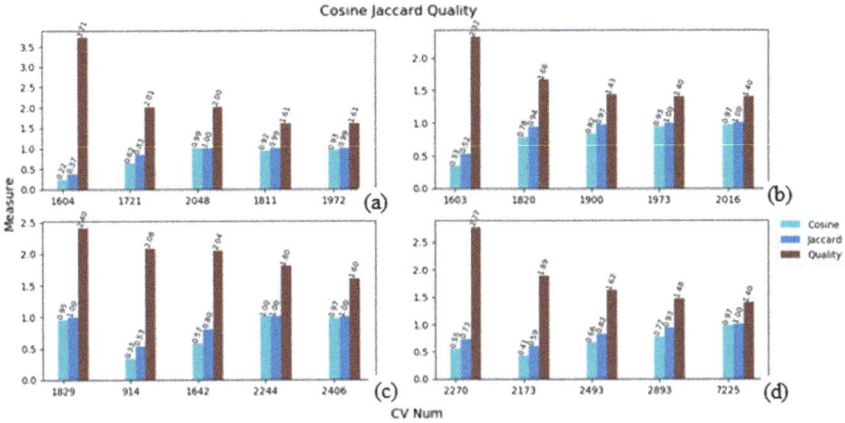

Fig. 5 Cosine Jaccard and quality measure for candidate profiles

next relevant is 1721 with the quality measure as 2.01. Similarly, in Fig. 5b–d, most similar candidate profiles 1603, 1829, and 2270, respectively. Thus, cluster formation and quality measures are useful in selecting the most relevant candidate profiles.

5 Conclusion

In the process of talent acquisition identifying more suitable candidate profiles according to job description more significant task. Usually, the candidate profile database consists of huge profiles, and manual searching is more time-consuming and causes errors. Ana incurs more cost. In this research, we have developed an AI-based model to select the candidate profiles that are most relevant to the job description. We preprocess both candidate profiles and job descriptions then create clusters of words. We also utilize NLTK to identify adjectives and adverbs from candidate profiles and JD. Using the clusters of skill words of candidate profiles and JD, the distance is measured using various techniques such as Jaccard, Cosine, Elken, LCSEQ. These distance measures provide the closeness of JD skills and candidate profile skills. The quality assessment of the candidate profiles is computed as reciprocal of Jaccard distance summed with a number of adjectives and adverbs. The quality measures of the most relevant candidate profile also highlight their attributes. The experiments are conducted on Kaggle dataset, and the most relevant candidate profiles are shortlisted.

References

1. Database KR (n.d.). No Title. https://www.kaggle.com/avanisiddhapura27/resume-dataset
2. Dixit VV, Patel T, Deshpande N, Sonawane K (2019) Ijresm_V2_I4_117. 4, 2–4
3. Faliagka, E., Ramantas, K., Tsakalidis, A., & Tzimas, G. (2012). Application of learning algorithms to online recruitment systems.pdf. In: ICIW 2012: the seventh international conference on internet and web applications and services application, pp 215–220. http://citeseerx.ist.psu.edu/viewdoc/download?doi=10.1.1.885.909&rep=rep1&type=pdf
4. Koyande BA, Walke RS, Jondhale MG (2020) Predictive human resource candidate ranking system 1:1–4
5. Loper E, Bird S (n.d.) NLTK : the natural language toolkit. 63–70
6. Nawaz N (2019) Artificial intelligence interchange human intervention in the recruitment process in Indian software industry. Int J Adv Trends Comput Sci Eng 8(4):1433–1442. https://doi.org/10.30534/ijatcse/2019/62842019
7. Nawaz N, Gomes AM (2019) Artificial intelligence chatbots are new recruiters. Int J Adv Comput Sci Appl 10(9):1–5. https://doi.org/10.2139/ssrn.3521915
8. Pandita D (2019, September) Talent acquisition : analysis of digital hiring in organizations. XVIII:66–72
9. Roy PK, Chowdhary SS, Bhatia R (2020) A machine learning approach for automation of resume recommendation system. Procedia Comput Sci 167(2019):2318–2327. https://doi.org/10.1016/j.procs.2020.03.284
10. Ruby Merlin PJ (2018) Artificial intelligence in human resource management. Int J Pure Appl Math 119(17):1891–1895
11. Sanyal S, Hazra S, Adhikary S, Ghosh N (2017) Research article 7(2):4484–4489
12. Sonar SC, Pawar NB, Raundal PV, Sonawane GU (2019) HR Assessment tool using AI abstract 2(5):899–903
13. Vedapradha R, Hariharan R, Shivakami R (2019) Artificial intelligence: a technological prototype in recruitment. J Serv Sci Manag 12(03):382–390. https://doi.org/10.4236/jssm.2019.123026
14. Wang Z, Tang X, Chen D (2016) A resume recommendation model for online recruitment. In: Proceedings—2015 11th international conference on semantics, knowledge and grids, SKG 2015, 256–259. https://doi.org/10.1109/SKG.2015.31

Chapter 38
A Survey on the ZigBee Protocol, It's Security in Internet of Things (IoT) and Comparison of ZigBee with Bluetooth and Wi-Fi

Malvika Gupta and Shweta Singh

1 Introduction

In today's scenario, internet plays a vital role and most importantly the wireless communication plays a vital role in gathering information or communication. Aim of wireless communication to gather information and analyzing the information in order to perform the specific tasks. As per today's scenario, the information is not gathered only from the specified areas but as researchers, we are always focused to retrieve or gather the information from different parts which includes the remote areas. So, this category of collecting information and devising the devices which is helpful in providing the help in remote areas through virtual connections comes under the term called Internet of Things (IoT).

In this paper, we are going to discuss about the collection of information virtually using a wireless protocol ZigBee, which a low power consuming, fast data transfer protocol. We are also going to discuss about the security issues of ZigBee and also its comparison with the other wireless protocols Bluetooth and Wi-Fi which are most commonly used.

2 Related Work

Many researchers and practitioners have given different reasons as why the ZigBee is most suitable protocol for the security of data in IoT. Few research works are discussed below.

M. Gupta (✉) · S. Singh
ABES Engineering College, Ghaziabad, India
e-mail: malvikagupta2008@gmail.com

S. Singh
e-mail: shweta.singh@abes.ac.in

© The Author(s), under exclusive license to Springer Nature Singapore Pte Ltd. 2021   473
X.-Z. Gao et al. (eds.), *Applications of Artificial Intelligence in Engineering*, Algorithms for Intelligent Systems, https://doi.org/10.1007/978-981-33-4604-8_38

Muthu Ramya, Shanmugaraj, Prabakaran [1], they discussed about the reason, ZigBee is suggested as it is low power, more reliable protocol. They discussed the ZigBee application areas, its stack, and security concerns.

Dini and Tiloca [2], they proposed some proposals to address the inefficiencies in handling both the network key securely as it is common and device certificates.

Bakhache, Ghazal, El Assad [3], there whole research revolves around the usage of AES-CTR algorithm for providing security, but there main concern was that this algorithm is time consuming and is not suitable for clinical or medicinal purposes. So, they presented a new robust and fast algorithm RFCA. They also gave comparisons among the RFCA, AES-CTR, and AES algorithm, so that it is easy to decide for the practitioners that which algorithm is best suited for the security in ZigBee.

Whitehurst, Andel, McDonald [4], they presented a survey on the vulnerabilities and security issues in the ZigBee and the issues faced during the exchange of network keys among the nodes. They presented these issues on the basis of their survey regarding the usage of ZigBee protocol mainly in home automation, door locks, etc., which requires high security.

Kulkarni, Ghosh and Pasupuleti [5], they proposed a secure routing protocol that uses message authentication code (MAC) and aggregate message authentication code for ZigBee networks, these are less expensive. They also compared the performance analysis among the proposed routing algorithms and the existing algorithms like 128 bit AES algorithm, which is considered to be the most suited algorithm for ZigBee network.

Al-alak, Ahmed, Abdullah, and Subramiam [6], they have suggested that AES is the best algorithm for the purpose of security, but they also suggested that due to symmetric key cryptography the key might be compromised in the future so they suggested ECC with AES to protect ZigBee with different types of attacks like replay attacks.

Sohel Rana, Halim, Kabir [7], they discussed about the different security threats in the ZigBee network and designed a security improvement framework (SIF) to take care the security threats in the network.

Khan et al. [8], they discussed about the comparison among the wireless networks like, Bluetooth, ZigBee, and Wi-Fi, they discussed the comparison on the basis of few parameters like range, data transfer rate, power consumption, network size, cost.

Pothuganti and Chitneni [9], they presented a comparative analysis of wireless networks like Bluetooth, Wi-Fi, and ZigBee, their study is more focused about the coding methods or algorithms used in these networks.

3 Protocol

3.1 ZigBee Introduction

ZigBee is an IEEE 802.15.4-based suite for a high level communication protocols used to create a personal area network, which is useful in home automation, information collection in different fields, etc. It is designed for the small scale applications which requires low power, low bandwidths, low data rate. Its low power consumption limits the transmission upto 10–100 m.

The reasons for using ZigBee are as follows:

- Reliable and self-healing
- Supports large number of nodes
- Easy to deploy
- Very long battery life
- Secure
- Low cost
- Can be used globally
- Industry support with large number of vendors according to the stipulated range are providing the products and services
- Protocols are open source, which means no license fee
- Chipsets available from multiple sources
- Remotely upgradeable firmware
- No new wires
- Low power consumption, can operate on batteries
- Low maintenance
- Standards-based security, 128 bit AES encryption Algorithm
- Ability to read gas meters.

All these above mentioned characteristics of ZigBee are the reasons of using this protocol in close proximity.

3.2 ZigBee Application Areas

- Remote Control
- Input Devices
- Home Automation
- Building Automation
- Health Care
- Telecom Services
- Retail Services
- Smart Energy.

3.3 *ZigBee Protocol Stack [10]*

Physical Layer: It is the bottom most layer of the ZigBee stack and also in the standard OSI model which works closely with the hardware. It is directly responsible for the communication, bit by bit data transfer, channel selection, quality of link management, and other activities which are to be dealt by the physical layer in the standard OSI model (Fig. 1).

MAC Layer: This layer works in between the physical layer and network layer or we can also say that it is the layer which acts as an interface between physical and network layer. This layer is mainly responsible for transmitting data packets to and from the network layer through the shared channels selected by the physical layer. This layer is also responsible for detecting errors or faults in the packets coming from the different nodes. It is also responsible for fixing these errors by using different error correction methods.

Network Layer: The main function of network layer is to enable the correct use of MAC layer and also provide the base to the above layer which is application layer. Its functionality is to provide routing and sets up the addresses of the devices that are contributing in the communication.

Application Layer: It is the top most layer which provides the interface to the end user to use the particular application. This layer is divided into two parts in a ZigBee stack, ZigBee device, and ZigBee Pro APS.

- *ZigBee Device Protocol*: This protocol is responsible for the security and privacy of the devices involved in the network. As in a ZigBee network, there are different devices, attached with the different roles so this protocol is responsible for deciding the role for a device and it is also responsible for the discovery of

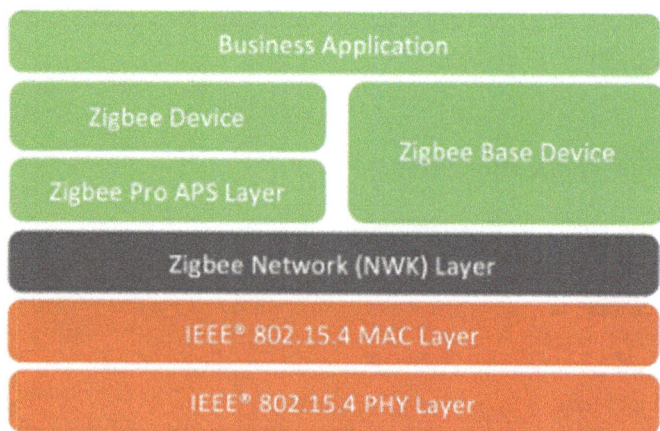

Fig. 1 ZigBee stack

new devices that can be included in the network taking care of secure link and connection.

- *ZigBee Pro APS Layer*: It is an application support sublayer, which acts in between the network layer and the components of application layer. As according to the stack, we can clearly see that it acts in between the network layer and ZigBee devices. This layer is responsible for binding a task with the appropriate device to carry out particular operation optimally.

3.4   ZigBee Network

ZigBee device protocol manages the devices as the coordinator, router, or end user. So, now, we will discuss about these devices which makes a ZigBee network [11] (Fig. 2).

Coordinator: It is responsible to initiate the communication in the network, there is only one coordinator in the ZigBee network which is responsible for the communication. It is like the head which initiates the process.

Router: It is useful for relaying the data among the devices connected in a network.

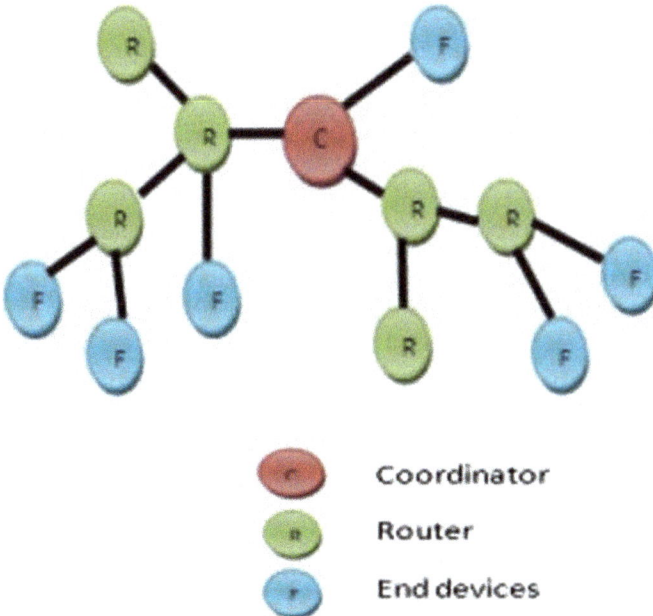

Fig. 2 ZigBee network

End devices: ZigBee end devices or ZED has the functionality to communicate with either coordinators or routers, they cannot relay data from other devices. It has long battery life and requires less memory hence it is less expensive than coordinator and router. It has the long battery life as the end device is not responsible for relaying of messages to other devices, so it will remain in sleep mode for maximum time hence power consumption is very less and battery life is more [12].

4 Security: Weakness and Methods to Secure the Protocol

As we know that ZigBee is an open source protocol, in which the changes can be made easily, it is a contradictory statement to what we have discussed till now in this paper, as earlier, we said that ZigBee provides strong security to the IoT. This strong security is provided by using the strong cryptography algorithms, like symmetric key algorithms which uses the same key for encryption and decryption and key is 128 bit. So, it is very difficult to decrypt the data during the transit time. In Sect. 3.1, we are going to discuss about the weaknesses of IoT, and in Sect. 3.2, we will suggest some methods that can strengthen the protocol.

4.1 Weakness in ZigBee

ZigBee is most secure and reliable protocol used for a range of 10–100 m. It is successful in communicating in this range with low power consumption and other important factors, but still there are some weaknesses which are to be dealt. These weaknesses are discussed below:

- In original version of ZigBee, there are nine exchange messages. This type of protocols require a number of operations directly proportional to the number of nodes in the system, which is not possible in wireless network, as it consists of large number of nodes and range is also higher in comparison of ZigBee.
- ZigBee faces a serious challenge in key distribution as it has only one master key which is shared by all the nodes.
- In the earlier version of ZigBee, no strong cryptography algorithms are used so, it is prone to attacks.

4.2 Methods Suggested Strengthening the Protocol

As in the above section, we have learnt about the different layers in the ZigBee stack. These layers are themselves responsible for the security. As we know that we have standard TCP/IP protocols that works for network layer, FTP, HTTP, and other protocols that works for the application layer. So, ZigBee in addition to these protocols of the particular layer, provides the secure passage or channel for the communication between different nodes by implementing strong encryption algorithms. These strong encryption algorithms require strong keys like 128 bit, which AES algorithm of cryptography provides. But as we have discussed earlier that ZigBee uses symmetric key cryptography, in which the same key is shared among the different nodes, so, to protect this, we require a trusted third party that will issue the keys to the authorized users. In ZigBee protocol, there is a huge trust level among the layers; so, the concerned part is only the transfer of data from nodes to nodes, which can be done using the following methodology:

- For any encryption algorithm, a key plays a major role, in this we have used a key of 128 bit which is shared among all the users or devices that are contributing to a network. In cryptography, it is very important that key should be shared securely, so there is always a trusted third party who is responsible to share the keys among the authorized users. These keys are used to perform encryption and decryption of data during the transit from network layer to application layer. In some places, these keys working in between network layer and application layer are also called network keys.
- Network keys are responsible for the security in between network and application layer, but in general, the communication takes place between the application layers of two different devices, so for that communication, we again need a key which is called link key. This key is used to establish a secure and authorized connection in between the two or more authorized users. Like in Bluetooth during the pairing of two devices, we need to accept or put a secret link in order to create a link and then an authorized connection is established.
- This link key is 128-bit as now the communication is in between the different application layers of different devices. So, now to secure this communication, a 128 bit key is shared among the communicating devices. The key is common as we are using symmetric key cryptography.
- AES encryption algorithm is used to strengthen the security.
- Timestamp or random values are used as a nonce, instead of rejecting the repeated frames. With the help of timestamp, the confirmation of receiving of particular data against a request happens only once.

5   Comparisons Between Bluetooth, ZigBee and Wi-Fi

There are different types of protocols that can be used to establish the connection among the devices. Most frequently used are Bluetooth, ZigBee, and Wi-Fi. But usage of these protocols completely depends upon the requirement of the companies and also the factors like power consumption, range, frequency, etc. In Table 1, we shown the comparison among Wi-Fi, ZigBee, and Bluetooth on the basis of mostly considered parameters. On the basis of this comparison, one can easily decide which protocol can be most suited as per the requirements.

Table 1 Comparison between Bluetooth, Wi-Fi, and ZigBee

S. No.	Features	ZigBee	Wi-Fi	Bluetooth
1	Range (m)	10–100	50–100	10–100
2	Network topology	Ad-hoc, Peer to peer, star or mesh	Point to Hub	Ad-hoc, very small networks
3	Frequency (GHz)	2.4	2.4 and 5	2.4
4	Complexity	Low	High	High
5	Power consumption	Very low	High	Medium
6	Security	128 bit AES + Application layer security	NIL	64 and 128 bit encryption AES and RSA

6 Conclusion

In this paper, we have discussed about the ZigBee protocol its stack, usage, features, security, and at last, we discussed about the comparison between the three most commonly used wireless protocols. As we know that nowadays IOT is the most common term coined in the world so to gather the information and analyzing the information, we need communication protocols and we have to be careful about the power consumption, data transfer rates, range, and most importantly security. So, in this paper, we have discussed all these points of ZigBee and also discussed the comparison between Bluetooth, Wi-Fi, and ZigBee. So, from this comparison, one thing is clear that the choice of protocol completely depends on the requirement of the user so as per the specifications of the applications the users can select either of the protocol. As far as security is concerned, ZigBee and Bluetooth are preferred, as they use maximum bit key for the encryption. But as we have seen the comparison, we can also draw a conclusion that ZigBee and Bluetooth are mostly useful for short range transferring of data. So, to choose either of the protocol requirement of the application has to be monitored or analyzed.

References

1. Muthu Ramya C, Shanmugaraj M, Prabakaran R (2011) Study on ZigBee technology, Research Gate, Conference Paper. Uploaded on 17 February 2019
2. Dini G, Tiloca M (2010) Considerations on security in ZigBee networks. In: 2010 IEEE international conference on sensor networks, ubiquitous, and trustworthy computing. Newport Beach, CA, pp 58–65. https://doi.org/10.1109/SUTC.2010.15.
3. Bakhache B, Ghazal JM, Assad SE (2014) Improvement of the Security of ZigBee by a New Chaotic Algorithm. IEEE Syst J 8(4):1024–1033. https://doi.org/10.1109/JSYST.2013. 2246011
4. Whitehurst LN, Andel TR, McDonald JT (2014) Exploring security in ZigBee networks. In: CISR '14: Proceedings of the 9th annual cyber and information security research conference April 2014. pp 25–28. https://doi.org/10.1145/2602087.2602090

5. Kulkarni S, Ghosh U, Pasupuleti H (2015) Considering security for ZigBee protocol using message authentication code. In: 2015 annual IEEE India conference (INDICON). New Delhi, pp. 1–6. https://doi.org/10.1109/INDICON.2015.7443625
6. Al-alak S, Ahmed Z, Abdullah A, Subramiam S (2011) Aes and ecc mixed for zigbee wireless sensor security. Computing 1:5
7. Sohel Rana SM, Halim MA, Kabir MH (2018) Design and implementation of a security improvement framework of Zigbee network for intelligent monitoring in IoT platform. Appl Sci 8(11):2305. https://doi.org/10.3390/app8112305
8. Khan MA, Khan A, Kabir M (2016) Comparison among short range wireless networks: Bluetooth, Zigbee, & Wi-fi. Adv Comput Sci Eng 4:19–28
9. Pothuganti K, Chitneni A (2014) A comparative study of wireless protocols: Bluetooth. UWB, ZigBee, and Wi-Fi. 4:655–662
10. https://www.digi.com/solutions/by-technology/zigbee-wireless-standard#:~:text=Zigbee%20is%20a%20wireless%20technology,900%20MHz%20and%20868%20MHz
11. https://www.assured-systems.com/uk/news/article/what-is-a-zigbee-network/
12. https://en.wikipedia.org/wiki/Zigbee
13. Fredeady Hands-on ZigBee: implementing 802.15.4 with microcontrollers
14. Yuksel E, Nielson HR, Nielson F (2008) ZigBee-2007 security essentials. Informatics and Mathematical Modelling, Technical University of Denmark Richard Petersens Plads Bldg 321, dk-2800 Kongens Lyngby, Denmark
15. Farahani S (2008) ZigBee wireless networks and transceivers. EBook, Published on December 2008
16. Sahni K, Kharb L (2019) ZigBee technology: a global standard for communication. Int Educ Appl Res J 3(3)
17. Baronti P, Pillai P, Chook V, Chessa S, Gotta A, Hu YF Wireless sensor networks: a survey on the state of the art and the 802.15.4 and ZigBee standards
18. Le KT (2004) Designing a ZigBee-ready IEEE 802.15.4-compliant radio transceiver. Chipcon
19. Protocols and architectures for wireless sensor networks holger karl university of paderborn, germany andreas willig hasso-plattnerinstitute at the university of potsdam, Germany, https://profsite.um.ac.ir/~hyaghmae/ACN/WSNbook.pdf, EBook.
20. Arrigault S, Zacharaki V Design of a ZigBee magnetic sensor node. Master of Science thesis
21. Part 15.4: wireless medium access control (MAC) and physical layer (PHY) specifications for low-rate wireless personal area networks (LR-WPANs). Sponsor LAN/MAN Standards Committee of the IEEE Computer Society
22. https://www.abiresearch.com/research/1003283
23. Kumar A, Gupta S (2013) Study on ZigBee technology. Int J Eng Sci Res Technol. October 2013. Impact Factor: 1.852, ISSN: 2277-9655
24. Razouka W, Crosby GV, Sekkakia A (2014) New security approach for ZigBee weaknesses, the UI international symposium on applications of Ad hoc and sensor networks (AASNET'14). Procedia Comput Sci 37:376–381
25. Khanji S, Iqbal F, Hung P (2019) ZigBee security vulnerabilities: exploration and evaluating. Conference Paper
26. https://www.ZigBee.org.
27. Borisov N, Goldberg I, Wagner D (2011) Intercepting mobile communications: the insecurity of 802.11. In: Proceedings of the 7th annual international conference on mobile computing and networking, ACM (2001). pp 180–189
28. Clavier C (2007) Passive and active combined attacks on AES. In: Fault diagnosis and tolerance in cryptography
29. Masica K (2007) Recommended practices guide for securing ZigBee wireless networks in process control system environments
30. Nabeel M, Zage J, Kerr S et al (2012) Cryptographic key management for smart power grids—approaches and issues
31. Saponara S, Bacchillone T (2012) Network architecture, security issues, and hardware implementation of a home area network for smart grid. J Comput Netw Commun 2012

32. Stelte B, Rodosek GD (2013) Thwarting attacks on ZigBee—removal of the killerbee stringer. pp 219–226
33. Li H, Jia Z, Xue X (2010) Application and analysis of ZigBee security services specification. In: 2010 Second international conference on networks security wireless communications and trusted computing (NSWCTC), vol 2. pp 494–497
34. Baig F, Beg S, Khan MF (2013) Zigbee based home appliances controlling through spoken commands using handheld devices. Int J Smart Home 7(1):19–26
35. Lee K, Lee J, Zhang B, Kim J, Shin Y (2009) An enhanced trust center based authentication in Zigbee networks. In: International conference on information security and assurance. Springer, pp 471–484
36. Dhillon P, Sadawarti H (2014) A review paper on ZigBee (IEEE 802.15.4) Standard. Int J Eng Res Technol (IJERT). ISSN: 2278-0181
37. Ali B, Awad DAI (2018) Cyber and physical security vulnerability assessment for IoT-based smart homes. Sensors 18:817
38. Betzler A, Gomez C, Demirkol I, Paradells J (2014) A holistic approach to Zigbee performance enhancement for home automation networks. Sensors 14:14932–14970
39. Kocher IS, Chow C-O, Ishii H, Zia TA (2013) Threat models and security issues in wireless sensor networks. Int J Comput Theory Eng 5:5
40. Kumar NVR, Bhuvana C, Anushya S (2017) Comparison of ZigBee and Bluetooth wireless technologies-survey. In: 2017 international conference on information communication and embedded systems (ICICES). Chennai, pp 1–4. https://doi.org/10.1109/ICICES.2017.8070716
41. Chhabra N (2013) Comparative analysis of different wireless technologies. Int J Sci Res Netw Secur Commun 1(5). ISSN: 2321-3256
42. Lee J-S, Su Y-W, Shen C-C A comparative study of wireless protocols: Bluetooth, UWB, ZigBee, and Wi-Fi. In: Information and Communications. Research Labs Industrial Technology Research Institute (ITRI) Hsinchu, Taiwan
43. Georgakakis E, Nikolidakis SA, Vergados DD, Douligeris C (2011) An analysis of Bluetooth, Zigbee and Bluetooth low energy and their use in WBANs. In: Lin JC, Nikita KS (eds) Wireless mobile communication and healthcare. MobiHealth 2010. Lecture Notes of the Institute for Computer Sciences, Social Informatics and Telecommunications Engineering, vol 55. Springer, Berlin, Heidelberg
44. Idris Y, Muhammad NA A comparative study of wireless communication protocols: Zigbee vs Bluetooth. IJESC 6(4). https://doi.org/10.4010/2016.867. ISSN: 2321 3361
45. Baker N (2005) ZigBee and Bluetooth: strengths and weaknesses for industrial applications. IEE Comput Control Eng 16(2):20–25
46. Mihajlov B, Bogdanoski M (2011) Overview and analysis of the performances of ZigBee based wireless sensor networks. Int J Comput Appl (0975–8887) 29(12)
47. ZigBee Alliance (2009) ZigBee technical document [Online]. Available: https://www.zigbee.org/Products/TechnicalDocumentsDownload
48. Lee J, Su Y, Shen C (2007) A comparative study of wireless protocols: Bluetooth, UWB, ZigBee, and Wi-Fi. In: The 33rd annual conference of the IEEE industrial electronics society (IECON). Taipei, Taiwan, 5–8 November 2007
49. Didla S, Ault A, Bagchi S (2008) Optimizing AES for embedded devices and wireless sensor networks. In: TridentCom '08: proceedings of the 4th international conference on testbeds and research infrastructures for the development of networks and communities. Article No.: 4, pp 1–10
50. Kwon Y, Kim H (2012) Efficient group key management of ZigBee network for home automation. In: 2012 IEEE international conference on consumer electronics (ICCE). IEEE, pp 378–379

Chapter 39
Application of Machine Learning in App-Based Cab Booking System: A Survey on Indian Scenario

Prerona Saha, Soham Guhathakurata, Sayak Saha, Arpita Chakraborty, and Jyoti Sekhar Banerjee

1 Introduction

The competition between traditional taxi companies and online cab services has become tougher over the years because of the huge number of companies that are entering the market worldwide. The companies which are new to the market are preferring cab booking via apps as this process is more transparent for customers. Day-by-day, the cab booking service has become online. To get a better understanding of the current situation, the market data [1, 2] from independent databases and third party are used here.

Various researchers have pointed out different drawbacks of existing systems and proposed some possible solutions for these problems in their papers [3–10]. The biggest stumbling block of the existing system is that all the systems use centralized servers [4]. So, if the server goes down, the whole booking system will go down inducing a massive amount of loss to the company.

P. Saha · S. Guhathakurata · S. Saha
Department of Computer Science Engineering, Bengal Institute of Technology, Kolkata 700150, India
e-mail: preronasaha0@gmail.com

S. Guhathakurata
e-mail: sohamguhathakurata1999@gmail.com

S. Saha
e-mail: sayak.sam98@gmail.com

A. Chakraborty · J. S. Banerjee (✉)
Department of Electronics and Communication Engineering, Bengal Institute of Technology, Kolkata 700150, India
e-mail: tojyoti2001@yahoo.co.in

A. Chakraborty
e-mail: chakraborty_arpita2006@yahoo.com

© The Author(s), under exclusive license to Springer Nature Singapore Pte Ltd. 2021   483
X.-Z. Gao et al. (eds.), *Applications of Artificial Intelligence in Engineering*, Algorithms for Intelligent Systems, https://doi.org/10.1007/978-981-33-4604-8_39

Thus, some of the researchers have suggested using a decentralized server to over-come this problem. Due to the usage of offline cab booking systems for a prolonged period of time, people are more used to bargaining before finalizing the ride. But the current system has no such facility. So, some of the authors have also proposed to give the user an option to negotiate [4] before confirming the trip. Currently, the cab systems do not have real-time updates about the demand for cabs in different regions. Thus, few authors have suggested providing real-time updates [6] to the system, so that cab availability never becomes an issue. Sometimes due to less avail-ability of cabs, users are forced to ride on a less suitable cab, which causes some serious inconvenience. To overcome this hindrance, Glaschenko et al. have proposed to assign the desired cab, which is about to complete its riding [6] over any less suitable cab. UBER has also recommended enhancing existing security [11] features by providing an option to record the audio [12] of the whole trip from both user and driver's end. All these features will reinforce the experience of the user and also make the ride a lot of secure using seamless mobile communication [13–32] services.

The letter is constructed as follows. Section 2 deals with the growth of the App–Cab industries in India. The study of existing problems of the App–Cab system and their possible solutions are presented in Sect. 3. Section 4 displays the mostly used algorithms in App–Cab-based system. Section 5 depicts the role of AI-ML. Section 6 presents the challenges faced by the App–Cab system accompanied by the conclusion in Sect. 7.

2   Growth of the App–Cab Industries in India

In this paper, authors are mainly focused on the different aspects of cab services in India specifically because densely populated this country demands heteroge-neous requirements. Cab services have transfigured the transportation system in metropolitan cities. India is a country known for its lowest car ownership because car ownership is burdensome for the urban dwellers. The mobility demands of this country got a much-needed thrust in the year 2013–2015 when online cab services ventured their online cab booking program.

Nowadays, cabs play a crucial role in bridging the gap between public and private transportation. The most interesting fact is that day-by-day the companies are hunting for new opportunities for the growth of their business. As OLA and UBER are the two largest companies of the ride-hailing market in India, the authors have restricted the comparison between these two cab services only in this paper (see Table 1).

Table 1 Comparison between OLA [2] and USER [1] based on some features

Features	OLA	UBER
Pin verification	Secured with OTP	OTP not required
Personal wallet	OLA Money	Don't have any personal wallet. Payment through e-wallet, debit cards and credit cards
Ride check	Not available	Latest trip recording facility launched by UBER for security purposes
Subsidiary	Foodpanda	Uber Eats
Availability in cities	250 cities since 2020	700 cities since 2019
Number of users in India	23.96 million MAU since 2019	5.5 million MAU (Monthly active users) since 2017
Share in Indian market	80% since 2016	50% since 2020
Number of drivers	2.5 million since 2020	3.9 million since 2019
Number of rides	1.5 million per day, since 2017	17 million per day, since 2019

3   Study of Existing Problems of App–Cab System and Their Possible Solutions

I.   ***Single Point Failure***: Currently, the cab service system is server-based. The database of the entire system is fed to the central server. Sometimes when a large number of people try to book a cab at the same time, the centralized server goes down. It will stop the entire system from working. During the downtime, this causes serious inconvenience to the users and a significant amount of loss to the company. It is known as a single point of failure [4] (Fig. 1).

Using a distributive approach, we can overcome this problem. In a distributive approach, the entire system is divided into multiple units. So, it can balance the load and does not create any hindrance. In vehicular ad hoc networks (VANET) [3], the distributed system consists of two parts: taxi booking and taxi de-blocking. The taxi booking part finds the nearest driver and assigns it to the passenger. On the other hand, the taxi de-blocking system ensures that no two passengers are booking the same taxi simultaneously. As many taxis are competing for a single passenger, so it is necessary to block the other taxis after assigning a taxi to a passenger.

II.   ***Bargaining Hindrance***: In offline taxi-hailing systems, people are accustomed to bargaining before confirming a ride. The existing online system does not allow the user and the driver to bargain the fare before confirming the trip. The fare decided by the system becomes the final fare, and the user or the driver cannot negotiate on that.

To overcome this problem, researchers have introduced a bargaining facility [4] where the user will be able to negotiate the fare. When a user will search

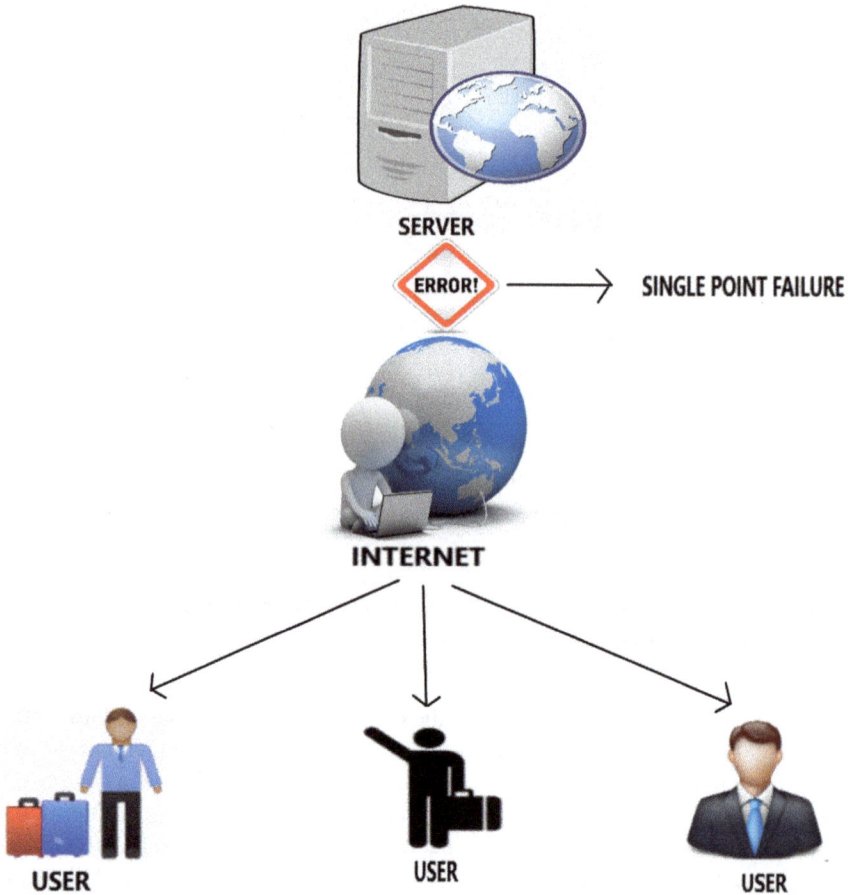

Fig. 1 Single point failure

for a cab, the nearest cab driver will receive a message asking if the cab is available or not. If the driver is available, then the driver will reply to that message with the desired fare. Then, the system will inform the user about the fare and ask the user to provide the user's desired fare. The fair will then be forwarded to the driver by the system. Then, the driver will send his final fare. The user will be informed about the final fare proposed by the driver. If the user agrees with the final fare proposed by the driver, the ride will be confirmed. Otherwise, the system will search for new drivers.

III. **_Enhanced Safety Measures_**: Existing online cab booking systems have some safety features for the travelers. They do not provide any safety measures for the Cab drivers if any mishap occurs with the driver who cannot report immediately through the system.

To enhance the safety of the driver as well as the travelers, the renowned cab service provider UBER has introduced a feature called audio recording [12]. If the driver or the travelers face any problem during the ride, they can start the audio recording of the trip. At the end of their trip, both of them will get a message asking about their feedback. They can submit the audio recording as a reference for their complaint. If they do not want to submit it instantly, it will be saved in their ride history in an encrypted format. The travelers or the driver will also be able to submit it later from the ride history. But they cannot play the audio recording on their device. After submitting the feedback only, they will be able to decrypt that recording, and they may take the necessary actions that are needed.

IV. ***Reacting to Real-Time Events***: Sometimes, the driver takes much more time than the estimated time of arrival (ETA) [10] to reach the pickup point because of traffic jams. Users prefer online cab booking over offline cab services to save their valuable time. Even after booking a cab online, if the user has to spend time waiting for the cab to arrive, then it will be resulting in a bad experience for the traveler.

To resolve this issue the system is designed to act according to real-time events [6]. According to Fig. 2, Cab-1 is stuck in a traffic jam, so the driver will give a message to the system. After receiving the message, the system will start looking for nearby drivers. If the system finds one, the system will assign it to the user. As the Cab-2 recently finished a ride near the location of the user, hence, now the Cab-2 is free for taking another ride.

V.   ***Increasing Availability***: In the current system, the driver's involvement is very less. But the cab driver is the one who is having better visibility and knowledge about the current scenario. Sometimes due to public gatherings or arrival of

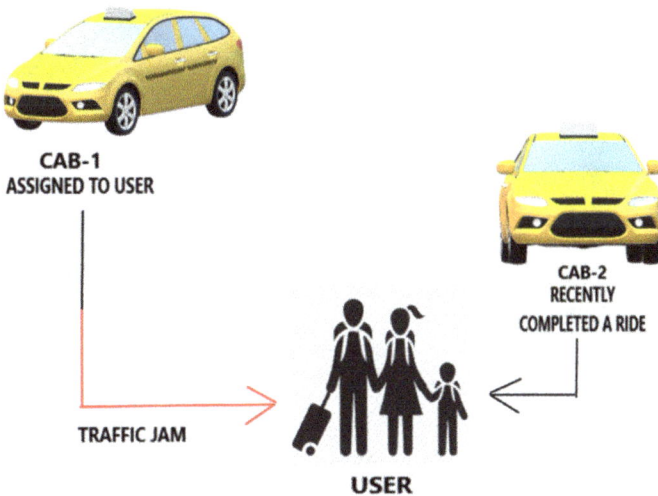

CAB-1
ASSIGNED TO USER

TRAFFIC JAM

USER

CAB-2
RECENTLY
COMPLETED A RIDE

Fig. 2 Reacting to real-time events

any VIP, certain places have high demands of cabs. But there is very little availability of cabs in comparison with the demand. However, cabs are easily available in some other locations with less demand.

To overcome such situations, the cab driver is playing the lead role in the described system as the driver is the person who aware of the real-time scenario. If the driver notices any gathering on the way or comes to know about any VIP arrival, then the driver will report the system about gathering so that the system can arrange some cabs around that location. Hence, travelers need not wait for a long time.

VI. ***Fleet Utilization***: While booking a cab, every user has some requirements. The system's job is to assign the best option to the user, fulfilling the traveler's requirements. But sometimes, it may happen that the available options do not accomplish the user's demands. Then, the system is forced to assign the user a less suitable cab.

To get rid of this problem, the suggested system has introduced fleet utilization [6]. In Fig. 3, the system will first search for nearby cabs that fulfill user's requirements. Cab-2 is free, but it does not fulfill all the requirements of the user. As any such cab is not free at that moment, the system will check the

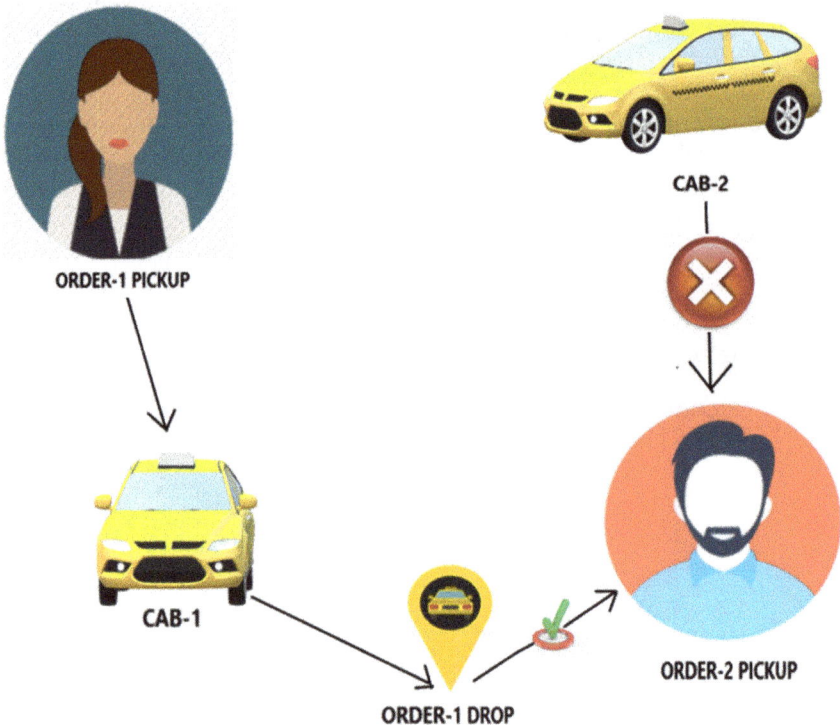

Fig. 3 Fleet utilization

nearby cabs which are about to finish their rides in a short time and fulfill the user's requirements. Cab-1 is about to finish its ride, and it completely satisfies the user's requirements. So, Cab-1 is assigned to the user.

VII.   ***Dynamic Order Dispatch***: In the current online cab system, mainly two approaches are used to assign a driver to any customer. One is a system assigning approach, where the system assigns a driver to the user based on the user's preference. The driver will get approximately 20 s to decide whether to confirm the ride or reject it. If the first driver rejects the ride, it is assigned to the next driver chosen by the service provider. This approach is used by UBER. The system assigning approaches mainly focus on the customer's preference. So, the driver's choice is ignored by the system. Another approach is the driver grabbing model. In this approach, the system broadcasts the user's request to all the nearby drivers. The first driver to respond to the request will continue the ride. This approach is used by DIDI, and it is more beneficial for the drivers, as it allows them to select the ride of their choice. Both of the two approaches have drawbacks because none of them brings benefits to both the customer and the driver.

The user and the cab driver, both of them are an integrated part of an online cab booking system. So, an ideal system should focus on the benefit of both of them. By keeping this fact in mind, the dynamic dispatch system [7] was implemented. This approach inculcates the advantage of both approaches. The nearby drivers are divided into sets based on their distance from pickup location. When the user requests a ride, the system broadcasts the request to the first set. The first driver to accept the request will continue the ride. If no one is ready to take the ride from that set, then the service provider broadcasts the request to the other set of drivers. It is beneficial for both the user and the driver.

VIII.   ***Detecting Driver's Attempt to Cheat***: The decision making regarding a trip should not only depend upon the driver. As they will take all the trips according to their personal preference, the user's preference will be ignored. Sometimes just to make some profit, the driver tries to mislead the system by giving false information to the system. Despite being at the starting of a trip, they inform the system that they will be free within 10 min for the next trip. So that they do not need to wait much to get their next trip. But this will increase the waiting time for the commuters. Another problem arises when the drivers mislead the system by repeatedly informing them that they are going home. So that they do not need to go far away from home, and they can get the rides only toward their home.

The previously mentioned reasons will cause an unequal distribution of drivers in all the regions. If everybody wants to roam around their home, then there will be insufficient cabs in some regions. To avoid this, the fleet can be distributed according to the flow forecasts, and the system should keep itself updated with the driver's schedule. So that the driver cannot make any unnecessary requests [6].

4   Mostly Used Algorithms in App–Cab Based System

4.1   Adaptive Scheduling Approach [6]

- The key feature of the new approach is it does not reschedule any disruptive event from scratch rather it only modifies the part affected by that event.
- Multi-agent technology [5] is the algorithm used for decision making in distributed events. Because of the distributive nature and object-oriented approach, it can easily deal with many complex events. In this approach, multiple decision-making nodes are cooperating with each other by exchanging messages or competing with each other by working on the same data space.
- The distributed paradigm partitions the software into self-contained parts. It represents the complex situations in the most normal form to the users.
- Clients and are represented in the system as agents. For computing each task, the system assigns an individual agent. According to this approach, order allocation managers are the most active agents. They keep track of available vehicles and perform the negotiation with vehicle agents. In this system, drivers are reactive agents. They only reply to requests from order agents and perform the action selected for them. There are some other agents like external event processors, region loading managers, etc.
- On a new order, the system automatically starts finding the best vehicle and books a resource. The system takes 9 min on average to provide a cab for urgent orders. When somebody schedules a ride, the system immediately allocates a cab and keeps reallocating the cabs if new suitable options appear. The system commits the final decision when the driver must start the journey in order to complete the journey.
- Authors have used multi-agent technology to apply all possible facilities of object-oriented designs. The proposed algorithm helps to increase the productivity and evolution of the system, and it makes it very easy to optimize those schedules where it is difficult to use formalized algorithms such as ant colonies algorithms or genetic algorithms. It reduces the time taken for implementing real-time scheduling software. So, the overall cost is also reduced significantly.

4.2   Distributive Approach [4]

- Distribution approach consists of multiple autonomous units. These units communicate with each other to coordinate their common goal.
- Multi-agent taxi dispatch architecture (MDTA) is used for bargaining features provided to the users.
- For cab gathering, it uses a multi-agent system (MAS)-based framework known as Java agent development framework (JADE).

- JADE has a flexible distributed infrastructure because of this flexible nature, easy and dynamic framework to design, debug, develop and test the agents-based applications. It allows easy extension with an add-on module.
- Java language is used to develop the JADE framework. So, any Java development framework, NetBeans, Eclipse, etc., can be easily plugged in with it.
- A unique name is provided to each agent in the platform using the agent management system (AMS). AMS represents the authority, and it can create/kill agents on remote containers in the platform.
- In order to achieve his goals, an agent needs another agent to fulfill his requirements. Yellow Pages services help the agent to find such agents. These Yellow Page services are provided by the Directory Facilitator (DF).
- JADE framework transmits messages asynchronously. JADE uses FIPA-ACL message structure specifications. Message formats consist of fields such as the sender, list of receivers, communicative act (REQUEST, INFORM, PROPOSE, etc.), content language and ontology.

4.3   PDPTW Approach [9]

- Pickup and delivery problem with time window (PDPTW) algorithm is a generalized form of a famous vehicle routing algorithm called the vehicle routing problem with time window (VRPTW) algorithm.
- PDPTW is an NP-hard algorithm. It was implemented to solve Singapore Taxi advance Reservation (STAR) problems using multi-vehicle pickup and delivery with time windows. To search the shortest time path from source to destination using the PDPTW approach, a link-to-link shortest path algorithm was embedded into the API program.
- Instead of starting from a central cab depot, it makes vehicles all over the street market. So, it becomes easier for the commuters to grab a cab anytime anywhere.
- As the customers are booking the cab in advance, they expect their cab to arrive at the pickup point at their desired time. PDPTW is using a hard and extremely narrow time window. In advance booking, the users specify the time point instead of the time window. So that the customers do not need to wait beyond their desired pickup time.
- This approach aims to minimize two factors. One is minimizing the number of vehicles, and another one is minimizing the distance traveled. These two factors can substantially decrease the fuel consumption and the costs of taxi operations. It mainly focuses on two types of move operations called exchange and reallocate.
- PDPTW approach is a two-phased method having the construction phase and improvement phase. In the construction phase, all the feasible roots are stored in an initial set. In this phase, a time window insertion algorithm was used to solve the 'ad hoc' problem of cab dispatching. And in the improvement phase, improvement of the existing solution is done by using tabu search.

5   Role of AI-ML

Artificial intelligence and machine learning (AI-ML) have powered machines with human intelligence [36–39]. Cab services have achieved a major milestone in the history of transportation with the help of AI-ML. The authors have discussed in this paper how the potential of AI-ML is used to enrich the online cab booking experience.

- **Demand Prediction:** From predicting cab demand on a special occasion to predicting the estimated time of arrival of a cab, demand prediction has created a new wave of mobility in the transportation system. Online cab booking systems mostly rely on forecasting algorithms for event prediction. Time series forecasting algorithms can easily predict the number of rides a system can get in a given location in a given time. So, it becomes easier for cab booking systems to reach the customer's demand.
- **Finding the Nearest Cab:** In offline taxi-hailing systems, people had to wait for hours by waving their hands to find a taxi. Most of the time customers used to be rejected by many taxi drivers. Online cab booking systems, having a high acceptance rate, rarely reject any hail. It uses GPS-based automatic vehicle location and dispatch system (AVLDS) [8] to assign the nearest ride to the user. Whenever a customer requests for a cab, the digital hail appears on the driver's cell phone. The driver gets 20 s to decide whether to accept the request or not.

It is almost impossible for a human being to allocate a cab keeping all the user's demands in mind. But with the help of ML, this can be done easily. The process of finding the nearest cab is discussed in Fig. 4. When the user requested a cab, the system started searching for nearby cabs. It found three available cabs. Among the three available cabs, Cab-2 needs to travel the least distance to reach the pickup point. So, the user was assigned Cab-2. ML uses all the data gathered from thousands of registered users and drivers to find the best and nearest ride for the user within a few seconds. Most of the online cab system uses the KNN, i.e., K-nearest neighbors algorithm to find all the nearest drivers and then apply regression on them to find the perfect match among them.

- **Fare Estimation:** Estimation of fare depends upon so many factors [33] like distance traveled, time of travel, day of travel, weather conditions, availability, etc. It is very hard for a person to estimate the fare keeping all these factors in mind.

  ML helps to calculate the fare very efficiently. It observes prior data and patterns to understand all the factors, including real-time changes like weather change, peak hours and escalate prices accordingly. There are different models [34] which are used to predict the fare. Machine learning algorithms like linear regression, random forest, light GBM are used to figure out these models.
- **Car Pooling:** Car pooling is one of the most innovative features of an online cab system. As it completes two rides at the same time, it improves traffic congestion. It is also an eco-friendly feature, as fewer cars will cause less pollution. But during drop off, it becomes difficult for the driver to decide whom to drop first.

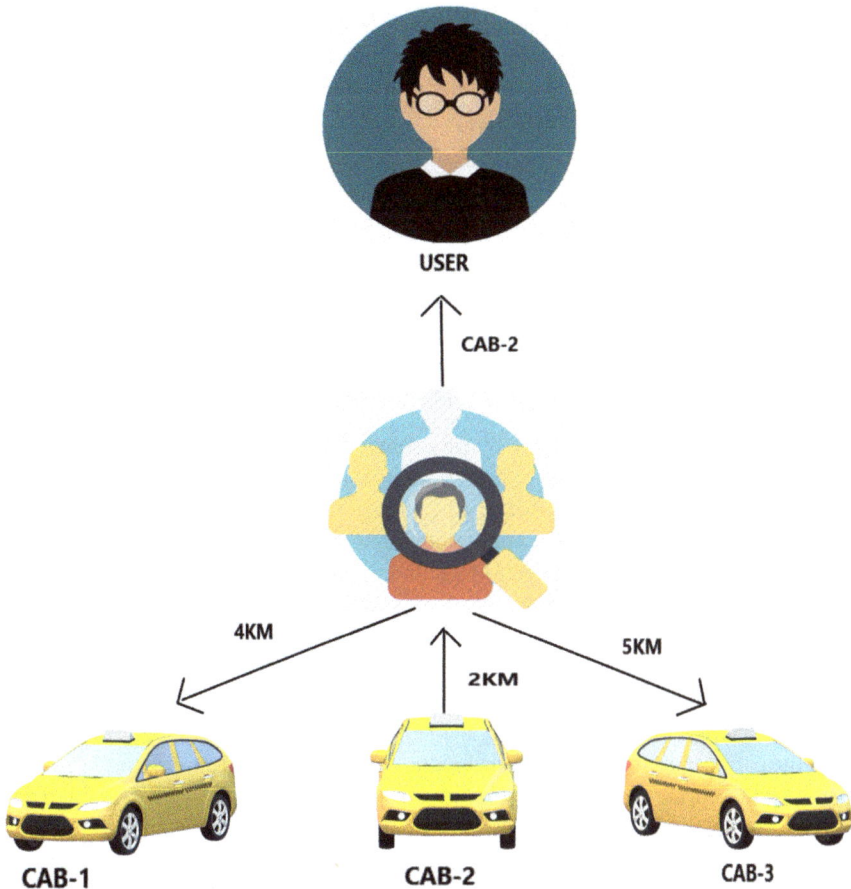

Fig. 4 Finding nearest ride

ML plays a vital role in this feature. With the help of data accumulated from the users, it can easily decide which rider to drop first.

- **Smart Chat:** Communication between driver and user is an important part of the trip. But most of the time, it is difficult for the driver to reply while driving the cab as it may lead to an accident.

  To respond effortlessly, drivers use an AI-based one-click-chat, which uses natural language processing and machine learning techniques to frame answers to common questions with a single click.

- **Optimized Route:** Taking a non-efficient route can result in delays in trip completion and can cause serious trouble for the user. It is also not possible for the driver to always identify the most efficient route. So, having a system that will be able to predict the shortest, and most efficient route will save the driver and the passenger a lot of time by avoiding traffic jams and other congestions on the road.

Uber uses a machine learning to update the possibilities in each route and suggest the quick-moving route to the driver. This process helps the driver to avoid crowded routes and make the journey a lot faster. This pleases the customer and offers the driver additional time for other trips.

6 Challenges Faced by App–Cab System

The demand for online cab services is increasing day-by-day. But with the increase in demand, the challenges faced by the cab industry are also increasing simultaneously. In this section, the authors have discussed some of the challenges encountered by the taxi booking companies, and how it is affecting the cab industry.

- **Increasing Competition**: Day-by-day many new companies are entering the market. It results in a cut-throat competition in the cab industry. Competition is an essential part of any industry be it a cab industry or any other industry. But, because of competition when a company tends to lose its customers, then it becomes a challenging situation for them. They need to come up with new features to stand out in this competitive market. Every company is trying its best to provide its customers with a comfortable ride. Some next-gen features like an extra discount, loyalty games, etc., will help them to make a stand to overcome this challenging situation.
- **Surge Pricing**: Surge pricing [35] is one of the most disadvantageous policies of online cab booking systems. Whenever the demand is high, it automatically results in a sudden price hike. The short-term effect of surge pricing may be beneficial as it significantly affects the rate of demand. But if we observe the long-term effect of surge pricing, it will tend to lose customers. Nobody will prefer to travel the same route at a higher price. As a result of this, customers will assume that making money is more important to the service providers than providing their customers a quality service. To serve the customers with quality service without hiking the price, service providers need to minimize the rush during peak hours by increasing the availability of cabs according to customer's demand.
- **Taking Too Long to Find a Ride**: Sometimes due to the unavailability of cabs, it takes a long time to find a ride. If the user is in a hurry, this can cause serious trouble to him. This bad user experience ends up in the user not using the service. To avoid this from happeningm the cab service providers have to increase the availability of their cabs so that the customers do not face such troubles.
- **Ride Cancelation Without Notification**: According to a survey mentioned in Financial Express, users face this issue almost daily [35]. Sometimes after accepting a ride, drivers take a very long time to arrive at the pickup location. At times, they even cancel the ride just before arriving at the pickup point without mentioning a valid reason. As per the standard cancelation policy of cab booking systems, the users have to bear some fine, if the ride is canceled after 10 min from the time of booking. Just because of the driver's negligence, the user has to pay

an unnecessary fine without even canceling the ride. In such cases, the user is left perplexed, and it becomes very difficult for them to find another ride straightaway.

7   Conclusion

Nowadays, in metropolitan cities, cab services have become an imperative part of everyday life. Day-by-day, App–Cab is replacing the other mode of transportations, with its comfort and hassle-free rides. Due to financial problems or maintenance problems, car ownership has become a burden to urban dwellers. So, they prefer to rent a car than having a car. In such scenarios, online cab booking systems have become a boon for the commuters. So, it is expected that the system should be mentored to provide an end-to-end service to its users. After analyzing some studies related to online cab booking systems, authors have tried to gather all the information in a single paper. So that it becomes easier for the readers to rectify the existing problems and find their solutions in the context of the existing system in India.

References

1. https://medium.com/analytics-vidhya/machine-learning-to-predict-taxi-fare-part-two-predic tive-modelling-f80461a8072e. Accessed June 2020
2. https://expandedramblings.com/index.php/uber-statistics/. Accessed May 2020
3. Sheu JP, Chang GY, Chen CH (2010). A distributed taxi hailing protocol in vehicular ad-hoc networks. In: 2010 IEEE 71st vehicular technology conference. IEEE, pp 1–5
4. Nath A, Khandelwal A, Kanojia A, Minocha I, Niyogi R (2017) Design and implementation of an intelligent cab service system. In: 2017 tenth international conference on contemporary computing (IC3). IEEE, pp 1–6
5. Awajan A (2013) An automated taxi booking and scheduling system. In: 2013 8th EUROSIM congress on modelling and simulation. IEEE, pp 502–505
6. Glaschenko A, Ivaschenko A, Rzevski G, Skobelev P Multi-agent real time scheduling system for taxi companies
7. Duan Y, Wang N, Wu J (2019) Optimizing order dispatch for ride-sharing systems. In: 2019 28th international conference on computer communication and networks (ICCCN). IEEE, pp 1–9
8. Liao Z (2003) Real-time taxi dispatching using global positioning systems. Commun ACM 46(5):81–83
9. Wang H, Lee DH, Cheu R (2009). PDPTW based taxi dispatch modeling for booking service. In: 2009 fifth international conference on natural computation, vol. 1. IEEE, pp 242–247
10. Lam HT, Diaz-Aviles E, Pascale A, Gkoufas Y, Chen B Grand challenge: real-time destination and eta prediction for maritime traffic
11. Banerjee J, Maiti S, Chakraborty S, Dutta S, Chakraborty A, Banerjee JS (2019) Impact of machine learning in various network security applications. In: 2019 3rd international conference on computing methodologies and communication (ICCMC). IEEE, pp 276–281
12. https://www.uber.com/en-IN/newsroom/new-safetyfeatures/#:~:text=We%20plan%20to% 20pilot%20audio,time%20through%20the%20Safety%20Toolkit. Accessed April 2020
13. Pandey I, Dutta HS, Banerjee JS (2019). WBAN: a smart approach to next generation e-healthcare system. In: 2019 3rd international conference on computing methodologies and communication (ICCMC). IEEE, pp 344–349

14. Paul S, Chakraborty A, Banerjee JS (2019) The extent analysis based fuzzy AHP approach for relay selection in WBAN. Cognitive informatics and soft computing. Springer, Singapore, pp 331–341
15. Paul S, Chakraborty A, Banerjee JS (2017) A fuzzy AHP-based relay node selection protocol for wireless body area networks (WBAN). In: 2017 4th international conference on opto-electronics and applied optics (Optronix). IEEE, pp 1–6
16. Chattopadhyay J, Kundu S, Chakraborty A, Banerjee JS (2020) Facial expression recognition for human computer interaction. In: Proceedings of ICCVBIC 2018. Springer (press)
17. Banerjee JS, Chakraborty A, Chattopadhyay A (2018) Relay node selection using analytical hierarchy process (AHP) for secondary transmission in multi-user cooperative cognitive radio systems. Advances in electronics, communication and computing. Springer, Singapore, pp 745–754
18. Saha O, Chakraborty A, Banerjee JS (2017) A decision framework of IT based stream selection using analytical hierarchy process (AHP) for admission in technical institutions. In: 2017 4th international conference on opto-electronics and applied optics (Optronix). IEEE, pp. 1–6
19. Saha O, Chakraborty A, Banerjee JS (2019) A fuzzy AHP approach to IT-based stream selection for admission in technical institutions in India. Emerging technologies in data mining and information security. Springer, Singapore, pp 847–858
20. Banerjee JS, Chakraborty A, Chattopadhyay A (2018) Reliable best-relay selection for secondary transmission in co-operation based cognitive radio systems: a multi-criteria approach. J Mech Continua Math Sci 13(2):24–42
21. Banerjee JS, Chakraborty A, Chattopadhyay A (2018) A novel best relay selection protocol for cooperative cognitive radio systems using fuzzy AHP. J Mech Continua Math Sci 13(2):72–87
22. Guhathakurata S, Kundu S, Chakraborty A, Banerjee JS (2020) A novel approach to predict COVID-19 using support vector machine. In: Data science for COVID-19. Elsevier (press)
23. Banerjee JS, Chakraborty A, Chattopadhyay A (2017) Fuzzy based relay selection for secondary transmission in cooperative cognitive radio networks. In: Proceedings of OPTRONIX. Springer, pp 279–287
24. Banerjee JS et al (2018) Relay node selection using analytical hierarchy process (AHP) for secondary transmission in multi-user cooperative cognitive radio systems. In: Proceedings of ETAEERE. Springer, pp 745–754
25. Banerjee JS, Goswami D, Nandi S (2014) OPNET: a new paradigm for simulation of advanced communication systems. In: Proc int conf contemp challenges manage, technol soci sci. SEMS, Lucknow, India, pp 319–328
26. Das D, Pandey I, Chakraborty A, Banerjee JS (2017) Analysis of implementation factors of 3D printer: the key enabling technology for making prototypes of the engineering design and manufacturing. Int J Comput Appl 8–14
27. Das D, Pandey I, Banerjee JS (2016) An in-depth study of implementation issues of 3D printer. In: Proceedings of MICRO 2016 conference on microelectronics, circuits and systems. pp 45–49
28. Banerjee JS, Chakraborty A (2015) Fundamentals of software defined radio and cooperative spectrum sensing: a step ahead of cognitive radio networks. In: Kaabouch N, Hu W (eds) Handbook of research on software-defined and cognitive radio technologies for dynamic spectrum management. IGI Global, USA, pp 499–543
29. Banerjee JS, Chakraborty A (2014) Modeling of software defined radio architecture and cognitive radio, the next generation dynamic and smart spectrum access technology. In: Rehmani MH, Faheem Y (eds) Cognitive radio sensor networks: applications, architectures, and challenges. IGI Global, USA, pp 127–158
30. Banerjee JS, Chakraborty A, Karmakar K (2013) Architecture of cognitive radio networks. In: Meghanathan N, Reddy YB (eds) Cognitive radio technology applications for wireless and mobile Ad Hoc networks. IGI Global, USA, pp 125–152
31. Banerjee JS, Karmakar K (2012) A comparative study on cognitive radio implementation issues. Int J Comput Appl 45(15):44–51

32. Biswas S, Sharma LK, Ranjan R, Banerjee JS (2020) Go-COVID: an interactive cross-platform based dashboard for real-time tracking of COVID-19 using data analytics. J Mech Continua Math Sci 15(6):1–15
33. https://medium.com/analytics-vidhya/machine-learning-to-predict-taxi-fare-part-one-explor atory-analysis-6b7e6b1fbc78. Accessed June 2020
34. https://www.mapsofindia.com/my-india/business/problems-faced-by-ola-and-uber-users. Accessed June 2020
35. https://expandedramblings.com/index.php/ola-statistics-facts/. Accessed May 2020
36. Guhathakurata S, Saha S, Kundu S, Chakraborty A, Banerjee, JS (2021) South Asian Countries are less fatal concerning COVID-19: a fact-finding procedure integrating machine learning & multiple criteria decision making (MCDM) technique. J Inst Eng (India): Series B, 2020, Springer (Press)
37. Guhathakurata S, Saha S, Kundu S, Chakraborty A, Banerjee JS (2021) A new approach to predict COVID-19 using artificial neural networks. In Cyber-Physical Systems: AI and COVID-19, Elsevier (press)
38. Biswas S, Sharma LK, Ranjan R, Saha S, Banerjee JS (2021) Smart farming & water saving based intelligent irrigation system implementation using IoT. In Recent Trends in Computational Intelligence Enabled Research, Elsevier (press)
39. Guhathakurata S, Saha S, Kundu S, Chakraborty A, Banerjee JS (2021) South Asian Countries are less fatal concerning COVID-19: a hybrid approach using machine learning and M-AHP. In Computational Intelligence Techniques for combating COVID-19, 2020, Springer (press)

Chapter 40
Churn Prediction in Telecom Industry Using Machine Learning Algorithms with K-Best and Principal Component Analysis

K. V. Anjana and Siddhaling Urolagin

1 Introduction

Churn rate is a metric that refers to the rate at which a business loses its customers over a chosen time frame. Churn is a problem that directly affect the revenue of any company. Churn applications are common in E-commerce companies and subscription business companies. Hence, the definition of churn varies from one company to another and depends on the chosen business model. Identifying potential churner in timely and accurate manner is very important for customer service providers (CSP) especially telecom companies. Customer churn prediction is detecting the customers who switches from one service provider to another. Phone number portability and rapid increase in telecom providers made switching between customer service providers much easier than before [1–3]. The prediction model helps the telecom industry to identify the list of customers at high risk of churn and factors influencing churning of customers. Individualized customer retention techniques are too expensive to implement, and hence if we could categorize the customers at risk of leaving, then customer retention efforts can be made more focused and effective [4, 5]. The cost of getting new customers make the companies invest in the area of retaining its customers. The data obtained from the telecom companies cannot be directly used in the classification model as a high level of imbalance exists between the non-churn data and the churn data. The huge volume of non-churn data that dominates the churn data, makes churn data unfamiliar to the system. Due to the huge volume of data, data mining techniques are the viable option to analyze the pattern among the churned customers and provide insight to strategically retain customers

K. V. Anjana (✉) · S. Urolagin
Department of Computer Science, BITS Pilani, Dubai Campus, PO Box 345055, Dubai International Academic City, UAE
e-mail: p20190005@dubai.bits-pilani.ac.in

S. Urolagin
e-mail: siddhaling@dubai.bits-pilani.ac.in

© The Author(s), under exclusive license to Springer Nature Singapore Pte Ltd. 2021
X.-Z. Gao et al. (eds.), *Applications of Artificial Intelligence in Engineering*, Algorithms for Intelligent Systems, https://doi.org/10.1007/978-981-33-4604-8_40

at high risk to churn. Decision trees, neural networks, regression models, Bayesian models are some of the most effective data mining techniques that have been used for churn prediction. Deep learning is also a field of machine learning that capacitate algorithms to learn, map and interpret given data without the need of human to specify any kind of information. The highly imbalanced deluge data including multitude of data types are the main challenges in customer churn prediction (CCP). Data analysis, feature extraction, identifying the important features that affect retention and appropriate choice of classification model are the important steps involved in CCP. Confusion matrix, sensitivity, specificity, receiver operating characteristics (ROC) curve, and lift charts are some of the most popular performance evaluation metrics for CCP algorithms. The data from telecom service providers mainly include customer demography, credit scores, usage pattern, billing, and payment details, value added services and customer care service details. Churn prediction can be considered as a supervised problem, i.e., predicting future churners from the data associated with each subscriber in the network.

2 Literature Review

In [6], the output of feature selection using Pearson correlation equation is fed to adaptive sampling (ADASYN) algorithm. Back propagation algorithm is adopted as the classification model. The combination of the two methods reduces unbalanced data between classes and adjust the threshold limits adaptively with difficult samples and thus improves the prediction performance. In [7], RFM features were considered to cluster the data. K-means clustering was used for customer profiling. The output is then fed to different algorithms like CART, CHAID, etc., and the result shows that hybrid methods gave better results when compared to single algorithms. In [8], the social network analysis (SNA) features were also added to the raw data and the model was tested on Spark environment using different tree algorithms. Area Under Curve (AUC) method was used for performance analysis, and XGBOOST algorithm was found to give better CCP results. In [9], Lasso regularization technique was used to prevent over fitting and H2o package of deep learning using K-fold cross validation method was used for CCP. The different parameters viz. epoch, number of hidden layers, drop out ratios, and activation functions were varied to achieve better sensitivity values.

In [10], fuzzy-based classification (Nearest Neighbor Classification) was used for CCP using real world telecom data sets. Churner severity and categorization were the two new approaches used in this paper. AUC method was used for performance analysis. In [11], Logit Leaf Model (LLM) hybrid algorithm is used for classification where different algorithms are used for different segments of data. The decision tree greedy algorithm is used in first phase to split data to small and homogenous subsets, and in second phase, logistic regression is applied to these subsets. The LLM regression algorithm adopts forward selection to identify important variables in each sub-set and has in-built feature election mechanism. Performance was evaluated using AUC and top decile lift (TDL) methods and was found to outperform the logistic models and random forests.

In [12], Churn scores, i.e., the probability of a customer leaving the company is predicted using decision trees and the scores are used to calculate the profit measure. "Expected Maximum Profit measure for customer Churn (EMPC)" is the performance metric used and ProfTree classifier uses evolutionary algorithm on EMPC measure. EMPC allocates benefits as well as cost to particular class, and thus, the profit driven decision trees help to detect the profitable customers who are worth retaining. In [13], the characteristics of publicly available telecom datasets and accuracy of different churn prediction models are analyzed in detail. Some of the performance metrics that can be used for both balanced and unbalanced datasets are also proposed.

In [11], a new hybrid algorithm consisting of two stages, a segmentation and a prediction phase is proposed. The method is called Logit Leaf Model that outperformed its building blocks logistic regression and decision trees in terms of both accuracy and interpretability. In [14], the exhaustive CHAID technique proved to be more efficient among all possible variants of decision tree for churn prediction in IBM SPSS data mining tool. In [15], the significance of F1 score as an evaluation metric for churn prediction with XGBoost classifier giving best results when compared to other algorithms such as KNN and RF classifiers.

3 Proposed Method

The historical data of telecom customers are stored and taken as the dataset for churn prediction. The classification problem has two class labels namely churn (denoted as 1) and non-churn (denoted as 0). The data has no missing values, and the shape is (3333, 17) out of which 2850 are non-churn customers, and 483 are customers likely to churn. Figure 1 shows the distribution of churn in dataset. Understanding the data is important before applying machine learning algorithm. Feature selection and data cleaning are the important methods in model designing. Heatmap helps to identify the features that are most related to the churn variable. Figure 2 shows the heatmap. The input data is pre-processed to remove the unwanted features based on the heatmap values. Standard scaler is used as the standardization method for this classification model. Feature selection and extraction are done using Select K-Best and Principal Component Analysis methods, respectively. Assuming that the first two columns in the dataset shouldn't affect churn viz. unique ID of the client, i.e., customer id and geographical information, i.e., state were dropped out as it gave minimum contribution for churn prediction. Out of the 14 remaining features, 4—most weighted features were identified using Select-K-Best method and PCA. Figure 3 shows the proposed method.

The output is then given to four classifiers for comparison. The tree-based methods chosen are XGBoost, and Random Forest Classifier (RFC). The other two methods are Support Vector Classifier (SVC) and Logistic Regression (LR). Confusion matrix

Fig. 1 Percentage of churn in dataset

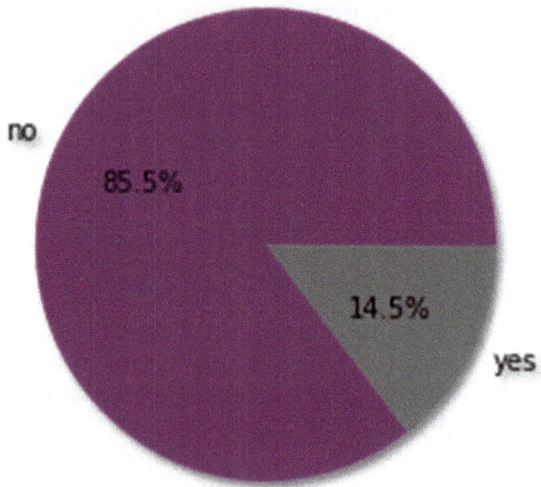

percentage of churn in dataset

and ROC curves are used for result analysis. Figure 3 shows the approach used for churn prediction.

4 Experimental Results

The dataset from BigML contained all customer information with 3333 user records. The output field, i.e., churn information was also available in the data set. Hence, supervised algorithms were used in the model for churn prediction. The model was build using Python by importing the appropriate Sklearn library algorithms. 70% of data was used for training the model, and 30% was used for the testing. Select K-Best method is chosen as the first method to identify the features that contribute most to the churn variable in the dataset. The value for K was chosen as 4 and the most weighted factors based on their scores are "international plan," "voice mail plan," "total day minutes," and "customer service calls." Figure 4 is graph between "Calls to customer care" and "churn." Principal Component Analysis was used for dimensionality reduction. Four principal components were identified. Figure 5 shows the contribution of each component to the churn variable. "0" indicates no–churn and "1" indicates churn.

The four algorithms XGBoost, Random Forest Classifier (RFC), Logistic regression (LR), and Support Vector Classifier (SVC) are applied on the outputs of Select K-Best and PCA methods. All the four methods are also directly applied on normalized data. The results are shown in the tables below. The outputs on normalized data

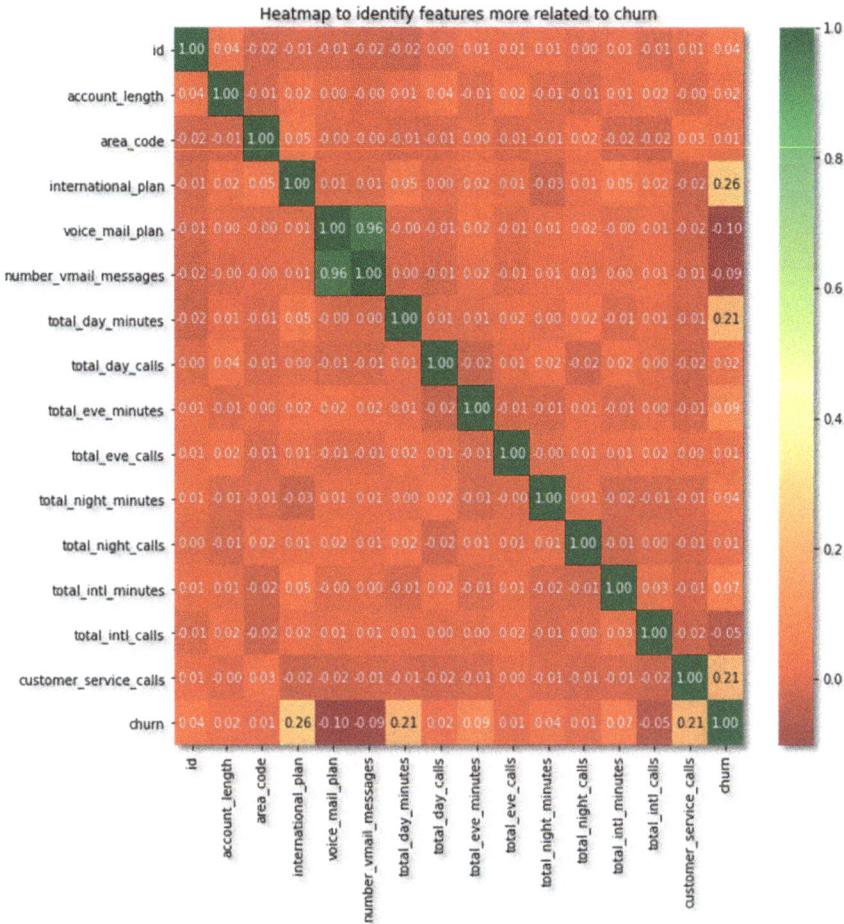

Fig. 2 Heatmap to measure correlation of features to target

are better than feature extraction or selection methods. However, in most of the real-life classification problems when the number of features is very large, it is important to consider only those features that are really important to reduce the training time, evaluation time, and overfitting.

Table 1 shows that XGBoost and SVC gave equal and best accuracy values on Select K-Best (SKB) and PCA outputs of 0.90 and 0.87, respectively. RFC and LR also gave equal accuracy values of 0.87 and 0.86 on Select K- Best and PCA outputs, respectively. Since accuracy values alone cannot define the performance of the algorithms, other evaluation metrics are also considered.

Table 2 shows that XGBoost gave best AUC scores of 0.861 and 0.773 on SKB and PCA outputs, respectively. SVC, RFC and LR gives better AUC scores in their

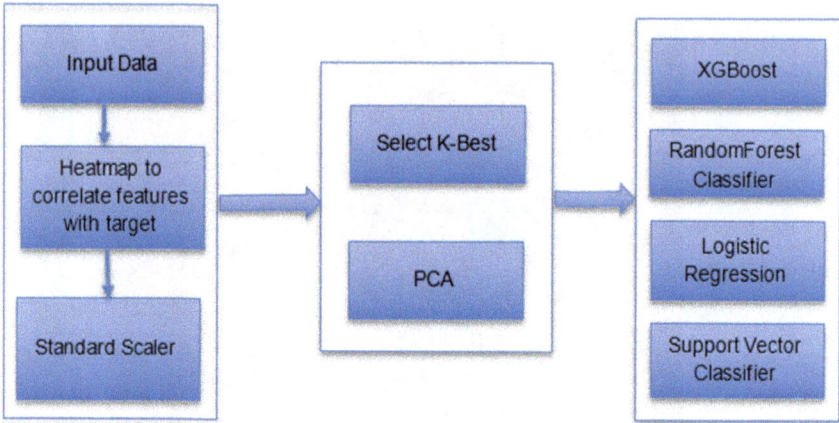

Fig. 3 Proposed model

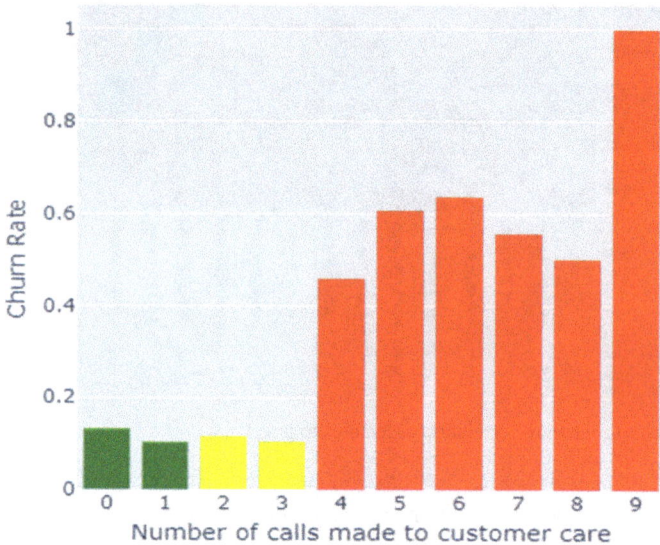

Fig. 4 Calls to customer care versus churn

order on SKB outputs whereas the order changes to LR, SVC, and RFC on PCA outputs.

Since an imbalanced class distribution existed in the data set and as false negatives are also crucial for telecom industry, F1 score was also considered as an evaluation metric. XGBoost, RFC, LR, and SVC are in the descending order of F1 scores for both SKB and PCA outputs (Table 3).

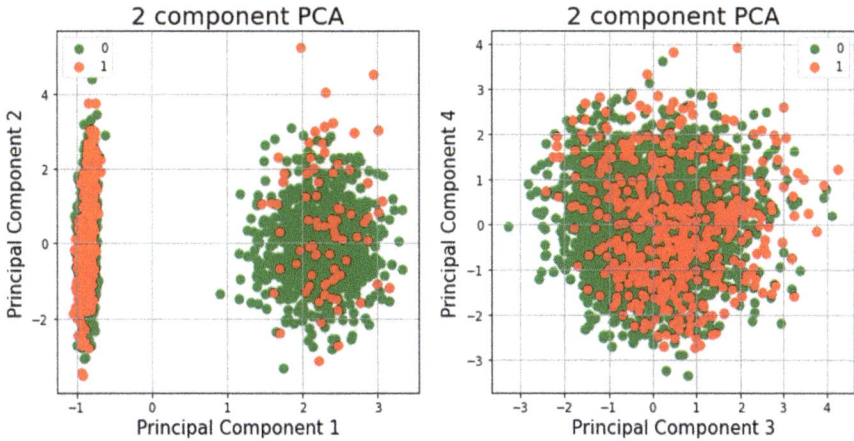

Fig. 5 Output of PCA on the telecom data

Table 1 Accuracy of different algorithms on the three data input datasets

Algorithm/data	Normalized data	Select K-Best	PCA
XGBoost	0.95	0.90	0.87
RFC	0.93	0.87	0.86
Logistic regression	0.87	0.87	0.86
SVC	0.91	0.90	0.87

Table 2 AUC score of different algorithms on the three data input datasets

Algorithm/data	Normalized data	Select K-Best	PCA
XGBoost	0.933	0.861	0.773
Random forest classifier	0.922	0.824	0.595
Logistic regression	0.820	0.797	0.695
SVC	0.918	0.842	0.667

Table 3 F1 score of different algorithms on the three data input datasets

Algorithm/data	Normalized data	Select K-best	PCA
XGBoost	0.826	0.597	0.333
Random forest classifier	0.730	0.547	0.325
Logistic regression	0.211	0.247	0.041
SVC	0.595	0.525	0.056

5 Conclusion

The study was conducted to help telecom companies to predict customer churn in advance and take necessary actions to make more profit. The prediction models need to achieve high AUC values. The sample data was divided to 70% for training and 30% for testing. Based on the applicability in prediction, four diverse classification algorithms are chosen. These are XGBoost, Random Forest Classifier, Logistic Regression, and Support vector classifier. XGBoost algorithm using Select K-Best gave better results than PCA in all AUC measurements, F1 scores as well as accuracy measurements using model score function. The accuracy of XGBoost on Select K-Best output was 0.90 were as on PCA output was only 0.87. The AUC score of XGBoost was 0.861 and 0.773 on Select K-Best and PCA outputs, respectively. For this particular scenario, feature selection method was proved to be better than feature extraction method. Further enhancements to this paper will focus on the combination of feature selection and feature extraction models to improve the accuracy of churn prediction that can help the telecom industries.

References

1. Ahmed AA, Maheswari D (2017) Churn prediction on huge telecom data using hybrid firefly based classification. Egypt Inform J 18(3):215–220
2. Amin A, Al-Obeidat F, Shah B, Adnan A, Loo J, Anwar S (2019) Customer churn prediction in telecommunication industry using data certainty. J Bus Res 94:290–301
3. Amin A, Shehzad S, Khan C, Ali I, Anwar S (2015) Churn prediction in telecommunication industry using rough set approach. New Trends Comput Collective Intell 83–95
4. De Bock KW, Van Den Poel D (2011) An empirical evaluation of rotation-based ensemble classifiers for customer churn prediction. Expert Syst Appl 38(10):12293–12301
5. De Caigny A, Coussement K, De Bock KW A new hybrid classification algorithm for customer churn prediction based on logis
6. Kurniawati YE, Permanasari AE, Fauziati S (2018) Adaptive synthetic-nominal (ADASYN-N) and adaptive synthetic-KNN (ADASYN-KNN) for multiclass imbalance learning on laboratory test data. In: 2018 4th International conference on science and technology (ICST). Yogyakarta, pp 1–6
7. Ullah I, Raza B, Malik AK, Imran M, Islam SU, Kim SW (2019) A churn prediction model using random forest: analysis of machine learning techniques for churn prediction and factor identification in telecom sector. IEEE Access 7:60134–60149
8. Ahmad ARK, Jafar A, Aljoumaa K (2019) Customer churn prediction in telecom using machine learning in big data platform. J Big Data. ISSN: 2196-1115
9. Nigam B, Dugar H, Niranjanamurthy M (2019) Effectual predicting telecom customer churn using deep neural network. Int J Eng Adv Technol (IJEAT), 8(5). ISSN: 2249-8958
10. Azeem M, Usman M (2018) A fuzzy based churn prediction and retention model for prepaid customers in telecom industry. Int J Comput Intell Syst 11(1):66–78
11. De Caigny A, Coussement K, De Bock KW (2018) A new hybrid classification algorithm for customer churn prediction based on logistic regression and decision trees. Eur J Oper Res 269(2):760–772
12. Hoppner S, Stripling E, Baesens B, Broucke SV, Verdonck T (2018) Profit driven decision trees for churn prediction. Eur J Oper Res

13. Umayaparvathi V, Iyakutti K (2016) A survey on customer churn prediction in Telecom insudtry: datasets, methods and metrics. Int Res J Eng Technol (IRJET) 3(4)
14. Saini N, Monika, Garg K (2017) Churn prediction in telecommunication industry using decision tree. Int J Eng Res Technol 6. https://doi.org/10.17577/IJERTV6IS040379
15. Raja B, Jeyakumar P (2019) An effective classifier for predicting churn in telecommunication. J Adv Res Dyn Control Syst 11:221–229

Chapter 41
A Hybrid Forecasting Model Based on Equilibrium Optimizer and Artificial Neural Network for Assessment of PM10 Concentration

Shalini Shekhawat, Akash Saxena, A. K. Dwivedi, and Rajesh Kumar

1 Introduction

Air pollution is a challenging and continuously growing problem in urban areas of developing and developed countries. Air pollution affects the environment in direct and indirect way, and the results of polluted air can be acute or chronical [1–5]. Densely populated and industrialized areas in metropolitan cities are on high risk. For the sake of future generations, control of air pollution is very essential. Also, some immediate actions must have also taken to face the worst period of air pollution in a year (i.e. smogy days in winter). These days are very critical for children, senior citizens and people suffering from any respiratory problem.

Air forecasting provides us the information about pollutant concentration, metrology and future prospective of different pollutants and help the decision maker to make new policies and accurate warning system [6, 7]. Air forecasting is a very vast and relevant subject; hence, a variety of methods have been applied to forecast different type of pollutants. These methods can be categorized in three types: first one is statistical models, which are commonly used due to their simple structure and easy to understand. Box model [8] and regression models [9] are some basic model under this category. As these models are based on linear relations, the accuracy

S. Shekhawat (✉) · A. Saxena
Swami Keshvanand Institute of Technology Management and Gramothan, Jaipur, India
e-mail: shekhawatshalini17@gmail.com

A. Saxena
e-mail: Aakash.saxena@hotmail.com

A. K. Dwivedi
Rajasthan Technical University, Kota, India
e-mail: akdrtu@gmail.com

R. Kumar
Malviya National Institute of Technology, Jaipur, India
e-mail: rkumar.ee@mnit.ac.in

© The Author(s), under exclusive license to Springer Nature Singapore Pte Ltd. 2021 509
X.-Z. Gao et al. (eds.), *Applications of Artificial Intelligence in Engineering*, Algorithms for Intelligent Systems, https://doi.org/10.1007/978-981-33-4604-8_41

of the forecast is not guaranteed. These models are replaced by non-linear models like artificial neural network [10] and least square support vector machine [11] and others. With less restrictions on input data better results obtained using NN models [12–14]. Second are the methods based on chemical properties of pollutants named as chemical transport methods, but their performance is not very good. The third category is models based on hybrid concepts in which two different methods are combined to get more accurate results. Some commonly used techniques in hybrid models are extreme learning machines (ELM), wavelet transforms-based neutral network, different data decomposition techniques (like empirical mode decomposition, variational mode decomposition, (VMD), complete ensemble empirical mode decomposition (CEEMD), optimization algorithms (like Harris hawks optimization, Grey Wolf Optimizer, Particle Swarm Optimization, and many more) [15, 16]. These hybrid models produce more accurate results and provide help for decision makers. However, these models also follow the drawback of the adopted method or algorithm and parameters. Hence, need of new forecasting models providing accurate results always exists.

1.1 Literature Review

In India, central pollution control board (CPCB) operates various air quality monitoring stations to measure various parameter concentrations including SO_2, NO_2, SPM, O_3, NH_3, and other toxic particles. Suspended particulate matter (PM) is one of the major pollutants based on guidelines of World health organization [17]. The mixture of solid dust particles emerged from industrial waste and fuel combustion and small liquid droplets present in the air are called suspended particulate matter. These are further classified in PM2.5 and PM10. In deeper sense, particles of more than 10^{-3} μm and less than 100 μm are known as PM2.5 whether particles having size smaller than 10 μm are called PM10 [18]. The long-term presence of these SPM's can lead to different respiratory and cardiovascular disease. By taking motivation, from the literature, we propose a hybrid architecture based on recently published metaheuristic algorithm EO and FFNN for predicting PM10 concentration of the Jaipur, Rajasthan. Following objectives are framed for this study:

1. To construct a time series based on rolling horizon and historical values of PM10 in Jaipur city. Further, with the help of this historical data construct a supervised architecture that comprises of FFNN.
2. To employ recently published EO for optimizing the weight of FFNN.
3. To derive comparative analysis of different forecasting engines tuned by different algorithm on the basis different error indices.

Rest part of the paper is organized as follows, in subsequent sections, metrological details of the city, details of employed algorithm EO, construction of forecasting engine, simulation results, and conclusions are presented.

2 Metrological Description of Study Area

For this study, we choose data of PM 10 concentration of Jaipur city. Jaipur popularly known as pink city is the capital of Rajasthan and one of the growing metropolitan cities of India. With the continuously increasing population and industrialization, Jaipur is also facing poor air qualities. It is situated in the eastern corner of Thar dessert at an altitude of 438 of sea level covering almost 470 km² area (Latitude 26.88°N, Longitude 75.83°E). CPCB has four air quality monitoring stations in Jaipur at Adarsh Nagar, Shastri Nagar, Police Commissionerate, and VK Industrial area. The dataset we have used here is taken from police Commissionerate AQI station. This station is situated in densely populated residential area and suffers from heavy traffic issue during day period.

In this paper, we propose a hybrid model of neural network and nature inspired algorithm named as EO-FFNN.

3 Equilibrium Optimizer

This algorithm is a recently given by Faramarzi et al. [19]. The inspiration source of this algorithm is the mass balance equation of physics, which states that the leaving and entering mass in a controlled volume is conserved. This is demonstrated through an ordinary differential equation of first order given by:

$$V\frac{dC}{dt} = RC_{eq} - RC + g \tag{1}$$

where

C—concentration inside controlled volume V.

R—flow rate into/out of the control volume.

$V\frac{dC}{dt}$—rate of change in mass in the control volume.

C—concentration in the control volume V.

C_{eq}—equilibrium state concentration.

g—generation rate of mass inside the control volume.

Equilibrium stage of control volume is a stage where generation is zero $\left(V\frac{dc}{dt} = 0\right)$. To solve above equation, we write it in the following way:

$$\frac{dc}{dt} = \frac{R}{V}C_{eq} - \frac{R}{V}C + \frac{g}{V} \tag{2}$$

Here $\alpha = \frac{R}{V}$ is known as turnover rate. Further it can be rearranged as:

$$\frac{dc}{\alpha C_{eq} - \alpha C + \frac{g}{V}} = dt \tag{3}$$

On integration between C_0 to C and t_0 to t it results in:

$$C = C_{eq} + (C_0 - C_{eq}) D + \frac{g(1 - D)}{\alpha V} \tag{4}$$

where $D = e^{[-\alpha(t - t_0)]}$, C_0 and t_0 are the initial concentration and start time, respectively, and both of these depend on integration interval. One can calculate the concentration if turnover rate α is known or can find the mean turnover rate if generation rate is known.

Above equations are the main inspiration and centre of the design of EO. The search particle is basically the solution, and position of the particle is represented by concentration. In EO, any particle updates its position in three steps, which can be expressed by three terms of Eq. (4). The first term randomly selected from equilibrium pool is also known as equilibrium concentration. It is considered as the best position till so far. The second term is a type of search mechanism represented by the difference between concentration of a particle at initial time and at equilibrium. This term can be considered as exploration stage where the particle searches the region globally. The third term is exploration phase which represents by generation rate.

1. Initialization: Alike most of the meta heuristic algorithms the optimization process in EO also starts with the initial population or initial concentration. A uniform and random population is constructed according to number of dimensions and particles following the below equation

$$C_{Initial}^i = C_{Min} + \beta(C_{Max} - C_{Min}), \quad i = 1, 2, ..., n \tag{5}$$

$C_{Initial}^i$ is the concentration of ith particle initially, and C_{min} are the concentrations at the maximum and minimum dimensions, $\beta \in (0, 1)$ is a randomly selected vector, n is total number of particles.

2. Selection of candidates and equilibrium state: Initially, the equilibrium which also represents as global optimum is unknown. The particles search towards their global positions through equilibrium candidates. The equilibrium candidates are taken as $\vec{C}_{eq(a)}$, $\vec{C}_{eq(b)}$, $\vec{C}_{eq(c)}$, $\vec{C}_{eq(d)}$, $\vec{C}_{eq(aveg.)}$. First four are the best identified particles under sufficient experiments and various other conditions. $\vec{C}_{eq(aveg.)}$ is the arithmetic mean of other four particles. These five equilibrium candidates formed a new vector known as equilibrium pool given by:

$$\vec{C}_{eq(p)} = \left[\vec{C}_{eq(a)}, \vec{C}_{eq(b)}, \vec{C}_{eq(c)}, \vec{C}_{eq(d)}, \vec{C}_{eq(aveg.)} \right] \tag{6}$$

The first four are responsible for exploration stage while the fifth one represents the exploitation stage. The number of particles (i.e. 5) is arbitrarily chosen but less number of particles does not guarantee good performance at multi model functions and composite functions. In each iteration the particle updates its concentration based on any one of these five equilibrium candidates and till the final global concentration each candidate is used to update the same number of particles.

3. Balancing term D: In EO, the exploration term D provides a balance between and exploitation phase. In real life problems, the turnover rate may vary with respect to time t so α is chosen between the interval [0, 1].

$$D = e^{[-\alpha(t-t_0)]} \tag{7}$$

$$t = \left(1 - \frac{\text{Iteration}}{\text{Iteration}_{\max}}\right)^{\left(h_2 \frac{\text{Iteration}}{\text{Iteration}_{\max}}\right)} \tag{8}$$

$$t_0 = \frac{1}{\alpha} \ln\left(-h_1 \, \text{Sign}(\vec{r} - 0.5)\left[1 - e^{-\alpha t}\right]\right) + t \tag{9}$$

where h_1 and h_2 are constant values and control the exploration and exploitation capacity, respectively, in EO. Iteration and Iteration$_{\max}$ are, respectively, the current number of iteration and maximum number of iterations. $\vec{r}$ is a random vector of interval [0,1] which affects the directions of exploration and exploitation stages. In this paper, we have taken $h_1 = 2$ and $h_2 = 1$ on the basis of testing on a set of test functions. These can be taken differently with other problems. After substituting values from Eqs. (8) and (9) and on further simplification, Eq. (7) can be represented as

$$D = h_1 \, \text{Sign}(\vec{r} - 0.5)\left[e^{-\alpha t} - 1\right] \tag{10}$$

4. Generation Rate (g): In EO, this term is very important as it is the control factor of exploitation stage. In various similar model's generation rate can be expressed as decreasing exponential function of time.

$$g = g_0 e^{[-k(t-t_0)]} \tag{11}$$

where B_0 is initial value and k is a constant which represents the decaying process. To limit the number of random variables and making the search more systematic, we have taken $k = \alpha$ here, so the search pattern is controlled and systematic.

$$g = \vec{g_0} e^{[-k(t-t_0)]} = \vec{g_0} \vec{D} \tag{12}$$

Here

$$\vec{g_0} = \vec{gcp}\left(\vec{C}_{eq} - \alpha \vec{C}\right) \tag{13}$$

$$\overrightarrow{gcp} = \begin{cases} 0.5\, r_1 & r_2 \geq gp \\ 0 & r_2 < gp \end{cases} \tag{14}$$

where $\overrightarrow{gcp}$ is a vector formed by the repetition of same values. This vector known as generation rate control parameter. gp is called generation probability which tells us about the number of particles updated with the help of generation rate. Every particle follows Eq. (14) and if gcp $= 0$ then using Eqs. (12) and (13), B is also zero, which means for that particular particle, positions are updated without any generation rate term. When gp $= 0.5$, the two phases exploration and exploitation set at balance. Hence, the final updating position is given by

$$\overrightarrow{C} = \overrightarrow{C}_{eq} + \left(C_0 - \overrightarrow{C}_{eq}\right)\overrightarrow{D} + \frac{g\left(1 - \overrightarrow{D}\right)}{\alpha\,\overrightarrow{V}} \tag{15}$$

The term $\overrightarrow{D}$ is same as in Eq. (10). Here, the first term tells us the concentration at equilibrium. In term second, we actually have the difference between concentration of any particle and equilibrium stage. Due to these concentration differences, this term helps in searching the global optima, and thus, it can be considered as exploration term. On the contrary, the third term of Eq. (15) depends on variations in $\left(1 - \overrightarrow{D}\right)$ which are comparatively small and thus it supports in exploitation and also helpful in achieving solution accuracy. According to the parameters available in both of these terms, two cases can arise: first if all parameters are of same sign, then the variations in concentration are large and helps in making the search area big, while in second case the signs of parameters are opposite to each other making the variations comparatively small which supports local search.

5. Memory saving ability: Like the best particle location saving concept of PSO, in EO also each particle follows a memory saving procedure to maintain the track position of its coordinates, which is shown by its fitness value. This procedure enhances the exploitation abilities of EO but with this quality the chance to be trapped in local minima also increases.

6. Exploration and Exploitation ability: There are some parameters and factors which controls or enhance these abilities of EO, some of them are:

   • h_1: This parameter helps in determining the new equilibrium particle location. It is recommended through experiments that higher values of h_1 improves the exploration but to avoid the final solution close to the boundary region its value should be less than 3.
   • h_2: Parameter h_2 is responsible for exploitation phase of EO, It helps in finding the optimum solution.
   • $(r - 0.5)$: This factor combinedly control both the phase as its sign can be either negative or positive.
   • Generation Probability gp: As defined earlier, gp provide the number of total updated particles through generation rate. If gp $= 0$, then generation rate is high which can lead the solution to be trapped in local minimum solution. On

the other side value of gp $= 1$ indicates that generate rate is not present in the process and hence the exploration of particles becomes scattered which can further generate no solution cases. If gp $= 0.5$, then both stages exploration and exploitation are balanced and provide a better solution.

- The equilibrium pool $\overrightarrow{C}_{eq(p)}$: As the pool consist of five particles, which are at a far distance from each other initially and updating their positions gradually. This process showcase that a larger area is explored through the particles. After some iterations, when they come nearer the search region becomes smaller and it leads to a good exploitation stage.

4   Construction of the Forecasting Engine

For constructing supervised architecture, we consider feed forward neural net topology for predicting the concentration of PM10. Feed forward neural network is a network that consists of three layers, input layer, output layer and hidden layers. These layers are connected with the help of weights. The neural network finds weights and biases by minimizing the following cost function:

$$
J = \min_{w_i} \left(\sum_{i=1}^{N} x_i w_i - y_t \right)^2
$$

where y_t is the required known output at instant t, w_i is the weights, and x_i is input.

For optimal tuning of weights, we employ different algorithms in order to minimize the error between predicted and actual values of PM10 concentration. For constructing the forecasting engine, we consider values of last four days as input to the forecasting engine and present-day value as the target value for the engine. Further, we choose rolling horizon of 4 days in this forecast. A rolling forecast is an add/drop process for predicting the future over a set period of time. In this work, we chose a rolling window of four days for predicting the PM10 concentration (Fig. 1).

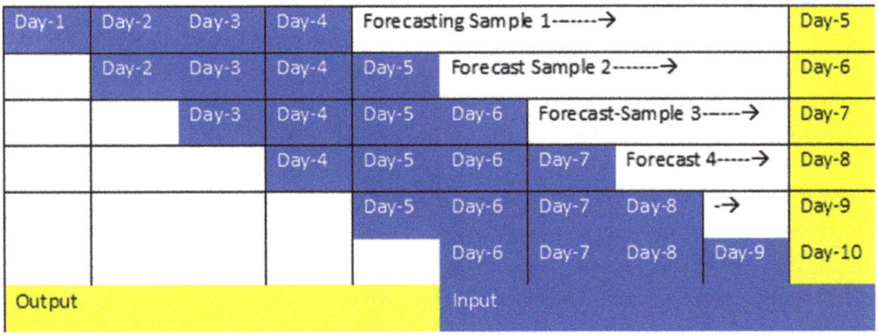

Day-1	Day-2	Day-3	Day-4	Forecasting Sample 1-------→					Day-5
	Day-2	Day-3	Day-4	Day-5	Forecast Sample 2-------→				Day-6
		Day-3	Day-4	Day-5	Day-6	Forecast-Sample 3-----→			Day-7
			Day-4	Day-5	Day-6	Day-7	Forecast 4-----→		Day-8
				Day-5	Day-6	Day-7	Day-8	-→	Day-9
					Day-6	Day-7	Day-8	Day-9	Day-10
Output				Input					

Fig. 1   Construction of forecasting engine

Algorithm 1: Equilibrium Optimizer (EO)

1: Begin

2: Initialize Population of Particles i = 1,.....,n

3: Chose a number for fitness of equilibrium candidates

4: Choose $h_1 = 2; h_2 = 1; C_1P = 0.5;$

5: While Iteration ; Iteration (Max)

6: for i=1; no. of particles (n) calculate fitness of ith particle

7: if fit $(\vec{C}_i) < fit(\vec{C}_{eq(a)})$

8: $\rightarrow exchange\ (\vec{C}_{eq(a)})\ with\ (\vec{C}_i)\ and\ Fit\ (\vec{C}_{eq(a)})\ with\ fit\ (\vec{C}_i)$

9: else if fit $(\vec{C}_i) > fit\ (\vec{C}_{eq(a)})\ \&\ fit\ (\vec{C}_i) < fit(\vec{C}_{eq(b)})$

10: $\rightarrow exchange\ (\vec{C}_{eq(b)})\ with\ (\vec{C}_i)\ and\ Fit\ (\vec{C}_{eq(b)})\ with\ fit\ (\vec{C}_i)$

11: else if fit $(\vec{C}_i) > fit\ (\vec{C}_{eq(a)})\ \&\ fit\ (\vec{C}_i) > fit(\vec{C}_{eq(b)})\ \&\ fit\ (\vec{C}_i) < fit(\vec{C}_{eq(c)})$

12: $\rightarrow exchange\ (\vec{C}_{eq(c)})\ with\ (\vec{C}_i)\ and\ Fit\ (\vec{C}_{eq(c)})\ with\ fit\ (\vec{C}_i)$

13: else if fit $(\vec{C}_i) > fit\ (\vec{C}_{eq(a)})\ \&\ fit\ (\vec{C}_i) > fit(\vec{C}_{eq(b)})\ \&\ fit\ (\vec{C}_i) > fit(\vec{C}_{eq(c)})\ \&\ fit\ (\vec{C}_i) > fit(\vec{C}_{eq(c)})\ \&\ fit\ (\vec{C}_i) < fit(\vec{C}_{eq(d)})$

14: $\rightarrow exchange\ (\vec{C}_{eq(d)})\ with\ (\vec{C}_i)\ and\ Fit\ (\vec{C}_{eq(d)})\ with\ fit\ (\vec{C}_i)$

15: End if

16: End for

17: $\vec{C}_{avg} = \frac{(\vec{C}_{eq(a)} + \vec{C}_{eq(a)} + \vec{C}_{eq(c)} + \vec{C}_{eq(d)})}{4};$

18: for m a equilibrium pool

19: $\vec{C}_{eq(p)} = (\vec{C}_{eq(a)}, \vec{C}_{eq(a)}, \vec{C}_{eq(c)}, \vec{C}_{eq(d)}, \vec{C}_{eq(avg)})$

20: Save the memory (if Iteration > 1)

21: $t = (1 - \frac{iteration}{iteration_{max}})(h_2 \frac{iteration}{iteration_{max}})$

22: For i=1: no of particles (n)

23: Select a candidate from the equilibrium pool vector

24: Generate random value of $\vec{\alpha}\ \vec{\gamma}$

25: for m D=$h_1\ Sign(\vec{\gamma} - 0.5)[e^{-\alpha t} - 1]$

26: Set $\vec{g}_{eq} = \begin{cases} 0.5r_1 & r_2 \geq GP \\ 0 & r_2 < GP \end{cases}$

27: Set $\vec{g}_0 = \vec{g}_{ep}(\vec{C}_{eq} - \vec{\alpha}\ \vec{\gamma})$

28: $g = \vec{g}_0, \vec{D}$

29: Update Concentration $\vec{C} = \vec{C}_{eq} + (\vec{C} - \vec{C}_{eq}), \vec{D} + g'(1 - \vec{D})\vec{\alpha}\ \vec{\gamma}$

30: END (FOR)

31: Iteration = Iteration +1

32: End While

---

5 Simulation Results

In this section, results obtained from EO-FFNN are presented. For making a fair comparison population size, search agent nos. and maximum iteration count have been fixed same. For analysis, we have taken latest set of algorithms and those are Grey Wolf Optimizer (GWO), Sine Cosine Algorithm (SCA), Harris Hawks Optimization (HHO) and EO (Equilibrium optimizer).

From Fig. 2, it can be concluded that the EO shows better convergence properties as compared to other algorithms. We observe a steep downfall in objective function value in the case of EO is optimizing ANN. However, performance of SCA is not as compared to other algorithms. On x axis, we have shown iteration count, and on y axis, logarithmic values of objective function (Mean Square Error) are reported.

Further, to judge the efficacy of the EO for potential solution provider for tuning weights of the ANN, we plotted the regression curve, and we can see that regression

Fig. 2 Convergence property analysis of different algorithms with equilibrium optimizer

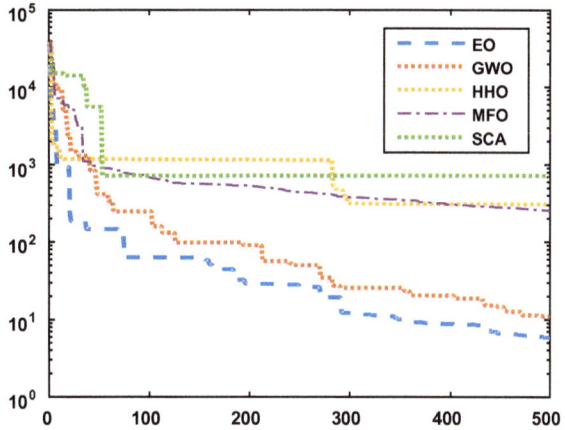

coefficient's value is high that indicates a good agreement in input and output data. Regression analysis is shown in Fig. 3.

As discussed earlier, the network is constructed through the series values of day-3, day-2 and day-1 and current day reading has been considered as the target value for the PM10 concentration. To test the efficacy of different ANNs, unknown sample concentrations of PM10 of five days have been chosen. These data are shown in Fig. 4.

After simulating ANNs tuned by EO, GWO, HHO, MFO and SCA, the prediction results are depicted in Table 1.

Following comments can be made with this prediction:

1. Prediction results of different ANNs are shown in table. From Table 1, it can be observed that the values obtained from EO tuned ANN are nearer to the original

Fig. 3 Regression results of EO

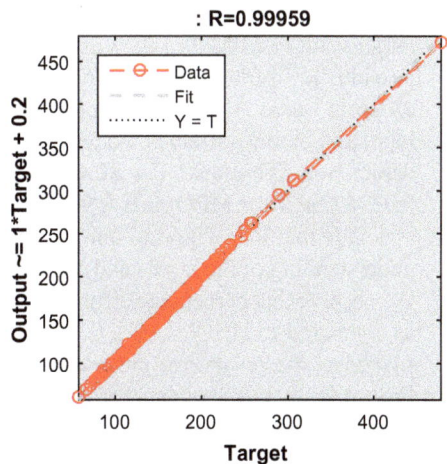

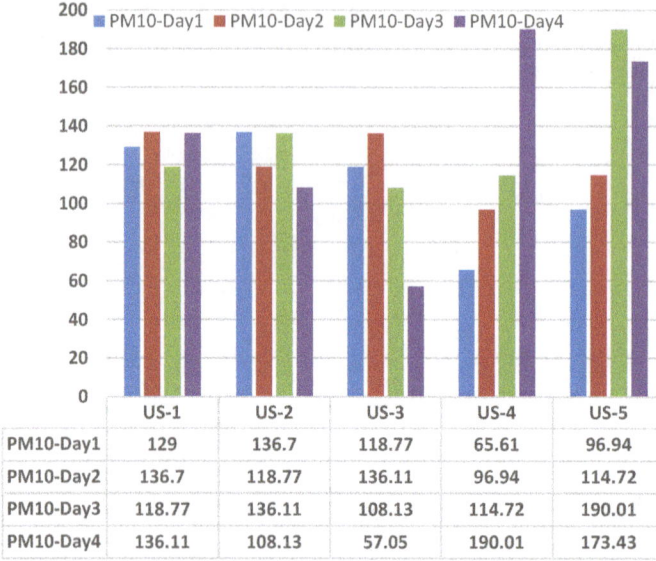

	US-1	US-2	US-3	US-4	US-5
PM10-Day1	129	136.7	118.77	65.61	96.94
PM10-Day2	136.7	118.77	136.11	96.94	114.72
PM10-Day3	118.77	136.11	108.13	114.72	190.01
PM10-Day4	136.11	108.13	57.05	190.01	173.43

Fig. 4 Data set of PM10 concentration for validation

Table 1 Results for unknown data samples to the simulated networks

Sample	Target	EO-NN	HHO-NN	GWO-NN	MFO-NN	SCA-NN
US-1	108.13	**106.8023**	105.1061	105.8523	104.8023	105.1909
US-2	57.05	**58.37613**	58.77703	55.85367	54.37613	53.01523
US-3	65.61	**68.0696**	59.93571	69.08563	72.0696	77.24561
US-4	173.43	**171.9831**	177.5766	170.0635	174.9831	185.3799
US-5	158.59	**155.1048**	153.0244	157.8559	159.1048	161.8458

target values of PM10. Likewise, it can also be concluded that SCS-NN shows pessimistic results as compared with other optimizers. Obtained R-values for different tuned ANNs indicate that EO exhibits better understanding between input and output variables. For EO, value of regression coefficient is 0.99959 as depicted in figure also. For SCA, it is 0.84, for GWO it is 0.98, for HHO it is 0.93243, and for MFO, it is 0.94663. From these observations, one can easily conclude that R is nearer to one for EO optimized network that indicates better understanding of the inputs and outputs. Optimal values are shown in bold faces.

2. We observe that percentage error obtained in sample-1 is optimal for EO (1.23%) as compared to HHO (2.8%), GWO (2.11%), MFO (3.08%) and SCA (2.72%). Likewise, the results for mean absolute percentage error is depicted in figure. From Fig. 5, it is evident that accuracy of the EO tuned ANN is far better as compared with other algorithms.

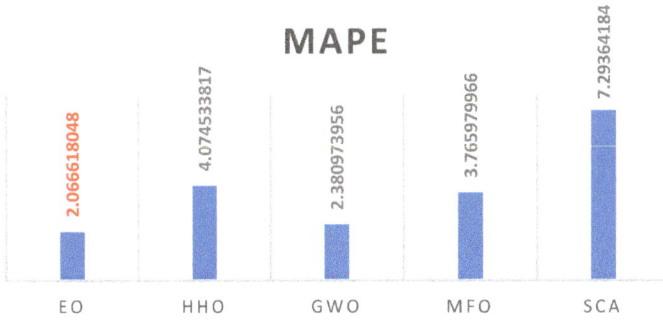

Fig. 5 Error analysis (MAPE)

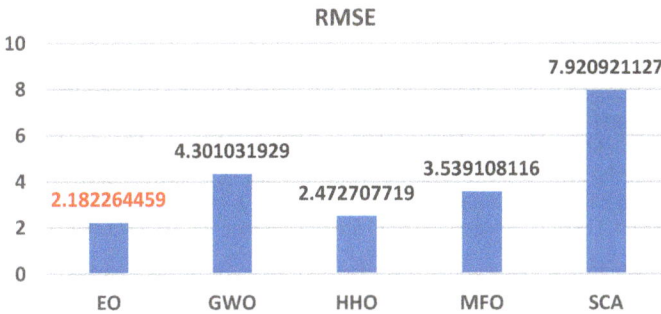

Fig. 6 Error analysis (RMSE)

3. As per the values of root mean squared error (RMSE) values, we observe that the values for EO is 2.182264 for HHO is 4.301032 for GWO is 2.472708, MFO 3.539108 and for SCA it is 7.920921 from these values it can be easily concluded, the values RMSE is optimal for EO. This result validates efficacy of EO. These values are depicted through Fig. 6.

6   Conclusion

The paper presents an application of newly developed algorithm EO for optimizing the neural networks microstructure by tuning the weights for obtaining higher accuracy in the predictions. For a case study, we choose the prediction of PM10 concentrations in Jaipur city. For analysis, we construct a time series-based inputs that are constructed by the data of PM10 of previous days and current day concentration is considered as the target values for the networks. A fair comparison is conducted between recently published algorithms that already have proven their optimization efficacy in different fields. We observe that for prediction of PM10 model.

Acknowledgements The Authors are thankful for the fully financial support from the CRS, RTU (ATU), TEQIP-III of Rajasthan Technical University, Kota, Rajasthan, India. (Project Sanction No. TEQIP-III/RTU (ATU)/CRS/2019-20/51).

References

1. Kolehmainen M, Martikainen H (2001) Ruuskanen J Neural networks and periodic components used in air quality forecasting. Atmos Environ 35(5):815–825
2. Wong C-M et al (2002) A tale of two cities: effects of air pollution on hospital admissions in Hong Kong and London compared. Environ Health Perspect 110(1):67–77
3. Diaz J et al (2004) Impact of temperature and air pollution on the mortality of children in Madrid. J Occup Environ Med 46(8):768–774
4. Afroz R, Hassan MN, Ibrahim NA (2003) Review of air pollution and health impacts in Malaysia. Environ Res 92(2):71–77
5. Katsouyanni K et al (1993) Evidence for interaction between air pollution and high temperature in the causation of excess mortality. Arch Environ Health: Int J 48(4):235–242
6. Kolehmainen M, Martikainen H, Ruuskanen J (2001) Neural networks and periodic components used in air quality forecasting. Atmos Environ 35(5):815–825
7. Hildén M, Marx A (2013) Evaluation of climate change state, impact and vulnerability indicators. in ETC CCA Technical Paper
8. Middleton DR (1998) A new box model to forecast urban air quality: BOXURB. Environ Monit Assess 52(1–2):315–335
9. Shi JP, Harrison RM (1997) Regression modelling of hourly NOx and NO2 concentrations in urban air in London. Atmos Environ 31(24):4081–4094
10. Bai Y et al (2019) An ensemble long short-term memory neural network for hourly PM2.5 concentration forecasting. Chemosphere 222:286–294
11. Hao Y, Tian C (2019) The study and application of a novel hybrid system for air quality early-warning. Appl Soft Comput 74:729–746
12. Agirre-Basurko E, Ibarra-Berastegi G, Madariaga I (2006) Regression and multilayer perceptron-based models to forecast hourly O3 and NO2 levels in the Bilbao area. Environ Model Softw 21(4):430–446
13. Kukkonen J et al (2003) Extensive evaluation of neural network models for the prediction of NO2 and PM10 concentrations, compared with a deterministic modelling system and measurements in central Helsinki. Atmos Environ 37(32):4539–4550
14. Russo A et al (2014) NO2, PM10 and O3 urban concentrations and its association with circulation weather types in Portugal. Atmos Environ 89:768–785
15. Wang D et al (2017) A novel hybrid model for air quality index forecasting based on two-phase decomposition technique and modified extreme learning machine. Sci Total Environ 580:719–733
16. Zhou Q et al (2014) A hybrid model for PM2.5 forecasting based on ensemble empirical mode decomposition and a general regression neural network. Sci Total Environ 496:264–274
17. World Health Organization (2006) Air quality guidelines: global update 2005: particulate matter, ozone, nitrogen dioxide, and sulfur dioxide. World Health Organization
18. Mohanraj R, Azeez PA (2004) Health effects of airborne particulate matter and the Indian scenario. Curr Sci 87(6):741–748
19. Faramarzi A et al (2020) Equilibrium optimizer: A novel optimization algorithm. Knowl-Based Syst 191:105190

Chapter 42
An Intelligent Path Estimation Technique for Transportation of Heterogeneous Droplets in Digital Micro Fluidic Biochips (DMFB)

Rupam Bhattacharya, Abhijit Sarkar, and Pranab Roy

1 Introduction

Droplet transportation in DMFB incorporating with advantageous features becomes more challenging in order to ensure the correctness of results for the bioassay. In DMFB architecture, droplet is transported from source electrode to target electrode through pre-estimated routes comprising of electrode configured cells. Droplets run in parallel can be heterogeneous. Heterogeneous droplet implies each net is constituted either for different bioassay execution or obtained from different subjects for execution of same bioassay schedule.

For example, within the same 2D DMFB layout multiple bioassays namely for glucose, urea and lactate detection are being executed. On the other hand, samples from different persons for execution of bioassay towards detection of blood glucose level within the same 2D layout have been accomplished simultaneously. Parallel execution of such variable types of bioassay refers to heterogeneous droplet routing procedures.

R. Bhattacharya (✉) · A. Sarkar
Institute of Engineering and Management, Salt Lake, Kolkata, India
e-mail: th_rup@yahoo.co.in

A. Sarkar
e-mail: abhijit.sarkar@iemcal.com

P. Roy
India Institute of Engineering Science and Technology, Shibpur, Howrah, India
e-mail: ronmarene14@gmail.com

1.1 Cross Contamination Issues

During droplet routing, there remains a possibility of unexpected mixing among droplets caused by crossover of route paths of two different nets. Occurrence of such phenomenon is termed as cross-contamination. Functional droplets often leave residue along configured electrode cells on their paths following which if other nets share any of these earlier cells along its path—cross-contamination occurs. Hence, during parallel routing of multiple droplets, arrival at the desirable location with intelligent avoidance of contamination, poses a major challenge.

The cross-contamination avoidance is made possible by avoiding contaminations altogether during droplet path estimation. Thereby, routing paths of different nets need ideally to be disjointing from each other in order to avoid cross-contamination. However, if no possible disjoint routing paths are available for certain net pairs— wash droplets are to be introduced to wipe out these residual contamination possibilities. Routing of functional droplets while taking an account of wash droplet routing, certainly brings about an additional overhead. Accordingly, an efficient and feasible routing scheme for DMFB architecture is required for optimal estimation of droplet routing path with minimum possible overlaps and minimal washing operations. Figure 1 shows droplet transportation with cross unavoidable cross contamination.

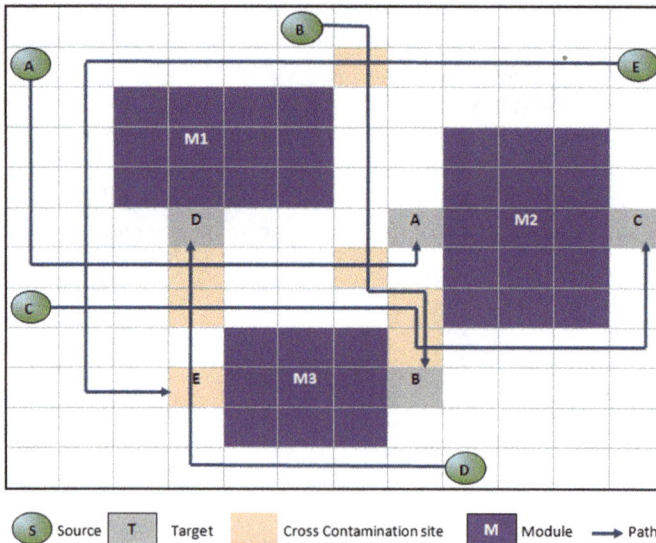

Fig. 1 Heterogeneous route path with unwanted crossovers

2 Related Work

Different researchers throughout the world are carrying out the research on routing of droplets on digital microfluidic biochip. Here we are mentioning some of the important work previously done by the researchers.

An algorithm for operation-variation-aware placement by utilising the real-time detection has been formulated in [1]. Experimental results on standard benchmarks show encouraging results. A deadlock and livelock avoidance routing technique has been proposed in [2]. An efficient wash droplet routing methodology has also been proposed. The proposed work shows encouraging results, particularly for resource-constrained chips. For the enhancement of routability in DMFB, a multi-level hierarchical method that works on reshaping and splitting of droplets has been proposed in [3]. The obtained results were compared with the recent similar works and significant reduction in latest arrival time and average arrival time have been observed. A heuristic for parallel droplet routing has been proposed in [4]. At first phase, interfering index of each net is calculated, then on the basis of that, parallel routing of nets is executed. A multi-level hierarchical approach was presented in [5]. Proposed technique is capable of taking proper decision regarding the droplet reshaping and splitting. This technique shows that this method, on an average, can reduce latest arrival time of around 18% and average arrival time of around 7%. An algorithm has been developed in [6] for efficient routing of heterogeneous droplets in digital microfluidic biochips. A complete routing simulator using JAVA has been developed. The method presented in [7] attempts to detect any possible error in routing and then tries to eliminate or minimise the problem. A heuristic for reducing the routing time and number of electrodes used in routing has been proposed in [8]. Some standard benchmarks have been used to test the proposed algorithm, and encouraging outcome has been observed. A routing scheme has been formulated in [9] for reducing multiple, redundant use of same resource to eliminate error or defect probabilities. Results obtain from experimental shows significant improvement over some existing approach. A technique of routing of test droplets has been proposed in [10]. The proposed technique activates the control pins of electrode based on the given methods of allocations, which allow test droplets to move within a digital microfluidic biochip. In [11], combined routing approach has been introduced where both homogeneous and heterogeneous droplet can be routed in parallel. In [12], authors proposed a homogeneous droplet routing technique that tries to maximise contaminations to improve cell utilisation. In [13], authors designed a complete routing simulator for graphically display and estimate their routing techniques.

3   Motivation and Objectives

As stated earlier, cross contamination is a major issue for heterogeneous droplet transportation. In heterogeneous droplet transportation, the major focus is to determine disjoint route paths to avoid unwanted contamination of samples or reagents due to crossovers. Such contamination may result in inaccurate analysis of the output—causing major error in the targeted diagnosis.

Hence, it becomes necessary to devise techniques that allow to estimate route paths for each net being disjoint to each other. This implies occurrence of no or minimal contamination.

4   Proposed Technique

4.1   Route Zone Based Path Exploration Technique

In a given 2D layout of DMFB the source, target and mixer locations of any net are pre-specified. The route zone for any 2 pin net is defined as the rectangle formed using all the cells with the respective source and target being its two diagonal points (Fig. 2). The significance of the routing zone is that it contains any stipulated Manhattan path for the given net. Any path explored out of this routing zone implies deviation of the specified net. This may in turn result in increase in number of cell utilisation as well as the arrival time for the net under consideration.

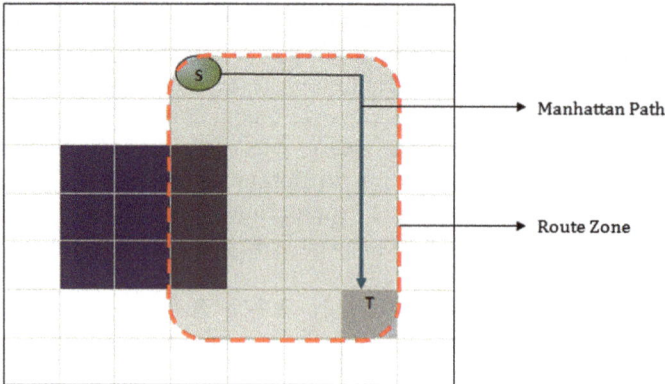

Fig. 2 Route zone for a given net with Manhattan path

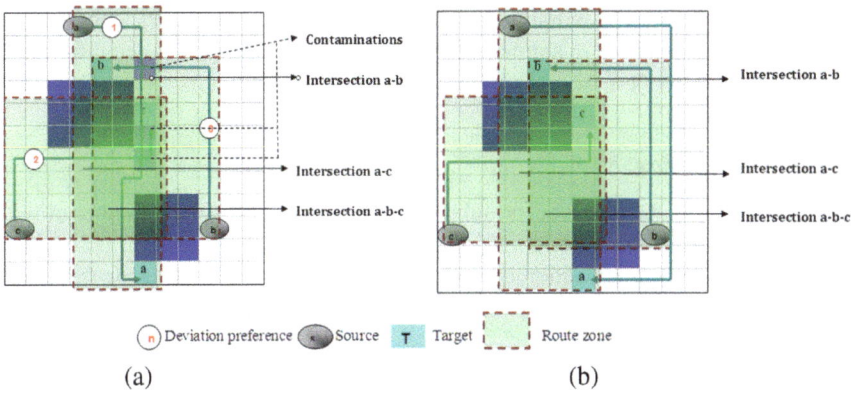

Fig. 3 **a** Nets with deviation preference. **b** Deviated path of net a with no contamination

4.1.1   Deviation Preference Based on Route Zones

Route path estimation process is a complex procedure where disjoint paths for multiple nets for a pre-specified layout are to be estimated. Initially, all the respective Manhattan paths are projected for each net. In the next step, necessary deviations to avoid hard blockages are estimated. Finally, the crossover sites are identified where same cell is shared by multiple nets. In such cases, paths for certain number of nets involved in crossover are to be eventually deviated.

In such cases, out of two or more nets, it becomes necessary to determine deviation preferences for one or more nets. Such nets once deviated may result in formation of disjoint paths—that in turn reduce or eliminate crossovers.

Such deviation preference for any nets can be determined using a route zone-based technique. In the proposed technique, route zone for each net is plotted. Intersection of the route zones are estimated. Such intersection with its cell coverage may provide an initial estimate of possibilities of cross over. So, any such route zone intersection will allow only one of the intersecting net to pass through. The rest needs to be deviated. In our proposed technique, we attempt to identify which of the nets have minimum intersecting area within its route zone. If this net involves in larger number of intersection with its route zone—this net becomes possible candidate for route path deviation. The proposed strategy for determination of deviation preference is stated in the next section (Fig. 3).

4.1.2   Route Path Estimation

Problem Formulation

We consider a 2D square layout of a DMFB with dimension $n \times n$. Number of preplaced two pin nets $= x$ and three pin nets $= y$. These are resolved into total number of two pin nets as $m = x + 3y$.

This in turn constitutes m number of routing zones (one for each two pin net).

Our goal is to estimate m number of route paths with optimal deviation to result in minimal number of crossovers together with possible shortest path length for each net.

Path Estimation Process

As stated earlier in the problem formulation section, we start with m number of two pin nets. One route zone is designated for each net. Hence, m number of route zones is formed. However, intersection of route zones of those nets composing three pin nets are not considered.

Input

An $n \times n$ 2D layout of DMFB with preplaced nets.
Number of 2 pin nets $= x$.
Number of 3 pin nets $= y$.
Number of microfluidic modules $= k$.

Objectives

To explore m number of paths one for each net such that:

- Number of cross overs are minimal or none
- Number of deviation for each path is optimal with minimal electrode utilisation
- Path length must be optimal if not shortest
- Estimated latest arrival time must be minimal for the overall route operation.

Assumptions

- Assume all route paths follow Manhattan distances
- Each hard blockage is treated as forbidden zone for droplet routes
- Transfer from one to adjacent unit cell takes timestamp of 1 unit
- Any movement for any given net beyond the route zone is treated as deviation.

Proposed Methodology

1. We assume there are s numbers of intersection regions among these route zones. We form a route zone graph G_{RZ} (D, E) with each node $d \,\varepsilon\, D$ represents Route zone and each edge $e \,\varepsilon\, E$ represents zone intersection.
2. Hence, there are m number of nodes and s number of edges in G_{RZ}.
3. A weight w is assigned to each edge e. Weight w represents number of cells constituting the corresponding intersection represented by e.
4. We plot Manhattan path from source to target for each net i.
5. For those paths passing through microfluidic modules /hard blockages- we initiate deviations using Hadlock's routing method to arrive at the target.
6. Such paths tentatively ensure shortest possible paths towards target with minimum deviations.
7. We compute respective length for route paths for each net.

8. Based on G_{RZ}, we define a table listing of each node of the graph. In other columns, we enlist respective node degree, computed path length and total edge weight associated with overall intersection for each net.
9. We identify the nets with maximum node degree. Net having maximum degree will be given higher deviation preference. For more than one nets with same degree—preference will be given to one having lower intersection area coverage represented in terms of total edge weight.
10. For nets with same degree and same total edge weight deviation preference goes to the net with lower path length.
11. Based on the preference, the route path for the net with highest deviation preference is plotted with possible minimum deviation.
12. Check for reduction of the number of contaminations.
13. Continue deviation with nets having next higher preference.
14. Terminate path estimation process if there remain no crossovers or deviation for all the nets (sequentially according to deviation preference) is over.
15. If any contamination still exists—it is considered as native contamination. These can be removed using wash droplets.

Output

1. Estimated path for each net n_i
2. Total number of contaminations
3. Estimated length of each path (n_i).

5   Illustrative Example

Here, we present a 16×16 layout of a 2D DMFB. It contains pre-scheduled placement of six droplets namely a, b, c, d, e, f—each representing one two pin net with respective source target pair as shown in Fig. 4a. The layout contains four microfluidic modules namely M1, M2, M3, M4 as shown in the Fig. 4a. Hence, six routing zones (A, B, C, D, E, F) are formed as shown in Fig. 4b.

Eventually this follows intersection between route zones and the regions of intersections are represented in Fig. 4c. They are indicated as A–B—representing intersection between route zones A and B and so on.

The corresponding route zone intersection graph G_{RZ} (V, E) is formed, where each vertex v represents the route zone and each edge e represents the corresponding intersection. The weight w assigned to each edge represents the number of cells involved in the intersection. Table 1 is formed to display degree of each vertex (intersection degree), net cells covered for intersection for each net and the respective Manhattan distance.

The route deviation preferences are assigned as described in the proposed technique as stated in proposed technique 4.1. If two zones have the same degree, the one with lower cell coverage are preferred for deviation. The reason being scope for finding alternate cells to avoid contamination within the intersection zone becomes less for the intersections comprising of lesser number of cells.

Now, we plot the estimated Manhattan path with deviation only to avoid hard blockages/modules. Figure 4e shows the initial plot for routes with possible contaminations (when stayed within the Route zones—deviated only to avoid hard blockages).

The corresponding path intersection graph G_{Int} (V, E) is shown in Fig. 4f. Here, each vertex v represents a two pin net with each edge e showing the path intersection. Each net is attached with a box representing estimated Manhattan distance for the respective path estimated so far.

Now, path estimation enters the final phase of contamination avoidance. Hence, route zone A becomes the most preferred candidate for deviation from its route zones. Next preferred deviation is carried out on F. The result is shown, in Fig. 4g. It has been found to have only one native contamination. The resultant path intersection graph is shown in Fig. 4g.

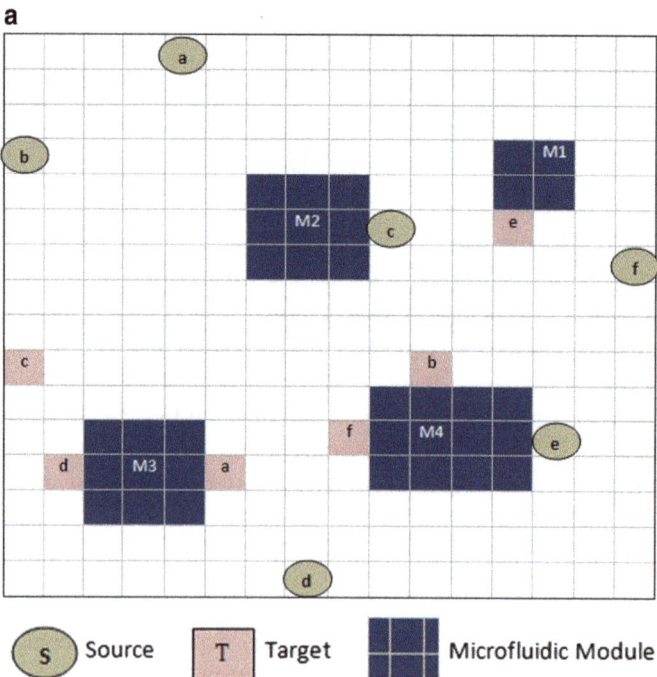

Fig. 4 **a** Pre-placed layout for a 2D DMFB. **b** Route zones for the given 2D layout. **c** Intersection between route zones. **d** The route zone intersection graph. **e** Initial plot of route paths with avoidable contaminations. **f** Path intersection graphs for the given layout. **g** Estimated route paths after deviation. **h** The path intersection graph after deviation. **i** Further deviation of net a with no improvement and increase in path length of a. **j** The final path estimation with computed deviation

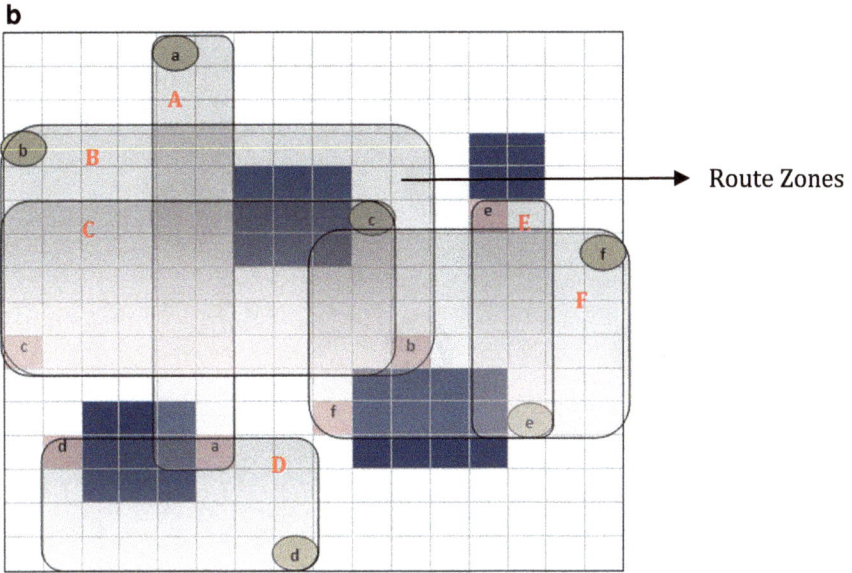

Fig. 4 (continued)

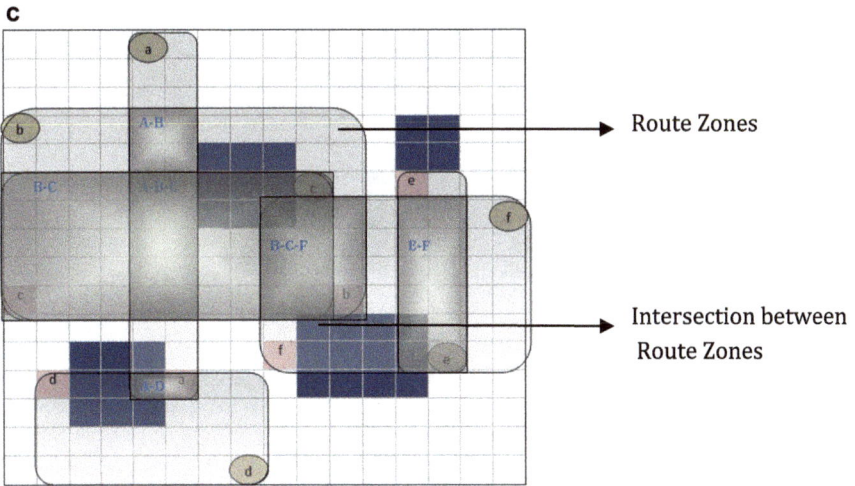

Fig. 4 (continued)

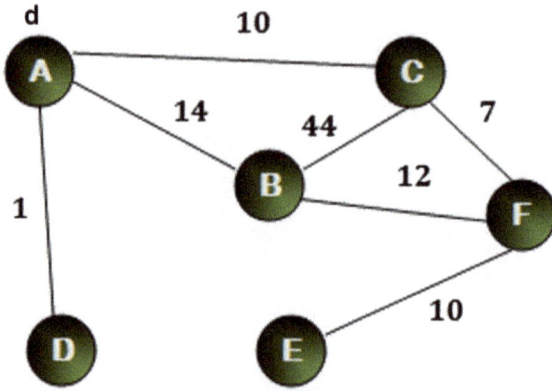

Fig. 4 (continued)

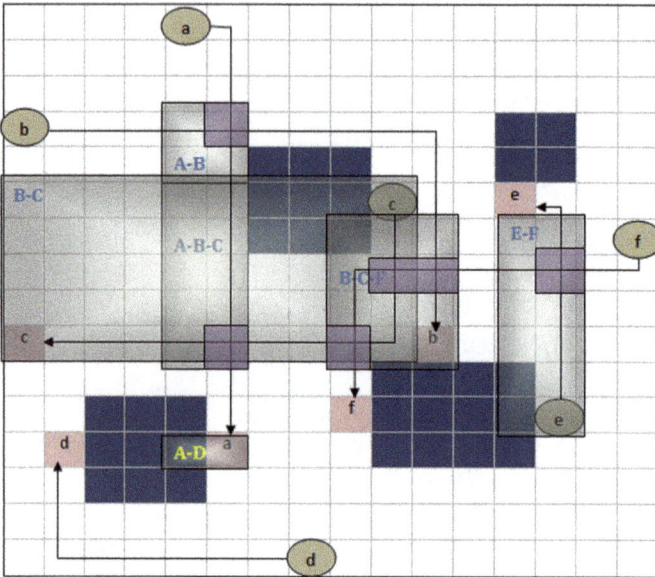

Fig. 4 (continued)

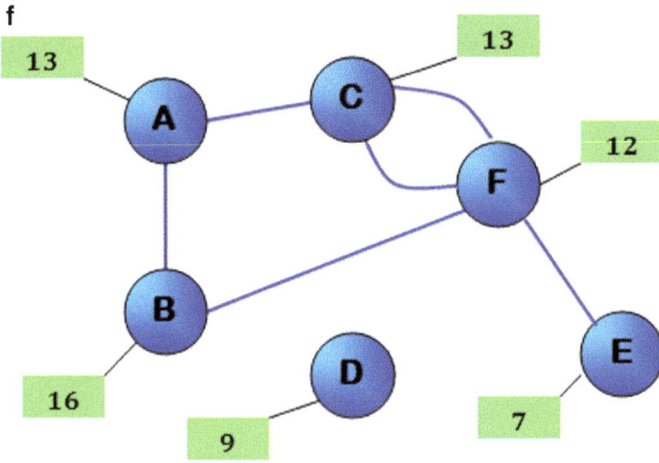

Fig. 4 (continued)

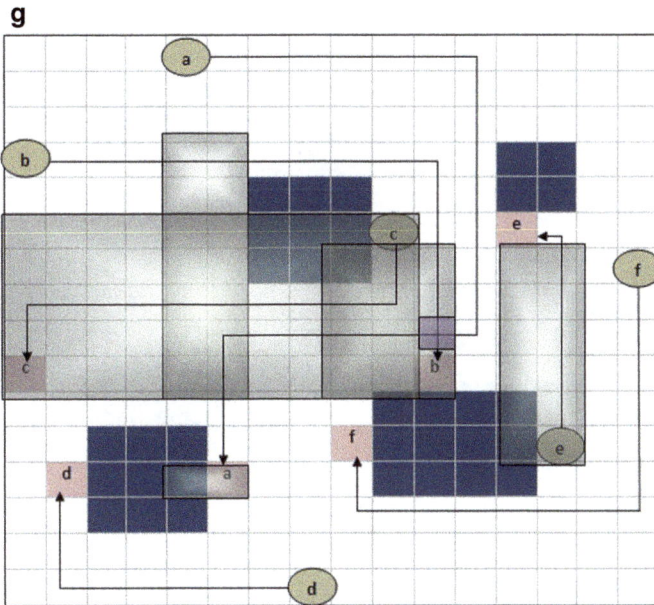

Fig. 4 (continued)

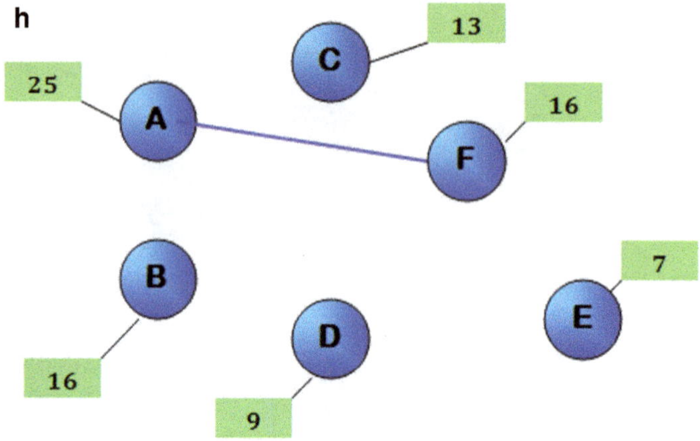

Fig. 4 (continued)

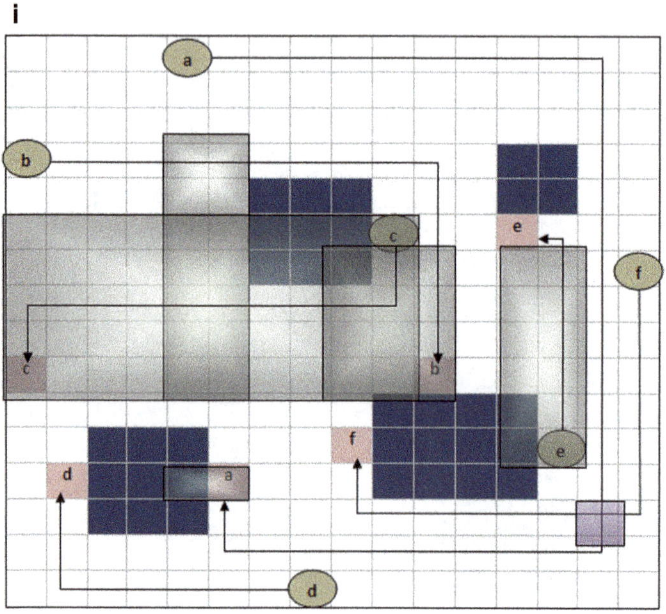

Fig. 4 (continued)

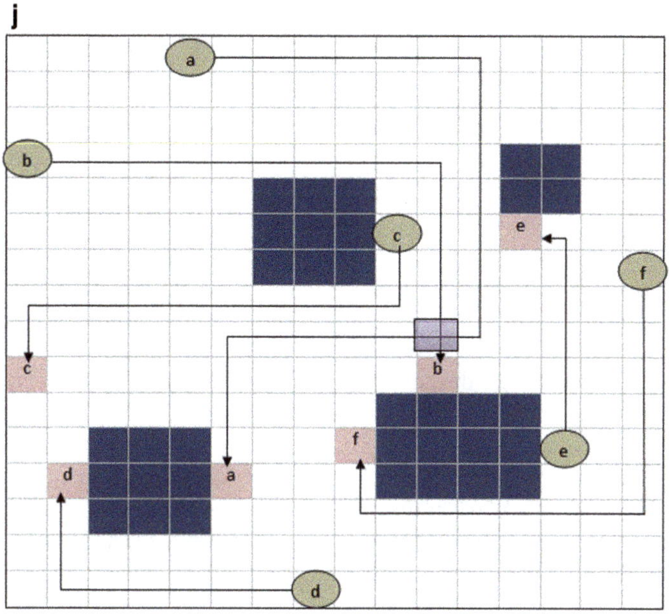

Fig. 4 (continued)

Table 1 Route zone intersection table

Net	Man dist	Int. zone Covg	Deviation preference	Intersection degree
A	13	15	1✓	3
B	16	48	4	3
C	13	44	3	3
D	9	1	5	1
E	7	10	6	1
F	12	7	2✓	3

Table 2 Simulation result for droplet transportation with benchmark suite III

Test bench				Heterogeneous droplets			
Benchmark suite	No. of subproblems	2 pin droplets	3 pin droplets	Max route time	Total route time	Contaminations	Cells used
In vitro 1	11	20	6	20	126.09	2	324
In vitro 2	15	29	6	15	132.25	2	244
Protein 1	64	170	8	24	705	20	1866
Protein 2	76	155	8	28	533.63	18	1127

We have explored further deviation for net a to avoid the single contamination so far as shown in Fig. 4i and the result worsens with single contamination and longer distance of a.

Thereby, the accepted path estimation result is clearly shown in Fig. 4j. The path implementation with arrival times and requisite stall is shown in Fig. 4k. The final result for the route performance is displayed in Table 2. The latest arrival time is computed as $= 25$ units, and the electrode/cell utilisation is found to be $= 83$. Number of native contaminations are found to be $= 1$.

6   Experimental Results

We have used benchmark suite III comprising of four testbenches for route simulation using the proposed route zone-based technique. The names of the testbenches are In vitro 1, In vitro 2, Protein 1 and Protein 2—each is comprised of a set of subproblems forming one bioassay each for a given testbench. Table 2 gives the results for the specified four testbenches representing parameters for route performances namely maximum route time, average route time, number of contaminations and the cells used. Table 3 displays comparison of the route performance using the proposed method and the contemporary contributions. It has been observed that in terms of both latest arrival time as well as cell utilisation parameters our proposed technique produces good improvement.

Total arrival time and cell utilisation for the four testbenches with two of the mostly cited contemporary works. Result shows marginal deterioration for the number of contaminations using the proposed method. This is because a in order to avoid crossovers, we did not compromise upon the other route performances namely latest arrival time and cell utilisation through optimal deviation using deviation preference techniques stated above.

Table 3 Comparison of route simulation results with contemporary contributions

Test bench	Zhao and Chakrabarty [14]				Huang, Lin, Ho [15]				Our method			
	Cotm	C.U	T_{la}	CPU	Cotm	C.U	T_{la}	CPU	Cotm	C.U	T_{la}	CPU
In vitro 1	4	621	268	0.06	21	351	193	0.58	2	324	126.09	0.25
In vitro 2	0	423	224	0.03	6	281	191	0.39	2	242	132.25	0.30
Protein 1	18	3215	1508	0.23	82	2213	1394	2.58	20	1866	705	2.98
Protein 2	11	1574	1287	0.14	51	1362	1108	1.49	18	1127	533.63	2.05

7 Conclusion

In this work, we proposed a method for routing of heterogeneous droplets. The main focus is to minimise the number of crossovers—that in turn reduces the number of wash droplet overheads necessary to handle the residual or native contaminations. The method forms routing zone for each net and determines the deviation preference in case of crossovers (contaminations) to optimise route performance in terms of contaminations, latest arrival time, cells used, etc. The simulation results shows improvement in route performance using benchmark suite III. However, the above method is not found to be as efficient for layouts with comparatively larger number of nets.

References

1. Windh S, Phung C, Grissom DT, Pop P (2017) Performance improvements and congestion reduction for routing-based synthesis for digital microfluidic biochips. IEEE Trans Comput Aided Des Integr Circuits Syst 36(1):41–54
2. Lu G-R, Bhattacharya BB, Ho T-Y, Chen H-M (2018) Multi-level droplet routing in active-matrix based digital-microfluidic biochips. In: 23rd Asia and South Pacific design automation conference, pp 46–51
3. Chakraborty S, Chakraborty S, Das C, Dasgupta P (2016) Efficient two phase heuristic routing technique for digital microfluidic biochip. IET Comput Digital Tech 233–242
4. Roy P, Bhattacharya R, Rahaman H (2012) A new algorithm for routing-aware net placement in cross referencing digital microfluidic biochips. In: 2012 IEEE computer society annual symposium on VLSI, pp 320–325
5. Lu G-R, Bhattacharya BB, Ho T-Y, Chen H-M (2018) Multi-level droplet routing in active-matrix based digital-microfluidic biochips. In: 2018 23rd Asia and South Pacific design automation conference (ASP-DAC), pp 46–51
6. Bhattacharya R, Rahaman H, Roy P (2016), A new heterogeneous droplet routing technique and its simulator to improve route performance in digital microfluidic biochips. In: 2016 International conference on microelectronics, computing and communications (MicroCom)
7. Singh A, Samanta T (2016) Vulnerability detection and error minimization in bioassay sample mixing and droplet routing for digital microfluidic biochips. In: 2016 sixth international symposium on embedded computing and system design (ISED), pp 66–71
8. Mukherjee S, Banerjee I, Samanta T (2014) Defect aware droplet routing technique in digital microfluidic biochip. In: 2014 IEEE International advance computing conference (IACC), pp 30–35
9. Das S, Roy S, Dey S (2014) Activation of control pins for routing of test droplets within a bi-partitioned digital microfluidic biochip. In: 2014 fourth international conference on advances in computing and communications, pp 121–124
10. Pan I, Samanta T, Voltage driven electrowetting based microfluidic operations for efficient droplet routing in digital microfluidic biochips. In: 2014 IEEE/ASME 10th international conference on mechatronic and embedded systems and applications (MESA)
11. Bhattacharya R, Roy P, Rahaman H (2018) A new combined routing technique in digital microfluidic biochip. In: 1st international conference on emerging technologies in data mining and information security, Springer, pp 441–450
12. Bhattacharya R, Roy P, Rahaman H (2018) A new homogeneous droplet transportation algorithm and its simulator to boost route performance in digital microfluidic biochips. In: 1st

international conference on emerging technologies in data mining and information security, Springer, pp 429–440
13. Bhattacharya R, Roy P, Rahaman H (2019) A complete routing simulator for digital microfluidic biochip. Int J Inf Syst Model Des 70–85 (Scopus indexed)
14. Zhao Y, Chakraborty K (2009) Cross contamination, avoidance of droplet routing in digital microfluidic biochips. In: Procedure of the design and automation test in Europe
15. Huang T-W, Ho TY (2009) A fast routability- and performance-driven droplet routing algorithm for digital microfluidic biochips. In: IEEE international conference on computer design (ICCD 2009), Lake Tahoe, pp 445–450, 4–7 October 2009

Chapter 43
MLP-WOA Neural Network-Based Automated Grading of Fruits and Vegetable Quality Detection for Food Industry Using Artificial Intelligence Techniques (Computer Vision—Image Recognition)

Syed Sumera Ershad Ali and Sayyad Ajij Dildar

1 Introduction

The quality of the food that obtained for the inspection is by imaging methods. The main aim is to set an image or several images employing a single camera or multiple cameras. The creation of the sensor with high quality gives an enhanced resolution as well as the quality of the image; an increment in power computed is necessitated to produce a lot of new best algorithms, and it uses the image tracking. Recently, the quality assessment of the food products includes utilizing some physical techniques (Calvoh et al. 2016). The inspection of the food quality can classify into two, non-destructive, and destructive, in which the non-destructive approach describes the reduction in quality while the product is normally weak and also it can be simply spoiled [1]. The non-destructive quality checking for the food factory is enhancing the need for the agro-food factory products is associated with the customer's physical condition and community encouragement. The enhanced quality, the protection, and the agro-food produce quality has attracted particular consideration worldwide [2]. The quality of the food ensured when the safety estimation motivates the methods to replace the conventional methods from the advanced methods by implementing them, which are less inconsistent and ineffective. The conventional methods examine the food once after long research. Through processing the quality of the food along with

S. S. E. Ali (✉)
Electronics and Telecommunication Department, CSMSS Chh. Shahu College of Engineering, Dr. B.A.M. University, Aurangabad, Maharashtra, India
e-mail: syed.sumera.ali@gmail.com

S. A. Dildar
Electronics and Telecommunication Department, MIT College of Engineering, Aurangabad, Maharashtra, India

© The Author(s), under exclusive license to Springer Nature Singapore Pte Ltd. 2021   539
X.-Z. Gao et al. (eds.), *Applications of Artificial Intelligence in Engineering*, Algorithms for Intelligent Systems, https://doi.org/10.1007/978-981-33-4604-8_43

protection, the advanced approaches are wanted that evaluate the resources used for the food and components at every stage. The enhancing agriculture food protective issues have been appeared in nowadays, such as excess addition of preservatives, toxic residues, and creating harmful chemicals through the procedures. The approaches that have been created for the quality estimation and measurement to conquer the drawbacks of the conventional approaches for spectroscopic and imaging approaches. The food structure evaluation employs spectroscopic approaches such as Raman Spectroscopy with Fourier transform nuclear magnetic resonance and near-infrared spectroscopy, and these approaches employed in the evaluation of food possessions such as salt, protein, fats, moistures, etc. and provides more accuracy and leads to present linear allocation of quality attributes that is more important for inspection of food protocols. The technique that usually works to achieve linear details; another method created called computer vision [3]. The linear details obtained from the food's digital image with texture, shape, size, and color.

2 The Motivation of the Research

In recent days, the quality of the food is based on the processing concepts and the need for the development of definite quality statements to the nature of the food as well as agricultural food items; it is necessary to maintain the food products in prolonged quality aimed at the increasing population.

The motivation of this study is to improve the quality of the food and also to classify the quality food as well as the defected food. However, to realize the defect detection more rapidly and accurately in-line, there are still many challenges to be overcome. But, the lack of solution for such drawbacks has motivated me to do the research work in this area.

3 Problem Identification

Recently, there is an enormous improvement in the food quality inspection on behalf of the technology. It is important to provide the quality food to the customers by inspect the quality of the food and also to classify the quality food as well as the defected food. There are some problems by using those techniques. The major problems are listed below:

- The noises are perfectly removed in the pre-processing by using the proper approach.
- The segmentation methods to detect the images are perfectly analyze.
- To enhance the scheme using the deep learning as well as the image processing technique, a light vision method developed.

- The real-time quality detection as well as the ranking of the food on the arranging lines is recognized.
- To inspect the quality of the food image, it may difficult to obtain large dataset.

4   Objectives of the Research Work

The main objective of research work is to separate the quality food and the damaged food also minimize the error and high accuracy with sensitivity, specificity.

- To separate the quality food and the damaged food.
- To minimize the error and high accuracy with sensitivity, specificity.
- To develop object recognition system in food products applications.
- To design a model or algorithm for an ideal image recognition.
- To develop a program in MATLAB based on the designed model.
- Extraction of quality from the databases are separated using the histogram pre-processing and segment the input products.
- Feature is extracted from the segmented output and classification is by using ANN classifier thereby achieving high accuracy, specificity, and sensitivity.
- Implementing neural networks and WAO process involved in inspection of food quality is to obtain the weights and bias by minimizing the error.

5   Research Methods and Tools

The quality of food products is very important for the human health. The large population and the increased requirements of food products makes it difficult to arrive the desired quality. Sorting tons of fruits and vegetables manually is a slow, costly, and an inaccurate process. Thus, computer vision is a rapid, economic, consistent, and objective inspection technique, which has expanded into many diverse industries. Its speed and accuracy satisfy ever-increasing production and quality requirements, hence aiding in the development of totally automated processes. Automation means every action that is needed to control a process at optimum efficiency as controlled by a system that operates using instructions that have been programmed into it or response to some activities. Automated grading system not only speeds up the time of the process, but also minimize error. Hence, the proposed method consists of four stages, such as pre-processing, segmentation of image, Feature extraction, and Classification.

Hence, after pre-processing, region growing-based segmentation is processed and then feature extraction will be carried out to extract the features to classify the output. Finally, based on extracted features, neural network classification is used for inspecting the food quality. Hence, this inspection method based on image analysis can be used in variety of different applications in the food industry. The developed model will be implemented in the platform of MATLAB, and the output will be compared with the existing technique to evaluate the performance.

Two phases namely food quality inspection by ANN and MLP-WAO were focused and presented to predict the accuracy and also to minimize the error by using various techniques

The first phases involved in the inspection of food quality using ANN (Phase—I). Whereas, the second phase involves inspection quality and also the weights and biases using MLP (Phase—II).

The steps involved in the research methodology of proposed ANN and MLP methods are of given as follows,

- Pre-processing
  - Histogram equalization
- Enhanced region growing segmentation
- Feature extraction
  - Color histogram features
  - GLCM features
- ANN—BPA neural network & MLP—WOA neural network.

Tools for this research work are as follows,

- MATLAB (R2013a & 2017a) or higher 2018.
- Window operating system as implemented in MATLAB. The work, however, is carried out on Window XP platform.
- Camera with 8MP or higher.
- Pentium 4 machine or higher.

6 Challenges in Object Recognition and Classification

The recognition as well as tracking the image from a sequence is a difficult job whenever the image has a difficult arrangement and able to alter the size, shape, or orientation over consequent frames [4]. The following Fig. 1 depicts several issues as well as the main challenges for food image recognition steps.

- **Low image quality**: The image that is captured by the device will result in the creation of a blurry or noisy image that causes false recognition. According to certain weather conditions such as fog, rain, etc. The images are noisy. The camera quality should be checked whenever using it in bad weather conditions.
- **Jitter**: The blur that occurred due to the moving camera that constraints the quality, and the resolution of the image in order to overcome by a new camera, called camera jitter that creates effective image capture effectively and employ any of the recognition processes be capable of conquering this limitation.

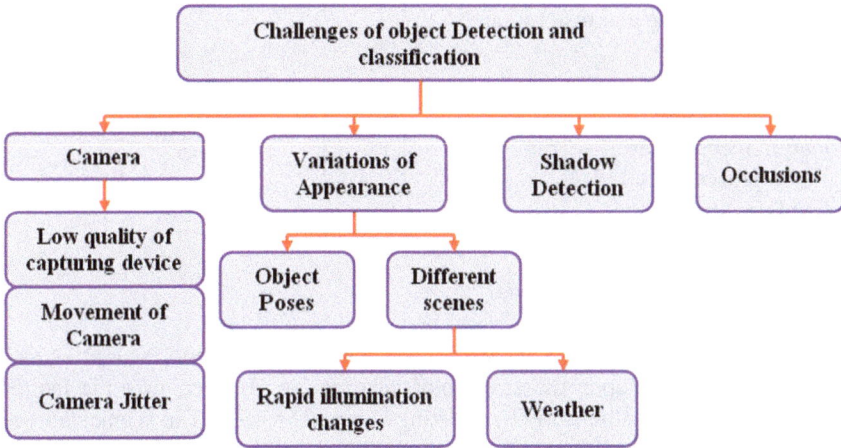

Fig. 1 Challenges in object recognition and classification

- **Moving camera**: The cameras that fixed on the top of the vehicles are another challenge to this. The moving camera could not produce accurate image recognitions and difficulty in image recognition with moving cameras, and some researchers investigate this.
- **Alterations in illumination**: Disappearance or presence of the source such as bulb, sun, tube light, etc. that lead to changes in the clarification of the images and causes false recognition in successive frames over various frames. The created solutions can be able to process on various stages of illuminations.
- **Weather**: Due to sudden background variation, the recognition process is difficult and causes motion in the background due to blowing of winds, submerged images, rapidly moving images, etc. and this also causes false recognition that is more common in the last recognized frame of the image.
- **Shadow recognition**: The recognition of shadows occurs from the background images or motion of images. Some of the researches overwhelmed this drawback by employing suitable techniques to take away the shadows from the exact image.
- **Occlusions**: The image that falls on the other image is called occlusion, and this is one of the major challenges in image recognition and results in the false computation of images count and the overlapping of recognized margins.

6.1 Applications of DIP

The digital image processing is an emerging technology in the agricultural as well as the food sector, which employed for online quality computation of different food items like fruits, vegetables, fishes, meats, grains, rice, canned food, etc. In recent years, more researches were done on food products. There are various application of food products in digital image processing listed.

- Image sharpening and restoration
- Medical field
- Remote sensing
- Color processing
- Pattern recognition
- Video processing
- Food processing.

6.1.1   Fruit Processing Methods

This describes the fruit quality inspection, i.e., defects in the fruits such as a crack in the melon depend upon the size, color, texture, etc. To recognize the faulted fruits and throw away the fruits by utilizing computer vision. The segmented and cracked portion of the melon recognized by employing the k-means clustering-based segmentation approach [5].

The fruit quality evaluation mainly depends on the size, color as well as weight, and k-means clustering is used to discard the defected fruit from the set of fruits in [6]. Then, the mass evaluation of mango fruits from the characteristics like length, thickness, width as well as area. These characteristics are further used as an input to the artificial neural network to calculate the mass (Utai et al. 2018).

6.1.2   Vegetable Processing Methods

Vegetable processing includes some of the vegetables like tomato, cucumber, asparagus, etc. For tomato, the quality inspected by collecting the sample, grading, cutting as well as lab tests. A computer vision system employs the enhancement of the classification of green asparagus and categorizes it according to the non-existence of white spots as well as the product width. CVS divided under the error percentage (Campos et al. 2019).

6.1.3   Grains Quality Processing Approaches

The grain quality processing system developed to check the quality of the rice grain, maize kernels, etc. It was generated to assess the appearance of rice quality. The grains classified into a whole as well as broken kernels.

6.1.4   Other Food Processing Approaches

For the meat quality processing approaches, one of the methods that arranging the chicken portions automatically, and its categories the portion into fillet, breast, leg, wings, and drumstick using classifiers such as ANN, LDA, PLSR. The test was carried out, and the accuracy is improved (Teimouri et al. 2018). The freshness of the fish identified by employing image processing approaches. The features that represent the degradation pattern for fish freshness [7].

7   Food Image Recognition

The challenging problems are overcome by the subsequent components of food image recognition. The following shows the image processing steps for object recognition. The food processing steps are depicted in the Fig. 2.

7.1   Pre-processing

The aim of pre-processing aims to be an enhancement of the image data that contains unwanted distortions or enhances some image aspects significant for additional processing. Figure 2 shows the pre-processing approaches are essential on grey-level, color, or binary document images that include graphics and graphics. The image enhancement comes under the pre-processing approaches and enhances the visibility of the image by the process of altering the digital images so that the results are more appropriate for the exhibit. The color spaces based on the RGB model and it consists of RGB, CIE RGB, and Adobe RGB.

The pre-processing technique performs the conversion of RGB to gray-level image reformation technique contains the histogram equalization process whereas the image contrast enlarged. The histogram equalization is the pre-processing technique used in the proposed method, and it enhances the contrast of the image. The histogram equalization converts the RGB to color level image in which the conversion improves the detectability of several image attributes because the contest expanded for various pixels of the image. This method of conversion enhances the gray level of the image (Fig. 3).

7.2   Segmentation

The process of dividing a digital image into several segments (sets of pixels, also known as image objects) is called image segmentation. The purpose of segmentation is to make simpler or to alter the demonstration of an image into some other thing, which is additionally significant as well as easier to evaluate. The approaches employed for the segmentation are clustering-based, edge recognition-based approaches, thresholding approaches, region-based approaches, artificial neural network-based approaches, watershed-based approaches, and partial differential equation-based, etc. Every pixel in a region is identical to features regarding

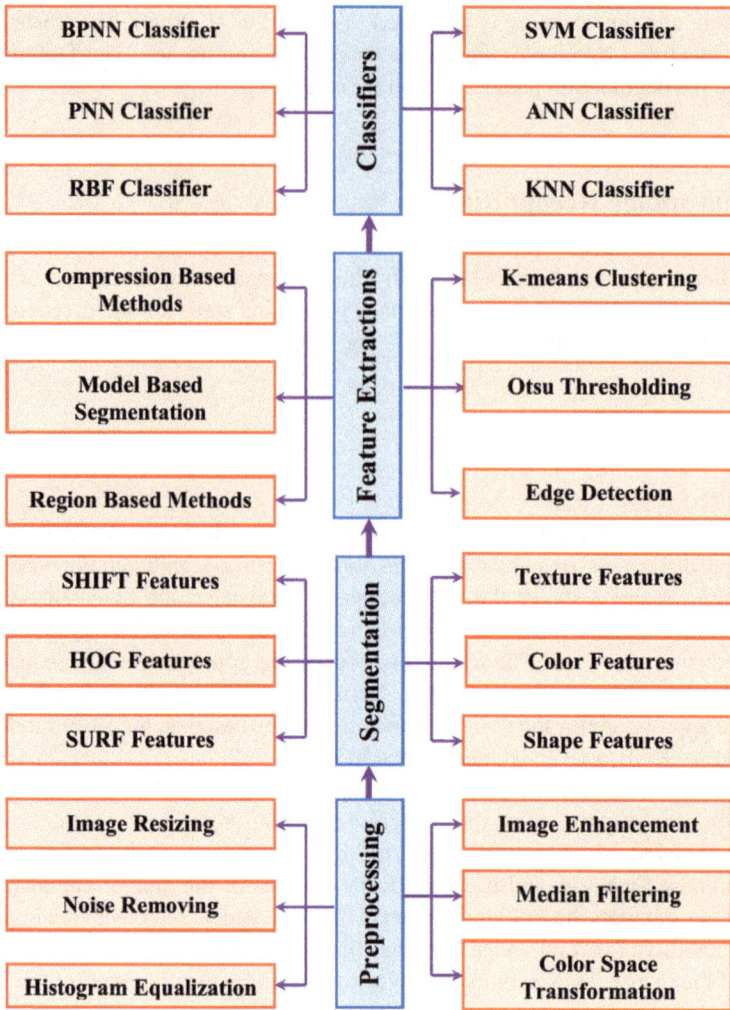

Fig. 2 Procedure in image processing

some other features or properties such as color, texture, or intensity and described in Fig. 4.

The MRG segmentation uses the proposed segmentation approach for this system. In this proposed fruit quality inspection, the limitations of region growing, such as noise addition, dimness of the image overcome by the utilization of MRG segmentation, i.e., modified region growing approach. The MRG approach has the following steps.

- Image block conversion
- Initialize seed points

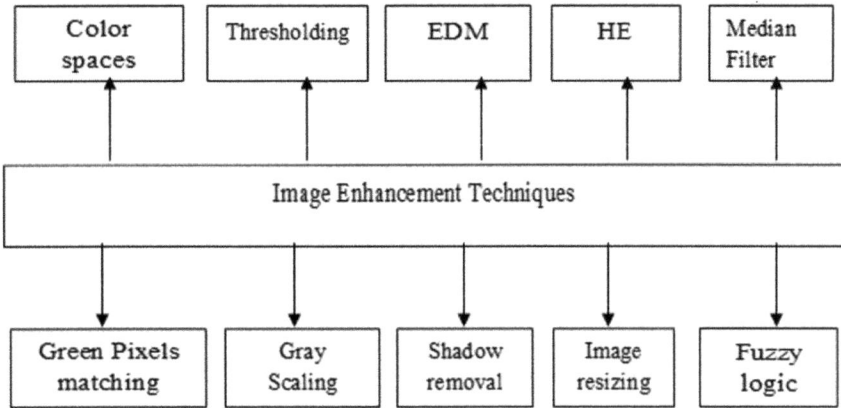

Fig. 3 Pre-processing techniques

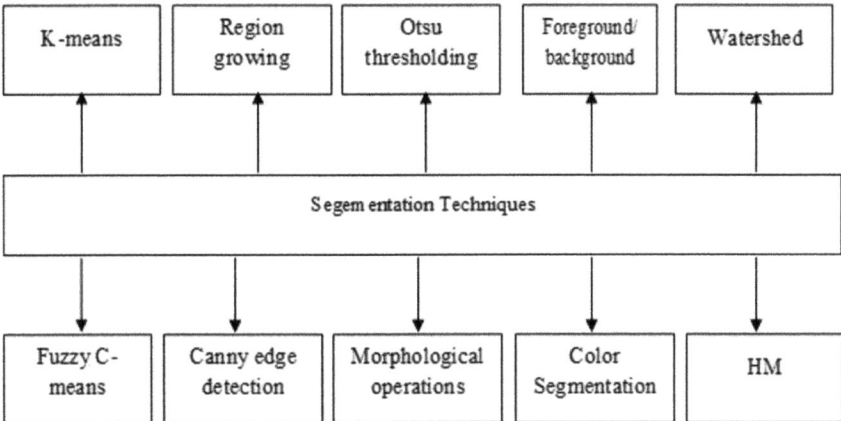

Fig. 4 Segmentation-based techniques

- Euclidean distance measurement
- Region growing constraint
- Image pruning
- Termination.

7.3 Feature Extraction

The feature extraction is that the feature is the information that resolves the evolutional risks evaluation task related to particular uses, and those features will be particular structures like edges, objects or points. In this, the data is diminished to

a dataset with a decreased number of variables that comprise the most discrimina-
tory information. There are many types of feature extraction in image processing
(Bhargava et al. 2018). They are image-based, camera-based, IR-based, ultrasonic
image-based, grayscale pixel values, the mean pixel of channels, multi-scale feature
extraction, etc.

The specific texture, as well as color related features, are discarded from the
segmentation of the image that is employed for the classification of the image is
explained in our proposed approach. The GLCM associated characteristics such as
cluster shade, entropy, autocorrelation, homogeneity, dissimilarity, contrast, energy,
correlation, maximum probability. The color histogram is not suitable for the change
in image as well as the resolution of the histogram, and the color histogram attributes
are evaluated using the color histogram approach is shown in Fig. 5.

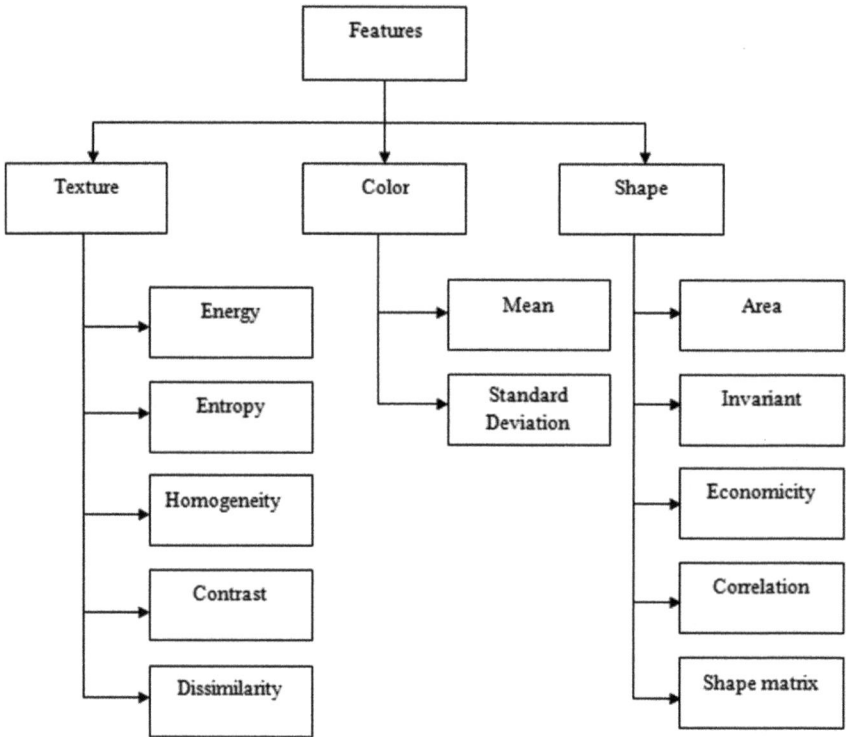

Fig. 5 Feature extraction techniques

7.4 Classification

The set of data is classified into various categories or classes, and it can be clearly understood by the link between the classes as well as the information. The classification techniques pose two phases of procedures: testing and training (Abid et al. 2018). The training stage is the beginning phases in which the characteristic properties of the image aspects are separated and depend on the separation a unique explanation of every classification group; *i.e., training phase* is generated. The feature space partitions are employed to classify the features of the image in the testing phase (Abid et al. 2018). The classification of images has many steps like efficiency, pattern recognition as well as accuracy based on the classification. The image classifiers are an artificial neural network, decision tree, radial basis function, K-NN, BPN, SVM, PNN, etc. (Fig. 6).

The classifiers that are used in Phase I are the ANN, in which the main aim is to cover the inputs into considerable outputs. The neural networks are trained by employing the back propagation method. The BPNN is used for the categorization of the dataset with an appropriate arrangement of training, transfer function as well as learning.

The classifiers that are used in Phase II is the MLP-WOA method in which it uses the feed-forward neural network, and it consists of input, hidden, and output layer. The input layer in the MLP architecture concerning neural networks includes correlation, contrast, energy, and homogeneity, and hidden layers are more in this model. The whale optimization is the other method that is combined with MLP to predict the accurate classification of the image and the whale produces the bubbles around the water to hold the smaller fishes.

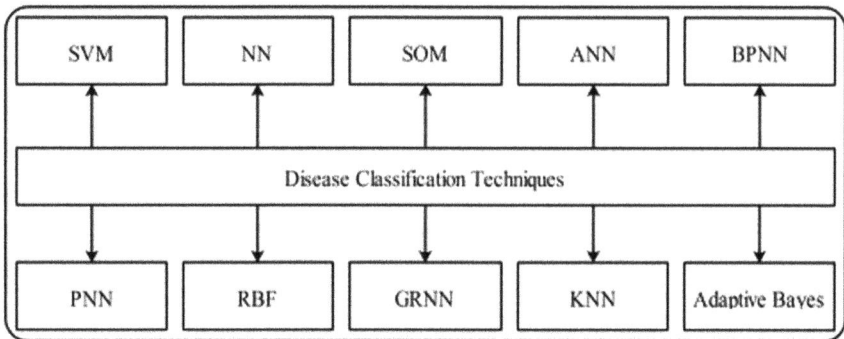

Fig. 6 Classification-based techniques

8 Overview

Literature survey is on food quality inspection relates the problem faced by the researches while the quality of the food detection by different approaches. A complete survey of such imaginative methods is done in this module to analyse every one of their disadvantages and effectiveness. Here, the limitations of various existing methods of quality inspection are clearly described. The primary goal of this is to overcome the limitation in the existing methods by proposing new techniques. The quality inspection thus helps to easily separate the defected food by using the classifiers like neural networks, support vector machine classifiers, k-nearest neighbors, genetic algorithm, etc. The major aim of this chapter is to analyse the drawbacks of several techniques, and also by overcoming these restrictions, proposed system is determined.

8.1 Taxonomy of Defects in Various Foods

In the agricultural industry, the quality of the food is about its ripeness; the customer accepts the quality of the fruit by its appearance in both internal as well as external. The visual appearance is to judge its ripeness that is calculated by its size, color, and shape. In this, the quality of the different of the food is evaluated in different procedures based upon the defects in each food product. Some of the defects in the food products are stated here. For apple, defects such as apple scab, apple rot and apple blotch; also, the apple leaf diseases include healthy leaves, rust, block rot as well as a scab. For tomato, the ripeness is evaluated based on the color, texture, size, as well as the shape of the food items, this ripeness of the fruit, calculates the quality of the tomato.

8.2 Food Defect Detection and Classification

The comprehensive depiction of the image processing techniques is employed for fruit recognition as well as classification of the images in this section. The food quality recognition comprised of the four main processes, such as pre-processing, feature extraction, segmentation as well as classification. Each process has its challenges, benefits as well as drawbacks and the overview of the literature review is shown (Fig. 7).

Fig. 7 Overview of our literature review

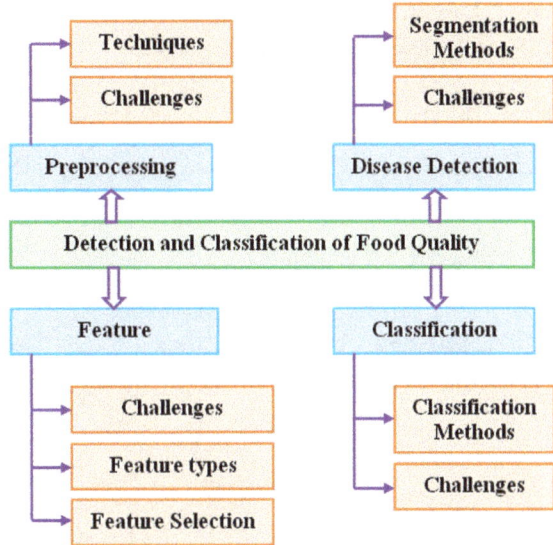

8.2.1 Pre-processing Based Approaches

The pre-processing is used to remove the unwanted noises from the signals, and also it is utilized to enhance the contrast of the image. The images that are captured from the cameras or sensors do not have a clear background, not having enough light so this will disturb the accuracy of the performances. Thus, the pre-processing works on it to enhance the quality of the image by using some of the pre-processing methods such as image enhancement, color space transformation, histogram equalization, etc. There are some of the filter methods such as Gaussian filter, mean filter, median filter, etc. are used in the pre-processing stage to remove the unwanted noises from the captured image (Hemapriya et al. 2018).

Challenges. The pre-processing stage makes the defected food image clearer when compared with the original input image. Some of the challenges in the pre-processing are (a) background of the image contains more noises (b) different images with same resolution (c) best contrast is to be obtained between the food image as well as the background (d) variations in lighting (e) input image intensity value is low.

The comparison of pre-processing-based approaches is described as shown in Table 1. In this, the quality food is detected through pre-processing approaches such as 3D box filtering, Gaussian filter, median filter, PCA, GLCM, ACO, SNV, Standard normal deviate, Savitzky–Golay algorithm, SNV, MSC, PLS, curve smoothing. These defects are identified by the above-stated methods and this converts images into grey-scale images for the segmentation process. The performance metrics that achieved using this process are accuracy, precision, sensitivity, specificity, etc.

Table 1 Comparison of pre-processing approach

S. No.	References	Food type	Purpose	Techniques	Performance measures
1	Khan et al. [8]	Apple	Blackrot, rust, and scab	3D box filtering, decorrelation, 3D Gaussian filter, and 3D median filter	SSIM
2	Li et al. (2019)	Green tea leaves	detection of volatile organic compounds	PCA, GLCM	PSNR, Accuracy
3	Khulal et al. (2017)	Chicken meat	Detect the total volatile basic nitrogen content	ACO, SNV	Better model performance
4	Sereshti et al. (2018)	Saffron	Identify the adulterants types	Asymmetric least squares approach, correlation optimization warping	Accuracy, Precision, Sensitivity, Specificity
5	Rady et al. (2015)	Potatoes	Prediction of glucose as well as sugar levels quickly	Standard normal deviate, weighted baseline, smoothing, multiplicative signal correction, Partial least squares regression	Classification error, accuracy
6	Włodarska et al. (2018)	Apple juices	Screening the quality parameter	Savitzky–Golay algorithm, SNV, MSC, PLS	Good calibration results
7	Mollazade et al. (2015)	Tomato	Predict the quality factors while storing the fruit	curve smoothing, Savitzky-Golay smoothing filter	Good correlation with modulus elasticity, SSC and TA
8	Canneddu et al. (2016)	Macadamia nuts	Nuts infected by Leucopteara coffeella as well as Ecdytolopha aurantiana	MSC, SNV, PLSR	SSIM, PSNR

8.2.2 Segmentation Based Approaches

The segmentation is defined as the process of dividing the digital image into multiple segments. The segmentation process is to simplify or alter the representation of the image into something that is easier to analyze. In this process, the image of the defected portion of the food is separated out from the pre-processed image by using anyone of the above approaches. The challenges along with their descriptions are described below.

Challenges. The image segmentation is one of the most challenging in various applications such as medical image processing, agricultural images as well as pattern detection. The image segmentation separates the defected portions alone from the image in the agricultural field as well as a background part of the image. There are various challenges in the segmentation of the defected part of the image like (a) variation in defect color while executing color-based segmentation; (b) actual color of the fruits is varied largely in the segmentation processes that are difficult to process (c) changes in lighting condition; (d) variation in size of defected region (e) number of fruits (f) region growing segmentation approaches that enormous amount of processing time (g) determination of defects as well as size of the fruit (h) variation in scale as well as origin and (i) defects texture these are the challenges that affect the accuracy as well as the performance of the system.

The segmentation-based techniques are used to segment the pre-processed image by the following approaches is shown in Table 2. The segmentation-based approaches such as canny edge detector, K-means clustering, histogram approach, color mapping approach, RGB color space approach, modified K-means clustering along with Otsu approach, bi-dimensional empirical mode decomposition, watershed segmentation for the purpose like recognition of the quality index of the fruit/vegetables, detection and counting of post-harvest fruits, defected region of fruits as well as vegetables, defected portion of the pome fruit, early decayed oranges by fungus identification, detects the quality of fruits. The limitations of these methods are consumption of time is more, complexity in computation. If the images have multiple edges, this approach does not work better, etc.

8.2.3 Feature Extraction Based Techniques

Feature extraction plays a major role in the computer vision as well as the machine learning field for the depiction of the object in an image. Each object has its shape, color, size as well as texture, while via feature extraction, the separated image is divided into their corresponding divisions. The features created by the structures that predict the defected part of the fruit image depends upon their size, color, shape as well as texture (Dai et al. 2015). Various feature extraction approaches are used for separating the color, size, texture, and shape but every feature has its challenges. The color features are very important since every defect has its color attributes that are utilized for the detection of the defects in the fruits.

Challenges. Every feature in the feature extraction has its benefits, as well as limitations and these limitations affect the overall accuracy of the process. There are a lot of challenges that exist in the feature extraction process such as features that are not appropriate, separate features that has high dimension, and calculation time is high, separated features has redundancy, variation in lighting, high correlation, these challenges straightaway affect the overall efficiency of the system.

Table 2 Comparison of segmentation techniques

S. No.	References	Food type	Purpose	Techniques	Performance measures
1	Prince et al. (2018)	Fruits/vegetables	Recognize the quality index of the fruit/vegetables	Canny edge detector	Size, shape and color detection are promising
2	Sethy et al. (2017)	Fruits	Detection and counting of post-harvest fruits	K-means clustering, histogram approach, HSI method, color mapping approach, RGB color space approach	Accuracy
3	Raut and Bora	Strawberry and cherry	Maturity of fruit assessment	Red channel conversion, green channel conversion, red and blue mask conversion, intermediate mask	Accuracy
4	Rozario et al. (2016)	Fruits such as apple, banana, etc. and vegetables such as tomato, potato, etc	Defected region of fruits as well as vegetables	K-means clustering, modified k-means clustering along with Otsu approach	Easy identification of defects
5	Yogesh et al. (2017)	Pome fruit, Pear fruit	Identify the defected portion of the pome fruit	Otsu method, Thresholding segmentation approach	Accuracy
6	Singhal et al. (2017)	Apple	Identify the defected portion of the fruit	Otsu, fuzzy c-means, k-means, texture, and watershed management	An efficient form of identification of defects
7	Nair et al. (2015)	Orange and apples	Defect recognition and classification	Improved k-means clustering approach	Accuracy

(continued)

Table 2 (continued)

S. No.	References	Food type	Purpose	Techniques	Performance measures
8	Li et al. (2019)	Orange	Early decayed oranges by fungus identification	Bi-dimensional empirical mode decomposition, watershed segmentation	Accuracy
9	Devi et al. (2017)	Fruits	Yield counting of fruits	Color and shape analysis, Canny edge detection, connected component labeling	Accuracy
10	Panda et al. (2017)	Carica Papaya fruit	Detects the quality of fruits	K-means clustering segmentation	Accuracy

The techniques such as SURF-based feature extraction approach, RGB features, grey-level co-occurrence matrix, SVR, MADM, ANN, NB, K-NN, SVM, BPNN, GLCM plus wavelet texture features, Wavelet transform using haar filter, SPA with SVR for the feature extraction from the segmentation portion of the image is shown in the Table 3. Normally, these feature extraction approaches were used to sort or grade the food quality from the image. Grading ratio, accuracy, recognition rates are the parameters achieved from this approach. There are some limitations in the above-stated approaches such as the color histogram needs spatial information, needs efficient resolution images, computation is high, etc. The approach that are used in the proposed method is grey-level co-occurrence matrix for feature extraction.

8.2.4   Classification Based Approaches

The classifiers-based approaches were employed for recognizing the images that depend on their extraction of features. There are various classification approaches that are examined in this part that includes SVM, PNN, RBF, K-NN, ANN as well as DT. The challenges along with their descriptions are described below.

Challenges. The image classification process is one of the important divisions that used to detect defected food from the different kinds of food products and it has its challenges. Some of the challenges in this classification process are (a) very slow training process, (b) number of images is required for training (c) time consumption is more for both training as well as testing phases (d) Complex network architecture (e) Scalability problem.

The approaches for the classification such as SVM, MLP, random forest, decision trees, k-nearest neighbors, SVM with linear kernel functions using RGB color space model, RBF-SVM classifier, CARS-SVM—AdaBoost approach, SVM for multiple

Table 3 Comparison of feature extraction techniques

S. No.	References	Food type	Purpose	Techniques	Performance measures
1	Dubey (2016)	Fruits	Detect the fruit quality	SURF-based feature extraction technique	Accuracy
2	Mohammadi et al. (2015)	Persimmon fruit	To detect the ripeness of the fruit	RGB features	Accuracy
3	Kumar et al. (2015)	Citrus fruits	Sorting and grading of citrus fruits	Grey-level co-occurrence matrix such as contrast, energy, correlation, homogeneity	Grading ratio
4	Nandi et al. (2016)	Mangoes	To grade the harvested mangoes	SVR, MADM	Accuracy, grading ratio
5	Sanaeifar et al. (2016)	Banana	To detect the organoleptic quality	SVR and ANN	Accuracy and computation time
6	Cen et al. (2016)	Cucumber	To detect the chilling injured cucumbers due to low temperatures	NB, K-NN, SVM	Accuracy
7	Olaniyi et al. (2017)	Banana	To sort quality banana	GLCM feature extraction, BPNN	Recognition rate
8	Semary et al. (2015)	Tomato	To grade tomato fruits based on the defects in the surface	GLCM plus wavelet texture features	Accuracy
9	Dutta et al. (2016)	Fish	To assess fish quality as well as freshness	Wavelet transform using haar filter	Freshness range, SD, mean
10	Zhang et al. (2015)	Eggs	To detect internal quality like freshness, bubble creation	SPA with SVR	Accuracy

class classification, optimum LS-SVM approach, deep CNN that includes AlexNet, MobileNet, GoogLeNet, VGGNet as well as Xception, partial least squares scheme, ANN for the classification defected food from the healthy food. The classification-based approach, such as ANN, MLP-WOA for the food quality inspection in our proposed system (Table 4).

Table 4 Comparison of classification techniques

S. No.	References	Food type	Purpose	Techniques	Performance measures
1	Araújo et al. (2019)	Vegetables like tomato, bell pepper, onion, lettuce	To evaluate the authenticity of organic as well as conventional vegetables	SVM,MLP, Random forest	Accuracy
2	Castro et al. (2019)	Cape gooseberry fruit	To visualize the ripeness level	SVM, decision trees, K-nearest neighbor approach	Accuracy, Precision
3	Varjão et al. (2019)	Citrus fruits	To detect the quality of the citrus fruits	SVM with linear kernel functions using RGB color space model	Accuracy
4	Ireri et al. (2019)	Tomatoes	To detect defect discrimination as well as the grading of tomatoes	RBF-SVM classifier with LAB color space pixel values	Accuracy
5	Li et al. (2019)	Soybean seed	To detect the viability of soybean seed	CARS-SVM—AdaBoost approach	Recognition accuracy
6	Mittal et al. (2019)	Rice grains	To categorize the grain samples into different classes	SVM for multiple class classification	Accuracy, Computational complexity
7	Zhao et al. (2019)	Beef	To detect the homologous adulteration in beef from the fresh minced beef	Optimum LS-SVM approach	Accuracy and Stability
8	Sustika et al. (2018)	Strawberry	To inspect the quality or grading system of strawberry	Deep CNN that includes AlexNet, MobileNet, GoogLeNet, VGGNet as well as Xception	VGGNet gives the best accuracy, GoogLeNet provides efficient computational architecture

(continued)

Table 4 (continued)

S. No.	References	Food type	Purpose	Techniques	Performance measures
9	Pacheco et al. (2019)	Tomatoes	To classify tomato variety such as Milano & Chonto	K-NN, MLP, K-means clustering	Accuracy, Precision, Sensitivity, Specificity
10	Tsouvaltzis et al. (2020)	Eggplant	To detect the eggplant fruit stored at chilling temperature earlier	Partial least squares scheme, SVM, K-NN approaches	Non-error rate, Accuracy
11	Liantoni et al. (2019)	Watermelon	To detect maturity classification depends on skin texture	K-NN	Accuracy
12	Hruska et al. (2018)	Grapevine leaves	To detect the anomalies in the grapevine leaves	K-means clustering approach, ANN	Accuracy

9 Performance Analysis

The performance evaluations of each image for the proposed MRG-based MLP-WOA image classification scheme is demonstrated. The convergence graph for the three cases namely best, worst and average values concerning error are shown in Fig. 8. The classification techniques such as K-NN, SVM, and CNN are compared with the proposed ANN and MLP-WOA approaches. The MLP-WOA approach gives the best accuracy when compared with other methods and other existing results.

The sensitivity, specificity, accuracy, FPR, and FNR values of the proposed and existing methods for the input image for segmentation results compares.

10 Result and Discussion

In this research, the proposed work is implementing in MATLAB, and the experiment is performed employing a system with 4 GB RAM and 2.10 GHz Intel i-3 processor.

The proposed MRG-based segmentation and MLP- WOA-based classification method inspects an efficient quality food product and identifies the products whether the food is defective or quality food products.

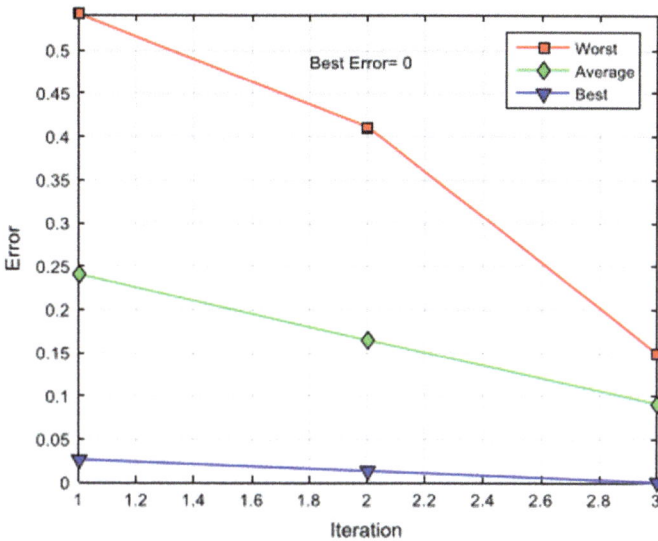

Fig. 8 Convergence of evolution

The main aim of this research work is to inspect the food quality from the input food databases in order to achieve high accuracy and low mean square error. In this phase, MLP-WOA technique used to classify the quality food from the image. This classification gives the high accuracy and low MSE.

The survey analysis for this phases involved in inspecting the quality food from the input image and the food quality with high accuracy, sensitivity, specificity, and minimal error.

In this, if two methods ANN and MLP are compared with the sensitivity, specificity, accuracy, and error, in which error is minimum for MLP-WOA compared with ANN, whereas accuracy, sensitivity, and specificity are high compared to ANN technique.

11 Conclusion

The various methods used for the detection of food quality. There are some limitations that happened by using this method for recognition.

Improved methods MLP-WOA have to be designed and implemented effectively to overcome the problems like noisy image, for real-time processing, to enhance accuracy, complexity, and computational time has to be reduced, mobile-based application shall be designed to automate on tree fruit detection as well as counting approach and also to obtain the healthy food from the defected food.

The aim of this research work is to inspect the food quality with image databases form the online. The quality food and the defected food are separated by means of the neural networks. There are different network classifiers used to extract the quality food from the input database.

The maximum accuracy with reduced error and high sensitivity and high specificity were obtained by means of MLP-WOA approach.

This proposed approach calculates the weights and based upon the weights it gives rankings for the food.

References

1. Ok G, Shin HJ, Lim M-C, Choi S-W (2018) Large-scan-area sub-terahertz imaging system for non-destructive food quality inspection. Food Control 96:383–389
2. Wang K, Sun DW, Pu H (2017) Emerging non-destructive terahertz spectroscopic imaging technique: principle and applications in the agro-food industry. Trends Food Sci Technol 67:93–105
3. Yaseen T, Sun DW, Cheng JH (2017) Raman imaging for food quality and safety evaluation: fundamentals and applications. Trends Food Sci Technol 62:177–189
4. Kaushal M, Khehra BS, Sharma A (2018) Soft computing based object detection and tracking approaches: State-of-the-Art survey. Applied Soft Computing 70:423–464
5. Bairam U, Green PJ (2018) Melon crack identification and classification using K-means clustering for quality inspection. In: 2018 Colour and visual computing symposium (CVCS), IEEE, pp 1–5, 19 September 2018
6. Jayanthi AN, Nareshkumar C, Rajesh S, Kumar S, Vazha Gurunathan K (2019) Fruit quality inspection system using image processing. Iconic Res Eng J 2:260–263
7. Sengar N, Gupta V, Dutta MK, Travieso CM (2018) Image processing based method for identification of fish freshness using skin tissue. In: 2018 4th international conference on computational intelligence and communication technology (CICT), IEEE, pp 1–4, 9 February 2018
8. Khan MA, Lali MI, Sharif M, Javed K, Aurangzeb K, Haider SI, Altamrah AS, Akram T (2019) An optimized method for segmentation and classification of apple diseases based on strong correlation and genetic algorithm based feature selection. IEEE Access 7:46261–46277
9. Ye H, Liu C, Niu P (2018) Cucumber appearance quality detection under complex background based on image processing. Int J Agric Biol Eng 11(4):193–199
10. Iqbal Z, Khan MA, Sharif M, Shah JH, ur Rehman MH, Javed K (2018) An automated detection and classification of citrus plant diseases using image processing techniques: a review. Comput Electron Agric 153:12–32
11. Chen S, Xiong J, Guo W, Bu R, Zheng Z, Chen Y, Yang Z, Lin R (2019) Colored rice quality inspection system using machine vision. J Cereal Sci 88:87–95
12. Arora M, Dutta MK, Travieso CM, Burget R (2018) Image processing based classification of enzymatic browning in chopped Apples. In: 2018 IEEE international work conference on bioinspired intelligence (IWOBI), IEEE, pp 1–8, 18 July 2018
13. Soto JN, Martínez SS, Gila DM, Ortega JG, García JG (2018) Fast and reliable determination of virgin olive oil quality by fruit inspection using computer vision. Sensors 18(11):3826
14. Uluişik S, Yildiz F, Özdemİr AT (2018) Image processing based machine vision system for tomato volume estimation. In: 2018 electric electronics, computer science, biomedical engineerings' meeting (EBBT), IEEE, pp 1–4, 18 April 2018
15. Tanwong K, Suksawang P, Punsawad Y (2018) Using digital image to classify phenotype of the rice grain quality under agricultural standards act. In: 2018 22nd international computer science and engineering conference (ICSEC), IEEE, pp 1–4, 21 November 2018

16. Aditya K, Saputro AH, Handayani W (2018) Enhancement of visible-NIR imaging prediction system using genetic algorithm: prediction of carotenoid content in Amaranthus sp. Leaf. In: International conference on electrical engineering and informatics (ICELTICs), IEEE, pp 106–110
17. Zhang B, Huang W, Li J, Zhao C, Fan S, Wu J, Liu C (2014) Principles, developments and applications of computer vision for external quality inspection of fruits and vegetables: a review. Food Res Int 62:326–343
18. Castro W, Oblitas J, De-La-Torre M, Cotrina C, Bazan K, Avila-George H (2019) Classification of Cape gooseberry fruit according to its level of ripeness using machine learning techniques and different color spaces. IEEE Access 7:27389–27400
19. Al-Marakeby A, Ayman A. Aly AA, Salem FA (2013) Fast quality inspection of food products using computer vision. Int J Adv Res Comput Commun Eng 2(11)
20. Soon JM (2019) Rapid food hygiene inspection tool (RFHIT) to assess hygiene conformance index (CI) of street food vendors. LWT 108304
21. Sahu D, Potdar RM (2017) Defect identification and maturity detection of mango fruits using image analysis. Am J Artif Intell 1:5–14
22. Thendral R, Suhasini A Automated skin defect identification system for orange fruit grading based on genetic algorithm. Curr Sci 112:1704–1711
23. Ganganagowdar NV, Gundad AV (2019) Intelligent computer vision system for vegetables and fruits quality inspection using soft computing techniques. Agric Eng Int: CIGR J 21:171–178
24. García M, Candelo-Becerra JE, Hoyos FE (2019) Quality and defect inspection of green coffee beans using a computer vision system. Appl Sci 9:1–18
25. Bairam U, Green PJ (2018) Melon crack identification and classification using K-means clustering for quality inspection. In: Colour and Visual Computing Symposium, pp 1–18
26. Benalia S, Cubero S, Prats-Montalbán JM, Bernardi B, Zimbalatti G, Blasco J (2016) Computer vision for automatic quality inspection of dried figs (Ficuscarica L.) in real-time. Comput Electron Agric 120:17–25
27. Kaur M, Sharma R (2015) ANN based technique for vegetable quality detection. IOSR J Electron Commun Eng 10:62–70
28. Ok G, Shin HJ, Lim M-C, Choi S-W (2018) Large-scan-area sub-terahertz imaging system for nondestructive food quality inspection. Food Control 1–27
29. Kaur M, Sharma R (2015) Quality detection of fruits by using ANN technique. IOSR J Electron Commun Eng 10:35–41
30. Wang J, Yue H, Zhou Z (2016) An improved traceability system for food quality assurance and evaluation based on fuzzy classification and neural network. Food Control 1–20
31. Pham VH, Lee BR (2015) An image segmentation approach for fruit defect detection using K-means clustering and graph-based algorithm. Vietnam J Comput Sci 2:25–33
32. Suganya V, Thilagavathi P (2019) A fruit quality inspection system using faster region convolutional neural network. Int Res J Eng Technol (IRJET) 6:6717–6720
33. Sustika R, Subekti A, Pardede HF, Suryawati E, Mahendra O, Yuwana S (2018) Evaluation of deep convolutional neural network architectures for strawberry quality inspection. Int J Eng Technol 7:75–80
34. Qin J, Kim MS, Chao K, Chan DE, Delwiche SR, Cho B-K (2017) Line-scan hyperspectral imaging techniques for food safety and quality applications. Appl Sci 7:1–22

Chapter 44
A Comparison of Sentiment Analysis Techniques on Movie Reviews

Brenden Carvalho and Siddhaling Urolagin

1 Introduction

Sentiment analysis is a rapidly growing field in data science. It deals with the classification of data into positive, negative, or neutral classes. Sentiment refers to the view or opinion that the author expresses in the text. A text with positive sentiment would mean that the author associates a positive emotion with the text subject. Similarly, for negative and neutral classes, the objective of this work is to develop a domain-specific system that can evaluate the sentiments expressed in movie reviews.

The huge growth in sentiment analysis might be due to the rise in social media use, for example, Twitter, Facebook, etc. These sites allow users to post their comments regarding any matter on the site and make it instantly available to all the other users. The comments might be related to any matter ranging from movies to politics. This results in a large collection of opinionated data from various different people. Mining this data will provide lots of useful information, which will be very helpful to various organizations. For example, during election season, the peoples' opinion regarding a certain law or policy can be gaged using sentiment analysis using data from social media sites; thus, helping the candidate to better understand the needs of their people.

In this paper, we intend to perform binary sentiment analysis of movie reviews. We obtained a movie review dataset [1] that contains 50,000 reviews. Each of the reviews are labeled into positive and negative classes. Any given movie doesn't have more than 30 reviews; further, the train and test sets contain a disjoint selection of movies; so, there is no performance gain from associating movie specific terms and class labels.

B. Carvalho (✉) · S. Urolagin
Department of Computer Science, BITS Pilani, Dubai Campus, Dubai International Academic
City, PO Box 345055, Dubai, UAE
e-mail: h20190009@dubai.bits-pilani.ac.in

S. Urolagin
e-mail: siddhaling@dubai.bits-pilani.ac.in

© The Author(s), under exclusive license to Springer Nature Singapore Pte Ltd. 2021　　563
X.-Z. Gao et al. (eds.), *Applications of Artificial Intelligence in Engineering*, Algorithms
for Intelligent Systems, https://doi.org/10.1007/978-981-33-4604-8_44

This paper will evaluate the performance impact of various text preprocessing techniques, as well as different learning algorithms including logistic regression and support vector machines and determine the best choice for this application. The paper will be structured as follows: In Sect. 2, we present some of the work related to the topic of this paper. In Sect. 3, we discuss the methodology for sentiment classification. In Sect. 4, we show the details of the experiments and discuss the results. In Sect. 5, we summarize our findings.

2 Literature Survey

Aloufi et al. [2] performed sentiment analysis on football specific tweets using different learning algorithms to compare their performance. Algorithms used were Support Vector Machine (SVM), Naive Bayes (MNB), and Random Forest (RF). The results showed that the BOW model (unigram) achieved the best performance and that SVM was most robust and consistent in performance as compared to MNB and RF. Traditionally sentiment analysis dealt with only three classes, positive, neutral, and negative. Bouazizi et al. [3] proposed an approach to sentiment analysis with more than the usual three classes. They experimented with classifying tweets into seven classes and achieved an accuracy of 60.2%. They further proposed a sentiment quantification approach [4], where they identify the different sentiments expressed and attribute a score to each sentiment. The sentiment with the highest score is selected. It averaged an F1-score of 45.9% for 11 different classes. Sentiment is usually assigned based on the information in the text; however, Li et al. [5] performed sentiment analysis using text data about a specific event from social media along with the time series of that data. It also took into account the identity and influence of the different users making the posts. Sentiment computation is traditionally based on supervised methods, but such methods are not scalable to big data like social media. Jiang et al. [6] proposed a method of sentiment analysis of news events using a word association emotion network which is modeled by the social media data of a news event. The network can be then used to compute the word and text emotions. Another issue with social media is that the data is constantly growing; Doan et al. [7] use an incremental learning approach to sentiment analysis. This is beneficial when the training dataset is constantly growing with newer texts. The system was implemented with a variant of random forests to perform sentiment analysis on customer reviews. An integral part of any sentiment analysis is text preprocessing; Jianqiang et al. [8] discuss the effects of different text preprocessing methods on sentiment analysis performance. The performance was measured using four different classifiers. The results revealed that removal of numbers, stop words, and URLs don't affect the classifiers performance; but it is good for reducing noise. On the other hand, replacing negation and acronyms greatly improve the classifiers performance. Sentiment can be calculated for other languages as well, Nanda et al. [9] proposed a system that performs sentiment analysis on movie reviews written in Hindi. It made use of a Hindi SentiWordNet, which is a dictionary that defines the polarity of Hindi words.

Sentiment is usually classified for an entire text and is based on the score contributed by each term. Sindhu et al. [10] proposed a system that processes the text and predicts an aspect and then specifies the sentiment of that aspect. This aspect-based opinion mining system was used for evaluation of faculty performance. It is based on a two layered LTSM model. The first layer predicts the aspect, and the second layer classifies the sentiment expressed. The system attained an accuracy of 91% in aspect extraction and 93% in sentiment detection.

3 Methodology

Sentiment analysis can be considered as a classification problem. Depending on the text data, the sentiment classification can be either positive or negative. The general process is as depicted in Fig. 1.

First the dataset [1] is retrieved from the source. The training and testing data are then read into memory. The data is then preprocessed using different techniques. Noise removal, this step removes irrelevant noise like html tags and punctuation marks, as these are not useful to the learning algorithm. Regular expressions are used to identify and remove the noise characters. Next, we perform stop words removal; these are common words like "he," "she," "we," "if," "but," etc. These words can be safely removed from the text as they do not contribute to the sentiment being expressed. Removal of stop words will also increase performance. Normalization is also an important part of the process; here, different forms of a given word are converted into a single form. Two methods can be used, stemming and lemmatization.

The next step to perform is feature extraction; learning algorithms cannot make direct use of raw text data; it needs to be converted into a numerical form. In this step, the raw text data is converted into feature vectors that can be used by different learning algorithms. Bag of words representation is used for this purpose. This representation consists of a matrix with one column for each token (word) in the corpus, and one row for each document instance. The process of converting raw text data into numerical feature vectors is referred to as vectorization. As most documents will only use a

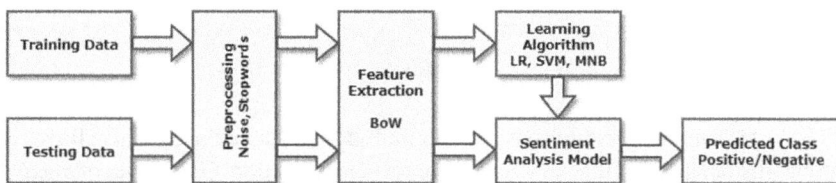

Fig. 1 Sentiment analysis and classification of reviews

small subset of the words in the corpus, most of the columns will be 0 and thus can be stored in a sparse matrix, to save memory and speed up algebraic operations.

The first type of vectorizer is the count vectorizer. This vectorizer counts the number of occurrences of a given token in the document and uses this value as its weight. Another type is the binary count vectorizer. It sets the value to 1 if the token is present and 0 otherwise. A more commonly used type is the TF-IDF vectorizer; it stands for term frequency-inverse document frequency. It represents how important a given word is in the corpus. The weight assigned to a token depends not only on the frequency in a given document, but on the number of documents that contain the token. The TF–IDF score (w) is given by,

$$w_{i,j} = \text{tf}_{i,j} * \log\left(\frac{N}{\text{df}_i}\right) \tag{1}$$

where $\text{tf}_{i,j}$ is number of occurrences of i in j. df_i is the number of documents containing i. N is the total number of documents.

n-grams are also used to increase the predicting power of the learning algorithms. An n-gram is a continuous sequence of n characters appearing in a document. Until now only unigram tokens were considered; however, bigram and trigram tokens can also be used.

After the feature vectors are prepared, we need to use a learning algorithm to train the model. The first learning algorithm discussed is, logistic regression. It can be used as a supervised classification algorithm, when combined with a decision threshold. This works very well in sparse datasets like the current one. Also, it tends to be much faster than the other algorithms. The cost equation is,

$$\min_{w,c}\frac{1}{2}w^T w + C\sum_{i=1}^{n}\log\left(\exp\left(-y_i\left(X_i^T w + c\right)\right) + 1\right) \tag{2}$$

Another important learning algorithm is the support vector machine. It is a supervised learning model. The goal of SVM is to choose a hyperplane that maximizes the margin between instances of the two closest classes. The optimization problem can be written as,

$$\text{argmin}_{w,b}\frac{1}{2}\|w\|^2 \tag{3}$$

$$\text{subject to } y_i(w.x_i + b) \geq 1, i \in [1, n]$$

The last learning algorithm considered is multinomial naive Bayes. Naive Bayes is a probabilistic classifier that works well for text classification. Given a set of classes c, documents d, and features f. The probability d_i belonging to class c is computed using Bayes rule given by,

$$P(c|d) = \frac{P(d|c) * P(c)}{P(d)} \qquad (4)$$

The probability that document d belongs to class c is calculated individually for each class c, and a class is assigned using the maximum a posteriori rule.

$$c_{(\text{map})} = \text{argmax}_{c \in C} P(c|d) \qquad (5)$$

4   Experimental Results

Dataset used contained 50,000 reviews. Each review was classified as positive or negative. The dataset was divided into two equal parts, for testing and training. Python 3.7 was used along with scikit-learn, nltk, and numpy libraries mainly. For each learning algorithm used, the classification report, confusion matrix and roc curve were recorded. First, we obtain baseline results using logistic regression. After removing the noise from the dataset, it was used to train a logistic regression model. The number of features is 92715. Binary feature vectors are used. Grid search was used to determine the value of regularization parameter C that provides the best accuracy. It was determined to be 0.05. Cross-validation was also done to prevent overfitting. Fivefold cross-validation was performed. The accuracy on the training set is 88.46%. The accuracy on the testing set is 88.152%. The results of the experiment are shown in Fig. 2. The confusion matrix shows us the false positive, and false negative values are quite low compared to the true positive and true negative. Figure 3 shows us the top five positive and negative words according to their sentiment scores. As expected, the generally accepted positive and negative words are classified accordingly.

Text preprocessing techniques were then applied in order to improve the accuracy of the model. Stop words were removed using the nltk library. Number of features

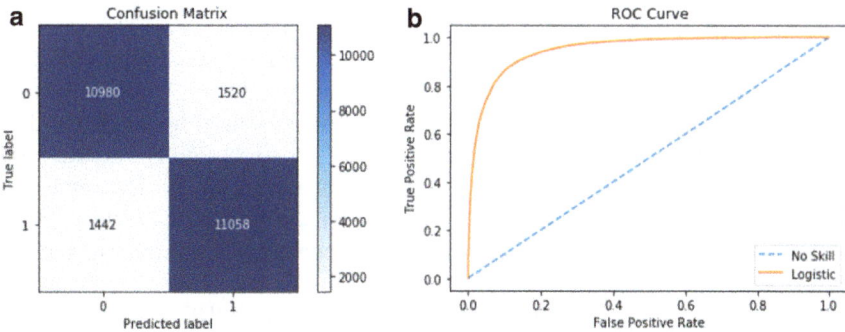

Fig. 2 Baseline logistic regression results. **a** Confusion matrix, **b** ROC curve

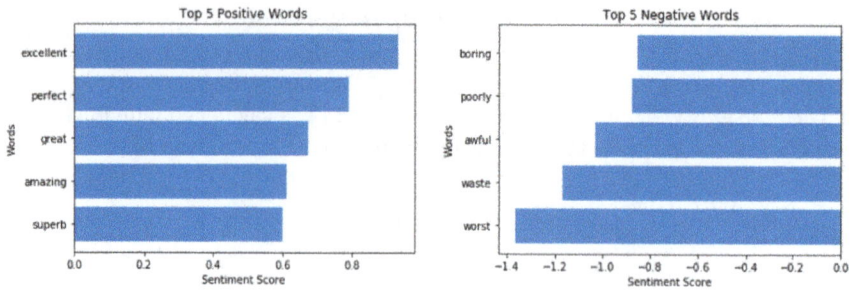

Fig. 3 Top five words' score as classified by logistic regression

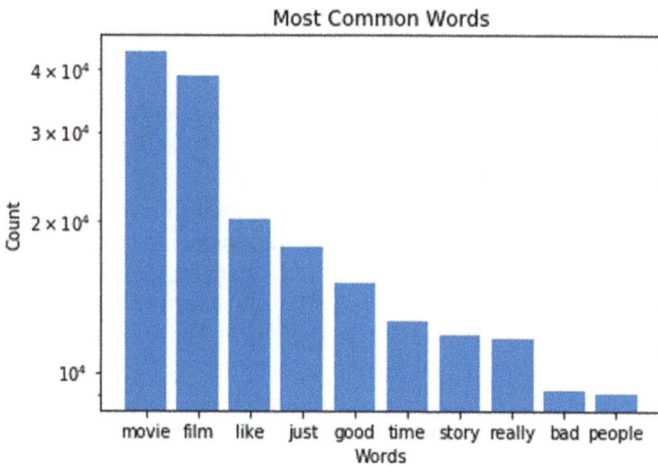

Fig. 4 Most common words

after removal is 92688. The accuracy on the test set was now 87.988%. In Fig. 4, common stop words are no longer the most common words as expected. Normalization was also performed, stemming removes affixes from words, leaving only the word stem. nltk library was used with the porter stemmer algorithm. The accuracy on the test set was 87.744%. Lemmatization is a different approach; it tries to identify part-of-speech of a given word and applies complex rules to get the word root. The accuracy on the test set was 88.004%. All features so far were single word or unigram. Feature vector with n-grams in the range $(1, 2)$ was tested. The accuracy on the test set was 89.759%. All feature vectors up to now were binary. The value was set 1 if the token was present and 0 otherwise. With word count, the feature vector stores the count of times the token appears in the document. The accuracy on the test set was 87.636%.

Term frequency-inverse document frequency statistic (TF-IDF) is calculated for each token, and that value is stored in the feature vector. The accuracy on the test set was 88.24%. Now, the various learning algorithms were tested. For the first test, the learning algorithm used was support vector machine along with n-gram and stop words. Support vector machine with linear kernel was used. Grid search was used to find penalty parameter C that provides best accuracy. It was determined to be 0.005. Fivefold cross-validation was also done to prevent overfitting. The model with best accuracy is selected. The accuracy on training set was 89.436%. The accuracy on testing set was 90.0279%. Figure 5 shows the corresponding results.

For the second test, learning algorithm used was logistic regression along with n-gram and stop words. The accuracy on training set was 89.312%. The accuracy on testing set was 89.96%. Figure 6 shows the corresponding results.

For the third test, learning algorithm used was multinomial naive Bayes with n-gram and stop words. The accuracy on testing set was 88.216%. Figure 7 shows the corresponding results.

The model trained using SVM with linear kernel on feature vectors with n-gram (1, 2) and stop words removed provided the best results as shown in Table 1. It reached an accuracy of 90.0279% on the testing set. Figure 8 shows the combined ROC curve of the three learning algorithms.

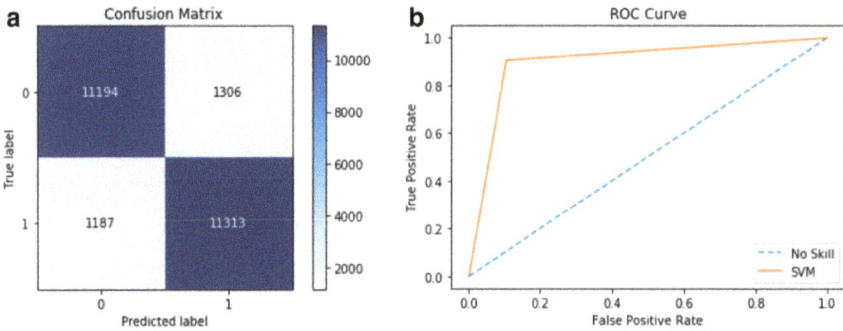

Fig. 5 SVM results. **a** Confusion matrix, **b** ROC curve

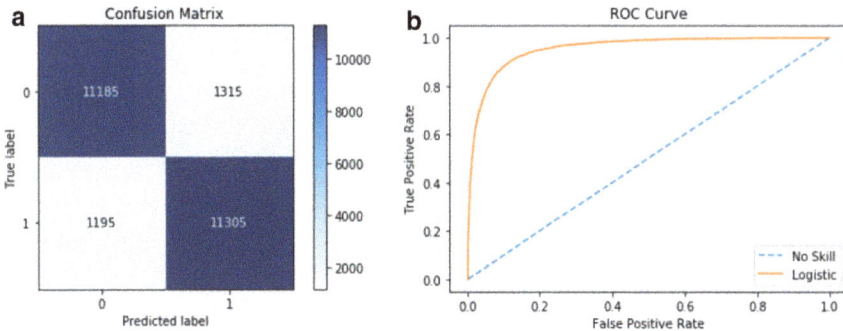

Fig. 6 Logistic regression results. **a** Confusion matrix, **b** ROC curve

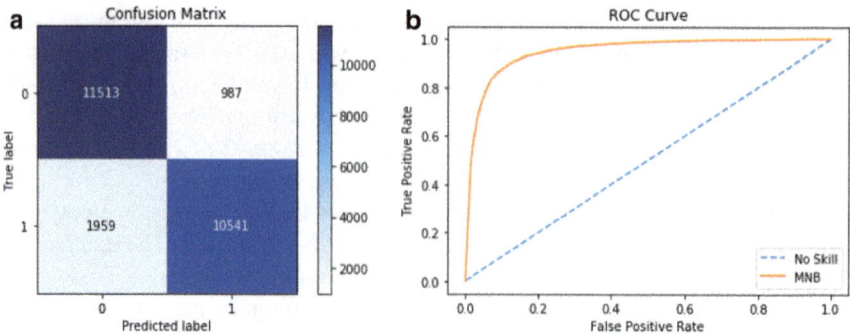

Fig. 7 Multinomial naive Bayes results. **a** Confusion matrix, **b** ROC Curve

Table 1 Overall learning algorithm accuracy results

Classifier	Accuracy (%)
Support vector machines	90.0279
Logistic regression	89.312
Multinomial naive Bayes	88.216

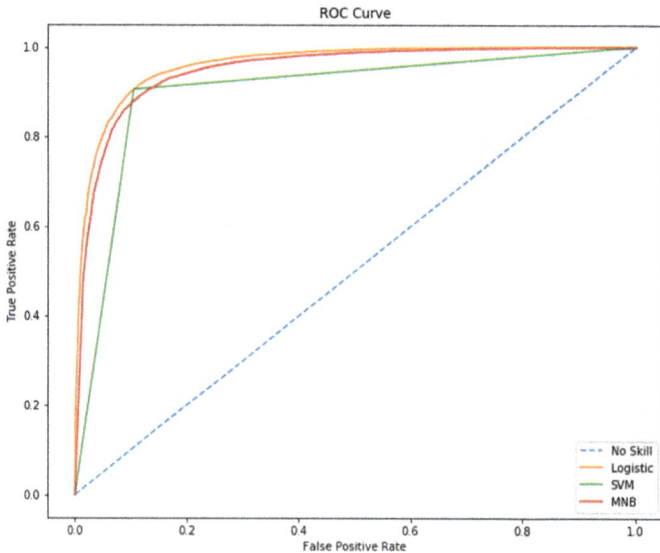

Fig. 8 Combined ROC curves

5 Conclusion

Multiple experiments were conducted to determine the performance of the different learning algorithms. Various text preprocessing techniques were also tested (Removal of stop words, normalization). Further, n-grams, word count feature vectors, TF-IDF were also tested.

Multiple experiments were performed to determine the optimal approach for the classification problem. Baseline results were obtained using logistic regression. Text preprocessing, stemming, lemmatization, n-grams, word counts were also used to improve accuracy. Three major learning algorithms were also evaluated, i.e., logistic regression, support vector machines, and multinomial naive Bayes. The model trained using SVM with linear kernel on feature vectors with n-gram (1, 2) and stop words removed provided the best results, achieving 90% accuracy on the test set.

References

1. Maas AL, Daly RE, Pham PT, Huang D, Ng AY, Potts C (2011) Learning word vectors for sentiment analysis. In: Proceedings of the 49th annual meeting of the association for computational linguistics: human language technologies, vol 1. Association for Computational Linguistics, pp 142–150
2. Aloufi S, Saddik AE (2018) Sentiment identification in football-specific tweets. IEEE Access 6:78609–78621
3. Bouazizi M, Ohtsuki T (2017) A pattern-based approach for multi-class sentiment analysis in Twitter. IEEE Access 5:20617–20639
4. Bouazizi M, Ohtsuki T (2018) Multi-class sentiment analysis in Twitter: what if classification is not the answer. IEEE Access 6:64486–64502
5. Li L, Wu Y, Zhang Y, Zhao T (2019) Time+User dual attention based sentiment prediction for multiple social network texts with time series. IEEE Access 7:17644–17653
6. Jiang D, Luo X, Xuan J, Xu Z (2017) Sentiment computing for the news event based on the social media big data. IEEE Access 5:2373–2382
7. Doan T, Kalita J (2016) Sentiment analysis of restaurant reviews on yelp with incremental learning. 2016 15th IEEE international conference on machine learning and applications (ICMLA). Anaheim, CA, pp 697–700
8. Jianqiang Z, Xiaolin G (2017) Comparison research on text pre-processing methods on twitter sentiment analysis. IEEE Access 5:2870–2879
9. Nanda C, Dua M, Nanda G, Sentiment analysis of movie reviews in hindi language using machine learning. In: 2018 International conference on communication and signal processing (ICCSP), Chennai, pp 1069–1072
10. Sindhu I, Daudpota SM, Badar K, Bakhtyar M, Baber J, Nurunnabi M (2019) Aspect-based opinion mining on student's feedback for faculty teaching performance evaluation. IEEE Access 7:108729–108741

Chapter 45
A Comprehensive Review of the Available Microgrid Protection Schemes and Their Challenges

Subhojit Paul, Nirmalya Maity, Samhita Sinha, Shounak Basu, Supriya Mondal, and Rohan Porel

1 Introduction

During the most recent decade, the idea of micro-grid has developed as a noteworthy method for incorporating maintainable vitality sources in the electric system. Its principle benefits lie in that it supplies power locally, diminishes network venture because of lower arrange limit prerequisites, lessens activity expenses and misfortunes, shaves the peak load, and thus, the reliability of the system increases. Be that as it may, alongside these benefits, micro-grids have likewise raised various challenges, among them is the issue of assurance.

Along these lines, there are two fundamental issues the micro-grid network has to deliver with respect to its assurance.

- Right off the bat, the assurance of when it ought to be islanded from the micro-grid, for example in light of strange conditions that the utility may experience.

S. Paul (✉) · N. Maity · S. Sinha · S. Basu · S. Mondal · R. Porel
Department of Electrical Engineering, University of Engineering and Management, Action Area-III, Plot No-3B/V, Newtown, Kolkata, West Bengal 700160, India
e-mail: subhojitpaul13@gmail.com

N. Maity
e-mail: nirmalya.maity18@gmail.com

S. Sinha
e-mail: samhitasinha15@gmail.com

S. Basu
e-mail: shounak99.sb@gmail.com

S. Mondal
e-mail: supriyomondal1080@gmail.com

R. Porel
e-mail: speak2rohanporel@gmail.com

- Also, furthermore, the arrangement of appropriately planned also, reliable protection system with the goal that it can dependably trip in case of a shortcoming inside it.

Among the distinctive security strategies of the micro-grid, we can examine some of them ahead of time and in extremely short so that in the hour of breaking down them we do not confront any issues (Fig. 1).

Adaptive protection: An adaptive protection system is a framework that can conceivably take care of issues by foreseeing an effect of micro-sources (DERs) and micro-grid setup on the relay performance and as needs be change the hand-off settings to guarantee that the entire micro-grid network is secured consistently. Adaptive protection is as 'an online action that adjusts the favored defensive reaction to an adjustment in framework conditions or prerequisites in an auspicious way by methods for remotely created signals or control activity.' The proposed adaptive protection gives a reasonable security to the micro-grid network for different fault conditions

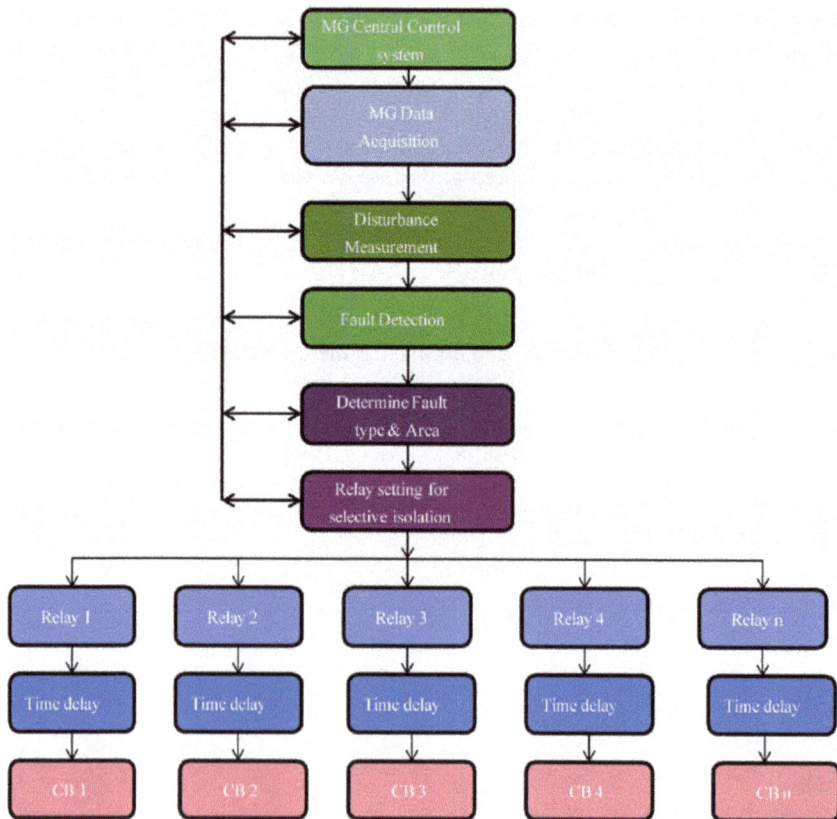

Fig. 1 Centralized protection scheme for microgrids

regardless of the working method of the micro-grid. The extraordinary part of the created adaptive protection is that it screens the micro-grid and in a split second updates hand-off deficiency current as indicated by the varieties that happen in the system. This is especially helpful for feeder moves and two-way power stream issues related with high-distributed energy resource.

Differential protection: Differential protection is a unit-type protection for a predetermined zone or bit of gear. It depends on the way that it is just on account of flaws inward to the zone that the differential current (contrast among input and output current) will be high. Be that as it may, the differential current can once in a while be significant even without an inside issue. Differential security plans are for the most part utilized for assurance against phase-to-phase fault and phase-to-earth fault. The differential protection utilized for power transformers depends on 'Merz-Prize Circling Current Principle'. This assurance conspire is chiefly utilized for the transformers in the microgrid network.

Distance protection: Separation protection is a non-unit system of protection, which gauges the impedance between the relay location and where the issue is occurrence and contrasts it and the set value. In the event that the deliberate impedance is not exactly the set value, the relay works and isolates the faulty section. Since, the line impedance is legitimately relative to line length, we get the specific location of the fault in kms. Since it secures a specific length of transmission line, it is known as a distance relay. On the off-chance that the deliberate impedance is not exactly the setting impedance, at that point the hand-off works. We utilize this distance protection method if there is any phase-to-phase or phase-to-earth fault occurs. What is more, once in a while it is utilized when the conductors are broken or the VT combine falls flat.

Voltage-based protection: Voltage-based protection procedure utilizes voltage estimations inside the framework to shield the micro-grids from various sorts of faults. The impact of various kinds of issues on the recreation center parts of the voltage is equipped for ensuring the micro-grids against 3 phase, 2 phase, 1 phase to ground faults.

Use of external devices for protection: In circumstances when fault current levels are radically extraordinary between the grid-connected and the islanded operation, the structure of a sufficient protection framework, which performs appropriately in the two circumstances, can be a genuine test. Right now, is a chance of applying an alternate methodology which effectively adjusts the fault current level when the microgrid changes from grid-connected to islanded operation and the other way around by methods for certain remotely introduced gadgets. These gadgets can either increment or reduction the fault level.

Overcurrent protection and symmetrical components: These recommendations attempt to upgrade the presentation of customary overcurrent protections, once in a while turning to estimation and figurines with balanced components. This assurance method depends on the investigation of current even segments which improves the performance and gives a ground-breaking framework to microgrids.

Cyber-protection schemes: Microgrids are progressively part of that recuperation plan since they can give an electric desert spring during a force blackout. Microgrids can provide power to a community's crucial administrations like law enforcement; fire security; medical care; conveyance of water, nourishment, and fuel; and correspondences. Some incorporate a public venue inside their impression, a safe house where the defenseless can assemble to accuse telephones and interface of cherished ones. These islands of power are made by utilizing utility disconnectable and independent force sources, for example, backup generators, spot generation, renewable, and batteries to control unavailable utility lines. Microgrids assume control over force conveyance during framework blackouts or voltage unsteadiness, or they can be set up as impermanent or portable force appropriation in crisis scenarios. More as of late, we have seen that it is not simply PC code that makes foundation powerless. Things being what they are, programmers may have an a lot more extensive playing field. Along these lines, a decent assurance plot is required so as to shield a microgrid from being exposed to anybody (Fig. 2).

In light of the attributes/properties of feeder(s) that establishes the foundation of a microgrid, it tends to be sorted as follows.

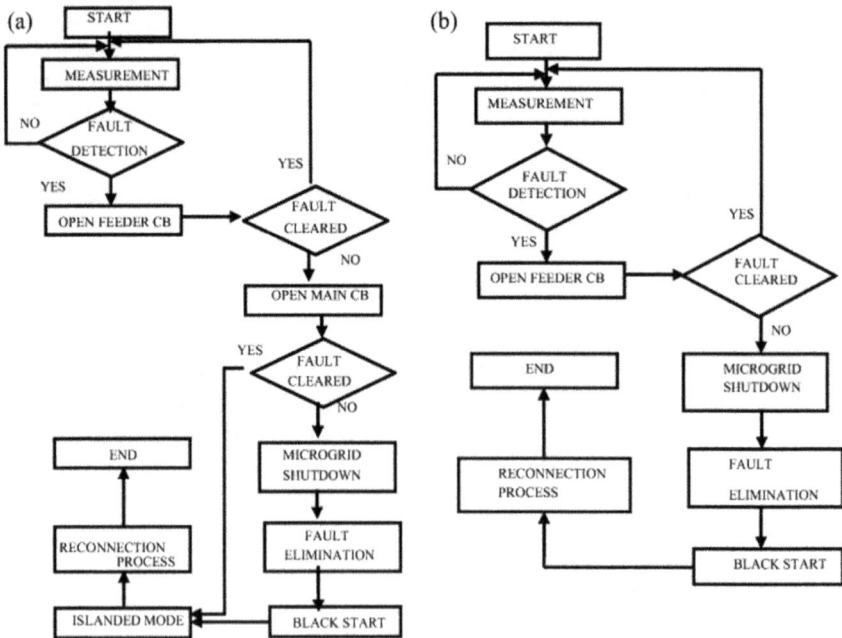

Fig. 2 **a** Flowchart of the protection scheme for grid-connected mode. **b** Flowchart of the protection scheme for islanded mode

Urban micro-grids: In this category, feeders are in a populated or a concentrated mechanical zone, the feeders are thickly stacked, the primary trunk and the laterals are genuinely short, and the level of op-sidedness is not high. Short out proportion of a urban smaller scale network at its PCC is regularly over 25. Along these lines, during the macrogrid-associated method of a urban microgrid, the voltage and recurrence are directed by the macrogrid, voltage profile is genuinely level, and transient quick voltage is well behaved. Thus, ECDG units can be synchronized with the microgrids no sweat, which is of significance for protection.

Rural micro-grids: In this category, feeders are situated in an inadequately populated zone, and along these lines, the heap is dispersed; the fundamental trunk and the laterals are recognizably long; the short out proportion is not really as high as that of the urban microgrid; furthermore, the voltage profile is not level. In a rural microgrid, voltage irregularity and vacillation can be critical. In this manner, DER units have sway on voltage, and whenever allowed, it can be controlled to aid the feeder voltage guideline.

Off-grid micro-grids: An off-grid microgrid is either geologically situated in a remote region with no opportunities for macrogrid connections, or encompassed by troublesome territory for transmission line association. By definition, an off-grid microgrid consistently works in islanded mode, and along these lines, it does not agree to the exacting meaning of 'microgrid,' i.e., does not have large-scale network-associated method of activity. Be that as it may, reconciliation of huge size DER units with off-grid microgrid is happening at quicker and higher pace than urban and provincial microgrid, created/a work in progress SC-EMS methodologies for off-grid microgrid are completely material to islanded method of urban and provincial microgrid, and there exist significant social, prudent, political, and ecological motivators/strain to coordinate sustainable power source based sources with off-grid microgrids. In this way, off-grid microgrids will possibly be the quickest developing microgrid type and stay as an individual from microgrid family as far as necessities and difficulties.

2   Existing Schemes for Microgrid Protection

The protection of microgrid ought to react to both utility grid and microgrid fault occurrences. On the off-chance that an issue happens on the utility grid, the ideal reaction is to confine the microgrid from the remainder of system. This prompts islanding operation of microgrid. On the off-chance that an issue happens inside the microgrid, the assurance framework ought to separate the littlest conceivable fault region of microgrid to clear the flaw. As of late, different procedures have been advanced to introduce a dependable assurance plot for microgrid that is abridged in Fig. 3. The methodologies will be delineated accurately in the accompanying subsections.

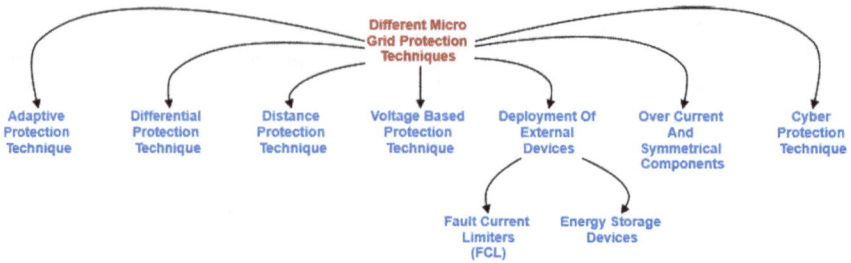

Fig. 3 Different microgrid protection schemes

2.1 Adaptive Protection: Available Schemes and Their Challenges

This can take care of the issues related with the micro-grids in both grid-connected and islanded method of operation. There is a programmed correction of relay settings when the microgrid adjusts from grid-connected mode to islanded method of operation and vice verse.

Adaptive protection is an online framework that alters the favored defensive reaction to change in framework conditions in auspicious way through outer produced signals or control actions. So as to give more assurance, a correspondence framework can be applied between individual relays and focal PCs with the goal that they can trade the data about the framework (Fig. 4).

Adaptive protection framework without the need of correspondence framework can be applied through the simulation of voltage response for both short-circuit and

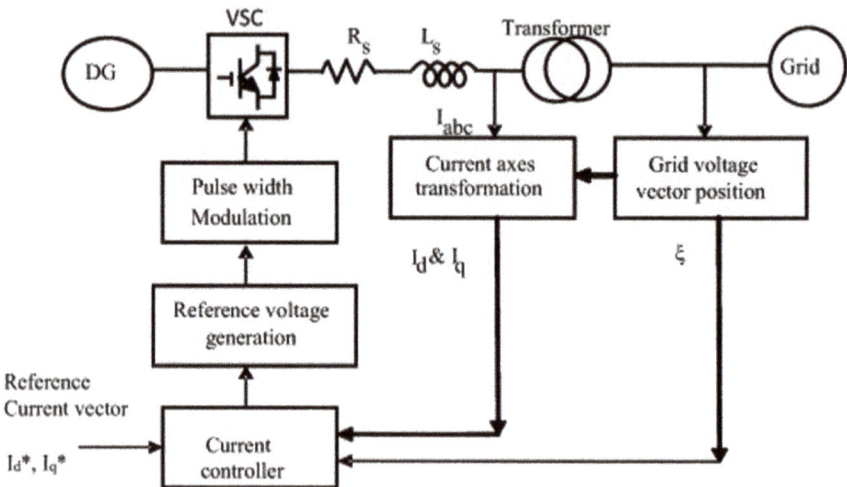

Fig. 4 Block diagram of adaptive protection system of a microgrid

over-load events. It is estimated from this re-enactment that the voltage drop coming about because of short circuits are essentially higher than the voltage drop coming about because of over loads. Consequently, a voltage-based fault detection strategy can be actualized to separate short-circuit and over-load events.

An epic protection scheme can be executed utilizing digital relays and propelled correspondence strategy, and the protection settings can be intermittently refreshed by the microgrid central controller as indicated by the microgrid states.

An adaptive fault current protection calculation can be executed by deciding the framework impedance from voltage and fault current parts. After that the current quick protection modifies the settings consequently by contrasting utility grid and microgrid impedances.

Likewise, a security framework can be executed by utilizing energy stockpiling and isolation transformers to ensure low voltage microgrids in both grid-connected and islanded method of operation. So as to perceive the grid's method of operation, the overcurrent assurance and dq0 voltage identification were utilized for grid-connected and islanded method of operation separately.

Extra protection schemes can be actualized in the utilization for broad correspondence for observing and refreshing the settings of relays as per the diverse method of operation of microgrid. There will be a 'Central Protection Unit (CPU)' which will speak with relays to refresh their working current.

Additionally, numerical relays in microgrid are utilized to adjust the settings of the relays to the status of the microgrid. At whatever point the microgrid is detached from the grid, the relay settings were consequently changes to the related gathering of settings.

Adaptive protection is basically helpful for feeder moves and two-way power stream issues related with high-distributed energy resource. The principle favorable position of utilizing this strategy is that by utilizing digital relays which have their own settings and attributes work we can change them according to our framework at whatever point we need (Fig. 5).

Challenges associated with adaptive protection schemes:

- The requirement for redesigning or refreshing the protection devices which are right now utilized in the power system network.
- It is important to know all the conceivable arrangement of the microgrids so as to execute these procedures.
- Making the entire correspondence structure can be exorbitant.
- Short-circuit calculations are confounded for a microgrid with various working modes.

Input Signal

Relay Output

ANALOG INPUT
SYSTEM

- Surge Filter
- Anti-aliasing Filter
- A/D Sample/Hold

Power
Supply

DIGITAL OUTPUT
SYSTEM

PROCESSOR

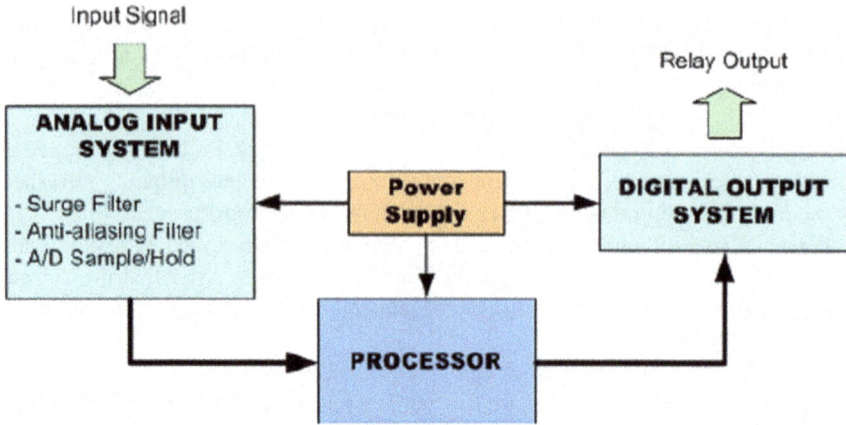

Fig. 5 Block diagram of digital relays

2.2 Differential Protection: Available Schemes and Their Challenges

This strategy is applied to ensure numerous components inside a power system network. It tends to be seen that differential relay is only a fundamental use of Kirchhoff's current law (KCL). For an external fault, the current in the two current transformers will be equivalent in magnitude and inverse in phase, and in this manner, no current will be flowing through the relay loop. For the short-circuit event, the current will be coursing through the operate circuit making the relay issue a trip output.

A consolidated method can be introduced to protect microgrid by differential protection and balanced components count against single line-to-ground (SLG) and line-to-line (LL) faults. Right now, the microgrid is partitioned into a few protection zones with relays. The differential current segments were sent to identify faults in the upper zone of protection (Fig. 6).

The balanced part is deployed to distinguish SLG faults in the downstream zone of protection and LL faults in all zones of protection. This all-out plan can secure a microgrid in islanded method of operation.

With respect to eventual fate of microgrids, two significant difficulties, for example voltage/frequency control and protection, can be mulled over. These two difficulties can be overwhelmed by utilizing relays intended to work in 50 ms which could secure the microgrid in both grid-connected and autonomous operation modes.

Another protection scheme can be used in microgrids containing synchronous-based and inverter-based DG. Additionally, there are three diverse protection systems accessible for recognizing phase to ground faults in isolated neutral microgrid.

An epic protection scheme can be built up dependent on a portion of the standards of synchronized phasor estimations and microprocessor relays to perceive all sort

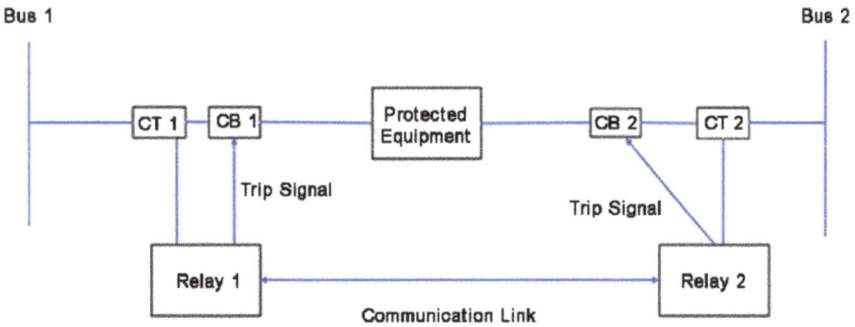

Fig. 6 Differential protection schemes in microgrids

of faults including high impedance faults (HIF). Here, if the supreme estimations of two examples were seen as over the pre-decided excursion limit, a trip signal will be produced by the microprocessor and it will be send to the tripping gadget.

It is discovered that a staggered approach would give the best type of system protection in a meshed microgrid. The idea is to build up an exceptionally solid correspondence design which can naturally recognize the faulted points in the framework and effectively confining them.

A novel protection system dependent on differential protection can be mulled over which will incorporate all the security difficulties, for example bidirectional power flow and fault current level in islanded mode of operation. here if the feeder protection is to be mulled over just as some arrangement identified with ensure buses and distributive generation sources inside the microgrid, must be finished.

One of the hugest advantages of actualizing differential protection plan is that they are not delicate to bidirectional power flow and decrease of fault current degree of microgrid in islanded microgrids (Fig. 7).

Challenges associated with differential protection schemes:

- On the off-chance that the communication system flops at any rate, at that point an auxiliary protection scheme is important.
- Making a communication structure is moderately costly.
- Unbalanced loads or systems may bring about certain challenges in the previously mentioned conspires as those are just material for balanced system or load.
- Transient steadiness during association and detachment of distributive generation may bring a few issues.

2.3   Distance Protection: Available Schemes and Their Challenges

A protection framework can be created utilizing a distance relay which is intended to work between the relay location and the chosen foreordained point. This plan

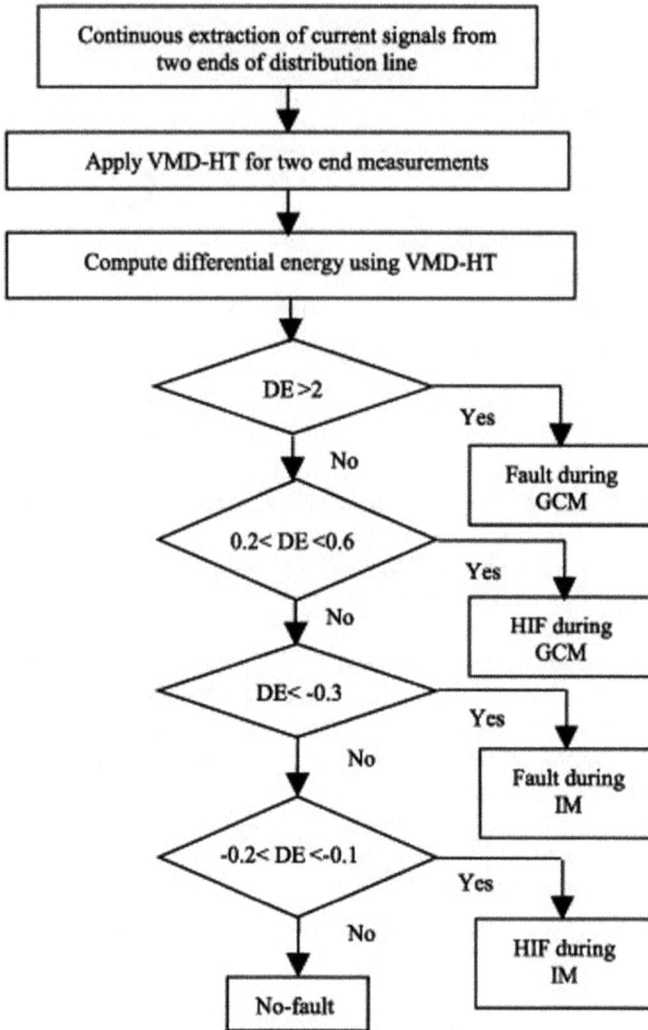

Fig. 7 Block diagram of differential protection system of microgrid

depends on the principle that impedance of a line is proportional to its length. So it is proper to utilize a relay which can quantify the impedance of a line up to a foreordained point (reach point) (Fig. 8).

Additionally, another protection plan can be created, in light of an admittance relay with reverse time tripping qualities. The relay could recognize and disconnect the faults for both grid-connected and autonomous microgrids. A while later by analyzing the fault qualities, the sequence current and voltages at the relay location can be determined. Despite the fact that the viability of this technique is still not approve.

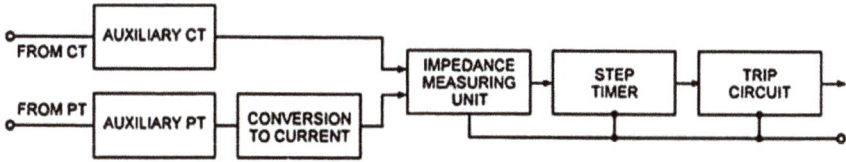

Fig. 8 Distance relay based on current comparison principle

As the distance relay has a quick operational time accordingly, this method may detect the fault part of the system quick and can expel that in extremely less time so as to protect the system. Likewise, this method may be helpful as the distance relay allows a high line loading (Fig. 9).

Challenges associated with distance protection schemes:

- Harmonics and transient steadiness may prompt a few issues in figuring the fundamentals.
- Fault resistance may impact in the count of the admittance.
- The estimation of admittance in short lines in distribution network is challenging.

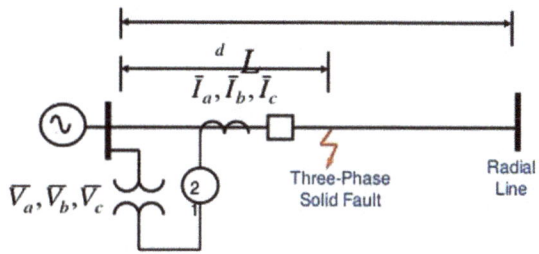

Suppose Relay Is Designed to Operate When:

$$|\bar{V}_a| \leqslant (0.8)|\bar{Z}_{L1}| \| \bar{I}_a|$$

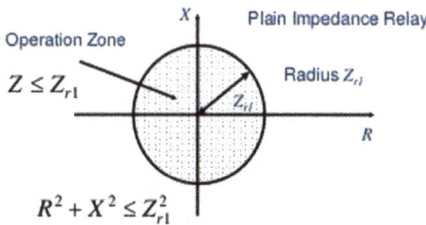

Fig. 9 Distance protection principle

2.4 Voltage-Based Protection: Available Schemes and Their Challenges

Voltage-based protection methods considerably utilize voltage estimations to ensure microgrids against various types of faults.

The plan, where output voltages of DG sources were observed and afterward changed into dc quantities utilizing the d-q reference outline, had the capacity to secure microgrids against in zone and out-of-zone faults. So as to ensure the microgrids, a communication interface has been built up during the process of transforming the DG sources into the dc quantities to recognize the in zone and out-of-zone faults (Fig. 10).

Additionally, another fault judgment strategy can be proposed dependent on distinguishing the positive sequence segment of the fundamental voltage to such an extent that it could give solid and quick identification to various sorts of faults inside the microgrid. Right now, waveforms of the three-phase voltages and the voltage magnitudes under balanced and unbalanced conditions were changed into the d-q reference

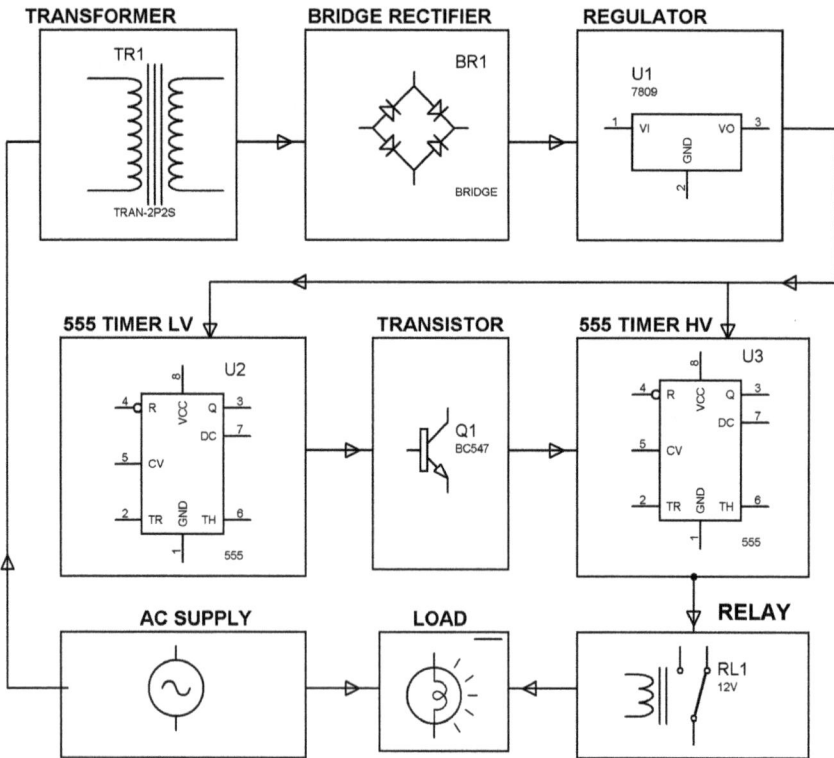

Fig. 10 Over voltage protection system

casing and contrasted with the adequacy of the crucial positive voltage sequence in the d-q coordinate system.

Another protection technique can be implemented which depended on the impact of various fault types on park parts of the voltage was equipped for ensuring microgrid against three phase, two phase and single phase to earth faults. The protection technique did not depend on the communication system during its activity; however, it could be enhanced through correspondence interfaces between various detection methods. The remarkable part of this plan contrasted with the one in was that the proposed procedure was intended for a specific microgrid as well as be utilized to ensure diverse microgrids with different setups.

Also, an extra assurance system can be implemented dependent on bus-bar fault direction to protect the microgrids in both grid-connected and autonomous operation modes. It is appeared that industrial personal computers (IPCs) are utilized for both hardware and software relay protection.

The principle advantage of utilizing this protection scheme which depends on the d-q change for the voltage waveform is that it can distinguish the nearness of a short-circuit fault and it can also separate between the faults which are available either inside or outside.

Challenges associated with voltage-based protection schemes:

- Any voltage drop inside the microgrids may prompt misactivity of the protection gadgets.
- High impedance issues cannot be recognized utilizing this protection framework.
- This protection strategy can be utilized distinctly for a particular setup of microgrids relying upon the quality and the state of the protection zones. Accordingly, it may not be reasonable for microgrids with various structures.
- This method is less reasonable in grid-connected method of activity.

2.5   Protection Schemes with the Organization of External Devices and their Challenges

2.5.1   Fault Current Limiters (FCLs)

As referenced before in the presentation, the fault current levels are essentially unique between grid-connected and the autonomous activity modes.

Hence, the plan of a sufficient protection system, which acts in the two methods of activity, can be a genuine test. Right now is a chance of utilizing an alternate methodology which adequately alters the fault current level when the grid-connected mode adjusts from the microgrid associated with the autonomous activity mode and the other way around through explicit remotely introduced devices. These devices can either increment or decline the fault level (Fig. 11).

To diminish the totaled commitment of numerous DG sources, which can change the fault current level enough to surpass the structure furthest reaches of different

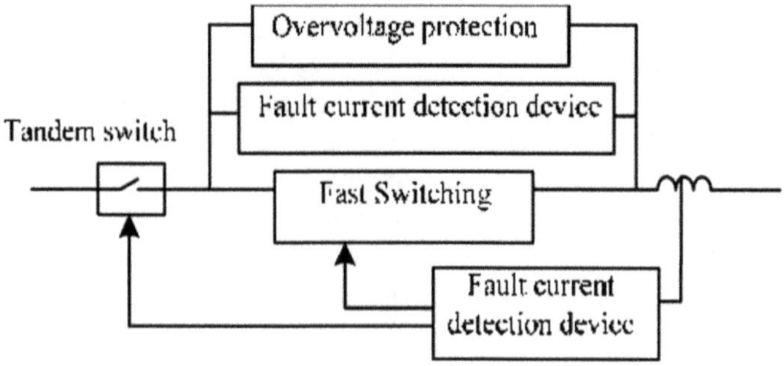

Fig. 11 Block diagram of the fault current limiter

gear parts, just as to ensure a sufficient coordination regardless of the taking care of impact of DG to fault current, fault current limiters (FCL) can be utilized. This impact is especially apparent with synchronous machine-based DG.

The fundamental advantage of this strategy is that it will take over the load for a brief time frame utilizing a reasonable grommet. Practically, for all intents and purposes, the fault current decrease procedure in a small-scale network like microgrids is performed by utilizing the fault current limiters (FCL). These gadgets are assessed in terms of their capacities like fault current restricting ability, power losses, and appropriate establishment areas.

2.5.2 Energy Storage Device

To level the fault current level in both grid-connected associated and in autonomous methods of activity, inferable from the constrained fault commitment by inverter interfaced DG sources. This can be accomplished in two distinct manners:

(a) By applying the vital energy storage gadgets like flywheels, batteries, and so on. In the microgrid, it will expand the shortcoming fault current to an ideal level, and permitting overcurrent assurance to work in a customary manner.

(b) By introducing explicit gadgets between the main-grid and the microgrid to lessen the commitment of fault current from the utility framework.

Energy storage gadgets have numerous natural advantages that can make them an important apparatus for meeting maintainability objectives. By improving the general proficiency of the power grid, storage quickens the more extensive selection of sustainable power source. Moreover, an energy storage gadget has no discharges, so it very well may be set at any place with no quick natural or air quality effects (Fig. 12).

Energy storage gadgets can move utilization of power from costly times of appeal to times of lower cost power during low interest. This permits facilities to take advantage of time-of-utilization valuing and diminish tariff structure change hazard to power cost.

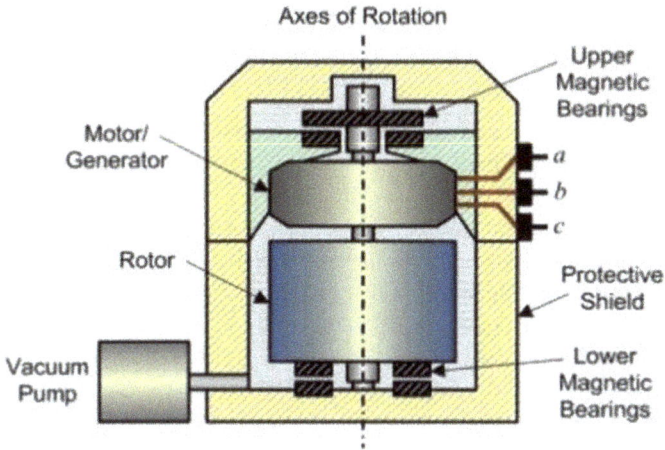

Fig. 12 Flywheel energy storage systems

Demand reaction for business and industries customarily includes tightening down use on occasion of peak demand. Energy storage can empower interest sought after reaction markets without affecting nearby energy use or activities. By reacting to utility cost signals, storage systems can increment money-related return, while profiting the grid network.

Making arrangements for crisis, backup power is a fundamental piece of the microgrid network. With advance storage arrangements, there are chances to redesign the framework to give crisis back-up as well as a host of other cash setting aside and cash making arrangements. Utilizing this strategy every day, its accessibility and reliability in case of a blackout can be expanded when contrasted with an independent battery framework and diesel generator that is just utilized during a blackout.

Challenges associated with the organization of external devices:

- Storage gadgets require enormous venture and need to coordinate the main grid's short-circuit level in order to ensure that faults are cleared in an auspicious way.
- The sending of plans dependent on the FCL innovation is just conceivable up to a specific measure of DGs associated. For elevated levels of DGs, it tends to be hard to decide the impedance estimation of the FCL because of the common impact of the DGs.
- Sources with high short out ability (flywheels and so forth.) require huge speculations, and their sheltered activity is reliant on the right support of the unit.
- The schemes dependent on an extra current source are exceptionally subject to the innovation of islanding recognition and the best possible activity of the present source.

2.6 Protection Schemes Based on Overcurrent Symmetrical Components and Their Challenges

These protection schemes which are mostly founded on the analysis of current symmetrical components endeavour to improve the performance of conventional overcurrent protection and give a powerful protection framework to microgrids.

Also, another protection system introduced a potential solution for perceive the fault in autonomous microgrid networks dependent on the estimations of current symmetrical components. To be exact, it is evident to utilize zero-sequence current detection in case of an upstream single line-to-ground fault (facilitated with unbalanced loads) and negative sequence current for line-to-line faults.

After few years, a three-stage communication helped selectivity scheme is proposed. In the plan, stage one recognized the issue occasion as per the local estimations. Stage two conveyed between communication breaker interchanges, and stage three adjusted the settings of the relays by means of supervisory controller (Fig. 13).

Furthermore, a microchip-based relay additionally works for this system related to the fault identification module. Unequivocal time relays are utilized beginning from the load side of auxiliary main and closure at the microgrid interface point expanding the time toward clear longer faults from the generation side of the grading way, not harming the microgrid framework. Cutoff points for the definite time delays at the generating sides are resolved dependent on certain constraints like sensitivity of the basic loads to voltage unsettling influences, the time over which the electronically coupled DGs can contribute to the fault current and the stability of the revolving machine-based DGs. In this technique, no direct communication link is introduced between the relays.

On the off-chance that the 3-phase (or multi-phase) framework is balanced then the phase segment for finding the impedance, voltage drops, currents, and so on are effortlessly decreased to single-phase arrangements since the arrangements of

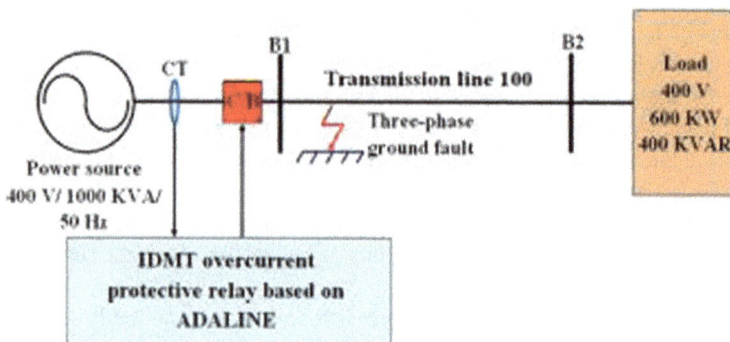

Fig. 13 Overcurrent protection

any 1-phase is a similar outcome for each phase. In this way, symmetrical components schemes for illuminating a balanced framework is time consuming and more involved.

Symmetrical component arrangements is a scientific development that takes the unbalanced 3-phase system esteems and changes over them into three balanced 3-phase systems significantly more smoothly than utilizing phase components.

Challenges associated with the protection schemes based on overcurrent symmetrical components:

- The main disadvantage of most of the mentioned protection strategies is identified with the need of communication frameworks. In such methods, the assurance coordination might be imperiled if there should be an occurrence of communication system failure.

2.7 Cyber-Protection in Microgrid

The smart grid energy system is a very efficient two-way delivery system of electrical energy between the generation and the distribution to customers.

As the smart grid uses renewable sources for generation and distribution, it can integrate multiple distributed energy resources (DERs), to its benefits such as reduced greenhouse gas emission and severely reduced transmission losses. Though being efficient, the smart grid is always prone to cyber-attacks due to its connectivity and advanced infrastructure. Reports from 2013 and 2014 show that the energy sector in the USA faced a loss of nearly 80 billion per year. However, this loss is recovered by increasing the tariff and other consumer expenses that does not act as a solution to the attacks (Fig. 14).

The incorporation of distributed energy resources into the grid system can provide the necessary solution to the cyber-attacks, as the smart grid includes many microgrids which control the distribution of power according to the needs of the consumer, the close monitoring time to place basis and feed important data required for energy management systems (EMS). The state of the system monitoring system can be used to collect data (such as stability analysis, fault detection, energy theft) for the microgrid so that it can be used for use later and studied later on for further improvement of the state of the system.

However, this communication infrastructure between the power generations in the smart grid to the distribution from the microgrids to the customer makes this infrastructure susceptible to cyber-attacks. Studies are continuously being carried on the state estimation of the cyber-attack on the smart grid.

Fig. 14 High-level
cyber-security objectives and
specific security
requirements

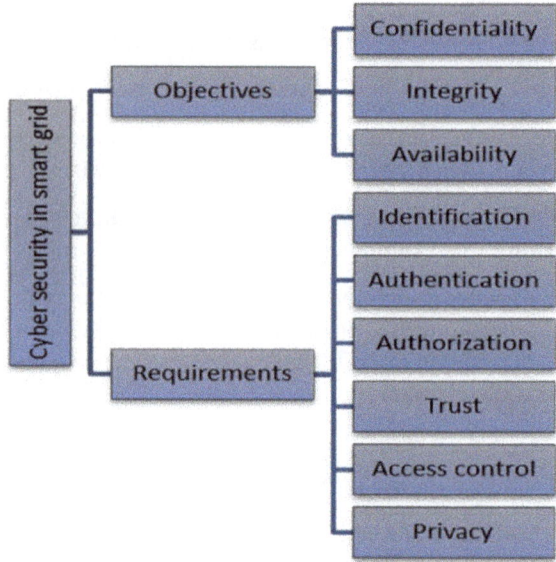

2.7.1 Software-Defined Network (SDN)

Due to the microgrids being more localized, power grid operates independently or along with smart grids via communication links with the smart grid.

Though there are lacks in the application such as generation control, state estimations are always prone to malicious cyber-attacks. The process of introducing a software-defined network (SDN) into the communication architecture enhances the cyber-security and also the microgrid resilience. SDN is an open-source programmable approach to designing, building, and managing communication networks. The SDN control layer is placed just in between the communication network layer and the application layer (such as monitoring, management, control which are microgrid applications). The SDN provides a self-healing network, verification into the network. This completes the multilayer architecture for the software-defined network. This separation of direct communication layer from the forward functioning control network by the SDN in the architecture is done by the abstractions and application programming interfaces (APIs); it consists of sensors, actuators, and networking devices (SCADA, supervisory control and data acquisition).

The SDN provides resilience against cyber-attacks; the SDN network can determine the location of the attack and then isolating the compromised device upon detection, not allowing malicious traffic in the communication network, and reconnecting the devices' sensors and meters back to the communication network to the preserve the working of the microgrid.

Application:

- **Self-healing Network**
  The self-healing network resets and reconfigures switches to establish new routes isolating malicious devices of the grid, triggers filtering switches which are located near the source of attack point and the control commands are made on demand. This minimizes the attackers' time window and the feature of global visibility enables self-healing on the point of attack not compromising the other network.
- **Real-Time Verification**
  The communication network being globally visible allows verification against security policies, network invariants to detect malware software and cyber-threats.
- **Intrusion Detection**
  SDN also incorporates the intrusion detection system (IDS) to continuously monitor the network activities and generate alarms against cyber-threats. As the workflow of microgrids is controlled by operators, a specification-based IDS is effective keeping in mind all the vulnerabilities of the microgrid protocols.

2.7.2   Kalman Filter in Cyber-Protection System of Microgrids

Kalman filter is a recursive filter which estimates the internal state of linear dynamic system from a series of noisy measurement. An X^2 detector easily detects discrepancies between the estimated data and the measured data and triggers the alarm instant. And, it can easily detect Denial of Service (DOS) attack and random attacks make the dynamic instability in power system, but not at the various time period. That is why Kalman filter is using the mathematical model. With the help of KF, it is very easy to detect the possible faults and the fault in the smart grid. Fault data injection (FDI) attack is also a big threat for a smart grid. Fault data injection (FDI) attacks in the target data integrity branches to make disturb a system. However, X^2 detector has a limitation and it is not detecting the fault data injection (FDI); so, Euclidean detector is used to be coupled with Kalman filter to solve the problem. Euclidean detector is based on cyber-attack detection in automatic generation control (AGC). Automatic generation control (AGC) is used in power system, to keep the system frequency with specified limits. Kalman filter (KF) is applied to estimate state variable system utilized by cyber-protection (Fig. 15).

The control system with the Kalman filter embedded for the estimation of the state vector and detector for the detection of attacks or faults. As shown in Fig. 15, $x(t)$ denotes the output of the state estimator that is fed to the controller and Z^{-1} is the control system feedback.

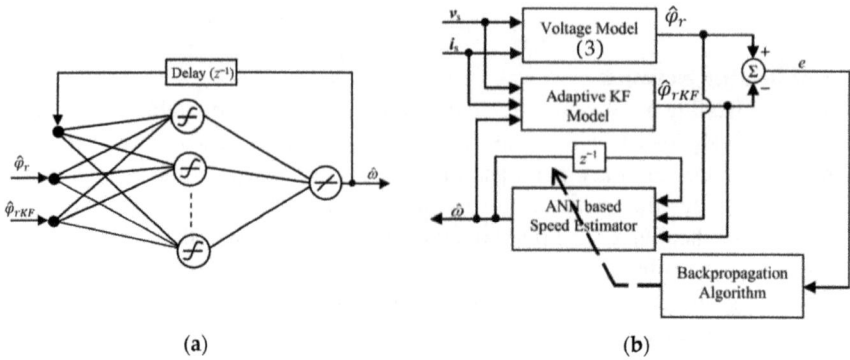

(a)   (b)

Fig. 15 Kalman filter using artificial neural network

2.8 Other Schemes Related to the Protection of Microgrids

Aside from the philosophies depicted in the past areas, there are additionally different methodologies that are less every now and again proposed. These have been separated into three gatherings.

- The main gathering is identified with high frequency treatment or transient data. The proposition in this field is essentially founded on wavelet transform, utilizing diverse wavelet capacities applied to high frequency calculations or voyaging waves.
- The subsequent gathering incorporates procedures dependent on artificial intelligence. Inside this gathering, the vast majority of the recommendations utilize the particle swarm optimization (PSO), for the most part for coordination improvement purposes. There are additionally a few creators that suggest the utilization of artificial neural networks (ANN), however, without having led any sort of study for microgrid protection.
- The third gathering incorporates the recommendations that depend on a simply numerical methodology. Inside this field, there is a wide assortment of approaches, for example,

  - Utilizing Dijkstra's algorithm (a graph search calculation) so as to acquire the relay progression of the system.
  - An alternate idea called brittleness entropy of a complex framework to identify deficiencies.
  - A state onlooker strategy as a fault finder, by partitioning the microgrid into various zones.
  - Utilizing the transient fault data (time and polarity highlights) remembered for the initial current voyaging waves and applying a changed numerical morphology innovation.
  - Utilizing time-coloured Petri nets for diagnosis of the faults in microgrids.

3   Technical Challenges in Microgrid Protection

The microgrid idea has to confront various challenges in a few fields, not just from the protection perspective, yet in addition from the control and dispatch point of view. All things considered, because of their particular characteristics and activity, microgrid protection frameworks need to manage new technical difficulties:

- Generation frameworks in both medium-voltage (MV) and low-voltage (LV) systems, making power stream bidirectional, and in this manner, planning a protection framework can be troublesome.
- With respect to protection frameworks, it needs to deal with the two operational methods of any microgrid, i.e., grid-connected and islanded/independent mode of operation.
- Topological changes in low-voltage (LV) system because of engagement/disengagement of generators, stockpiling frameworks and loads may make issues for the protection system.
- Irregularity in the generation of a few micro-sources associated in the microgrid.
- Expanding penetration of revolving machines, which may cause fault current that surpass ratings of the equipment and accordingly it could be hard for the framework to ensure the microgrid.
- Inadequate degree of short-circuit current in the islanding activity mode, because of the application of power electronics interfaced distributed generation (DG).
- Decrease in the allowable tripping times when any fault happen in MV and LV frameworks, so as to keep up the dependability of the microgrid.
- Aggravation tripping of protection because of faults on adjoining feeders.

As referenced over, the truth of the matter is that numerous micro-sources are commonly associated with the microgrid by methods for a power electronic inverter, either on the grounds that their output is not good for the grid voltage (photograph voltaic boards, micro-turbines, and so forth) or as a result of the adaptability gave by power electronics in the management of extracting of energy (wind turbines, etc.). Also, due to the reason that semiconductor switches have very less thermal inertia, inverters are effectively current restricted, and, as a result of their little fault current commitment, they lead unavoidably to various issues that must be considered by the protection framework.

- Attributes of the inverters under any fault conditions may not be steady with the current assurance gadgets.
- Throughout the entire microgrid, there might be various inverters with various attributes.
- Even on account of individual inverter, its fundamental qualities may contrast contingent upon its structure or application.
- There might be troubles in describing inverter behavior for cut off, since it relies upon the control methodology.
- Significantly diminished fault current level when the microgrid is changed to islanded mode of operation from the grid-connected mode.

The last point especially is one of the key issues which have been under research for as far back as years. The qualities of most defensive gadgets utilized in microgrids are normally like those utilized in distribution systems, and the traditional distribution network's protection depends on huge fault currents. Be that as it may, under islanded activity, the utility grid cannot contribute to the issue, and, accordingly, its extent is constrained to what the micro-sources can give. Thus, traditional overcurrent protection plans might be not at this point appropriate because of the current constraints of most inverters.

4   Comparative Analysis

The area of research for microgrids has confronted various challenges in fields of protection, control, and other additional point of view. Microgrids have been designed to improve the power quality and reliability of the customers. It is to be mentioned that the challenges have always been the key area of research to put forward an adequate protection technique in the grid system.

The method of choosing the most reliable technique for protection depends on selectivity, reliability, speed, cost as well as the simplicity of the method of protection.

Adaptive protection: Voltage is restrained, and the overcurrent relay/numerical directional relay is used, while the grid is connected in operation and the part of fault is islanded. The cost of this technique is reasonable and affordable. The distributed generation is rotating based and inverter based.

Differential protection: The protection device used is a digital relay, while the grid remains connected and only the part of the fault is islanded. The use of digital relays makes this protection technique costly. The distributed generation in this case is also rotating based and inverter based, and the communication link exists during the time of fault.

Distance protection: Distance relay is used for the protection technique, and the grid remains connected, and the operation in the part of the fault is islanded. The distributed generation is inverter based, and the communication link is broken off or cut off during the time of fault. The cost efficiency is quite affordable.

Voltage-based protection: The protection technique is different than using relays; here, the voltage measurements are used to protect the microgrid from the faults. The microgrid operation is islanded until the fault is cleared, and there is communication link available during the occurrence of the fault. The cost efficiency is reasonable.

Deployment of external devices: Mainly, overcurrent relay is used for the protection of these micro-grids. The total grid is islanded during the fault period, and the distributed generation is inverter based. The communication link is cut off during the fault in the grid. The deployment of the external devices makes it expensive.

Overcurrent and Symmetrical components: The relay used in this protection technique is a digital relay. Though the grid remains connected the part of the fault is islanded. The distributed generation is rotating and inverter based. The communication is based on the technique used. The cost is quite reasonable in this protection technique.

5   Suggestion

Though the publications are limited on this vast domain of microgrids and smart grids, the area for research and suggestions is still open in this domain, and with time, it will keep on coming up with suggestions for improvement. Though most of the ideas proposed to date are yet to be applied in practical scenarios. The other part of importance is the network architecture, which plays a big role in all the protection schemes, most importantly a network architecture that defines the control of the power grid. The layer of cyber-protection also works mainly based on the network architecture of the grid. The network architecture hence also calls for a room for improvement to be capable of protecting from various cyber-attacks. Most of the ideas proposed work for radial feeders but cannot work for the microgrids with the looped feeders.

Final suggestion maybe, the grid system can be protected from any kind of faults by optimizing a combined action of all the available protection schemes.

6   Conclusion

These days' clients need greater dependability and power quality. Microgrids have intended to remain upon all the necessities of clients. In any case, with huge difficulties, the development of microgrids has been gone with. In late decades, to introduce a sufficient protection strategy for micro-grids, various methodologies have been advanced. A strong protection plan ought to have the option to ensure the miniaturized scale framework against various kinds of faults and guarantee its protected and secure activity in both grid-connected and autonomous mode. Moreover, an endeavor was made to organize these movements into explicit gatherings; lastly, some discourse and practical ascriptions are getting from the analyzed references. The principle conclusion and suggestions concerning the assurance of microgrids can be consolidated as follows:

- Despite the protection methodology, it seems to be expected that some kind of transmission is going to be necessary, either centrally operated or dispensed.
- In appears to be clear that there is an even necessity to enhance many protection devices (*fuses*, etc.) presently used in power systems, mainly in low voltage (LV) systems. To supply new potentialities for the solicitation of new unconventional

protection systems of micro-grids, some other type of enhanced protection devices is to be introduced.

- Even though the originators that are most related to protection manufacturing companies seem to esteem adaptive protection systems, a universal trend in this direction has not been remarked.
- To concern adequately with the protection problems equated with bidirectional power flow, the need for a directional feature is clear.

References

1. Buigues G, Dysko A, Valverde V, Zamora I, Fernandez E Microgrid protection: technical challenges and existing techniques
2. Mirsaeidi S, Said DM, Mustafa MW, Habibuddin MH, Ghaffari K An analytical literature review of the available techniques for the protection of micro-grids.
3. Manandhar K, Cao X, Hu F, Liu Y Detection of faults and attacks including false data injection attack in smart grid using Kalman filter

Chapter 46
Personalized Recommender Systems: An Empirical Analysis

Poonam Tijare, S. Athreya Uppili, M. Ajay, Anisha Rao, and K. K. Chaithra

1 Introduction

With the overload of information over the recent years, recommender systems have grown to become an important part of people's everyday lives. Online platforms are an absolute essential in this digital age. Increasing the utility of recommendation systems on these platforms has increased user interaction and is also a cost-effective method of the same. Consumers expect a personalized experience and sophisticated recommendation systems to find relevant products and content, all to save consumers time and money. Recommendation technologies are widely used to help people identify relevant products or services or information. YouTube, Amazon, Netflix and many other such web services are the famously known recommendation systems. The system can suggest a set of items to the users and recommend top items to the user. People are always provided with too many options to choose from; the recommendation system focuses on the user's best interest and suggests the best options by learning from the users. There are several algorithms that have been developed over time to help improve the efficiency of these systems. Systems mainly can be categorized into content based and collaborative filtering based [1].

Content-based recommender systems aim mainly to recommend items based on their similarity metrics. For example, if a movie recommender system where a user has recently seen a movie of horror genre is considered, a simple content-based system might recommend similar horror movies or movies by the same director or having the same actors, etc. There are various measures available to compare and

P. Tijare (✉)
CMR Institute of Technology (VTU RC), Bengaluru, Karnataka, India
e-mail: poonamtijare@gmail.com

S. Athreya Uppili · M. Ajay · A. Rao · K. K. Chaithra
Department of Computer Science & Engineering, CMR Institute of Technology, Bengaluru 560037, Karnataka, India

© The Author(s), under exclusive license to Springer Nature Singapore Pte Ltd. 2021
X.-Z. Gao et al. (eds.), *Applications of Artificial Intelligence in Engineering*, Algorithms for Intelligent Systems, https://doi.org/10.1007/978-981-33-4604-8_46

compute the similarity of two items including but not limited to cosine similarity, Pearson's coefficient and Euclidean distance.

Collaborative filtering deals with a user-item matrix in most cases and tries to compute the rating that a user will give to an item he has not yet interacted with. This can be done by finding other users with similar tastes. Till now, our discussion was focussed on explicit ratings; i.e. the ratings take on numbers or values in a given domain. There is also the existence of implicit ratings, which can take the form of whether the intended user has viewed, clicked or seen a particular item rather than explicitly providing a rating.

2 Evaluation Metrics

There are many different methods and metrics to evaluate the accuracy of a recommender system [2]. Traditional accuracy and error methods for collaborative filtering systems include mean absolute error (MAE), mean square error (MSE) and root mean square error (RMSE). Decision support methods are used in cases where we want to understand how useful the system was in helping a user take a better decision. Two common metrics are precision and recall. These are generally used to evaluate the accuracy of content-based recommender systems. However, other methods like feedback from real-time users are also used to evaluate content-based recommendation systems. The accuracy of the algorithms implemented is measured using MAE and RMSE as these allow simple calculation and subsequent interpretation of the results. The RMSE and mean error formulas are described using Eq. 1 and 2:

$$\text{RMSE} = \sqrt{\sum_{i=1}^{n} \left(\text{predicted Rating}_i - \text{actual}_i\right)^2 / n} \tag{1}$$

$$\text{Mean Error} = \sum_{i=1}^{n} \left|\text{predicted Rating}_i - \text{actual}_i\right| / n \tag{2}$$

where n denotes the total number of samples in the test set, actual denotes the real rating value corresponding to that sample of the test set, predicted denotes the rating that the proposed model predicts.

3 Datasets

One of the datasets [3] used for movie recommendations is the classical MovieLens dataset which contains 100 k ratings between 1 and 5 aggregated from 943 users on 1682 movies. There is also a latest version of the above dataset containing 100,836 ratings and 3683 tag applications across 9742 created by 610 users between March 29, 1996, and September 24, 2018. This dataset was generated on September 26, 2018. The secondary dataset used for movie recommendation is the Kaggle's TMDB 5000 movies dataset [4]. For restaurant recommendation systems, a variety of datasets including the Yelp Challenge dataset [5] and the Zomato Restaurants Dataset [6] have been incorporated. An overview of other common datasets typically used for recommender systems research purposes is specified in Table 1. Table 1 provides a list of various datasets used in different recommender systems.

Table 1 Overview of various datasets for recommender systems

Name	Summary
MovieLens [3]	Collection of movie ratings. Comes in 1M, 10M and 20M ratings. The largest set uses 140,000 users and spans across 27,000 movies
Jester [7]	List of various jokes and their rating
Book Crossing [8]	A Book rating and metadata dataset
Last.fm [9]	Dataset for songs and includes the top song in a playlist and the amount of times that song has been listened to
Yelp Dataset (Kaggle)	A subset of Yelp's businesses, reviews and user data. It was originally put together for the Yelp Dataset Challenge for students to conduct research or analysis on Yelp's data and share their discoveries
Restaurant Data (Kaggle) [10]	This dataset was used for a study where the task was to generate a top-n list of restaurants according to the consumer preferences
TMDB-5000(Replacement by Kaggle)	Contains movie information relating to the metadata and credits
Zomato Restaurants Data (Kaggle)	All metadata, rating and location information about restaurants fetched via Zomato's API
OpenStreetMap	Contains map related data. Objects in the dataset includes roads, buildings, points-of-interest, etc.
Wikipedia	This contains encyclopaedia data written by its users
Python Git Repositories	This contains all the python code found in Gitrepositories
Stanford Large Network Dataset Collection (SNAP)	Contains a wide variety of datasets from different sources and of varying size

Table 2 Overview of various research papers

Research author and year	Dataset	Approach	Evaluation Metric
Zhang et al. [11]	MovieLens 100 K and MovieLens 1M dataset from GroupLens	Scalable CF algorithm called Weighted KM-Slope-VU is proposed	RMSE (100 K) —0.95062 RMSE (1 M) —0.94676
Li et al. [12]	MovieLens 100 K MovieLens 1M and MovieLens 10M datasets from GroupLens	Deep Neural Networks	MAE 100 K—0.722 1M—0.673 10M—0.655
Xinchanang et al. [13]	MovieLens 100 K dataset from GroupLens	Social Network Analysis along with Collaborative Filtering Technique Collaborative Filtering Technique to solve Cold start problem	MAE—3.55
Nagamanjula and Pethalakshmi [14]	MovieLens 100 K Dataset from GroupLens	Opinion Mining and User Similarity Analysis. Opinion detection is performed using naïve Bayes-based Support Vector Machine (NbSVM) Similarity among users is computed using weight adjusted cosine scoremetric	Precision—91% Recall—89% f-measure—89% Accuracy—93.45% Execution time—7.84 s
Hossain and Nazimuddin [15]	Merged required datasets from GroupLens (20million ratings)	Neural Networks that recognizes user's behavioural patterns	MAE—1.97 MRE (%)—6.06 MSE—4.75
Kuppani et al. [16]	Unspecified restaurant dataset	Conopy algorithm, Hierarchical algorithm, Filtered clustering algorithm, Farthest first algorithm and simple KMeans method	Time taken to build model- Conopy: 0.05 Hierarchical Clustering: 0.01 Filtered Clustering: 0.04 Farthest First:0.03 Simple KMeans:0.02

(continued)

Table 2 (continued)

Research author and year	Dataset	Approach	Evaluation Metric
Fakhri et al. [17]	Zomato Restaurants Data (Kaggle)	User-based collaborative filtering method	MAE—1.492
Basudkar et al. [18]	Zomato Restaurants API	Content-based and collaborative filtering method	Recommendation score is calculated with normalized distance values
Logesh et al. [19]	Yelp Dataset and Trip Advisor	Content-based approach with location data	Yelp: RMSE: 1.0643 Coverage: 94.73 F-measure: 0.7283 Trip Advisor: RMSE: 1.0755 Coverage: 92.38 F-measure: 0.7254
Katarya and Varma [20]	MovieLens 100 K dataset from GroupLens	KMeans cuckoo-based collaborative filtering framework KMeans as clustering algorithm and cuckoo search as the optimization algorithm	MAE—0.754 RMSE—1.266
Bhojne et al. [21]	Historical Restaurant Customer Database	Hybrid approach with content-based and collaborative filtering method, naïve Bayes and sentiment analysis	Ranking of restaurants is calculated using SentiWord-Net analysis
Zhang et al. [22]	MovieLens 100 K from GroupLens	Level filling method to predict the non-rated items Time weighted-based recommendation prediction formula is adopted	Accuracy of the proposed algorithm is higher than the traditional algorithms mentioned

(continued)

Table 2 (continued)

Research author and year	Dataset	Approach	Evaluation Metric
Ravinarayana et al. [23]	Clustered Restaurant Database	Content-based approach using KMeans method with clustered database	Results with clustered database are compared with non-clustered database; time taken for comparisons has reduced by 94%
Sawant and Pai [24]	Yelp Dataset	naive baseline, singular value decomposition, weighted bipartite graph, hybrid cascade of kNN clustering, weighted bipartite graph projection, clustered weighted bipartite graph projection, multi-step random walk weighted bipartite graph projection and cascaded clustered multi-step weighted bipartite graph projection	Cascaded clustered multi-step weighted bipartite graph projection was considered to have the highest accuracy: RMSE: 1.09263 MAE: 0.67548

4   Related Work

Table 2 provides a brief overview of various research papers. The table describes the contribution by research authors and year, dataset used and the approach used and the metric used for evaluation.

As mentioned in [25], there are a few challenges that are faced while implementing recommender systems. Some of them include:

1. Cold Start—This problem can occur when new users enter the recommendation system and the user's preferences are not known. This can be solved by asking the user to indicate his/her preferences the first time he/she signs up for the service.
2. Grey Sheep—This occurs when one person's tastes differ from that of the group thereby rendering the recommendations provided to him/her useless. By perusing collaborative filtering systems, this can be avoided as they provide recommendations based on the personal interests and profile of the user.
3. Synonymy—This problem occurs when two words or items have different ways of expression, but they point to the same entity. For example, action movie and action film have the same meaning however a rote learning or memory-based approach to filtering systems will not be able to capture this semantic similarity. Using methods like SVD, this error can be averted.
4. Shilling attacks—This attack happens when a malicious user starts providing false ratings intentionally in order to sabotage the recommender system and lower the trustworthiness/relevance of the items recommended. The remedy to this attack involves identifying prediction shift, hit ratio, etc.
5. Sparsity—Users tend to rate a very small subset of all the available items in a dataset leading to a sparse rating matrix. This makes it difficult in cases where we are using algorithms such as kNN due to the few available ratings. Algorithms like SVD and some content-based collaborative algorithms provide relief.

5   Algorithms

5.1   Movie Recommendation

5.1.1   Baseline Metrics

Surprise module is a specialized module built to provide users complete control over their recommender systems research and experiments as well a way to evaluate such systems. A few datasets like MovieLens and Jester are built-in along with implementations of popular algorithms.

A few of the built-in popular algorithms like SVD, kNN were run on the Movie-Lens 100 k dataset to establish baseline metrics. Fivefold cross-validation was conducted, and the mean is documented. Table 3 shows the error estimates observed in the datasets using algorithms SVD, kNN and normal predictor.

Table 3 Evaluation results of standard algorithms

Built-in algorithm	Root mean square error (RMSE)	Mean absolute error (MAE)
SVD	0.9364	0.7385
kNN	0.9790	0.7730
Normal Predictor	1.5233	1.2232

Table 4 Algorithms with the mean of all ratings and random value scores

Algorithm	RMSE	MAE
Mean of all ratings	1.1257	0.9447
A Random rating value (1–5)	1.8869	1.5129

5.1.2 Mean and Random Measures

The surprise module provides a simple to implement and executes one's own algorithms by implementing two of the methods fit and estimate derived from AlgoBase. Basic process followed for implementing an algorithm using the Python surprise module can be presented as:

1. Define the name of the algorithm being developed.
2. Implement the fit method (called once on training data).
3. Implement estimate method, i.e. the prediction for a given user ID and item ID, given the knowledge gained from training.

A dummy baseline model that returns 0 for any given user-/item ID combination was first implemented to get familiarized with the intricacies of surprise. It can be seen that such a method would produce the most deviation as ratings range from 1 to 5. The mentioned model was built to serve the purpose of proving that with surprise; building models are very simple and also provide an example of extreme RMSEs to be avoided. This method achieves an RMSE of 2.7690 and MAE of 2.5299 as expected. To summarize, in order to build algorithms in surprise, we must load a dataset, implement our own algorithm/use built-in ones, and it provides us with the error estimates.

In case the need arises to test implemented model, the fit and estimate methods must be implemented. The fit method is called once on the training set, and the estimate method is called for every row of the test set and expects the predicted rating for given user id (u) and item ID (i) taken as parameters (Table 4).

5.1.3 Age-Based Clustering

People are classified according to their age into certain age groups, and ratings are provided to a particular movie based on the cluster in which they belong. Exploratory data analysis on the data reveals that the youngest person in the dataset is 7 years old and the oldest person is 73 years old. Based on this, the age group clusters

Table 5 Analysis of proposed approach based on age

Technique	RMSE	MAE
Age-based	1.254	0.9436
Age-based along with movie	1.26	1.00
Most common rating	1.2199	0.8942

identified are 7–17, 18–29, 30–40, 41–50, 51–60 and 61–73. The reasoning behind this approach is that few movies are targeted towards certain age groups. Taking the average of the ratings of other users in a similar age groups for a particular movie, the rating can be computed. However, this is not the best method as choice may differ from the group/cluster allotted, movies for all age groups or fewer outliers during clustering. Post-analysing the data, it is found that the most common rating users provided to a movie was 4.

A few variations of this method were implemented, one involving taking just the average of all ratings of users in the same cluster as the one we are trying to predict irrespective of the rating. For example, in order to predict the rating of a user who falls in cluster 3 (30–40), the average of all ratings in that age group is returned. The second method also takes into consideration the movie while predicting the rating of a user. In order to predict the rating of a user for a movie and that user falls in cluster 2 (18–29), the average rating of that movie in that age group is returned. Table 5 shows the accuracy of the age-based clustering algorithm in each variation.

5.1.4   Genre-Based Recommendation

In accordance with the MovieLens dataset, the user's previous ratings on the movies are grouped into three groups, i.e. ratings 3, 4 and 5, respectively. To predict how the user would rate an unprecedented movie, the genres of the movies the user rated previously is stored and the frequency of each genre occurring in each group is calculated. For instance, if the user rated 5 for action movies 10 times, rated 4 three times, then the user is most likely to rate the action movie 5 again. Considering the genres comprised in a movie, the mean of ratings cohering with the frequency is computed. The error metrics evidently proves a decent score for a genre-based proposed approach. Table 6 provides an analysis of the proposed approach based on genre.

Table 6 Analysis of proposed approach based on genre

Technique	RMSE	MAE
Genre	1.45199	1.14648
Genre with supplemented rating	1.24133	0.90754

Table 7 Evaluation of cosine similarity for collaborative filtering

Metric	RMSE	MAE
Similarity of plot (giving director weight)	1.2323	0.9433
Mixture of plot, genres, actors, etc.	0.2208	0.9355

5.1.5 Cosine Similarity

The hybrid approach implemented by utilizing two data sets—the Kaggle TMDB replacement dataset which has information regarding 5000 movies and the existing MovieLens data set. The cosine similarity measure is used in order to identify similar movies based on how similar the plot vectors of two movies are providing more weight to the "director" factor thereby getting movies with similar plot along with good movies from the same director. The other similarity measure was used where the metric of comparison was not weighted in terms of director but was considered to be a mixture of actors, genre and plot. So, the idea involved fetching similar movies (in the other data set) to the one currently trying to predict (in MovieLens). The estimated predicted movies were the weighted average of these (Table 7).

5.1.6 K-Nearest Neighbours

This machine learning algorithm is used to find clusters of similar users based on common movie ratings and make predictions using the average rating of top-k nearest neighbours. Using the MovieLens 2M dataset from GroupLens, ratings were presented in a matrix, with the matrix having one row for each movie and one column for each user. The k item that has the most similar user engagement vectors was found. This algorithm uses Brute force implementation to compute the nearest neighbours and cosine metric to calculate the cosine similarity between rating vectors.

Next, the closeness of instances is determined by calculating the distance. Then, the algorithm classifies an instance by finding its nearest neighbours and picks the most popular class among the neighbours.

Approach for implementing kNN:

1. Load the dataset.
2. Initialize the value of k.
3. For each example in training data,

   (a) Calculate the distance between test data and each row of training data using cosine similarity as the distance metric.
   (b) Sort the calculated distances in ascending order based on distance values.
   (c) Pick the top k entries from the sorted list.

The most important feature of kNN algorithm is its simplicity and easy-to-implement feature. The drawback is that the computation time is more when a larger dataset is used.

5.1.7 Singular Value Decomposition

The inspiration was drawn from Nicolas' article [26] on matrix factorization and tried to implement SVD from scratch with stochastic gradient descent for optimization. The two matrices (p and q) were found whose product will give the rating matrix (R). The minimising function is given by Eq. 3.

$$\text{Function to minimise: } \sum_{r_{ui} \int R} (r_{ui} - p_u q_i)^2 \tag{3}$$

where p_u the row vector of the matrix p and q_i the column matrix at position i. A stochastic gradient approach is used to minimise the above expression by starting off with random values for p and q and for a given number of epochs updating those parameters by subtracting the product of the derivative and learning rate. One of the defining characteristics of the rating matrix is that it is most of the time sparse.

It is assumed that SVD will help to identify the latent factors and the corresponding strength of each factor. In a simple sense in the case of movies, it can be stated as how much a user is prone to some factors like action, comedy, etc., and how much the strength of that factor is in each movie in the dataset. Thus, the multiplication of the two factored matrices would provide a way of constructing back the matrix and guessing the values that are not known.

Table 8 shows the error metrics based on the number of latent factors. In the above experiments, learning rate was fixed to 0.01 and the number of epochs at 10. Next, the error estimates when factor is 20 and learning rate is slowly varied. This is shown in Table 9. This takes a lot of time if the number of epochs is very large as it is implemented in the fit method and for each iteration the entire training set is processed. Large nudges to the learning rate are avoided so as to prevent overshooting the minima. The best learning rate was found to be 0.01.

Table 8 Evaluation for various number of latent factors

No. of factors	RMSE	MAE
10	0.9598	0.7521
15	0.9600	0.7534
20	0.9587	0.7522
50	0.9696	0.7614
100	0.9807	0.7696

Table 9 Error estimates for varying learning rates

Learning rate	RMSE	MAE
0.001	1.5000	1.1771
0.01	0.9602	0.7501
0.02	1.0000	0.7776

Table 10 Neural network evaluation using RMSprop as optimiser

No. of epochs	RMSE
50	0.9754
100	0.9409
150	0.9419
200	0.9450

5.1.8 Neural Networks/Auto-encoders

The available files in the MovieLens dataset for cross-validation (u1.base, u1.test, … u5.base, u5.test) have been used. A stacked auto-encoder with Pytorch was constructed and trained it on the u4.base and test it on u4.test. If SVD built-in is run on the mentioned files, an RMSE of 0.9337 is obtained.

The auto-encoder results are as follows (Table 10).

Other optimizers such as Adagrad and SGD have been considered. RMSprop provides the best output with the least hassle.

5.1.9 Restricted Boltzmann Machine (RBM)

Restricted Boltzmann machine is a probabilistic graphical artificial neural network model that learns from probability distribution over a set of inputs. This algorithm uses the movie lens 100 k ratings dataset and is trained and tested on the train-test splits (u1.base, u1.test, …, u5.base, u5.test).

The algorithm performs binary classification; i.e. if user likes the movie, it returns 1 else 0. The ratings dataset is preprocessed to achieve a matrix having users as rows and movies as columns. If the movie is rated below 3, it implies the user did not enjoy the movie, whereas if he/she rated 3 and above, they enjoyed it. Hence, ratings of 1 and 2 are made 0 in the matrix, while the ratings 3, 4 and 5 are given 1. A rating of -1 is used if the user has not rated the movie.

Approach for implementing RBM Model:

1. Initialize weights(W), bias for hidden nodes(a) and bias for visible nodes(b) \
   a: = randNumber(1,nh).
   b: = randNumber(1,nv).
2. Function sample_h(x)
   /* To calculate probability of activation of hidden nodes given visible nodes */
   wx: = MatrixMultiplication of x and W(transposed).
   activation: = wx + a p_h_given_v: = sigmoid(activation).
   return p_h_given_v, bernoulliDistribution(p_h_given_v).
3. Function sample_v(y)
   /* To calculate probability of activation of visible nodes given hidden nodes */
   wy: = MatrixMultiplication of y and W(transposed).
   activation: = wy + b p_v_given_h: = sigmoid(activation).

Table 11 RBM evaluation on various batch sizes

Batch size	RMSE	MAE
1	0.44339	0.25367
20	0.42073	0.23402
50	0.41354	0.22840
75	0.41669	0.22786
100	0.41304	0.22756
200	0.41952	0.23083

return p_v_given_h, bernoulliDistribution(p_v_given_h).

4. Function train(v0, vk, ph0, phk)
    /*v0 initial visible node; vk— visible node after k-step contrastive divergence; ph0—probability sampling of hidden nodes; phk— probability sampling of hidden nodes after k-step contrastive divergence. */
    W: = (MatrixMultiplication of v0 and ph0) - (MatrixMultiplication of vk and phk).
    a: = sum((ph0 - phk)).
    b: = sum((v0 - vk)).

This model is used to sample the hidden and visible nodes using Gibbs sampling technique. Pytorch libraries have been used for sampling probability distributions. K-step contrastive divergence is performed to minimize the overall cost function. Computing gradients are computationally expensive, and hence, techniques like contrastive divergence have been employed to approximate the likelihood gradient (Table 11).

Batch size of 100 works out well to produce the best results as seen. From the results, 75% of the times, i.e. 3 out of 4 times given a movie, the algorithm is to predict whether the user will like the movie or not. This algorithm finds correlations purely based on user ratings. Hence, it can be used as a good filtering technique on some other base algorithm (e.g., auto-encoders).

5.2   Restaurant Recommendation

5.2.1   Content-Based Approach Using Cosine Similarity

Content-based algorithms use features of items and recommend similar items to users based on previous actions or explicit feedback. Algorithm uses the Zomato Bangalore which has been modified to include restaurant IDs as well. Restaurants with similar cuisines are to be recommended, so feature extraction on cuisines is performed.

Table 12 Restaurant recommendation using content-based approach

ID	Name	Similarity score	Rating
8269	Truffles	0.93	3.4/5
3147	Here & Now	0.88	3.9/5
1196	Brew Point	0.73	0
104	A Hole Lotta Love Cafe	0.73	3.7/5
5812	Pizbur Trap	0.55	0
675	BLOW	0.47	0
1937	Cravy wings	0.47	3.2/5
3705	Java City	0.42	3.7/5
1082	Bistro Claytopia	0.41	3.5/5
2521	Fatso's	0.41	4.4/5

A maximum of 5000 features is extracted using TF/IDF in the cuisines columns across all the restaurants. The sporadic terms that occur in less than three documents are eliminated to have rich and meaningful features.

Cosine similarity is used to find pairwise similarity between the restaurants using the computed. The similarity scores are used to find the top ten scores similar to the given restaurant and recommend them. The recommendations are searched on "The Hole in the Wall Café". The results are shown in Table 12.

The next implementation of a Content-based recommendation system considers the containing restaurants across the world. It includes features such as city, locality, cuisine, aggregate rating and the total number of votes for each restaurant in the. In order to provide more personalized recommendations to the user, the top, similar restaurants are recommended for a user by taking into consideration their location as well as the name of the restaurant. Cosine similarity was used to compute the similarity score between a pair of restaurants. This score will be used to determine the top ten most similar restaurants with respect to the user's location and cuisine of the restaurant specified. The results on search query is shown in Table 13.

Search query: restaurant_recommend_func('Connaught Place', 'Sbarro').

In both cases, cosine similarity as the similarity metric is chosen. Cosine similarity is proven to be better when compared with other metrics like Euclidean distance and Pearson's correlation as it measures the similarity between items irrespective of their size. It measures the cosine of the angle between two vectors (or documents) projected in a multi-dimensional space. Therefore, even if two documents are far apart with respect to their Euclidean distance, there may be a chance they could have a smaller angle between them as smaller the angle, greater its similarity [27].

Content-based systems do not require any data about the other users using the system. For this reason, it is easier for the system to scale to a large number of users. The recommendations made to a user are specific to that user only so content-based systems work well if a user has an interest that is not very common among all users. However, implementing such a system requires an extensive knowledge of that particular domain. These systems only make predictions based on the user's

Table 13 Restaurant recommendation using content-based approach using location

ID	Restaurant name	Aggregate rating	Cosine similarity
63	Pizza Hut	3.5	0.86
26	Caffe Tonino	3.9	0.79
32	Domino's Pizza	3.7	0.70
91	Ovenstory Pizza	0.0	0.70
24	Cafe Public Connection	3.7	0.33
12	The Immigrant Cafe	3.2	0.33
1	Attitude Kitchen & Bar	2.9	0.33
112	Smoke on Water	4.1	0.33
95	Ardor 2.1	4.1	0.28
94	Ambrosia Bliss	4.0	0.28

current, existing likes and dislikes; it does not provide any way of expanding the user's interests beyond their currents interests.

5.2.2 Item-Based Collaborative Filtering Using k-Nearest Neighbours

Collaborative filtering is a machine learning technique that is used to make predictions based on the past behaviour of users. It can be either item based or user based. The bottleneck in traditional collaborative filtering algorithms is searching for neighbours among a large user population of potential neighbours. Item-based filtering algorithms avoid this bottleneck by exploring the relationships between items' ratings instead of the relationships between users as user behaviour tends to be dynamic in nature, whereas items' ratings tend to remain static. Item-based algorithms provide recommendations for users by finding items that are similar to other items the user has liked in the past [23].

An item-based collaborative filtering technique using k-nearest neighbours algorithm is the algorithm proposed. kNN algorithm is a supervised machine learning technique that uses labelled input data for learning in order to make predictions on new, unlabelled data. The advantage of using a kNN algorithm-based approach is that it does not require any training before making predictions and adding new items does not affect the accuracy of the algorithm. Another reason as to why kNN is a popular approach is that it is relatively simpler to implement as it considers

Table 14 Evaluating RMSE, MAE of algorithm KNNBasic on two splits

	Fold 1	Fold 2	Mean	STD
RMSE	0.8218	0.8324	0.8271	0.0053
MAE	0.6584	0.6636	0.6610	0.0026
Test time	0.0	0.0	0.0	0.0
Fit time	0.03	0.05	0.04	0.01

only two parameters: the value chosen for 'k' and the distance measured between items. For the implementation, the Restaurant Dataset (Kaggle) which contains user ratings for particular restaurants is chosen. A 'popularity threshold' is considered in order to recommend restaurants having more number of ratings than the 'popularity threshold'.

A pivot table is created containing the ratings given by each user for a particular restaurant. The table has restaurant names which are the indices and the user ID as the columns. This table is then converted into an array matrix by importing csr.matrix library from scipy.sparse package. The metric chosen for calculating the distance between items is cosine similarity.

Recommendations output for the restaurant 'puesto de tacos' shown below:

1: Tortas Locas Hipocampo with distance of 0.4173017144203186.
2: La Cantina Restaurante with distance of 0.5863252878189087.
3: Gorditas Doa Gloria with distance of 0.6199256777763367.
4: Cafeteria y Restaurant El Pacifico with distance of 0.7170524597167969.

Surprise module is used to calculate the RMSE and the MAE by importing KNNBasic.

This is shown in Table 14.

6 Conclusion

An attempt has been made to study existing literature surveys and research papers to obtain knowledge about the various types of recommender systems, their uses, the algorithms implemented and the challenges faced in each of them. Using this knowledge, implementation of some of the known algorithms as well implementation of our own techniques was carried out. The errors obtained for each algorithm and approach have been documented extensively.

This paper will provide an overview to those entering the field of recommendation systems and arm them with more than sufficient intuition to select the best algorithm suited for their task. Table 15 provides a view of the error metrics obtained in the approaches implemented.

Table 15 Overall results

Approach	RMSE	MAE
Mean	1.1257	0.9447
Random	1.8869	1.5129
Age cluster	1.2199	0.8942
Genre cluster	1.24133	0.90754
Cosine similarity	1.2208	0.9355
SVD	0.9587	0.7522
Auto encoders	0.9409	–
Restricted Boltzmann machine	0.40373	0.21915

7 Future Work

One of the main goals is developing a recommendation algorithm that would decisively beat the existing baseline measures. Apart from that a few ideas for future include:

1. Exploring the possibility of probing genetic algorithms for selecting population size and features and adopting at each iteration the sets with most potential.
2. How to be more informed while predicting the rating of movies? Trying out ensemble algorithms, a mix of different algorithms taking the majority vote in case of a mismatch in case of collaborative and content filtering algorithms.
3. Is it possible to recognize enough outliers (people who always give high ratings, people who like one type of movie, people who are swayed by trends) to radically change the performance accuracy?
4. Explore the accuracy of graph-based models. If all the users are represented as nodes in a graph and the differences between the rating matrices as edges, will it help? How to intelligently find similar users? Can giving weight to different parameters play a role?

References

1. Beel J, Gipp B, Langer S et al. (2016) Research-paper recommender systems: a literature survey. Int J Digit Libr 17:305–338. https://doi.org/10.1007/s00799-015-0156-0
2. Jalili M, Ahmadian S, Izadi M, Moradi P, Salehi M (2018) Evaluating collaborative filtering recommender algorithms: a survey. IEEE Access 6:74003–74024. https://doi.org/10.1109/ACCESS.2018.2883742
3. Harper FM, Konstan JA (2015) The MovieLens datasets: history and context. ACM Trans Intell Syst. https://doi.org/10.1145/2827872
4. Kaggle TMDB 5000 movies dataset. https://www.kaggle.com/tmdb/tmdb-movie-metadata. Last Accessed 06 May 2020
5. Yelp Challenge Dataset. https://www.yelp.com/dataset. Last Accessed 15 May 020
6. Zomato restaurant dataset. https://developers.zomato.com/api. Last Accessed 15 May 2020

7. Goldberg K, Roeder T, Gupta D, Perkins C (2001) Eigentaste: a constant time collaborative filtering algorithm. Inf Retrieval 4(2):133–151
8. Ziegler C-N, Mcnee SM, Konstan JA, Lausen G (2005) Proceedings of the 14th international world wide web conference (www '05). ACM
9. Bertin-Mahieux T, Ellis DP, Whitman B, Lamere P (2011) The million song dataset. In: Proceedings of the 12th international conference on music information retrieval (ISMIR 2011). ACM
10. Restaurant & consumer data, recommender systems domain. https://www.kaggle.com/uciml/restaurant-data-with-consumer-ratings. Last Accessed 09 May 2020
11. Zhang J, Wang Y, Yuan Z, Jin Q (2020) Personalized real-time movie recommendation system: practical prototype and evaluation. Elsevier
12. Li Q, Choi I, Kim J (2020) Evaluation of recommendation system for sustainable E-commerce: accuracy, diversity and customer satisfaction. Preprints
13. Xinchanang K, Vilakone P, Park D (2019) Movie recommendation algorithm using sentiment analysis to alleviate cold-start problem. J Inf Process Syst 15(3):616–631
14. Nagamanjula R, Pethalakshmi A (2019) A novel scheme for movie recommendation system using user similarity and opinion mining. Int J Innov Technol Explor Eng (IJITEE) 8(4S2):318–322. ISSN: 2278-3075
15. Akter Hossain Md, Uddin MN, A neural engine for movie recommendation system. In: 4th international conference on electrical engineering and information & communication technology. https://doi.org/10.1109/CEEICT.2018.8628128
16. Sathish K, Ramasubbareddy S, Govinda K, Swetha E (2019) Restaurant recommendation system using clustering techniques. Int J Recent Technol Eng (IJRTE) 7(6S2):917–921. ISSN: 2277-3878
17. Fakhri A, Baizal A, Setiawan E (2019) Restaurant recommender system using user-based collaborative filtering approach: a case study at Bandung Raya Region. J Phys: Conf Ser 1192:012023. https://doi.org/10.1088/1742-6596/1192/1/012023
18. Basudkar B, Bagayatkar S, Chopade M, Darekar S (2018) Restaurant recommendation system using customer's data analysis. Corpus ID: 21250386
19. Logesh R, Subramaniya Swamy V, Vijayakumar V (2018) A personalised travel recommender system utilising social network profile and accurate GPS data. Int J Electron Government 14:90. https://doi.org/10.1504/EG.2018.089538
20. Katarya R, Varma OP (2017) An effective collaborative movie recommender system with Cuckoo search. Egypt Inf J 18. https://doi.org/10.1016/j.eij.2016.10.002
21. Bhojne NG, Deore S, Jagtap R, Jain G, Kalal C (2017) Collaborative approach based restaurant recommender system using Naive Bayes. Int J Adv Res Comput Commun Eng 6(4)
22. Zhang Z, Xu G, Zhang P (2016) Research on E-commerce platform-based personalized recommendation algorithm. Hindawi Publishing Corporation, Applied Computational Intelligence and Soft Computing, Article ID 5160460
23. Ravinarayana R, Pooja MC, Raghuveer K (2016) Using clustered database for food recommendation system. IEEE, Seoul
24. Sawant S, Pai G (2013) Yelp food recommendation challenge
25. Khusro S, Ali Z, Ullah I (2016) Recommender systems: issues, challenges, and research opportunities. Springer, Singapore. https://doi.org/10.1007/978-981-10-0557-2_112
26. Hug N (2017) Understanding matrix factorization for recommendation. https://nicolas-hug.com/blog/matrix_facto_1. Last Accessed 12 May 2020
27. Prabhakaran S (2020) Cosine similarity—understanding the math and how it works. https://www.machinelearningplus.com/nlp/cosine-similarity/. Last Accessed 14 May 2020
28. How to build a Recommendation Engine quick and simple. https://towardsdatascience.com/how-to-build-a-recommendation-engine-quick-and-simple-aec8c71a823e?gi=e8b74e514cdd. Last Accessed 06 May 2020

Chapter 47
Feature Selection on Linked Data: A Review

Tanjina Das, Srikanta Paitnaik, and Smita Prava Mishra

1 Introduction

The modern and fast development of technology produces a large amount of data in every fraction of second. These data can be either video, text, images, audio, etc. The data here can be characterized as data with high dimensions. Data collected from the Internet and other sources is considered as raw data which comprises noise (error) and is also of very high dimension. Processing the data of high dimensions also encounters challenges in data analysis, data selection, decision making, data retrieval, storage, etc. [1]. In the direction of solving the above issues along with reducing dimension of the data, data mining and machine learning [2–5] techniques come to our rescue.

Dimensionality reduction has high efficiency, and it is a simple process to reduce the redundant attribute along with the noise from the data. It can be broadly classified into two types, such as feature selection and feature extraction [6–8]. Feature selection is one such technique which helps us to deal with high-dimensional data in a much easier way.

Feature selection is the subset of original dataset [9, 10]. The subset is formed from the original high-dimensional data on the basis of certain criteria. Among the aforesaid methods, feature selection is preferred over feature extraction, as it can

T. Das (✉)
Department of Computer Science and Engineering, Institute of Technical Education and Research, Siksha'O' Anusandhan University, Bhubaneswar, India
e-mail: tanjinadas@gmail.com

S. Paitnaik · S. P. Mishra
Department of Computer Science and Information Technology, Institute of Technical Education and Research, Siksha'O' Anusandhan University, Bhubaneswar, India
e-mail: srikantapatnaik@soa.ac.in

S. P. Mishra
e-mail: smitamishra@soa.ac.in

be easily understood and interpreted since there is not any change in the data as that in the extraction method. Feature extraction can be characterized as univariate and multivariate. Univariate evaluation is one where each entity is evaluated independently without having any connection with the other entity in the feature space, whereas multivariate is the one which considers multiple features together to evaluate and reduce redundancy.

There are several categorizations of the feature selection technique, depending on various criteria for selection of the features:

- Based on the learning methods used for feature selection, they may be classified as supervised, unsupervised and semi-supervised methods.
- Based on the selection criteria of features, the methods may be categorized as filter [11], wrapper [12] and embedded methods [13].
- According to the search strategies used, feature selection can be classified as forward selection, backward deletion, random and hybrid model.
- According to evaluation criteria, feature selection can be carried out on the basis of Euclidean distance, consistency, dependency and information measure.
- Based on the output obtained, the feature selection is represented through feature rank and/or feature subset.

The performance of these feature selection methods is evaluated with the help of machine learning techniques. The method having the highest accuracy is considered to be the best available method for the respective dataset along with the one with less computational overhead for obtaining the selected feature [14, 15].

2 Feature Selection

The major applications for feature selection are image recognition and retrieval, bio-informatics, data mining, data analysis, data clustering, etc. It eliminates the redundant and/or irrelevant data and selects the best data for further process.

2.1 Generic Feature Selection Method

The generic process of feature selection is depicted as in Fig. 1 [16].

The various types of data which are considered for feature selection are represented taxonomically in Fig. 2.

Data can be broadly classified into:

Streaming data and static data. Streaming data is that which changes with respect to time, whereas static data does not change with respect to time.

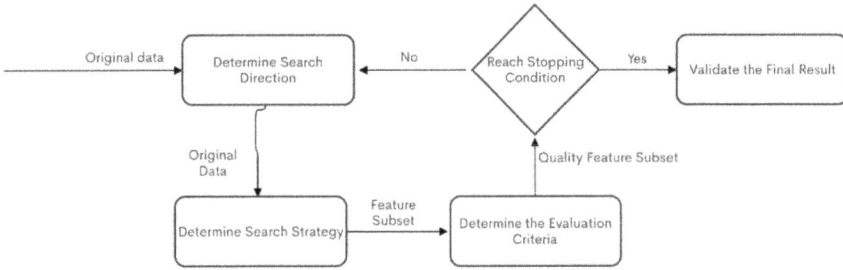

Fig. 1 Stages of feature selection

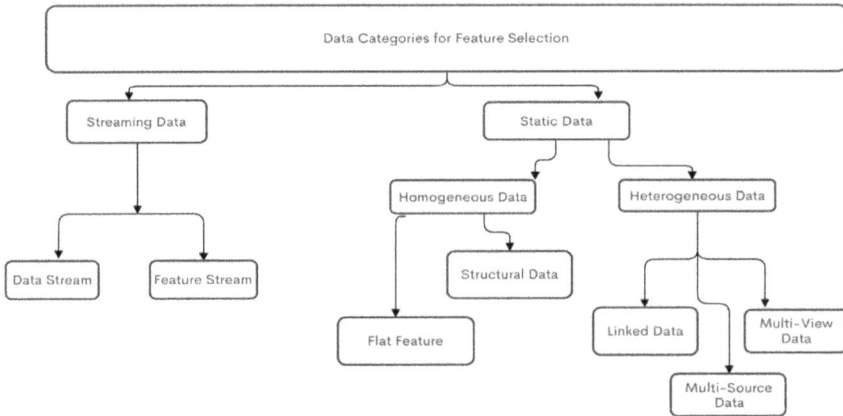

Fig. 2 Categories of data for feature selection

- Streaming data can further be categorized into data stream and feature stream. Data stream is generally sequential flow of digital signals from sender to receiver, and feature stream is the flow of features.
- Static data comprises homogeneous data and heterogeneous data.
- Homogeneous data is that data which consists of similar type of data and can be classified as flat feature [17] and structural data. Structural data is also known as organized data.
- Heterogeneous data is that data which does not consist of similar type of data and can be further divided into linked data, multi-score data and multi-view data [17].

Filter Method

Filter method is also known as "open loop method." In this method, the irrelevant data is filtered out using some criteria [18]. Every single entity is considered, and thereafter, it is evaluated on the basis of the criteria, and then, the entity with the highest computed score is selected in the feature selection process. This method does not obey any specific algorithm. Examples of such method usage are feature dependency [16], entropy-based distance, Laplacian score, etc.

Wrapper Method

Wrapper method is also known as "close loop method" [18]. This method uses clustering algorithms to find the suitable subset of features from the feature space, after which it further evaluates to find out the best possible features as per requirement. Heuristic approach is used for search space reduction since calculation of all the subsets of high-dimensional data is costly as well as time-consuming.

Though filter method is computationally cheaper than wrapper method, still wrapper method has the ability to produce better and efficient subset clustering than that of the filter method. This method has a tendency to generate optimal solution as the algorithm continuously repeats itself in evaluating the generated subset for better results.

Hybrid Method

Hybrid method comprises both wrapper method and filer method [16]. It considers the best features among both the methods that use different algorithms at different scenarios to get the optimal and accurate outcome with very few overheads. It initially filters out the candidate subsets; thereafter, it evaluates each of them individually. The best among them is eventually considered.

Embedded Method

This method automatically considers the best features from the feature space during the training process. This method does not evaluate continuously as that of wrapper model, thus making it all the more efficient. It is the most complex method among the three discussed methods as the feature selection is embedded in the learning algorithm itself [19].

2.2 Feature Selection for Linked Data

The feature selection can be carried out on generic data, stream data, linked data and text data [19, 20]. Algorithms used for generic data are also applicable to text, streaming as well as linked data [21]. Below discussion is the algorithm used exclusively for linked data.

Linked Data

Linked data is structured data and is interlinked with each other thus helpful in providing information. Linked data in this twenty-first century is present everywhere, like the tweets from Twitter, posts from Instagram or Facebook or any other social media websites [1]. The linked data is also of high dimensional. If the linkage among the data could be established, then they serve as important sources of information. The linked data is generally unlabeled in nature. Due to high dimensionality of data,

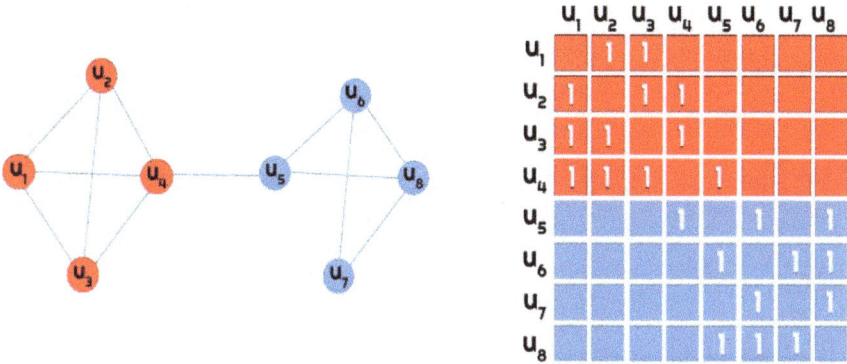

Fig. 3 Linked data with its adjacency matrix

it is expensive to label such data. The data uses social and semantic contexts for the existing channels. Figure 3 depicts an example of linked data in social media and its representation along with its adjacency matrix. This figure consists of eight linked instances (U1-U8) which are grouped together. The instances within a group are further connected to each other. The groups in the figure are denoted in red and blue. The interconnection between the instances among the groups is represented in the adjacent matrix.

Feature Selection Methods

Feature selection methods used for social media data or linked data are as follows (Table 1):

2.3   Challenges for Linked Data

- Linked data is generally heterogeneous data. The dimensions of these data are high; thus, performance challenges and the methods used for implementation are complex.
- To reach usable solution, generalization is required which is yet another performance issue in linked data feature selection.
- Methods like NetFS and LUFS loose the link information, thus generating partial order methods to quickly work on the link information so that the link is not lost [20].
- Text analysis is an issue for feature selection of linked data. Due to the performance issues, streaming methods which deal with real-time analysis are slowly coming into the picture and overshadowing the static methods [21].
- The nature of linked data is basically dynamic and heterogeneous; thus, feature selection for these types of data as pre-processing is not mandatory.

Table 1 Different feature selection method with description

Sl no.	Method	Description
01	Feature selection in network (FSNet) [22]	• It was introduced in the year 2011 • It is a supervised regression type feature selection method • It is used when traditional feature selection cannot be applied • It uses two regularization methods after which FSNet is applied to get the best features • Graph regularization is applied to bring out the necessary information from linked structures. The directed graph can be calculated as: $$tr\left(W^T X L X^T W\right) \text{ (1)}$$ where $L = \sqcap - \frac{1}{2}\left(\sqcap P + P^T \sqcap\right)$, $\sqcap$ is the diagonal matrix, P is the transition matrix, and X is the set of features The undirected graph can be calculated as: $$tr\left(W^T X L X^T W\right) \text{ (2)}$$ where $L = D - A$, A is the symmetric matrix, and L is the diagonal matrix • Laplacian regularization least square (LapRLS) [22] is used to minimize the empirical error It can be calculated as: $$W = \left(XX^T + \lambda_A I + \lambda_I X L X^T\right)^{-1} XY \text{ (3)}$$ $\lambda_A, \lambda_I > 0$ are regularization parameters • To achieve optimization, accelerated proximal gradient descent [22] is used
02	Linked feature selection (LinkedFS) [23]	• It was introduced in 2012 • It is a semi-supervised feature selection method • It extracts data and then integrates using generic feature selection method • LinkedFS is a framework of social media which integrates different relations of linked data • There are four different relations along with their respective hypothesis [23]: • Single user having multiple post—CoPost Hypothesis Formulation for optimization problem is $$\min_W tr\left(W^T B W - 2EW\right) + \alpha\|W\|_{2,1} \text{ (4) where}$$ $E = XX^T + \beta F L_A FT$ and $E = Y^T X^T$ L_A is Laplacian matrix • Two users follow another user—CoFollowing Hypothesis Formulation for optimization is $$\min_w tr\left(W^T B W - 2EW\right) + \alpha\|W\|_{2,1} \text{ (5)}$$ where $B = XX^T = \beta F H L_{F1} H^T F^T$ and $E = Y^T X^T$ [23] • A third user follows two other users—CoFollowed Hypothesis Formulation for optimization is $$\min_w tr\left(W^T B W - 2EW\right) + \alpha\|W\|_{2,1} \text{ (6)}$$ where $B = XX^T = \beta F H L_{FE} H^T F^T$ and $E = Y^T X^T$ [23] • A single user follows another user—Following Hypothesis $$\min_w tr\left(W^T B W - 2EW\right) + \alpha\|W\|_{2,1} \text{ (7)}$$ where $B = XX^T = \beta F H L_S H^T F^T$ and $E = Y^T X^T$ L_S is the Laplacian matrix defined over matrix S [23]

(continued)

Table 1 (continued)

Sl no.	Method	Description
03	Linked unsupervised feature selection (LUFS) [24–26]	• It was introduced in the year 2013 • It is an unsupervised method • It works of linked data (social media data) • It uses social dimension reduction along with "pseudo-class" [25] • The data points are assigned with the class in the matrix mapping as [24] $W \in R^{m*c}$ (8) where m refers to the number of original features present, and c refers to the number of pseudo-class using in the technique. • Y is referred as the label indicator of the class as $Y = W^T X \in R^{m*c}$ (9) • LUFS gathers the information related to pseudo-class by fetching constraints from different types of data, i.e., linked and attribute. The constraints can be obtained by social dimensional regularization [26] • The dimensional regularization improves the performance of learning; it further computes the dependency of linked data [24]
04	CoSelect [27]	• It was introduced in the year 2013 • It is unsupervised in nature • Uses social correlation theory [27] in extracting relevant instances from the data with feature concurrency • The process of evaluation is • Initially applies different co-relation theories to extract link information • Applies instance selection • Selects relevant instance and/or features • This method tries to solve two main challenges of exploiting the link information and incorporating the instance and thereafter finds out the best possible features [27] • The optimizing problem of CoSelect can be written as $$\min_{W,B} \left\| W^T X - B^T - Y \right\|_F^{22}$$ $$+ \alpha \|W\|_{2,1} + \beta \|B\|_{2,1} + \gamma tr\left(W^T X L X^T W \right) \quad (10)$$
05	Generative feature selection (GFS) [20]	• It was introduced in the year 2016 • It is an unsupervised method • Used for specific and high-quality network • Applies information from both linked structures and network attributes • Pseudo-class with the help of clustering summarizes the data within a class • Inappropriate summarization is a cause of losing important data • To avoid such loss, direct connection is made between the original dataset and the features • Projection matrix W can be finally obtained using this method is: $$W = \left(\operatorname{diag}(s) X X^T \operatorname{diag}(s) + \beta I_D \right)^{-1} \operatorname{diag}(s) X X^T \quad (11)$$ where X is the set features, and s is the diagonal matrix I_D is the $D*D$ matrix [20]

(continued)

Table 1 (continued)

Sl no.	Method	Description
06	DetectFS [28]	• It was introduced is the year 2018 • It is an unsupervised method • Overcomes the quadratic errors which are generated from the CoSelect method
07	Unsupervised feature selection framework (UFFS) [29]	• It is an unsupervised static method • It is used to overcome restrictions of missing link information used • It considers inter-user relationship formation and features of the users • Reduces feature space in lower dimension to obtain subset of the features • It uses link information to partition them into known labels • It brings out definite cluster with the help of discrete features in a subset • To update the mapping matrix M, using scaled partitioning matrix G, the equation can be given as: $$M_{t+1} = \left(XX^T + \gamma X\left(I_n - GG^T\right)X^T - \lambda D_{Mt}\right)^{-1} XP^T$$ (12) where X is centered matrix, γ is discrimination control parameter, and λ is sparsity control parameter [29]
08	Robust unsupervised feature selection framework (NetFS) [30]	• It was introduced in 2016 • It is an unsupervised method • Reduces noisy data • Does not provide any information about the interconnected link in the network • It is used to apprehend all the interchanges that takes place between different networks • This method works on two phases • Latent representation: Used to uncover hidden attributes of the network • Noise reduction: Reduces the negative effect of the noise • The transformation parameter matrix W can be obtained by: $$W = (X^T X + \alpha D)^{-1} X^T U \ (13)$$ where X is the set of all the features, D is the diagonal matrix, U is the low dimensional latent space, and α is the parameter to control sparseness [30]

- With the availability of various methods, choosing the appropriate method becomes a challenge.
- The data obtained from the several tweets, posts, images and videos is huge, and thus, it has many problems.
- The major problem here is collecting authentic data from authentic posts, as social media is filled with fake news and post. Non-theoretical methods are generally used to collect the data for authenticity purpose [21].
- The technical issues of such data are primarily volume, velocity, variety and veracity. It is having the same issues like big data [21].

3 Conclusion

Linked data uses unlabeled heterogeneous data; thus, unsupervised feature selection method is highly recommended. The data obtained as linked data generally comes from social media and various other dynamic networks and is extremely huge and complex; thus, it is impossible to group them into certain classes. The linked data uses embedded model, and scholars are looking how to use linked data in filter and wrapped model. Unsupervised techniques are now in practice but they are soon likely to get replaced with the increasing requirement of the real-time methods. Different techniques of feature selection for linked data are present out of which mostly all come under unsupervised learning. Scholars are working on linked data as the data available is huge and effective methodology is yet to come to efficiently bring the feature space clearing all the issues as discussed in the paper.

References

1. Alelyani S, Tang J, Liu H (2013) Feature selection for clustering: a review. Data Cluster Algorithms Appl 29(1)
2. Zheng W, Zhu X, Wen G, Zhu Y, Yu H, Gan J (2020) Unsupervised feature selection by self-paced learning regularization. Pattern Recogn Lett 1(132):4–11
3. Gao L, Guo Z, Zhang H, Xu X, Shen HT (2017) Video captioning with attention-based LSTM and semantic consistency. IEEE Trans Multimedia 19(9):2045–2055
4. Song J, Gao L, Liu L, Zhu X, Sebe N (2018) Quantization-based hashing: a general framework for scalable image and video retrieval. Pattern Recogn 1(75):175–187
5. Zhang S, Li X, Zong M, Zhu X, Wang R (2017) Efficient knn classification with different numbers of nearest neighbors. IEEE Trans Neural Netw Learn Syst 29(5):1774–1785
6. Crowley JL, Parker AC (1984) A representation for shape based on peaks and ridges in the difference of low-pass transform. IEEE Trans Pattern Anal Mach Intell 2:156–170
7. Sun ZL, Huang DS, Cheun YM (2005) Extracting nonlinear features for multispectral images by FCMC and KPCA. Digital Signal Process 15(4):331–346
8. Sun ZL, Huang DS, Cheung YM, Liu J, Huang GB (2005) Using FCMC, FVS, and PCA techniques for feature extraction of multispectral images. IEEE Geosci Remote Sens Lett 2(2):108–112
9. Zhang H, Zhang Y, Huang TS (2013) Pose-robust face recognition via sparse representation. Pattern Recogn 46(5):1511–1521
10. Zhu X, Li X, Zhang S, Ju C, Wu X (2016) Robust joint graph sparse coding for unsupervised spectral feature selection. IEEE Trans Neural Netw Learn Syst 28(6):1263–1275
11. Gu S, Cheng R, Jin Y (2018) Feature selection for high-dimensional classification using a competitive swarm optimizer. Soft Comput 22(3):811–822
12. Gutlein M, Frank E, Hall M, Karwath A (2009) Large-scale attribute selection using wrappers. In: 2009 IEEE *Symposium on computational intelligence and data mining* Mar 30, IEEE, pp 332–339
13. Savić M, Kurbalija V, Ivanović M, Bosnić Z (2017) A feature selection method based on feature correlation networks. In: *International conference on model and data engineering* Oct 4, Springer, Cham, pp 248–261
14. Zhang JR, Zhang J, Lok TM, Lyu MR (2007) A hybrid particle swarm optimization–back-propagation algorithm for feedforward neural network training. Appl Math Comput 185(2):1026–1037

15. Huang DS, Du JX (2008) A constructive hybrid structure optimization methodology for radial basis probabilistic neural networks. IEEE Trans Neural Netw 19(12):2099–2115

16. Yu L, Liu H (2003) Feature selection for high-dimensional data: a fast correlation-based filter solution. In: *Proceedings of the 20th international conference on machine learning* (ICML-03), pp 856–863

17. Li J, Cheng K, Wang S, Morstatter F, Trevino RP, Tang J, Liu H (2017) Feature selection: a data perspective. ACM Comput Surveys (CSUR). 50(6):1–45

18. Ang JC, Mirzal A, Haron H, Hamed HN (2015) Supervised, unsupervised, and semi-supervised feature selection: a review on gene selection. IEEE/ACM Trans Comput Biol Bioinfo 13(5):971–989

19. Saeys Y (2004) Feature selection for classification of nucleic acid sequences (Doctoral dissertation, Ghent University)

20. Wei X, Cao B, Philip SY (2016) Unsupervised feature selection on networks: a generative view. In: Thirtieth AAAI conference on artificial intelligence, Mar 2

21. Cherrington M, Airehrour D, Lu J, Xu Q, Wade S, Madanian S (2019) Feature selection methods for linked data: limitations, capabilities and potentials. In: Proceedings of the 6th IEEE/ACM international conference on big data computing, applications and technologies, Dec 2, pp 103–112

22. Gu Q, Han J (2011) Towards feature selection in network. In: Proceedings of the 20th ACM international conference on Information and knowledge management, Oct 24, pp 1175–1184

23. Tang J, Liu H (2012) Feature selection with linked data in social media. In: Proceedings of the 2012 SIAM international conference on data mining, Apr 26, Society for Industrial and Applied Mathematics, pp 118–128

24. Tang J, Liu H (2012) Unsupervised feature selection for linked social media data. In: Proceedings of the 18th ACM SIGKDD international conference on knowledge discovery and data mining, Aug 12, pp 904–912

25. Tang J, Liu H (2014) An unsupervised feature selection framework for social media data. IEEE Trans Knowled Data Eng 26(12):2914–2927

26. Tang L, Liu H (2009) Relational learning via latent social dimensions. In: Proceedings of the 15th ACM SIGKDD international conference on knowledge discovery and data mining, Jun 28, pp 817–826

27. Tang J, Liu H (2013) Coselect: Feature selection with instance selection for social media data. In: Proceedings of the 2013 SIAM International conference on data mining, May 2, Society for Industrial and Applied Mathematics, pp 695–703

28. Benkabou S-E (2018) Détection d'anomalies dans les séries temporelles: application aux masses de données sur les pneumatiques. PhD diss

29. Hoseini E, Mansoori EG (2016) Selecting discriminative features in social media data: An unsupervised approach. Neurocomputing 12(205):463–471

30. Li J, Hu X, Wu L, Liu H (2016) Robust unsupervised feature selection on networked data. In: Proceedings of the 2016 SIAM international conference on data mining, Jun 30, Society for Industrial and Applied Mathematics, pp 387–395

Chapter 48
NAARI: An Intelligent Android App for Women Safety

Shreya Chakraborty, Debabrata Singh, and Anil Kumar Biswal

1 Introduction

Women are frequently worked across racial, sacred, opinionated, and intellectual divides to encourage serenity. So that we are so conscious about the security of women, but we must know that they should be well secured and freely traveled everywhere [1], where a woman is not so much strong as compared to the physical strength of men, that's why needs a helping hand to relieve them from a crisis or any difficult situation? When a victim of violent crime (robbery, sexual assault, rape, domestic violence) is identified, then it needs to minimize and use some resources to protect them from unsafe situations [2]. Whether a woman is in any critical situation like got separated from friends at night and does not know how to travel alone to home as well as feeling insecurities, at that moment, by using this app on phone can reduce the cause of risk and provide support whenever it is required. In the recent crime in Delhi which shivered the nation and made us more aware of the safety issues for our daughters, that's why the public is awaked and ready to fight back in different ways. A smart app has been designed to provide security features to women on their android phones. Here, we introduce a smart android app that can aware of the safety of women from any unavoidable circumstances. This app also minimizes the possibility of risk and helps us to identify the location, where the person is in danger [3].

S. Chakraborty · D. Singh (✉)
Dept of CSIT, SOA Deemed to Be University, Bhubaneswar, Odisha, India
e-mail: debabratasingh@soa.ac.in

S. Chakraborty
e-mail: shreyac22@gmail.com

A. K. Biswal
Dept of CSE, SOA Deemed to Be University, Bhubaneswar, Odisha, India
e-mail: anil.biswal123@gmail.com

Android is the free and robust platform that is widely used mobile OS motorized by Linux kernel, due to that gain much popularity [4]. Especially, it was designed by Google team and also managed its functionality with the help of the Java language. The advanced safety app is based on an android platform that utilizes the proper way of memory management for a custom virtual machine and hardware resources in an android phone [5]. So any application is built that will contain equal access to a phone's capabilities which are providing users with a lot of applications and services [6, 7].

1.1 Problem Definition

In the world, women are unsafe to travel alone at night. To provide this woman safety how to take them out from any unsafe situation to protect them from being a victim to any violent action is to identify all resources in a hand and use them. In today's generation, almost all use smartphones. Therefore, the wisest option is to have a safety application on our phones. Hence, this application has been built. This app, unlike other apps that work only at the time of emergency, works also to take preclusive measures as it is rightly said that "Precaution is better than cure."

This paper we made revolves around the concept of making phone apps for women safety that would fetch the real-time location using GPS and also the vehicle no using QR code scanner. It would then send a text message to the emergency contacts added by the user. It is our try to make each woman using our app to not to feel alone and unsafe. "NAARI we care for you."

1.2 Project Overview

Unfortunately, the safety of women is in doubt, and security is not concerned. Many headlines still coming across against women indicates that increasing trends of such sexual assault, rapes still happening in today's generation [8]. Around 80% of women are losing confidence and have fear toward the realization of freedom [9]. So we are trying to provide little efforts toward women for their safety, respect and also have rights to grow equally like men. This mobile application is very much helpful for anyone. This application will help user through scanning the QR code of app which will be nothing but scan and extract the vehicle details with also fetched current address through GPS. Then, it sends to any contact depending on user. Here, the user can take precautions before coming to the actual danger [10].

2 Literature Survey

Akram et al. [11] developed a safety device to protect women from vulnerable activity and lonely traveling on the road using IoT based on the fingerprint security method. This device alerts closure people as well as a police station, if any wrong or criminal movements with that women. Saikumar et al. [12] proposed a gadget to locate a risky place and many threats for women through IoT modules like Arduino UNO controller. So this controller is also integrated with Bluetooth device, teaser, and Android app. But the overall system is tracked the risky spot of the women with the help of GSM and GPS module. Ullah et al. [13] have designed an IoT-based smart system and android application that can track the nearest location of the bus and police station using the GPS module. Similarly, the IR sensor provides data about seat availability. But any women harassment situation occurred then press an alert button that available in a seat. When the button pressed, it forward the message to the nearest police station with the current spot of the bus.

Roy et al. [14] have constructed a wearable sensor band that protects women from various threats like lonely walking on road and harassment, whereas that band is designed through body sensor, Bluetooth module, GPS, SMS, and mobile database system that works followed by a supervised method of machine learning.

Sogi et al. [15] implemented a smart security wearable ring for women based on IoT modules that connected with Raspberry Pi controller, Pi camera, and buzzer. This system is activated by pressing the button; then, the buzzer is enabled and the camera has captured a picture of abuser or attacker. So that related information forwards to police through a smartphone for taking protecting action. Khandoker et al. [16] build an android application to reduce the criminal and harassment movements through passing voice command or pressing the SOS key. So it also facilities the continuous location tracking of victim spot that provides useful features in offline mode. Soman et al. [17] designed a multipurpose smart wrist band for the security of women from various hazardous situations by observing through heartbeat movement, shivering, and sweating through various sensors. In addition to locating the risky zone as well as the health parameter of users or women, forwards automatically alert messages to an emergency station to provide security.

2.1 Existing System

When we think about the safety of a woman/girl in the city, we considered 354 women candidates living in the city about our survey. From that survey, we represent our analysis in the depicted Fig. 1 and concluded that 54% of women realize unsafe at night time, whereas 46% of women do. Which puts us with a conclusion that more women feel unsafe at night time? After finding this percentage wreaked the women about their feeling when they have police patrolling in their vicinity. To this 68% of women replied that they have seen police patrolling at night, whereas rest

Break-Up of Women respondants who feel safe at Night

Action taken by Respondents   **Time Duration when Women Respondent's avoid going Out**

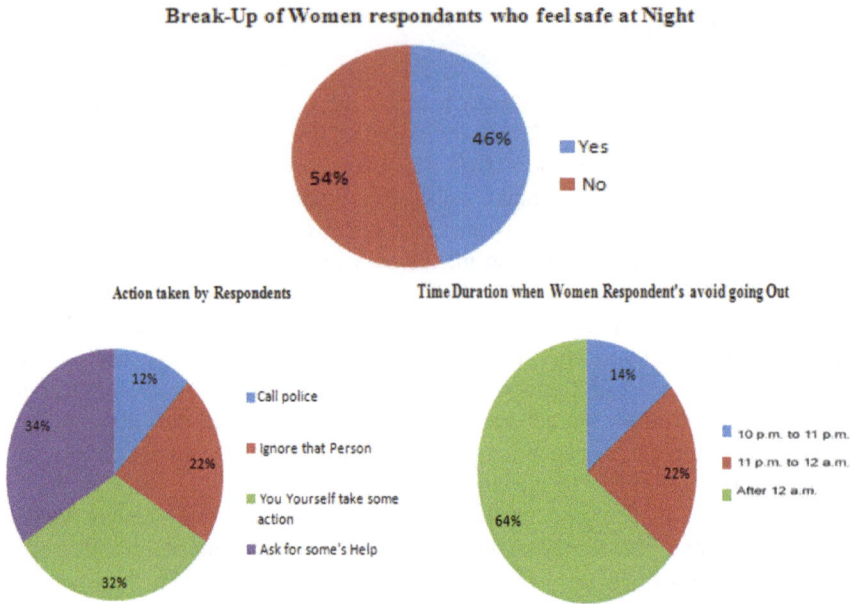

Fig. 1 Women respondents at night and action taken care, when they are outside home

32% of women have never seen police patrolling at night. Those who have seen cops patrolling they said that they feel safe at night.

Almost around 64% of women populations said that they do not find going out at night, especially between 11 and 12 pm a good option as most of the crimes and mishaps occur during that time. Therefore, women take to finishing their work before 10:00 pm. There were almost 22% of women who stated that stepping out after 10 even in the presence of cops is not a wise choice. After all this analysis the result revealed that girls experience teasing by male counterparts, and they take not being out during that time as a safety measure.

But at the same time, it was very pleasing to hear that almost 36% of women do not need help from anybody or cops and said that they can deal in such situations by themselves. There was also a very less percentage of girls who stated that they always have either a birthday or family member with the more nearby them at night. Therefore, feel safe as they do not find any exceptional action taken against the mat night.

3 Proposed System

In the proposed system can implement QR scanning and GPS location using Android, where the current address of the user will be retrieved through GPS. After that, we are going to scan the QR code to find details of the vehicle by pointing the camera at the QR code that automatically changed into text format. Finally, this system can send an SMS to any of the authorized users.

Product function:

Our summary of the major high-level function that the application can perform

- Registration page: User register himself by providing name, email, password, emergency number.
- Login page: User login by valid email id and password.
- Scanning page: A page where users can scan the QR code and fetch vehicle information.
- GPS tracker page: A page where the user can fetch the location.
- Safety tips page: A page with safety tips and log out button.

3.1 Feasibility Study

As we know that the rate of crimes has readily increased in today's world. With the increasing crime rate the women of our society have taken to certain ways, one of the ways to deal with this is to have a safety application in their smartphone [18, 19]. This safety application ensures that there is always an eye on her of her mentioned emergency contact numbers.

The emergency contact number always receives an updated location and the vehicle number which she is traveling in. It reduces the risk factor.

4   Methodologies

4.1 Design and Test Steps

The application is built to run on devices having Android KitKat (API 29), and above and during testing, all parameters and functions were tested successfully [20]. Some dummy data were built and tested, and it was found that the app is working properly. For the login purpose, the application only accepts valid password as login credential, and it was found working properly [21, 22].

The Front Ends of the Project are Designed Using XML (Extensible Markup Language)

- activity_main.xml: Image background with a launch button to get into using the app.
- activity_registration.xml: User register himself by providing name, email, password, emergency number.

- activity_second.xml: User login by valid email id and password.
- activity_functionality.xml:
- activity_scan.xml: A page where user can scan QR Code and fetch vehicle information.
- activity_mylocation.xml: A page where user can fetch the location.
- activity_safety_tips.xml: A page with safety tips and log out button.

4.2 Algorithm and Code

Step 1:   Go to launch page.

Step 2:   Login providing valid email address and password. If not registered user, then "NEW USER? SIGNUP."

Step 2.1: In case is not a registered user, then register by entering name, valid email address, password, emergency contact number. Then, after registering click on "ALREADY REGISTER? LOGIN"

Step 3:   From the login page by entering valid email and password go to the functionality page.

Step 3.1. You can click SCAN QR AND BARCODE to fetch vehicle information.

Step 3.2: Then, "CLICK ON THIS TPO FETCH LOCATION" button present right below the message "Done fetching vehicle information? Great!" to fetch the location.

Step 4:   After fetching the information go to the safety tips page which has a set of safety tops written in it.

Step 5:   In case you want to logout, then click on the "LOG OUT" button present at the right bottom corner of the safety tips page.

4.3 Sequence Diagram

we execute our programs with these three sequence diagrams depicted in Figs. 2,3 and 4, respectively.

4.4 Testing Process

Step 1:   The application installed successfully as represented in Figs. 5, 6 and 7 with their design phases. After building the *apk* file, we can install it on our device. After the installation is complete, we have the app ready to use. In between the process of building and installing the *apk* file is scanned for any malware presence.

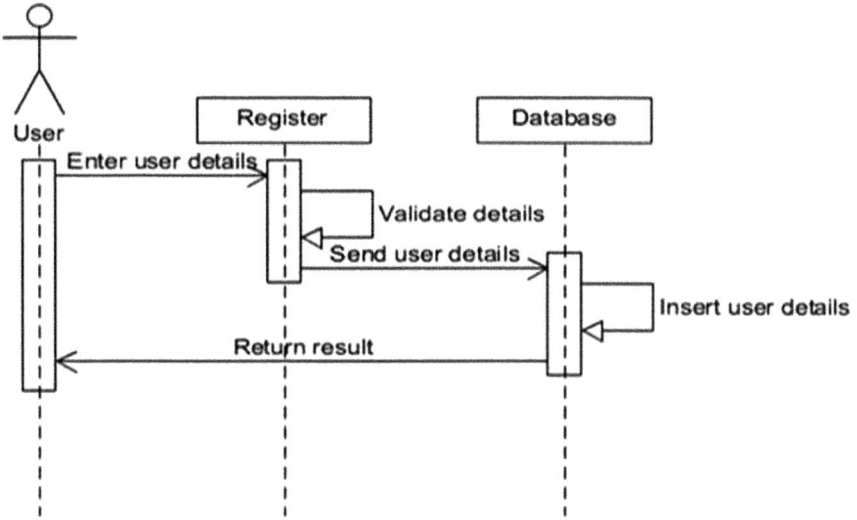

Fig. 2 Sequence diagram 1

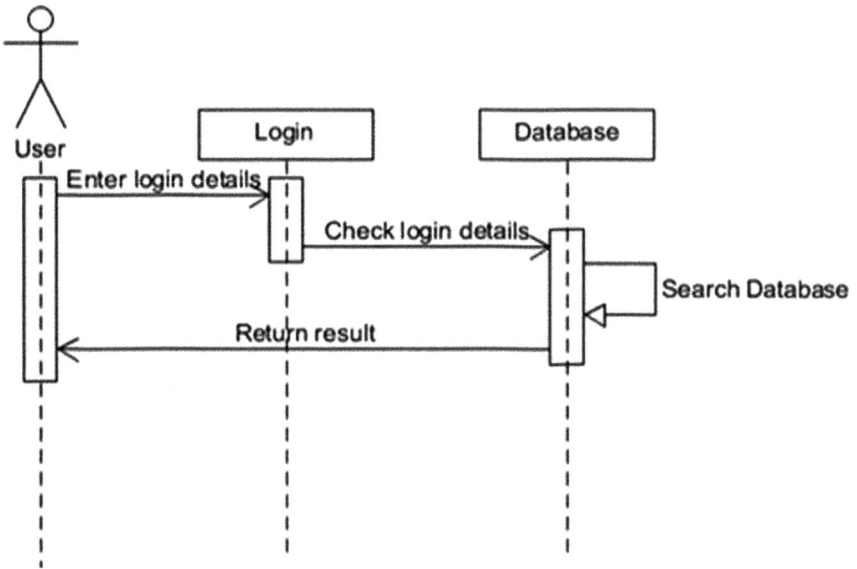

Fig. 3 Sequence diagram 2

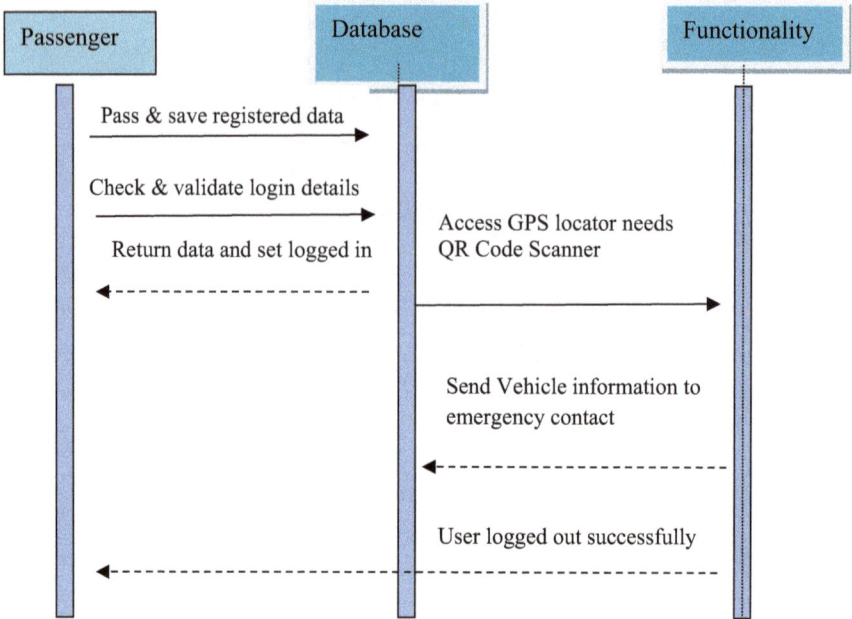

Fig. 4 Sequence diagram 3

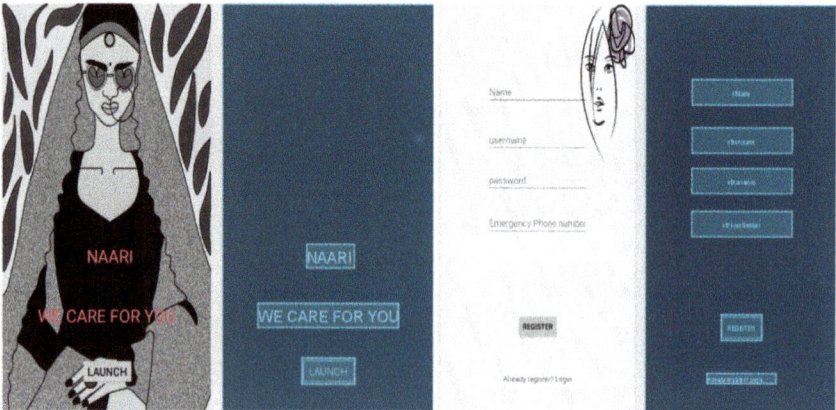

Fig. 5 Designing of launch page and designing of registration page

Step 2: "NAARI Safety App" in the application list or menu on mobile screen. After installing the *apk* file in android, it shows in the ion of the menu bar. We can access the application by clicking on the icon.

Step 3: Main welcome page with launch option to proceed to login page. This is the launch or first page that we get by clicking on the application. On this page, we have to click on Launch to go to the login page.

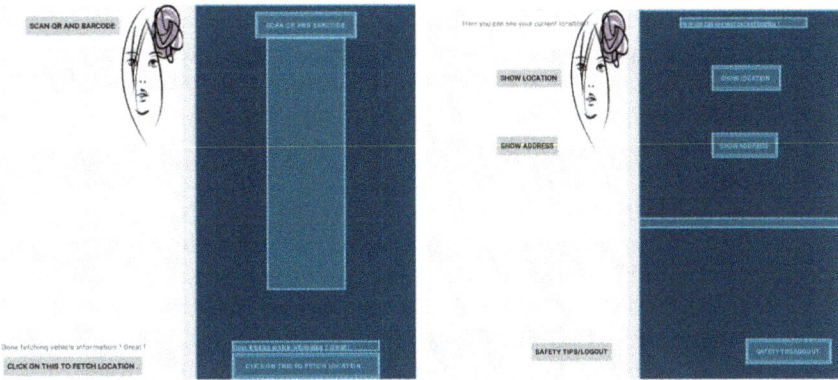

Fig. 6 Designing of QR code scanning page and designing of location fetching page

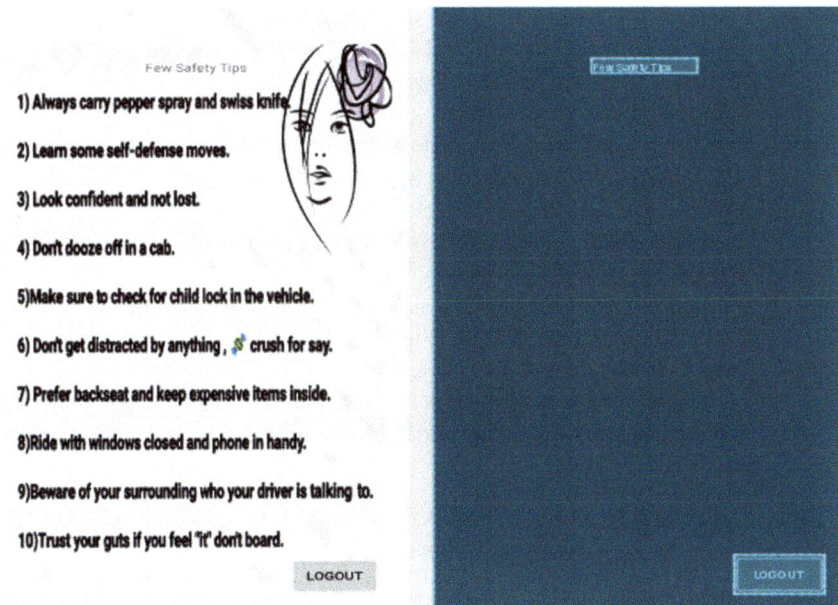

Fig. 7 Designing of safety tips page

Step 4:   Login page with maximum allowed three chances of logging in and new
          user register button. After clicking on the launch, we are directed to the
          login page which has maximum three numbers of attempts of successfully
          logging in. We have to entered valid username and password matching the
          design pattern and then click on "LOGIN" button to login. If not registered,
          then click on "NEW USER? SIGN UP" button which will take you to the
          registration page.

Step 5: Registration page with apt details and button to take back to login page. This is the page that asks to the new user to enter valid bio data for registering himself. The details asked to the new user are: Name, email, password, emergency phone number. After registration, he have to click on the "ALREADY REGISTER! LOGIN Again" to go back within the login page.

Step 6: "LOGIN UNSUCCESSFUL" message is displayed if username or email ID doesn't match pattern. In case the user entered invalid email id or password, then he received an error message. The error message is as followed "LOG IN UNSUCCESSFUL. The email address is badly formatted."

Step 7: "USER CREATED SUCCESSFULLY" message is displayed after with correct pattern of data. After entering the correct bio data (Name, email, password, emergency contact) the user directed to the login page. On the login page, the new user receives the message "USER CREATED SUCCESSFULLY." He has to enter the username and password to login.

Step 8: Functionality page with options of "SCAN QR AND BARCODE" and "CLICK ON THIS TO FETCH LOCATION." This is the main functionality page of the application. This page has two options (Scanning vehicle information and fetching user location)

Step 9: Scanner to scan vehicle information from QR Code. By clicking on "SCAN AND BARCODE" button you will be directed to this page. On the scanning page the user has to correctly place the QR code or Bar code in order to fetch the vehicle information. After scanning the vehicle information, he is redirected back to the functionality page

Step 10: Fetched vehicle information. It shows the vehicle information that the user got by scanning the QR code (Fig. 8).

Step 11: Page to fetch address: After clicking on "CLICK ON THS TO FETCH LOCATION" button the user is redirected to this page. On this page, the user can click on show location to fetch the address in latitude and longitude number.

Then, the user has to click on "SHOW ADDRESS" button to convert the longitude and latitude number into actual text address.

Step 12: Fetched address: This is the fetched address in latitude and longitude and actual text address.

Step 13: Safety tips with log out button at bottom left corner. This page has numerous safety tips that the user can take to avoid unforeseen situation. The users can also logout from this application by clicking on "LOGOUT" button.

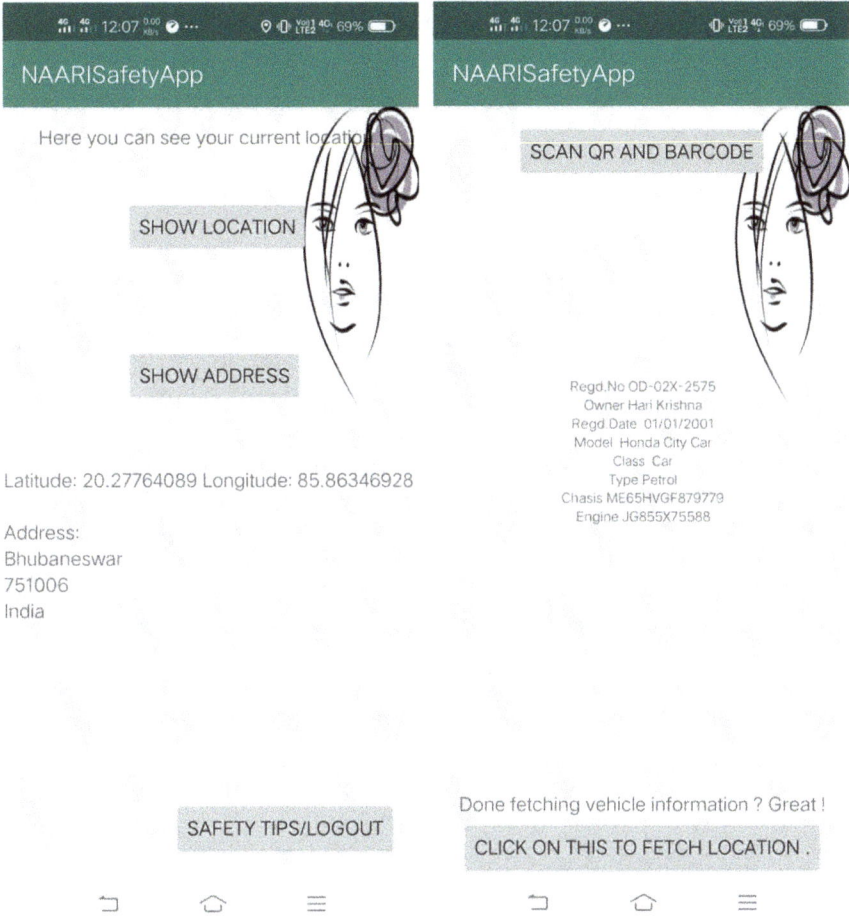

Fig. 8 Current location & fetched vechile information

5 Result and Analysis

This is the fetched address in latitude and longitude and actual text address, and also it shows the vehicle information that the user got by scanning the QR code. After the vehicle information is fetched, the user can fetch the location by clicking on the "CLICK ON THIS TO FETCH LOCATION."

Data stored in database. The bio data of the new user is stored in the firebase database. For storing the data in database, firebase is given the authentication, and as soon as, a user signs up by providing his details in correct pattern (name, email id, phone no, emergency phone no.) her details get stored in the data base [23]. This information's are used when the user wants to sign in into her account, or even the mobile number of hers is used to geo tract her location, and the emergency contact number is used to send her updated location all the while she is traveling and also send the vehicle information she is traveling by. The admin can access to delete an user's account in case of rule invalidation.

6 Conclusion

Unfortunately, the safety of women is in doubt, and security is not concerned. Many headlines still coming across against women indicates that increasing trends of such sexual assault rapes still happening in today's generation. Around 80% of women are losing confidence and have fear of the realization of freedom. So we are trying to contribute little efforts toward women which will ensure the safety and respect for women so that she canal so have the right to grow equally like men. This mobile application is very much helpful for anyone. This application will help the user by scanning the QR code which will be nothing but she can attach the vehicle detail send through GPS the current address which will be fetched and send it to any contact depending on the user. Here, the user can take precautions before coming to the actual danger.

It is to let every NAARI is now safe to travel alone as someone is getting their updated location and also has vehicle information. For the future, we have in mind to extend this app, where she can also contact nearby police patrolling vans in case of need. This project that I have made is small scale but has a large development scope, and I look further to the day it can be extended and used by all common people so in totality, this project is an initiative taken by the youth community to contribute to the betterment of the society in whatever way we can.

References

1. Gupta M, Thakur S, Singh L, Rana V (2016) In: Design of women safety system using RFID and GSM technology
2. Varade S, Itnare T, Parande H, Sonawane P, Bhardwaj R (2017) Advanced women security system based on IOT. Int J Recent Innov Trends in Comput Commun 12:57–61
3. Harini R, Hemashree P (2019) In: Android app for women security system
4. Kadkol RJ, Aman K, Keerthi M, Neha K (2017) GPS Based android application for women security. Int J Eng Sci 11016
5. Lehman WE, Pankow J, Rowan GA, Gray J, Blue TR, Muiruri R, Knight K (2018) StaySafe: a self-administered android tablet application for helping individuals on probation make better decisions pertaining to health risk behaviors. Contemporary Clinical Trials Commun 10:86–93
6. Mane IA, Babar JR, Patil SS, Pol SD, Shetty NR (2016) Stay safe application. Int Res J Eng Technol (IRJET), SJ Avenue 3(5):2157–2160
7. Miriyala GP, Sunil PVVNDP, Yadlapalli RS, Pasam VRL, Kondapalli AT, Miriyala A (2016) Smart intelligent security system for women. Int J Electron Commun Eng Technol (IJECET) 7(2):41–46
8. Paradkar A, Sharma D (2015) All in one intelligent safety system for women security. Int J Comput Appl 130(11):33–40

9. Cohn C, Kinsella H, Gibbings S (2004) Women, peace and security resolution 1325. Int Feminist J Politics 6(1):130–140
10. Yarrabothu RS, Bramarambika T (2015) Abhaya: an android app for the intelligent safety safety of women. In: 2015 Annual IEEE India conference (INDICON), IEEE, pp 1–4
11. Akram W, Jain M, Hemalatha CS (2019) Design of a smart safety device for women using IoT. Proc Comput Sci 165:656–662
12. Saikumar P, Bharadwaja P, Jabez J (2019) Android and bluetooth low energy device based safety system. In: 2019 3rd International conference on computing methodologies and communication (ICCMC) March, IEEE, pp 1180–1185
13. Ullah A, Hossain MA, Zaman N, Dey M, Kundu T (2019) Enhanced women safety and well-suited public bus management system in Bangladesh using IoT. Adv Internet Things 9(4):72–84
14. Roy S, Sharma A, Bhattacharya U (2015). Move free: a ubiquitous system to provide women safety, In: Proceedings of the third international symposium on women in computing and informatics August, pp 545–552
15. Sogi NR, Chatterjee P, Nethra U, Suma V (2018) SMARISA: a raspberry pi based smart ring for women safety using IoT. In: International conference on, inventive research in computing applications (ICIRCA), IEEE, pp 451–454
16. Khandoker RR, Khondaker S, Nur FN, Sultana S (2019) Lifecraft: an android based application system for women safety. In: 2019 International conference on sustainable technologies for industry 4.0 (STI), IEEE, pp 1–6
17. Soman S, Sreelakshmi G, Asok A, Embrandhiri S (2017) Intelligent multipurpose safety wrist band for women using arduino. Int J Innov Implement Eng 1
18. Bhanushali P, Mange R, Paras D, Bhole C (2018) Women safety android app
19. Yarrabothu RS, Thota B (2015) Abhaya: an android App for the safety of women, In: 2015 Annual IEEE India conference (INDICON), IEEE, pp 1–4
20. Alani MM (2017) Android user's privacy awareness survey. Int J Interactive Mobile Technol (iJIM) 11(3):130–144
21. Ismail NA (2015) In: Stay safe mobile application, IRC
22. Mareeswari V, Patil SS (2018) Smart device for ensuring women safety using android app. In: Advanced computational and communication paradigms, Springer, Singapore, pp 186–197
23. Singh D, Das A, Mishra A, Pattanayak BK (2017) Safety and crime assistance system for a fast track response on mobile devices in bhubaneswar. In: Computational intelligence in data mining, Springer, Singapore, pp 1–12

Chapter 49
Performance Benchmarking of GPU and TPU on Google Colaboratory for Convolutional Neural Network

Vijeta Sharma, Gaurav Kumar Gupta, and Manjari Gupta

1 Introduction

Deep learning algorithms [1] have been emerged as the most popular field of artificial intelligence research. This has various network architectures with multiple layers such as deep neural network, recurrent neural network, deep belief network. Convolutional Neural Network (CNN) [2] is one of the widely used algorithm of deep learning for image classification tasks. CNN mainly consists of multiple layers of simple neural network. CNN classifies images under certain categories or objects based on the input images. Usually, computer interprets given input image as an array of pixels, and it depends on the image resolution. During the training process of CNN, each input image (pixel's numeric value) processes via multiple convolution layers with filters called Kernels, then pooling layer and afterward via fully connected layers (FC). Finally, softmax function is applied to classify an object with probabilistic values between 0 and 1. Figure 1 shows the complete flow of CNN to classify handwritten digits.

To train the CNN model on large and complex dataset of images, huge computational power is required for processing. This is also true that traditional computers with CPU are not capable enough to train CNN in sequential manner. For CNN's training, large amount of floating point operations (mathematical calculation) are required; also CNN has parallel I/O architecture that needs parallel processing device for faster processing. Graphics Processing Unit (GPU) is the revolution for deep

V. Sharma (✉) · G. K. Gupta · M. Gupta
Computer Science, DST-Center for Interdisciplinary Mathematical Sciences, Institute of Science, Banaras Hindu University, Varanasi, India
e-mail: vijeta.it@gmail.com

G. K. Gupta
e-mail: gauravkumarg068@gmail.com

M. Gupta
e-mail: manjari@bhu.ac.in

© The Author(s), under exclusive license to Springer Nature Singapore Pte Ltd. 2021
X.-Z. Gao et al. (eds.), *Applications of Artificial Intelligence in Engineering*, Algorithms for Intelligent Systems, https://doi.org/10.1007/978-981-33-4604-8_49

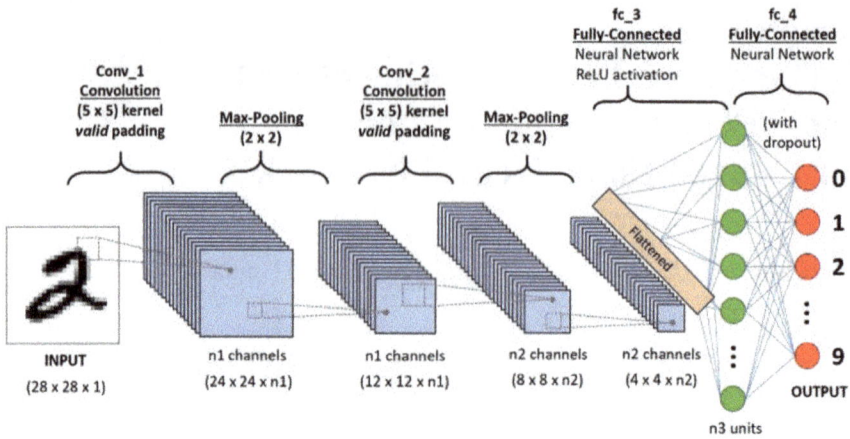

Fig. 1 A typical flow of CNN to classify handwritten digits. *Source* https://miro.medium.com/max/700/1*uAeANQIOQPqWZnnuH-VEyw.jpeg, last accessed 2020/08/06

learning algorithms. GPU can perform mathematical calculations much faster than CPU. GPU has thousands of Arithmetic Logic Units (ALUs) in single processor, which means that it can perform thousands of multiplication and addition parallelly, which allows GPU to execute up to 50 times faster than traditional CPU. Image itself is a matrix, and GPUs are good at performing matrix operations because of its parallel architecture. So, this is the reason why CNN performs well on GPU for image processing task. Earlier, GPUs were used mainly for 3D images and video rendering such as gaming software. Later, Nvidia launched its GeForce series [3], which became quite popular. They also developed Compute Unified Distributed Architecture (CUDA) [4], as an API for parallel computing platform to use Nvidia GPU.

Since GPU performs parallel computation on thousands of ALU, it spends more energy on memory access. To resolve this problem, Google designed a Tensor Processing Unit (TPU) [5]. TPU loads the parameter from memory into the matrix of multipliers and adders. After that, it loads the data from memory. Result of each multiplication execution get passed to next multipliers simultaneously performing summation too. So, the output is sum of the result of all the multiplication between data and parameter. In the entire process of complex calculations and data transfer, access of memory is not required. This is why TPU can achieve high computational throughput on neural network algorithms. Deep learning algorithms, which are developed on TensorFlow software only, are used for training and inference on TPU as it is meant to deal with tensors

only. Usually, a tensor is n dimensional matrix. Google has provided this TPU on "Google Colaboratory" [6] for the research purpose. It is a free online cloud-based Jupyter notebook environment that allows researchers to train complex neural network like models on CPUs, GPUs, and TPUs. Table 1 shows the specifications of GPU and TPU offered by Google Colab:

Table 1 Specifications of GPU and TPU on Google colaboratory

GPU	TPU
Nvidia K80s, T4s, P4s, and P100s with 12 GB of GDDR5 VRAM	Cloud TPU with 180 teraflops of computation
Intel Xeon processor with two cores @2.20 GHz and 13 GB RAM	Intel Xeon processor with two cores @2.30 GHz and 13 GB RAM

2 Related Work

Only few researches have done on usage or performance benchmarking of Google colaboratory GPU and TPU. One of them is [7], where researcher had trained U-NET-based eye fundus images segmentation. It observed that training speed was much faster on TPU than training on GPU because of it's highly parallelism nature. In another paper, author Carneiro et.al [8] did performance analysis of Google Colaboratory GPU for the applications of object detection, classification, localization and segmentation. They compared the hardware under the accelerated runtime with a mainstream workstation and a robust Linux server equipped with 20 physical cores. It is found that Google colaboratory GPU accelerated runtime which is a faster to train a CNN than using 20 physical cores of a Linux server. Author of [9] used Google colaboratory GPU to train CNN and deep neural network model for Intrusion Detection System (IDS), heart disease prediction, and skin lesion classification.

A systematic benchmarking of deep learning technique on Google cloud TPU, NVIDIA's V100 GPU, and Intel Skylake CPU platform has been done [10] by introducing a new architecture called ParaDnn. This deep learning networks architecture consists of a fully connected (FC), convolutional (CNN) and recurrent (RNN) neural network. In cross-platform comparison (TPU, GPU, and CPU) authors concluded that TPU is highly optimized for large batches, and CNN has the highest training throughput, whereas GPU shows better flexibility and programmability for irregular computations, such as small batches and non-matrix multiplication computations. Since CPU has the best programmability, thus, it achieves the highest FLOPS utilization for RNNs.

While there is more focus on the development of domain specific hardware, researchers of Google [11] have shown the datacenter performance analysis of TPU. They revealed that TPU accelerates the inference phase of neural networks, which offer a peak throughput of 92 TeraOps/second (TOPS) by using 65,536 8-bit MAC matrix multiply unit and a large (28 MiB) software-managed on-chip memory. In this study, TPU was compared with a server class Intel Haswell CPU and a Nvidia K80 GPU, which were deployed in the same datacenters. The NN applications (MLPs, CNNs, and LSTMs) with TensorFlow framework were used to measure the efficiency of NN inference. As a result, it was observed that 95% of datacenter's NN inference demand were fulfilled by TPU, and on average, it is about 15X -30X faster than its contemporary GPU or CPU, with TOPS/Watt about 30X - 80X higher. Moreover, using the GPU's GDDR5 memory in the TPU, it raises TOPS/Watt to nearly 70X the GPU and 200X the CPU.

This paper highlights the feature of GPU and TPU in terms of processing speed for the different batch sizes of dataset when a model is built on convolution neural network. For research, we have taken an application of handwritten digits recognition with popular MNIST dataset. The use case of digit recognition system is — read bank cheque, recognize vehicle's number plate, numeric entries in forms filled up by hand, and so on. The Modified National Institute of Standards and Technology (MNIST) [12] dataset is a large database of handwritten digits that are commonly used for training various image processing systems. This study is also an effort to show the intersection between HPC and artificial intelligence [13].

The structure of the paper is as follows: Sect. 2 explains the methodology, Sect. 3 summarizes the result of benchmarking of GPU and TPU, Sect. 4 concludes, and Sect. 5 explores about future research.

3   Methodology

Below steps has been followed to build the CNN model for handwritten digit recognition system and for a comparative study for performance benchmarking.

3.1   Choosing Dataset

MNIST dataset contains 70,000 images of handwritten digits, in which 60,000 images are in training set and 10,000 images for in testing set. Both the training and testing sets are labeled images from digits 0 to 9. These images are in the form of 28*28 gray scale intensity of each digit. We have used the "mnist_train.csv" for training, where first column represents label values ranging from 0 to 9 for every image. The same has been opted for the cases of the testing dataset "mnist_test.csv" as 10,000 images with a label of 0 to 9. Each column represents as $1 \times 1, 1 \times 2, 1 \times 3, 1 \times 4 \ldots 28 \times 7, 28 \times 28$.

3.2   Building CNN Model

The CNN model for MNIST handwritten digit recognition system for the performance benchmarking of GPU and TPU has been developed by using Keras [14] which has TensorFlow version 2.2.0 [15] at the backend. Data has been handled using tf.data.Dataset API. For CNN, batch size is 1024, epochs are 10, and LEARNING RATE is 0.01.

3.3 *Training CNN Model*

We have used the cooperative iPython notebook development environment Google colaboratory [16] to train the CNN model. The environment has very good support for Keras, with the possibility of implementing and training network's based on GPUs and TPUs in Google cloud. In Google colaboratory, there is an option to run multiple CPU, GPU, and TPU instances simultaneously, and the resources are shared between these instances. Training on GPU and TPUs allowed us to test wider and deeper architectures that always show out of the memory limits on many of current single CPU systems.

The higher training speeds also allows us to prune the network to make it lighter with small effects in prediction efficiency. We use the version of Keras included in TensorFlow. This is necessary to be able to execute it on TPUs. Firstly, we to train CNN model on GPU and then TPU. To do so, colaboratory has an easy option to change the runtime environment. That we have chosen from the runtime menu and then "Change runtime type," where we chose runtime type as "GPU" among the three options MNIST GPU, TPU, none [17].

To test whether our program is running on TPU, we used the following steps mentioned in Algorithm 1.

Algorithm 1 Check TPU device

1. Import package os
2. Check "COLAB_TPU_ADDR"
3. if COLAB_TPU_ADDR Not in os environ

   Display not connected to TPU
   Else
   Display connected to TPU

Similarly, to confirm the running environment of GPU, we used the following steps mentioned in algorithm 2.

Algorithm 2 Check GPU device

1. Import package TensorFlow
2. Test gpu_device_name
3. If GPU connected

   Display/device:GPU:0
   Else
   Display nothing

Google colaboratory has very good provision to import the data from Google drive or directly clone from github [18]. Dataset has been uploaded directly to Google drive and mounted the Google drive on colaboratory.

After completing training on GPU, we noted down the complete duration and started training on TPU by changing the runtime environment and with few modifications in CNN code.

On successful CNN model training on TPU, we have observed the time of each step and epoch against different batch sizes. It has also been observed that TPU setup initially took some time when it started to compile the model. So, the first epoch took few longer times. So, we have reported only the time for the later epochs. We have calculated the average time across different epochs. In all the experiments, training time was very less for both the GPU and TPU but comparatively different which is shown in result section.

4 Result and Discussion

In our training experiments, we trained for 10 epochs with 1024 batch size per epoch. As we trained CNN on 60,000 images, the result was very interesting to see; epoch time for different batch size was less with the TPU. It was also notable in results that with the batch size 1024, epoch time of TPU is half than the GPU (Table 2).

While Training, steps time of GPU and TPU also shows the clear difference in the results (Table 3).

Here, if we see, then the training step time for batch size 1024 is again remarkable for TPU which took almost half time than GPU. Also, here, we did not expect the outstanding performance by TPU on a dataset because MNIST is not a big image dataset.

Training on Google cloud TPUs has allowed us to test many different configurations, training them in a time almost independent of the network architecture. The

Table 2 Epoch time for different batch size of CNN training

Batch size	GPU (s)	TPU (s)
256	6	6
512	5	3
1024	4	2

Table 3 Training step time for different batch size of CNN training

Batch size	GPU	TPU
256	94	97
512	82 μs	58 μs
1024	79 μs	37 μs

speedup obtained with TPUs which are 2–3 times faster than GPU makes this environment ideal for other large neural networks such as deep neural network, recurrent neural network, etc. It has also been observed that the performance of the GPU and TPU made available by colaboratory may be enough for several researchers to train the CNN for other image processing task.

One thing we have faced during this experiment is lack of documentation of Google's TensorFlow TPU support. However, due to user friendly interface of Google colaboratory, we were able to do performance benchmarking experiments efficiently.

5 Conclusion

However, handwritten digit recognition application is not as complex and as huge to observe the big difference between the performance of Google GPU and TPU but definitely, this research work would be an interesting point to start with the complex and huge application oriented research to leverage the benefits of fastest environment like TPU for neural network training. Also, performance benchmarking on compute intensive and data intensive applications such as large-scale video data processing, cancer research, and natural language processing by RNN, etc. will definitely give big difference in their epoch time and training step time.

6 Future Research

There are many possibilities for expanding this work in future. Other possible topics related to deep learning architecture can apply on various application to be train and validate on Google colaboratory TPU and GPU in fully parallel environment.

References

1. Deep Learning (2020) https://en.wikipedia.org/wiki/Deep_learning. Last Accessed 04 Aug 2020
2. Convolutional Neural Network, http://deeplearning.stanford.edu/tutorial/supervised/ConvolutionalNeuralNetwork, last accessed 2020/08/03
3. NVIDIA GPU (2020). https://www.nvidia.com/en-in/geforce. Last Accessed 05 Aug 2020
4. CUDA Toolkit (2020). https://developer.nvidia.com/cuda-toolkit. Last Accessed 03 Aug 2020
5. Google Cloud Tensor Processing Units (2020). https://cloud.google.com/tpu/docs/tpus. Last Accessed 03 Aug 2020
6. Google Colaboratory (2020) https://colab.research.google.com. Last Accessed 04 aug 2020
7. Civit-Masot J, Luna-Perejón F, Vicente-Díaz S, Rodríguez Corral JM, Civit A (2019) TPU cloud-based generalized U-Net for eye fundus image segmentation. IEEE Access 7: 142379–142387

8. Carneiro T, Medeiros Da NóBrega RV, Nepomuceno T, Bian G, De Albuquerque VHC, Filho PPR (2018) Performance analysis of google colaboratory as a tool for accelerating deep learning applications. IEEE Access 6:61677–61685
9. Syam R, Marapareddy R (2019) Application of deep neural networks in the field of information security and healthcare. In: SoutheastCon, Huntsville, AL, USA, pp 1–5
10. Wang YE, Wei G-Y (2019) David brooks: benchmarking TPU, GPU, and CPU platforms for deep learning. arXiv:1907.10701v4 (cs.LG)
11. Norman PJ, Cliff Y et.al (2017) In-datacenter performance analysis of a tensor processing unit. In: ACM/IEEE 44th International symposium on computer architecture (ISCA), Toronto, Canada, pp 1–12
12. MNIST Dataset (2020) https://www.kaggle.com/oddrationale/mnist-in-csv. Last Accessed 07 Aug 2020
13. Intersection of HPC and Machine Learning (2020). https://www.nvidia.com/en-us/data-center/resources/intersection-of-hpc-and-machine-learning. Last Accessed 06 Aug 2020
14. Keras (2020). https://keras.io. Last Accessed 06 Aug 2020
15. Tensorflow (2020) https://www.tensorflow.org. Last Accessed 06 Aug 2020
16. iPython Jupyter Notebook (2020). https://ipython.org/notebook.html. Last Accessed 05 Aug 2020
17. Runtime Environment (2020). https://research.google.com/colaboratory/local-runtimes.html. Last Accessed 04 Aug 2020
18. Import External Data (2020). https://colab.research.google.com/notebook/io.ipynb. Last Accessed 05 Aug 2020

Chapter 50
A Study on Steam Cleaning Devices and Its Impact on Health and Hygiene

Amrita Dey and Sasthi Charan Hens

1 Introduction

One has witnessed many traditional cleaning techniques, the most common being the one with a bucket and mopping cloth. The one where some detergent is added to a water-filled bucket followed by mopping. But over the years with advancement in the research, proper hygiene is given utmost importance and the comfort of the people has been prioritized. Various studies throw light on many disadvantages associated with traditional mopping. Like it is unable to sterilize the surface in essence unable to kill the microbes completely. Then, the labor one has to devote to carrying those large vessels especially for an old person or someone with the knee joint, and back pain it is a strenuous job. So, the need arises to develop a device that facilitates proper and more efficient sanitation. Most microbes are not resistant to heat so the cleaning which left a hot surface came into light. That is where steam cleaning finds its importance in today's world. We know steam is a natural substance it does not cause any harmful reaction to those who are suffering from allergies or asthma. The high temperatures are generated by the steam which in turn will kill most bacteria and allergens. Using steam mops makes the floor quickly dry as compared to conventional cleaning with a mop and bucket. Steam cleaning is eco-friendly, along with being safer and more effective.

A. Dey (✉) · S. C. Hens
Department of Mechanical Engineering, University of Engineering and Management, Jaipur, Rajasthan 303807, India
e-mail: dey.amrita93@gmail.com

S. C. Hens
e-mail: hens.sasthi@gmail.com

2 Literature Review

2.1 Steam Cleaning Apparatus

In several inventions, the portable machine is made for steam cleaning carpets, rugs, automotive engines, and soiled areas. An electrolytic generator is used for substantially instant production of the steam; a positive type displacement pump is basically used for the delivery of a water and chemical solution from a tank to the electrolytic generator. Then, an electrolytic generator forms the steam. Discharge of steam takes place via hose and nozzle on the soiled area. The machine includes a solenoid and air pump for the delivery of detergent from a tank to the nozzle when it is adapted carpet cleaning or the rug and for picking up excess moisture and dirt particles released by the steam solution achieved via vacuum motor with the nozzle. It provides a dual operating system in essence the vacuum cleaner is also there but the biggest disadvantage stays as it focuses on mixing some sort of chemical or detergent along with it. The diagrammatical representation is given below [1] (Fig. 1).

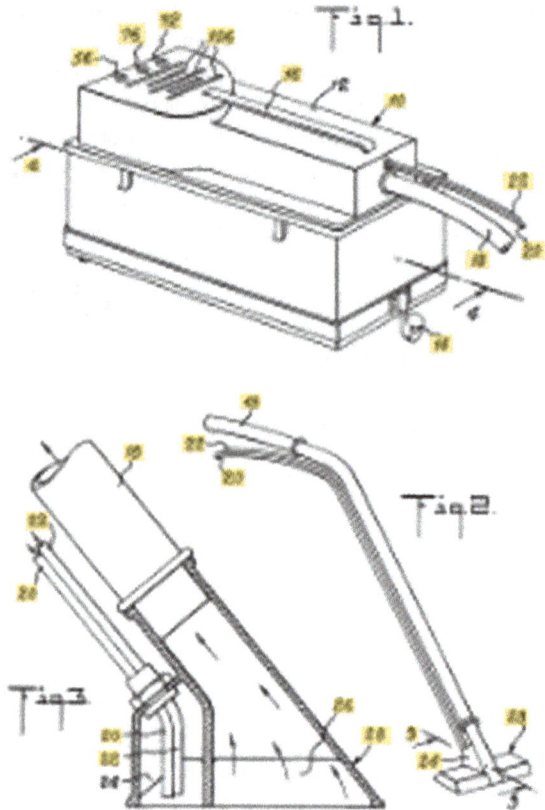

Fig. 1 Schematic representations of combined steam and vacuum unit

In other similar inventions, a steam mop has a water pump for selectively injecting water from a reservoir to a boiler is provided. The pump is actuated by the user's movement to pump water to the boiler for distribution to a steam pocket frame attachment. A fabric steam pocket actually is mounted on the steam pocket having frame to provide a cleaning surface. The mop includes a housing with an electric boiler and a water pump coupled to a water tank [2]. Some inventions provide apparatus in which the clean water is subjected to super atmospheric pressure utilizing some fan or blower which is employed to establish the vacuum cleaning. The invention also tries to provide an organization of elements that accomplishes the feeding of clean water by the application of super atmospheric pressure. It is held in a reservoir within the apparatus to establish a siphoning action from the reservoir to the floor tool and interrupts the siphoning action when the desired amount of clean water is removed from the apparatus [3].

Inventions are there to implement a filter to the steam cleaner for removing the impurity of the water to prevent the impurity from blocking the nozzle. A safety switch is added to the steam cleaner such that the steam producing equipment can automatically turn off when overheats [4].

A steam formation device, which has a water inlet and a steam outlet, the steam outlet are communicated with the steam outlet of the decontamination surface of the shell and a steam formation cavity is provided between the water inlet and the steam outlet for communicating with the both. A heating device is provided for heating the steam formation cavity. A water pump is implemented between the water tank and the steam formation device and used to drive the water in the water tank to flow into the steam formation device. Here, the water pump comprises a synchronous motor and reciprocating piston rod and is connected to the piston rod of the water pump to drive the water pump to actuate [5].

An upright bare floor cleaner with a handle assembly is mounted on the base in one of the inventions. The water tank is embedded in the handle to store water and there is a fluid distribution system that includes a heater with spray nozzle. Steam is generated by the heater, and then, it is distributed on a removable cleaning cloth thus it applies the steam to the surface to be cleaned. An auxiliary hand tool for steam cleaning above-floor surfaces is further included by the distribution system. Here, we see that that the steam mopping can be done even at the upper surface. Which is unique concerning other mopping systems as it can be used for upper surfaces as well. The steam mop of the invention provides both floor and upper-floor steam cleaning as depicted in Fig. 2 [6].

Then, other inventions include a wet mop with an interchangeable scrubbing pad and cloth wipe is provided, which comprises a flat rigid plate. The invention also provides a mop which can accommodate an abrasive scrubbing pad. The plate has an elongated handle pivotally connected to it. The bottom has a deformable cushion attached to the plate grippers countersunk within the foam pad, and it has bristles that extend downwards. A porous scrubbing pad is sometimes attached to the bottom of the foam pad where the bristles are engaged. A terry cloth wipe can also be fitted easily over the foam cushion and around the plate. An elastic band around the perimeter of the wipe helps it in place. Therefore, it is an objective of the invention

Fig. 2 Upright steam mop
with the auxiliary hand tool

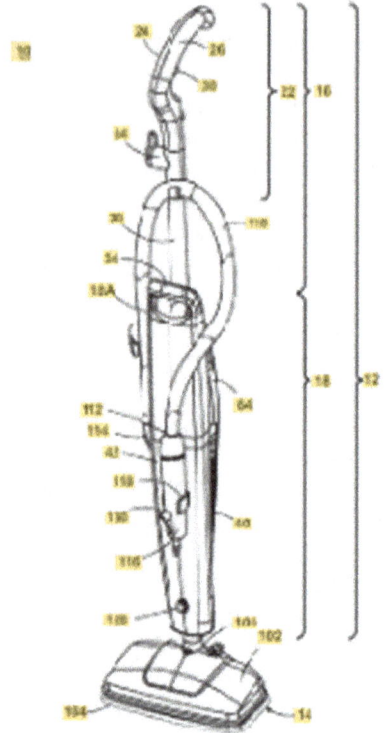

to provide a mop that is adaptable for damp mopping as well as scrubbing the floors, respectively. The invention also aims to provide a mop which is deformable and have cushioned head covered with a cloth wipe. It also emphasizes to provide a mop with convenient interchangeability of cleaning media. A further objective of the invention is to provide a scrubbing pad that does not interfere with the use of cloth wipes.

It is depicted in the figure below. Here, basically, a wet mop for in combination with interchangeable scrubbing pad and cloth wipe is provided. Then, with a flat rigid plate, the elongated handle is connected pivotally to the plate. The deformable cushions are then attached along the bottom of the plate. The pad gripper is provided for having a foundation connected to plate and the countersunk, respectively [7] (Fig. 3).

On a steam pad frame, a multilayer fabric steam pad is mounted in some device. The fabric steam pad consists of substantially planar backing layers and a fabric layer with at least one fastener region secured to the backing layer for attaching the pad to the steam frame. Velcro fastener, a conventional hook-and-loop is used commonly. In an embodiment, a mesh layer is joined to the fabric backing layer on the outer surface [8].

Hooks and the loops fastener or the Velcro tape has a hook portion that is directly connected to the towel cleaning cushion. Then, we see additional features in some

Fig. 3 Wet mop with an interchangeable scrubbing pad and cloth wipe

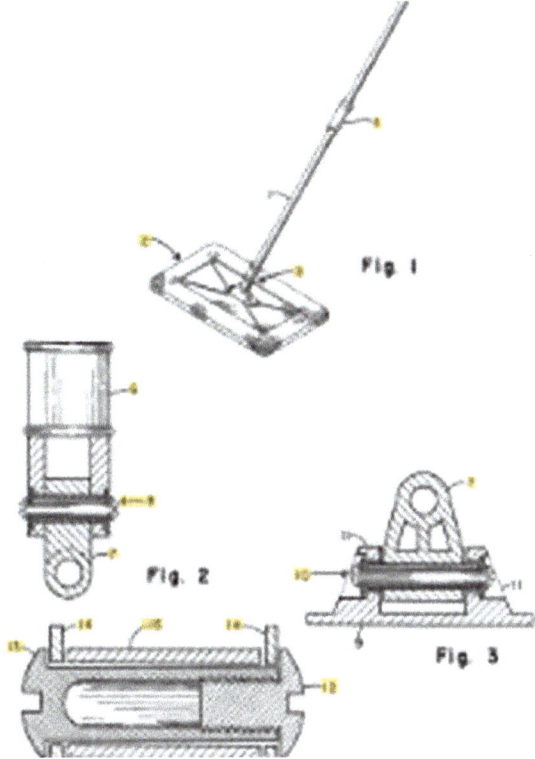

inventions that aim to produce a burst of steam therefore the reservoir includes a mechanism for dispensing additional water to the heating element on demand. A cloth that is of absorbent type is attached to basically a relatively stiff perimeter frame that is adapted to fit around the bottom of a housing. The cloth assembly can be easily removed for cleaning [9] (Fig. 4).

Some steam mop also having a main body consist of a boiler, a water container, and a mechanical water pump having one sidearm connecting the boiler steam outlet to a fabric steam pocket frame. To send water to the boiler, the water pump is actuated by movement of the mop when cleaning to send water to the boiler to distribute cleaning steam to the surface by replaceable fabric pocket which fits snugly over the frame. The steam pocket frame is rectangular with baffles to able to distribute steam disposed of substantially perpendicular to a stream channel that has openings

Fig. 4 Hooks and loops fastener or Velcro tape has a hook portion directly connected with the towel cleaning cushion

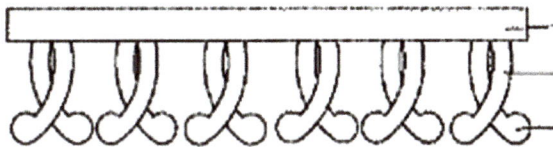

to distribute steam between the baffles. In a steam mop, a mechanical water pump is actuated via the movement of a user to pump water from container having water to a boiler for generating steam to be distributed to a steam pocket applied to a surface to be cleaned and the invention relates generally to a steam mop. We see steam mop having a water pump for selectively injecting water from a reservoir to a boiler is provided in this invention. To achieve a cleaning surface, a water-pump is coupled to a water tank and the pump is actuated by the movement of the user to pump water to the boiler for the steam distribution. To impart an improved cleaning surface, the steam pocket frame is connected to the boiler by at least one sidearm. Water is stored in a water tank formed as part of the handle in one embodiment. Water is pumped to the boiler when a user push the handle for generating the steam, and the steam is then fed to the steam pocket frame through the sidearm, respectively.

Therefore, we see, unlike other steam mops where the heated water generates steam, which may be directed toward the destination through a nozzle which controls the application of the steam here the user itself regulates the flow of steam. The steam pocket frame moves from sidewall to another sidewall. The sidearm is connected to the central steam channel. The handle is telescopic. The pocket frame is covered by two opposed side arms connecting the steam outlet of the main body to the steam pocket frame [2] (Fig. 5).

Fig. 5 Steam mop

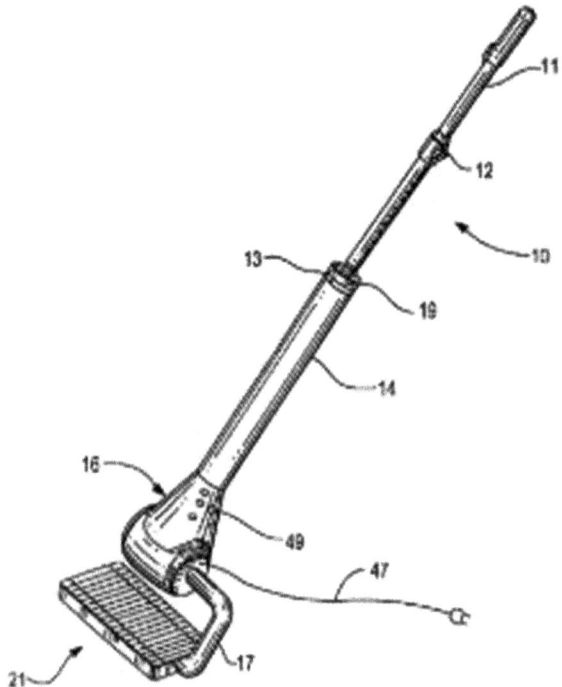

2.2   Steam Mop

In another invention, stain on a floor surface is treated with a cleaning apparatus that consists of a steam generator which comprises a cleaning composition that has a per oxygen and surfactant mixture which is applied to a stain on the floor surface, and cleaning composition is heated with the applied steam. Thus, the transfer of energy takes place from the steam to the cleaning composition to enhance the cleaning property. That is having much greater efficacy as compared to the cleaning without steam. Under another aspect of the invention, for treating a stain on a floor surface, a steam generating appliance is employed, a cleaning pad that is attached in such a way that it can be removed. A cleaning composition comprising a surfactant and per oxygen for application on the stain on the floor surface and is then carried by the cleaning pad or by a second solution reservoir provided on the surface cleaning apparatus wherein the steam generator is embedded to apply steam to the cleaning composition at the time of delivery of the cleaning composition to the stain in some steam mopping units. Here, we see apart from cleaning, it heats the surface, thus, in a way softens the surface with stains first then the steam is applied, and the mopping is done after that [10].

2.3   Surface Cleaning Apparatus

Another invention deals with an upright bare floor cleaner with a handle assembly pivotally mounted to a base assembly. The base assembly has incorporated a sweeper that includes a rotatable brush which is adapted to sweep the dust and dirt particles into the removable dirt receptor. The handle includes a water tank to store a quantity of the water. A fluid distribution system is present that includes the heater along with the nozzle for spraying steam generated by the heater and is distributed to a removable mop cloth through which the steam reaches the surface to be cleaned. Here, a rotatable brush has been used and the liquid is stored in the handle itself. So, obviously, a large quantity cannot be filled here because the operator cannot hold bulk and move. An upright handle is mounted to the base housing that includes handle housing, and a sweeper is mounted to the base housing in the brush chamber and adapted to contact the surface to be cleaned by the opening to remove dirt and dust particles. A dirt receptacle is positioned in the housing to receive the dust, and dirt particles are swept from the surface that is to be cleaned by the sweeper. A steam generator is mounted in the housing of the fluid distributor so the steam is distributed on the surface to be cleaned. Then, a water tank is mounted to one of the handle housing and the base housing which is designed in such a way. For holding quantity of water, a fluid distribution system between the fluid distributor water tank and for distributing fluid from the water tank to the surface which has to be cleaned. A heating element within the fluid distribution system is employed, and the steam is dispensed to the surface to be cleaned by the fluid distributor. Then, a mop cloth is

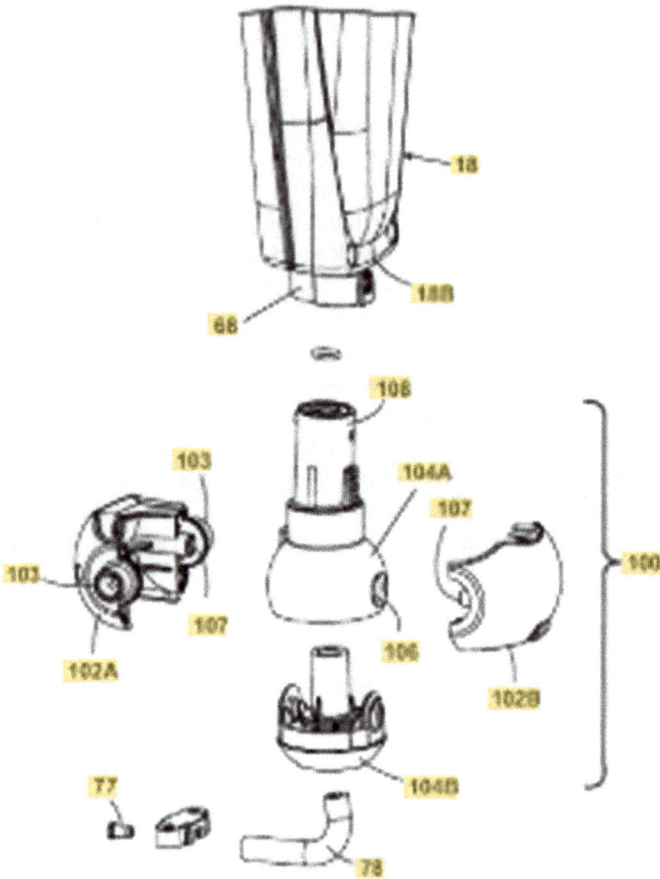

Fig. 6 Cleaning robot roller processing

affixed to the base plate of the base housing and positioned in such a way that it come in contact with the surface to be cleaned [11] (Fig. 6).

2.4 Upright Steam Mop Sweeper

A more specific object of another invention is to provide apparatus of the type in which the clean water is subjected to super atmospheric pressure employing the same blower or fan means which is employed to establish the cleaning vacuum. The objective of several inventions is to provide an organization of elements which possess feeding of clean water by the application of super atmospheric pressure to the surface of the clean water when it is held in a reservoir within the apparatus to thus it somehow establish a siphoning action from the reservoir to the floor tool and to interrupt the siphoning action when the desired amount of clean water is removed from the apparatus. Floor cleaning apparatus for supplying clean water leads to a handled floor tool and for exhausting dirt or water-laden air from second flexible hose leads to said floor tool and for containing the dirt and water which is to be exhausted, the apparatus comprising means for separating from the air so exhausted the dirt and water. A reservoir for clean water, second walled means defining a path of communication from the output side of the mentioned fan to the top of the reservoir, damping means for the closing of air exhaust aperture to thereby subject the surface of water held in a reservoir to super atmospheric pressure sufficient to establish a siphoning action through said water conduit means and first flexible hose means, bleed valve means in water conduit and above the reservoir, and control means for opening bleed valve means when damping means is opened, where siphoning action may be interrupted positively when super atmospheric pressure is no longer applied to the surface of water held in the reservoir [3] (Fig. 7).

The features which are desirable for damp mopping, a deformable, smooth surface, are not for scrubbing the floor. For this purpose, a stiff abrasive pad is required. It is more beneficial if the scrubbing pad is porous so that it does absorb the cleaning solutions if any being applied to the floor. It increases the performance when a cloth

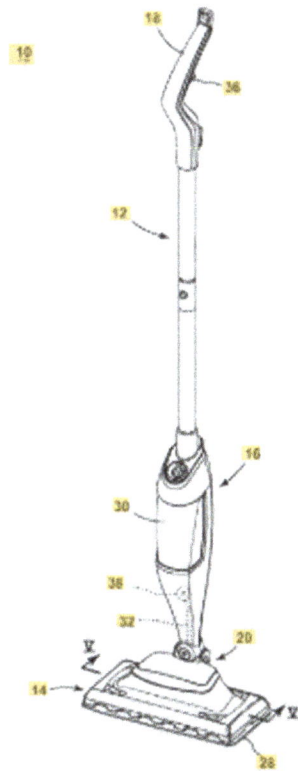

Fig. 7 Surface cleaning apparatus

wipe is used [7]. A steam mop having an improved mop housing containing a water tank with a water inlet having a funnel-shaped entry to facilitate filing a reservoir in the mop housing is implemented. The housing also includes at least one transparent window to the reservoir allowing a user to view the level of water in the reservoir. An LED in the tank provides improved visibility of the water level in the tank. In a preferred embodiment, the housing is easily disconnected from an adapter on the steam frame [12].

Many inventions provide a steam mop including a mop body, a connection member connecting the towel cleaning cushion to the mop body, and a towel cleaning cushion. The connection member used in the steam mop is a special Velcro tape which is easy to directly adhere to the towel cleaning cushion. The ejection of steam cannot be stopped by the loop strip thus improves the utilization ratio of the steam, thereby enhancing the cleaning effect. This relatively simplified manufacturing process in which raw materials are saved, and it is helpful to save cost. A loop strip attached to the towel cleaning cushion is not needed here no loop strip is attached to the towel cleaning cushion [8].

Improved steam cleaning devices, comprises: a steam generator, a pole, a platform cleaning polygon housing, and at least a cleaning tool, wherein the platform cleaning polygon housing could be triangular or tetragonal, the cleaning tool could be jetting nozzles, a circular brush, a triangular brush, a glass scraper, or an implanted brush, and said steam generator could be selectively used alone, or adapted to the pole, the platform cleaning polygon housing or cleaning tool [13].

A steam appliance having a water pump for selectively pumping water from a reservoir to a boiler for generation of steam in response to a user action is provided. The user actuates the pump by the movement of the appliance to activate a motion switch or micro-switch, in response to either a forward movement, a downward pressure, or a backward movement. The steam appliance includes a housing with a user handle and a towel frame or cleaning pad where the generated steam is distributed for steaming. Water is pumped by an electric pump, a mechanical or systolic pump closing a motion switch timer or circuit, or by mechanical movement of a pump piston or wheel. A steam frame for distributing steam generated in the boiler may be connected to the base of the appliance, such as a steam mop, steam brush, or steam iron. A fabric steam pad is mounted on the steam frame to distribute steam to the fabric to provide an improved steaming surface [14].

2.5   Steam Appliance with Motion Switch

Off late, the present invention is to provide a c when the pressure of the boiler increases. The invention also provides a steam cleaner that automatically powers off when a boiler has reached a certain temperature. When a certain temperature is reached it automatically switches off. It includes a machine body having first and second protection provided therein. The objective of the present invention is to

Fig. 8 The artificial
intelligence robot cleaner

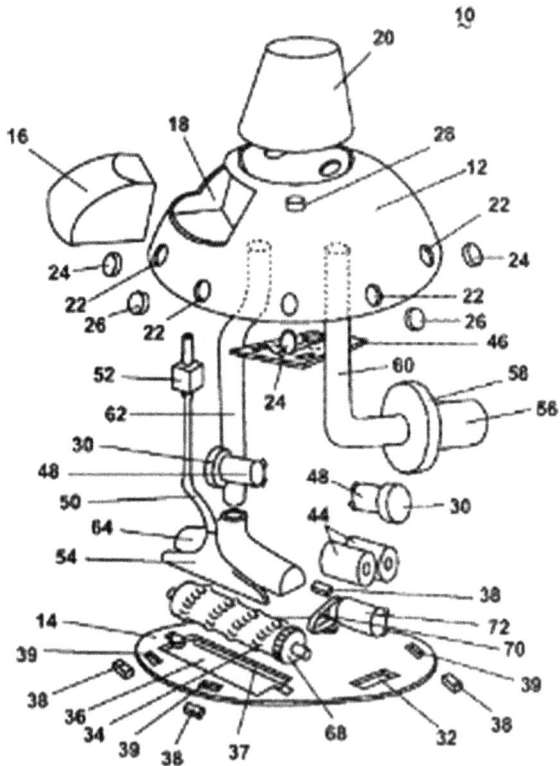

provide a steam cleaner that automatically powers off when the internal pressure and temperature of the boiler exceeds a preset value.

To achieve the above and other objectives, the steam cleaner of the present invention includes a machine body having first and second protection provided therein. The body includes a boiler, a power control unit, and a water storage unit; the first protection includes a pressure controller and a relief valve, and the second protection includes a temperature controller and a fuse unit. A power control unit is also present. A boiler and a water storage unit and a relief valve are also present. It is a steam cleaning machine with many protection layers. One of the second protections includes a temperature controller and a fuse unit and the first protection includes a pressure controller [15] (Fig. 8).

2.6   Steam Cleaner with Multiple Protections

Robotic extraction cleaner with a dusting pad. In this, an autonomously movable home cleaning robot is there that consist of a computer processing unit for storing, receiving and transmitting data, respectively. It is attached to base housing. A dusting assembly is operatively associated with the base housing and is designated selectively rest on the surface to be cleaned. The recovery tank is then mounted on the base housing, is in fluid communication with the suction nozzle. A power source is connected to the drive system and the computer processing unit. The computer processing unit controls the horizontal movement. A drive system mounted to base housing wherein the drive system is adapted to autonomously move the base housing on a substantially horizontal surface having boundaries. A suction nozzle is mounted on the base housing for withdrawing dirt and debris from the surface to be cleaned, and a recovery tank is mounted on the base housing and is in fluid communication with the suction nozzle. A power source is connected to the drive system and the computer processing unit. The cleaning robot actually includes the floor conditioning sensors for detecting the dirtiness of the floor. It generates a control signal that forms a part of the input data to the computer processing unit. Further, the computer processing unit controls at least one of the agitators, the delivery of fluid by the fluid delivery system, the suction source, and the drive system in response to the control signal. In the embodiment, proximity sensors are mounted on the base housing for detecting the boundaries of the surface to be cleaned. Then, it generates a second control signal that forms the input data to the computer processing unit [16].

2.7 The Artificial Intelligence Robot Cleaner

Then in advanced-level research, an artificial intelligence robot cleaner is invented. The artificial intelligence robot cleaner includes the driving unit is made in such a way that it removes the contaminant. The rear camera configured to photograph a rear area of the artificial intelligence robot cleaner, and a processor configured to determine whether cleaning of an already cleaned area is completed by using the image of the rear area of the artificial intelligence robot cleaner. Suppose if the cleaning is not completed properly, the driving unit and the cleaning unit to re-clean the area for this purpose the invention has been made [17] (Fig. 9).

2.8 Robot Cleaning System

A robot cleaning system is greatly capable of performing the cleaning automatically and manually with lesser number of devices. The first cleaning unit basically performs the automatic cleaning. It moves itself in an area to be cleaned. The second cleaning unit performs the manual cleaning and is coupled to the first cleaning unit. The user moves it in the area to be cleaned. Each of the first and second cleaning units contains a blower and dust collector to vacuum. The first cleaning unit has a dust outlet to deliver dust to the second cleaning unit when the first cleaning unit is coupled to the second cleaning unit via the dust outlet of the first cleaning unit, a connector, and the connection port of the second cleaning unit [18] (Fig. 10).

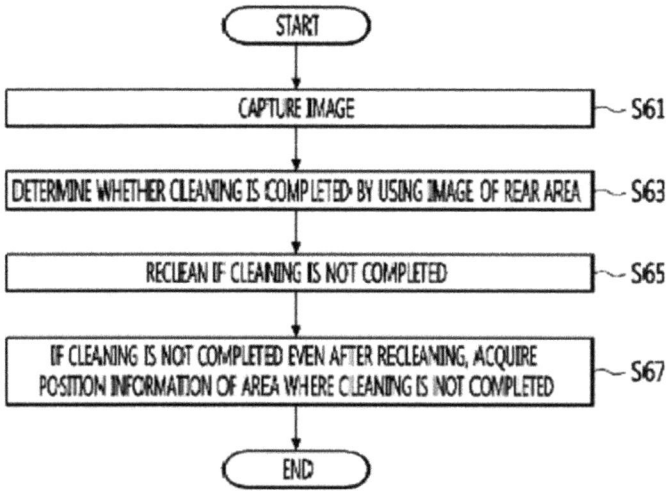

Fig. 9 Robot cleaning system

Fig. 10 Robot cleaning system

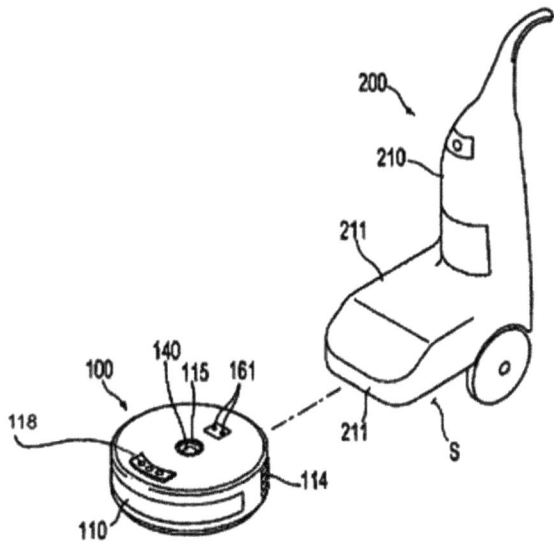

2.9 Robot Cleaning System.

A robotic cleaner includes a cleaning assembly employed for cleaning the surface. The cleaning assembly is located in front of the drive system, respectively. The width of the cleaning assembly is greater than the width of the main robot body. The movement of the robotic cleaner is caused by the drive system. A microcontroller is present to control the movement of the robotic cleaner. The robot cleaner consists of

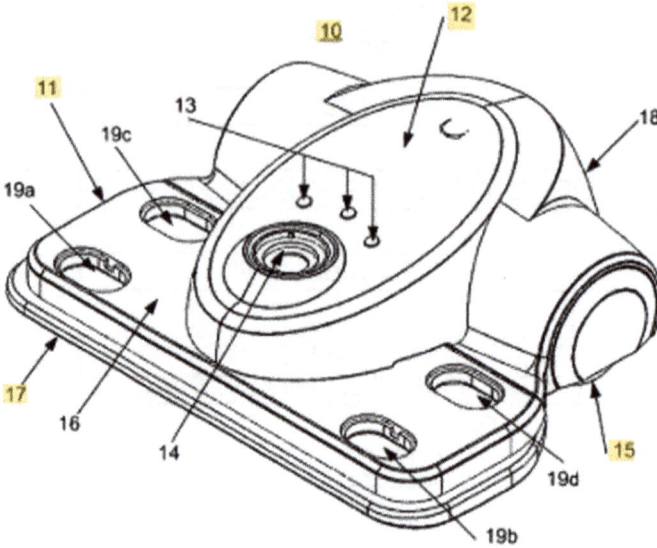

Fig. 11 Mobile robot for cleaning

a plurality of cleaning assemblies for cleaning the surface. The cleaning assembly is located in front of the drive system and each of the cleaning assemblies is detachable from the main robot body. And, all of them has a unique cleaning function [19] (Fig. 11).

2.10 Mobile Robot for Cleaning

In the robotic floor-cleaning apparatus with shell connected to the cleaning assembly and suspended over the drive system, it has the same functioning as that of the mobile robot for cleaning [20].

A robot cleaner consists of a pivot able brush that pivots following a condition of a floor surface to be cleaned. The pivotal's brush main function is to prevent the overload of a suction motor caused by excessive contact of the brush and floor surface to be cleaned. A hinge receiving portion, and a brush frame with a hinge protrusion pivotally connected to the hinge receiving portion is given to the robot cleaner. There is a suction port is connected with the dust suction portion of the robot body. The brush cover locking mechanism made it possible to detach the brush frame. Then, the rotatable brush is disposed between the brush frame and the brush cover in a rotatable manner [21].

2.11   Robot Cleaner with an Adjustable Brush

An autonomous floor-cleaning robot comprises of a chassis, so as to provide the energy to power the autonomous floor- cleaning robot. A power subsystem is employed, then a motive subsystem operative to propel the autonomous floor-cleaning robot for cleaning operations. Then comes a command and control subsystem which operates the control of the autonomous floor-cleaning robot to effect cleaning operations. A vacuum assembly disposed in combination with the deck and powered by the motive subsystem to ingest particulates during cleaning operations. The deck, and the chassis that is automatically operative in response to an increase in brush torque in said brush assembly to pivot the deck concerning. The autonomous floor-cleaning robot includes a side brush assembly that is mounted in combination with the chassis that is powered by a motive subsystem. That results in the particulates to enter outside the periphery of the housing infrastructure followed by directing the particulates toward the leaning head that is self-adjusting subsystem [22].

2.12   The Gutter Cleaning Robot

It is equipped with a debris auger at a front end that contacts and ejects the debris and has a drive system for propelling the gutter cleaning robot along the rain gutter. The debris auger includes a spiral screw sometimes or various other forms of the auger and may be interchanged by the user to increase the effectiveness of the gutter cleaning robot in various environments and the modes of operation. By diverting roof runoff away from the walls of a building, rain gutters also reduce structural damage that would otherwise be caused by the flow of rainwater onto the walls. Substantial amounts of debris such as leaves, tree branches, silt runoff from roof shingles, and accumulate in rain gutters over time, which can eventually constrict or prevent any rainwater from flowing properly. A gutter cleaning robot can traverse rain gutters to agitate and remove the debris [23] (Fig. 12).

An autonomous floor-cleaning robot consists of a transport drive and control system that is arranged for autonomous movement of the robot over a floor for performing cleaning operations. The robot chassis actually carries the first cleaning zone comprising the cleaning elements arranged in such a way that it sucks the loose particulates up from the surface. Then, a second cleaning zone comprising cleaning elements arranged to apply a cleaning fluid onto the surface. Though the cleaning fluid is taken up from the surface, after it has been used to clean the surface. The robot chassis actually carries a supply of cleaning fluid. There is a waste container for storing waste materials that are collected up from the cleaning surface [24].

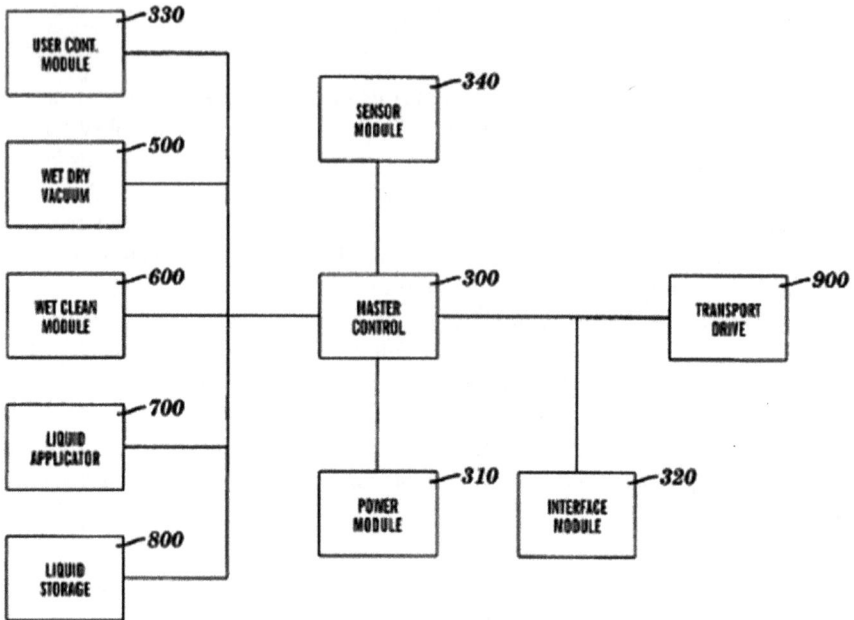

Fig. 12 Robotic wash cell using recycled pure water

2.12.1   Robotic Wash Cell Using Recycled Pure Water

A robotic wash cell actually includes the six-axis robotic arm and end effector equipped with nozzles that is able to spray unheated, solvent-free, pure water at high pressure for cleaning or removing objects by maintaining the nozzles, respectively. The robotic cell wash is particularly useful for cleaning oil and grease from complex shaped items. The nozzles actually produce a multi-zone type spray pattern with the continuous effective cleaning zone. A water recycling and the pressurizing system very effectively collects the used water, separates the oil and grease contaminants to a level of about 5 ppm, and pressurizes the pure water to about 3000 psi for washing operations and 6000 psi for debarring operations [25]. A coverage robot consists of a chassis, a controller, a drive system, and a cleaning assembly. The cleaning assembly includes a housing and driven cleaning roller. The coverage robot basically includes a roller cleaning tool that is carried by the chassis employed for longitudinally traverse sing the roller to remove the accumulated debris. Here, the cleaning assembly includes housing. There is one driven cleaning roller which includes an elongated core with end mounting features that defines a central longitudinal axis of rotation, multiple floor-cleaning bristles extending radially outward from the core. The compliant flap actually is designed to extend radially outward from the core to for sweeping the floor surface [26].

Cleaning robot roller processing (Fig. 13).

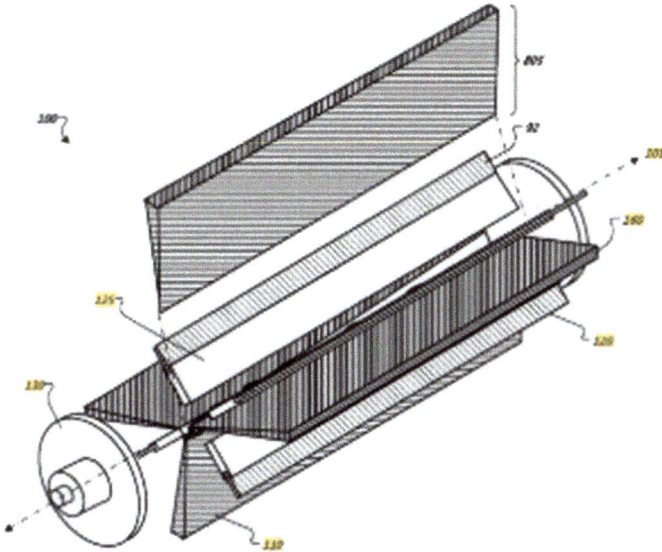

Fig. 13 cleaning robot roller processing

3   Results and Discussions

The efforts taken by the various group of researchers in implementing a device which works on steam generation is overwhelming. As we have seen previously in some of the inventions along with steam generation, the detergent or cleaning agent has been imparted. But again it's not feasible. Anyway, the use of external agents along with steam can lead to pollution or allergies in children and even in pets and adults. Then, light upright steam cleaners were produced which were able to clean the upper surfaces as well. The biggest advantage of light weight steam cleaners is that it can be easily used by even an old person or someone with back or knee problems. Therefore, the labor one has to give in carrying those large vessels is completely removed. Then, other features were added like abrasive on cleaning pads for better cleaning which would be able to clean tough stains or damp surfaces. For proper cleaning, rotatable brushes were also introduced. Then, a porous scrubbing pad can be attached with the bottom of the foam-pad that engages the bristles. It is an objective of the invention to provide a mop that is adaptable for damp mopping as well as scrubbing the floors, respectively. Later on, the focus also shifted to the safety of the user and total power consumption. A steam cleaner that automatically powers off when the pressure of the boiler increases is one of the best inventions. When a certain temperature is reached, it automatically switches off. Thus, there will be no chance of overheating and much safer for the operator. When the device reaches a preset pressure, it will again switch off. It will not only save electricity or power consumption, but will be able to prevent or protect the operator from any sort of accident. Then, there are robot where a microcontroller is present to control the movement of the robotic cleaner. The robot cleaner consists of a plurality of cleaning assemblies for cleaning the surface, then the coverage robot includes a roller cleaning tool carried by the chassis employed to longitudinally traverse the roller to remove accumulated debris from the cleaning roller. A robotic wash cell is provided and includes a six-axis robotic arm and end effector equipped with nozzles that spray unheated, solvent-free, pure water at high-pressure to clean or debar objects by maintaining the nozzles nearby and substantially normal to each surface being cleaned or edge being debarred. A cleaning unit is designed to remove the dirt. Then, there is rear camera configured to photograph a rear area of the artificial intelligence robot cleaner. A processor determines whether cleaning of an already cleaned area is completed by using the image of the rear area. This is achieved with the help of artificial intelligence. All these inventions are greatly promising and can bring potential positive improvements when health and hygiene are concerned. There is a growing need for properly sanitizing our nearby environment. Many studies have been carried out for using water in an appropriate way along with keeping it in mind the comfort of the people who are involved in the cleaning process. So in earlier days, the traditional mopping consisted of a bucket and a mopping cloth. It has many disadvantages associated with it. Firstly, it was not at all comfortable for the person to carry those large vessels of water and then changing the dirty water time and again. All this was in vain only as it was not able to sterilize the floor in fact, observations showed it was leaving the stresses of chemical used along with water which ultimately resulted in allergies and inappropriate cleaning. Hence, the need is felt to think of other alternatives. The steam cleaning thus found its application. So, steam cleaning has many advantages associated with it. There must be awareness among people to adopt steam cleaning in the coming days for better sanitation which will ultimately benefit the society by reducing the rate of diseases and as it also takes into consideration the comfort of the operator and sustainable use of water.

4 Research Gap

1. There is still scope in the modification in the steam cleaning devices for its better functioning where the steam energy can be completely utilized. Here, in the existing steam cleaning devices, the steam is not directly dispensed to the floor instead it was reaching the surface of concern via a mopping cloth thus the steam energy somehow drops by a certain level as the temperature drops when it comes in contact with the mopping cloth.
2. Then, most of the apparatus is bulky, thus difficult for the user to handle.
3. The ones which are lightweight again cannot accommodate a large amount of water.
4. A dual-purpose device can be made which contains both the vacuum cleaner and steam cleaner.

Reference

1. Gilbert JJ (1982) U.S. Patent No. 4,327,459. Washington, DC, U.S. Patent and Trademark Office
2. Rosenzweig M, Vrdoljak O (2008) U.S. Patent Application No. 11/496,143
3. Noble JW (1962) U.S. Patent No. 3,056,994. Washington, DC, U.S. Patent and Trademark Office
4. Chen SY (2002) U.S. Patent No. 6,490,753. Washington, DC, U.S. Patent and Trademark Office
5. Tang L (2013) U.S. Patent Application No. 13/821,989, vol 16
6. Nolan MP, Krebs AJ, Ashbaugh KE (2014) U.S. Patent No. 8,850,654.Washington, DC, U.S. Patent and Trademark Office
7. Krajicek SW (1989) U.S. Patent No. 4,852,210. Washington, DC, U.S. Patent and Trademark Office
8. Chow O (2012) U.S. Patent Application No. 13/014,529
9. Shaw RR (2003) U.S. Patent No. 6,584,990. Washington, DC, U.S. Patent and Trademark Office
10. Kellis JM, Haley K, Scholten JA, White JM (2016) U.S. Patent No. 9,420,933. Washington, DC, U.S. Patent and Trademark Office
11. Kasper GA, White JM, Jansen JL, Griffith AP (2013) U.S. Patent No. 8,458,850. Washington, DC, U.S. Patent and Trademark Office
12. Rosenzweig M, Vrdoljak O, Wall E, Xu D (2009) U.S. Patent Application No. 12/163,537
13. Tsai MH (2008) U.S. Patent No. 7,380,307. Washington, DC, U.S. Patent and Trademark Office
14. Rosenzweig M, Vrdoljak O (2013) U.S. Patent No. 8,402,597. Washington, DC, U.S.Patent and Trademark Office
15. Chen CM (2006) U.S. Patent No. 7,103,270. Washington, DC, U.S. Patent and Trademark Office
16. Huffman EC, Miner JL (2008) U.S. Patent No. 7,320,149. Washington, DC, U.S. Patent and Trademark Office
17. Jonghoon CHAE (2020) U.S. Patent Application No. 16/563,561
18. Lee JH, Joo JM, Kwak JY, Shin GW, Wee H (2008) U.S. Patent No.7,412,748. Washington, DC, U.S. Patent and Trademark Office
19. Romanov N, Johnson CE, Case JP, Goel D, Gutmann S, Dooley M (2015) U.S.Patent No. 8,961,695. Washington, DC, U.S. Patent and Trademark Office
20. Romanov N, Dooley M, Pirjanian P (2017) U.S. Patent No. 9,725,013. Washington, DC, U.S. Patent and Trademark Office
21. Kim KM (2007) U.S. Patent No. 7,200,892. Washington, DC, U.S. Patent and Trademark Office
22. Jones JL, Mack NE, Nugent DM, Sandin PE (2009) U.S. Patent No. 7,636,982. Washington, DC, U.S. Patent and Trademark Office
23. Lynch J (2013) U.S. Patent No. 8,453,289. Washington, DC, U.S. Patent and Trademark Office
24. Ziegler A, Gilbert D, Morse CJ, Pratt S, Sandin P, Dussault N, Jones A (2008) U.S. Patent No. 7,389,156. Washington, DC, U.S. Patent and Trademark Office
25. Laski SJ (2008) U.S. Patent No. 7,321,807. Washington, DC, U.S. Patent and Trademark Office
26. Kapoor DR, Dubrovsky ZA (2014) U.S. Patent Application No. 14/067,119

Chapter 51
Video Description Based YouTube Comment Classification

**Asha Shetty, Bryan Abreo, Adline D'Souza, Akarsha Kondana,
and Kavitha Mahesh Karimbi**

1 Introduction

YouTube has been recognized as the second most popular website in the world by
Alexa Internet[1] [2]. One of the most important and helpful features of YouTube
is its commenting system which allows users to post comments on videos that are
uploaded. This commenting system allows the user to share their ideas about videos,
ask questions regarding the videos, clarify all the doubts pertaining to the video
posted. Users are also free to complement the video author for providing good video
or express displeasure otherwise. However, one of the concerns with the use of commenting system is that, malicious users employ it as a medium for sharing irrelevant
comments, often referred to as 'spam'. The comments are termed 'spam' as they
represent content that are irrelevant in the context of the uploaded video and are
often generated by automated bots disguised as a user [2].

The commenting system in video sharing websites such as YouTube, also allows
users to react or comment on the posts of other users. However, it is quite common

---

[1]https://try.alexa.com/competitive-website-analysis.

A. Shetty · B. Abreo · A. D'Souza · A. Kondana · K. M. Karimbi (✉)
Department of Computer Science & Engineering,
St Joseph Engineering College, Mangaluru, India
e-mail: kavitham@sjec.ac.in

A. Shetty
e-mail: shettyasha417@gmail.com

B. Abreo
e-mail: bryanabreo@gmail.com

A. D'Souza
e-mail: adlinedsouza59@gmail.com

A. Kondana
e-mail: akarshakondana@gmail.com

© The Author(s), under exclusive license to Springer Nature Singapore Pte Ltd. 2021   667
X.-Z. Gao et al. (eds.), *Applications of Artificial Intelligence in Engineering*, Algorithms
for Intelligent Systems, https://doi.org/10.1007/978-981-33-4604-8_51

for people to post comments which are in no way related to the topic of the video. The comments can turn irrelevant either accidentally due to typing errors or intentionally when posts involve comments that has no relevant meaning to the video content. Of the two concerns, second is more serious. Many a times we observe that users post unnecessary comments which includes malicious links, opinions on controversial topics or hate comments which are in no way related to the video itself. Further, in such scenario, it has also been observed that a substantial rise in off-topic comments causes the people involved in discussions or commenting to completely deviate from the topic of interest, thereby inducing lot of outrage and consequently triggering hatred among genuine participants. Subsequently, lot of unnecessary comments gets generated causing difficulty in identifying those comments that are relevant or important, as for instance, comments which are significant to the author of the video, or comments for which the author of the video might want/have to respond. Thus, a commonly observed problem in the YouTube commenting system is posting comments which are totally irrelevant[2] to the video content.

When a user comments by seeking clarifications or posing questions on a particular video, it might be required and important for the author of that video to reply to those comments. Hence, as the irrelevant comment increases, it becomes difficult for author to reply to only those specific comments that are relevant. Also, outrage caused due to hate comments distracts user's attention in watching the video further. Hence it is essential to identify the comments that are totally irrelevant to the topic. In YouTube Commenting System, it is also observed that if the uploaded video is of good quality, majority of the users issue complementary comments appreciating the author. Similarly, if the video quality is bad, users post negative comments. Consider a scenario where a large number of users had watched a video of good quality. In this case, all those comments containing content analysing or referring to the video gets hidden among the 'positive' or 'negative' comments. We refer to such comments as 'relevant comments'. Locating 'relevant comments' from among other comment types is also tedious. Therefore, it is important to identify all the 'positive' and 'negative' comments and separate it from the 'relevant' comments. In this paper, we present a classifier based tool which automatically identifies the 'positive' and 'negative' comments in a video uploaded. Further, the comments which are not relevant to the video are also separated as 'irrelevant' comments. Thereby, entire collection of video comments is classified into four categories, namely 'relevant', 'irrelevant', 'positive' and 'negative'. In our classification experiments, we assume that any video posted has an associated description pertaining to the video uploaded. In the said context, the objective of our work is to develop a classifier-based tool for segregation of YouTube Comments with respect to video description into one of four categories: relevant, irrelevant, positive and negative.

---

[2]https://www.youtube.com/watch?v=KzYW6EPMK7w.

2 Related Work

Currently, YouTube provides features that helps its users to sort the comments based on top comments and newest first. The newest first method uses date of post, that is, date on which a video is posted by the author. Top comments method on the other hand sorts the comments based on three features namely the like-dislike ratio, the number of replies, and the author who had posted that particular video. By default, YouTube does not categorize the comments based on the importance and relevance of the subject. Consider a scenario, where the comments that are totally irrelevant to the video content getting high like-dislike ratio. As per the available features, these irrelevant comments are placed on the top. In our current work, we focus on categorizing the YouTube comments with specific preference to the subject of the video considering video description. Below we discuss the related work for Spam Detection exclusively in the context of YouTube Comments.

One of the earliest reported works provide a characterization of social and content attributes for distinguishing users from spammers, promoters and legitimates [7]. The authors report correct identification of majority of the promoters except for miss-classifications in a small percentage of legitimate users. Further, the work points towards the difficulty in distinguishing spammers from legitimate users.

The spam detection technique proposed by Alex et al. uses a language and tokenization independent metric namely content complexity to identify features for comment spam detection. Content complexity is estimated over groupings of messages with same author, sender IP, included links and so forth [14]. Network motif profiling has been recommended for tracking spam campaigns over time [18]. Two different spam campaign strategies are detected along with the associated user accounts using discriminating motifs identified from motifs of up to five nodes.

By conducting a content analysis of 66,637 user comments on YouTube videos, Amy et al. [17] created a classification schema to categorise the types of comments users leave. The schema reveals ten broad categories, and 58 subcategories reflecting the wide-ranging use of the YouTube comments facility. Comment types that do not appear in the scheme outlined could appear as YouTube continues to evolve and requires to be dealt. Microsoft SQL Server Data Mining Tools (SSDT) were used to provide a heuristic for classifying an arbitrary video as either spam or legitimate [9].

Severyn et al. focus on opinion mining on YouTube comments by pre-modeling classifiers for predicting the polarity of opinions and the type of YouTube comments. The authors report of using tree kernel technology to automatically extract and learn features. To improve model adaptability, they augment it with robust shallow syntactic structures [22].

A survey on various techniques to analyse user opinions about a particular YouTube video is presented in the review paper by Asghar et al. [4]. Uysal et al. [3] provide a comparative analysis taking into account the performance of five state-of-the-art text feature selection methods for spam filtering on YouTube using two widely-known classifiers namely Naïve Bayes (NB) and decision tree (DT). The author reports distinguishing feature selector (DFS) and Gini Index (GI) feature selection methods to provide best classification results. Also, the study indicates that performance of DT classifier is better than NB classifier in most cases for spam filtering on YouTube.

Naïve-Bayes based multi-label classifier is used to understand the behaviour and response of individuals on seeing a particular YouTube video [16]. Poche et al. [19] focus on the link connecting programmers creating coding videos with their audience. The results show that Support Vector Machines can detect useful viewers' comments on coding videos with an average accuracy of 77%. Further, the authors advocate the use of SumBasic, an extractive frequency-based summarization technique with redundancy control in capturing the main concerns reflected in viewers' comments. Savigny et al. exclusively focus on emotion classification on Indonesian Youtube comments. The authors report that word embedding with CNN provide best accuracy of 76.2% for the said task [21]. In sorting out the spam comments on YouTube videos from the legitimate one, Sharmin et al. [23] analyse the performance of ensemble classifier over single classifier algorithm.

A comparative analysis of common YouTube comment spam filtering techniques show that high filtering accuracy (more than 98%) can be achieved with low-complexity algorithms, further hinting towards a suitable browser extension to alleviate comment spam on YouTube [1]. This study uses features based on the Edge Rank algorithm. The authors report highest accuracy (i.e average) of 98% and the lowest of 74% based on experiments employing nine different learning classifiers such as decision trees, Bayesian and function-based. Aiyar et al. [2] proposed a methodology to detect spam comments on YouTube by applying standard machine learning algorithms such as Random Forest, Support Vector Machine (SVM), Naïve Bayes along with N-grams. Ezpeleta et al. [10] focus on mood analysis, among various content-based analysis techniques. The approach uses a new dataset created from a labeled dataset consisting of Youtube comments, including spam, by adding mood feature to each comment followed by application of best spam filtering classifiers. Authors report that incorporating the mood feature can help to improve social spam filtering results. Yet another work for YouTube Spam Comment Detection employs SVM and K-Nearest Neighbour [5]. Burhanudin [8], in classification of spam comments on YouTube, use several methods such as Support Vector Machine, K-Nearest Neighbour, Naïve Bayes and the model evaluation shows that Support Vector Machine with Linear approach provide highest accuracy of 91.92%.

Samsudin et al. [20] present a YouTube Spam detection framework using Naive Bayes and Logistic Regression run in Weka and RapidMiner. The reported results suggest that Naive Bayes enables higher precision in comparison to Logistic Regression. Kaur et al. [15] focuses on the sentiment analysis of Hinglish comments on cookery channels by identifying different comment patterns employing DBSCAN. The authors report accuracy of about 75% using Logistic regression with the term frequency vectorizer. A detailed analysis of the behavioural characteristics of two video categories, namely, phone spam and link spam, based on metadata content and graph based features is discussed by Tripati et al. [24]. The study reveals that such videos are highly related to each other, and specific motifs discriminating such videos exist. In segregating the comments as fake, meta-fake and genuine reviews, Jawaid et al. [13] employ Sentiment Analysis, Negative Ratio Checking and Cosine Similarity. Fernando et al. [11] use the bag-of-words representation with Multi-nomial Naive-Bayes method in classifying 2019 Indonesian Election YouTube comments. In improving the relationship of coding lesson video creators and their viewers, Jain et al. [12] proposed an approach that allows classification and summarization of YouTube comments of only genuine viewers. In a recent work, Bansal's study provide performance-based comparison of deep learning approaches employed in detection of offensive YouTube comments [6].

The proposed tool purely relies on categorizing comments based on content relevance to video description and hence focuses majorly on textual data. In the sections that follow, we present the approach involved in video description based YouTube comment classification and provide the experimental results with discussion.

3   Approach

The steps involved in YouTube comment classification task is outlined in the subsections that follow.

3.1   YouTube URL and Description

The URL for the video provided by the user is first verified for correctness. The video id following '?=v' in the URL should contain exactly 11 characters. YouTube video should have an associated description. As the proposed approach compares user comments with the video description, the description about the video is mandatory and this requirement is verified from the client side.

3.2   Bag of Words Model for Description Based Comment Classification

Following URL verification as specified in the earlier subsection, the sub tasks involved in comment classification are depicted in Fig. 1 and are as elaborated below:

1. *Data Extraction*: For the valid URL entered by the user, the comments, description of the video and author details are extracted from the YouTube. This is achieved by making use of the YouTube API.[3]
2. *Pre-processing*: The Pre-processing techniques performed involves Tokenization and Stop word Removal.

---

[3]https://developers.google.com/youtube/v3/docs/comments.

Fig. 1 YouTube comment
classifier

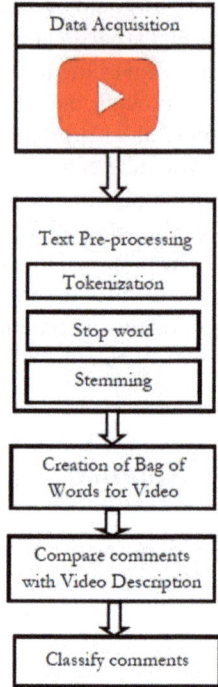

- *Tokenization*: The YouTube comments are tokenized by splitting each comment into individual words with space as the delimiter.
- *Stopword Removal:* We use the NLTK toolkit[4] to eliminate stop words such as 'the', 'to', 'a', 'an' and so forth. NLTK supports stop word elimination for multiple languages, however in our experiments despite the fact that comments occasionally involved multi-language words, we considered elimination of stop words specific to English only.

3. *Comment Classification*: The comments are classified into one of the four classes namely 'relevant', 'irrelevant', 'positive' and 'negative' by representing the comment data considering the Bag of Words model.
   Relevant Comments: Comments which are closely related or appropriate to what is being discussed in the video. For instance, for a video from the domain 'MS Word Document', the comment 'how to merge two column in table' is a 'relevant' comment.
   Irrelevant Comments: Comments which are not related or inappropriate to what is being discussed in the video. For the video mentioned in the previous paragraph, the comment 'I am going to Mumbai' is considered as 'irrelevant'.
   Positive Comments: Comments which are characterized by positive emotion

---

[4]https://www.nltk.org/.

words. The comments 'I like your video.. Awesome' is referred to as 'positive' comment.

Negative Comment: Comments which are characterized by negative emotion words. For instance, the comment 'Bad Video' is referred as 'negative' comment.

We used Bag of Words model for classifying user comments into relevant categories. The general approach is summarized below:

1. Create two separate lists of positive and negative sentiment words using the dataset from Keenformatics.[5] This vocabulary serves as features for positive and negative class.
2. Score the word occurrence in test comment based on its presence or absence in the sentiment word lists.
3. Label the test comment as 'positive' or 'negative' based on the scoring.
4. Create a bag of words using the video description.
5. Extract features by choosing the most frequent words in the description.
6. Score the word occurrence in test comment based on its presence or absence in the feature set.
7. Label the comment as 'relevant' based on the scoring. This will override the classification done in step 3.
8. If the test comment does not satisfy the scoring conditions in steps 3 and 7, label it as 'irrelevant'.

4 Results and Discussion

4.1 Data Set

For testing the tool, video from "MS Word" domain with 46 comments were used. The video consisted of 22 relevant, 12 positive, 1 negative and 11 irrelevant comments. Experiments were also conducted using comments for videos taken from multiple domains. Comments corresponding to the videos from "Natural Language Processing", "Sports" and "Movie" domains were specifically employed and used in testing.

4.2 Experimental Setup

The comments from the YouTube is collected using the video URL and are further manually categorized into the four classes for testing. Precision and Recall are the

---

[5]http://keenformatics.blogspot.com.

metrics employed in evaluating the classified comments and are estimated as shown in equation below.

$$\text{Precision} = \text{TP}/(\text{TP} + \text{FP}) \tag{1}$$

$$\text{Recall} = \text{TP}/(\text{TN} + \text{FN}) \tag{2}$$

$$\text{Accuracy} = (\text{TP} + \text{TN})/(\text{TP} + \text{FP} + \text{TN} + \text{FN}) \tag{3}$$

where, Precision is the ratio of the number of true positives to the total number of comments classified to be true; Recall is the ratio of the number of true positives to the total number of comments that are true and Accuracy is the total comments that are classified as true to the total number of comments that are classified. For relevant comments, TP indicates the number of relevant comments that are correctly classified as relevant, FP indicates the number of non-relevant comments that are incorrectly classified as relevant, TN represents the number of non-relevant comments that are correctly classified as non-relevant comments and FN represents the number of relevant comments that are incorrectly classified as non-relevant comments.

Experiments were conducted using the data set mentioned in Sect. 4.1. Experimental results showed that of the 22 relevant, 12 positive, 1 negative and 11 irrelevant comments, 24 were classified as relevant, 10 as positive, 1 negative and 11 were classified as irrelevant comments respectively.

Table 1 shows the experimental results for videos having subjective description. Columns TP, TN, FP, FN shows the list of true positives, true negatives, false positives and false negatives respectively. The table shows that the approach allows classification of comments based on the video description with an accuracy of 97.75%.

The relationship between the expected and current outcome for the classification of comments using Bag of Words based feature representation scheme is as shown in Fig. 2. The Y-axis shows the number of comments.

Also, we conducted a set of three experiments using comments for video taken from multiple domains. Table 2 shows the comment classification results obtained for videos belonging to different domains. Experiment 1 (Exp 1) was conducted using comments from "Natural Language Processing" domain. Similarly, Experiment 2 (Exp 2) and Experiment 3 (Exp 3) involved videos obtained from "Sports" and "Movie" domains respectively.

Table 1 Classification results using bag of words model

Category	TP	FP	TN	FN	Precision (%)	Recall (%)	Accuracy (%)
Relevant	20	4	2	2	83.33	90.90	86.95
Positive	10	0	34	2	100	83.33	95.65
Negative	0	1	44	1	0	0	95.65
Irrelevant	10	1	34	1	90.90	90.90	95.65

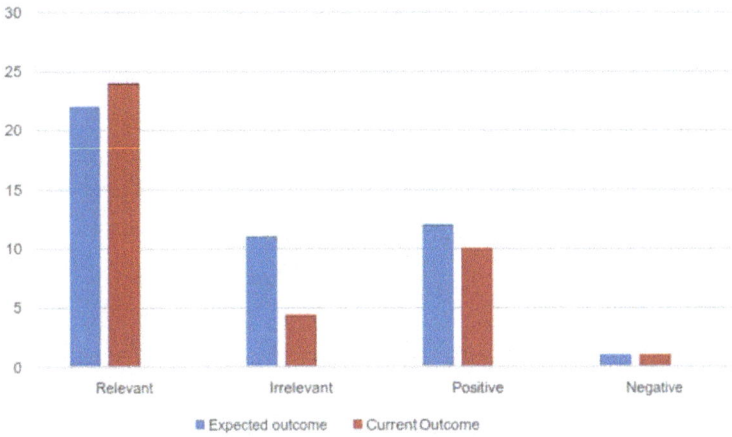

Fig. 2 Comparison between expected and actual outcome

Table 2 Comment classification results for multiple videos from various domains using bag of words based comparison

	Category	TP	FP	TN	FN	Precision (%)	Recall (%)	Accuracy (%)
Exp 1	Relevant	84	1	27	5	98	94.38	94.8
	Irrelevant	16	2	98	1	88	94	97.4
	Positive	11	3	103	0	78.60	100	97.40
	Negative	0	0	117	0	0	0	100
Exp 2	Relevant	20	4	20	2	83.30	90.90	86.95
	Irrelevant	10	1	34	1	90.90	90.90	95.65
	Positive	10	0	34	2	100	83.33	95.65
	Negative	0	1	44	1	0	0	95.65
Exp 3	Relevant	27	4	18	1	87	96	90
	Irrelevant	4	0	45	1	100	80	98
	Positive	11	1	36	2	91.70	84.60	94
	Negative	3	0	46	1	100	75	98

Figure 3 shows the comparative analysis of the precision obtained for the experiments conducted with comments from chosen domains. X axis shows the various experiments that were conducted and Y axis shows the precision obtained in percentage. While there were no negative comments for first two experiments, we observe a sharp increment with respect to the third experiment.

Figure 4 shows the comparative analysis of the accuracy obtained for the experiments conducted with comments taken from different domains. X axis shows the various experiments that were conducted and Y axis shows the accuracy obtained in percentage. For all experiments, we were able to achieve accuracy greater than 85% using Bag of Words approach.

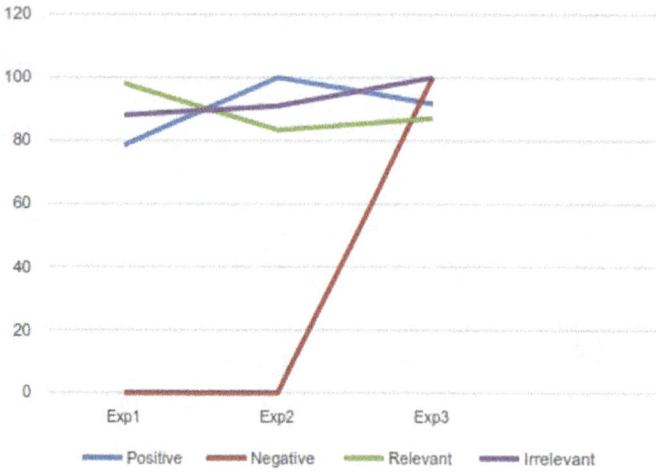

Fig. 3 Comparison of precision for classification of comments for videos from various domain using bag of words feature

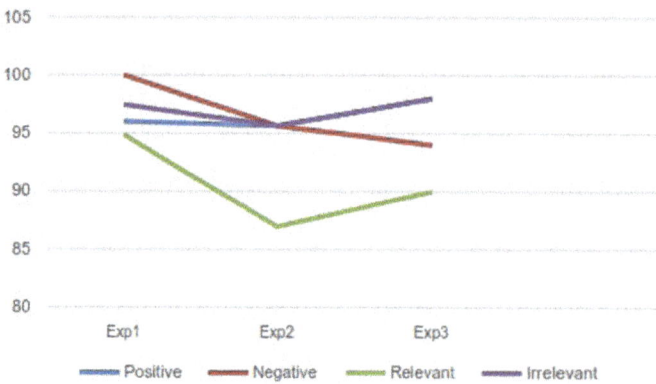

Fig. 4 Comparison of accuracy for classification of comments for videos from various domain using bag of words feature

5 Conclusion and Future Work

In this paper, we have discussed a classifier-based tool, for segregation of comments in YouTube video sharing website into one of the four categories namely, relevant, irrelevant, positive and negative based on the relevance of the comment to the video content given by the description associated with the video posted. The current work uses Bag of Words approach for feature extraction. Association list based representation could be explored in future to handle data sparsity. The work can be extended further to incorporate n-grams as features. Further, to handle negative emotions observed in messages such as, 'The movie was not as good as expected.', non-contiguous phrases needs to be explored as features.

References

1. Abdullah AO, Ali MA, Karabatak M, Sengur A (2018) A comparative analysis of common youtube comment spam filtering techniques. In: 2018 6th international symposium on digital forensic and security (ISDFS), pp 1–5
2. Aiyar S, Shetty NP (2018) N-gram assisted youtube spam comment detection. Proc Comput Sci 132:174–182 (2018 international conference on computational intelligence and data science, ICCIDS 2018; Conference date: 07-04-2018 Through 08-04-2018)
3. Alberto TC, Lochter JV, Almeida TA (2015) Tubespam: Comment spam filtering on youtube. In: 2015 IEEE 14th international conference on machine learning and applications (ICMLA), pp 138–143
4. Asghar MZ, Ahmad S, Marwat A, Kundi FM (2015) Sentiment analysis on youtube: A brief survey. arXiv: 1511.09142
5. Aziz A, Foozy CFM, Shamala P, Suradi Z (2018) Youtube spam comment detection using support vector machine and k–nearest neighbor
6. Bansal P (2019) Detection of offensive youtube comments, a performance comparison of deep learning approaches
7. Benevenuto F, Rodrigues T, Almeida VAF, Almeida JM, Gonçalves MA, (2009) Detecting spammers and content promoters in online video social networks. IEEE INFOCOM Workshops 2009:1–2
8. Burhanudin YM, Wihardi Y (2018) Klasifikasi komentar spam pada youtube menggunakan metode Naïve Bayes, support vector machine, dan k-nearest neighbors
9. Chowdury R, Adnan Md, Mahmud GAN, Rahman RM (2013) A data mining based spam detection system for youtube. In: Eighth international conference on digital information management (ICDIM 2013), pp 373–378
10. Ezpeleta E, Iturbe M, Garitano I, de Mendizabal IV, Zurutuza U (2018) A mood analysis on youtube comments and a method for improved social spam detection. In: HAIS
11. Fernando JR (2019) Udayawibawamukti. Klasifikasi spam pada komentar pemilu 2019 indonesia di youtube menggunakan multinomial naïve-bayes
12. Jain S, Patel M (2019) Analyzing user comments of learning videos from youtube using machine learning
13. Jawaid A, Dev S, Sharma R (2019) Predilection decoded : Web based spam detection and review analysis for online portals
14. Kantchelian A, Ma J, Huang L, Afroz S, Joseph A, Tygar J (2012) Robust detection of comment spam using entropy rate. In: Proceedings of the ACM conference on computer and communications security, pp 59–70
15. Kaur G, Kaushik A, Sharma S (2019) Cooking is creating emotion: a study on Hinglish sentiments of youtube cookery channels using semi-supervised approach. Big Data Cogn Comput 3(3):37
16. Khan AUR, Khan M, Khan MB (2016) Naïve multi-label classification of youtube comments using comparative opinion mining. Proc Comput Sci 82:57–64 (4th symposium on data mining applications, SDMA2016, 30 March 2016, Riyadh, Saudi Arabia)
17. Madden A, Ruthven I, McMenemy D (2013) A classification scheme for content analyses of youtube video comments. J Documentation
18. O'Callaghan D, Harrigan M, Carthy J, Cunningham P (2012) Network analysis of recurring youtube spam campaigns. ArXiv, abs/1201.3783

19. Poché E, Jha N, Williams G, Staten J, Vesper M, Mahmoud A (2017) Analyzing user comments on youtube coding tutorial videos. In: 2017 IEEE/ACM 25th international conference on program comprehension (ICPC), pp 196–206
20. Samsudin NM, Foozy CFM, Alias N, Shamala P, Othman NF, Din W (2019) Youtube spam detection framework using Naïve Bayes and logistic regression. Indonesian J Electr Eng Comput Sci 14:1508
21. Savigny J, Purwarianti A (2017) Emotion classification on youtube comments using word embedding. In: 2017 international conference on advanced informatics, concepts, theory, and applications (ICAICTA), pp 1–5
22. Severyn A, Uryupina O, Plank B, Moschitti A, Filippova K (2014) Opinion mining on youtube
23. Sharmin S, Zaman Z (2017) Spam detection in social media employing machine learning tool for text mining. In: 13th international conference on signal-image technology & internet-based systems (SITIS), pp 137–142
24. Tripathi A, Bharti KK, Ghosh M (2019) A study on characterizing the ecosystem of monetizing video spams on youtube platform. In: Proceedings of the 21st international conference on information integration and web-based applications and services

Chapter 52
Internet of Things: Current Research, Challenges, Trends and Applications

Dipankar Debnath and Sarat Kr. Chettri

1 Introduction

The Internet of Things (IoT) is referred to as a network of physical objects or "Things" embedded with sensors for the purpose of connecting and exchanging data with other devices and networks over the Internet. Although the idea of IoT has been around for a long time, recent advances in various supporting technologies such as low-cost low-power sensors cloud and advanced computing, machine learning and data analytics have made it the most important technology of twenty-first century. The paper is organized as follows: Sect. 2 provides background and current research trends followed by key enabling technologies in Sect. 3. Section 4 outlines a variety of application areas and Sect. 5 discusses issues and challenges. Section 6 highlights some of the open research problem, and comments on the conclusion are presented in Sect. 7.

2 Background and Current Research

Radio-Frequency identification (RFID) [1] technology is considered to be the key building block of IoT. A number of IoT-based applications have been developed to understand, manage and control personal RFID data. The emergence of several key technologies such as intelligent sensors, low-energy wireless communication, Wireless Sensor Network (WSN), and cloud computing among others have provided

D. Debnath (✉)
Department of Computer Science, St. Mary's College, Shillong, Meghalaya, India
e-mail: d.debnath@smcs.ac.in

S. Kr. Chettri
Department of Computer Applications, Assam Don Bosco University, Guwahati, India
e-mail: sarat.chettri@dbuniversity.ac.in

© The Author(s), under exclusive license to Springer Nature Singapore Pte Ltd. 2021   679
X.-Z. Gao et al. (eds.), *Applications of Artificial Intelligence in Engineering*, Algorithms for Intelligent Systems, https://doi.org/10.1007/978-981-33-4604-8_52

ambient intelligence and independent control in IoT [2]. The International Telecommunication Union (ITU) [3] presents the enabling technologies, future markets, and
emerging challenges and the implications of the IoT.

Currently, researchers are fusing different technologies with the internet of things
in order to develop machine to machine (M2M), Machine to Human (M2H) and
Human to Machine (H2M) interaction [4]. Broadly, the study identifies an uptrend
in the following research areas:

- Intelligent IoT

Artificial intelligence, particularly machine learning and deep learning are incorporated in IoT applications. The association of AI and ML with IoT provides the ability
to automatically identify patterns, detect anomalies and make better decisions. Intelligent IoT enables the creation of smart cities, smart homes, smart transportation,
and smart outdoor monitoring (SOM) [5] in the near future.

- Green IoT

The power consumption of sensors is a big concern and limitation. Saving energy is
a critical design goal for IoT devices. Researchers are focusing on energy efficient
techniques such as green energy harvesting [6] and wireless charging methodologies.

- Cloud IoT

IoT Cloud [7] is a platform designed to store and process data from millions of
globally dispersed devices to deliver solutions in real time. In the last few years,
Internet of Things, Cloud computing, Edge computing, and Fog computing have
gained a considerable attention in industry and academia.

- Social Internet of Things (SIoT)

Another emerging interdisciplinary field is autonomous connectivity between social
networks and the Internet of Things, which is a promising model called the Social
Internet of Things (SIoT). In SIoT [8], objects may have mutual ties between people-
and-things and between things-and-things that serve as social circles.

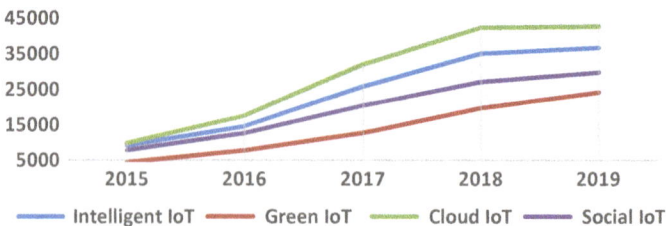

Fig. 1 Number of the IoT-based research over the period 2015–2019 (*Source* Google Scholar)

Figure 1 shows the trend in the number of paper publications indexed in Google Scholar over the period (2015–2019). The spending on IoT devices and applications is also on the rise. According to International Data Corporation (IDC) [9], the worldwide spending on IoT is expected to surpass $1 trillion by 2022.

3   Enabling Tools and Technologies

3.1   IoT Objects

IoT objects mainly consist of hardware platforms, mechanical, electrical components and embedded software. These objects handle key tasks and functions such as sensing, actuation, control, communication, security, identification, tracking and monitoring activities. The hardware platforms include Arduino, Raspberry Pi, BeagleBone, pcDuino, Netduino, etc. along with mechanical and electrical components. Sensors are the primary hardware devices used to detect, sense certain characteristics of the surroundings and measure some events such as temperature, light, proximity, air pressure, motion, water quality etc. Sensors typically measure something of interest and provide output for further processing. In fact, the features of many smart sensors are driven by the use of modern microcontrollers [10] such as ESP32, ESP8266, SMART SAM G54/G55, and so on. IoT's key enabling technology RFID (Radio Frequency Identification) [11] is being integrated with wireless sensor network (WSN) to provide real-time identification, tracing and tracking of physical objects within a short range. Similar technologies are NFC (Near Field Communication) [12] and Ultra-Wide Bandwidth (UWB). NFC is based on the ISO/IEC 18092:2004 standard and supports bi-directional communication; card emulation and peer-to-peer, used in electronic payment systems.

3.2   Communication Technologies and Networks

Generally, communications in IoT takes place between humans and devices or direct communication between devices without any human intervention. IoT-related communication can broadly be classified on the basis of its range of communication; long-range communication such as WiMAX, 2G/3G/4 G, NB-IoT, NWave, Satellite Communication, etc. and short-range communication such as RFID, Bluetooth, Z-Wave, NFC, 6LowPAN, etc.

Table 1 provides a comparison of some of the long-range (SigFox, LoRaWan, and NB-IoT) and short-range (BLE, ZigBee, 6LowPAN, and Wi-Fi) communication standards [13] used in IoT applications. Communication requirements for IoT applications are specific in terms of area coverage, security, power consumption, data

Table 1 Comparison of long range and short range communication standards

Wireless technology	Frequency band	Wireless standard	Transmission range	Data rate and power requirement
Bluetooth smart (BLE)	2.4 GHz	Bluetooth (Formerly IEEE 802.15.1)	10–150 m	1 Mbps, 1 W
ZigBee	2400–2483.5 MHz	IEEE 802.15.4	10–100 m	250 Kbps, 1 mW
6LowPAN	2.4 GHz	IEEE 802.15.4	20–100 m	20 Kbps–250 Kbps, 1 mW
Wi-Fi	2.4 GHz, 5–60 GHz	IEEE 802.11 a/c/b/d/g/n	20–100 m	1 Mbps–6.75 Gbps, 1 W
SigFox	Unlicensed ISM bands—868 MHz (Europe), 915 MHz (North America), 433 MHz (Asia)	Sigfox	10 km (urban), 40 km (rural)	100 bps, N/A
LoRaWan	Unlicensed ISM bands—868 MHz (Europe), 915 MHz (North America), 433 MHz (Asia)	LoRaWAN	5 km (urban), 20 km (rural)	50 kbps, N/A
NB-IoT	Various—Licensed LTE frequency bands, re-framed GSM bands	3GPP	1 km (urban), 10 km (rural)	200 kbps, N/A

rate, and cost effectiveness. To facilitate the communication between the connected things gateways are used.

The four layered IoT protocol stack based on TCP/IP reference model are:

1. Application layer with the most commonly used protocols are MQTT (Message Queuing Telemetry Transport), CoAP (Constrained Application Protocol), AMQP (Advanced Message Queuing Protocol), MQTT-SN (MQTT for Sensor Networks), etc.
2. Transport layer has UDP (User Datagram Protocol), DTLS (Datagram Transport Layer Security) protocol providing secured communication for data-gram based applications.
3. The Internet layer consists of address and routing protocols. The IPv6 (Internet Protocol version 6) [14] used for addressing (128 bits) has become a de facto standard as the new identity for IoT devices. The RPL (Routing Protocol for Low Power and Lossy Networks) is mostly used for IoT-based applications in low-energy wireless networks. Likewise, 6LoWPAN, an acronym for IPv6

over low power wireless personal area network enables communication between 6LoWPAN devices with all other IP-based devices over the Internet.

4. The Network/Link layer is a repository of communication protocols and standards that are interoperable to meet the IoT requirements. For example, the IEEE 802.15.4 protocol is designed to allow communication between devices that require a long battery life. Likewise, Zigbee [15] based on IEEE 802.15.4 communication protocol standard is designed to allow communication between devices in a personal area network that require a long battery life and supports multi-hop routing opposed to BLE (Bluetooth Low-Energy).

3.3   Software and APIs

Software (including OS) and API (Application Programming Interface) are required to enable the functionality of IoT applications. Some of the common embedded operating systems used are Contiki, LiteOS, Android, Riot OS, Snappy, TinyOS, and Windows IoT etc. Contiki is an open source IoT OS mainly used for low power microcontrollers and other IoT devices. Android Things is a lightweight IoT operating system developed by Google to simplify IoT development and support Bluetooth and Wifi technology. Riot is also a free and open source IoT operating system built using C and C++ and provides content-centric networking and network protocols like TCP, CoAP, and UDP, etc. The SDKs (Software Development Kit) for these operating systems support the programming of IoT applications using various programming languages such as C, C++, C#, Java, etc. APIs are tools for building software applications and act as logical connectors allowing IoT applications to communicate with IoT devices such as JML, Web GL, and RAML, etc. REST (Representational State Transfer) APIs are required to truly achieve the potential of IoT. Many IoT applications rely on REST or RESTful APIs to offer interoperability between different network technologies.

3.4   Data Storage and Data Processing Mechanisms

The International Data Corporation (IDC) [16] has estimated that by 2025, there will be 41.6 billion connected IoT devices generating 79.4 ZB of data demanding larger data storage and analytics. Data storage facilities include document-oriented database systems such as MonGoDB, SimpleDB, CouchDB or column-oriented databases such as Cassandra, HBase, BigTable and HyperTable, etc. These data platforms make it convenient to store a variety of heterogeneous IoT-generated data [17]. The data analytics component processes and displays the information collected through IoT. It includes analytical tools, artificial intelligence (AI), machine learning, and data visualization capabilities. There are also cloud-based data processing and open source platforms including Apache Kafka, Storm, Cloud Datalab, and Apache Hadoop, etc. Some other software platforms that provide data collection, data processing and management services are AWS IoT Platform, Microsoft Azure, ThingWorx, IBM Watson IoT, DynThings and Xively among others. These software platforms facilitate secure and bi-directional communication between IoT devices connected via the Internet, enabling telemetry data collection from multiple devices, storage and analysis. The ThingWorx IoT platform is specifically designed to develop powerful applications with augmented reality (AR) experience for industrial IoT (IIoT).

3.5 Supportive Technologies

Supporting technologies include cloud computing, fog computing, web technology, edge computing, Software-Defined Networking (SDN), grid computing, blockchain, distributed computing, Information-Centric Networking (ICN) and Cyber-Physical. Systems (CPS) among others. These supporting technologies make the smart ecosystem of IoT possible. For example, SDN [18] provides orchestration for network management of IoT devices connected over the Internet by providing flexible and programmable architectures and scalable communication methods. Blockchain technology [19] integrated with IoT provides access management for specific scalable IoT scenarios. The CPS [20] has tremendous potential for next generation smart systems that integrate cyber technology into physical processes. The ICN [21] has the potential to address the deficiencies of current IP-address-based networking and thus overcome the lack of address-space in the IoT environment. The adoption of the Cloud-IoT paradigm [22] is significant in designing platforms that need accurate, secure and real-time processing of sensor data.

4 Application Areas of IoT

IoT is applied in several diverse fields practically in all areas of the daily life of individuals, organizations and society as a whole. Some of the important application domains are presented in the following:

Healthcare

There are many applications of IoT in the field of healthcare and medicine, ranging from detection to monitoring of several critical diseases such as diabetes, cancer, coronary heart disease, stroke, chronic obstructive pulmonary disease, cognitive impairments, seizure disorders and Alzheimer's disease. Table 2 summarizes different approaches, tools and techniques applied in the field of smart healthcare.

Agriculture and Irrigation

IoT adoption has enhanced the monitoring and management of firms in the field of agriculture, the sharing of knowledge between producers and consumers and the

Table 2 A summary of IoT-based applications in healthcare

References (Year)	Tools and techniques	Approaches
[23] (2016)	Wireless sensors, Cloud IoT, Deep learning	Designed a deep learning model to detect mood fatigue by using user's physiological information, external environment conditions and user behavior
[24] (2018)	Wireless body area network (WBAN) and Block Chain	The proposed remote patient monitoring system uses real time log transaction metadata of medical sensors using block chain technology
[25] (2019)	IoT and wireless sensor network (WSN)	Implemented a health care system for cancer care.
[26] (2019)	Machine learning, decision tree, random forest	Developed a prediction model to detect several diseases such as heart diseases, breast cancer, diabetics, thyroid and liver disorder
[27] (2020)	Modified deep convolutional neural network (MDCNN)	Proposed an IoT framework to evaluate heart disease using MDCNN with an accuracy of 98.2%

efficient use of available resources. Table 3 presents a description of the different methods and tools and techniques applied in the field of smart agriculture present in the literature.

Smart City

The smart city concept encompasses better service management of energy, electricity, traffic, waste management, urban planning, air pollution and so forth. Table 4 summarizes the various approaches and applied tools and strategies incorporated in the literature of the smart city domain.

Manufacturing and Supply Chain

Intelligent network supporting infrastructures, optimization of production processes, smart labels to track and trace, supply chain monitor are some of the key benefits of IoT in manufacturing industry. Table 5 provides a summary of various methods and applied tools and techniques present in the literature in smart industry, manufacturing and supply chain.

Media and Entertainment Industry

IoT in media and entertainment covers everything about telecommunications: film, television, games and sports. Table 6 provides a summary of various approaches and applied tools and techniques in the domain of media and entertainment.

IoT is fuelled by an explosion of Internet-based business applications. As a result, there are several issues and challenges that will be addressed in the section below.

Table 3 A summary of smart agriculture and irrigation applications

References (Year)	Tools and techniques	Approaches
[28] (2018)	Agricultural cyber-physical social system (CPSS)	Proposed an agricultural CPSS for better agricultural production management, with a case study on the solar green house
[29] (2018)	Neural Network and fuzzy logic-based weather model	Implemented a fuzzy logic weather model to provide uniform farm irrigation in almost all the weather conditions
[30] (2018)	Wireless sensor network (WSN) and cloud IoT	Proposed a prototype for potato late blight disease prevention decision support system based on sensor network and cloud
[31] (2019)	Block chain	Designed architecture for IoT based blockchain to enhance security and data transparency in smart agriculture
[32] (2019)	LoRAWaN network	Proposed customizable model architecture to provide data analytics solution for a large-scale real time data and support decision-making to monitor and control the irrigation system

Table 4 A summary of IoT-based applications in smart city

References (Year)	Tools and Techniques	Approaches
[33] (2016)	Black networks, software defined network (SDN) controller	Developed a secure IoT Smart City model that adds privacy, identity management, authentication and secure key management
[34] (2017)	Deep learning, long short term memory (LSTM)	Proposed a model for predicting the air quality of a city using the long short term memory (LSTM) networks and deep learning techniques
[35] (2018)	Low Power wide area network (LPWAN), Narrowband IoT (NB-IoT)	Implemented a NB-IoT smart grid to provide secure and reliable communications between the different components of the power system
[36] (2019)	Edge computing, deep reinforcement learning	Proposed an IoT-based energy management system that provides efficient energy scheduling scheme based on edge computing infrastructures with deep reinforcement learning
[37] (2019)	Augmented reality (AR), Virtual reality (VR)	Discusses AR and VR technologies used for smart policing

Table 5 A summary of IoT-based applications in industry, manufacturing and supply chain

References (Year)	Tools and techniques	Approaches
[38] (2017)	Smart manufacturing system (SMS), Industry 4.0	Elaborates industry 4.0 intelligent manufacturing system and associated technologies
[39] (2017)	RFID, supply chain for perishable food with two-echelon supply hubs (SCPF-2ESH)	Presented a conceptual model for the IoT-enabled perishable food supply chain with two-echelon supply hubs
[40] (2018)	Convolutional neural networks and fog computing	Implemented a product inspection system for smart industry
[41] (2018)	Neutrosophic decision making trial and evaluation laboratory (N-DEMATEL) and analytic hierarchy process(AHP)	Proposed a supply chain management framework which integrates (N-DEMATEL) technique with analytic hierarchy process (AHP)
[42] (2020)	Industry 5.0, absolute innovation management (AIM)	Presented a new innovation management framework called as "AIM" to make innovation more understandable, implementable as a part of the day-to-day routine of organization

Table 6 A summary of IoT-based applications in media and entertainment

References (Year)	Tools and techniques	Approaches
[43] (2017)	Deep belief network framework (DBNF)	Developed a real time internet cross-media retrieval method based on DBNF
[44] (2018)	Reinforcement learning (RL), Markov decision process (MDP)	Proposed a traffic scheduler based on IoT using reinforcement learning to generate adaptive controllers that can dynamically adapt traffic variations
[45] (2018)	Deep learning object recognition (DLOR)	Implemented an edge-based system for efficient and large-scale video stream analysis using DLOR
[46] (2019)	Intrusion detection system (IDS), Machine learning (ML)	Developed an intelligent, real-time intrusion-detection system using deep-learning algorithm to detect malicious traffic in IoT networks
[47] (2019)	Machine learning, deep learning cyber forensics (DLCF)	Presented a prototype to help digital forensic investigators to manage investigation process using DLCF Framework

5   Issues and Challenges in IoT

The challenges of IoT technologies and applications are mostly related to information security and privacy, IoT standardization, power and energy consumption, legal challenges, environmental issues, IoT system modeling, data storage and analysis, and other developmental issues [48]. Broadly, three major issues and challenges are discussed in the following subsections.

5.1   *Technical Challenges*

Notwithstanding advancements in IoT technology, there are still numerous technical challenges [49] which require further study. With the growing number of connected IoT devices, there are scalability problems in terms of data transmission, networking, and service delivery, etc. The growing size of the vast amount of data generated by IoT applications makes it difficult to gather any meaningful information from its analysis. In this direction, there is a need to develop more effective context-awareness IoT middleware solutions to gain valuable insights into the big IoT data analytics.

5.2   *Security and Privacy*

Given the technical and societal benefits and IoT's tremendous economic potential, the protection and privacy issue [50] is a major concern. The problem is complicated due to the inherent complexity of its deployment and mobility [51]. It is one of the main reasons why many innovations are applicable to consumer use but not implemented for industrial applications which have strict safety and security requirements. Some of the open IoT security questions [51] are: How do we effectively identify and evaluate the IoT device security features? How are we assessing the efficacy of the initiative and countermeasures taken for IoT security? Should there be any legislation to penalize the device manufacturers for selling software or hardware for IoT with security flaws?

Data privacy protection in the IoT setting is a serious concern due to the prevalence of various types of attacks on IoT entities [50], lack of transparency due to inadequate authentication and authorization, absence of strict rules and regulations against the collection and use of information etc. For example, IoT devices are used in eHealth services [52] to capture patient information, such as blood pressure, blood sugar level, and then transfer it over the network so that health care professionals can track subjects' health-related data anywhere at any time. Patient data may however be stolen or compromised. Another example is that bio-sensor [53] is used in the food industry to detect target bacteria early for healthy food delivery and to prevent contamination of the food. Such data shall be transmitted over a network and kept strictly confidential to protect the food company's reputation.

5.3 Interoperability and Standards

Interoperability refers to the ability of various IoT devices and systems to interact whatever their hardware architecture and software systems. One can see the IoT interoperability [54] as:

- Interoperability between high-end, middle-end and low-end devices [55].
- Interoperability in dynamic and heterogeneous networks, for seamless communication among IoT systems.
- Interoperability of the format of message exchange between IoT systems.
- Interoperability between IoT platforms and IoT domains (for example healthcare, agriculture, transportation, etc.).

The interoperability issues arise due to the adoption of various solutions and a range of standards and technologies used to build IoT systems by different vendors. The proliferation of communication infrastructure and protocols such as IPv6, IPv4, 6LoWPAN, Thread, CoAP, ZigBee, AMQP, JSON-LD, EPC, UDP, NanoIP, and RFID among others has added to the challenging issue of technical interoperability among IoT systems. The established internet infrastructure does not help the heterogeneous devices' maximum connectivity and therefore there are complex integration problems. The different IoT standards such as security standards, identification standards and communication standards and the lack of standard configurations for interfacing large numbers of IoT devices also give rise to technical and semantic interoperability problems.

6 Open Research Issues

Some of the open research areas (see Fig. 2) which require further investigations and research based on the above-mentioned issues and challenges are listed below:

Technical Challenges

- Secured data transfer over the cloud and data storage.
- Connecting the ever-increasing number of IoT devices for seamless communication.
- Better data storage facility and analysis of Big-IoT data.
- Having a proper scalability in terms of IoT data transfer over the network service delivery.

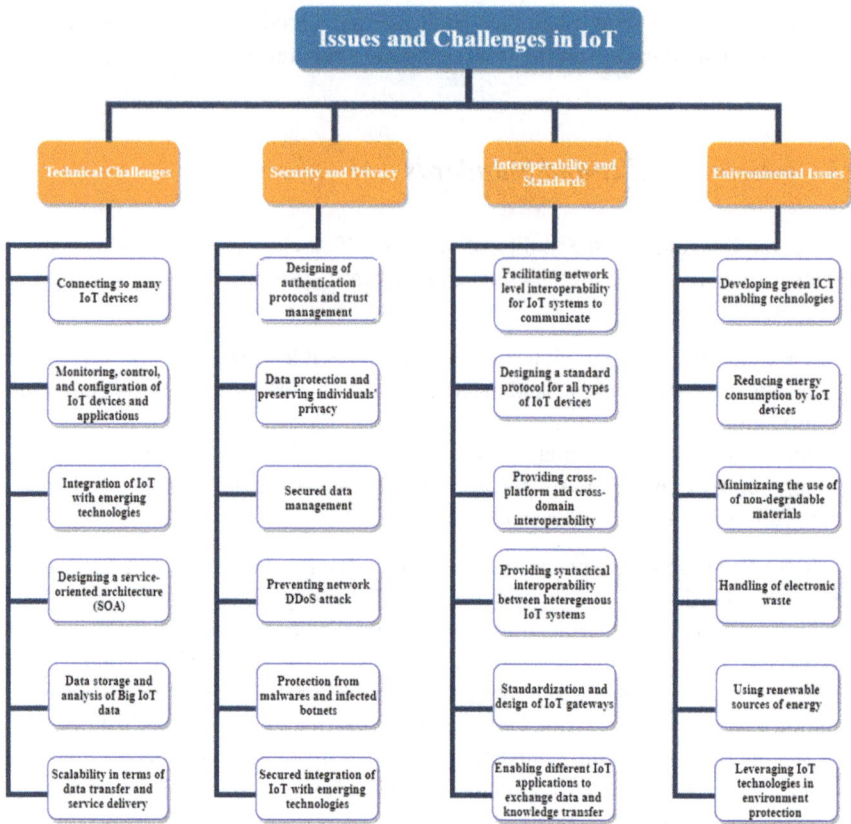

Fig. 2 Open research issues and challenges in IoT

Security and Privacy Issues

- Developing effective techniques to overcome the network DDoS attack.
- Securing the communications that takes place between IoT devices.
- Developing solutions for maintaining data transparency and provide full rights to users to set their own privacy preferences.
- Creating common data protection rules and regulations to have a standard security policy for all IoT devices of different vendors.
- Developing effective techniques for data protection in IoT-based applications.
- Build solutions to protect IoT devices from malwares and compromised botnets.
- Developing authentication protocols and authorization schemes.
- Secured integration of IoT and computing technologies such as cloud, fog and edge computing.

Interoperability and Standards Issues

- Developing technologies and APIs for providing cross-platform interoperability among IoT systems.
- Developing a standard IoT application protocol for all types of devices (low-end, middle-end and high-end IoT devices).
- Building practical scalable solutions for multiple platforms with the flexibility to incorporate new platforms as they arise.
- Designing a strategy to automate the testing of technologies and standards for interoperability and make them less labor intensive.
- Making a gateway-free interoperability solution to be scalable for device to device communication in IoT systems.
- Making a standard configuration to interface a large number of IoT devices.
- Developing solutions to facilitate seamless communication between IoT systems across different networks.

Environmental Issues

- Putting more stress on IoT systems with low power consumption.
- Developing green ICT enabling technologies.
- Applying IoT devices to protect the environment like reducing air pollution, noise emissions and so on.
- Designing a comprehensive strategy to reduce the e-waste problem caused by discarded IoT devices.
- Promoting the use of renewable sources of energy for IoT devices.

7   Conclusion

IoT involves devices ranging from basic electronics to mobile devices and industrial systems where communication between humans and machines takes place or direct communication between systems without interference by humans. The words IoT and IoE (Internet of everything) are interrelated; however, IOE has a more detailed version than IoT, including people, items, data and processes. IoT has tremendous technical and social benefits with significant economic growth, thanks to the rapid developments in technology, capital and infrastructure. This paper reviews the current research and focuses primarily on IoT applications in different fields, discussing the issues and challenges and potential research opportunities for future IoT work.

References

1. Welbourne E et al (2009) Building the internet of things using RFID: the RFID ecosystem experience. IEEE Internet Comput 13(3):48–55

2.	Khalil N, Abid MR, Benhaddou D, Gerndt M (2014) Wireless sensors networks for Internet of Things. In: IEEE 9th International conference on intelligent sensors, sensor networks and information processing (ISSNIP), IEEE, pp 21–24
3.	Trappey AJ, Trappey CV, Govindarajan UH, Chuang AC, Sun JJ (2017) A review of essential standards and patent landscapes for the internet of things: a key enabler for industry 4.0. Adv Eng Info 33:208–229
4.	Da Xu L, He W, Li S (2014) Internet of things in industries: a survey. IEEE Tran Indus Info 10(4):2233–2243
5.	Lemayian JP, Al-Turjman F (2019) Intelligent IoT communication in smart environments: an overview. In: Artificial intelligence in IoT, Springer, pp 207–221
6.	Liu X, Ansari N (2019) Toward green IoT: energy solutions and key challenges. IEEE Commun Mag 57(3):104–110
7.	De Donno M, Tange K, Dragoni N (2019) Foundations and evolution of modern computing paradigms: cloud, IoT, edge, and fog. IEEE Access 7:150936–150948
8.	Sciullo L, Aguzzi C, Di Felice M, Cinotti TS (2019) WoT Store: enabling things and applications discovery for the W3C web of things. In: 16th IEEE Annual consumer communication and networking conference, IEEE, pp 1–8
9.	Torchia M, MacGillivray C, Bisht A, Siviero A (2020) Worldwide internet of things spending guide. Int Data Corpor [Online]. https://www.idc.com/getdoc.jsp?containerId=prUS44596319. Last accessed 2020/6/10
10.	Maier A, Sharp A, Vagapov Y (2017) Comparative analysis and practical implementation of the ESP32 microcontroller module for the internet of things. In: Internet technology and applications (ITA2017), IEEE, pp 143–148
11.	Akgün M, Çalayan U (2015) Providing destructive privacy and scalability in RFID systems using PUFs. Ad Hoc Netw 32:32–42
12.	He D, Kumar N, Lee J (2015) Secure pseudonym-based near field communication protocol for the consumer internet of things. IEEE Trans Consum Electron 61(1):56–62
13.	Mekki K, Bajic E, Chaxel F, Meyer F (2019) A comparative study of LPWAN technologies for large-scale IoT deployment. ICT Expr 5(1):1–7
14.	Ziegler S et al (2013) IoT6 - moving to an IPv6-based future IoT. In: Future internet assembly, Springer, Berlin, pp 166–172
15.	Ndih EDN, Cherkaoui S (2016) On enhancing technology coexistence in the IoT era: ZigBee and 802.11 case. IEEE Access 4:1835–1844
16.	International Data Corporation [Online] (2020). https://www.idc.com/getdoc.jsp?containerId=prUS45213219. Last Accessed 2020/6/10
17.	Kang YS, Park IH, Rhee J, Lee YH (2015) MongoDB-based repository design for IoT-generated RFID/sensor big data. IEEE Sens 16(2):485–497
18.	Tayyaba SK, Shah MA, Khan OA, Ahmed AW (2017) Software defined network (SDN) based internet of things (IoT): a road ahead. In: Proceeding of international conference on future networks and distributed systems, ACM, pp 1–8
19.	Novo O (2018) Blockchain meets IoT: an architecture for scalable access management in IoT. IEEE Internet Things 5(2):1184–1195
20.	Yao X, Zhou J, Lin Y, Li Y, Yu H, Liu Y (2019) Smart manufacturing based on cyber-physical systems and beyond. J Intell Manuf 30(8):2805–2817
21.	Arshad S, Azam MA, Rehmani MH, Loo J (2018) Recent advances in information-centric networking-based internet of things (ICN-IoT). IEEE Internet of Things 6(2):2128–2158
22.	Darwish A, Hassanien AE, Elhoseny M, Sangaiah AK, Muhammad K (2019) The impact of the hybrid platform of internet of things and cloud computing on healthcare systems: opportunities, challenges, and open problems. J Ambient Intell Humaniz Comput 10(10):4151–4166
23.	Shi X et al (2016) Cloud-assisted mood fatigue detection system. Mobile Netw Appl 21(5):744–752
24.	Griggs KN, Ossipova O, Kohlios CP, Baccarini AN, Howson EA, Hayajneh T (2018) Healthcare blockchain system using smart contracts for secure automated remote patient monitoring. J Med Syst 42(7):1–7

25. Onasanya A, Elshakankiri M (2019) Smart integrated IoT healthcare system for cancer care. Wireless Netw 1:1–16
26. Kaur P, Kumar R, Kumar M (2019) A healthcare monitoring system using random forest and internet of things (IoT). Multimedia Tools and Appl 78(14):19905–19916
27. Khan MA (2020) An IoT framework for heart disease prediction based on mdcnn classifier. IEEE Access 8:34717–34727
28. Kang M, Fan XR, Hua J, Wang H, Wang X, Wang FY (2018) Managing traditional solar greenhouse with CPSS: A just-for-fit philosophy. IEEE Trans Cybernet 48(12):3371–3380
29. Keswani B et al (2019) Adapting weather conditions based IoT enabled smart irrigation technique in precision agriculture mechanisms. Neural Comput Appl 31(1):277–292
30. Foughali K, Fathallah K, Frihida A (2018) Using Cloud IOT for disease prevention in precision agriculture. Proc Comput Sci 130:575–582
31. Devi MS, Joshi AS, Suguna R, Bagate R (2019) Design of IoT blockchain based smart agriculture for enlightening safety and security. In: International conference n emerging technologies in computer engineering, Springer, pp 45–54
32. Davcev D, Mitreski K, Trajkovic S, Nikolovski V, Koteli N (2018) IoT agriculture system based on LoRaWAN. In: 14th IEEE-international workshop on factory communication systems, IEEE, pp 1–4
33. Chakrabarty S, Engels DW (2016) A secure IoT architecture for smart cities. In: 13th IEEE annual consumer communications and networking conference, IEEE, pp 812–813
34. Kök I, Şimşek MU, Özdemir S (2017) A deep learning model for air quality prediction in smart cities. In: 2017 IEEE International conference on big data, IEEE, pp 1983–1990
35. Li Y, Cheng X, Cao Y, Wang D, Yang L (2018) Smart choice for the smart grid: narrowband internet of things (NB-IoT). IEEE Internet of Things 5(3):1505–1515
36. Liu Y, Yang C, Jiang L, Xie S, Zhang Y (2019) Intelligent edge computing for IoT-based energy management in smart cities. IEEE Netw 33(2):111–117
37. Bang J, Lee Y, Lee YT, Park W (2019) AR/VR based smart policing for fast response to crimes in safe city. In: IEEE International symposium on mixed and augmented reality (ISMAR-Adjunct 2019), IEEE, pp 470–475
38. Zhong RY, Xu X, Klotz E, Newman ST (2017) Intelligent manufacturing in the context of industry 4.0: a review Engineering 3(5):616–630
39. Zhang Y, Zhao L, Qian C (2017) Modeling of an IoT-enabled supply chain for perishable food with two-echelon supply hub. Ind Mgmt Data Sys 117(9):1890–1905
40. Li L, Ota K, Dong M (2018) Deep learning for smart industry: efficient manufacture inspection system with fog computing. IEEE Trans Industr Inf 14(10):4665–4673
41. Abdel-Basset M, Manogaran G, Mohamed M (2018) Internet of Things (IoT) and its impact on supply chain: a framework for building smart, secure and efficient systems. Future Gener Comput Syst 86:614–628
42. Aslam F, Aimin W, Li M, Ur Rehman K (2020) Innovation in the era of IoT and industry 5.0: absolute innovation management (AIM) framework. Information 11(2):1–25
43. Jiang B, Yang J, Lv Z, Tian K, Meng Q, Yan Y (2017) Internet cross-media retrieval based on deep learning. J Visual Commun Image Rep 48:356–366
44. Chinchali S et al (2018) Cellular network traffic scheduling with deep reinforcement learning. In: 32nd AAAI conference on artificial intelligence, pp 766–774
45. Ali M et al (2018) Edge enhanced deep learning system for large-scale video stream analytics. In: IEEE 2nd International conference on fog and edge computing, IEEE, pp 1–10
46. Thamilarasu G, Chawla S (2019) Towards deep-learning-driven intrusion detection for the internet of things. Sensors 19(9)
47. Karie NM, Kebande VR, Venter HS (2019) Diverging deep learning cognitive computing techniques into cyber forensics. Forens Sci Int Syner 1:61–67
48. Kumar NM, Mallick PK (2018) Blockchain technology for security issues and challenges in IoT. Proc Comput Sci 132:1815–1823
49. Banafa A (2017) Three major challenges facing IoT-IEEE internet of things. In: IEEE Internet of things, no. March 2017. pp 18–21. [Online] Available https://iot.ieee.org/newsletter/March-2017/three-major-challenges-facing-iot

50. Sun G et al (2017) Efficient location privacy algorithm for internet of things (IoT) services and applications. J Netw Comput Appl 89:3–13
51. Rose K, Eldridge S, Chaplin L (2015) Internet of things: an overview. Internet Soc (ISOC) 80:1–50
52. Farahani B, Barzegari M, Aliee FS, Shaik KA (2020) Towards collaborative intelligent IoT eHealth: from device to fog, and cloud. Microprocess Microsyst 72:102938
53. Massad-Ivanir N et al (2016) Porous silicon-based biosensors: towards real-time optical detection of target bacteria in the food industry. Sci Report 6:1–12
54. Noura M, Atiquzzaman M, Gaedke M (2019) Interoperability in internet of things: taxonomies and open challenges. Mobile Netw Appl 24(3):796–809
55. Ojo MO, Giordano S, Procissi G, Seitanidis IN (2018) A review of low-end, middle-end, and high-end iot devices. IEEE Access 6:70528–70554

Chapter 53
The Possibilities of Artificial Intelligence in the Hotel Industry

Sunil Sharma and Yashwant Singh Rawal

1 Introduction

In precise provisions it means building augmented things intelligent. Dissimilar to human, who have unusual intelligence, the abilities to imagine and to take decisions, machines are just dumb systems with no intelligence. Therefore construction of a machine to work as a person, feel like humans and have decision-making capability is termed as artificial intelligence. Artificial Intelligence is a cram of preparation the machines and the other procedure to perform jobs as humans do.

Artificial intelligence (AI) as a scholar ability was identified in the era of 50 s. In fact the "AI" dialect was think up by "**John McCarthy**", who was an "American Computer Scientist" (ACS), at the time of 1956 at "**The Dartmouth Conference**". According to John McCarthy, AI was defined "**The science and engineering of designing intelligent machines, especially intelligent computer programs**" [1] (Fig. 1).

However it was not in anticipation of newly, it became part of everyday life. Thanks to advances in big data accessibility and inexpensive high computing control. AI works at its top by combining huge amounts of data sets by means of rapid, iterative exemption and smart algorithms. This offers the AI software to be skilled unwillingly from prototypes or attributes in that massive datasets.

AI is actually an extensive idiom and to some extent this also causes each company to maintain their product as AI has these day [1]. The Machine Learning (ML) is a compartment of Artificial intelligence (AI), and it consists of supplementary advanced systems and models that assist computers to summarize possessions out

S. Sharma
Rajasthan Technical University, Kota, India
e-mail: ersharma.sunil@gmail.com

Y. S. Rawal (✉)
Amity University, Jaipur, India
e-mail: yashwantr84@gmail.com

The **Turing test** is developed by Alan Turing to test whether a machine is capable of human intelligent behaviour or not.

The First Robot is introduced on an assembly line at General Motors.

IBM's Deep Blue - a chess playing computer - beats then chess world champion, Garry Kasparov.

1950   1955   1961   1965   1997

John McCarthy, an American computer scientist, coined the term '**Artificial Intelligence**'.

Eliza, the first chatbot is created by MIT AI Laboratory based on Natural Language Processing (NLP).

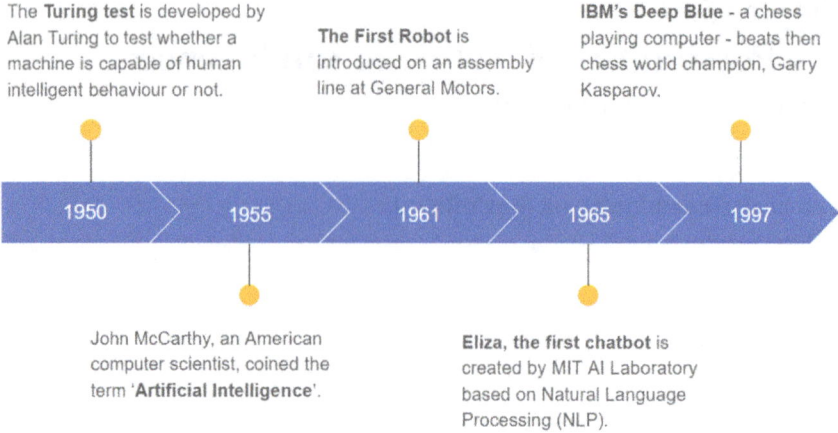

Fig. 1 Dr. Alok Aggarwal, Domains in Which Artificial Intelligence is Rivaling Humans, 2018

from the statistics and allocate AI applications. The Machine Learning (ML) is the order of getting computers to execute without being unequivocally designed.

2 Impacts and Role of AI in Hotel Industry

The well defined customer service plays an essential element of the travel business, with hotels frequently living and fading depends on the mode they extravagance their regular and other customers. By mean of artificial intelligence, the potential for humanizing these characteristics are almost eternal, which is ranging from improved personalization to customized proposals. Even though the applications of artificial intelligence contained by the hotel business are still in its comparative immaturity [2], it previously has abundant convenient applications, of which some are summarize here in more detail:

1. **In-Person Customer Service**

Artificial intelligence can be widely used in the hospitality business to convey **"in-person customer services"**. It has already been considered the improvement of robots by mean of artificial intelligence and the latent for this technology to develop is huge. Previously, it was capable to covenant with essential customer-facing conditions (Fig. 2).

The most suitable example at this position can be considered as an **AI robot** sometime called as '**Connie**', and it was initially implemented by Hilton. This robot 'Connie' is capable to supply traveler in sequences to clientele who act together with it [3]. Most remarkably, it is intelligent to be trained from personal communication

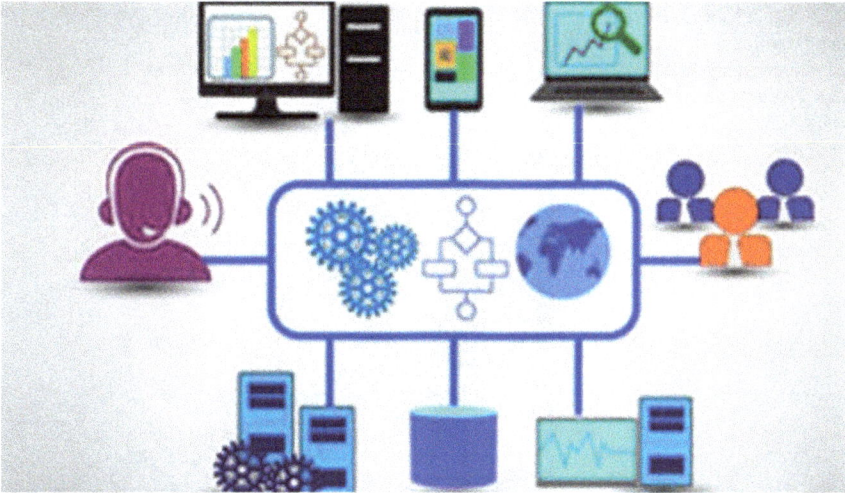

Fig. 2 Brian Frank, Deliver smarter service with artificial intelligence built right into the world's service platform, 2020

and acclimatize to individuals. Eventually, this ways the more clientele converse to it, the superior it will acquire.

2. **Chat-Bots and Messaging**

This will possibly be the most noticeable way of artificial intelligence that can be set up by those who fall within the generosity region to provide front-desk customer services [4]. In general, this technology has been revealed to be tremendously effectual when it approaches to direct messaging and online chat services and in return responding to straightforward queries or requirements (Fig. 3).

3. **Data Analysis**

Data Analysis plays an important role in different way in which AI is being utilized within the hotel business [5] and which is far away from conventional customer service. In this competence, this technology can be used to swiftly organize through huge amounts of statistics and sketch significant conclusions concerning to customers, or prospective customers (Fig. 4).

3 The Technology for Straight Smart Hotels

Here we will observe some of the customs in which AI is enhancing and reshaping hospitality industry all over the world [6]. It includes:

Fig. 3 Pavel Kordik, How
Deep Learning and
Recommender Systems
make Chatbots useful and
more Intelligent , Chatbots
life, 2016

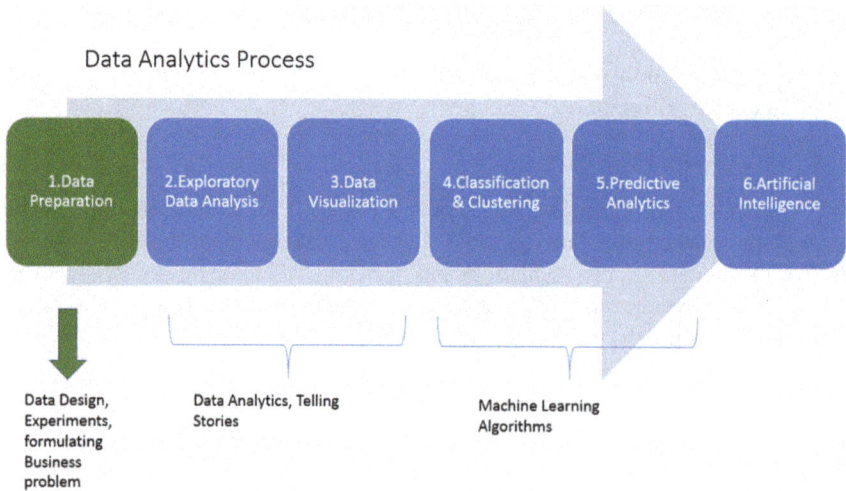

Fig. 4 Self Designed

1. **Personalization**

Personalized understandings boost up the revenue and reliability [7], and this is a
region where hospitality has long been in advance of the bend.

2. **Habitation and Rate Optimization**

Since hotel property is pricing-optimized for ski season. Accepting the contact of
these on occupancy levels is demanding.

3. **Booking and Staff Interaction**

Bookings are a foremost abrasion point for hotel industries. Many customers book by mean of aggregator websites, and those who land on an individual hotel page often bounce without booking [8]. If a user does wish to book, isolating preferred dates and check-in details can be time-consuming and complex, and opportunities to actually communicate with hotel staff are limited.

4. **Updates and Maintenance**

Every hotelier wish to deliver the best guest experience they can, but you also need to strategically channel their resources for the best return on investment.

5. **Informing Competitive Intelligence**

AI's ability to parse multiple databases to provide valuable context can help drive the future direction of your business. For example, it can map population density data against circulation data to identify hotel "deserts" or areas where demand is exceeding availability [9].

It has been observed that by utilizing an intelligent or smart hotel wills improve the customer potential and it is used as a strategy by many contemporary generosity leaders and service associates conjure up. There are convinced indispensable fundamentals which make hotels smart and intelligent. Some of the characteristics are mentioned below:

1. Concierge robotics
2. Digital support and assistance
3. Voice-controlled or instruction-activated services
4. Travel incident enhancers
5. Automatic data processing (ADP)

It has been investigated that the AI robots not only moderate the individual contribution by mean of its voice—controlled or voice activated assistants but also redecorate the system for guest services which is to be delivered to customer's contentment [10]. From customer's individual selection to their negligible requirements, AI-equipped hotel concierge bots can propose diminutive services and vigilant assistance too.

4 Hotels Leveraging Artificial Intelligence

Progressively, more the hotel business is inundated by mean of AI robots just about the globe. At the moment, the gesticulation of robotic butlers with cognitive proficiency looks like extraordinarily on the escalate Take a stare at the consequent example of the hotel implementation the encouragement of AI robots:

AI has dislodged the generosity business in inestimable traditions [11]. The most indispensable way in which AI has unaffected the tourist considerate is commencing from side to side the exercise of 'chat-boat' messaging services.

It has been identified that the grand Hotel named "**Hilton**" which is outfitted by mean of the world's foremost AI powered-operated by "AI robot hotel" which named as **Connie**,

The technology named AI in accumulation has get in touch with the hospitality administration by data analysis method [12]. "**The Caesars Entertainment Corporation** is very well known hospitality conglomerate which has worn out data analysis to an extra bump up within its defined characteristics.

5   Advantages and Disadvantages of AI

Parallel to all expertise AI also have several advantages and disadvantages. But somewhat we require preferring this technology as it will offer best services to the clients in hotel business.

Advantages

1.   **Responding Guest's Queries in Hotel Premises**

After having survey on various hotels we have come to know that some hotels have been using an AI based robot named "**Sepp**" to respond practical queries of customers and providing cooperative information to visitors. **Sepp's** awareness comes as of IBM's Watson Conversation Services [13], which assist it with accepting customer's desires and culture.

Another hotel is there who has also established an advanced robot in the form of **Connie**, an AI-oriented caretaker. This small robot acquire advantage of IBM's Watson Conversation Services and assist guests by finding their path in the order of the hotel and responds various other questions.

2.   **Hotel Supervision**

European and the other countries are not the only target places where AI-enabled robots are utilizing to create a disparity. Singapore-based hotels properties also have been using related technologies since 2018. Various service deliverance robots are performing their responsibility as taking facilities to visitor rooms and providing surveillance the hotel buildings [14].

3.   **The capability to allow travelers to investigate tour options with a Chat-bot**

Advantages:	Disadvantages:
Reduces in human error	Cost overruns
Helps in lessening repetitive work	Dearth of talent
Provides Digital assistance	Lack of practical products
Faster and more accurate decisions	Potential for misuse

Fig. 5 Advantages and disadvantages of AI

This advantage falls in the category about personalization. A current study which have been done by "**booking.com**" exposed that around 78% of clients favored self-service preference to get answers for straightforward requirements, which compose chat-bots a wonderful alternative in this scenario (Fig. 5).

Disadvantages

1. **AI is Under Developing**

After having fabulous improvement in topical years, the technology is pretty much in its immaturity. Various changes need to be done in this technology, which will make it best.

2. **Data Privacy Issues**

This is one of the main imperative apprehensions that all businesses trying to receive benefits of AI will face. Data collected by the technology during interactions with users can be potentially used for other purposes [15]; therefore, the risk of data privacy violation is involved.

6   Conclusion

Still while AI is experiential and dealing as potential technology expansion, the hospitality business has by currently competent unbelievable profit and extensive propel on their traffic impulsion. It is experiential that from the frontage segment where AI-equipped robotic doorkeeper alter stands in the hotel rooms where voice-activated subsidiary momentum to positive customer understanding, Artificial Intelligence is identified efficiently bouncing the hotel business at the moment. AI demonstrates an indisputable guarantee in getting higher feasibility, eye-catching personalization, as well as elevating gratification and commitment for hotel guests. Conversely, there are some hotels like "**Alibaba's FlyZoo**" Hotel individually and extremely powered and fully equipped through diverse smart technologies including **AI robots** and **facial recognition**, hoteliers is required to commence to locate the means of a stability linking human being and AI. Consequently, we can come to an end that Artificial Intelligence (AI) is carrying out extremely as remote to the opinion for hotel organizations.

On or after back up hotel workforce or staff members by mean of answering guests' questions to given that tailored proposals via steadfast search engines, AI indisputably has an prospect in the hospitality business. On the other hand, in normalize for the smart technology to extend into the innovative unsurpassed companion for both hoteliers and explorers, some extensive face ups have to be resolute.

References

1. Agarwal N (2018) Top 5 ways chatbots are revolutionizing the hotel industry
2. Bethesda (2017) Marriott International's. AI-Powered chatbots on facebook messenger and slack, and Aloft's ChatBotlr. Simplify travel for guests throughout their journey, Marriott International Newscenter
3. Biron B (2019) Chinese E-commerce giant Alibaba has a hotel run almost entirely by robots that can serve food and fetch toiletries. Take a Look Inside, Business Insider
4. Davis L (2016) Hilton and Pilot IBM 'Connie. The world's first watson-enabled hotel concierge robot, IBM
5. Kohlmayr K (2019) In: AI and the inevitable future of hospitality, Travel Daily Media
6. Kotler P (2017) In: Marketing for hospitality and tourism, Pearson Education
7. Marr B (2015) Big data at caesars entertainment "A one billion dollar asset", Forbes
8. Mest E (2017) Marriott launches facebook chatbot to assist in career development hotel management
9. Discover the Story of English More than 600,000 Words, over a Thousand Years, Oxford English Dictionary
10. Rosenbaum E (2013) What happens in vegas, stays in vegas as data, CNBC
11. Trejos N (2016) Introducing Connie, Hilton's new robot concierge, USA Today
12. Kang B, Brewe, KP, Bai B (2007) Biometrics for hospitality and tourism: a new wave of information technology. FIU Hospitality Tourism Rev 25(1):1–9
13. Kelsey J (2017) The risks for ASEAN of new mega-agreements that promote the wrong model of E-Commerce
14. Kelsey J (2017) TiSA Foul play. UNI Global Union
15. Möller K, Halinen A (2017) Managing business and innovation networks from strategic nets to business fields and ecosystems. Ind Mark Manage 67:5–22

Chapter 54
Automated Vehicle Emergency Support Using Edge Computing Concept

Anonyo Sanyal, Pijush Das, Pratik Gon, and Sutirtha Kumar Guha

1 Introduction

We all like a little get-away from the hustle and bustle of our daily lives of routine. Travelling is fun until some unexpected and unforeseen happens. In those dire situations what we need is easily, accessible and reliable help. Services that are available in typical scenario are server based support. Request is sent to the server and solution or support is provided by the server.

The assistance system that we use is highly dependent on the availability of service. The travelers in fact at times are not even aware of the available services nearby. The assistance provided to the travelers is nowadays achieved using GPS system. The traveler locates its location and finds the nearby assistance system using various Navigation Applications using GPS Maps.

It is obvious that large amount of requests results traffic congestion in server end. An amount of requests would be analyzed and served at the edge of the network. A significant amount of time would be saved and prompt remedial measure would be initiated.

A. Sanyal · P. Das · P. Gon · S. K. Guha (✉)
Meghnad Saha Institute of Technology, Kolkata, West Bengal, India
e-mail: sutirthaguha@gmail.com

A. Sanyal
e-mail: anonyosanyal1@gmail.com

P. Das
e-mail: daspijush.jgm@gmail.com

P. Gon
e-mail: gon.pratik@gmail.com

2   Related Work

Edge computing is a variation of distributed computing that reduces the burden at cloud server by analyzing and filtering amount of data at the edge of the network [1]. Multi directional research area are already found by the researchers for implementing edge computing as suggested in [2]. Usage of edge computing to resolve scalability issues in IoT is suggested in [3]. Offline services at the edge of the internet for mobile devices is discussed in [4]. A proposal of handling different challenges such as high latency, low spectral efficiency etc. in cloud environment is suggested in [5]. Research work have been focusing on vehicle assistance on road traffic. Support for the vehicle in case of emergency is not primarily focused by the researchers. An ontology based driver assistance system has been proposed in [6]. Car assistance system for indicating wrong lane selection, safety distance maintenance has been proposed in [7]. Different level and types of automated support for the driver in modern day road traffic scenario is proposed in [8, 9].

3   Proposed Work

3.1   Proposed System

In this work an automated emergency vehicle support system is introduced. Unexpected breakdown of vehicles would be entertained in most optimal way in highway. In our proposed system request from effected car would be generated first and then it would be attained at different levels. We consider three types of entities for solving the issues namely, Car, Service Car and Service Station.

The existing system needs to be automated for the benefit of the user. We propose a 3-phase system for supporting the cars in case of emergency. Initially effected car will go through phase-1 for support, if it does not get the help it will go to next phase.

In phase-1 neighbor car will be communicated. Phase-2 comprises of service car support. Phase-3 needs intervene of service station.

Phase-1

Cars would be connected with each other using bus topology concept so that each car will be connected to each other as shown in Fig. 1.

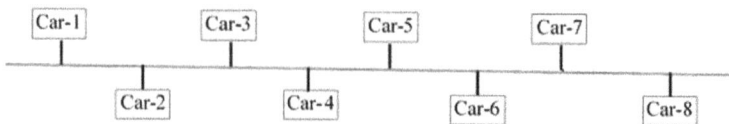

Fig. 1 Proposed phase-1 network

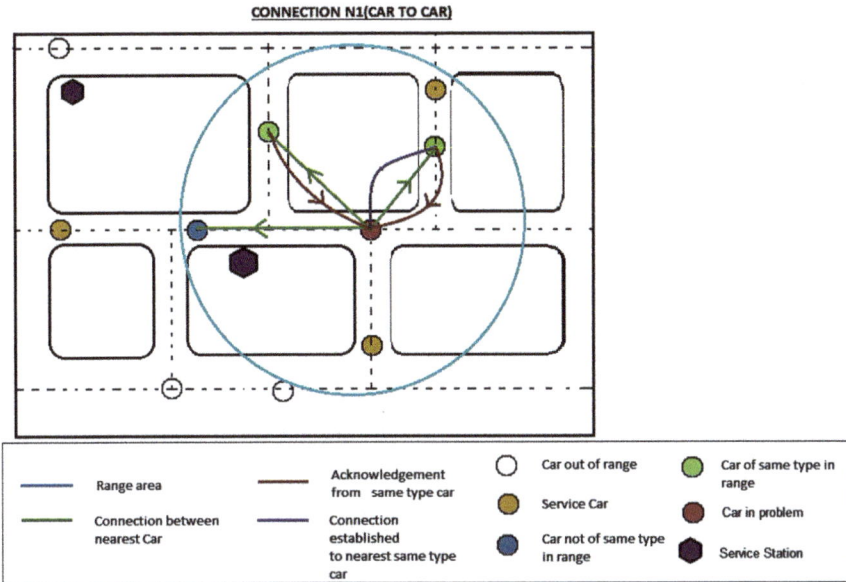

Fig. 2 Custer network among all the cars

Generated network of neighbor cars would be small in size for better communication and support to each other in case of emergency.

The car that needs help will act as a hub or head of the cluster. In other words, the bus network of cars turns into a cluster network with the effected car as hub as shown in Fig. 2. Hub car now triggers distress signals to the nearby cars. Process will be successfully terminated when neighbor car in the cluster is found to be able to provide the assistance to the effected car.

In case of unsuccessful process or in other words no neighbor car is capable enough to solve issue, Phase-2 will be activated. In that case a cluster among the effected car and the service cars would be constructed.

Operational activity is depicted in Algorithm 1.

Algorithm 1: Case_1

Input: Distress signal from affected vehicle
Output: Connection between vehicles.

Step 1. Distress signal is activated to all the
fellow cars in the network.
Step 2. Available supporting vehicles would be
filtered based on the feature extraction and resource
availability.
Step 3. If then problem would not be handled then go
to Case 2. Otherwise go to Step 4.

Fig. 3 Proposed phase-2
network

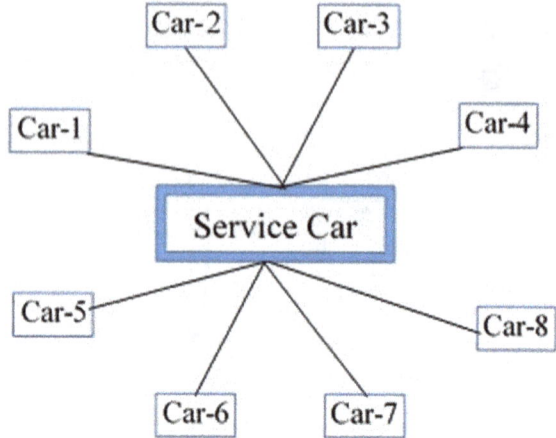

Step 4. Request would be transmitted to the filtered
cars and receive the acknowledgement.
Step 5. Supporting vehicle would be selected based
on the positive acknowledgement and shortest distance
from the affected vehicle.
Step 6. Connection would be established between
service provider and service receiver vehicles.

Phase-2

In Phase-2 service car would be introduced in the network. Each car of the network
would be connected with the service car using star topology concept as shown in
Fig. 3.

The network would be converted into a cluster as shown in Fig. 4. In the cluster
stressed car will act as cluster head and will produce distress signal. Process will be
successfully terminated when service car in the cluster is found to be able to provide
the assistance to the effected car.

In case of unsuccessful process or in other words service car is not able to solve
issue, Phase-3 will be activated. In that case a cluster among the effected car and the
service stations would be constructed.

Operational activity is depicted in Algorithm 2.

Algorithm 2: Case_2

Input: Distress signal from affected vehicle
Output: Connection between service car and vehicle.

Step 1. Distress signal is activated to all the
available service cars in the network.
Step 2. Available service cars would be filtered
based on the resource availability.

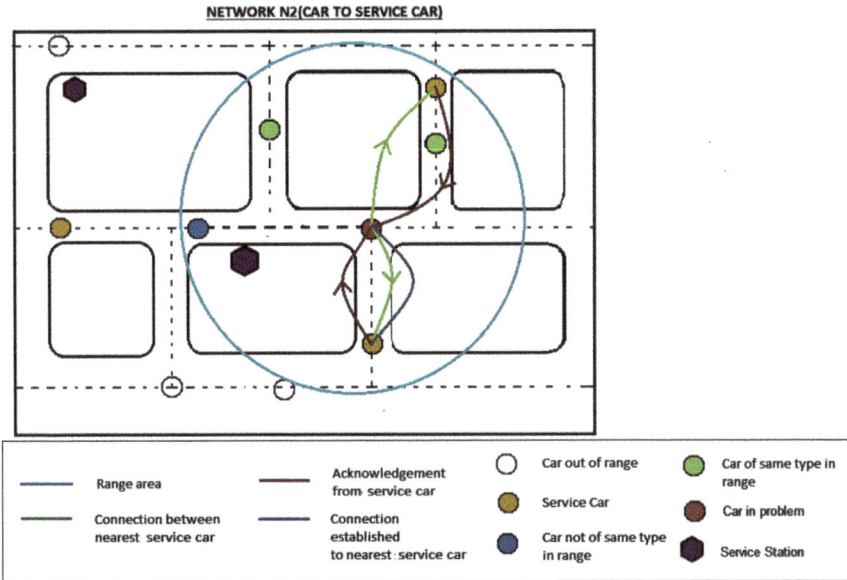

Fig. 4 Cluster network of cars and service car

Step 3. If then problem would not be handled then go to Case 3. Otherwise go to Step 4.
Step 4. Request would be transmitted to the filtered service cars and receive the acknowledgement.
Step 5. Service car would be selected based on the positive acknowledgement and shortest response time.
Step 6. Connection would be established between service car and the vehicles.

Phase-3

In Phase-3 service station would be accessed by the affected car. As the issue remains unsolvable by the neighbor cars or service car, it would be escalated to the service station. A cluster would be constructed with affected car as the cluster head as shown in Fig. 5. Service station would be selected based on the possible quick response time.

Car issues would initially be handled by the passing by cars. Unsolved issues would be directed to service cars. Finally service station would be contacted for the issues that are not resolved at initial two phases. This proposed method reduce the load on work station, minimize the time delay for work station response even for a repair that would easily be handled by passing cars.

Issues are analyzed and filtered at the network level without sending the entire issues into the 'cloud' in terms of service station. Hence the concept of edge computing is implemented here to minimize the cost in terms of human resource and fuel consumption.

NETWORK N3 (CAR TO SERVICE STATION)

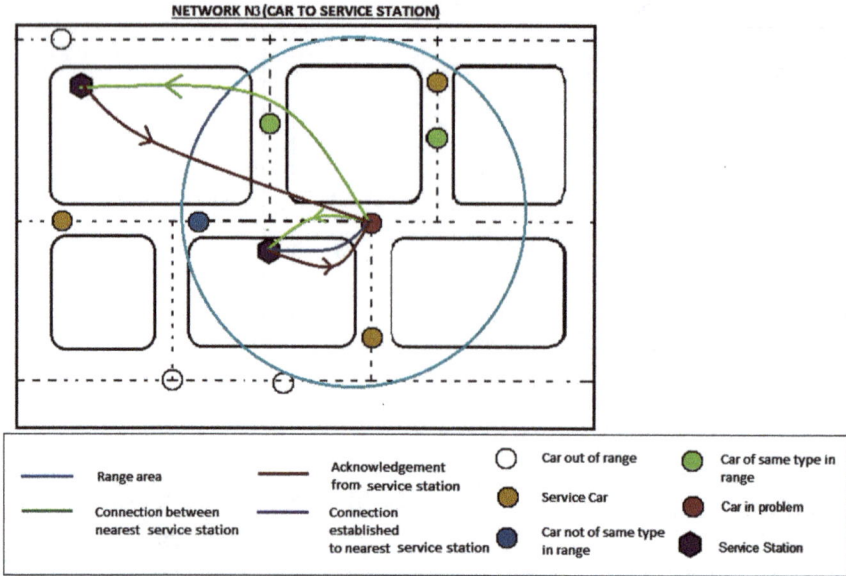

Fig. 5 Cluster network of car and service station

3.2 Analytical Study of Load Handling

Phase-1

Let,

N = no. of cars in trouble in a radius of r kms. at a time instant T.

f = total no. of fellow car present at a time instant T.

$p1$ = percentage of fellow car who can provide assistance to some or all of the N cars in trouble.

Hence, the no. of problems solved by fellow cars = $\left(\frac{p1}{100}\right) \times f$

Phase-2

Let,

s = no. of service cars presents at time instant T.

r = maximum radius allowed for each car in trouble.

$p2$ = percentage of service vehicles who can provide assistance to some or all of the N cars in trouble.

So, no. of problems solved by service cars = $\left(\frac{p2}{100}\right) \times s$

Phase-3

It is obvious that unsolved issues would be handled in this phase.

Hence,

No. of problems solved by service stations =

$$N - \left(\left(\frac{p1}{100} \right) \times f \right) - \left(\left(\frac{p2}{100} \right) \times s \right)$$

In case of typical system entire problem is likely to be handles by the service station.

Hence,

No. of problem solved by the service station $= N$

It is observed that the no. of car issues served by the service station by implementing our proposed method would be less as,

$$\left\{ N - \cdot \left(N - \left(\left(\frac{p1}{100} \right) \times f \right) - \left(\left(\frac{p2}{100} \right) \times s \right) \right) \right\}$$

$$= \left(\left(\frac{p1}{100} \right) \times f \right) + \left(\left(\frac{p2}{100} \right) \times s \right)$$

3.3 Analytical Study of Fuel Consumption

Fuel consumption is directly proportional to the distance travelled.

Phase-1

Let,

$N =$ no. of cars in trouble at time T.

$f =$ no. of fellow cars presents at time instant T.

$r =$ maximum radius allowed for each car in trouble.

$p1 =$ percentage of fellow vehicles who can provide assistance to some or all of the N cars in trouble.

Now, we note that the fellow cars which will provide assistance will actually travel no extra distance because fellow cars accept requests only if they find that the assistance has to be provided in to a vehicle which is on their desired route they are travelling in.

Hence, distance travelled by fellow cars to provide assistance $= d_1 = \left(\frac{f \times p_1}{100} \right) \times 0 = 0$

Phase-2

Let,

$s =$ no. of service cars presents at time instant T.

$r =$ maximum radius allowed for each car in trouble.

$p2 =$ percentage of service vehicles who can provide assistance to some or all of the N cars in trouble.

davg $=$ average up-down distance from service car to car in trouble. It is obvious that davg<=r.

So, distance travelled by service cars to provide assistance $= d_2 = \left(\frac{s \times p_2}{100}\right) \times d_{\mathrm{avg}}$

Phase-3

Unsolved issues would be directed to the service station. They are either towed or reached there by its' own. Hence,

Davg = average up-down distance from service station to car in trouble.

So, distance travelled to reach service station $= d_3 = \left(N - \left(\frac{f \times p_1}{100}\right) - \left(\frac{s \times p_2}{100}\right)\right) \times D_{\mathrm{avg}}$

Hence, Total distance travelled would be,

$$\mathrm{dedge} = d1 + d2 + d3 = \left(\frac{s \times p_2}{100}\right) \times d_{\mathrm{avg}} + \left(N - \left(\frac{f \times p_1}{100}\right) - \left(\frac{s \times p_2}{100}\right)\right) \times D_{\mathrm{avg}}$$

In typical cases total distance travelled by affected cars would be dnon-edge $= N \times D_{\mathrm{avg}}$

where,

N = no. of cars in trouble at time T.

Davg = average up-down distance from service station to car in trouble.

As all the affected cars would likely to be directed to service stations.

Fuel consumption would be significantly improved by implementing our proposed method as (Fuel Consumption $\propto$ Distance Covered).

$$\mathrm{dimproved} = \mathrm{dnon} - \mathrm{edge} - \mathrm{dedge} = \left(N \times D_{\mathrm{avg}}\right)$$

$$- \left(\left(\frac{s \times p_2}{100}\right) \times d_{\mathrm{avg}} + \left(N - \left(\frac{f \times p_1}{100}\right) - \left(\frac{s \times p_2}{100}\right)\right) \times D_{\mathrm{avg}}\right)$$

$$= \left(\left(\frac{f \times p_1}{100}\right) + \left(\frac{s \times p_2}{100}\right)\right) \times D_{\mathrm{avg}} - \left(\frac{s \times p_2}{100}\right) \times d_{\mathrm{avg}}$$

It is found that implementation of Edge Computing in car assistance in a remote highway would be advantageous in many aspect. Amount of issues directed to the service station would be significantly less that yields man power savings. Fuel consumption would also be significantly less as a substantial amount of issues would be resolved by the fellow vehicles or service cars.

4 Experimental Result

In this section, our proposed edge computing based assistance system is implemented on a dataset [10]. The result is compared with conventional assistance method where issues are directed and processed at corresponding service centre in terms of time and distance travelled. Our proposed method is implemented using C language on a Notebook PC with CPU 2 GHZ.

Different types of issues are considered in a remote highway. More than 600 cases are considered and are segregated based on our proposed method. Case wise segregation is shown in Fig. 6.

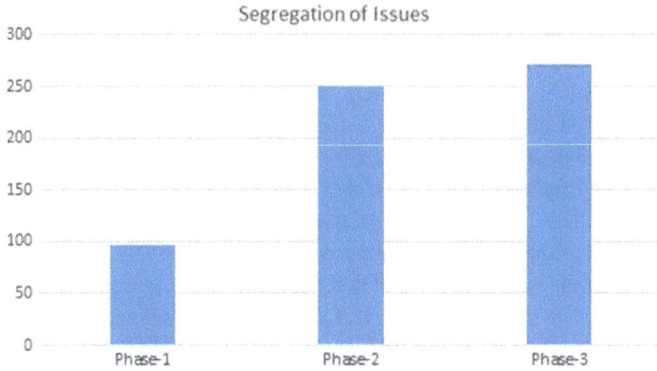

Fig. 6 Case wise segregation

It is found more than 50% issues could be resolved without intervention of service centre in terms of resource sharing. It could ensure cost effectiveness in terms of fuel consumption and human resource involvement. Estimated issue wise expenditure by implementing our proposed approach is shown in Fig. 7 based on the dataset [10].

It is obvious that in typical system issues would be forwarded to the service centre and remedial measure would be taken by the centre that results higher expenses in terms of human resource involvement and distance traversed. In our proposed method issues are analysed and filtered and then sent to the service station. Thus concept of edge computing is used here as issues are analysed at network edge.

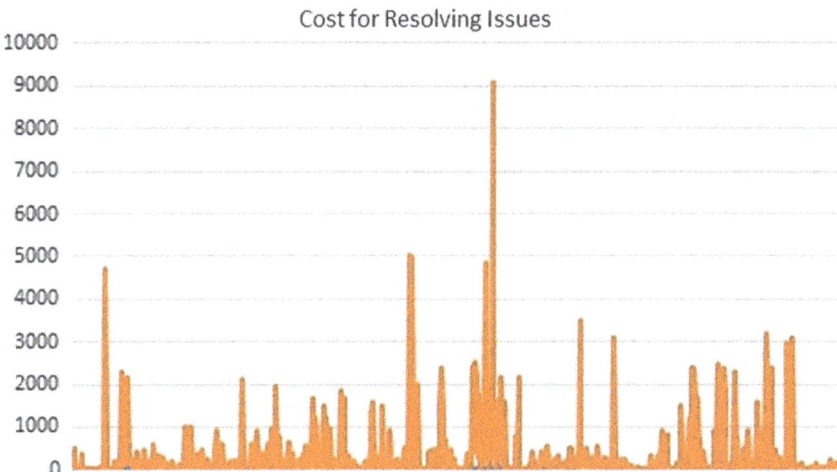

Fig. 7 Estimated cost for resolving issues

5   Conclusion and Future Scope

An automated vehicle assistance system is proposed in this paper to reduce the expenditure in terms of human resource and fuel consumption. Here vehicle issues are initially communicated and taken care by the passing by vehicles. In the next phase issues would be redirected to the service cars directly. In both cases it is presumed that supporting vehicles need to travel minimum distance. Hence fuel consumption would be minimum. Rest pf the unsolved issues would be handles by the service station. It is observed based on the real time data set that more than 50% issues would be resolved by the fellow vehicles or service cars. It is definitely advantageous over the typical system that need the entire issues at service station and then analyze and decide centrally. Hence implementation of edge computing concept would be beneficial. Edge computing makes the entire process faster by filtering the entire process into multiple levels and reduces unnecessary pressure on the upper layers. Requirement and remedial measure selection procedure would have been made automated in future by implementing similarity index based cluster. Selection of available service from affected vehicle needs to be agent based in future to minimize human delay and to provide more optimality in terms of resource selection.

References

1. Eric H (2018) What is edge computing: the network edge explained. cloudwards.net
2. Tran TX, Hosseini M-P, Pompili D (2017) Mobile edge computing: recent efforts and five key research directions. IEEE COMSOC Commun-Front 12(4):29–33
3. Sun X, Ansari N (2016) EdgeIoT: mobile edge computing for the internet of things. IEEE Commun Magazine 54(12):22–29
4. Hu W, Gao Y, Ha K, Wang J, Amos B, Chen Z, Pillai P, Satyanarayanan M (2016) Quantifying the impact of edge computing on mobile applications. APSys'16, Hong Kong
5. Ai Y, P M, Zhang K (2018) Edge computing technologies for Internet of Things: a primer. Digital Commu Netw Sci Direct 4(2):77–86
6. Kannan S, Thangavelu A, Kalivaradhan RB (2010) An intelligent driver assistance system (I-DAS) for vehicle safety modelling using ontology approach. Int J UbiComp 1(3):15–29
7. Zhou Y, Wang G, Xu G, Fu G (2014) Safety driving assistance system design in intelligent vehicle. In: IEEE International conference on robotics and biomimetrics (ROBIO 2014), Bali, Indonesia
8. Kukkala VK, Tunnell J, Pasricha S, Bradley TH (2018) Advanced driver-assistance system: a path toward autonomous vehicles. IEEE Consum Electron Magaz 7(5):18-25
9. Park K, Kwahk J, Han SH, Song M, Choi DG, Jang H, Kim D, Won YD, Jeong IS (2019) Modelling the intrusive feelings of advanced driver assistance systems based on vehicle activity log data: case study for the lane keeping assistance system. Int J Automot Technol 20:455–463
10. https://catalog.data.gov/dataset/vehicle-repairs

Chapter 55
Majority Voting Machine Learning Approach for Fault Diagnosis of Mechanical Components

Priyanka S. Patil, Mahadev S. Patil, S. G. Tamhankar, Sangram Patil, and Faruk Kazi

1 Introduction

Fault diagnosis of mechanical equipment is a crucial task for maintenance personal. Therefore, in recent years Internet of Things (IoT) has changed the scenario of industries to automate processes in every sector of our economy and daily life because of its improved monitoring and remote operating and controlling facility. Data collected from equipment by using a data acquisition system and given to edge devices for edge analytics. Further, data pushed and stored at cloud environment via IoT gateway, and data would use for analytics purpose [1]. Several researchers demonstrated the use of machine learning algorithms in predictive maintenance for fault diagnosis and prognosis. In industrial equipment, mainly rotating machinery fault diagnosis is carried out with vibration analysis techniques because vibration signals contain a massive amount of information about the machine condition [2].

Machine learning-based fault diagnosis has carried out by several researchers and conclude about its benefits in automated condition monitoring. Kankar et al. [3] have used Support Vector Machine (SVM) and Artificial Neural Network (ANN) for

P. S. Patil (✉) · M. S. Patil
Rajarambapu Institute of Technology, Sangli, Maharashtra, India
e-mail: priyapatil22055@gmail.com

M. S. Patil
e-mail: mahadev.patil@ritindia.edu

S. G. Tamhankar
Walchand Collage of Engineering, Sangli, Maharashtra, India
e-mail: sunil.tam@gmail.com

S. Patil · F. Kazi
Veermata Jijabai Technological Institute, Mumbai, Maharashtra, India
e-mail: sspatil_p14@me.vjti.ac.in

F. Kazi
e-mail: fskazi@el.vjti.ac.in

© The Author(s), under exclusive license to Springer Nature Singapore Pte Ltd. 2021   713
X.-Z. Gao et al. (eds.), *Applications of Artificial Intelligence in Engineering*, Algorithms for Intelligent Systems, https://doi.org/10.1007/978-981-33-4604-8_55

fault diagnosis of rolling element bearing with its raw time series data and observed that SVM is more accurate than ANN for classification. Further, Kankar has used wavelet coefficient features for training the machine learning approaches such as SVM, ANN and SOM [4]. Murlidharan and Suguraman [5] has proposed tree-based classifier, i.e. Decision Tree (DT) classifier for fault diagnosis of the centrifugal pump. Results show that classification potential of DT to classify different faults like cavitation, impeller fault, bearing faults etc. k-nearest neighbour (k-NN) based method is presented by Dong wang [6]. Fault diagnosis of gear in a mechanical transmission system using the k-NN based method. The author also highlights the impact of factor 'k' of k-NN in analysis with statistical features from temporal vibration signals. Lei and Zau [7] used k-NN classifier for the gear crack level identification. Vibration-based feature extraction and feature selection for avoiding redundant data with k-NN give satisfactory result in terms of gear crack level identification. Further, k-NN for life grate recognition of rotating machinery is implemented by Li et al. [8]. Recently Gohari and Eydi [9] utilize k-NN, and Decision Tree (DT) for modelling of shaft unbalance and compare its results with each other. k-NN outer forms DT for the same task. Duan et al. [10] used Naïve Bayes (NB) and SVM for infrared image classification for rotating machinery fault. Zoungrana et al. [11] demonstrated an automatic fault classification method using SVM, DT, k-NN and NB classifiers. The main work of this design is to analyze the vibration signals of a rotating machine depending on historical data of the sensors. Some researcher used ensemble machine learning technique for diagnosis, as well as prognosis of industrial equipment and components ensemble. Patil and Phalle [12] described the use of AdaBoost classifier for bearing fault classification and investigated the effect of estimators and ranked features are described. Further, they demonstrate three ensemble technique, i.e. Extra tree, Gradient boosting and Random forest for fault diagnosis rotating machinery. These three classifiers are tree-based ensemble classifiers and reported their potentials in antifriction bearing fault classification [13]. Recently, Majority Voting type of ensemble techniques used for fault diagnosis of some systems such as Refrigeration [14], Transformers [15] and preserved its capability to identify the faults in systems.

Above literature explained the use of different machine learning algorithms and the importance of ensemble classifiers for better performance and majority voting classification is part of the Ensemble technique in which several low-performance base classifiers clubbed together for classification. Therefore, in this paper, the majority voting algorithm is used for fault diagnosis of mechanical components such as gear and bearing. Five base classifiers such as SVM, k-NN, DT, NB and MLP are used for Majority voting technique. Statistical Features are calculated from time-domain vibration signals and used as training and testing data for machine learning algorithms.

Table 1 Base classifiers and its hyper-parameters

Sr. No.	Classifier	Hyperparameters
1.	Support vector machine	C, γ and activation function
2.	K-Nearest neighbors	K
3.	Naive byes	alpha
4.	Multilayer perceptron	Number hdden of layers and activation functions
5.	Decision tree	Number of splits, criteria and depth

2 Majority Voting Machine Learning Technique

Majority Voting Classifiers are the ensemble techniques wherein the predictions of various models are aggregated to give a single prediction. These techniques are generally used to create custom ensemble models. In Majority Voting Classifiers, the models predict the probabilities of each class, after which the outputs of each model are averaged to provide the final probability of each class. The class with the highest probability is given out as a definitive prediction [14].

In this paper, Majority Voting techniques is made-up with five base classifiers, i.e. Support Vector Machine, K-Nearest Neighbor, Naive Bayes, Multilayer Perceptron and Decision Tree. Output results have been carried out with individual base classifiers with hyperparameter tuning by grid search algorithm [16]. Further, tuned hyperparameters are used in the majority voting base classifiers. Table 1 shows hyperparameters tuned in base classifiers.

3 Methodology

Methodology for fault diagnosis of mechanical components, i.e. gear and bearing with Majority Voting technique is as shown in Fig. 1. For experimentation in machine learning techniques, data is an important factor. For this purpose, in this study, online available data set of gear and bearing are used, and details are as follows.

3.1 Gear Data

For the gearbox fault diagnosis, online available data has been used in this study [17]. This dataset consists of 2 conditions of the gearbox, i.e. healthy gearbox and broken tooth fault. There are four vibration sensors placed in 4 different directions on the gearbox. These sensors are used to capture the data from gearbox for both the conditions with varying load from 0 to 90% (Total ten txt files) with 30 Hz frequency. For the analysis, these ten txt files are divided into four parts each. So rearranged the

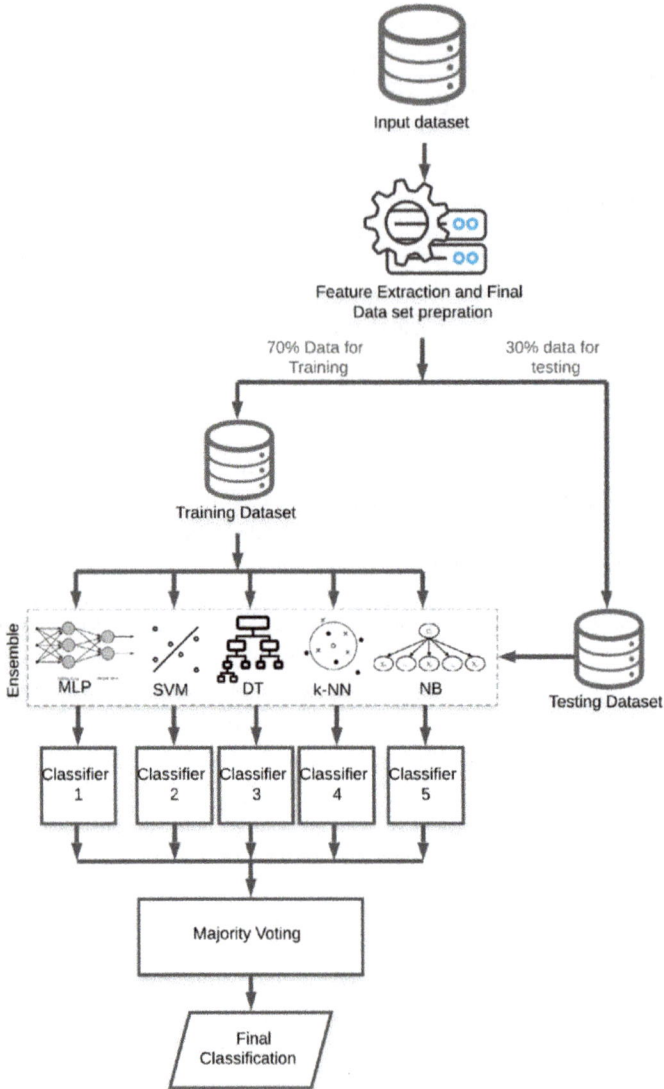

Fig. 1 Fault diagnosis methodology: majority voting technique

dataset includes 40 samples of healthy and 40 samples of broken teeth. Statistically features (see Table 2) are extracted from 80 samples for further analysis.

Table 2 Statistical features

Sr. No	Features
1	RootMeanSquare $(a_{rms}) = \sqrt{\frac{\sum_{n=1}^{N}(a(n))^2}{N-1}}$
2	Kurtosis $(K_a) = \frac{\sum_{n=1}^{N}(a(n)-\mu_a)^4}{(N-1)\sigma_a^4}$
3	PeaktoPeak $(a_{p-p}) = \max(a(n)) - \min(a(n))$
4	ShapeFactor $(SF_a) = \frac{a_{rms}}{\frac{1}{N}\sum_{n=1}^{N}|a(n)|}$

3.2   Case Western Reserve University (CWRU) Bearing Data Centre

For the analysis of bearing faults, the dataset was downloaded from CWRU bearing data centre respiratory [18]. The experimental test facilities consist of 2HP drive, three-phase induction motor, torque transducer and a dynamometer as shown in fig A. Experimentation for collecting signals are done on drive end ball bearing with 12 kHz sampling frequency. There are four bearing condition, i.e. Healthy, Inner race fault, outer race fault and ball fault. In this paper, the analysis smallest fault size, i.e. 0.007" for inner race and outer race fault and 0.028" for rolling element fault is used. Four signals of each are available and those converted into 40 samples each by splitting the data and used for further analysis.

As per the above description, vibration data are input data to the algorithm. Four mostly used statistical features such as kurtosis, RMS, peak to peak and shape factor are extracted from the signals (see Table 2) and used that feature set as an input vector to machine learning technique.

For the preparation of training and testing datasets, the ratio of 70–30% of the total feature set is used. Both the feature sets are normalized for further experimentation. For tuning the hyperparameters of the individual base classifier, Grid search algorithm was deployed. Testing results of individual classifiers are calculated with similar hyperparameters. A Majority Voting Classifier is created from the tuned base classifiers and trained on the training data. This basically forms an ensemble of all base classifiers and its testing results are achieved from the testing data. Results of all the classifiers are discussed in the result and discussion section.

4   Result and Discussion

Experimentation for performance analysis of individual base classifier and Majority Voting ensemble technique is carried out with the same input dataset, i.e. gear and bearing. Gear dataset classifies into two class and bearing dataset classifies into four class. The results of six classifiers applied on the gearbox and bearing dataset is tabulated in Table 3. Also, the ROC curves of these six classifiers are shown in Figs. 2 and 3 for Gearbox data and Bearing data, respectively. In plots of ROC curves were in, True Positive Rate is plotted against False Positive Rate, the area under the

Table 3 Summary of classifier performance

Case	ML techniques	Training accuracy	Testing accuracy	Recall	Precision	F1 Score	Mean absolute error	R2
Gearbox	SVM	0.88	0.8994	0.9	0.9	0.9	0.1005	0.60
	DT	0.79	0.8125	0.81	0.82	0.81	0.1875	0.25
	MLP	0.82	0.8208	0.87	0.87	0.87	0.1284	0.48
	KNN	0.78	0.75	0.75	0.77	0.75	0.25	0.22
	NB	0.74	0.7187	0.72	0.77	0.7	0.2812	0.12
	Majority Voting	**0.95**	**0.9375**	**0.94**	**0.94**	**0.94**	**0.073**	**0.82**
Bearing	SVM	0.86	0.8912	0.89	0.89	0.89	0.1991	0.65
	DT	0.86	0.8657	0.87	0.87	0.87	0.2245	0.62
	MLP	0.83	0.8379	0.84	0.84	0.84	0.26851	0.55
	KNN	0.72	0.7361	0.74	0.74	0.73	0.449	0.28
	NB	0.71	0.7291	0.73	0.73	0.73	0.4444	0.31
	Majority voting	**0.94**	**0.9583**	**0.96**	**0.96**	**0.96**	**0.081**	**0.85**

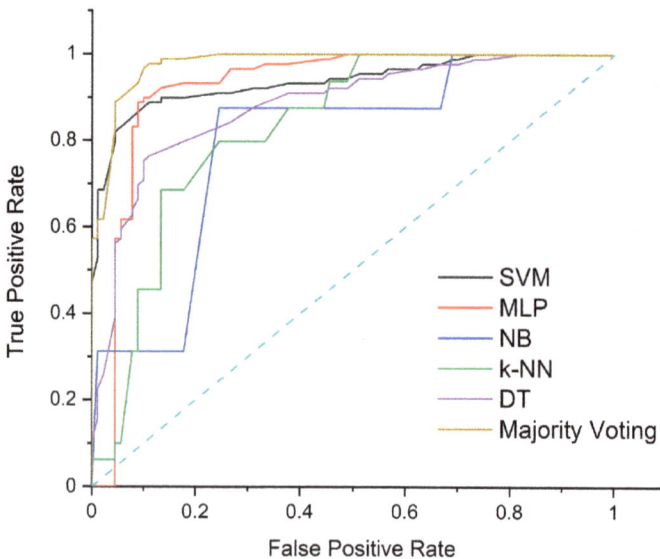

Fig. 2 ROC curve with gear data

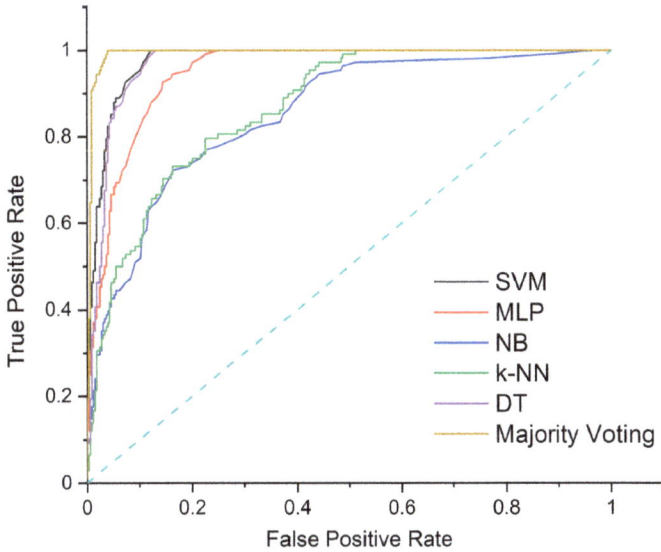

Fig. 3 ROC curve with bearing data

curve corresponds to the accuracy of the classifier. Here, in both RoC curve figures (see Figs. 2 and 3) the curves of Majority Voting Classifier have a higher area under the curve in both the cases. In contrast, Navies Bayes Classifier has the lowest area under the curve in both cases.

Finally, Majority Voting Classifier, which consists of all other classifiers, having a testing accuracy of 95.83% for Gearbox data and 93.75% for Bearings data, outperforms the individual corresponding accuracy of other algorithms. The best confusion matrix with Majority voting classifier is presented in Tables 4 and 5 for gear

Table 4 Confusion matrix for gearbox dataset with 95.83% accuracy

		Predicted class	
		Healthy (%)	Broken teeth (%)
Actual class	Healthy	*91.66*	**8.33**
	Broken teeth	**0**	*100*

Table 5 Confusion matrix for bearing dataset with 93.75% accuracy

		Predicted class			
		Healthy (%)	Outer race (%)	Inner race (%)	Ball (%)
Actual class	Healthy	**100**	*0*	*0*	*0*
	Outer race	*0*	**100**	*0*	*0*
	Inner race	*0*	*8.33*	**91.66**	*0*
	Ball	*8.33*	*0*	*8.33*	**83.33**

and bearing data set respectively. Not only the AUC score, but even other accuracy metrics of Majority Voting Classifier are far better than those achieved individually by different classifiers.

5 Conclusion

Ensemble machine learning-based Fault diagnosis of mechanical components such as gears and bearings are discussed in this paper. For the experimentation purpose, available online data is used, and results are obtained for individual base classifiers, i.e. SVM, MLP, DT. k-NN and NB. Hyperparameters of these classifiers are tuned using Grid search technique. Those tuned and trained base classifiers are used in Majority Voting Ensemble technique. Results are obtained for various factors such as classification accuracies, mean absolute error, RoC curves etc. As this small input data, SVM gives best results amongst individual base classifiers, but Majority voting techniques outperforms all individual base classifiers and gives the best result. Best classification accuracy is obtained with the Majority voting for both the cases, i.e. 95.83% and 93.75% for gear and bearing respectively. Thus the aggregation of all the five classifiers applied is proven to be more effective than individual classifiers. This study shows that the possible implementation of the Majority voting classifier and its potential in automatic fault detection systems of mechanical equipment.

References

1. Wang K (2016) Intelligent predictive maintenance (IPdM) system–industry 4.0 scenario. WIT Trans Eng Sci 113:259–268
2. Liu R, Yang B, Zio E, Chen X (2018) Artificial intelligence for fault diagnosis of rotating machinery: a review. Mech Syst Signal Process 108:33–47
3. Kankar PK, Sharma SC, Harsha SP (2011) Fault diagnosis of ball bearings using machine learning methods. Expert Syst With Appl 38(3):1876–1886
4. Kankar PK, Sharma SC, Harsha SP (2011) Rolling element bearing fault diagnosis using wavelet transform. Neurocomputing 74(10):1638–1645
5. Muralidharan V, Sugumaran V (2013) Feature extraction using wavelets and classification through decision tree algorithm for fault diagnosis of mono-block centrifugal pump. Measurement 46(1):353–359
6. Wang D (2016) K-nearest neighbors based methods for identification of different gear crack levels under different motor speeds and loads: revisited. Mech Syst Signal Process 70:201–208
7. Lei Y, Zuo MJ (2009) Gear crack level identification based on weighted K nearest neighbor classification algorithm. Mech Syst Signal Process 23(5):1535–1547
8. Li F, Wang J, Tang B, Tian D (2014) Life grade recognition method based on supervised uncorrelated orthogonal locality preserving projection and K-nearest neighbor classifier. Neurocomputing 138:271–282
9. Gohari M, Eydi AM (2020) Modelling of shaft unbalance: modelling a multi discs rotor using K-nearest neighbor and decision tree algorithms. Measurement 151:107253
10. Duan L, Yao M, Wang J, Bai T, Zhang L (2016) Segmented infrared image analysis for rotating machinery fault diagnosis. Infrared Phys Technol 77:267–276

11. Zoungrana WB, Chehri A, Zimmermann A (2020) Automatic classification of rotating machinery defects using machine learning (ML) algorithms. In: Human centred intelligent systems, Springer, Singapore, pp 193–203
12. Patil SS, Phalle VM (2019) Fault detection of anti-friction bearing using adaboost decision tree. In: Computational intelligence: theories. Applications and future directions. vol 1, Springer, Singapore, pp 565–575
13. Patil S, Phalle V (2018) Fault detection of anti-friction bearing using ensemble machine learning methods. Int J Eng 31(11):1972–1981
14. Zhang Z, Han H, Cui X, Fan Y (2020) Novel application of multi-model ensemble learning for fault diagnosis in refrigeration systems. Appl Therm Eng 164:114516
15. Zhang L, Zhai J (2019) Fault diagnosis for oil-filled transformers using voting based extreme learning machine. Cluster Comput 22(4):8363–8370
16. Zhang X, Zhou J (2013) Multi-fault diagnosis for rolling element bearings based on ensemble empirical mode decomposition and optimized support vector machines. Mech Syst Signal Process 41(1–2):127–140
17. Data.World (2020). https://data.world/gearbox/gear-box-fault-diagnosis-data-set. Last Accessed June 2020
18. Loparo KA (2020) Bearing vibration dataset, Case Western Reserve University. Available at: https://csegroups.case.edu/bearingdatacenter/home. Last accessed May 2020

Chapter 56
Simulation of Colour Image Processing Techniques on Verilog

A. Shaikh Abuzar and S. Patil Mahadev

1 Introduction

Visual applications have become widely known in many fields, including image processing, personalization or object. Image processing is the process of performing certain functions in a duplicate, toward get an improved image or to abstract approximately useful information from it. It is a type of signal correction where the contribution is an image and the output can be an image or features associated with that image. Image processing among the wildest rising technologies. Build a world-class research laboratory. Image processing contains the subsequent three steps:

- Introducing photos with image gaining tools.
- Analyze then analyze the image.
- An effect where the result can be changed to an image or report constructed on an image evaluation.

Two types of approaches used for image processing are, digital and analogue image processing. Analogue image operation is used as firm copies such as printouts and pictures. Graphic forecasters use the basics translation although using these visualization techniques. Digital image processing techniques help in digital photography using computers. The three general categories of all types of data to be used while using the digital method are pre-processing, development, and data visualization. Common steps in image processing are image scanning, storing, editing, and translation. Digital image processing is about building a digital system that works on a digital image. Digital cameras usually include special digital printable. It is used

A. Shaikh Abuzar (✉) · S. Patil Mahadev
Department of Electronics and Telecommunication, Rajarmbapu Institute of Technology,
Islampur 414405, India
e-mail: abushaikh7972@gmail.com

S. Patil Mahadev
e-mail: mahadev.patil@ritindia.edu

© The Author(s), under exclusive license to Springer Nature Singapore Pte Ltd. 2021   723
X.-Z. Gao et al. (eds.), *Applications of Artificial Intelligence in Engineering*, Algorithms for Intelligent Systems, https://doi.org/10.1007/978-981-33-4604-8_56

to operate the image by the custom if algorithm. Digital image dispensation delivers a stage to accomplish various operation like image ornamental, processing of digital and analog signals, image indications, voice indications.

2 Literature Survey

Chiuchisan et al. potorac has explained the disadvantages of phase-even phase processing, it can dramatically reduce the number of multiple complex multipliers and complex processors used. Thirumarai Selvi and Amudha have explained the procedure used to help people who are facing the challenge of crossing a road near a traffic signal or pedestrian with the help of automatic video monitoring. Route analysis: According to the Indian road organization it is determined that a minimum of 1.8 m (width) × 2.2 m (height) is reserved for the travel area. The 1800 mm width is reserved for wheelchair movement for pedestrians. Transportation of stable materials requires a high degree of security. Ravikumar and Arulmozhi have defined image is one of the most visible sources in image printing applications. Image processing will dramatically change people's computer interactions in the future. A large number of photo applications, tools and techniques help to extract complex image features. While today the processing of images is about more than just diversity and see what's actually in the picture. Radhika Kharade et al. have describe the lump is made by uncontrolled multiplication of cell division. Several strategies were developed to find and classify the neoplasm to exploit multiple classification formulas such as a fluid algorithm, a k-group, methods, Fuzzy c-means Fuzzy c. This is an effective formula wherever the part of the tumor is distributed and its options as a center of mass, circumference and area are calculated from the differentiated tumor [1–4].

Karthikeyan and Vanitha have explained text recognition is a method that recognizes text from a paper document in the format you want. Process the text recognition involves a number of ladders, including pre-processing, segmentation, feature extraction, categorization, and previous editing. The split section is used to split the image provided online and to use each part of the split line. Feature extraction is to incorporate skills into image types. This document describes strategies for converting paper text content into machine-readable format. This paper analyzes and compares the technological challenges, methods, and practice of finding text and the lessons adopted in color photographs. It summarizes the basic issues and lists the things that should be considered when dealing with them. Dixit has explained the image indicate edges of the borders. Reduce the filters and amount of data out of idle information. Maintains image structure possessions. Discovery of edge is useful to subdivision, its registering and identification. Therefore, algorithms are defined. The original state-of-the-art imagery was tested using the VLSI architecture. The disease-affected and standard picture population are associated and the proportion of the sickness has attempted. Deyy et al. have explained the fundamental of Euler number have a feature of the model's environment. A computational algorithm for calculating Euler's binary image future grounded on the divide-and-conquer paradigm, which is most applicable to the traditional techniques used in image printing tools. Yun Yang has described in this paper, a new RAM/ ROM renaming system for three-dimensional (3D) VLSI memory design. In order to enable the flexibility of dynamic image data, direct input/ input control is essential for high quality image printing software. A high-speed VLSI system also requires control of the operation of the pipeline data. New builds of RAM/ ROM integration codes are available for specific editing via RAM, ROM, pin and connection. Flip-Flop control, temporary buffer installation and sensitive signal] improved the overall system speed. The memory system is also proposed to enable faster acceleration and chip system technology. 3D VLSI image editing can also be enhanced by the maintenance of appropriate data and pipeline control flow in the RAM/ ROM rename system [5–8].

Yao et al. have described the design and implementation of a new automated white and color converter in the FPGA. This white balance method can effectively detect global gray faults and more gray-retina alternatives by calculating Cr and Cb deviation values. This color-changing approach can significantly reduce effective artefacts by interpreting edges by comparing the differences of neighboring pixels with predefined edges. Chiuchisan et al. have described simulation produces logic results, the natural step consist of extending the use of hardware simulators in the filed of signal processing. This paper presents a new modeling model followed by immediate implementation. The current HDL method is used for image processing and as a result degradation of the basic and principles, as well as the algorithms widely used for image development are described. The paper focuses on image development in the local domain, with specific reference to measurement technique. Hirak Kumar Maitya and Santi P. Maityb have explained that in Reversible Contrast Mapping (RCM) and many of its widely modified models used to convert water-marking (RW) to embed confidential information into digital content. The RC-based version includes a simplified version made of pixels and their composite fragments (LSB) are used to embed information. It is completely unresponsive even if the modified LSB pixels are lost during the data embedding. RCM offers a high degree of embedding with low viewing rates (crazy embedding). In addition, the low cost of integrating and simplifying Hardware monitoring makes it a real focus. This paper proposes a VLSI-based gateway (VLSI) architecture based on the RCM-RW digital portal gateway that can be used for real world coverage. Dinu Coltuc and Jean-Marc Chassery have explained that RCM is a lightweight version that works in pairs with pixels. At two pixels, RCM cannot be underestimated, or large fragments (LSBs) of changed pixels are lost. The data space occupied by LSBs is suitable for data encryption. The minimum admission rates for the proposed domain renewal scheme are very close to the highest rates reported to date [9–12].

Nithiya and Imtiaz have explained the concept of image capture system is designed to work in the FPGA (Field Programmable Gate Array). FPGAs have speed advantages and DSP regeneration, which are required directly for video applications. The research work done on image resolution techniques using the FPGA is limited. The purpose of this study is to produce and apply color printing techniques such as pseudo color analysis, youth filtering, medium filtering and edge detection that are used to detect throat burnout and dementia. Boscaro et al. have explained Internal electrical measurements are an important step in troubleshooting and troubleshooting. Due to the many metal layers, the active zones of the chip are only accessible from the die back. The strength of nonlinear communication methods using the silicon substrate and the few sample preparations required have greatly contributed to maximizing it as an unstoppable tool for distorting the fault zone. The problem of time or irregular use can be identified by static and dynamic analysis of the emission photon and Electro Optical Selecting. In paper 14 Huang et al. have explained large accounting resources are often required for SA operations. In this paper, the corresponding construction of the SA imaging algorithm can be identified in the Field Programmable Gate Array (FPGA), for light gain and minimum construction. The proposed design was validated by MATLAB and was employed to create an ultra-high-resolution icon from the raw data of the 128 array transducer data. In paper 16 Iuliana and geman have describe a real-time digital image processing program using Verilog description language that can be tracked to achieve faster hardware performance. The image enhancement algorithm includes in the system described are used in the ultrasound image. The paper focused on the method for image extension such as contrast and light scaling, invert image and pseudo colors, which are described and synthesized using a word-word definition [13–16].

Poorani and Vasantha Kalyani David has explained A soft computer is a combination of methods designed to modulate and enable a solution to real-world problems, which cannot be measured or are too difficult to simulate. A soft computer is an integral mode of interconnecting systems and provides a unique, flexible programming experience for managing real-time situations. A soft computer is a field that contains something that comes with the right mind, neural computer and natural integration. Soft computer software has found application in many areas. Another very important application is image processing. Nandhagopal et al. have explained medical image processing is a research challenge because photographic images are noisy and a bad contrast. The efficiency of the processing of medical images depends on the quality oh the medical images. The major features of small variations in medical images are the age of the requirements of the scan, the light conditions and the inexperience of medical personnel. Comparative algorithms for adding an algorithm are a powerful tool for developing a comparison graphic hidden by a very small range of gray or color levels. There are various ways to enhance image contrast. Pooja et al. have explained classification and labeling are still the weakest step in many medical imaging applications. This paper presents a water-based approach designed to solve the common problem encountered in various applications. Also, that can be changed by changing their parameters. Two of these modules are presented with lung cancer detection, a method for the classification of cancer incidents from CT images, a full-blown brain imaging algorithm for the detection of neurological MRI images. Mahavir Sigh and Gitanjali Pandove have defined whenever a single image is transformed into another form, the occurrence occurs in original image quality. Therefore, to improve the image's visibility, the image stabilization process is used optimally. These capabilities are distinguished by stretching, light control, cluster performance, bandwidth etc. Image development can be done with the help of software launch and hardware implementation [17–20].

Fig. 1 Brightness operation

3   Objectives

Image processing techniques in accumulation to improvement techniques, is practical to improve the data practicality. The image dispensation methods are as follow:

- Brightness operation.
- Threshold operation.
- Invert operation.

3.1   Brightness Operation

Figure 1 shows the image flexibility. Brightness and Comparison Adjustment. The image should have a clear and unique brightness for easy viewing. Brightness means high brightness or darkness of an image. The difference is the difference of light between objects or regions. Brightness is an unbalanced term, so light might be defined as quantity of energy productivity a source. In some cases, we can easily say that the picture is cheerful in other cases, it is not informal to understand. Image brightness is defined by high gray matter values. For example, if the image has gray values like 175, 200, 255 etc. it can be said that the image brightness is higher. That is, the higher the gray scale the higher the brightness of the image. Therefore, image brightness with higher values of gray levels.

3.2   Threshold Operation

Image threshing is a simple, yet effective way to separate an image into a frame and a layer. Method of image analysis is a type of image classification which separates objects by changing grayscale images into the binary images.

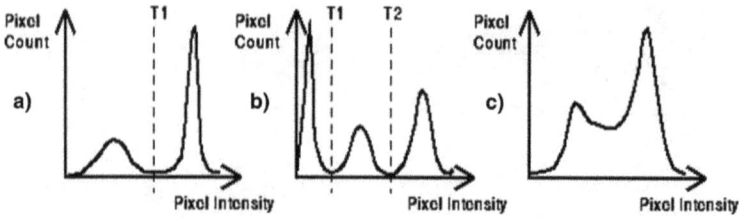

Fig. 2 Histogram equitization in threshold operation

Fig. 3 Thresholding process

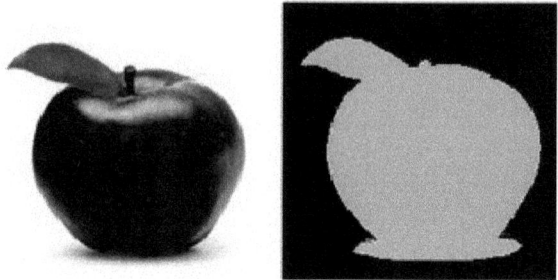

Figure 2 shows the fit of the image histogram. Histogram Equalization is a computer-based image analysis technique that is used to enhance contrast in images. It works this by effectively extending the most common values, e.g. Stretching Distance for image Stabilization. This approach often amplifies the differences in country imagery where its data used is represented by the closest values. This allows different geographical areas to get the most difference. In multiple view applications, it is useful to be able to distinguish image regions that correspond to our preferences, from image regions corresponding to layers. Empowerment often provides a simple and easy way to make these distinctions on the basis of the size or colors that are unique to the front parts of the image. In addition, it is useful to be able to see which areas of the image contain pixels whose values lie within a certain distance, or band thickness.

Thresholding operation shows in the Fig. 3. The left side of the image is an original image and the right-side of the image shows the threshold operation on original image.

3.3 Invert Operation

Rotating image awake and miserable, as left and right that is, an object is need to rotating 180° approximately the from line object to object; such images are made up of many telescopes. Also known as a modified image. Therefore, the logical use of NOT in a binary image changes its shape. It can also be used with a gray level image

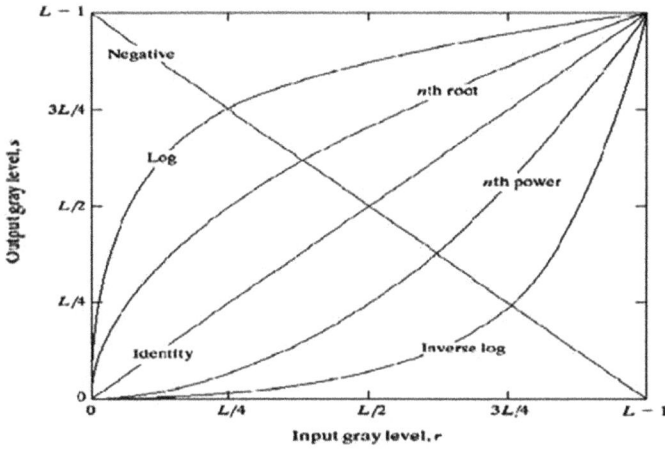

Fig. 4 Grey level conversion

stored in pixel format by using it in a very subtle way the value of a pixel is the input value subtracted from 255.

Gray level transformation shown in Fig. 4. After compresses of the log transformation the dynamic variety and upscales the image.

4   Methodology

Block diagram of the total operation is shows in Fig. 5. This image is converted to a bitmap file (.bmp) and this image is provided by MATLAB to convert a bitmap (.bmp) file to hexadecimal image values. This hexadecimal code is provided by the VIVADO simulation tool. A single image is converted in different way. According to the application, the difference, the light, the negative image is made.

For image conversion, code is written in MATLAB. After the code conversion hexadecimal image values are displayed in the command window. These values are converted according to pixel values (8-min is 0–255) the hexadecimal values of the

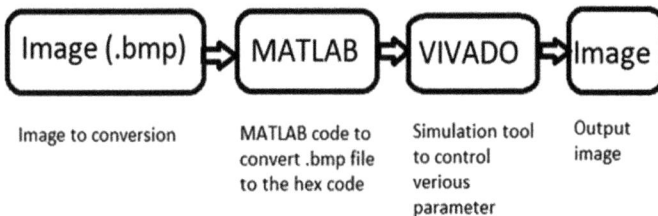

Fig. 5 Block diagram

image are changes. These hexadecimal ones are provided by VIVADO for further calculation. And for a different version and a negative symbol set the simulation parameter according to it. After all available steps a new image in which the image parameter changes.

5   Result and Discussion

5.1   *Convert the JPG to Hex*

Verilog cannot read image file such as BMP, JPJ, TIF, etc. directly, image is converted into binary text file so that Verilog can read using the TEXTIO VHDL package.

The hexadecimal values after conversion is shown in the Fig. 6. The values specified to the VIVADO for the next calculation.

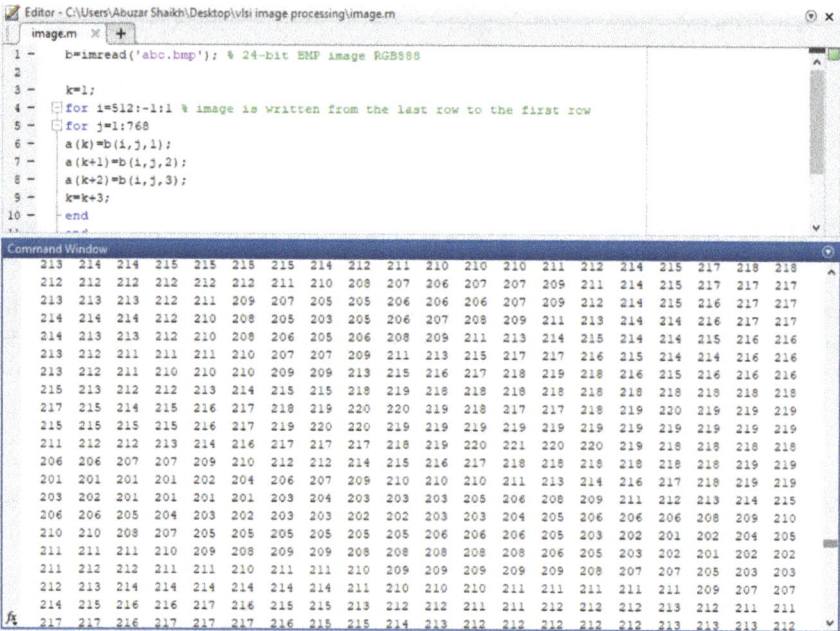

Fig. 6 Hexadecimal standards

5.2   Image Read

In this function the data is get from the hexadecimal file (image. Hex). For the reading purpose different parameters of the image is need to set. Which is nothing but the height, input file name, start up and horizontal synch delay.

HEIGHT = 512,
INFILE = "image.hex",
START_UP_DELAY = 100,
HSYNC_DELAY = 160,
VALUE = 100,
THRESHOLD = 90,
SIGN = 0

Output is stored into the 8-bit register. These registers of content value are red and green and blue pixel values. Two different files are stored is for even part and the odd part.

output reg [7:0] Data _R,
output reg [7:0] Data _G,
output reg [7:0] Data __B,
output reg [7:0] Data __R,
output reg [7:0] Data __G,
output reg [7:0] Data __B,

5.3   Parameter

To select the different parameter which task has to perform. Input file name is INPUTFILENAME (image. Hex) and the output file is OUTPUTFILENAME (output.bmp).

 ¸´define INPUTFILENAME ´´image.hex´´   // Input file name
´define OUTPUTFILENAME ´´output.bmp´´   // Output file name'
/ * ´define BRIGHTNESS_OPERATION
' ´define INVERT_OPERATION
´define THRESHOLD − OPERATION*/

5.4   Image Write

In the image write function the initial parameters are need to set. Output file (output.bmp) then the most important part is bmp header.

Table 1 Header

Offset	Size	Description
0	2	Signature, must be 4D42 hex
2	4	Size of BMP file in bytes (unreliable)
6	2	Reserved, must be zero
8	2	Reserved, must be zero
10	4	Offset to start of image data in bytes
14	4	Size of BITMAPINFOHEADER structure, must be 40
18	4	Image width in pixels
22	4	Image height in pixels
26	2	Number of planes in the image, must be 1
28	2	Number of bits per pixel (1,4,8, or 24)
30	4	Compression type (0 = none, 1 = RLE-8, 2 = RLE-4)
34	4	Size of image data in bytes (including padding)
38	4	Horizontal resolution in pixel per meter (unreliable)
42	4	Vertical resolution in pixel per meter (unreliable)
46	4	Number of colors in image, or zero
50	4	Number of important colors, or zero

HEIGHT = 512,   // Image height

INFILE = "$output$.bmp",

After all this process the output is stored into a output.bmp file. The input file is hexadecimal and the output file is the bitmap.

Table 1 shows the header format of the image. The BMP is diffident raster visuals image file arrangement premeditated to stock bitmap digital image self-sufficiently of a display device.

5.5 Output Images

Invert operation

Figure 7 shows the Invert operation. After run the simulation for 6 ms while selecting the invert as output this image will appears. In the invert operation the white part is converted into the dark and the dark part is converted into the bright.

Threshold operation

Threshold operation shows in Fig. 8. The contrast value is set to the 100. After the threshold operation the pixel value which is less than 100 is converted into a 0 and the pixel value grater than 100 is converted into a 255.

Fig. 7 Invert operation

Fig. 8 Threshold operation

Brightness operation

Figure 9 shows the Brightness operation. In the image brightness we either increase or decrease the brightness. In the above figure the brightness of an image is decreases. After run the simulation for 6 ms while selecting the invert as output this image will appears.

Fig. 9 Brightness operation

5.6 *Flow Chart*

Figure 10 shows the flow graph of the project. In flow diagram first, original image remains converted into a bit map file (64-bit) color image.

6 Conclusion

Digital image startup techniques mean inserting a sample of the area and increasing the size. The algorithms associated has been developed in VIVADO. The processing of image is cost effective and faster. No fixing or processing chemical are wanted to take and process digital image. The displayed result shows the image performance of Brightness, Threshold and Invert function. An Image enhancement approaches provides a large diversity of methods to improve the visuality of the image. Here the operation of the image enhancement is described by using the Verilog HDL. Verilog HDL cannot operate with the image format that's why first converting the image into binary form. In this image performance enhancement techniques are applied using point performance such as light intensity, performance input and specific grip.

Fig. 10 Flow chart

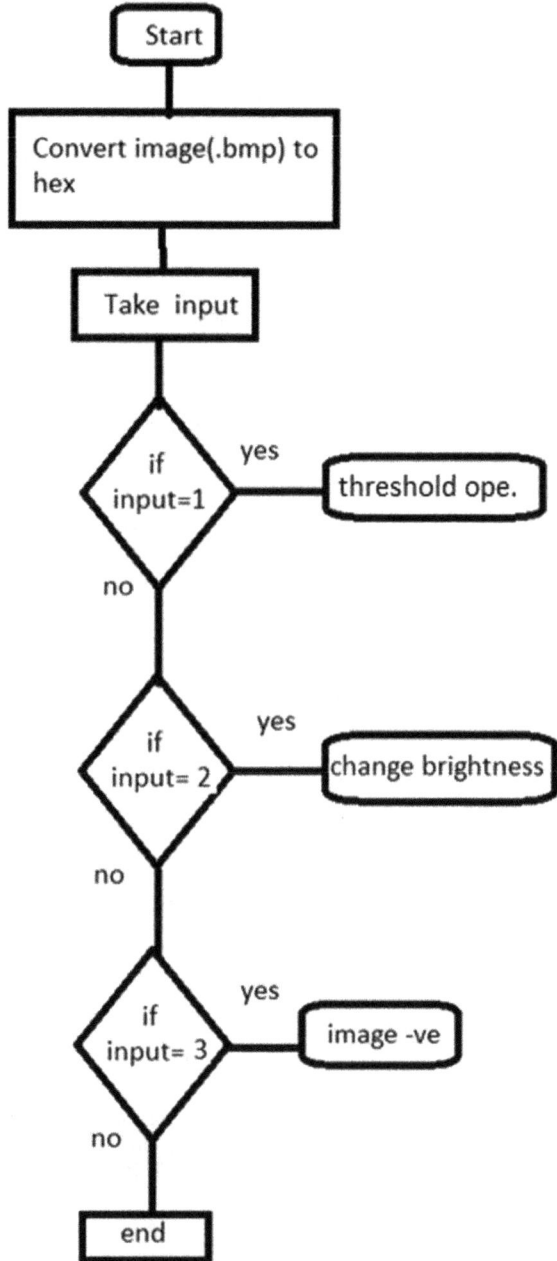

References

1. Chiuchisan I, Cerlinca M (2018) Alin-Dan Potorac. Digital Image Processing, Adrian Graur
2. Thirumarai S (2019) Automatic video prediction is used to process digital image processing. 12(2)
3. Ravikumar R, Arulmozhi V (2019) Digital image processing-a quick. Review 2:16–24
4. Khan S, Kharade R, Lavange V (2019) Segmentation in digital image processing. 2:292–294
5. Karthikeyan U, Vanitha M (2019) Research on text recognition using image processing with mining data techniques, pp 1–5
6. Dixit SR (2013) Edge design techniques used for image representation with VLSI. In: 6th international conference emerging trends in engineering and technology, pp 115–116
7. Deyy S, Bhattacharya BB, Kundu MK (2012) A fast algorithm for computing the euler number of an image and its VLSI implementation. Intel Corporation, USA, pp 1–6
8. Yang Y (2011) Three-dimensional VLSI image editing and possible memory retrieval and RAM/ROM composing. IEEE 134–137
9. Zhang Y, Yao S, Zhang N, Xu J (2008) Design of two DM Image image sensor processing procedures based on FPGA. In: Supported by national natural science foundation of China, pp 132–141
10. Chiuchisan I, Cerlinca M, Potorac A-D, Graur A (2012) Image enhancement techniques using verilog hardware definition language. In: 11th World conference on development plans and applications, Suceava, Romania, pp 145–149
11. Maitya HK, Maityb SP (2014) Design and implementation of an effective graphic design framework using handel-C
12. Coltuc D, Chassery J-M (2007) very fast watermarking by reversible contrast mapping. IEEE Signal Process Lett 14(4):255–258
13. Nithiya N, Imtiaz R (2017) Focused color image-based implementation of FPGA
14. Boscaro A, Jacquir S, Sanchez K, Perdu P, Binczak S (2016) Signal and image processing techniques for VLSI failure analysis. In: International conference on informatics and computing, pp 1–4
15. Huang HJ, Yu Y, Wang JJ, Meng W, Pun SH, Mak PU, Vai MI (2015) Equal design of ultrasound synthetic aperture imaging FPGA. In: State key laboratory of analog and mixed-signal VLSI, University
16. Chiuchisan I, Geman O (2014) A review of the HDL-based system of real image performance used for plant testing. In: 11th International conference on development and application systems, Suceava, Romania, pp 144–148
17. David VK, Poorani S (2019) A survey of soft computing techniques on bio medical image processing. J Netw Commun Emerg Technol (JNCET) 9(1):6–9
18. Nandhagopal N, Naveethan S, Ramesh M, Kandasamy R (2019) Image enhancement using adaptive histogram equalization for medical image processing. Cikitusi J Multidisciplinary Res 6(2)
19. Supel PV, Bhgat KS, Bhgat KS, Chandhari JP (2019) Chandhari: image segmentation classification for medical image processing. Int J Future Comput Sci Commun Eng 5(1):45–52
20. Singh M, Pandove G (2018) An implementation of image enhancement on real time configurable system using HDL. IJARECE 7(3):202–208

Chapter 57
Integrated Multi-biometric Template Security Based on Hybridization of Feature Transformation and Image Transformation

Sonali Patil and Pallavi Dhade

1 Introduction

Biometric technology is a process of user proof of identity comprising biological data, or indeed the screening or examination of some body part [1]. Biometric Systems offer automated techniques to validate or recognize a living individual's appearance mostly on bases of some of these physiological characteristics, including a Fingerprint or facial template, or even certain aspects of activity, such as handwriting.

Because of multiple modalities, multi-biometric based authentication methods are superior to unimodal single biometric systems. This has huge benefits compared with traditional password authentication schemes [2]. Such approaches provide an alternative solution to password-based security schemes because biometrics templates of the user cannot be stolen and those are really hard to copy. But for password authentication, we need to remember password and there are high chances of hacking the password. Unimodal biometric authentication systems have several drawbacks, as it experiences from several problems [1] like Noise in grasped data, differences in intra-class, Non centrality etc. Hence there is need of multi-biometric for authentication.

Biometric data is private and sensitive data of a person, so privacy preservation is very much essential in such systems. Though multi-biometric authentication systems are more secure, few issues have been posed because of the great specificity of biometric details that contributes to privacy implications, massive size repositories and an unified structure that may generate risks to safety and privacy. Any information leakage may cause critical privacy and security issues. The Proposed system uses hybridization of feature transformation and image transformation for

S. Patil · P. Dhade (✉)
Pimpri Chinchwad College of Engineering, Pune, Maharashtra, India
e-mail: pallavi.dhade@gmail.com

S. Patil
e-mail: sonalimpatil@gmail.com

multi-biometric template protection. It accepts two biometric traits (Fingerprint and Iris) and it performs two phases on the input data. In first phase feature transformation method is applied where Discrete Cosine Transformation (DCT) technique is used on biometric templates feature transformation. And polynomial based Secret Sharing scheme (SSS) are used for hybridization which results in protection of multi-biometric template. This results in not revealing any information of the original biometric trait. It uses secret sharing for secure biometric authentication.

1.1 Why Multimodal Biometric System?

Multimodal Biometric System offers significant benefits, as shown below:

a. **Accuracy**: Multimodal biometrics possesses information from multiple or more biometric features—(e.g., fingerprint and face; or face and iris) while biometric unimodal systems collect the information from one biometric feature [3, 4].
b. **Increased and Reliable Recognition**: A multimodal biometric approach empowers a higher degree of confidence for efficient match of identification and verification methods [3, 4].
c. **Enhanced Security**: In the multimodal biometric system, the use of several identification and recognition methods enables a system to achieve a higher degree of threshold recognition and a system administrator can make a decision on the precise security level that is essential [3, 4].
d. **User Acceptance**: As mentioned above, multimodal biometric systems are much more reliable, accurate, have strong security choices available and many countries support this sort of infrastructure more extensively [2]

1.2 Biometric template protection

ISO/IEC 24745 [2] suggests following 3 important characteristics to be achieved while using biometric templates for authentication systems.

a. **Irreversibility**: Irreversible transforms should be applied on input biometric data before storage.
b. **Unlinkability**: There should not be any link between stored biometric with applications or databases.
c. **Renewability**: While issuing new template (in case if required), it shuld not match with previous template.

2 Literature Survey

The survey on Multi-biometric Template Protection shows there are three main types Feature transformation, Biometric cryptosystem, Image transformation. The proposed system uses the hybridization of Feature transformation and Image transformation.

In paper [5] proposed framework by the author that has been assessed for finger-print identification in which the attribute retrieval method is conducted that used discrete cosine transformation (DCT).The proposed system's mission is to deliver incredibly sealed authentication by the use of unrevealed distribution. Transformation is effective in strength densification and ending in vector pruning of the size of the function [6] explains various features associated with secrete sharing. In paper [7] visual cryptography is used for secure biometric authentication.

In paper [2], the authors report a user identity problem solving method. Biometrics has a leading role in identifying and verifying identity, but some issues have increased due to the high rise in biometric input vulnerability. Extreme threats to privacy arise from the exposure of data. Authors suggested that only secure assured models should be saved and sent back for detection purposes in order to help out in this matter. Homomorphic stochastic encoding is used to make private information-protected conservation network more secure. Because the encryption algorithm it's indeed computation time is powerful.

The authors in paper [4] **presented** a template protection algorithm based on feature transformation method for multimodal biometric recognition. This uses combination highlights by blending DCT coefficients in face and palm print images. Based on Discrete cosine transform (DCT), retrieval of low frequency features which have high specificity is done. The hybrid feature vector is then pushed to the eigenvector which generates low-spatial hybrid feature vector that pull facial and palm print images private information.

In paper [3], introduced a template security calculation dependent on DNA encoding for multimodal biometric acknowledgment. The proposed conspire changes over the biometric characteristics of the multimodal trait into a DNA arrangement by DNA encoding. At first, it produces a clamorous succession and afterward it will be changed over it to a DNA grouping. In the wake of getting DNA Sequence, it performs the DNA inclusion activity on the two DNA arrangements. Finally, it changes over the aggregate to decimal numbers that is the encrypted format.

In paper [8] proposed characteristics level merging method. It is utilized for making the structure of multimodal biometric attributes, for example, finger print, retina and finger vein. It uses RSA technique which secures numerous traits. By making use of fingerprint, finger-vein and retina, real-time biometric modalities cryptosystem also underwent a security review. This provides a significant increase in the efficiency of RSA-based multimodal biometric cryptosystem.

The authors of paper [9] proposed form of authentication with vein and signature of the finger. For several authentication schemes, finger-vein is used as the most effective fool proof form for automatic personal identification. By using the definition of cross numbers and the compound analysis principle, the finger-vein and signature image are processed and characteristics obtained. The visual cryptographic scheme on the biometric models is then applied to produce the shares. The shares are held in different repositories, such that no information is disclosed.

The authors of paper [10], proposed unified network improved protection framework. Two modules, Enrollment and Authentication, form the framework. The system administrator gathers the users' iris image and fingerprint image in the first module, and then characteristics retrieval is performed on enrolled images. The shares copies would be produced for both the enrolled images. Control is only granted unless the functionality match else the verification is denied.

[11, 12] and [13] also proposes methods for secure biometric authentication. The literature shows a need of more robust multi biometric template protection method.

In paper [14], the study presented the multimodal biometric premise of Biohashing-facial and fingerprint features along with privacy conserving. The results of the proposed scheme are strengthened, which use facial features and fingerprints, and tested utilizing two paradoxical bi-modal datasets. These are built from the facial and fingerprint databases which are publicly accessible.

In paper [15], the authors suggested secrecy rules for the treatment of multi-biometric authentication schemes which itself magnifies data leakage worries. It addresses proposed scheme in the sense of applications focused on the Smartphone and cloud. The scholars have introduced super secure, multi-biometric strong authentication retaining secrecy that delivers better accuracy. In paper [16, 17] privacy preserving authentication systems for Iris and biometric fusion are proposed respectively.

3 Proposed Technique

Biometric data is private data of a person, so privacy preservation is very much essential in such systems. Unimodal biometric authentication systems suffer various issues. Hence there is need of multi-biometric for authentication. Multiple factor authentication suffers following difficulties

1. User privacy preserving.
2. Achieve the ISO 24547 standard biometric template protection.
3. Centralized Database.

Hence, this work aims to provide Privacy-Preserving Authentication System using Multi-Biometric Template Protection. In this proposed system, fusion of Feature Transformation with Image Transformation is used to enhance security and user privacy. For Feature Transformation DCT [4] and for Image transformation Polynomial Based Secret Sharing [18] is used.

Feature Transformation

We employ discrete cosine transform (DCT) that is the most appreciated compression algorithm but is most widely used for the image and video processing that compact the data by packaging the Image data into different frequency ranges. DCT approach is amongst the most commonly used evaluation retrieval methods. This transforms spatial domain data into frequency domain features [4].We must derive DCT coefficients through images in two dimensional Fingerprint and Iris. And that we're extending two dimensional-DCT forth to derive information from its frequency domain. The 2D DCT transform with matrix form is given by Eq. 1. However, we use DCT coefficients for feature transformation of Fingerprint and Iris template; those are required in a hybridization process. 2D DCT transform computation is given by the Formula which is as:

$$F(u, v) = \alpha(v)\alpha(u) \sum_{x=0}^{N-1} \sum_{y=0}^{N-1} f(x, y) \ldots \cos\left[\frac{u\pi(2x+1)}{2}\right] \cos\left[\frac{v\pi(2y+1)}{2N}\right]$$

(1)

where,

$0 \leq u \leq N - 1,$

$0 \leq v \leq N - 1$

$\alpha(u, v) = \left\{ 1/\sqrt{N} \text{ for}(u, v) = 0 \right.$

$\sqrt{2/N} \text{ for } 1 \leq (u, v) \leq N - 1$

Image Transformation In this work, for each of template T_i we have used unique secret S that is obtained from the user. The templates and secrets can then be represented as:

$$T_1, T_2, T_3, \ldots T_{i+1} : S$$

where S is the user's secrete key value and I is really the total of templates considered by the user to obtain the secure template for a single person. Equation 2 is polynomial based can be formed by following Lagrange interpolation formula with the various models, their respective cost and the user control key is as follows:

$$P(x) = a_0 + a_1 x + a_2 x^2 + \cdots a_{i-1} x^{i-1}$$

(2)

For which a_i are the coefficients obtained accordingly, so by putting value of respective coefficient by referring the corresponding templates transformation matrix, we can generate the new secrete template for the user. Equation (2) came from well-known method of Lagrange interpolation

Lemma 1 One and just one polynomial curve $f(x)$ of degreel-1 could be defined with l points on the plane $(x1, y1), (x2, y2), \ldots (xl, yl)$. Hence, we get Eq. 3 as:

$$f(x) = a_0 + a_1 x + a_2 x^2 + \cdots a_{i-1} x^{i-1}, \{f(x_i) = y_i\}._{i=1}^{l} \tag{3}$$

The sensitive data of the user can be kept private by the Lagrange interpolation formula with secret S is shown through Eq. 4. Therefore, based on the l shares received, the recovery of the system secret S is done [18].

$$S = (-1)l \sum_{i=1}^{l+1} S_i \coprod_{j=1, j\neq i}^{l+1} \frac{F_j}{F_i - F_j} \tag{4}$$

Only one curve can be given by the subsequent polynomial (Lemma 1) formulation of Lagrange, hence specific S is obtained.

In the proposed system, from multi-biometric templates (fingerprint and Iris), the feature vector is generated, and then fusion of that feature vector is done. Next step generates shares and instead of storing whole image feature template at one place, it is stored in the form of shares using secret sharing. One share is saved to the database and another share stored into authentication server. Thus, the privacy of biometric traits is maintained. The authentication process has two phases involved.

1. Enrolment
2. Authentication

Biometric characteristics are taken in the enrolment phase using a sensor or camera. After collecting the fingerprint and iris images, normalization is carried out on the data to eliminate the noise. The important features are derived from the images using the transform domain, and the function vector is created. The data base of feature vector is vulnerable to centralized database attack. The enrolment phase is shown in Fig. 1.

Feature Transformation

Image Transformation

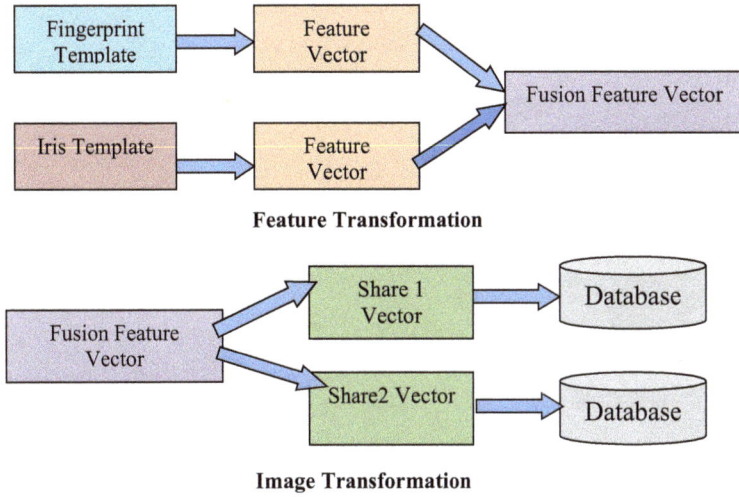

Fig. 1 Enrolment process

Steps of Enrolment Process

1. Accept biometric templates—Fingerprint Template and Iris Template.
2. Apply Feature Transform—DCT to individual Fingerprint and Iris template.
3. Do fusion of these two biometric templates.
4. Apply polynomial secret sharing technique to generate two shares.
5. Store these two shares in two separate databases viz. Database server and authentication server.

A. Steps for Authentication Process:

Using the sensor the fresh fingerprint and Iris are collected during the user authentication. Features are retrieved after image capture using the transform domain. The obtained feature pattern is evaluated by comparing with pattern from the share data base. Resemblance tests are being used to match the two templates, and that's checks whether or not the user is authorized. The authentication process is shown in Fig. 2.

1. Reconstruct the fusion vector by applying Secret reconstruction algorithm of polynomial secret sharing on shares stored in the databases.
2. Accept Iris and Fingerprint template from user and apply DCT on each template.
3. Apply Feature level fusion on obtained two DCT templates.
4. Compare this fusion multi-biometric vector with the formed fusion vector generated in step 1.
5. If it matches, user is authenticated else rejected.

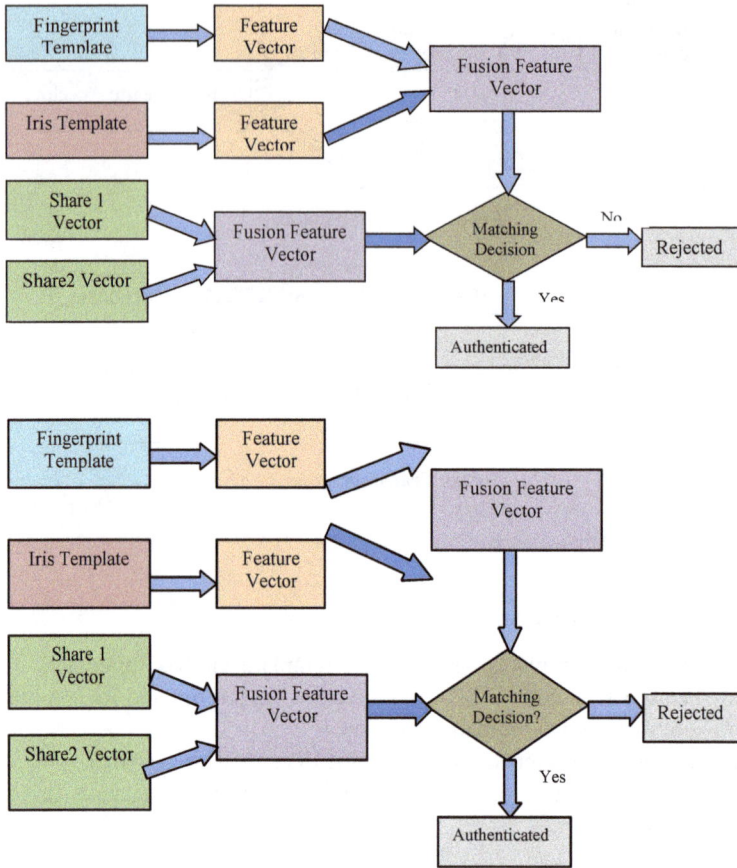

Fig. 2 Authentication process

4 Results

The implementation of proposed technique is done in python. Test bed of 500 samples of fingerprint template and iris template of 100 individuals are used to evaluate the proposed technique. The sample image of fingerprint template and iris template is shown in Fig. 3.

The tests are evaluated on feature vectors of different sizes, such as 4×4, 8×8, 16×16 and 32×32. Discrete Cosine Transformation (DCT) is used on each template of size 512×512 to convert it into 4×4, 8×8 and 16×16 size feature vector.

To decide threshold for matching criteria SD

The standard deviation (*SD*) is a function of how often a range of values differs or disperses. If the values are much closed to mean then it is a low standard deviation

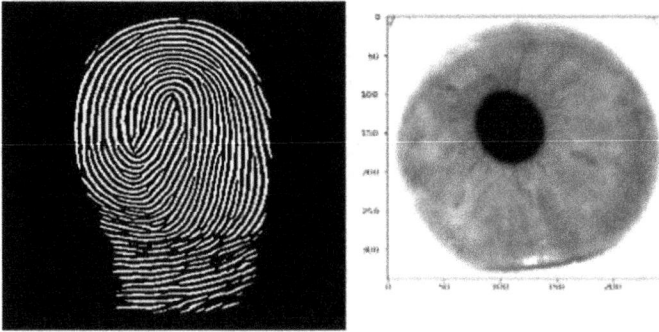

Fig. 3 Original sample fingerprint template and iris template

and if the values are dispersed over a wide range, the it is a high standard deviation. The standard deviation of a dataset is the square root of its variance which is given by Eq. 5.

$$SD = \sqrt{\frac{\sum |x - \mu|^2}{N}} \tag{5}$$

N - Size of population
x - each value from population
μ- Population mean

Performance Measures

Genuine Acceptance Rate (GAR): That is described as a percentage of legitimate users that the system agrees. GAR=100-FRR is given on it.

False Rejection Rate (FRR): Due to the poor quality of acquired biometric signal, the registered user is continuously getting denied access and rejected for authentication. This occurs because if image signal is out of the intra-class variation. Unless the biometric characteristics of a user fail to fit a prototype generated during enrolment, the user will be falsely denied. It's known as a percentage of legitimate users that the biometric system rejects. FRR is defined by the formula and give by Eq. 6. In this equation FR implies the amount of false refusals and P signifies the maximum amount of identity verification.

$$FRR = \frac{FR}{P} * 100 \tag{6}$$

Genuine Acceptance Ratio (GAR) is calculated for separate vector of template of Fingerprint and Iris as well as fusion vector of Fingerprint and Iris.

Reduce the size of fingerprint image of 8 × 8 is taken on which the Discrete Cosine Transform (DCT) is applied. From the transformed image, and compress size of original image. MSE and RMSE of original image divided in 8*8, 32*32, 64*64, 124*124, 256*256 and compare the results of all divided images.

Results for fingerprint and iris based biometric authentication using DCT and secret sharing method. Here, fingerprint, iris image of 8×8, 16×16, 32×32, 64×64, 256×256 is taken on which the Discrete Cosine Transform (DCT) is applied. From the transformed image, original image applies the DCT compression and generate the new compressed image. MSE and RMSE of original image with the approximate compressed image are computed. The calculated GAR and FRR values for feature vector size 4 x 4 of fingerprint, iris and fusion are illustrated in Table 1.

Standard Deviation:

The results for accuracy in authentication for feature vector with size 4×4. Fusion achieving maximum accuracy 87.58. Fingerprint GAR 77.77 and FRR is 22.22. For iris GAR is 83.00 and FRR is 16.99. The calculated GAR and FRR values for feature vector size 8 x 8 of fingerprint, iris and fusion are illustrated in Table 2.

Table 1 GAR and FRR values for 4×4 feature vector size

Threshold	Fingerprint GAR	Fingerprint FRR	Iris GAR	Iris FRR	Fusion GAR	Fusion FRR
Mean + SD	33.33	66.66	35.29	64.70	37.25	62.74
Mean + 2*SD	36.60	63.39	37.25	62.74	38.56	61.43
Mean + 0.75* (Max-Min)	71.24	28.75	69.28	30.71	73.20	26.79
Mean + 3*SD	66.013	33.98	68.62	31.37	67.32	32.67
Mean + 4*SD	73.85	26.14	76.47	23.52	79.08	20.91
*Mean + 5*SD*	**77.77**	22.22	**83.00**	*16.99*	**87.58**	*12.41*

Table 2 GAR and FRR values for 8×8 feature vector size

Threshold	Fingerprint GAR	Fingerprint FRR	Iris GAR	Iris FRR	Fusion GAR	Fusion FRR
Mean + SD	38.56	61.43	35.94	64.05	41.17	58.82
Mean + 2*SD	39.86	60.13	37.9	62.09	43.79	56.20
Mean +0.75* (Max-Min)	75.81	24.18	71.89	28.1	79.08	20.19
Mean + 3*SD	72.54	27.45	69.93	30.06	76.47	23.52
Mean + 4*SD	80.39	19.6	77.78	22.23	85.62	14.37
*Mean + 5*SD*	**86.92**	*13.07*	**84.31**	*15.69*	**93.46**	*6.53*

Table 3 GAR and FRR values for 16 × 16 feature vector size

Threshold	Fingerprint GAR	Fingerprint FRR	Iris GAR	Iris FRR	Fusion GAR	Fusion FRR
Mean + SD	43.13	56.86	38.56	61.43	42.48	57.51
Mean + 2*SD	45.09	54.90	41.83	58.16	45.75	54.24
Mean + 0.75* (Max–Min)	81.04	18.95	75.17	24.83	81.69	18.30
Mean + 3*SD	77.77	22.22	73.85	26.14	79.08	20.91
Mean + 4*SD	86.92	13.071	80.39	19.6	90.84	9.15
*Mean + 5*SD*	**92.15**	*7.84*	**86.92**	*13.07*	**96.07**	*3.92*

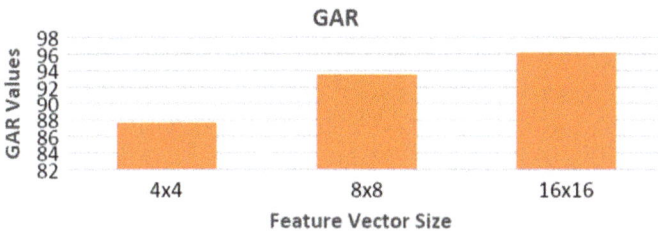

Fig. 4 GAR values for feature vector size 4 × 4, 8 × 8 and 16 × 16

The authentication accuracy results for feature vector size 8 × 8. Fusion achieving the highest accuracy 93.46. Finger-print GAR 86.92 and FRR is 13.07. For iris GAR is 84.31 and FRR is 15.69. The calculated GAR and FRR values for feature vector size 16 x 16 of fingerprint, iris and fusion are illustrated in Table 3.

The authentication accuracy results for feature vector size 16 × 16. Fusion achieving the highest accuracy 96.07. Fingerprint GAR 92.15 and FRR is 7.84. For iris GAR is 86.92 and FRR is 13.07. The comparative graph of GAR values for feature vector size of 4 × 4, 8 × 8 and 16 × 16 is shown in Fig. 4.

In this graph 16 × 16 Feature vector size gives highest accuracy. As compare to feature vector size 4 × 4 and 8 × 8, the size vector 16 × 16 and fusion of Fingerprint and Iris gives best results.

5   Conclusion

In this paper multi-biometric template protection technique is proposed using hybridization of feature transformation and image transformation. Two biometric templates viz; Fingerprint and Iris are used. Feature transformation is done using Discrete Cosine Transform (DCT) and Image transformation is done using polynomial Secret Sharing Scheme (SSS). Proposed system is validated using test bed of 500 samples of Fingerprint template and Iris template. The original 512×512 size biometric template is reduced to various size feature vectors. The size 16×16 vector is giving best results in fusion of both the template vectors. While storing in database the size of vector is reduced to vector size 16×8 due to image transformation phase. The image transformation provides protection to database by decentralization of database. The database is not revealing any information about original template which proves high security of private biometric template. We use Genuine Acceptance Ratio (GAR) and False Rejection Rate (FRR) to properly assess the effectiveness and performance of the proposed scheme. The obtained high average value 96.07% of GAR and low average value 3.92% of FRR validates high accuracy. Overall, the proposed technique is a secure authentication system which provides template protection, high accuracy, reduced size database and avoidance of single point database failure. In future the system can be tested with more number of biometric templates.

References

1. Ahlawat M (2015) An introduction to multimodal biometric system: an overview. IJSRD- Int J Sci Res Develop 3(2):1150–1154. http://www.ijsrd.com/articles/IJSRDV3I2690.pdf
2. Gomez-Barrero M, Maiorana E, Galbally J, Campisi P, Fierrez J (2017) Multi-biometric template protection based on homomorphic encryption. Pattern Recogn 67:149–163. https://doi.org/10.1016/j.patcog.2017.01.024
3. Dong J, Meng X, Chen M, Wang Z (2017) Template protection based on DNA coding for multimodal biometric recognition. In: 2017 4th International conference on systems and informatics (ICSAI). https://doi.org/10.1109/icsai.2017.8248565
4. Ahmad MI, Mohamad N, Md Isa MN, Ngadiran R, Darsono AM (2017) Fusion of low frequency coefficients of DCT transform image for face and palmprint multimodal biometrics. In: 2017 3rd IEEE International conference on cybernetics (CYBCONF). https://doi.org/10.1109/cyb conf.2017.7985778
5. Patil RS, Patil SD, Thepade SD (2017) Performance evaluation of fingerprint trait authentication system. In: Advances in intelligent systems and computing, pp 143–151. https://doi.org/10.1007/978-981-10-5520-1_14
6. Patil S, Deshmukh P (2012) An explication of multifarious secret sharing schemes. Int J Comput Appl 46(19):6–10. http://citeseerx.ist.psu.edu/viewdoc/download?doi=10.1.1.736.3421&rep=rep1&type=pdf
7. Patil S, Tajane K, Sirdeshpande J (2013) Secret sharing schemes for secure biometric authentication. Int J Sci Eng Res 4(6):2890–2895. https://www.ijser.org/researchpaper/secret-sharing-schemes-for-secure-biometric-authentication.pdf
8. Jagadiswary D, Saraswady D (2016) Biometric authentication using fused multimodal biometric. Proced Comput Sci 85:109–116. https://doi.org/10.1016/j.procs.2016.05.187
9. Nandhinipreetha A, Radha N (2016) Multimodal biometric template authentication of finger vein and signature using visual cryptography. In: 2016 International conference on computer communication and informatics (ICCCI). https://doi.org/10.1109/iccci.2016.7479963

10. Patil S, Bhagat K, Bhosale S, Deshmukh M (2015) Intensification of security in 2-factor biometric authentication system. In: 2015 International conference on pervasive computing (ICPC). https://doi.org/10.1109/pervasive.2015.7087058
11. Patil RS, Patil S, Thepade SD (2015) Secret sharing based secure authentication system. Int J Comput Appl 118(22):8–11. https://doi.org/10.5120/20875-3613
12. Kumar G, Bhatia PK (2014) A detailed review of feature extraction in image processing systems. In: 2014 Fourth international conference on advanced computing and communication technologies. https://doi.org/10.1109/acct.2014.74
13. Meraoumia A, Chitroub S, Bouridane A (2015) Robust multimodal biometric identification system using finger-knuckle-print features. In: 2015 3rd International conference on control, engineering and information technology (CEIT). https://doi.org/10.1109/ceit.2015.7233103
14. Anzaku ET, Sohn H, Ro YM (2011) Privacy preserving facial and fingerprint multi-biometric authentication. Digital Watermarking 239–250. https://doi.org/10.1007/978-3-642-18405-5_20
15. Natgunanathan I, Mehmood A, Xiang Y, Hua G, Li G, Bangay S (2017) An overview of protection of privacy in multibiometrics. Multimed Tools Appl 77(6):6753–6773. https://doi.org/10.1007/s11042-017-4596-y
16. Morampudi MK, Prasad MVNK, Raju USN (2020) Privacy-preserving iris authentication using fully homomorphic encryption. Multimedia Tools Appl. https://doi.org/10.1007/s11042-020-08680-5
17. Toli C-A, Aly A, Preneel B (2016). A privacy-preserving model for biometric fusion. In: Cryptology and network security, pp 743–748. https://doi.org/10.1007/978-3-319-48965-0_54
18. Ometov A, Bezzateev S, Mäkitalo N, Andreev S, Mikkonen T, Koucheryavy Y (2018) Multi-factor authentication: a survey. Cryptography 2(1):1–31. https://doi.org/10.3390/cryptography 2010001

Chapter 58
Classifying Chromosome Images Using Ensemble Convolutional Neural Networks

Muna Al-Kharraz, **Lamiaa A. Elrefaei**, and **Mai Fadel**

1 Introduction

Medical imaging processing has become an important diagnostic and therapy assistant tool. Chromosomes carry the genetic information of the human and the healthy human cell contains 23 pairs of chromosomes, 22 pairs of autosomes, and one pair of sex chromosomes (XX for female and XY for male) [1]. Changes in chromosomes number or structure do occur on rare occasions and maybe a sign of a genetic disorder or cancer. Hence, cytogenetics is used to confirm or aid in diagnosing disease and plan for treatment. A series of steps are involved in the cytogenetics laboratory to diagnose the sample case. It includes culturing, harvesting, slide-making, banding, staining, karyotyping, and finally chromosome analysis [2]. Figure 1 illustrates the process of preparing chromosomes from a peripheral blood sample.

Cytogeneticists perform karyotyping which involves separation and classification of human chromosomes from a cell image [3]. The classification categorizes each chromosome into one of the 24 chromosome classes based on various chromosomes features like length, area, banding pattern, centromere position, etc. However, performing this task is done manually in the cytogenetic laboratory and this process required effort, domain experience, and spend a considerable amount of time [4, 5].

M. Al-Kharraz (✉) · L. A. Elrefaei · M. Fadel
Computer Science Department, Faculty of Computing and Information Technology, King Abdulaziz University, Jeddah, Saudi Arabia
e-mail: malkharaz0001@stu.kau.edu.sa

L. A. Elrefaei
e-mail: laelrefaei@kau.edu.sa; lamia.alrefaai@feng.bu.edu.eg

M. Fadel
e-mail: mfadel@kau.edu.sa

L. A. Elrefaei
Electrical Engineering Department, Faculty of Engineering at Shoubra, Benha University, Cairo 11629, Egypt

© The Author(s), under exclusive license to Springer Nature Singapore Pte Ltd. 2021   751
X.-Z. Gao et al. (eds.), *Applications of Artificial Intelligence in Engineering*, Algorithms for Intelligent Systems, https://doi.org/10.1007/978-981-33-4604-8_58

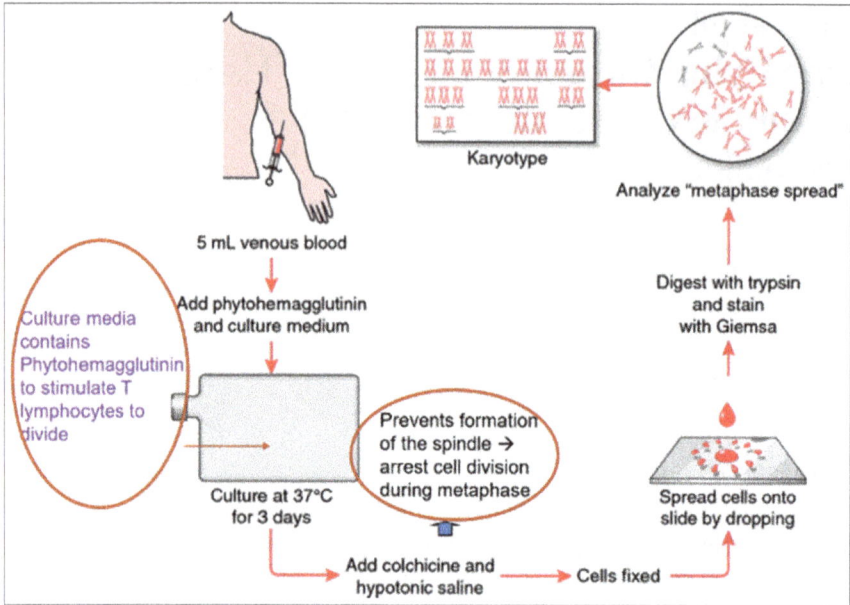

Fig. 1 Chromosome preparation steps from peripheral blood [2]

To reduce the burden of this process and to assist laboratory technicians, many computer-aided systems for chromosome classification have been developed to automate the chromosome classification and help cytogeneticists to perform this task due to the unpredictable shapes and appearances caused by non-rigid nature of chromosomes. These systems follow two steps: feature extraction and classification. In developing such an automated system, the chromosome's features extraction as well as the selection of the classifiers are two most significant challenges. A good chromosome classification system is mainly related to feature extraction; where it allows an appropriate classification.

Deep learning has proved to be the state-of-art technique for a wide range of computer vision tasks [6, 7], such as image classification within the recent past. Convolutional Neural Network (CNN) is a class of deep learning that was used in the medical image field due to its capability in dealing with complex features [8, 9]. Recent researches have increasingly replaced traditional techniques on manually designed features and feature selection methods with learning-based features where features extracted automatically using CNN.

CNNs are multi-layered neural networks, where their architecture is like the human visual system. They use restricted receptive fields and a hierarchy of layers that progressively extract more abstracted features [10]. However, CNN requires a large amount of labeled data. Therefore, classifying images with several class labels using only a small number of labeled data is a difficult task; hence, the transfer learning can be used in this situation. The transfer learning idea is utilizing the learned knowledge for one problem to solve a new task rather than training from scratch [11–13]. VGG19, ResNet50, and MobileNetv2 models were trained on more than a million images and can classify them into 1000 classes. From these images, they learned huge feature representations [7].

A classifier may perform poorly on the test data and not able to well learn all patterns and as a result, make different mistakes. To avoid this problem is to train a set of classifiers on the same dataset [14]. Ensemble classifier combines different classifiers trained on a dataset in a supervised classification problem to improve the predictive performance. Generally, combining the predictions from different classifiers is more accurate than the prediction from an individual classifier [14, 15]. Various ways of combining predictions from different classifiers such as average voting, where the average of the softmax class probabilities is calculated from the different networks, weighted average, where each classifier has a weight that is proportional to its performance, and majority voting in which the prediction is the most common class in all predictions [16, 17].

We attempt in this paper to automate the classification step of karyotyping to overcome the effort that spent on manual chromosome classification by feeding the chromosome images into CNN to classify them into 24 classes. The main contributions are summarized as follows:

1. Integrating pre-trained CNNs (VGG19, ResNet50, and MobileNetv2) into an ensemble learning utilizing the decision-level fusion via average voting and majority voting.
2. We benchmark the models against state-of-the-art methods on the biomedical imaging laboratory dataset [18].

The remainder of this paper is organized as follows: Sect. 2 reviewed various classification systems. In Sect. 3, classification models based on CNN and ensemble methods are explained. In Sect. 4, the conducted experiments, evaluation, and result discussion are presented. Section 5 presents the conclusion and future work.

2   Related Work

Chromosome classification considered the most challenging issue in karyotyping, and different researches have been made to automate the classification stage. In recent years, various researches focused on applying deep learning to automate extracting features, where the network includes several layers to learn and extract features while training and they showed very promising results. The number of created features can be practically infinite and without any human bias.

Monika et al. [19] built their own CNN for feature extraction and classification. It includes four blocks; every block has two convolutional layers, one dropout, and one maxpooling layer. The classification part of CNN has two fully connected layers and a softmax layer with 24 units. They have been classified 18,400 chromosomes acquired from the hospital. Before the classification, they applied some preprocessing such as straightening of chromosomes, finding bending orientation, and normalizing chromosome length. They evaluated their proposed CNN before and after preprocessing and obtained 68.5, 86.7% accuracy, respectively.

Siamese Network was proposed by Swati et al. [3] which consisting of twin neural networks and used CNN as the base network. This CNN has two convolutional layers, maxpooling layer, and a fully connected dense layer followed by the computation of an energy function. They trained the embeddings obtained from Siamese network by Multi-layer Perceptron (MLP) as the second training step. They straightened chromosomes using two approaches: via medial axis extraction and crowdsourcing and via projection vectors. They got an accuracy of 84.6% when they straightened chromosomes using projection vectors.

Sharma et al. [20] utilized Biomedical Imaging Laboratory (BioImLab) dataset that publicly available and contains 5474 chromosome images. They extracted features from these images using ResNet-50 then fed these features into long short-term networks followed by an attention block and fully connected softmax layer. They achieved an accuracy of 90.42%.

Zhang et al. [21] constructed a CNN contained five types of layers: convolution, pooling, dropout, flatten, and dense layers. They acquired 10,304 vertical chromosome images from a local company and got an accuracy of 92.5% and the proportion of well-classified karyotype of 91.3%. They will work on different chromosomes orientations in future.

Also, Somasundaram [22] constructed a CNN for feature extraction and classification. It consisted of five types of layers: convolution layer, pooling layer, flatten layer, dense layer, and dropout layer. The dataset includes 5000 normal chromosomes and 3000 abnormal chromosomes. They performed the augmentation technique and achieved an accuracy of 98.9%.

Global features like chromosome's length, size, and shape have been extracted by Qin et al. [8] via the G-Net, and for extracting local features like texture patterns of local parts, they applied L-Net; then, they fused these features and applied two MLP classifiers with augmentation technique. A dispatch strategy was utilized to assign each chromosome to its type. They gained an accuracy of 98.9% and F1-score of 98.7% on 87831 segmented chromosome images collected from the Xiangya Hospital of Central South University, China.

Chromosomes length on BioImLab dataset was normalized by Swati et al. [23]; then, the resolution of these images was converted from low to high by applying a convolutional super-resolution network. For classification, they utilized Xception pre-trained model and got an accuracy of 92.36%. They plan in future to detect chromosome structural abnormalities.

From the above related work, we observed the automation of feature selection and extraction achieved promising results. However, in this paper, we propose to automatically extract features from images based on pre-trained CNN and then ensembles their predictions.

3 Methods

In this paper, the feature extraction and classification are based on fine-tuning pre-trained CNN. The evaluated pre-trained models are:

Visual Geometry Group (VGG19): includes 19 layers (16 convolutional with 3*3 convolution filters, 3 fully connected) as seen in Fig. 2 [24]. It accepts images in RGB color scheme with size 224-by-224 pixels.

Residual Networks (ResNet50): deeper than VGG19 and includes 50 residual layers. It has global average pooling layers instead of fully connected layers, and it accepts RGB images in size 224-by-224 pixels as shown in Fig. 3 [24].

Mobilenetv2: includes 54 layers organized into two blocks; one is residual block with stride of 1, and another one is block with stride of 2 for downsizing. Also, it accepts RGB images in size 224-by-224 pixels [25] as illustrated in Fig. 4.

We performed an online data augmentation technique on the training samples that suitable in case of having a small dataset to help preventing the model from memorizes the exact details of the training data and therefore preventing the model from over-fitting. During training, the training dataset has a random combination of transformations in the minibatch, and in this way, each epoch uses different datasets without changing the number of training samples [26, 27]. The training images augmented by scaling vertically up and horizontally to 10%, flipping along the vertical axis, and translating up to 30 pixels.

We fine-tuned the above models by freezing the weights of earlier layers to speed up the training process and due to the differences between our dataset and the dataset, where the above models trained on [12, 28]. We replaced the last fully connected layer and classification layer of the above models with the new layers according to the number of chromosomes classes which are corresponding to 24. Figure 5 demonstrates the idea of the classification process on VGG19 model. We trained the models using stochastic gradient descent (SGD) optimizer with a learning rate of 10^{-3}, 10 min-batch size, and 60 epochs.

Then, we ensembled the decisions produced by the pre-trained models into a single decision using average voting by computing the average of the softmax class probabilities to generate posterior labels, and using majority voting to get the most frequent class as obvious in Fig. 6.

4 Results and Discussion

In building this system, we used the following tools:

- **Hardware**: The models are trained and tested on a computer with the following specification: Core i7-6700 K CPU @ 4.00 GHz processor, 16 GB RAM, 2T HDD, and NVIDIA GTX 1070 graphics card.

Fig. 2 VGG19 architecture
[24]

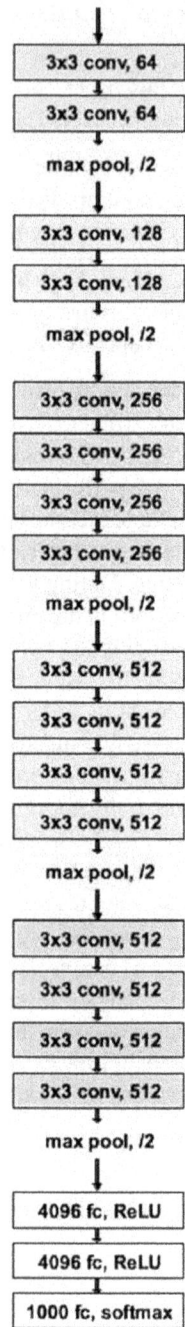

3x3 conv, 64

3x3 conv, 64

max pool, /2

3x3 conv, 128

3x3 conv, 128

max pool, /2

3x3 conv, 256

3x3 conv, 256

3x3 conv, 256

3x3 conv, 256

max pool, /2

3x3 conv, 512

3x3 conv, 512

3x3 conv, 512

3x3 conv, 512

max pool, /2

3x3 conv, 512

3x3 conv, 512

3x3 conv, 512

3x3 conv, 512

max pool, /2

4096 fc, ReLU

4096 fc, ReLU

1000 fc, softmax

Fig. 3 Resnet50
architecture [24]

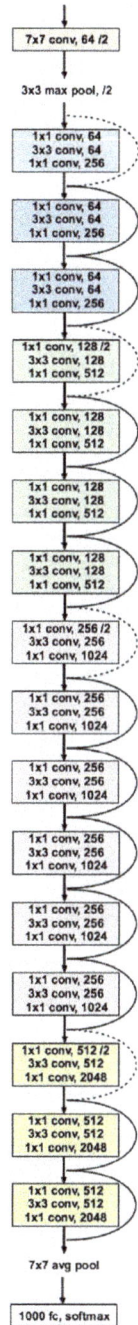

Fig. 4 Mobilenetv2
architecture

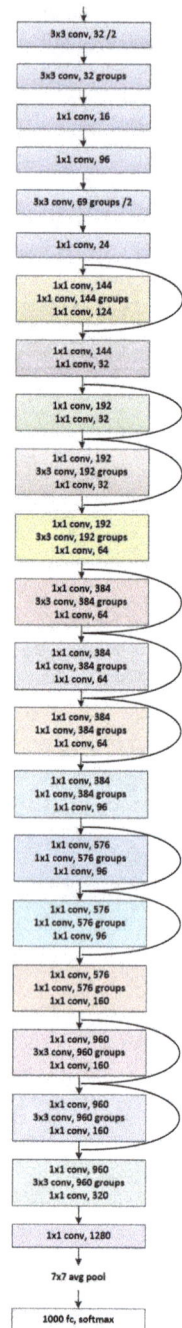

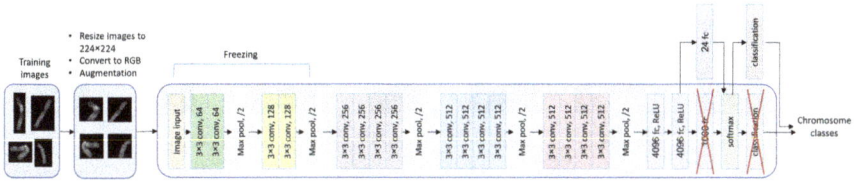

Fig. 5 Classification using VGG19

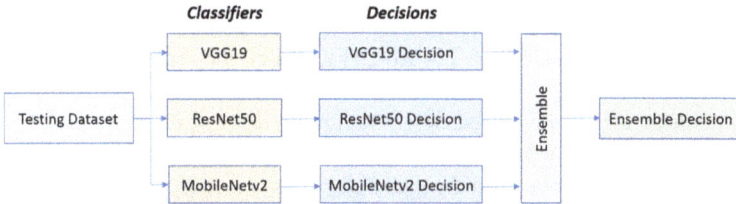

Fig. 6 Ensemble decisions of VGG19, ResNet50, and MobileNetv2 models

- **Software**: The proposed system implemented using MATLAB R2019b with installed deep learning toolbox.

In our experiments, we evaluated our methods in two datasets:

- **Diagnostic Genomic Medicine Unit (DGMU) dataset**: From the center of excellence in Genomic Medicine Research at King Abdulaziz University [29]. This dataset segmented using thresholding and Otsu's technique by Bashmail et al. [30]. It produced 6011 segmented chromosome images, and theses images have different sizes like $38 \times 125, 30 \times 61, 45 \times 146, 44 \times 87$, and 209×126. Figure 7 shows sample of images for DGMU dataset.
- **Biomedical Imaging Laboratory (BioImLab) dataset**: available online [18] and contains 5474 segmented chromosome images. Also, these images have different sizes. Figure 8 shows sample of images for this dataset.

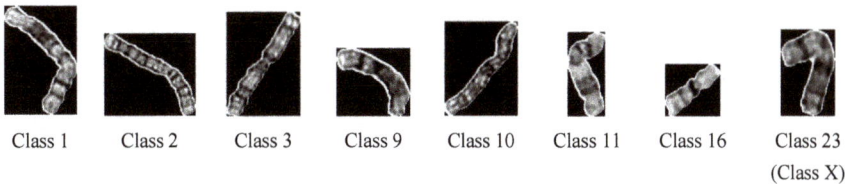

Class 1	Class 2	Class 3	Class 9	Class 10	Class 11	Class 16	Class 23
							(Class X)

Fig. 7 Sample of chromosome images DGMU dataset

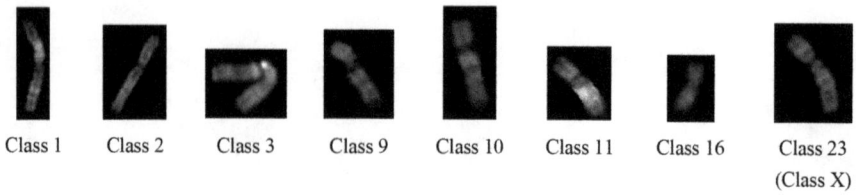

| Class 1 | Class 2 | Class 3 | Class 9 | Class 10 | Class 11 | Class 16 | Class 23 (Class X) |

Fig. 8 Sample of chromosome images BioImLab dataset

We split each dataset to 80% training and 20% testing, and all images were converted to RGB and resized to 224-by-224 to match the requirements of the first layer.

The methods are evaluated based on accuracy, recall, precision, and F1-score performance metrics (Table 1). Accuracy is the ratio of the number of correctly classified chromosome images to the total number of chromosome images as defined in (1).

$$\text{Accuracy} = \frac{\text{Number of correct classified images}}{\text{Total number of images}} \times 100 \qquad (1)$$

Equation 2 represents the recall that measures the percentage of correctly classified positive chromosome images to all images in the actual class.

Table 1 Pre-trained models and ensemble results on BioImLab and DGMU datasets

Dataset	Model	Accuracy (%)	Recall (%)	Precision (%)	F1-score (%)
BioImLab	VGG19	91.67	91.20	91.80	91.50
	Resnet50	94.11	94.30	94.41	94.35
	Mobilenetv2	93.30	93.10	92.92	93.01
	Ensemble (majority voting)	95.47	94.87	95.36	95.11
	Ensemble (average voting)	**97.01**	**96.34**	**96.70**	**96.52**
DGMU	VGG19	91.49	90.01	92.02	91.01
	Resnet50	92.67	92.49	92.94	92.71
	Mobilenetv2	92.89	92.27	93.47	92.87
	Ensemble (majority voting)	93.63	92.71	94.16	93.43
	Ensemble (average voting)	**94.97**	**94.57**	**95.29**	**94.93**

$$Recall = \frac{TP}{TP + FN} \tag{2}$$

where TP is the true positive, and FN is the false negative. Precision measures the percentage of correctly classified positive chromosome images to the total classified positive chromosome images as defined in (3).

$$Precision = \frac{TP}{TP + FP} \tag{3}$$

where FP is the false positive. F1 score is calculated in (4) as the weighted average of precision and recall.

$$F1\ score = \frac{2*(Recall*Precision)}{(Recall\ + Precision)} \tag{4}$$

As seen from Table 1, the BioImLab dataset got different results on the base classifiers. For VGG19, it got an accuracy of 91.67%, recall of 91.20%, precision of 91.80%, and F1-score of 91.50%. While for Resnet50, it got an accuracy of 94.11%, recall of 94.30%, precision of 94.41%, and F1 score of 94.35%. An accuracy of 93.30%, recall of 93.10%, precision of 92.92%, and F1 score of 93.01% were obtained when using Mobilenetv2. Therefore, Resnet50 outperformed other classifiers on the BioImLab dataset. On the other hand, the best accuracy of 97.01%, recall of 96.34%, precision of 96.70%, and F1 score of 96.52% are achieved when ensemble classifiers using average voting.

Even for the DGMU dataset, it has different results on the base classifiers. VGG19 achieved an accuracy of 91.49%, recall of 90.01%, precision of 92.02%, and F1 score of 91.01%. Resnet50 got an accuracy of 92.67%, recall of 92.49%, precision of 92.94%, and F1-score of 92.71%. Mobilenetv2 achieved results near to Resnet50 results, and it got an accuracy of 92.89%, recall of 92.27%, precision of 93.47%, and F1-score of 92.87%. Mobilenetv2 outperformed other classifiers on the DGMU dataset. On the other hand, the best accuracy of 94.97%, recall of 94.57%, precision of 95.29%, and F1-score of 94.93% obtained when ensemble classifiers using average voting.

Based on the result achieved from ensemble using majority voting, it is clear that it slightly improves the accuracy over the best result from the base classifiers (1.36% for BioImLab dataset and 0.47% for DGMU dataset) but average voting proved successful in improving the accuracy over the best result from the base classifiers (2.9% for BioImLab dataset and 2.08% for DGMU dataset). Average voting outperformed majority voting in both datasets' experiments.

Comparison with the state-of-the-art methods

Table 2 compares our methods which give the best result on BioImLab dataset experiments with the state-of-the-art methods that used the same dataset. All of them utilized deep learning for feature extraction and classification but our method obtained better

Table 2 Comparing our method with state-of-the-art methods on BioImLab dataset

Research paper	Feature extraction	Classifier	Accuracy (%)
Sharma et al. [20]	ResNet50	LSTM followed by an attention block	90.42
Swati et al. [23]	Super-Xception network		92.36
Our method	Ensemble VGG19, ResNet50, and MobileNetv2 (average voting)		97.01

accuracy of 97.01% when we ensembled pre-trained models VGG19, ResNet50, and MobileNetv2 using average voting.

5   Conclusion

The need for automating chromosome classification is increased in recent years to save cytogeneticists time and effort. We proposed in this paper chromosomes classification methods using CNN transfer learning in pre-trained models (VGG19, ResNet50, and MobileNetv2) for feature extraction and classification then ensemble their predictions to obtain the highest prediction accuracy via average voting and majority voting. We find that the average voting leads to a higher accuracy than majority voting. Our method achieved the best accuracy of 97.01% on BioImLab dataset, and it outperformed the state-of-the-art methods that used the same dataset also for the DGMU dataset, the best accuracy obtained was 94.97%. For the future, we will extract hand-crafted features and classify chromosomes based on both hand-crafted and learning-based features and try to use another kind of ensemble.

References

1. Munot MV, Joshi PM, Kulkarni P, Joshi MA (2012) Efficient pairing of chromosomes in metaphase image for automated karyotyping. In: 2012 IEEE-EMBS conference on biomedical engineering and sciences, pp 916–921
2. Faiyaz-Ul-Haque M (2011) Human Chromosomes Human Karyotype. Available https://slidep layer.com/slide/7949123/
3. Swati, Gupta G, Yadav M, Sharma M, Vig L (2017) Siamese networks for chromosome classification. In: 2017 IEEE international conference on computer vision workshops (ICCVW), pp 72–81
4. Lerner B (1998) Toward a completely automatic neural-network-based human chromosome analysis. IEEE Trans Syst Man Cybernet Part B (Cybernet) 28(4):544–552
5. Wang X, Zheng B, Li S, Mulvihill JJ, Wood MC, Liu H (2009) Automated classification of metaphase chromosomes: optimization of an adaptive computerized scheme. J Biomed Info 42(1):22–31
6. Krizhevsky A, Sutskever I, Hinton GE (2012) Imagenet classification with deep convolutional neural networks. In: Advances in neural information processing systems, pp 1097–1105

7. Simonyan K, Zisserman A (2014) Very deep convolutional networks for large-scale image recognition. arXiv preprint arXiv:1409.1556
8. Qin Y et al (2019) Varifocal-net: a chromosome classification approach using deep convolutional networks. IEEE Trans Med Imag 38(11):2569–2581
9. Abid F, Hamami L (2018) A survey of neural network based automated systems for human chromosome classification. Artif Intell Rev 49(1):41–56
10. Kheradpisheh SR, Ghodrati M, Ganjtabesh M, Masquelier T (2016) Deep networks can resemble human feed-forward vision in invariant object recognition," (in eng). Sci Rep 6:32672–32672
11. Yosinski J, Clune J, Bengio Y, Lipson H (2014) How transferable are features in deep neural networks?.In: Advances in neural information processing systems, pp 3320–3328
12. Marcelino P (2018) Transfer learning from pre-trained models. Available https://towardsdatascience.com/transfer-learning-from-pre-trained-models-f2393f124751
13. Pan SJ, Yang Q (2010) A survey on transfer learning. IEEE Trans Knowled Data Eng 22(10):1345–1359
14. Rahman A, Tasnim S (2014) Ensemble classifiers and their applications: a review. arXiv preprint arXiv:1404.4088
15. Sagi O, Rokach L (2018) Ensemble learning: a survey. WIREs Data Mining and Knowledge Discovery 8(4):e1249
16. Pingel J (2019) Ensemble learning. Available https://blogs.mathworks.com/deep-learning/2019/06/03/ensemble-learning/
17. Yazdizadeh A, Patterson Z, Farooq B (2019) Ensemble convolutional neural networks for mode inference in smartphone travel survey
18. Poletti E, Grisan E, Ruggeri A (2008) Automatic classification of chromosomes in Q-band images. In: 2008 30th Annual international conference of the IEEE engineering in medicine and biology society, pp 1911–1914
19. Sharma M, Saha O, Sriraman A, Hebbalaguppe R, Vig L, Karande S (2017) Crowdsourcing for chromosome segmentation and deep classification. In: Proceedings of the IEEE conference on computer vision and pattern recognition workshops, pp 34–41
20. Sharma M, Vig L (2018) Automatic chromosome classification using deep attention based sequence learning of chromosome bands. In: 2018 International joint conference on neural networks (IJCNN), IEEE, pp 1–8
21. Zhang W et al (2018) Chromosome Classification with convolutional neural network based deep learning. In: 2018 11th international congress on image and signal processing, biomedical engineering and informatics (CISP-BMEI), IEEE, pp 1–5
22. Somasundaram D (2019) Machine learning approach for homolog chromosome classification. Int J Imaging Syst Technol 29(2):161–167
23. Swati S, Sharma M, Vig L (2019) Automatic classification of low-resolution chromosomal images. In: Computer vision–ECCV 2018 workshops, Cham, Springer International Publishing, pp 315–325
24. Leonardo MM, Carvalho TJ, Rezende E, Zucchi R, Faria FA (2018) Deep feature-based classifiers for fruit fly identification (Diptera: Tephritidae). In: 2018 31st SIBGRAPI conference on graphics, patterns and images (SIBGRAPI), pp 41–47
25. Sandler M, Howard A, Zhu M, Zhmoginov A, Chen L-C (2018) Mobilenetv2: inverted residuals and linear bottlenecks. In: Proceedings of the IEEE conference on computer vision and pattern recognition, pp 4510–4520
26. Raj B (2018, 28 Feb 2020). Data Augmentation| how to use deep learning when you have limited data—Part 2. Available https://medium.com/nanonets/how-to-use-deep-learning-when-you-have-limited-data-part-2-data-augmentation-c26971dc8ced
27. Shorten C, Khoshgoftaar T (2019) A survey on image data augmentation for deep learning. J Big Data 6
28. (15/02/2020) Train deep learning network to classify new images. Available https://www.mathworks.com/help/deeplearning/examples/train-deep-learning-network-to-classify-new-images.html

29. C. o. E. I. G. M. R. (CEGMR) (2015) Cytogenetic Service Unit. Available https://cegmr.kau. edu.sa/Pages-Cytogenetic-Service-Unit-en.aspx
30. Bashmail R, Elrefaei LA, Alhalabi W (2018) Automatic segmentation of chromosome cells. In: Proceedings of the international conference on advanced intelligent systems and informatics 2018, Cham, Springer International Publishing, pp 654–663

Chapter 59
Power Loss Sensitivity and GWO-Based Approach for Optimal Capacitor and DG Allocation in Distribution System

Kingshuk Roy, Laxmi Srivastava, and Shishir Dixit

1 Introduction

In the current situation, power demand is increasing promptly. Power loss in network, decrement in system voltage profile, system power quality and stability have become relatively common due to large growth in power demand. Before for producing electrical energy, only conventional resources were used. These conventional resources are coal, nuclear, hydro, etc. These traditional generator supplies the widely connected load through long transmission and distribution lines. These conventional power generating stations are weakening due to shortage of conventional resources and also environmental issues, etc. As a result of this, transmission and distribution costs are also progressively increasing. These above challenges shifts the attention of researchers towards small source generation connected towards load side known as distributed generation. Distributed generation allocation is the advance way to solve the many of these above-mentioned challenges. DG in simple terms can be termed as generation in small or dispersed generation that is neither centrally planned nor dispatched. DGs are used nearer to the distribution side or load side, and they use energy sources that are eco-friendly and its capacity is also not more than 100 MW. There are basically four types of DG [1].

Type 1. DG supplies real power only (i.e. solar photovoltaic, fuel cell).
Type 2. DG supplies reactive power only (i.e. capacitors, synchronous condenser).
Type 3. DG supplies both real and reactive power. (i.e. synchronous generator).

K. Roy (✉) · L. Srivastava · S. Dixit
Madhav Institute of Technology and Science, Gwalior, India
e-mail: kingshukroy52@gmail.com

L. Srivastava
e-mail: srivastaval@hotmail.com

S. Dixit
e-mail: shishir.dixit1@gmail.com

© The Author(s), under exclusive license to Springer Nature Singapore Pte Ltd. 2021   765
X.-Z. Gao et al. (eds.), *Applications of Artificial Intelligence in Engineering*, Algorithms for Intelligent Systems, https://doi.org/10.1007/978-981-33-4604-8_59

Type 4. DG supplies real power but consumes reactive power (i.e. wind turbine).

Authors in [2] have used analytical approach for allocating DG at optimal location on radial distribution system (RDS). PSO algorithm has been implemented by authors to calculate location and sizes of multiple types of DG units. Authors in [3] have done a brief overview of previous works on DG placement topic. Combination of bacteria foraging optimization algorithm (BFOA) and CPLS has been proposed for optimal DG placement problem on 33-bus system [4]. In [5], author has implemented back-tracking algorithm for optimal DG placement problem by taking active power loss reduction and enrichment in system voltage profile as the objective function (O.F) on 33-bus and 94-bus RDS. Multiobjective PSO algorithm has been executed for optimal DG and capacitor banks allocation on IEEE 33 and 94 bus and Portuguese RDS by taking reduction of system real power loss, voltage profile advancement and current balancing in system section as the O.F of the problem [6]. Combination of analytical approach and PSO algorithm has been used to calculate optimal sizing and position of multiple types of DG units in 33-bus and 69-bus RDS [7]. GA has been implemented by the authors to calculate adequate size and position of capacitor and DG units while taking costs minimization as the system objective function [8]. Moth-Flame optimization algorithm has been used to calculate size and location of DG unit in IEEE 33-bus system. Index vector methods have also been used by the authors to select the sensitive buses for DG allocation [9]. Real and reactive power loss minimization have been taken as O.F for optimal DG and capacitor allocation problem by a decomposition-based multiobjective evolutionary algorithm on 33, 69 and 119 radial distribution system [10]. Hybrid GMSA and SSA has been implemented to calculate optimal locations and sizes of DG and capacitor units on 33 and 69 bus system in [11, 12], respectively. In [13], authors have used voltage stability index and modified GA method to calculate DG and capacitor locations and sizes, respectively. SSA has been implemented for optimal DG and capacitor placement problem in 33-bus system by taking real power loss reduction, system voltage profile advancement and generation costs reduction as the O.F of the problem [14].

2   Problem Formulation

The simultaneous DGs and capacitors placement problem have been formulated to minimize the real power loss of the system. The advantages associated with DGs and capacitors placement basically depend on how optimally they have been placed in the distribution system. For an N-bus distribution system, formulation of the O.F has given below.

2.1 Objective Function

$$P_l = \sum_{k=1}^{N} \left(|I_k^2| \times R_k \right) \tag{1}$$

$$\mathcal{F} = \min(P_l) \tag{2}$$

2.2 System Constraints

The objective function has minimized subject to various system operational constraints. These constraints are.

Power flow constraints

For each and every bus of the distribution system, power flow equations must be satisfied. These equations are as following

$$P_{Gk} - P_{Dk} = \sum_{k=1}^{N}\sum_{m=1}^{N} V_k V_m (G_{km} \cos(\Theta_k - \Theta_m) + B_{km} \sin(\Theta_k - \Theta_m)) \tag{3}$$

$$Q_{Gk} - Q_{Dk} = \sum_{l=1}^{N}\sum_{m=1}^{N} V_k V_m (G_{km} \sin(\Theta_k - \Theta_m) - B_{km} \cos(\Theta_k - \Theta_m))$$
$$\forall_k = 1, 2, 3, \ldots, N \tag{4}$$

$$P_k = P_{Gk} - P_{Dk} \tag{5}$$

$$Q_k = Q_{Gk} - Q_{Dk} \tag{6}$$

P_k and Q_k are the net active and reactive power at ith bus, respectively. Q_{Dk} and Q_{Gk} are reactive power demand and generation at kth bus. P_{Dk} and P_{Gk} are active power demand and generation at kth bus. V_m and Θ_m are the voltage amplitude and voltage angle at mth bus.

Voltage limit Constraints

Voltage at each and every bus must be within the permissible limits.

$$V_{\min} < V_k < V_{\max} \quad \forall_k = 1, 2, 3, \ldots, N \tag{7}$$

DG and capacitor power limit constant

Generated power DGs and capacitors must be within permissible minimum and maximum limits.

$$P_{DG}^{min} < P_{DG} < P_{DG}^{max} \tag{8}$$

$$Q_{QC}^{min} < Q_{QC} < Q_{QC}^{max} \tag{9}$$

3 Combined Power Loss Sensitivity

Integration of DGs in the system not only influences the real power loss of the system but also system voltage profile and reactive power loss of the system. For these reason, combined power loss sensitivity (CPLS) [15] is calculated for both active and reactive power loss. Combined power loss sensitivity factor for kth bus are calculated as follows.

$$\frac{\partial P_{loss}}{\partial Q_k} = \frac{2 * Q_k * R_{lk}}{V_k^2} \tag{10}$$

$$\frac{\partial Q_{loss}}{\partial Q_k} = \frac{2 * Q_k * X_{lk}}{V_k^2} \tag{11}$$

CPLS with respect to reactive and real power can be defined as follows

$$\frac{\partial S_{loss}}{\partial Q_k} = \frac{\partial P_{loss}}{\partial Q_k} + j\frac{\partial Q_{loss}}{\partial Q_k} \tag{12}$$

$$\frac{\partial S_{loss}}{\partial P_k} = \frac{\partial P_{loss}}{\partial P_k} + j\frac{\partial Q_{loss}}{\partial P_k} \tag{13}$$

Loss sensitivity factor matrix (LSM) can be defined as follows

$$\text{LSM} = \begin{bmatrix} \frac{\partial P_{loss}}{\partial P_k} & \frac{\partial Q_{loss}}{\partial P_k} \\ \frac{\partial P_{loss}}{\partial Q_k} & \frac{\partial Q_{loss}}{\partial Q_k} \end{bmatrix} \tag{14}$$

where
P_k, Q_k = real and reactive power at kth bus, respectively.
P_{loss}, Q_{loss} = system total real and reactive power loss respectively.
S_{loss} = system total apparent power loss.
R_{lk} = resistance of the branch connected between bus l and bus k.
X_{lk} = reactance of the branch connected between bus l and bus k.

V_k = voltage at kth bus.

Normalized voltage (N.V) for kth bus can be calculated as follows

$$V(k) = V(k)/0.95 \tag{15}$$

All these parameters can be calculated from running the base case load flow programme. After calculating the LSM, they are organized in descendant order. Buses having higher value of CPLS and N.V less than 1.01 are selected for DG and capacitor allocation in the distribution system. This approach can significantly reduce the simulation time.

4 Optimization Algorithm

Grey Wolf Optimization (GWO) Algorithm is developed by Ali Mirjalili in 2014 [16]. GWO is a population-based algorithm; it mimics the hunting and group hierarchy behaviour of grey wolves. Grey wolves basically lives in a group that contain around 5–12 member. Group of grey wolves contain four types of member. First-level member of the group is known as alpha, they are responsible for taking decision, place of living. The next-level member of the group is known as beta. Betas are responsible for helping the alpha wolf in taking decision or other actions. The next-level member after betas is known as delta. They are basically plays a role of scout, hunters or caretaker of the groups. The elder member of the group also belongs to this category. The lowest level members of the group are omega. This activity shows that in a group or pack organization and discipline is more important than its strength. In this GWO algorithm, α, β and δ represents the first three best solutions. Other less fittest solution in denoted by ω. Along with this group hierarchical behaviour the hunting is also an exciting behaviour of grey wolves. Hunting of grey wolves is categorized into three phases which are as follows:

The following operators are used to exhibit the GWO algorithm mathematically.

i.   Following and approaching the prey
ii. Encircling around the prey and troubling the prey until the prey stops moving
iii. Attacking towards prey.

4.1 Group Hierarchy of GWO

In modelling of GWO algorithm, α is considered as the fittest solution. Next fittest solutions are β and δ, respectively. Rest of the solutions are ω.

4.2 Encircling of Prey

Grey wolf encircle the prey in their second stage of hunting. This behaviour can be modelled mathematically using the following Eqs. (16) from to (19).

$$\vec{H} = \left| \vec{G}.\overrightarrow{L_p(t)} - \overrightarrow{L(t)} \right| \tag{16}$$

$$\overrightarrow{L(t+1)} = \overrightarrow{L_p(t)} - \vec{F}.\vec{H} \tag{17}$$

$\vec{L}$ is the position vector of grey wolf. $\overrightarrow{L_P}$ is the position vector of prey. Here t is the current iteration. F and G are the coefficient vector. These F and G coefficient vectors are calculated as follows Eq. (18) from to (19).

$$\vec{F} = 2.\vec{p}.\vec{x}_1 - \vec{p} \tag{18}$$

$$\vec{G} = 2.\vec{x}_2 \tag{19}$$

Here in this algorithm p linearly decrease from two to zero over the iteration and x_1 and x_2 are two random vector uniformly distributed between [0, 1].

4.3 Hunting

Grey wolves hunt theirs prey directed by α wolf. B and δ also participate in the hunting of prey. To model these hunting activities of grey wolves, it is presume that α, β and δ have the better knowledge about prey (optimal solution in algorithm). Other ω wolves need to update their position according to α, β and δ. The following formulas are developed for updating position in accordance with Eqs. (20) to (26).

$$\vec{H}_\alpha = \left| \vec{G}.\overrightarrow{L_\alpha} - \overrightarrow{L_o} \right| \tag{20}$$

$$\vec{H}_\beta = \left| \vec{G}.\overrightarrow{L_\beta} - \overrightarrow{L_o} \right| \tag{21}$$

$$\vec{H}_\delta = \left| \vec{G}.\overrightarrow{L_\delta} - \overrightarrow{L_o} \right| \tag{22}$$

$$\overrightarrow{L_1} = \overrightarrow{L_\alpha} - \vec{F}.\left(\overrightarrow{H_\alpha} \right) \tag{23}$$

$$\overrightarrow{L_2} = \overrightarrow{L_\beta} - \vec{F}.\left(\overrightarrow{H_\beta} \right) \tag{24}$$

$$\vec{L_3} = \vec{L_\delta} - \vec{F}.\left(\vec{H_\delta}\right) \tag{25}$$

$$\vec{L_o}(t+1) = \frac{\left(\vec{L_1} + \vec{L_2} + \vec{L_3}\right)}{3} \tag{26}$$

Here L_α, L_β and L_δ are position vector of alpha, beta and delta wolves, respectively. $\vec{L_o}$ is the position vector of one omega wolf. $\vec{L_1}$, $\vec{L_2}$ and $\vec{L_3}$ are the distance of that omega wolf from alpha, beta and delta wolves, respectively.

4.4   Searching for Prey

Exploration of the optimization algorithm has done by this phenomenon. Grey wolf diverge from each other when searching for prey. This is denoted by $|F| > 1$. To model this phenomenon mathematically, $\vec{F}$ operator has been utilized. Here $\vec{F}$ operator value has been taken as either more than 1 or less than -1. $\vec{G}$ operator also has been used for exploration by taking its value between $[0, 2]$.

4.5   Attacking Towards Prey

Grey wolves converge to each other for attacking the prey. This is denoted by $|F| < 1$. This phenomenon has done the exploitation of the optimization problem.

5   Computational Procedure

The GWO algorithm has been executed on a 33 bus radial distribution system. The entire connected load of the test bus system is 3.715 MW and 2.3 MVAr [2]. MATLAB R2013a software has been used for executing the programme. Computational steps for performing the GWO algorithm have been described in flow chart of Fig. 1.

6   Numerical Results

Backward forward sweep (BFS) power flow method [17] has been used to calculate the 33 bus system profile before placing the DGs or capacitors in the system. After that combined power loss sensitivity has been used to calculate the most sensitive buses out of all the 32 load buses. GWO algorithm has been implemented on

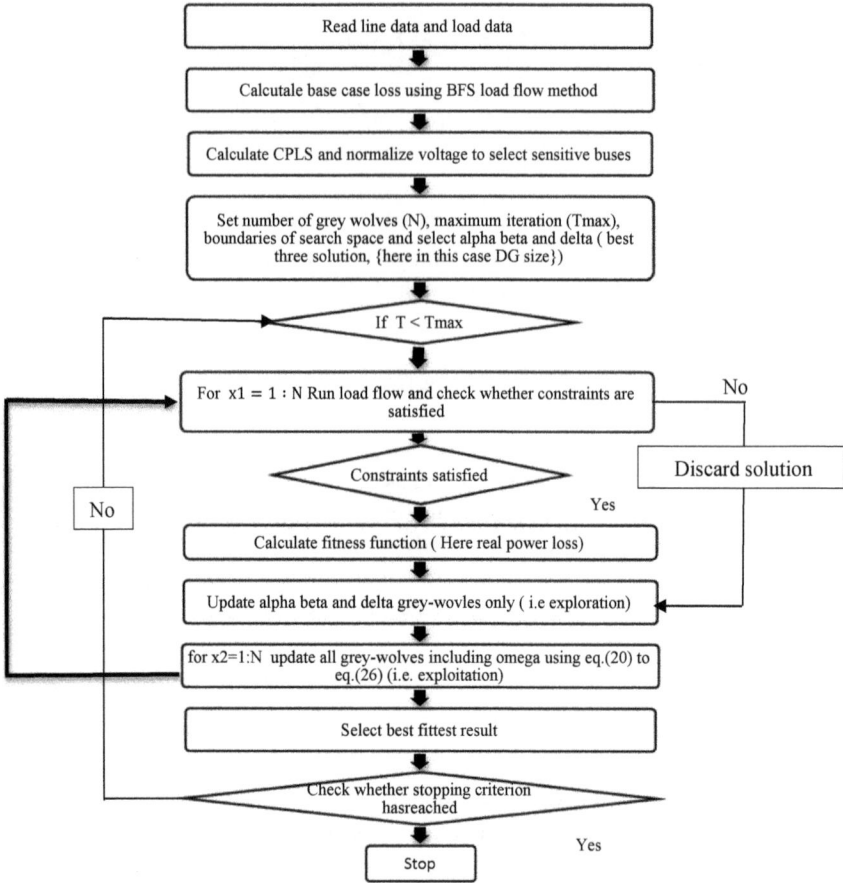

Fig. 1 Computational steps of the problem

these sensitive buses for optimal sizing and placement of DGs and capacitors units. Table 1, described the summary of results obtained for different cases after GWO implementation to the system. From Table 1, it can clearly be observed that before connecting any devices to the system active power loss was 210.9856 kW. After placement of two DGs units, system active power loss has reduced by 58.69% that of base case. Further, allocation of two capacitors into the system has reduced the real power loss by 32.736% that of base case. Simultaneous capacitor and DG placement have reduced the power loss of the system by 82.94% as compare to base case. Table 1 also described that the advancement in system minimum voltage is evident when simultaneously two DG and one Qc are connected to the system. Minimum voltage of the system was 0.9038 p.u at bus 18 before installation of DG or Qc. After two DG allocation, minimum bus voltage of the system becomes 0.9684738 p.u. at bus 33. An 7.15% increment in minimum bus voltage has observed after two DG allocation. After two capacitors placement, minimum voltage of the system has increased

Table 1 Summary of results obtained

33 bus system			
Device	Parameters		
	Real power loss (kW)	Minimum voltage (at bus number)	Location (bus no.) and Size (kW/kVAr)
Without DG	210.9856	0.9038 (18)	–
Two DGs	87.16592509	0.9684738 (33)	1. 13 and 849.4651114 (kW) 2. 30 and 1157.9404133 (kW)
Two capacitors	141.9184742	0.9304106 (18)	1. 10 and 489.4939319 (kVAr) 2. 30 and 1053.0759449 (kVAr)
Simultaneous DGs and Qc	35.9841935	0.9793 (18)	1. 13 and 837.6351762 (kW) 2. 30 and 1146.8612738 (kW) 3. 30 and 1239.8685488 (kVAr)

by 2.94%. In this case, minimum voltage becomes 0.9304106 p.u at bus 18. Simultaneous DG and capacitor placement in the system results in rise in minimum voltage upto 8.35% of the base case. Figure 2 shows the system voltage profile for all the cases, i.e. without DG or capacitor, with two DGs, with two capacitors, simultaneous two DGs and one capacitor.

Table 2 provides the comparison of obtained results for only DG allocation from GWO with other reported results, i.e. analytical approach, PSO, hybrid PSO, backtracking algorithm, Moth-Flame algorithm, etc. Tables 3 and 4 provide the comparison of obtained results for only capacitors and simultaneous DG and capacitor allocation, respectively, with other reported results, i.e. analytical, hybrid PSO, modified GA, MOPSO, BFOA, etc.

It can clearly be observed from the above table that the objective function has minimized best by implementing GWO algorithm under among other compared algorithm. Among all three cases the objective function has minimized best when simultaneous two DGs and one capacitors are installed to the system. In the case of simultaneous DG and capacitor real power loss has minimized by 83.22% and minimum voltage has improved by 8.35%.

Figure 3 shows the convergence characteristics for two DGs placement. This results show that it has been converged in 29 iterations. In Figs. 4 and 5, illustrate convergence characteristics of two capacitors unit placement and simultaneous DGs

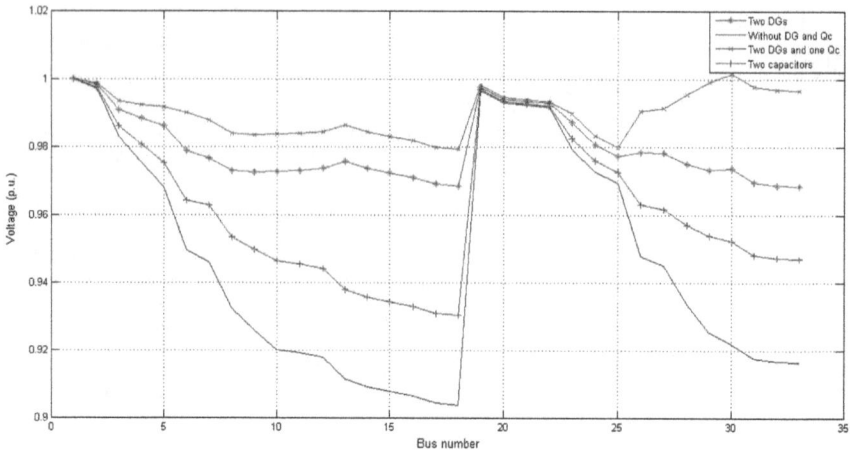

Fig. 2 Comparison of voltage profile for different cases

Table 2 Comparison of results for only DGs allocation

Algorithm	DGs locations (at bus no.) and sizes (MW)	Real power loss (Kw)	Minimum voltage (p.u.) at bus no	% Reduction in real power loss
Analytical [2]	(18) and 1	142.34	0.9311 at 33	33.29
PSO [3]	1. (13) and 0.85 2. (30) and 1.16	87.28	–	58.64
Hybrid PSO [8]	1. (13) and 0.83 2. (30) and 1.11	87.17	0.9680 (18)	58.69
BFOA [6]	1. (17) and 0.6335 2. (18) and 0.0908 3. (33) and 0.9470	98.30	0.9645	53.41
Back-tracking [7]	1. (13) and 0.88 2. (31) and 0.924	89.24	0.9665	57.62
Moth-Flame [12]	1. (13) and 0.8516 2. (30) and 1.1575	87.166	0.9685 (33)	58.68
GWO (Proposed)	1. (13) and 0.8494651114 2. (30) and 1.1579404133	87.16592509	0.9684738 (33)	58.69

and Qc placement have been shown respectively. For two capacitors, placement characteristics have been converged in 32 iterations, while for simultaneous DGs and Qc placement characteristics have been converged in 35 iterations.

Table 3 Comparison of results for only capacitors allocation

Algorithm	Capacitors locations (at bus no.) and sizes (MVAr)	Real power loss (Kw)	Minimum voltage (p.u.) at bus no	% Reduction in real power loss
Analytical [2]	(33) and 1	164.6	0.91654 at 18	22.83
Hybrid PSO [9]	1. (12) and 0.43 2. (30) and 1.04	141.94	0.93 (18)	32.73
BFOA [6]	1. (18) and 0.3496 2. (30) and 0.8206 3. (33) and 0.2773	144.04	0.9361	31.72
GWO (Proposed)	1. 10 and 0.4894939319 2. (30) and 1.0530759449	141.9184742	0.9304106 (18)	32.736%

7 Conclusion

In this paper, a population-based nature-inspired algorithm known as grey wolf optimization has been used for DG and capacitor allocation in 33-bus radial distribution system. The BFS power flow method has been used for determination of the base case power loss. After calculation of base case system parameters (i.e. real power loss, system voltage profile, etc.), CPLS has been calculated for all load buses. Sensitive buses have been selected after calculation of CPLS from all the load buses. GWO algorithm has been applied on these sensitive buses to calculate optimal locations and sizes of DGs and capacitors. The presence of simultaneous DGs and Qc unit in distribution network causes best minimization in active power losses and advancement in the system voltage profile. The comparison of obtained results from GWO has been compared with other recently reported results, i.e. PSO, hybrid PSO, backtracking algorithm, BFOA, etc., to demonstrate effectiveness of implemented GWO algorithm.

Table 4 Comparison of results for simultaneous DG and capacitor allocation

Algorithm	DG/capacitor locations (at bus no.) and sizes (MW/MVAr)		Real power loss (Kw)	Minimum voltage (p.u.) at bus no	% Reduction in real power loss
Analytical [2]	DG	(18) and 1	96.7	0.96003(30)	54.66
	Capacitor	(33) and 1			
PSO [3]	DG	(6) and 2.5317	58.45	0.9570 (18)	72.29
	Capacitor	(30) and 1.2258			
BFOA [6]	DG	1. (17) and 0.5424 2. (18) and 0.1604 3. (33) and 0.8955	41.41	0.9783	80.37
	Capacitor	1. (18) and 0.1632 2. (30) and 0.5410 3. (33) and 0.3384			
Hybrid PSO [9]	DG	(6) and 2.483	58.51	–	72.27
	Capacitor	(30) and 1.223			

(continued)

Table 4 (continued)

Algorithm	DG/capacitor locations (at bus no.) and sizes (MW/MVAr)		Real power loss (Kw)	Minimum voltage (p.u.) at bus no	% Reduction in real power loss
Modified GA [16]	DG	1. (6) and 0.764 2. (3) and 1.930 3. (28) and 1.047	43.6947	0.9641 (18)	79.29
	Capacitor	1. (6) and 0.550 2. (3) and 1.050 3. (28) and 0.750			
MOPSO [8]	DG	1. (9) and 0.911 2. (23) and 0.669 3. (30) and 1.423	80.8	–	65.24
	Capacitor	1. (10) and 1.05 2. (21) and 1.2			
GWO (Proposed)	DG	1. 13 and 0.8376351762 2. 30 and 1.1468612738	35.9841935	0.9793 (18)	83.22
	Capacitor	1. 30 & 1239.8685488 (kVAr)			

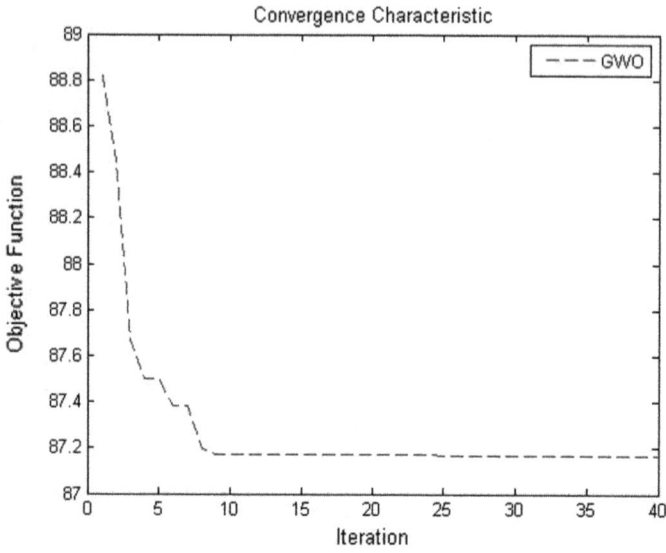

Fig. 3 Convergence characteristics for two DGs

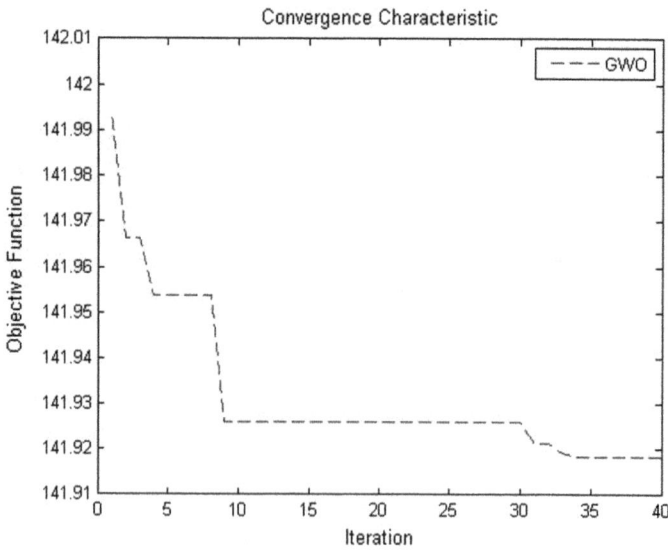

Fig. 4 Convergence characteristics for two capacitors

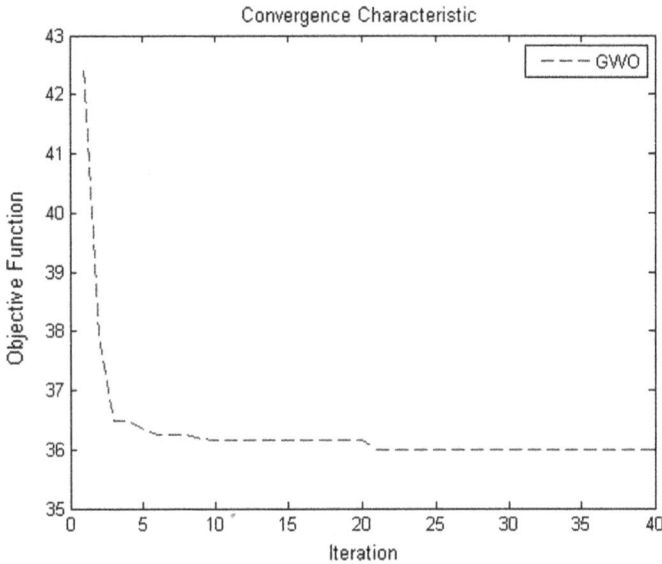

Fig. 5 Convergence characteristics for simultaneous DG and capacitor

References

1. Kansal S, Kumar V, Tyagi B (2013) Optimal placement of different type of DG sources in distribution networks. Int J Electr Power Energy Syst 53:752–760
2. Gopiya Naik S, Khatod DK, Sharma MP (2013) Optimal allocation of combined DG and capacitor for real power loss minimization in distribution networks. Electr Power Energy Syst 53:967–973
3. Yadav A, Srivastava L (2014) Optimal placement of distributed generation: an overview and key issues. In: International conference on power, signals, controls and computation, Thrissur, January, 2014
4. Kowsalya M, Mohamed Imran A (2014) Optimal distribution generation and capacitor placement in power distribution networks for power loss minimization. In: International conference on advances in electrical engineering, Vellore, 2014
5. El-Fergany A (2015) Optimal allocation of multi-type distributed generators using backtracking search optimization algorithm. Electr Power Energy Syst 64:1197–1205
6. Zeinalzadeh A, Mohammadi Y, Moradi MH (2015) Optimal multi objective placement and sizing of multiple DGs and shunt capacitor banks simultaneously considering load uncertainty via MOPSO. Electr Power Energy Syst 67:336–349
7. Kansal S, Kumar V, Tyagi B (2016) Hybrid approach for optimal placement of multiple DGs of multiple types. Int J Electr Power Energy Syst 75:266–235
8. Rahmani-Andebili M (2016) Simultaneous placement of DG and capacitor in distribution network. Electr Power Syst Res 131:1–10
9. Das A, Srivastava L (2017) Optimal placement and sizing of distribution generation units for power loss reduction using moth-flame optimization algorithm. In: International conference on intelligent computing, instrumentation and control technologies, July, 2017
10. Biswas PP, Mallipeddi R, Suganthan PN, Amaratunga GAJ (2017) A multiobjective approach for optimal placement and sizing of distributed generation and capacitors in distribution network. Appl Soft Comput 60:268–280

11. Mohamed EA, Ali Mohamed A-A, Mitani Y (2018)Hybrid GMSA for optimal placement and sizing of distributed generation and shunt capacitors. J Eng Sci Rev 11:55–65
12. Smbaiah K, Jayabarathi T (2019) Optimal placement of renewable distribution generation and capacitor banks in distribution systems using Salp swarm algorithm. Int J Renewab Energy Res 9(1): 96–107
13. Das S, Das D, Patra A (2019) Operation of distribution network with optimal placement and sizing of dispatchable DGs and shunt capacitoes. Renewab Sustain Energy Rev 113
14. Dehghani M, Montazeri Z, Malik OP (2020) Optimal sizing and placement of capacitor banks and distributed generation in distribution system using spring search algorithm. Int J Emerg Electr Power System 21(1): 1–7
15. Murty VVSN, Kumar A (2013) Comparison of optimal DG allocation methods in radial distribution systems based on sensitivity approaches. Int J Electr Power Energy Syst 53:246–256
16. Mirjalili S, Mirjalili SM, Lewis A (2014) Grey-wolf optimizer. Adv Eng Software 69:46–61
17. Ravi Teja B, Murthy VVSN, Kumar A (2016) An efficient and simple load flow approach for radial and meshed distribution networks. Int J Grid Distribut Comput 9(2): 85–102

Chapter 60
Real-Time COVID-19 Detection and Prediction Using Chest X-rays and CT Scan: A Comparative Study Using AI

Dhwani Trivedi, Meha Dave, Rutvik Patel, Vatsa Dave, and Ghansyam Rathod

1 Introduction

An acute transmissible respiratory disease widely referred to as COVID-19 spread all over the globe in 2020 like wildfire, leaving people helpless. It was initiated by a virus SARS-CoV-2 and affected the mass in mammoth figures, precisely millions. The ex-ponentially increasing death toll, the lack of any efficient cure or vaccine coupled with the prolonged incubation duration amid the virus affecting a human to the development phase of their symptoms has led to concerns over how the patient can get admitted earlier and flinch with the prevailing medication to curb the disease.

Individuals infected by the virus may leave droplets on surfaces and entities, referred to as fomites when they touch somewhere, cough, or sneeze. These surfaces might be touched by others, then touching themselves before washing their hands thereby falling prey to the transmitted infection. Small droplets containing the virus known as aerosolized droplet nuclei or aerosols can remain suspended in the air. If inhaled by anyone who is not wearing appropriate personal protective equipment, then it can infect them as well. Aerosol transmission, predominantly in enclosed

D. Trivedi (✉) · M. Dave · R. Patel · V. Dave · G. Rathod
Birla Vishwakarma Mahavidyalaya, Anand, Gujarat 388120, India
e-mail: dhwanitrivedi999@gmail.com

M. Dave
e-mail: davemeha60@gmail.com

R. Patel
e-mail: rutvikpatel8@gmail.com

V. Dave
e-mail: vatsa2331@gmail.com

G. Rathod
e-mail: ghansyam.rathod@bvmengineering.ac.in

places which are crowded or inadequately ventilated spaces, where infected persons spend long periods of time with others, can cause local transmission and recklessness could cost one their life.

The situation in developing countries such as India and Brazil has led researchers to believe that nowhere in the near future, the case curve would flatten. Testing is quite difficult, considering resources and the enormous population. When this coincides with a developing country, where hygiene might be lesser than the fellow developed ones, chances of the disease spread escalate.

Research, as stated in [1] observed the clinical characteristics in COVID-19 positively tested close associates of COVID-19 patients. About 30% of those COVID-19 acquaintances who showed positive never developed any symptoms or changes on PCR tests. The remainder showed changes in CT, but around 20% reportedly developed symptoms during their hospital course, none of them developed the severe disease, suggesting that a high percentage of COVID-19 carriers are asymptomatic. This has led to concerns regarding the detection of the reliability of these tests.

Despite the fact that PCR tests are very specific, they have a subordinate sensitivity of 65–95%, indicating that the test can depict a negative result even when the patient is diseased. One more problem is that one has to wait for the test results, which end up taking more than 24 h, while CT results are available on the spot. COVID-19 infected personnel diagnosed in laboratories have been found to exhibit a decreased lymphocyte count and a high-sensitivity C-reactive protein (CRP) level [2–13].

In recent times, numerous cases in India claimed to get infected by coronavirus again, within a week of recovering from it. Research on this matter indicated that the patients never recovered from the previous infection, the PCR-test gave a false negative. This false negative could be due to manual errors while taking the throat or nasal swab, or mishandling of test samples during transportation, etc. These errors are eliminated while detecting COVID-19 through X-Ray or CT scan, as shown by the authors in [14]. Hence, X-Rays and CT scans can be used for early diagnosis and during discharge, respectively. We use these two methods and use deep learning algorithms using CNN and transfer learning approach to get these results with commendable accuracy (Fig. 1).

2 Methodology

COVID-19 test kits are scarce and are not available easily. Even if they were accessible, we do not have enough of them in quantity to suffice the demand. With the view that there are insufficient test kits, we need to look out for other diagnostic techniques. As it is known to us that the epithelial cells which are found in our respiratory tracts are being assailed by COVID-19, we can use CT scans and X-rays to assess the health of a patient's lungs [15, 16]. In this preemptive study for the COVID-19 detection using CT scans and X-rays, a deep learning model was developed using neural networks to identify indications of COVID-19 from an abundance of CT scans

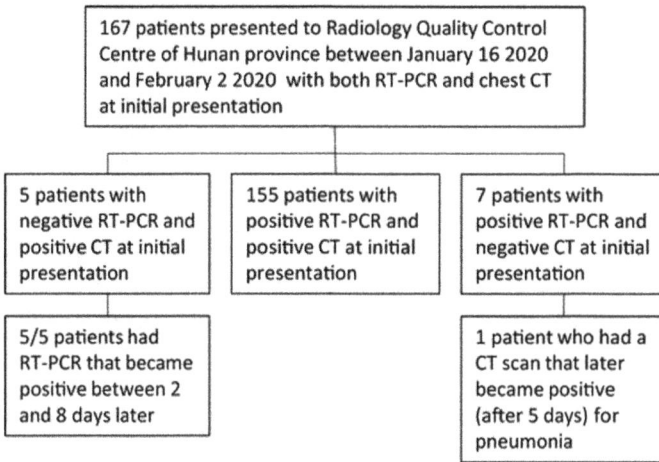

Fig. 1 Results of a study conducted on 167 patients (*Source* https://pubs.rsna.org/doi/10.1148/rad iol.2020200343)

and X-rays community-obtained scans of specific patients having COVID-19 as well as the CT scans and X-rays of people having no ailments.

2.1 Dataset Used for the Study

To conduct the experiment and to prepare a model of our system, various CT scans and X-rays from verified sources were obtained. The samples of the CT scans and X-rays were obtained which were classified as COVID-19 patients and non-COVID patients (normal patients). The dataset that has been compiled contains 840 images of CT scans and chest X-rays which include 420 images each of the people affected by COVID-19, and the ones which were not influenced, respectively, [17–20] (Fig. 2).

They were further labeled as CT_covid and CT_noncovid for the CT scans. Similarly, the chest X-rays were labeled as X_covid and X_noncovid. The images

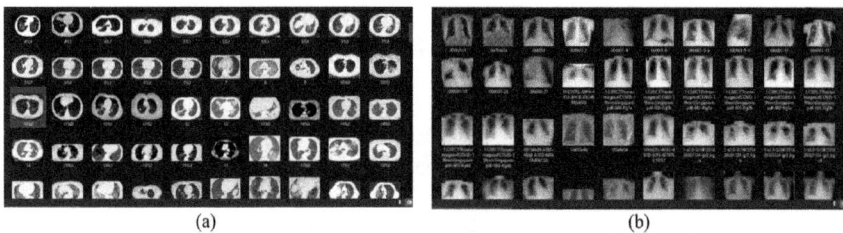

Fig. 2 Dataset used for the study: **a** CT scan and **b** X-ray

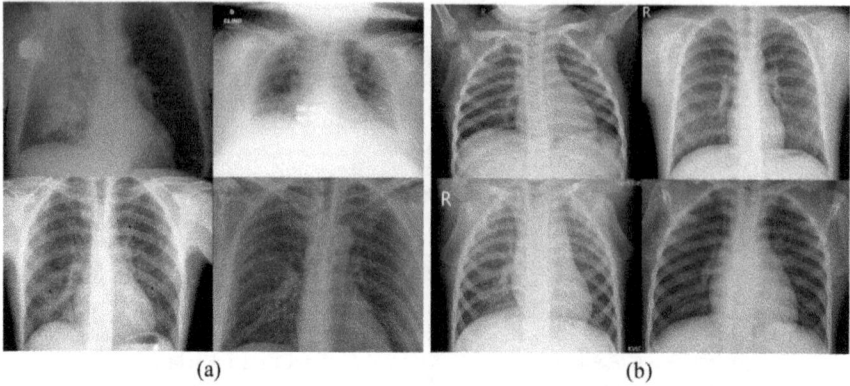

Fig. 3 Lung X-Ray images of **a** COVID-19 patients and **b** healthy patients

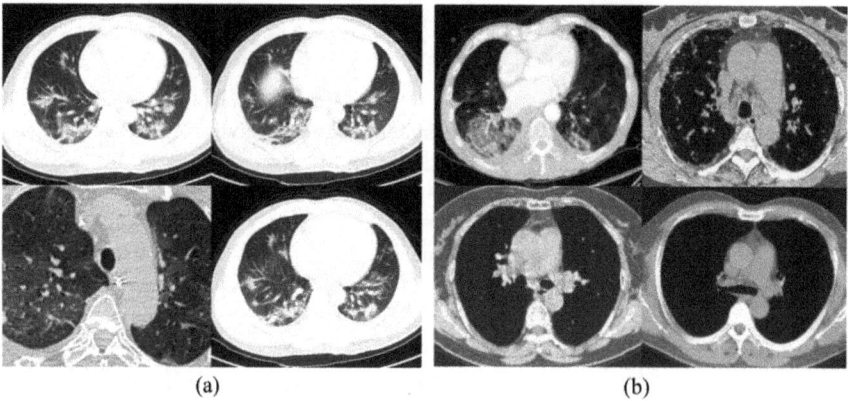

Fig. 4 Lung CT Scans of **a** COVID-19 patients and **b** healthy patients

procured from the Internet for our dataset were not all similar. The orientations of some of the images obtained were not precise. A few were tilted, and a few were found in the inverted position. All the images were regulated and set according to our requirements. Some images were marked or not perfectly noticeable. Those images were removed as they would wither our system's accuracy. Improper and blurred images were also removed to increase the precision of our system (Figs. 3 and 4).

2.2 Typical COVID-19 Lung Patterns

As per the inspections of many confirmed COVID-19 patients, it can be perceived that distinct patterns are developed inside their lungs. These patterns can be indexed as follows:

Ground Glass Opacification (GGO) Ground glass opacification is a term defined to cite a particular area of increased attenuation in the lung. In the preliminary stage of the disease, it may be present as a unifocal lesion; otherwise, they are usually peripheral, bilateral, and multifocal. GGO pattern is most frequently found in the inferior lobe of the right lung of COVID-19 patients.

Crazy paving: In crazy paving, we usually spot thickened interlobular and intralobular lines in a blend with a ground glass pattern.

Vascular dilatation A standard finding in the region of ground glass is the broadening of the vessels (as visualized in the figure).

Traction bronchiectasis Traction bronchiectasis is a twisting of the aviation routes auxiliary to the mechanical footing on the bronchi from fibrosis of the encompassing lung parenchyma.

Subpleural bands and architectural distortion In some cases, there is architectural distortion with the formation of thin curvilinear opacities, 1–3 mm in thickness, lying under 1 cm from the pleural surface (Fig. 5).

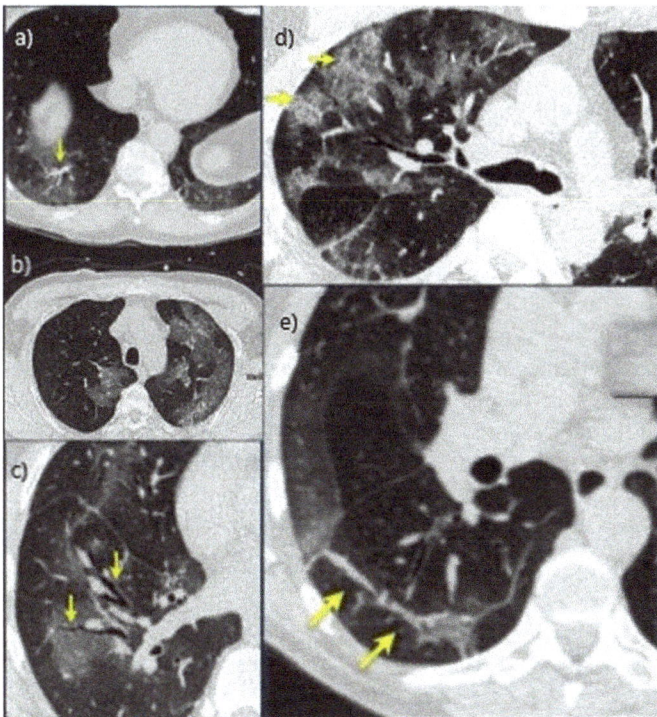

Fig. 5 **a** Vascular dilation, **b** ground glass opacification, **c** traction bronchiectasis, **d** crazy paving, **e** subpleural bands

Table 1 Common patterns and distribution on initial CT images of 919 patients COVID-19 [21]

Lung pattern in COVID-19	Distribution percentage (%)
Ground glass opacification	88
Bilateral involvement	88
Posterior distribution	80
Multilobar involvement	79
Peripheral distribution	76
Consolidation	32

These patterns can be noticed as they change over a period of time in a patient's lung. In the early stage (0–4 days), we can notice GGO, partial crazy paving with a low number of involved lobes. In the progressive stage (5–8 days), we can see the extension of GGO and an increased crazy paving pattern. In the peak stage (10–13 days), the locale of ordinarily compressible lung tissue loads up with fluid rather than air (Table 1).

2.3 CNN Aided Transfer Learning

Transfer learning is a procedure, wherein the information mined by a CNN from given information is transferred to comprehend a different, however, related assignment, including new information, which for the most part is of a littler populace to prepare a CNN from the beginning. As of late, artificial intelligence (AI) utilizing deep learning innovation has exhibited extraordinary achievement in the clinical imaging space because of its high capacity for feature extraction.

In this study, libraries like Keras, NumPy, and Panda were used for image classification. The dataset was trained using CNN. Common hyper-parameters are shared in CNNs here, and the convolution neural layers are activated by sigmoid; here, a dropout layer is added to prevent overfitting. The inception model is used which is a pre-trained model by Google, where more than 3 lakh images have been trained. Hence, it will be functional when the dataset is small, i.e., less data. Because of this, the accuracy will increase and will give reliable output. The dataset was divided into three categories: (a) Training dataset (b) Validation set (c) Test dataset to prevent overfitting or underfitting.

3 Metrics

With respect to characterization undertaking of the CNNs, explicit measurements were recorded as follows: (a) correctly recognized COVID-19 cases (True positives, TP), (b) incorrectly recognized COVID-19 cases (False negatives, FN), (c) correctly

		ACTUAL	
		Positive	Negative
PREDICTED	Positive	These values are our positive tests AND they have been predicted as positive too. This is True Positive (TP)	These values are our negative tests BUT they have been predicted as positive. This is False Positive (FP)
	Negative	These values are our positive tests BUT they have been predicted as negative. This is False Negative (FN)	These values are our negative tests AND they have been predicted as negative too. This is True Negative (TN)

Fig. 6 After undertaking CNN, the values are categorized as predicted and actual (*Source* https://medium.com/fintechexplained/classification-evaluation-for-data-scientists-28d84d30dbb)

recognized non-COVID-19 cases (True negatives, TN), and (d), incorrectly recognized non-COVID-19 cases (False positives, FP). It is important to mark that TP indicates the accurately speculated COVID-19 cases by the CNN; FP indicates the pneumonia cases that were speculated as COVID-19; TN indicates pneumonia or normal cases that were speculated as non-COVID-19 cases, while FN indicates the COVID-19 cases speculated as pneumonia or as normal cases. This characterization was carried out both for CT scans and X-ray datasets (Fig. 6).

Items from the test dataset were labeled as actual positive and actual negative. Then, the model started to produce the results. The results produced by the model were labeled as predicted positive and predicted negative. After getting various parameters such as TN, TP, FP, and FN, the confusion matrix was generated separately for both the datasets (X-ray and CT scan), where we can find the actual COVID-19 data and the non-COVID-19 data likewise with the normal and pneumonia cases.

In view of the above measurements, we evaluate the specificity, accuracy and sensitivity of the model. The equations defining the previously mentioned measurements are Eq. 1, 2, and 3, respectively:

$$Specificity = TN/(TN + FP) \tag{1}$$

$$Accuracy = (TP + TN)/(TP + TN + FP + FN) \tag{2}$$

$$Sensitivity = TP/(TP + FN) \tag{3}$$

Sensitivity means that it tells us the probability of a patient to test positive given that person has any condition or symptom. So, a better way to calculate these values there are two other probability methods that can be used for this problem.

Positive Predictive Value (PPV) is the probability that means when the patient tests positive when a patient has a disease.

Negative Predictive Value (NPV) is the probability that means when the patient tests negative when a patient does not have a disease.

$$PPV = TP/(TP + FP) \qquad (4)$$

$$NPV = TN/(FN + TN) \qquad (5)$$

4 Results

The model was trained, validated, and tested rigorously throughout our research for both the datasets (CT scan and X-ray). The confusion matrix for each dataset for CNN Inception v3 is depicted in Table 2. In Table 3, the sensitivity, specificity, and accuracy for each dataset are illustrated. Table 4 represents the Positive Predictive Value (PPV) and the Negative Predictive Value (NPV) for the CT scan and X-ray dataset.

Table 2 Confusion matrix for CT scan and X-ray dataset for Inception v3 CNN

CNN (Dataset)	True Positives (TP)	True Negatives (TN)	False Positives (FP)	False Negatives (FN)
Inception v3 (CT Scan)	105	82	10	13
Inception v3 (X-ray)	90	63	20	37

Table 3 Accuracy, sensitivity, and specificity for CT scan and X-ray dataset for Inception v3 CNN

Dataset	Accuracy (%)	Sensitivity (%)	Specificity (%)
CT scan	89.04	88.98	89.13
X-ray	72.37	70.21	63.03

Table 4 Measurement of Positive Predictive Value (PPV) and Negative Predictive Value (NPV) for CT scan and X-ray for Inception v3 CNN

Dataset	Positive Predictive Value (PPV) (%)	Negative Predictive Value (NPV) (%)
CT scan	91.34	86.31
X-ray	81.81	63.00

After examining the results encapsulated in Table 3, it implies that the CT scan had better accuracy (89.04%) in distinguishing COVID-19 cases from pneumonia and normal cases than the X-ray (72.37%) for the same Inception v3 CNN.

5 Conclusion

In light of the outcomes, it is shown that transfer learning with CNNs may have significant effects on the autonomous extraction and recognition of crucial highlights from X-ray as well as CT scan images, associated with the prognosis of COVID-19. The proposed work adds to the odds of automatic and spontaneous analysis of the COVID-19. Besides the fact that the pertinent treatment is not deduced exclusively from an X-ray or CT scan image, prior screening of the coronavirus cases would be valuable, not for the purpose of treatment, but for the early implementation of isolation measures in the case of positive patients, until a progressively holistic assessment and a certain treatment are established. Another major advantage of the proposed approach is the curtailment of exposure of front-line medical workers to the pandemic.

The accuracy, sensitivity, and specificity of the exemplified model can be increased by carrying out an extensive research based on a much more substantial COVID-19 patient dataset in future. Furthermore, the model can be enhanced to differentiate COVID-19 cases from other respiratory diseases like MERS, SARS, etc.

References

1. Early Epidemiological and Clinical Characteristics of 28 Cases of Coronavirus Disease in South Korea, https://pubmed.ncbi.nlm.nih.gov/32149037. Last accessed 21 July 2020
2. Apostolopoulos I, Mpesiana T (2020) Covid-19: automatic detection from X-ray images utilizing transfer learning with convolutional neural networks. Phys Eng Sci Med 43(2):635–640
3. Xu Z, Shi L, Wang Y, Zhang J, Huang L, Zhang C, Liu S, Zhao P, Liu H, Zhu L, Tai Y, Bai C, Gao T, Song J, Xia P, Dong J, Zhao J, Wang F (2020) Pathological findings of COVID-19 associated with acute respiratory distress syndrome. Lancet Respiratory Med 8(4):420–422
4. Li L, Qin L, Xu Z, Yin Y, Wang X, Kong B, Bai J, Lu Y, Fang Z, Song Q, Cao K, Liu D, Wang G, Xu Q, Fang X, Zhang S, Xia J, Xia J (2020) Using artificial intelligence to detect COVID-19 and community-acquired Pneumonia based on pulmonary CT: evaluation of the diagnostic accuracy. Radiology 296(2):E65–E71
5. Greenspan H, van Ginneken B, Summers R (2016) Guest editorial deep learning in medical imaging: overview and future promise of an exciting new technique. IEEE Trans Med Imaging 35(8):1153–1159
6. Deng L (2014) Deep learning: methods and applications. Found Trends® Sign Process 7(3):197–387
7. Ronneberger O, Fischer P, Brox T (2015) U-Net: convolutional networks for biomedical image segmentation. In Navab N, Hornegger J, Wells W, Frangi A (eds) Medical image computing and computer-assisted intervention—MICCAI 2015. MICCAI 2015. Lecture notes in computer science, vol 9351, pp 234–241. Springer, Cham

8. Krizhevsky A, Sutskever I, Hinton G (2017) ImageNet classification with deep convolutional neural networks. Commun ACM 60(6):84–90
9. Simonyan K, Zisserman A, Very deep convolutional networks for large-scale image recognition, https://arxiv.org/abs/1409.1556. Last accessed 24 June 2020
10. Weiss K, Khoshgoftaar T, Wang D (2016) A survey of transfer learning. J Big Data 3(1):124–139
11. Bahrami-Motlagh H, Sanei Taheri M, Abbasi S, Haghighimorad M, Salevatipour B, Alavi Darazam I (2020) Accuracy of low-dose chest CT scan in detection of COVID-19. Radiol: Cardiothoracic Imag 2(3):e200256
12. Dangis A, Gieraerts C, Bruecker Y, Janssen L, Valgaeren H, Obbels D, Gillis M, Ranst M, Frans J, Demeyere A, Symons R (2020) Accuracy and reproducibility of low-dose submillisievert chest CT for the diagnosis of COVID-19. Radiol: Cardiothoracic Imag 2(2):e200196
13. Ai T, Yang Z, Hou H, Zhan C, Chen C, Lv W, Tao Q, Sun Z, Xia L (2020) Correlation of chest CT and RT-PCR testing for coronavirus disease 2019 (COVID-19) in China: a report of 1014 cases. Radiology 296(2):E32–E40
14. Ozturk T, Talo M, Yildirim E, Baloglu U, Yildirim O, Rajendra Acharya U (2020) Automated detection of COVID-19 cases using deep neural networks with X-ray images. Comput Biol Med 121:103792
15. Tackling the coronavirus pandemic. https://uwaterloo.ca/stories/eng-news-tackling-corona virus-pandemic. Last accessed 25 Apr 2020
16. CT Scans provide best diagnosis for covid-19. https://www.itnonline.com/content/ct-provides-best-diagnosis-novel-coronavirus-covid-19. Last accessed 15 May 2020
17. CT Scan dataset about COVID-19. https://github.com/UCSD-AI4H/COVID-CT. Last accessed 21 May 2020
18. Chest X-Ray Dataset for COVID-19. https://github.com/ieee8023/covid-chestxray-dataset. Last accessed 27 May 2020
19. COVID-Net. https://github.com/lindawangg/COVID-Net. Last accessed 30 Apr 2020
20. Kaggle X-ray dataset. https://www.kaggle.com/andrewmvd/convid19-X-rays. Last accessed 05 May 2020
21. COVID-19 Imaging Findings. https://radiologyassistant.nl/chest/covid-19/covid19-imaging-findings. Last accessed 06 May 2020

Chapter 61
Heart Disease Prediction—An Online Consultation Software

Monalisa Dey, Anupam Mondal, and Darothi Sarkar

1 Introduction

Heart diseases are considered as one of the most prevalent fatal diseases. Different surveys have been conducted by medical practitioners to gather important information on heart disease, their types and symptoms. The term "heart disease" includes various diseases and the symptoms they show which affect the heart severely. Report form World Health Organization states that the number of people suffering from heart disease is on the rise [1–3]. The database of the healthcare centres and the hospitals has a huge pool of data which if processed properly can be proved to be a boon in the field of medical science. The hospital management system keeps a track of patient's data which are rarely used to make any medical analysis. The system is neither very intelligent nor user friendly. The information systems of the hospital generate huge numbers of data, texts, charts or images. Using machine learning algorithms on proper datasets, an automated system can be generated, which can predict the symptoms of heart disease. In this paper, we propose a web application, from which users can get instant guidance on their heart disease. The system uses machine learning algorithms to predict the result with maximum precision and extract hidden relation between data and disease, using which the system enhances its intelligence [4]. The paper proposes to use data mining in the field of medical science using which prediction of heart diseases can be assisted [5–7]. New approach to find hidden patterns among various data can be provided, and human biasness can also be avoided. Using

M. Dey (✉) · A. Mondal (✉) · D. Sarkar
Department of Computer Science & Engineering, Institute of Engineering & Management, Kolkata, India
e-mail: monalisa.dey@iemcal.com

A. Mondal
e-mail: anupam.mondal@iemcal.com

D. Sarkar
e-mail: darothi.sarkar@iemcal.com

this system, costs of medical tests can be reduced and unwanted medical errors can be avoided which in turn enhances the precision of the result generated and safety of the concerned patient.

2 Methodology

In this system, we are implementing effective heart disease prediction system, using various machine learning algorithms. The dataset contains 13 features and 1 target value. The target value depicts if the patient has a heart disease or not. As this is an end-user application, so the users can input their data in the website which are necessary for diagnosis. Exploratory data analysis (EDA) has been done to get familiar with our data set and perform the necessary data processing part before training our model. Once, this has been done and necessary data processing is completed, we have trained our model with various machine learning algorithms. The algorithms that were used were logistic regression, K-nearest neighbour classifier, decision tree, random forest classifier, support vector machine (SVM) and naive Bayes algorithm. This work mainly encompasses two major fields of computer science, namely machine learning and web development. The subsections of this paper are:

1. Importing the required libraries.
2. Exploratory data analysis.
3. Data processing.
4. Training the model.
5. Testing the model.
6. Website development.

Let us sequentially explore the subsections one by one.

Importing the required libraries—The first thing while implementing machine learning is to import the necessary libraries. The libraries imported were NumPy, Panda, Matplotlib, Seaborn, Scikit-learn. Each library has its own features and requirements. NumPy was used to process multidimensional arrays and perform mathematical operations. Panda was used to handle data structures, numerical tables and axis indexing. Matplotlib was used to plot 2D graphs which were further visualized using the features of seaborn library. Sci-kit learn provides all the basic machine learning algorithms which were used to train the model.

Exploratory Data Analysis—EDA was performed to gather as many insights from our data as possible before training our model. The dataset consisted of 14 columns and 303 rows. Out of these, 14 columns 13 were features and one was the target attribute (target) depicting if the person has a heart disease or not. Among these 13 features, sex, cp, fbs, restecg, exang, slope, ca, thal have categorical values, and age, trestbps, chol, thalach, oldpeak have continuous values. Our dataset has neither got any null values nor any string values. After having a basic "data visualization",

Fig. 1 Dataset used

importing the data set

```
ds = pd.read_csv('heart.csv')
ds.info()

<class 'pandas.core.frame.DataFrame'>
RangeIndex: 303 entries, 0 to 302
Data columns (total 14 columns):
age         303 non-null int64
sex         303 non-null int64
cp          303 non-null int64
trestbps    303 non-null int64
chol        303 non-null int64
fbs         303 non-null int64
restecg     303 non-null int64
thalach     303 non-null int64
exang       303 non-null int64
oldpeak     303 non-null float64
slope       303 non-null int64
ca          303 non-null int64
thal        303 non-null int64
target      303 non-null int64
dtypes: float64(1), int64(13)
memory usage: 33.2 KB
```

we have plotted the correlation matrix to know the relationship among the available features, followed by plotting histogram, pairplot, factorplot and countplot (Fig. 1).

On plotting correlation matrix, it was found that three features cp, thalach and slope have the highest correlation with the target, with a value of 0.433798, 0.421791, 0.345877, respectively. So, we can explore these features further to find the interdependencies of these variables with the target.

Histogram was plotted next, in order to have an estimate of the range of distribution of data. In our dataset, age, trestbps, chol, thalach, oldpeak have continuous values where age ranges from 20 to 80 while oldpeak ranges from 100 to 400. On the other hand, the categorical data like sex, fbs has values only 0 and 1. Thus, after plotting the histogram, it is quite evident that we need to scale our data, so that our model can be trained properly.

Using pairplots, we can easily find out which features helps to form a distinct cluster of our target. Hence, we can use these features to train our model. In our dataset, we have 14 features. So $^{14}C_2$ numbers of plots were plotted using pairplot. We found that the plots of thalach versus ca, thalach versus sex, fbs versus ca and oldpeak versus sexform a distinct cluster of the target value and separates the target value (0 and 1). Next, we have explored these pair of attributes further using scatter plots.

Factor plot was used to visualize how the count of each variable uniquely related to our target, using a parameter "col" which separates the two target values 0 and 1 with separate colours. In our factor plot, we have plotted the features in the x-axis and count in the y-axis. The factor plots uniquely show how many males and females suffer from heart attack. It also shows the number of people (both male and female) suffering from heart attack based on their fasting blood sugar (fbs) value. Similarly, age, trestbps, thalach, oldpeak were also plotted to have an idea of the count of the people suffering from heart disease based on the specific feature attributes.

Lastly, countplot was done to get an idea whether our dataset is imbalanced or not. Imbalanced datasets are those where there is a severe skew in the class distribution, such as 1:100 or 1:1000 examples in the minority class to the majority class. This bias in the training dataset can influence many machine learning algorithms, leading some to ignore the minority class entirely. The count plot was plotted with "target" in the x-axis, to know whether our dataset is imbalanced or not. On plotting, we found that our dataset has a good ratio of positive and negative classes (target = 1 and target = 0, respectively) and thus can be used to train our model without under sampling or oversampling.

Data Processing—Preprocessing of data was done to ensure that the model is trained with proper data on which the algorithms can run effectively. First, the categorical variables were converted into dummy variables, so that no extra priority is given to any of the features having a higher value (1). 8 categorical values "sex", "fbs", "cp", "restecg", "exang", "slope", "ca" and "thal" were converted into dummy variables. After dummy variables were created, feature scaling was performed to normalize or standardize the independent features present in the dataset in a fixed range. If feature scaling is not done, then a machine learning algorithm tends to weigh greater values, higher and consider smaller values as the lower values, regardless of the unit of the values. So, we use feature scaling to bring all values to same magnitudes and thus,] tackle this issue. In our work, we have used standardization to scale our data. Standardization is a scaling technique where the values are centred on the mean with a unit standard deviation. This means that the mean of the attribute becomes zero and the resultant distribution has a unit standard deviation.

Here is the formula for standardization:

$$X' = \frac{X - \mu}{\sigma}$$

μ is the mean of the feature values and σ is the standard deviation of the feature values. In this case, the values are not restricted to a particular range. We avoided using normalization as it gave a lower scaling accuracy compared to standardization. We have used standard scalar to scale our data. Once, data-scaling is done, splitting the dataset into training and testing was required to avoid over-fitting and under-fitting, which is a common problem for many machine learning algorithms. We have used 67% of our data to train our model and 33% of our data were kept for testing the dataset. The "target" column is taken as the y-value of the testing and training data. The rest 13 features are taken as the X-value of the training and testing data. These are obtained after dropping the "target" column.

Training the model—At first the model was trained with six machine learning algorithms separately to find out which algorithm generates the highest accuracy. logistic regression, K-nearest neighbour, support vector machine, decision tree, random forest classifier and naïve Bayes classifiers were used. Among them K-nearest neighbour (KNN) gave the highest accuracy. KNN is a non-parametric algorithm which may be used both for classification and regression. The input consists of K closest training examples in the feature space, and the output depends on whether the problem is classification or that of regression. In classification, the output falls to that class which is most common among its K-nearest neighbours. In case of regression, the output is the average of the values of the K-nearest neighbours.

The KNN algorithm is as follows:

1. Load the data.
2. K is to be initialized to a chosen number of neighbors.
3. For each example in the data:

   3.1 The distance between the query example and the current example from the data is to be calculated.

   3.2 The distance and the index of the example is to be added to an ordered collection.
4. The ordered collection of distances and indices from smallest to largest (in ascending order) by the distances is to be sorted next.
5. The first K entries from the sorted collection is to be picked.
6. The labels of the selected K entries should be fetched then.
7. If regression, the mean of the K labels is to be returned.
8. If classification, the mode of the K labels is to be returned [8].

Testing our model—For Testing our model and to have a proper idea about the accuracy of each model, we have used classification report, confusion matrix and plotted the ROC curve, where KNN had the highest area of 0.87.

Website Development—The website has been made using Bootstrap as the front-end and SpringBoot Framework as the back-end. We have employed an online MySQL database for storing the user details, and those details can be fetched by the admin to predict whether the user is having a heart disease or not. HTML, Bootstrap, Rest Template, Thyme leaf, Ajax and jQuery have been used to make the front-end of the website. Java Database Connectivity (JDBC), Java Persistence API (JPA), GetMapping, PostMapping, Regex, Exception Handler and Object Mapper have been used to make the back-end of the website. In the back-end, the concept of microservice has been implemented for the users/patients so that they can submit their health details through the UI (front-end), and the data gets pushed into the database. Also, while checking their report, users can use the same UI to see their results.

3   Results

The above figure shows that there is hardly any imbalance between the positive and the negative class. Thus, the model can be trained without doing the necessary under sampling or oversampling (Figs. 2 and 3).

From the above image, it is quite evident that KNN gives the maximum accuracy. Thus, KNN was chosen to train the model for further use and to make the system more intelligent. KNN gave an accuracy of 87% followed by random forrest classifier and logistic regression with 84%.

```
In [473]:  fig= plt.figure(figsize=(5,5))
           sns.countplot(x='target',data=ds)
           plt.title("Count of Target", fontsize=18)
```

Out[473]: Text(0.5, 1.0, 'Count of Target')

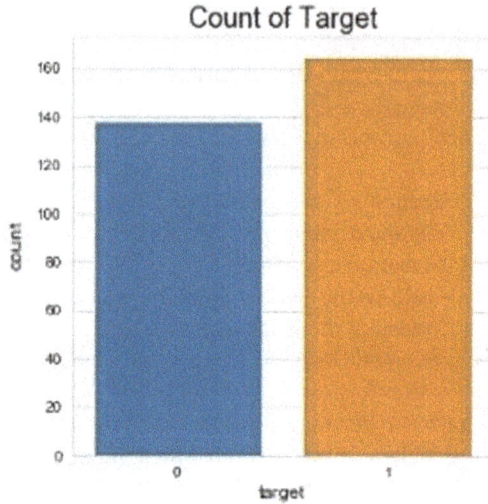

Fig. 2 Count plot showing that the data set is balance

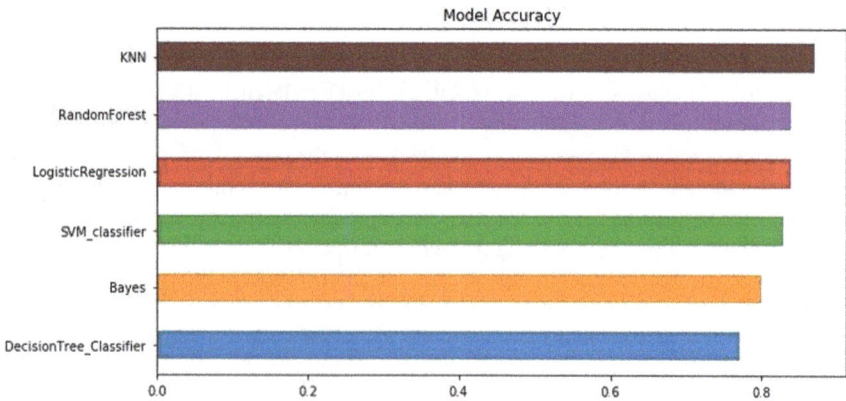

Fig. 3 Model accuracy comparison

KNN gave an accuracy of 87% when 8 neighbours were chosen. This accuracy was the highest among other neighbours and other algorithms also.

The above image depicts the confusion matrix and ROC curve of KNN. ROC curve and confusion matrix were plotted to test our dataset with 33% of the data available (Figs. 4, 5 and 6).

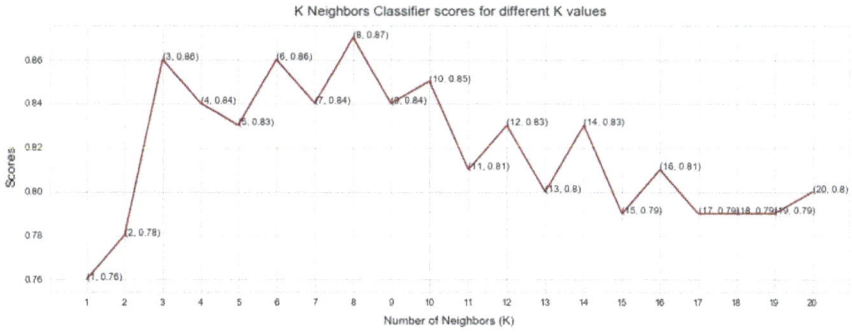

Fig. 4 Plotting the KNN curve

Fig. 5 Confusion matrix of KNN

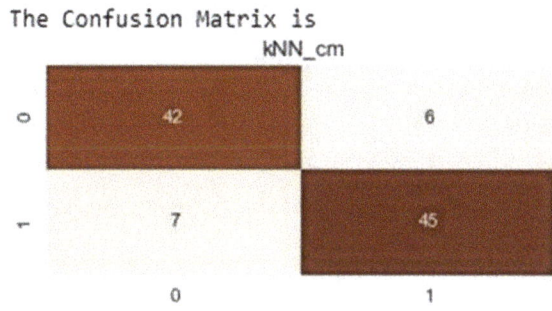

Fig. 6 ROC curve of KNN

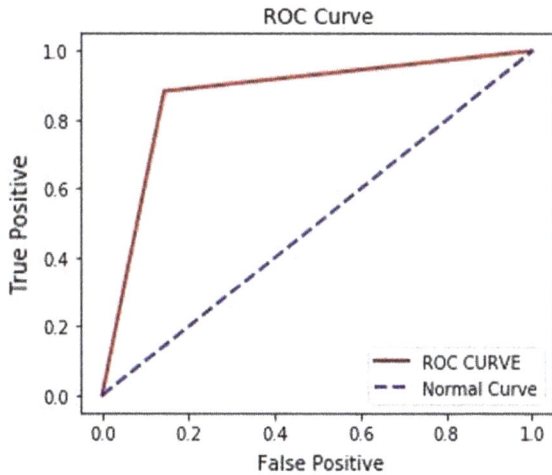

The above diagrams are the implemented UI. Figure 7 is the registration page of the user. The user will have to create ANA count in the website after which he or she will be redirected to a login page (Fig. 8). Upon providing valid credentials, the user will be allowed to access the website he or she will be asked to provide certain information about user's heart (Fig. 9). Once, the user has given all the details, the system which is artificially trained, will be predicting the result which will be finally shown in the result page (Fig. 10).

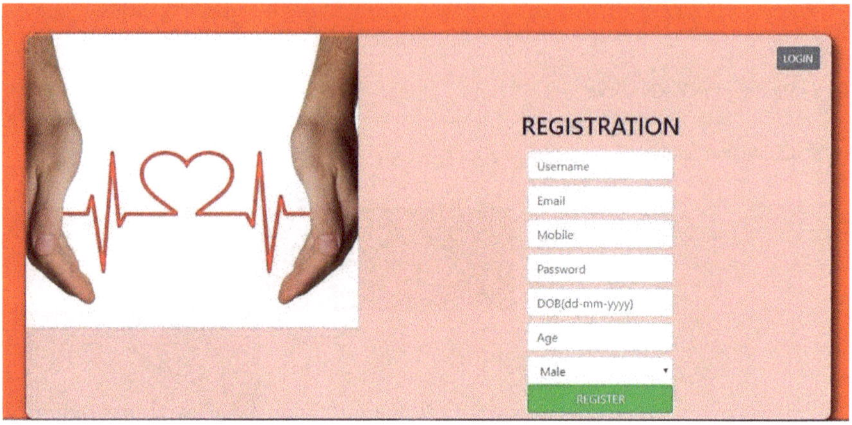

Fig. 7 Registration page

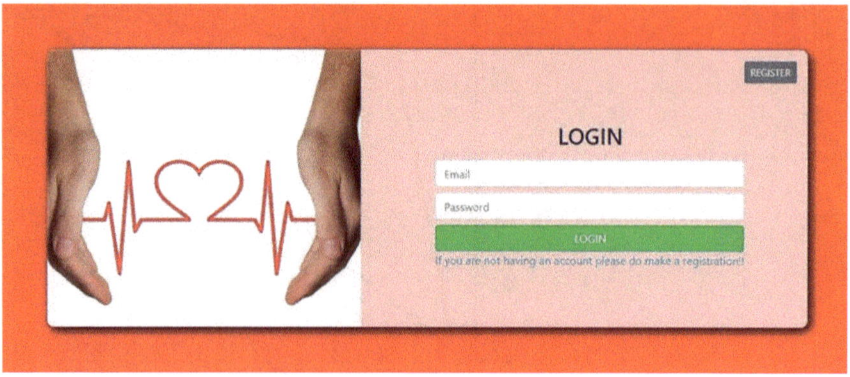

Fig. 8 Login page

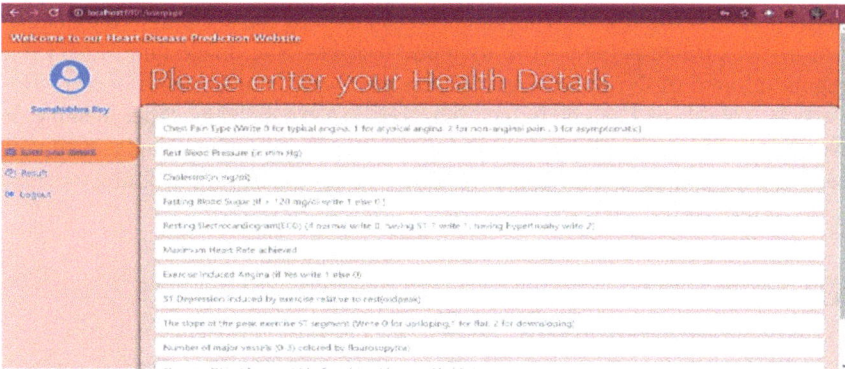

Fig. 9 Heart details page

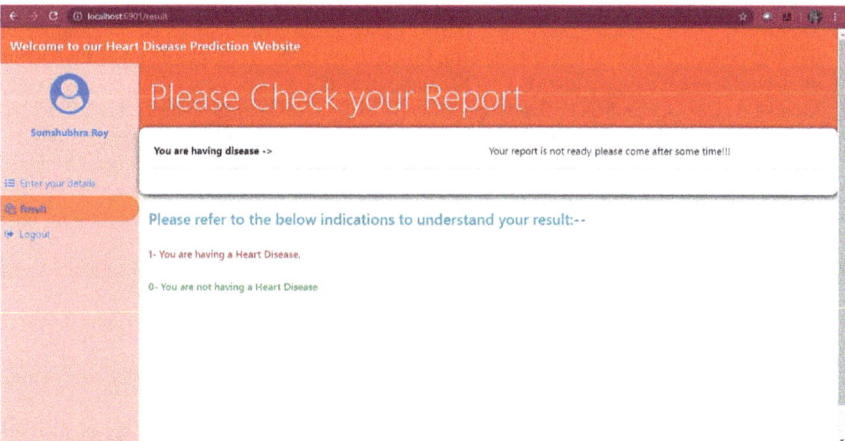

Fig. 10 Result page

4 Conclusion

The heart disease prediction is a necessary burden in medical offices, healthcare facilities and wellness centres. It can be completely automated through an efficient and intelligent online software programme. The benefits of implementing this technology are that everyone is involved in the scheduling process, as administrators and users can conduct their tasks more efficiently and accurately. The system extracts hidden knowledge from a historical heart disease database. The hidden information and patterns enable the model to re-learn on their own and increase its efficiency from time to time. The main advantage of this system is that it is a cost-effective, labour-saving model which reduces unforced human errors. The proposed model can be used as a training tool for medical students and a soft diagnostic tool for general physicians and cardiologists. This model can be used by the doctors for initial treatment of the cardiac patients. The user interface has been made quite interactive keeping in mind of the wide range of users who will be using this software. This system is scalable and can be further enhanced and expanded by experimenting on several other dataset. Implementation of augmented neural network and convolution neural network will make the proposed model more efficient in the future.

References

1. World Health Organization (2010) The world health report 2008. https://www.who.int/whr/2008/whr08_en.pdf. Accessed Nov 2010
2. Mondal A, Chaturvedi I, Das D, Bajpai R, Bandyopadhyay S (2015) Lexical resource for medical events: a polarity based approach. In: 2015 IEEE international conference on data mining workshop (ICDMW). IEEE, New York, pp 1302–1309
3. Mondal A, Das D, Bandyopadhyay S (2017) Relationship extraction based on category of medical concepts from lexical contexts. In: Proceedings of the 14th international conference on natural language processing (ICON-2017), pp 212–219
4. Mondal A, Das D, Cambria E, Bandyopadhyay S (2018) WME 3.0: an enhanced and validated lexicon of medical concepts. In: Proceedings of the 9th global WordNet conference (GWC)
5. Kirmani M (2017) Cardiovascular disease prediction using data mining techniques. Oriental J Comput Sci Technol 10(2):520–528. https://doi.org/10.13005/ojcst/10.02.38
6. Sai PP, Reddy C (2017) Heart disease prediction using ANN algorithm in data mining. Int J Comput Sci Mob Comput 6(4):168–172. Retrieved from www.ijcsmc.com
7. Wilson P, D'Agostino R, Levy D, Belanger A, Silbershatz H, Kannel W (1998) Prediction of coronary heart disease using risk factor categories. Circulation 97:1837–1874
8. Harrison O (2018) Machine learning basics with the k-nearest neighbors algorithm. Towards Data Science. 10 September
9. Science C, Faculty GM (2009) Heart disease prediction using machine learning and data mining technique. IJCSC 0973–7391(7):1–9
10. Ali AA, Hassan HS, Anwar EM (2020) Improve the accuracy of heart disease predictions using machine learning and feature selection techniques. In: Bhattacharjee A, Borgohain S, Soni B, Verma G, Gao XZ (eds) Machine learning, image processing, network security and data sciences. MIND 2020. Communications in computer and information science, vol 1241. Springer, Singapore

Chapter 62
Evaluating Snort Alerts as a Classification Features Set

Anas I. Al Suwailem, Mousa Al-Akhras, and Kareem Kamal A. Ghany

1 Introduction

Nowadays, if a specified entity in the organization is not able to access certain data that is considered to be important, then this is an incident and need to be identified and mitigated. Also, this particular threat will be increased by the number of devices connected to the same network topology. This kind of attack is hard to be detected and traced back. This type of attack is known as Disturbed Denial of Services (DDoS). No one will imagine that the service will be shut down and it will not be available to be used. This type of attack needs a sophisticated and smart method that can understand the deep packets that run through the network as packets of flow. As we already know that, the Intrusion Detection System (IDS), also, a famous IDS is called Snort, it is open source and easy to be configured [1].

The IDS (Snort) is detecting based on the misuse approach as the application will depend on a signature predefined through an organization that calls Checkpoint. This type of approach is not strong enough to understand and act on himself as a dynamic

A. I. Al Suwailem (✉)
Saudi Information Technology Enterprise (SITE), Riyadh 12382, Kingdom of Saudi Arabia
e-mail: aialsuwailem@site.sa

A. I. Al Suwailem · M. Al-Akhras · K. K. A. Ghany
College of Computing and Informatics, Saudi Electronic University,
Riyadh 11673, Kingdom of Saudi Arabia
e-mail: mousa.akhras@ju.edu.jo; m.akhras@seu.edu.sa

K. K. A. Ghany
e-mail: kareem@bsu.edu.eg; k.abdrabouh@seu.edu.sa

M. Al-Akhras
King Abdullah II School of Information Technology,
The University of Jordan, Amman 11942, Jordan

K. K. A. Ghany
Faculty of Computers and Artificial Intelligence, Beni-Suef University, Beni-Suef, Egypt

system. Due to the need for a smart application, the authors decided to engage the Artificial Intelligence (AI) as a Machine Learning tool that learning the attributes and imply testing on a pre-configured percentage of the same data that will have a training on. Machine Learning is a concept and it does introduce as applications such as Anaconda and Weka. This method of (AI), will be used as a subpart of the IDS preprocessing task, where the IDS (Snort) will process the flow packets through the network as first preprocessing, then the (AI) machine learning will come as second preprocessing to make sure that there is no true false rate in the alerts generated from snort. Also, there is another method that has been induced already, this type is the upset of misuse IDS approach where it calls anomaly detection system. This method of detection is working to detect any changes in the behavior of the network such as malicious activities. within the researches that have been done on the subject of anomaly detection based. The Intrusion Detection System (IDS) has not been approved yet to be strong enough for the industrial environment network.

IDS is a general name of a security system where the most popular IDS is called Snort. Snort is an open source application where it can be used as IDS and IPS as well by enabling the inline mode. Snort is a misuse intrusion detection system based were the system is performing an analysis process on the network traffic based on a signature that already been setup from a well-known organization that calls Checkpoint. Also, Snort is popular enough to be deployed worldwide with around 1 million downloads and 300,000 registered. Snort is using libPcap/winPcap to capturing the traffic from the TCP/IP and UDP as a flow. The traffic is flowing through an engine that calls decoder in the session layer. According to Snort functionality, we will define the preprocessing phase in Snort compared with Machine Learning (Decision tree Algorithms). For the Machine Learning processes of Decision Tree, we will conduct an analysis based on J48 as well as Random forest as a preprocessing to detect the high amount of abnormal traffic with minimizing the rate of positive false alarms [2]. Weka (Waikato Environment for Knowledge Analysis) is a great tool to be used as a machine learning function, it is open source application (GUI based) contain a plentiful number of integrated algorithms. Also, Weka has a feature of pre-processing as well as a classification clustering and association rules. Weka is not limited to data mining, also it can be used to perform developments through the Java language as well.

In this paper, a comparison is conducted between the two techniques that used to be detecting malware through the network. The two techniques are known as Intrusion Detection System (IDS) particularly using Snort application and the other technique is depending on the Artificial Intelligent method by using Machine Learning applications such as Weka. The main purpose is to minimize the rate of True Negative alarms generated from the IDS and increasing the rate level of the precision on the prediction of unknown malware this will be achieved by defining the highest rate of accuracy results that granted from the two techniques used. The two techniques used the same dataset fetched from the dataset as PCAP file extension. For Weka, the PCAP file needs to be converted to CSV file as well as adding 80 plus attributes with labels for each instance by using the application CICFlowMeter to be readable over Weka. Multiple algorithms such as decision tree (J48 and random forest), are applied to calculate the results. The results are compared with Snort results by injecting the same dataset as PCAP to be read by Snort and exporting the alerts generated by Snort as CSV file so we can have the ability to conduct the same experiment to generate the results of True Positive and True Negative for both techniques. Those two techniques can be merged to be work together as preprocess to be working as layered IDS system that can detect both misuse and anomaly intrusion detection systems.

The rest of the paper is organized as follows: Section II discusses a background about Literature review. While section III shows the Methods and the Materials. Last but not least is section IV showing the Research results and discussion. Finally, section V is the conclusion and future work.

2   Literature Review

2.1   Overview

Machine learning is a process to provide cybersecurity to protect network-based devices both on a personal and organizational scale. The system was invented to protect efficient data from world complexity and technical challenges contemporary. Initially, the SNORT system was divided into two parts—IPS and IDS to enable internet and public security from harm. The invention of this SMART system was quite facilitative to users as threats of cyberspace reduced effectively by its introduction in the market. The logical operation of IDS is categorized in two terms to detect cybercriminals and ensure security. Signature-based IDS and Anomaly-based IDS are efficient to define variability and detection of behavior trafficking for future incorporation and quality advancement. Traditional approach of machine learning is divided into two categories such as supervised approach and unsupervised approach. Both of them are prone to detect known and unknown attackers for better protection of the machine. Therefore, it can be said that digitalization in machine learning is required to provide sufficient security of data as shown in Fig. 1.

SNORT is termed as real time packet sniffer to detect trafficking efficiently for reducing the viability of cybercrime. It finds suspicious players readily for better protection of system and its behavior. SNORT can scrutinize closely to detect library

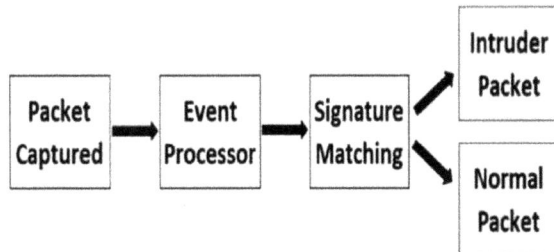

Fig. 1 Supervised approach of machine learning

capture dangerously, a tool widely used by traffic sniffers and data analyzer. Recently, efficiency of this viable system is under several challenges because of numerous reasons listed down.

2.2   Previous Studies

Article Topic: Using Classification Techniques for Creation of Predictive Intrusion Detection Model [3]. Authors: Abdulrazaq Almutairi and David Parish. The Method used: Machine learning. Algorithm used: Naive Bayse (Decision Tree). Data traffic: KDD'99, Labeled dataset. Also, it has 22 classes of attacks. Each instance considered as one class. Attacks are categorized into four attacks as the following: (DoS, Probing, User to root Attack, and remote to user attack). Research Methodology: The author attempted to prove the concept of machine learning accuracy by detecting attacks that already been recognized as classes. The authors have used the Decision tree algorithm to classify the known classes (True Negative and False Positive) and unknown classes (True positive and False Negative). The authors used the dataset of KDD'99. The data set consists of (41) discrete and continuous attributes and has 22 attack classes and 1 normal class, where each instance in the data set has been categorized as one class. And then, the authors compared the results with the process made in machine learning. The goal of conducting that is to find an alternative method of Snort that it is not adaptable to detect new attacks where that will lead snort to generate false alarms detection as shown in Fig. 2.

Article Topic: Bayesian Classifier and Snort based network Intrusion Detection System in Cloud Computing [4]. Authors: Chirag N. Modi, Dhiren R. Patel, Avi patel, and Rajarajan Muttukrishnan. The Method used: Integration between Machine learning and Snort. Algorithm used: Bayesian classification algorithm. Data traffic: KDD'99, Labeled dataset. Also, it has 22 classes of attacks. Research Methodology: The authors attempted to demonstrate a new method of Intrusion detection system used in Cloud Computing. The purpose of that is to minimize the rate of false alarms generated by using Snort only as IDS. The authors decided to integrate the machine learning system with the Bayesian algorithm as another method of preprocessing to decrease the rate of the false alarm that might occur from Snort. So here, the authors integrated the machine learning with Snort to be in the same preprocessing method block. Snort will stand as Misuse IDS where it detects only the attacks that have a known signature, the machine learning will detect the attacks that have no signatures or what it is called anomaly attacks as shown in Fig. 3.

Article Topic: Improved Signature Based Intrusion Detection Using Clustering Rule for Decision Tree [5]. Authors: Phuong Do, Ho-Seok Kang, and Sung-Ryul Kim. The Method used: Performing clustering rules by dividing the dataset into two groups of subsets and processing that through machine learning. Algorithm used: ID3 algorithm. Research Methodology: In this proposed article, the authors tried to demonstrate a method that can strengthen the process of detecting malicious traffic that runs through the network flow. The authors tried to integrate the rules used in

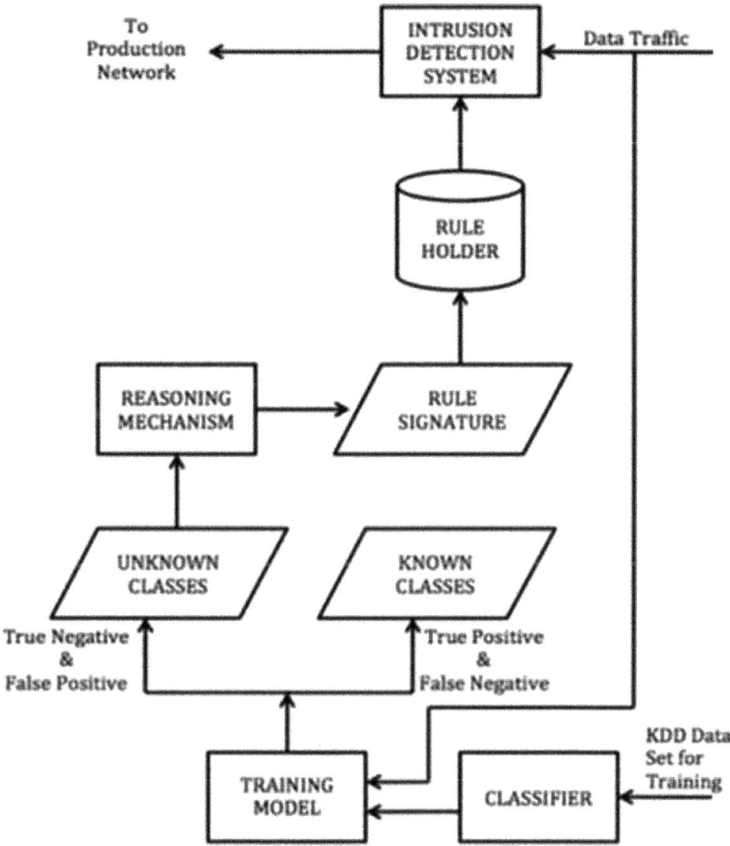

Fig. 2 HLD: research methodology [3]

Snort to be used to detect attacks in machine learning by using the decision tree algorithm (ID3). The authors used a classified dataset, and he was facing an issue by applying the rules to machine learning (ID3) where that will cause a problem by optimizing the tree with unspecified features and that will make the tree unmanageable. The authors used a great method to avoid this problem by dividing the dataset rules into different subsets and building a separated tree to each subset. The subsets that have been divided are two, the first one is having a rule that having a similar rule, the second one has the rest rules. The authors did the experiment by specifying the feature with (Protocol, Source IP Source Address, and Destination IP, and Destination Address), these features have been used because in Snort there are many features that are not being able to specify. The experiment has made on three experiments.

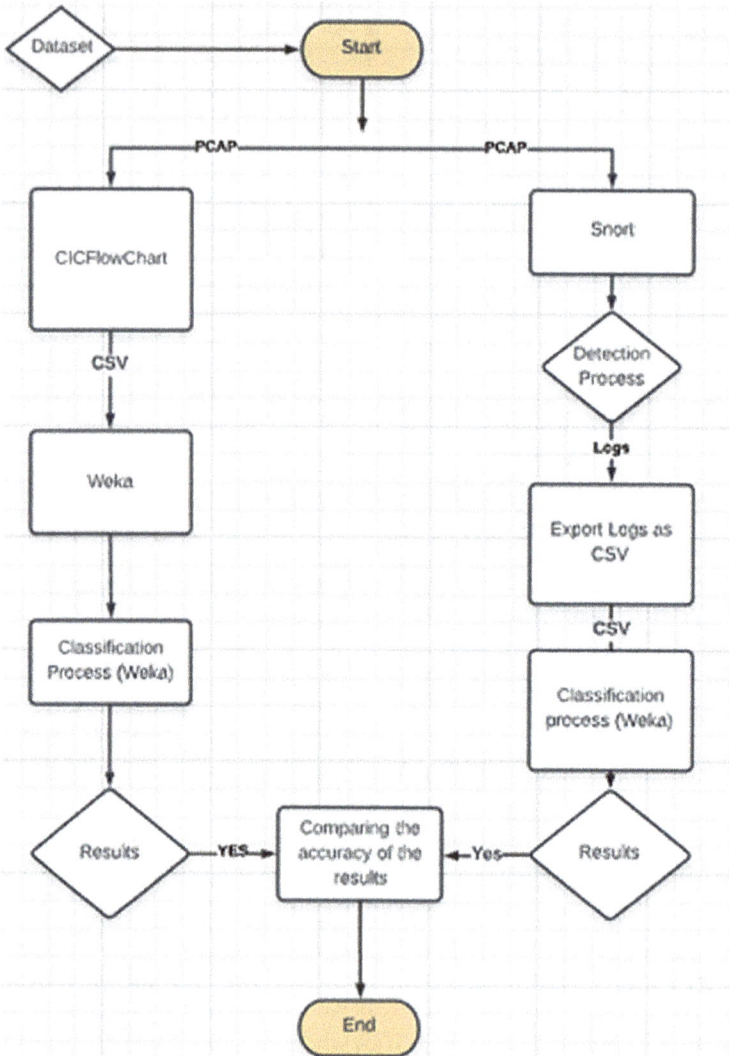

Fig. 3 Workflow of the proposed NIDS module [4]

3 Methods and the Materials

In this paper, an approach of evaluating Snort alerts as classification features set will be applied through two phases, the first phase, is using Data set that acquired from Amazon Cloud as a PCAP file extension captures through Wireshark application (Libpcap).

The data set is considered as a raw data where that will enable Snort to read the data set and applying Snort rules on the data set. Snort is generating alerts based on the rules set up already where these rules will be compared on the Protocol type, source IP, source port, destination IP, and destination port after that alerts will be generated and logged on a text file. The problem here is Snort is generating a huge number of alerts where that causing Snort to generate false-positive where that considered as a major issue [6]. When the IDS system (Snort) generating numerous numbers of positive false alerts, the accuracy of detecting malware will be decreased and the possibility to be breached will increase. For this purpose, an approach of having a raw data set that has the data state is followed, where each flow will contain multiple protocols and attacks such as DoS, Ping Scans, Probs, Remote user access, and DDoS. An application through Python language is generated where this application will be responsible for generating the number of alerts for each attack in a vector table where that will show the frequency for each attack as well. For example, DDoS has 270 of alerts, Probs have 7900 of alerts, and DoS has 3344 alerts, these alerts will put in three blinks in a vector table and this table will be used as a classification feature to be used in machine learning application. The machine learning application that I will use called Weka, Weka is an open source machine learning tool where it is a GUI based. Weka is used to examine the data set accuracy by using the data set as a CSV extension file, the Data set is having 83 number of features and over 4 million instances. After generating Snorts alerts as a classification features set, the data is inserted over Weka and the results between machine learning method (Weka) are compared through the algorithm decision tree (ID3 and J48) in order to have the high level of results of malicious activity detection and eliminating the possible rate of false-positive alerts. For a better view of the work, please refer to the following flowchart listed below where it gives you a wide understanding, as shown in Fig. 4.

4 Research Results and Discussions

In this section, we will discuss the results that we come along with for the process of analyzing the traffic of the network to determine whether if they are compromised or normal traffic of the network. The purpose of these results is finding a new way to mitigate the abnormal network traffic that is not easily captured by either misuse or anomaly intrusion detection systems. First of all, we will start with the results achieved by the Wireshark application. Wireshark is a network sniffer application that used to sniff and dig deep to the network packets in order to analyze the protocol, port source, IP source, port destination, IP destination. Wireshark has defined a number of multiple uncompleted three TCP handshake by sending SYN from a particular source where we considered that source by the intruder. The sniffed traffic has been extracted in two files, one as PCAP and the other is CSV by an application that calls CICFlowMeter in order to use it for the purpose of machine learning analysis through Weka application as shown in Figs. 5 and 6.

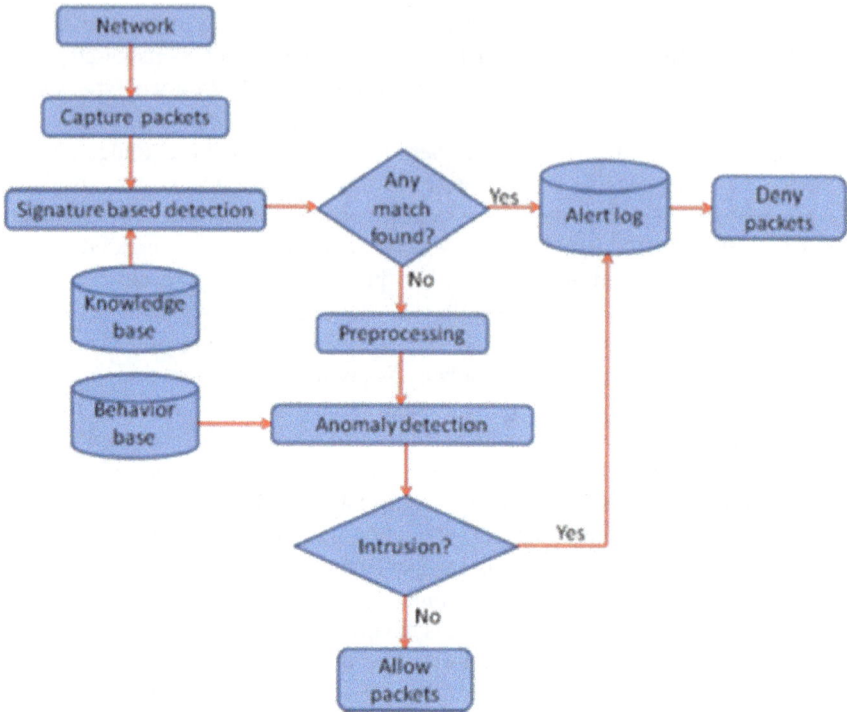

Fig. 4 Workflow of the proposed NIDS module

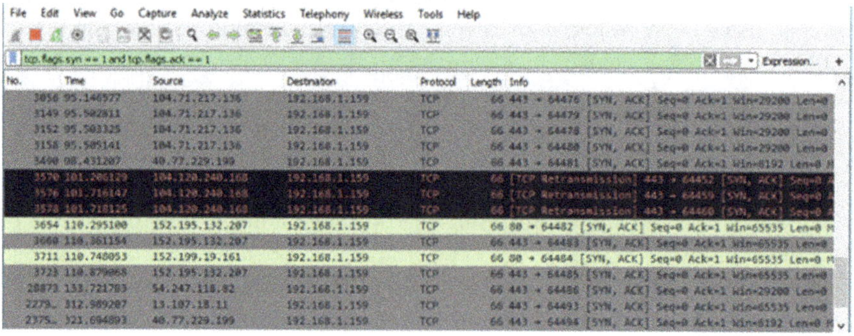

Fig. 5 Wireshark SYN attack—DoS

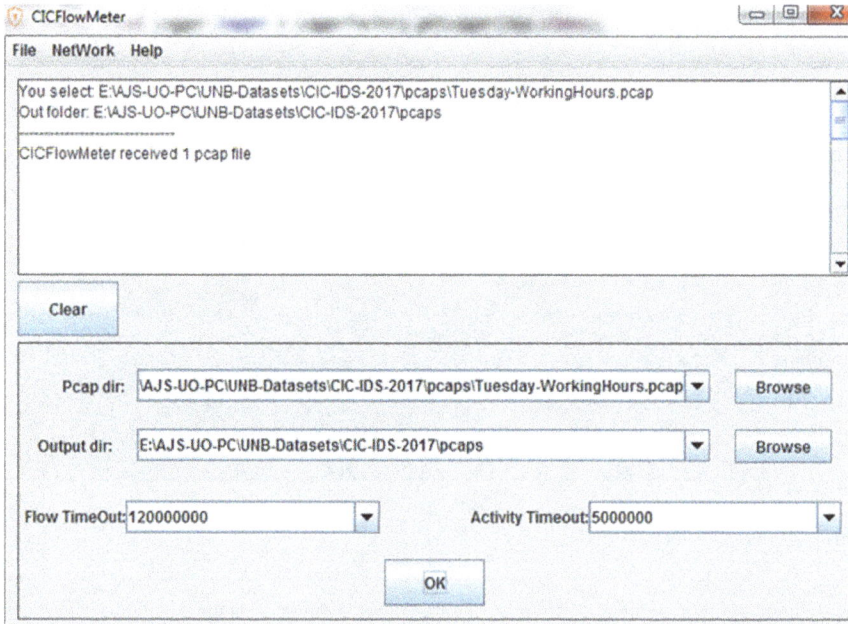

Fig. 6 CICFlowMeter

Using Snort intrusion detection system (IDS) application analyzing the data traffic that received and captured from Wireshark application by applying the rules that preconfigured on Snort application (IDS) produced results. The results that will be generated by Snort will be on an alert form because we will not activate the inline mode as an intrusion prevention system (IPS). The results will be listed as logs on the directory of Snort/logs. The logs generated will give us an overview sight about the nature of the attack as well as the source of the attack whether if it performed internally or externally. At the same time, we will ask Snort (IDS) to generate and export the alerts logs as CSV file extension in order to use it in a sophisticated and advanced analysis process in order to define the accuracy of the alerts generated and defining if they are real alerts or fake as shown in Fig. 7.

Machine learning (Weka) is an application that learns data and then applying a test as per the data that feed with. This kind of application will be nearest to the human being's minds where it learns and predict the upcoming traffic and how it will look like and how it is going to decide. The results of the machine learning (Weka) will present the accuracy of the data captured from Wireshark and converted through CICFlowmeter. Also, the analysis will be performed on the network traffic generated from Wireshark and feed to Snort (IDS) and exported as a CSV file in order to be readable from the machine learning (Weka).

After the analysis of the two traffics that belong to CICFlowMeter and Snort (IDS), we come with two results presenting the accuracy of the precision of the prediction

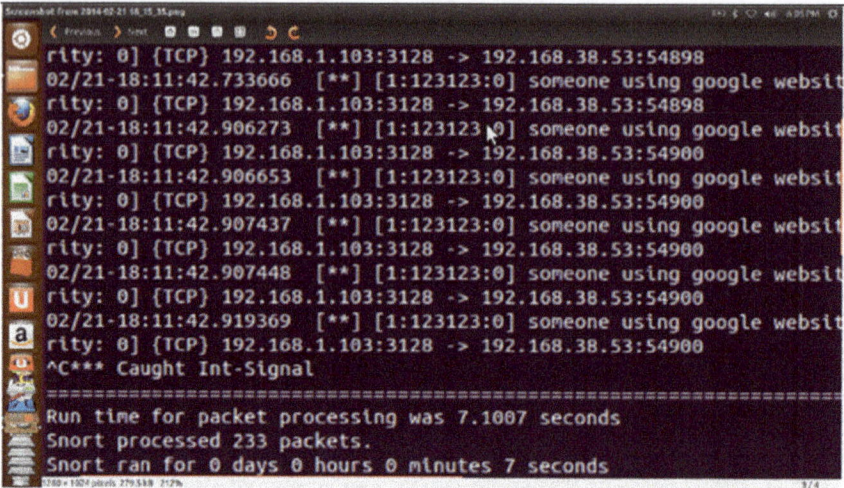

Fig. 7 Snort alerts captured—results

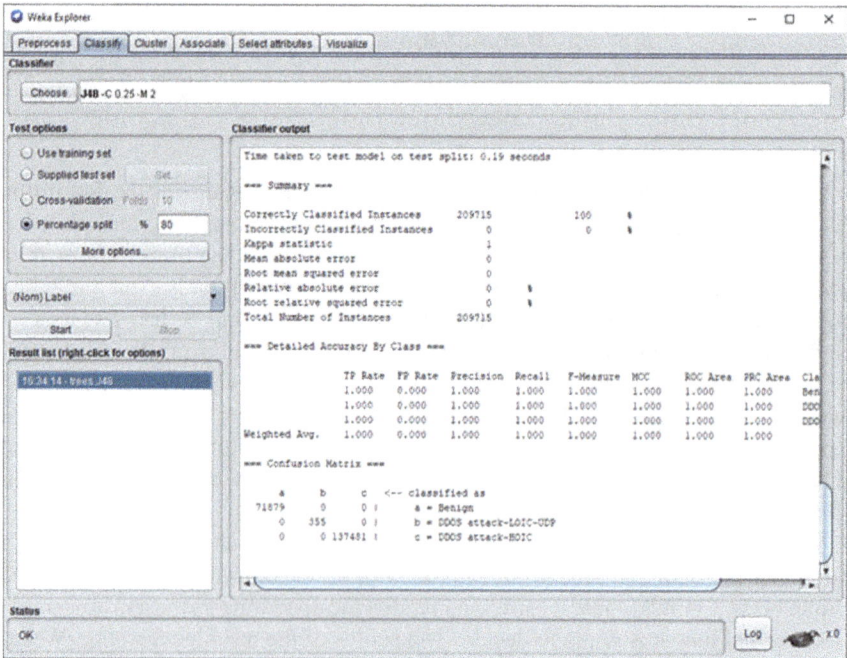

Fig. 8 CICFlowMeter Weka analysis—results

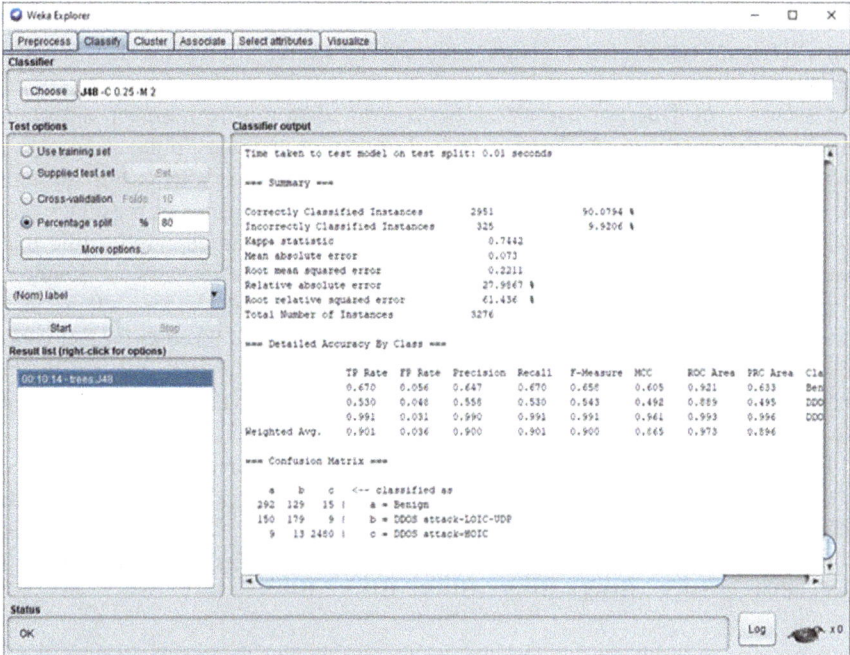

Fig. 9 Snort Alerts.CSV Weka—results

as well as the true positive and true negative results. The result of traffic generated from CICFlowMeter giving better accuracy and a high rate of prediction level with a high percentage of true positive alerts as shown in Fig. 8.

The results granted for CICFlowMeter are perfect unlike the results of Snort as CSV. Snort CSV file has been generated and inserted to the machine learning in order to determine the accuracy of the alerts generated from Snort (IDS), but we notice that the results were weak and have a low rate of prediction, true-positive, and true negative, as shown in Fig. 9.

In the end, Snort (IDS) and machine learning (Weka) can both work together as preprocessors in order to have a better and strong layer of analysis detection to misuse and anomaly detection methods.

5   Conclusion and Future Work

In conclusion, cyber-attacks are growing up rapidly and they become smart to be detected as well. To conduct a sophisticated cyber-attacks analysis, Wireshark is a suitable tool that can be used to capture traffic and used for the experiment of cyber-attacks detection. Such threat like that, a customized IDS is needed to have a better visibility and detection rate to attacks. The customized IDS solution consists of two parts, IDS application sensor (Snort) and machine learning tool (Weka). Combined the two applications as a preprocessing phase gave a high level of detection process and accuracy of True positive alerts. Snort worked as misuse intrusion detection system to detect known attacks with defined signatures and rules, if the attacks have not shown as a threat and there is no alert shows up, then the traffic passed through the machine learning (Weka) to deg deeper through the attributes of the attack and classified as anomaly attack. Algorithms are applied on the captured traffic that has been converted through CICFlowMeter such as, decision tree (J48 and Random forest) to classify the instances of traffic and learning the attributes for each instance for the sake of determining the anomaly activity that move through the network. Also, the two algorithms have different methods of classification and that might provide different results such as, (True positive, False positive, and the precision of the prediction).

For the future work we will try using machine learning technique and applying the Artificial Neural Network (ANN) algorithm instead of depending on decision tree only. Additional, the researchers could divide the dataset that will be experimented to two parts, similar instances will be gathered in one group and the second part will contain the all other instances.

References

1. Stampar M, Fertalj K (2015) Artificial intelligence in network intrusion detection, pp 1318–1323. https://doi.org/10.1109/MIPRO.2015.7160479
2. Fang X, Liu L (2011) Integrating artificial intelligence into Snort IDS. In: 2011 3rd international workshop on intelligent systems and applications, Wuhan, pp 1–4. https://doi.org/10.1109/ISA.2011.5873435
3. Almutairi A, Parish D (2014) Using classification techniques for creation of predictive intrusion detection model. In: The 9th international conference for internet technology and secured transactions (ICITST-2014), London, pp 223–228
4. Modi CN, Patel DR, Patel A, Muttukrishnan R (2012) Bayesian classifier and snort based network intrusion detection system in cloud computing. In: 2012 third international conference on computing, communication and networking technologies (ICCCNT'12), Coimbatore, pp 1–7. https://doi.org/10.1109/ICCCNT.2012.6396086
5. Do P, Kang H-S, Kim S-R (2013) Improved signature-based intrusion detection using clustering rule for decision tree. In: Proceedings of the 2013 research in adaptive and convergent systems (RACS '13). ACM, New York, NY, USA, pp 347–348. https://doi.org/10.1145/2513228.2513284
6. Rani AS, Jyothi S (2016) Performance analysis of classification algorithms under different datasets. In: 2016 3rd international conference on computing for sustainable global development (INDIACom), New Delhi, pp 1584–1589

Chapter 63
A System Design for Combined Approach of WCID and Wavelet Transformation to Optimize the Underwater Image Enhancement

Shrinivas Shirkande and Madhukar Lengare

1 Introduction

Underwater imaging plays an essential role in describing the life submerged in studies of ocean. Underwater images are taken to organize submersed studies and to research about aquatic features and life. But it is critical to acquire clear objects images submerged due to lamentable overtones. As the light enters the medium of water, it gets scattered by suspended elements and withal a component of light is absorbed by the medium. Image enhancement techniques and denoising are particularly very important in case of improving the picture quality of the underwater images. The amount of light decreases as it moves deeper under water. As an outcome, the underwater images seem to be dark and also a part of light from objects meets the camera and the remaining light gets distributed while traversing the path through the medium of water. Due to the features of medium of water the components. Of light having larger wavelengths are acquired easily at the surface, and those with smaller wavelengths manage to pass deeply. This is why several underwater images possess bluish or greenish cast color. The surface water effects are depicted Fig. 1.

From Fig. 1 certain incident light meets through water and the remaining light gets reflected at the surface. Due to the selective features of absorption, the component of red color of light faces serious attenuation from surface. So the light involving deep into water mainly comprises of green and blue color elements. The light diffusion indicates the light which distributes through the medium of water by scattering. Image enhancement is a technology of developing the image quality by developing their RGB values and their features.

The image processing of underwater images has acquired substantial attention revealing essential accomplishments. The image processing methods are capable of expanding the underwater image processing range, increasing the image resolution

S. Shirkande · M. Lengare (✉)
Pacific University, Udaipur, India
e-mail: madhukarlengare@gmail.com

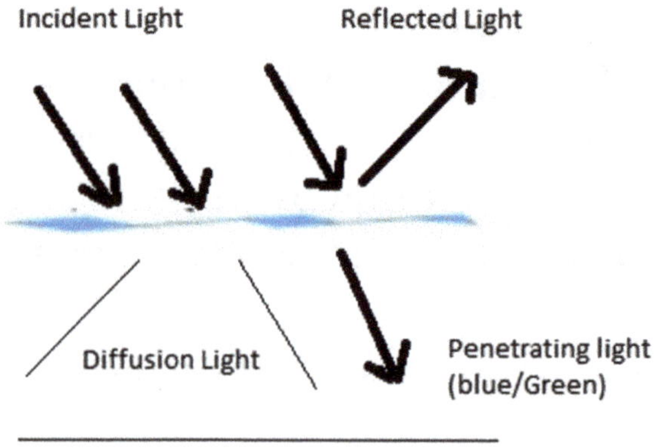

Fig. 1 Surface of water effects

quality and contrast level. The process of light reduction is affected by the spreading and absorption. According to Maini and Aggarwal, image enhancement is usually developing the perception or interpretability of data in images for viewers of human [1] being and offering good input for other automated techniques of image processing. The major aim of image enhancement is to change image attributes to make it much applicable for a specific observer and a given task. During this method, more than one attributes of the image are changed. The attributes option is specific to a particular task.

2 Literature Review

The images seized under water are generally degraded due to the impacts of scattering and absorption. The underwater images which are degraded reveal certain limitations when they are used for analysis and display. This chapter explores several works related to optimization of underwater image enhancement technique by combining wavelength compensation and image dehazing (WCID) and wavelet transformation technique. Iqbal et al. as per this paper to enhance underwater images in HSI and RGB color models, there is use of slide stretching algorithm. For the demonstration purpose, there is development of interactive software tool for underwater image enhancement. For RGB color model, there is use of contrast stretching, and for HSI color model, there is use of saturation and intensity stretching. With the benefit of color contrast equalization in the images also lightning issue is solved. The quality of the images was statistically illustrated through the histograms. As per paper, future work includes. The future work will include a better comprehension and comparison of these methods with summarization of image enhancement and restoration. There was complete analysis on the basis of quantitative and quality metric for assessment. Nowadays, leading advancements in optical imaging technology and the use of sophisticated sensing techniques was rapidly increasing the ability to image objects in the sea. According to Chiang et al., there is proposal of algorithm called WCID for restoring [2] the color balance and haze effect of image. To the best of our knowledge, no existing techniques can handle light scattering and color change distortions suffered by underwater images simultaneously. The experimental results demonstrated superior haze removing and color balancing capabilities of the proposed WCID over traditional dehazing and histogram equalization methods. Prabhakar et al. give preprocessing technique [3]. for enhancing the quality of degraded underwater images as The proposed technique included four filters such as anthropomorphic filtering, wavelet denoising, bilateral filtering and contrast equalization, which were applied sequentially. The main contribution of this research was inclusion of bilateral filter for smoothing in addition to existing other filtering techniques. Karthigal R. et al. derived an incipient submersed optical model to describe the formation of an submersed image, and then proposed an efficacious submersed image enhancement algorithm with this model [4]. In future work will further amend the adaptability and flexibly of this algorithm. Bharal need reviewed a portion systems from claiming picture upgrade [5]. WCID is utilized to upgrade those submerged pictures adequately. This algorithm need the proficiency will adjust those conflicts from claiming debilitating along those span of transmission. Difference keeping stretching, experimental mode decomposition, anthropomorphic filtering, an-isotropic filtering, wavelet denoising toward Mundane filter, red channel strategy; histogram adjustment Furthermore contrast set multifarious histogram adjustment (CLAHE) need avail a few of the strategies to underground picture upgrade. These systems bring postulated a proficient part with diminishing commotion issue accessible in the methods from claiming present picture change. Writers additionally have discovered out that the wavelet denoising Eventually Tom's perusing Normal channel which provides for obliged comes about as far as crest indicator will commotion proportion (PSNR) Also mean square lapse (MSE). Jaiswal Furthermore Padole [6] bring reviewed an enhanced calculation to submerged picture upgrade Furthermore denoising Similarly, as for every researches, no existing systems need took care of the diffusing about light what's more shade transform distortions endured Eventually perusing submerged pictures. Writers need pointed out that preprocessing of submerged picture appears to be on a chance to be all the more necessary because of the picture personal satisfaction caught under water. Additionally, creators need recommended those wavelets convert and furthermore Weber theory to upgrading the submerged picture characterization. In fact, those wavelets change need assumed an effective part should uproot the picture denoising progressively. Wavelet convert may be acknowledged likewise those powerful guidelines behind an estimation technobabble of nonlinear wavelet built sign. Dhar et al. proposed a method to segment an underwater image which has some intrinsically perturbations. The method utilized the multi-scale and multi-directional M-band wavelet transform of the submerged images to capture the texture of the objects present in it. The proposed method judiciously chooses the congruous sub-band depending on the human psycho-visual phenomenon. Kumar et al. proposed a technique to eliminate the presence of blurriness and recuperation of submersed images that were affected by color distortion that occur [3] due to the dust-like particles in the water. Submerged images or videos always suffer from the back scattering effect. A component of the light may be scattered by the floating particles due to which light reflected from objects and propagates toward the camera. This resulted in the haze-like effect for submersed images and greatly reduced the scene contrast. Submerged image were fundamentally described by their poor deceivability in light of the fact that the light was enervated through the water and the image results ineffectively differentiated.

In this technique to eliminate the presence of blurriness and restoration of underwater images that were affected by color distortion that occur due to the dust-like particles in the water. Underwater images or videos always suffer from the back scattering effect. This resulted in the haze-like effect for underwater images and greatly reduced the scene contrast. Submerged image were basically described by their poor deceivability in light of the fact that the light was weakened through the water and the image results ineffectively differentiated. The retention and diffusing procedures of the water impact the general execution of submerged imaging framework.

3 Research System Design

Under water image quality can be improved in the form of contrast, brightness, visibility in low-resolution image by using image enhancement technique. A new enhancement technique called Wavelength compensation and image dehazing (WCID) in addition with wavelet discrete transform (WDT) is used here for amplifying the contrast adjustment, color modification and edges concentration. These wavelets contain different data about the image. LL corresponds to low-frequency data of smooth region of images like region of less brightness and less contrast. And the other three wavelets contain the high-frequency data of the image. The WCID algorithm has been employed for distortion mitigation caused due to light scattering and color change. In the first state, the image is incorporated into the image processing system as input, and the map of the light transmission for a specific image is evaluated. The medium transmission is reevaluated, and the wavelength of the atmospheric light is calculated. Based on depth of image object from the source, segmentation of background and foreground within the image is formulated. In image acquiring process to determine an artificial light scattering effect, the foreground and background light intensities are compared with each other. So luminance is removed by detecting artificially.

System design flow:

In the first step, single underwater image is selected from database and to be processed for the further process of image processing. Then apply DWT for which wavelets are discretely sampled; a key advantage it has is Fourier transform which has temporal resolution, so it captures both frequency and location of information. DWT mainly is used to capture image information and its locations (Fig. 2).

Further apply wavelength compensation image dehazing (WCID) algorithm. This algorithm is mainly used to restore the superior quality of images by minimizing the

Fig. 2 System design

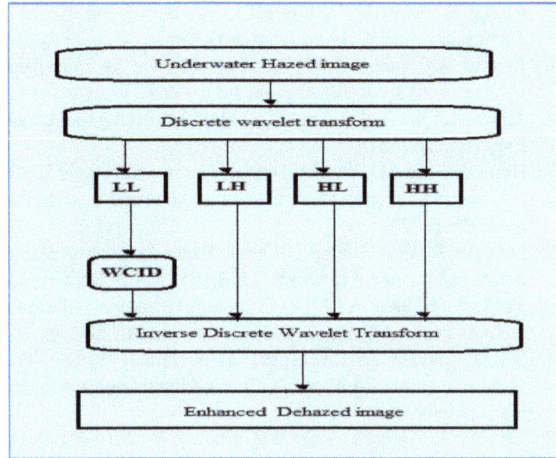

haze effect.Its mainly combination of dehazing algorithm and energy compensation. The dark medium is used to evaluate the distance here. In this WCID algorithm, the background and segmentation of object foreground are computed within the image by using the depth of object. Further, the IDWT is performed for WCID-enhanced image and other high-frequency components of input image, which results an enhanced dehazed image as our final image. The given technique was observed by MSE and PSNR values, and simulation was done using MATLAB.

4 Conclusion

The given system design work is combined approached of WCID and wavelet transformation technique to optimize the underwater image enhancement technique.

It will help to get the resultant enhanced dehazed image by using above research system design.

References

1. Maini R, Aggarwal H (2010) A comprehensive review of image enhancement techniques. J Comput 2(3)
2. Chiang J, Chen Y (2012) Underwater image enhancement by wavelength Compensation and image dehazing
3. Prabhakar CJ, Praveen Kumar PU (2011) An image based technique for enhancement of underwater images. Int J Mach Intell 3(4):217–224
4. Karthiga R et al (2013) Illuminance estimation in underwater using color constancy. Care J Appl Res 1(1)

5. Bharal S, Amritsar GNDU (2015) A Survey on Various Underwater Image Enhancement Techniques. Int J Comput Appl 5(4):160
6. Jaiswal AP, Padole BV (2016) Review on an improved algorithm for underwater image enhancement and denoising. Int J Adv Res Electron Commun Eng (IJARECE) 5(1)
7. Trucco E, Olmos-Antillon AT (2006) Self-tuning underwater image restoration. IEEE J Ocean Eng 31(2):511–519
8. Hou DJ, Gray AD, Weidemann G, Fournier R, Forand JL (2007) Automated underwater image restoration and retrieval of related optical properties. In: Proceedings of IGARSS, 2007, vol 1, pp 1889–1892
9. Schettini R, Corchs S (2010) Underwater Image processing: state of the art of restoration and image enhancement methods. EURASIP J Adv Sign Process 2010:01
10. Thakare SS, Sahu A (2014) Comparative analysis of various underwater image enhancement techniques. Int J Comput Sci Mob Comput 3(4):33–38
11. Sahu P, Gupta N, Sharma N (2014) Underwater image color enhancement using color stretching technique and USM filters. In: International conference on research in science, engineering & management (IOCRSEM)
12. Chang YT, Wang JT, Yang WH, Chen XW (2014) Contrast enhancement in palm bone image using quad-histogram equalization
13. Priyadharsini R et al (2015) Underwater acoustic image enhancement using wavelet and K-L transform
14. Alex Raj SM, Abhilash S, Supriya MH (2016) A comparative study of various methods for underwater image enhancement and restoration. IOSR J VLSI Sign Process 6(2):30–33
15. Shirkande S, Lengare MJ (2017) A survey on various underwater image enhancement techniques. Int J Innov Res Comput Commun Eng 5(7):13701–13710
16. Natrajan K (2017) A review on underwater image enhancement techniques. Int Res J Eng Technol 4(4):780–782
17. Ashwin S, Srinivasan S, Pan KH, Kalimuthu KK (2017) Underwater image enhancement: an integrated approach. Int J Electron Data Commun 5(6):102–107
18. Patil SV, Patil AV, Suryawanshi SS (2018), Underwater Image Enhancement, 13th International Conference on Recent Innovations in Science, Engineering and Management, pp 394–400.

Chapter 65
Daily Plant Load Analysis of a Hydropower Plant Using Machine Learning

Krishna Kumar, Ravindra Pratap Singh, Prashant Ranjan, and Narendra Kumar

1 Introduction

The monitoring and maintenance of a hydropower plant are important for efficient energy generation. Hydropower generation depends on the discharge and head of water. For proper planning and monitoring, machine learning techniques can be used. Machine learning is a technique to make machines to think like a human. Machine learning uses supervised and unsupervised learning techniques to predict future outcomes and to better insight of data.

(i) *Supervised Learning*

Supervised machine learning constructs a model in the presence of uncertainty which makes predictions based on proof. A supervised learning algorithm takes a known set of input data and known data (output) responses and trains a model to produce appropriate predictions for new data response.

K. Kumar (✉)
Hydro and Renewable Energy Department, Indian Institute of Technology, Roorkee, India
e-mail: kkumar@ah.iitr.ac.in

R. P. Singh
Bipin Tripathi Kumaon Institute of Technology, Dwarahat, Uttarakhand, India

P. Ranjan
University of Engineering and Management, Jaipur, India

N. Kumar
School of Computing, DIT University, Dehradun, Uttarakhand, India

© The Author(s), under exclusive license to Springer Nature Singapore Pte Ltd. 2021
X.-Z. Gao et al. (eds.), *Applications of Artificial Intelligence in Engineering*, Algorithms for Intelligent Systems, https://doi.org/10.1007/978-981-33-4604-8_65

(ii) *Unsupervised Learning*

Unsupervised machine learning is used to find hidden patterns in data or inherent structures. It is used to extract inferences from datasets that consist of input data without labeling answers.

The self-organizing map (SOM) is a popular unsupervised machine learning algorithm [1]. It provides vector quantization and topology-preserving mapping from a high dimensional input space to usually two-dimensional map space. In this paper, we have provided daily plant load data of MB-II hydropower plant of one-year duration and classified the data for the elimination of outliers, and to extract the useful data which can be used for the future prediction and monitoring of the plant.

2 Research Review

Many researchers have already used machine learning techniques for future predictions based on past data. Hosein et al. [2] proposed a data-driven deep learning system for load forecasting, and the peak load variables are computed from the Copula models. A comparative study was performed between the support vector regression machine and the classical neural networks. Vantuch et al. [3] demonstrated how predictive accuracy decreases with time scale increase due to the impossibility of using all variables. Khantach et al. [4] examined the accuracy of electricity consumption.

Sarhani et al. [5] investigated the importance of the feature selection approach for removing the irrelevant factors of electric load. Scott et al. [6] compared several regression models that can be used for accurate predictions of energy load. Wee et al. [7] proposed reinforcement learning for load-prediction. The suggested solution checked and used the best models and recalibrates them iteratively by comparing the estimation of the model against the real load data in order to address the varying accuracy of the forecast models on loads of different substations.

He et al. [8] forecasted a one-day ahead hourly load based on deep learning and implemented a new scalable architecture incorporating multiple types of input features where inputs are processed using different types of neural network components based on their common characteristics. Zuniga et al. [9] developed a load-prediction method using ANN. Tian et al. [10] presented a deep neural network architecture incorporating the CNN model and the LSTM model to improve predictive accuracy. Mosavi et al. [11] reviewed the deep learning application areas based on the popularity and effectiveness of the models. Yang et al. [12] proposed a SOM clustering method and concluded that the algorithm can get as good clustering results when the U-matrix is well separated.

Riese et al. [13] introduced the supervised SOM framework to perform unsupervised, supervised, and semi-supervised classification. Xayyasith et al. [14] presented a machine learning technique for predictive maintenance of a water cooling system in the hydropower plant. The classification learner was used to train model and concluded that SVM and decision trees are better for predictive analysis. Kumar et al. [15] analyzed the oscillations by the synchronous generator based on historical data with a conventional PSS regulator to maintain the constant stator current.

From the research review, we have concluded that SOM is popularly used in many applications for data clustering. In this paper, the SOM algorithm is used for the daily plant load data classification. The 304 MW MB-II hydropower plant is selected for data collection. This power plant is severely suffering from erosion and cavitation problems, and most of the time running at partial load due to low discharge and erosion. The prediction of future power generation is important for the load scheduling of the plant and to provide more support to grid stability in terms of reliable power generation.

3   Methodology

The daily power generation data of 304 MW hydropower plant is shown in Fig. 1. From the figure, it is clear that the power plant is running at half-capacity in January, February, and no power generation due to the lean discharge period in first two weeks of March. Power generation from mid of March is increasing and from the last week of May to September machines are running at their rated capacity, and after that power generation again decreases till December end.

MATLAB's machine learning application is used to cluster daily plant load data through the self-organizing map. A hex top topology of 10×10 nodes was selected with 200 training epoch. An unsupervised SOM clustering technique is used to understand the pattern of the data. The details of the training window data are shown in Fig. 2. A 366 days (due to leap year) plant load data has been taken as input with 100 layers and 100 outputs for 200 iterations.

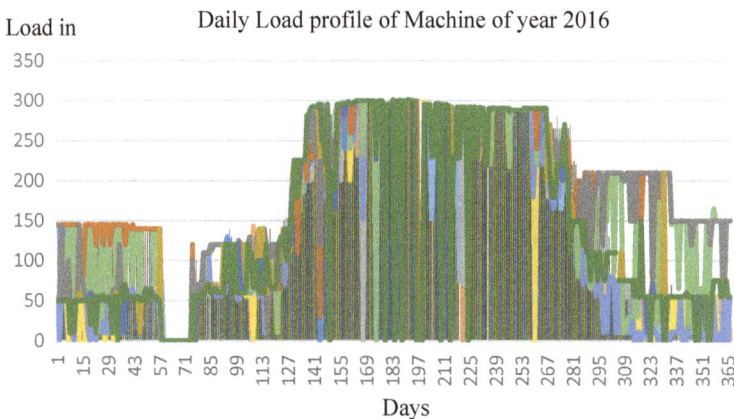

Fig. 1 Daily plant load (in MW) of 304 MW MB-II hydropower plant

Fig. 2 Neural network
training window

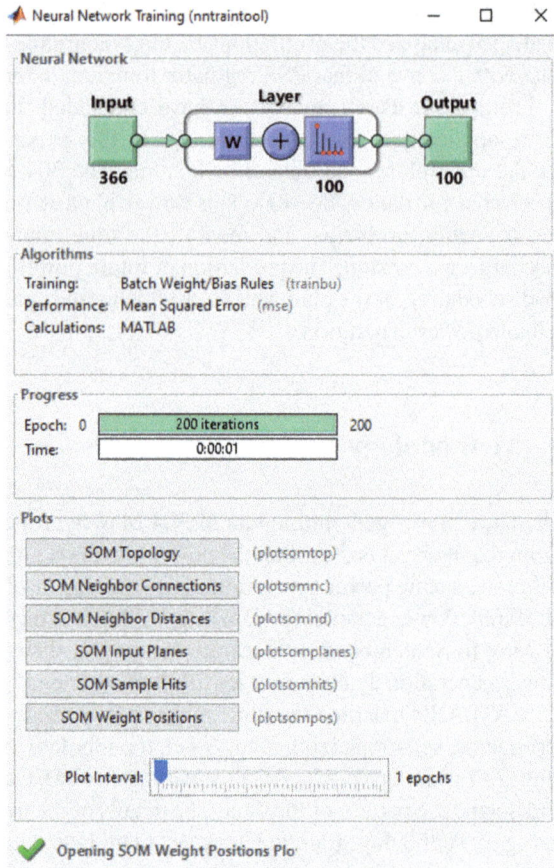

4   Results and Discussion

The weight vector associated with each neuron progresses toward being the center
of an input vector cluster. Through topology, neurons adjacent to each other can also
pass in the input space next to each other. This topology shows the hexagonal and
two-dimensional grid of 100 neurons arranged in a 10×10 matrix.

In Figs. 3 and 4, the neurons are represented in blue hexagons. In the neighborhood,
the red lines connect neurons. The colors showing the neuronal distances in the
regions which include the red lines. The darker shades are larger distances, and
smaller distances are the lighter colors. SOM neighbor weight distance matrix shows
one large cluster, with small distances between the member records, is present. Two
small and one very small clusters are also present, with relatively large distances
between the member records.

A sequence of subplots displays the weights from the input nth to the neurons
of the layer. The most negative connections display black, zero connections display

Fig. 3 SOM neighbor
weight distances

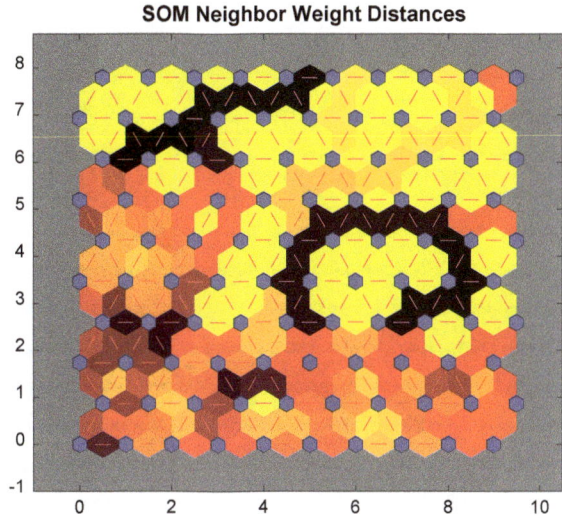

SOM Neighbor Weight Distances

Fig. 4 SOM neighbor
connections

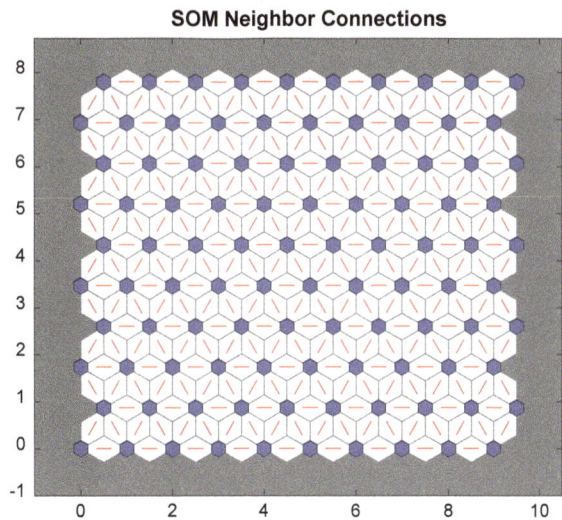

SOM Neighbor Connections

red, and the most positive connections display yellow. The sections of input weight planes display the weight on each day power generation for each of the 100 neurons. Weight plane which looks identical with strongly dependent indicated characteristics; although identical planes indicate independent characteristics as shown in Fig. 5.

SOM sample hits plot, as shown in Fig. 6, counts the number of data records associated with each neuron. The plot of sample hits shows how many of the training samples falls to each cluster. In an ideal situation, relatively even distribution across

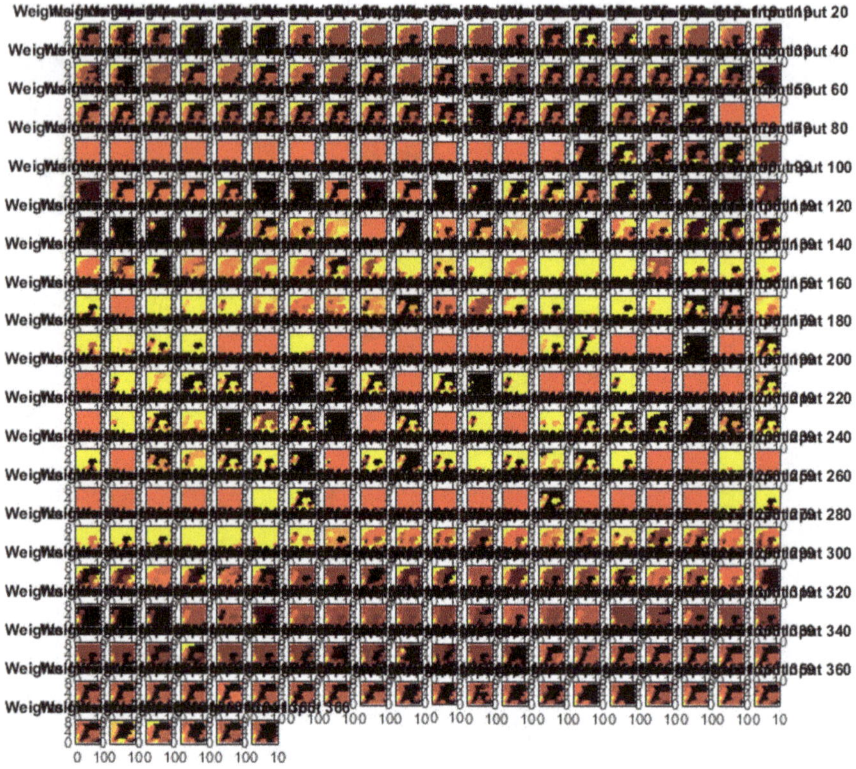

Fig. 5 SOM input planes

Fig. 6 Hits

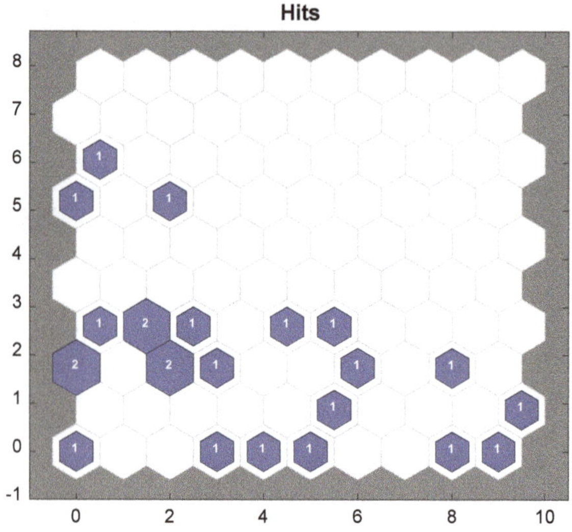

Fig. 7 SOM weight
positions

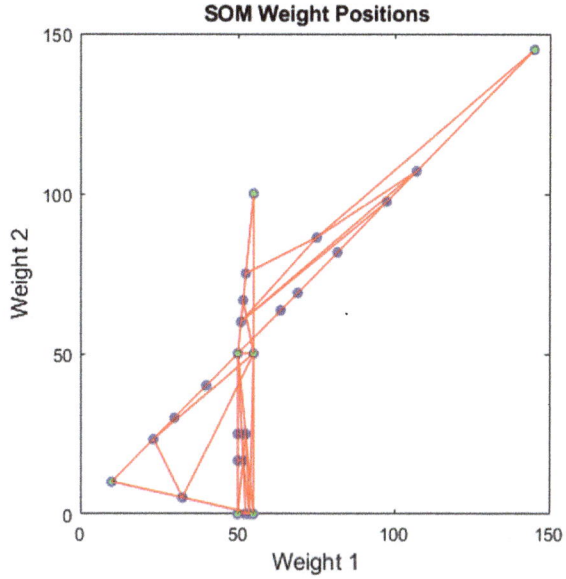

the neurons is desired. In this paper, the distribution is clustered in four groups, which indicates very similar data without much separation.

Topology for the standard SOM is hexagonal. Figure 7 shows the positions of the neuronal topology and the number of training data is correlated with each of the neurons. The mean number of hits corresponding to each neuron is 02. Therefore, 02 input vectors are in the cluster. Sections of the weight region show the sample data region as green dots in terms of their weights, and then the neurons are repressed as blue dots by weight. The red lines indicate which neurons are neighbors.

5   Conclusions

In this paper, power plant load data on daily basis is analyzed using self-organizing map machine learning visualization technique. It has been concluded that everyday power plant load is varying in nature due to variation in daily discharge and availability of the machines. In this paper, it is clear that the nearest neighbor approach to be especially useful for map with a large number of units compared to the number of data points. It detects the outliers which cannot be used for the prediction of plant load. Based on the reliable data, the next-day energy generation schedule can be given to the grid monitoring authorities which will help to maintain the grid stability.

References

1. Kohonen T (2001) Self-organizing maps, 3rd ed. Springer, Berlin
2. Hosein S, Hosein P (2017) Load forecasting using deep neural networks. In: 2017 IEEE power energy society innovative smart grid technology conference ISGT 2017, 2017, pp 1–8. https://doi.org/10.1109/ISGT.2017.8085971
3. Vantuch T, Vidal AG, Ramallo-Gonzalez AP, Skarmeta AF, Misak S (2018) Machine learning based electric load forecasting for short and long-term period. In: IEEE world forum internet things, WF-IoT 2018—Proc 2018; 2018-January, pp 511–516. https://doi.org/10.1109/WF-IoT.2018.8355123
4. El khantach A, Hamlich M, Belbounaguia NE (2019) Short-term load forecasting using machine learning and periodicity decomposition. AIMS Energy 7:382–394. https://doi.org/10.3934/ENERGY.2019.3.382
5. Sarhani M, El Afia A (2015) Electric load forecasting using hybrid machine learning approach incorporating feature selection. CEUR Workshop Proc 1580:1–7
6. Scott D, Simpson T, Dervilis N, Rogers T, Worden K (2018) Machine learning for energy load forecasting. J Phys Conf Ser 1106. https://doi.org/10.1088/1742-6596/1106/1/012005
7. Wee CK, Nayak R (2019) Adaptive load forecasting using reinforcement learning with database technology. J Inf Telecommun 3:381–399. https://doi.org/10.1080/24751839.2019.1596470
8. He W (2017) Load forecasting via deep neural networks. Proc Comput Sci 122:308–314. https://doi.org/10.1016/j.procs.2017.11.374
9. Zuniga-Garcia MA, Santamaría-Bonfil G, Arroyo-Figueroa G, Batres R (2019) Prediction interval adjustment for load-forecasting using machine learning. Appl Sci 9:1–20. https://doi.org/10.3390/app9245269
10. Tian C, Ma J, Zhang C, Zhan P (2018) A deep neural network model for short-term load forecast based on long short-term memory network and convolutional neural network. Energies 11. https://doi.org/10.3390/en11123493
11. Mosavi A, Salimi M, Ardabili SF, Rabczuk T, Shamshirband S, Varkonyi-Koczy AR (2019) State of the art of machine learning models in energy systems, a systematic review. Energies 12. https://doi.org/10.3390/en12071301
12. Yang L, Ouyang Z, Shi Y (2012) A modified clustering method based on self-organizing maps and its applications. Proc Comput Sci 9:1371–1379. https://doi.org/10.1016/j.procs.2012.04.151
13. Riese FM, Keller S, Hinz S (2020) Supervised and semi-supervised self-organizing maps for regression and classification focusing on hyperspectral data. Remote Sens 12. https://doi.org/10.3390/RS12010007
14. Xayyasith S, Promwungkwa A, Ngamsanroaj K (2019) Application of machine learning for predictive maintenance cooling system in Nam Ngum-1 Hydropower Plant. In: International Conference on ICT Knowledge Engineering 2019; 2018-November 3–7. https://doi.org/10.1109/ICTKE.2018.8612435
15. Kumar K, Singh AK (2020) Power system stabilization tuning and step response test of AVR:a case study 482–485

Chapter 66
Enhanced Medical Monitoring Wireless Sensors Networks Using Proposed Greedy Multipoint Relays Protocol Algorithm

C. Naveeth Babu and V. S. Prakash

1 Introduction

In future years, wireless communication in all area will be embedded in everyday stuffs like agriculture, home-based applications, roads, toys as well as Humans. The integration of a wireless sensor module is not just aiding a way to communicate but it is a measure to make objects smarter and conceding those new abilities.

Wireless sensor networks will aid a wide range of new concepts and usages. The most important one of the applications is in body sensor networks by which the patients can able to take treatment with any doctors though out worldwide. This computing will help on enhancing the aspects of quality of life of human being and change the way of any individuals to perceive the world [11].

Newly, the WSN-based telemedical treatment is considered as a new model or a pattern for supporting remote clinical health cares. It permits smoothing initial and periodically health diagnostics through a database source history of the patient's record. This standard was fully successfully deployed in several studies and also commercial applications [12]. In telemedicine area, the wireless sensor networks are been composed of many devices, i.e., nodes dispersed on the patient body as wearable devices. These devices are stimulated by the physical sensors that are responsible for measuring target metrics [16]. They will collaborate together to send the data to one or more nodes or sinks, i.e., doctors and medical staffs for diagnostic treatment [13].

The aim of this research work is to perform and achieve a simulation result for large scale wireless sensor networks which implemented on a medical diagnosis. The main problem in this type of environment focuses on the density of nodes and how

C. Naveeth Babu (✉) · V. S. Prakash (✉)
Department of Computer Science, Kristu Jayanti College (Autonomous), Bengaluru 560077, India
e-mail: naveeth@kristujayanti.com

V. S. Prakash
e-mail: vsprakash@kristujayanti.com

to guarantee their connectivity despite their connectivity and within different radio choices [7]. Our approach helps doctors or medical staffs to diagnose the disease in an automated way. In this framework, each patient is equipped by a set of sensors [17]. Each sensor is connected to a transducer device which dedicates to know the health parameters.

In this paper, we organized the research works as follows. Section 2 briefly discusses about the related works to medical monitoring. Section 3 draws the content of the proposed work and introduces the evaluation work. Sections 4 and 5 describe the about the simulation processes and the analysis results, respectively. Finally, in Sect. 6, we conclude the research paper and derives some future works.

2   Related Works

Authors [1, 15] discuss how wireless technologies, reducing the healthcare cost, are often used for medical applications and the way well they perform during a health-care and hospital environment. This work considers low-rate Wireless Personal Area Network (WPAN) technology as laid out in the IEEE 802.15.4 standard and evaluates its suitability to the medical environment. They specialize in scalability issues and multi-disciplinary of communicating devices during a patient's room. In [2, 10], authors represent the scheduling, implementation, and demonstration of a sensible medication system, which employs the wireless sensors network (WSN) technologies. The starting function of the system includes the medication reminder and tracking for the patients with chronic diseases. The systems are often easily deployed during a home which accommodates many elderly people. Also [3, 9] presents and demonstrates a wireless heart condition monitoring hardware/software system supported unplanned interconnection of small ECG sensors called medical sensor wireless networks.

Researchers [4, 14] have reliable in replication to authenticate their ideas and methods implemented to the problem of organizing real systems to deploying tens or many sensor nodes within the physical environment, code them, execute the nodes, and monitor their performance, behavior, state because the algorithm applying the study idea unfolds. While several authors [5, 6, 8] present the actual experimental implementation with the real equipment for monitoring.

3   Proposed Model

Applications for wireless telemedicine were attainable as a very small bit rate application delay sensitive with related to truncated cost and robustness topology. Researches outline always the flexibility fact of patients which to needs a self-organized network. While the protocols and procedures for the self-organization process presented for

Fig. 1 Medical wireless sensor network-based network

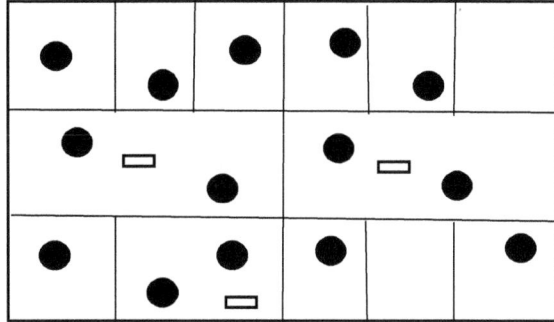

the wireless sensor networks is still inadequate face mobility, some procedures of source dependents just like the multipoint relay appears promoted.

In this paper, we aim to assess the performance of mobile medical wireless sensors networks used to monitor the numerous various healthcare patient parameters. Each patient presenting an emergency situation should have the necessary treatment at the exact time. The delay is very complex, especially for the heart diseases.

Sometimes, patients need many doctors; consequently, network should be able to send message for numerous destinations including the sink node, where the data are been stored. So, the doctors can react and even give better treatment with specific device like injecting sugar dosage in patient physique body by vibration.

Filled circles represent the patients, and rectangles represent doctors

While one patient is transmitted each time the problem of collusion and delay access are been avoided about the problem of persistent at emergency cases while more patients need care on treatment. In this paper, we analyze the performance behavior of the networks which are shown on Fig. 1, where all the nodes, i.e., patients are moving, the filled circle represents the patients, and the rectangle represents the doctors.

4 Simulation

The simulation work is very important step to improve the performance and ability, efficiency of our proposed proposal work to monitor the vibrant parameters in telemedicine environment. However, if the network in the node is to be analyzed is yet to be deployed on the different environment and to be made operational, so the simulation modeling is used. It is the most important method for evaluating the network performance analysis.

In this research work, the network topology is constructed using 30 patients sensing nodes with three sensors for each one, 4 nodes for doctors, and one for the sink node. Nodes will exchange Hi messages to complete the neighbor table. Each node will measure their own relays with the Multipoint relays protocols (MRP)

Table 1 Simulation parameters

Parameter	Value
field_x	200
field_y	200
numNodes	30
TxOutputPower	0
constantDataPayload	20, 100, 1000, 2400
packetHeaderOverhead	5
startupDelay	0
speed	0.2
maxNetFrameSize	5000
maxMACFrameSize	5000
maxPhyFrameSize 2500	5000

presented in this paper as a proposed technique to reduce the number of terminated retransmissions in the wireless sensor nodes by selecting a special specific node set to cover the network based on three-hops neighbor knowledge. Several rules and methods are proposed for this calculation. In this paper, we deployed to the proposed greedy MPR set computation described as the algorithm below.

Proposed greedy MPR algorithm

i.   Start the procedure with an empty multipoint relay set.
ii. Add nodes which are the only neighbor of some nodes in the three-hop neighbors.
iii. If still there will be some three-hop nodes which are not yet covered, so computes the two-hop nodes degrees.
iv. Two-hop nodes degree chooses the maximum one.
v.   Repeat step 3 until all three-hop neighborhoods are covered.
vi. The simulation study was carried out to estimate the act of the network based on the metrics.

Media access delay and packet delivery ratio simulation parameters are listed in Table 1.

5 Results

5.1 Media Access Delay

Media access delay is the time that are needed for a device to gain access to the media for primary transmissions. This parameter represents the total queuing concept and delay of the data frames which are transmitted. This delay is defined as the interval from when the frame is inserted into the broadcast queue until when it is sent to

Fig. 2 Media access
delay—greedy MRP
algorithm versus modified
greedy MRP algorithm

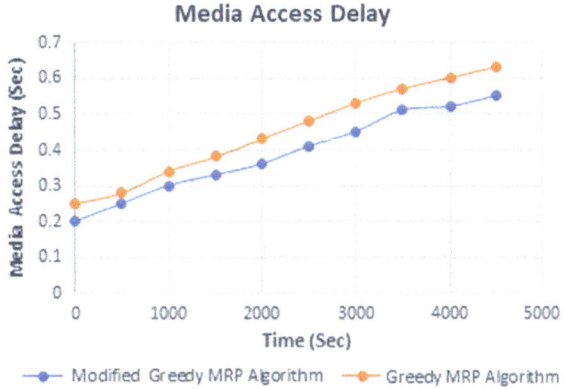

the physical layer. The queuing delay which includes the data packets arrival time
from higher layer and frame creation time. Figure 2 shows the media access delay
for which is less for the proposed greedy MRP Algorithm

5.2   Packet Delivery Ratio

Packet Delivery Ratio (PDR) can be measured as the ratio of number frames delivered in total to the total number of frames that sent from source node to destination
node in the network. It measures the loss rate as realized by the transport protocols
and as such it characterizes both the accuracy and efficiency of WSN routing protocols, where high packet delivery ratio is desired in any network. Figure 3 shows the
packet delivery ratio for the proposed greedy MRP algorithm is higher compared to
the existing normal multipoint relay protocol algorithm

Fig. 3 Packet delivery
ratio—greedy MRP
algorithm versus modified
greedy MRP algorithm

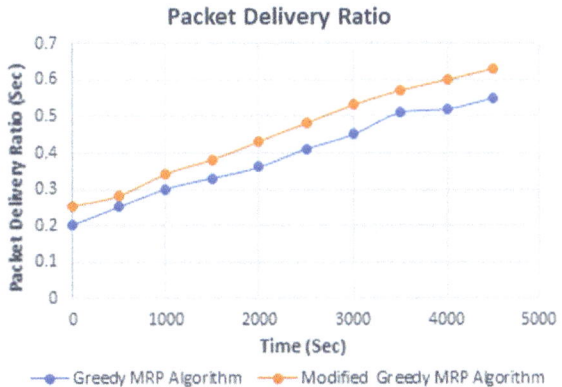

6 Conclusion and Future Scope

In this paper, the performance of the proposed greedy MPR protocol for medical sensor network was analyzed using the NS2 simulator and has simulate the media access delay time and the packet reception rate by changing the number of traffic sources and sinks in a full mobile network to define the possible scenarios of emergencies cases. Results show that the proposed greedy MPR protocol enhances the behavior of the medical network, and still it is challenging to improve the performance of the device. Also, security is an important aspect should not be avoided in the multi-user devices atmospheres to grant patients privacies and to protect data from unwanted actions of unauthorized users. In future, modified greedy MRP algorithm will be implemented in various medical monitoring wireless sensor devices, and further, more improvements will be derived.

References

1. Ahmed A, Bakar KA, Channa MI, Haseeb K, Khan AW (2016) A trust aware routing protocol for energy constrained wireless sensor network. Telecommun Syst 61(1):123–140
2. Akyildiz IF, Su W, Sankarasubramaniam Y, Cayirci E (2002) Wireless sensor networks: a survey. Comput Netw 38(4):393–422
3. Alwan H, Agarwal A (2013) MQoSR: A multiobjective QoS routing protocol for wireless sensor networks. ISRN Sensor Netw 1–13
4. Aswale S, Ghorpade VR (2015) Survey of QoS routing protocols in wireless multimedia sensor networks. J Comput Netw Commun 2015:1–29
5. Chang CC, Le HD (2016) A provably secure, efficient and flexible authentication scheme for ad-hoc wireless sensor networks. IEEE Trans Wirel Commun 15(1):357–366
6. Chen D, Varshney PK (2004) QoS support in wireless sensor networks: a survey. Int Conf Wirel Netw 233:1–7
7. Devisri S, Balasubramaniam C (2013) Secure routing using trust based mechanism in wireless sensor networks (WSNs). Int J Sci Eng Res 4(2):1–7
8. Gaurkar S, Ingole PK (2013) Access control and intrusion detection for security in wireless sensor network. Int J Sci Tech Res 16(2):63–67
9. Hamid Z, Hussain FB (2014) QoS in wireless multimedia sensor networks: a layered and cross-layered approach. Wirel Pers Commun 75(1):729–757
10. Jayshree A, Biradar GS, Mytri VD (2012) Energy efficient prioritized multipath QoS routing over WMSN. Int J Comput Appl 46(17):34–39
11. Kim HS, Lee SW (2009) Enhanced novel access control protocol over wireless sensor networks. IEEE Trans Consum Electr 55(2):492–498
12. kumar p, gurtov a, iinatti j, sain m, ha ph (2016) access control protocol with node privacy in wireless sensor networks. IEEE Sens J 16(22):8142–8150
13. Li M, Lou W, Ren K (2010) Data security and privacy in wireless body area networks. Wirel Commun 17:51–58
14. Safa H, Artail H, Tabet D (2010) A cluster-based trust aware routing protocol for mobile ad-hoc networks. Wirel Netw 16(4):969–984
15. Vamsi PR, Kant K (2016) Trust and location-aware routing protocol for wireless sensor networks. IETE J Res 62(5):634–644

16. Chi J, Wang Y, Chu H (2020) Mean shift-based multisource localization method in wireless binary sensor network. J Sens 1–8
17. Wang B, Yu J, He J (2020) An optimal backoff time-based internetwork interference migration method in wireless body area network. J Sens 1–12

Chapter 67
Streetlight Management and Control System Using IOT

A. R. Aswatha and J. Shwetha

1 Introduction

IOT is a widespread global interconnection of sensors (the major component), evaluating, and digital equipment connected through the Internet to share and transfer data with each device identified by a unique identifier (UID). With the immense growth in various commercial platforms and societies, various automation ideas have been evolving to improve these premises which in turn led to the competitive healthy environment. This interest toward technology made everyone to focus on the things which can be improved and creating a way for new inventions which are more efficient and advanced smart systems. User nowadays utilizes more convenient Web application and mobile or laptop-based monitoring and control system which is connected to the IoT cloud server for optimized energy conservation and quick detection upon any kind of error in the system. The modern era requires smart cities to invest in flawless energy-based systems like, the technologies which makes use of less resources in developing traffic systems and streetlighting appliance. Therefore, the attempt is made to give users a genuine and easy to use application according to their convenience for monitoring an array of electrical devices.

Kevin Ashton is the earliest individual to propose the concept of IoT in the year of 1999, and he mentioned that the IoT is a differently identifiable connected objects with radiofrequency identification and detection (RFID) technology [1]. A significant requirement of an IoT is that objects in the network must be interconnected. The factors like networking, communication, processes, and some others are involved in the process of IoT designing. Most important thing to remember is to consider

A. R. Aswatha · J. Shwetha (✉)
Dayananda Sagar College of Engineering (Affiliated to VTU, Belgaum), Bangalore, Karnataka, India
e-mail: shwethagowdaj@gmail.com

A. R. Aswatha
e-mail: ashwath.ar@gmail.com

operability, extensibility, and the scalability while designing the IoT's architecture, as they play a vital role.

2 Applications of Internet of Things

IoT is used in various fields, and some of them are listed as below:

2.1 IoT in Day-to-Day Activities

Consider a traffic camera that can monitor the streets for accidents, weather conditions, and communicates this data to a standard gateway. The traffic management system can analyze data acquired and get routes around the area to avoid bottlenecks. The system could also convey line instructions to diverse through smart devices or radio channels.

2.2 IOT in Smart Home Applications

IoT plays a vital role in the home mechanization framework where various household tasks or works are initiated by utilizing a string of melodic notes, and it can also be used in security cameras and door unlocks modes.

2.3 IOT in Biometric Systems

Usually, fingerprint scanning systems or other scanning systems are used in the biometric system, which depends on the organization. Considering the fingerprint scanner, it will extract the unique pattern from the fingerprint scanner and stores it on the database. When every time the fingerprint is scanned, it will compare with the stored data with present data and gives the results.

3 Literature Survey

The total power or the energy consumption is minimized by partitioning the IoT demands optimally among the available processing nodes [2]. A case where the requests of IoT are minimum and less amount of demand, the power consumption is up to significant savings of about 93%. And when the requests of IoT number increases, still 32% of total saving can be obtained. The application should be maintained with a smaller number of IoT requests as possible and make use of all the processing nodes for the application.

Time which is being spent on each task by all the resources are not available or it will be too difficult to collect all the details by human observation, for example, obtaining information about the time spent by each student on each question. It helps to assess the exact time required for the specific application or task. It also helps in providing valuable information like the progress of the task and efficiency of the outcome over time. The system is designed in a way that it can be easily used by all kind of users, what is more, as the measurements will be consequently done by the system, just changes have to be applied to the upcoming tasks. The framework acquires the information with capacitive sensors and moves respective information to a Web application with the help of a Wi-Fi module [3].

Indoor and outdoor lightings consume more amount of energy than one can think, which leads to lowering the efficiency. So, it must be controlled, as an example in this section assume the public controlled smart streetlights. Where it uses an almost familiar technology called light-emitting diode (LED) lights which are controlled electronically. These are used to achieve an energy-efficient smart lighting system for monitoring and for adaptable illumination [4]. Using this technology, very good improvement in the reliability is obtained with more user satisfaction, and this is of low-cost network. In addition to the above advantages, this system can also save energy up to 68–82%, and of course, it is depending on the daylight hours fluctuation in between summer and winter. Not only the increased energy saving but also this reduces the amount of greenhouse gases in atmosphere and improves system reliability and decreases the major maintenance problem as it is implemented based on smart controlling idea.

For indoor lighting systems, the distance or area occupied will be less so one can make use of wireless technology where the available devices will be connected to a shared Wi-Fi network. However, in case of outdoor lighting systems, the number of appliances will be more, and if they are in one direction, the wired configuration can be used to avoid the range problems [5].

Approximately, 6.7 TWh of electric energy is being consumed by the public lighting system as per the Central Electricity Board of India, which costs almost $500 M per annum [6]. Saving energy is the most critical challenge in the world today. An effective solution that can be undertaken to this problem is making use of Internet of Things. In the case of a streetlight control system, by controlling the intensity of the lights based on the surrounding brightness, one can save energy, and it must be done using a protocol. And interesting fact here is that this technique can be used for both outdoor and indoor lighting systems.

Not only for indoor or outdoor purposes, but the streetlights can also be controlled based on night or day. This can be done based on the detection of the sunlight. Streetlights gets ON automatically during the night times and OFF during the daytime. By placing a camera on the top of streetlight, the actions performed on the road can be tracked completely, where the footage's will be stored in a specific server [7]. Every streetlight pole will have a button for emergency use of passengers, if a person faces

Fig. 1 Node MCU module

any danger, he/she can press this button, so that the respective footage will be sent to the nearby police stations,, and an alarm gets ON. It is also better to use LED system rather than using HPS system to save energy. They are proven as best mode for general lighting systems and roadside lighting systems concentrating on energy consumption [8].

The Node MCU is shown in Fig. 1. Any network with 2.4 GHz supporting Wi-Fi module can get the access by TCP/IP protocol. The Node MCU can either connect to an available wireless connection or it can also host an application over the HTTP (Hypertext transfer) protocol [5].

In the work, which is being explained, the Raspberry Pi is used as the master controller for another electrical device which are being used, as shown in Fig. 2. Specifications of the Raspberry Pi Model B are given as following, operating voltage is at 5 V, it consists total of 40 GPIO pins, it has flash memory of 32 Kilobytes, and 40 mA of DC. Powering of Raspberry Pi can be done either through external power supply or it can also be done by USB connection, with the range 7–12 volts. It also has input and output pins which can be used as power source for several devices. Serial communication can be performed on the input/output digital pins using a software serial library. An integrated development environment which is provided by Arduino can be used for programming the Raspberry Pi board which has support for every product of the Arduino component. It is available license-free from the official Web site of Arduino.

Arduino mega specification are as follows, it has 5 V of operating voltage with a USB port and a DC power jack, Input voltage of 7–12 V, microcontroller, SRAM, and LED. But it does not have any Wi-Fi module in it.

Fig. 2 Raspberry Pi-microcontroller development board

4 Methodology

The information about the status of LED (ON or OFF) must be fetched to the raspberry pi server, so here the Wi-Fi module in node MCU is being used. Turn ON the LED's and measure the current flowing across them. And based on the present estimated, the status of LED is decided. Figure 3 shows the block diagram of streetlight management system, the basic methodology used here is that the current measurements values are sent to the server, and later based on those values from the server, the LED's are made OFF or ON, in order to save the energy The steps followed to measure current is as depicted in Fig. 4.

The schematic representation to measure current across the LED's is as shown in Fig. 4. Here, the resistors are used to measure the flow of current, and they are known as shunt resistance. From the point in between resistor and diode, current

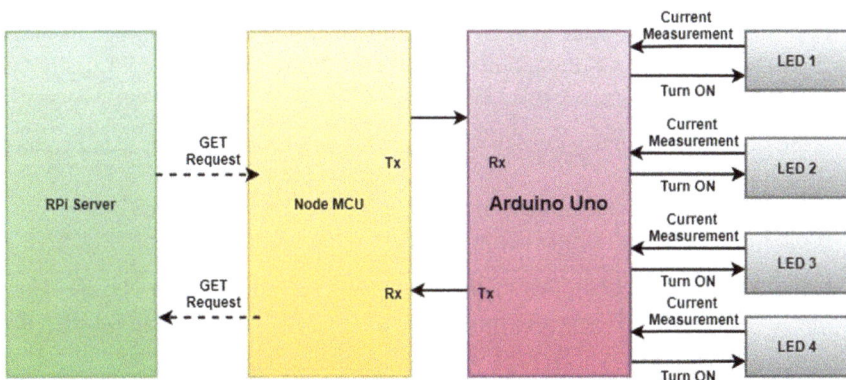

Fig. 3 Block diagram for streetlight management

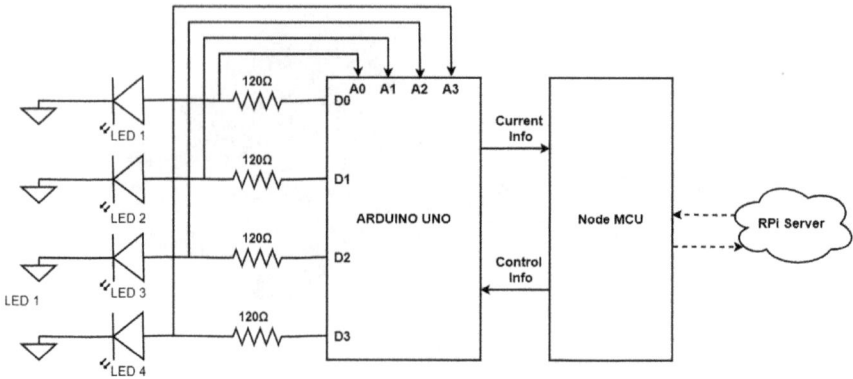

Fig. 4 Schematic representation for streetlight management

is measured by connecting it to the analog pins of the Arduino, later these current values are fed to the server using Wi-Fi module through the node MCU where the data can be displayed using database tables, and according to the status of the current, the LED's are made OFF to save the electric current. As all of us know that the Ohms law formula,

$$V = I \times R \tag{1}$$

$$R = VI \tag{2}$$

Node MCU gives output of 3.3 V, but here the Arduino is being used, so that the output voltage will be 5 V, and as known,, the rated current value for Arduino is 12 mA. Calculating the value for resistor from available values and Eq. (2) gives,

$$R = 416 \,\Omega \tag{3}$$

Choosing the nearest value of the resistor to use in the circuit. Give this connection to the analog pins in the Arduino; this is where the current is measured. Conditions for measuring current are as explained below,

- If the LED is ON and working the analog, value will be less than 1024 but greater than 200.
- If the LED is ON and it is open circuited, value will be 1024 or 1023.
- If the LED is ON and short circuited, value will be less than 200.

From these above conditions taking LEDs as load, measure the current. Now update the status of LED's to the cloud, whether they are working or not. Using the Tx pin of Arduino, transmit it to the cloud across the node MCU. Sensing this data, the status of LED is identified, and it is indicated at the output side.

5   Flowchart

See Fig. 5.

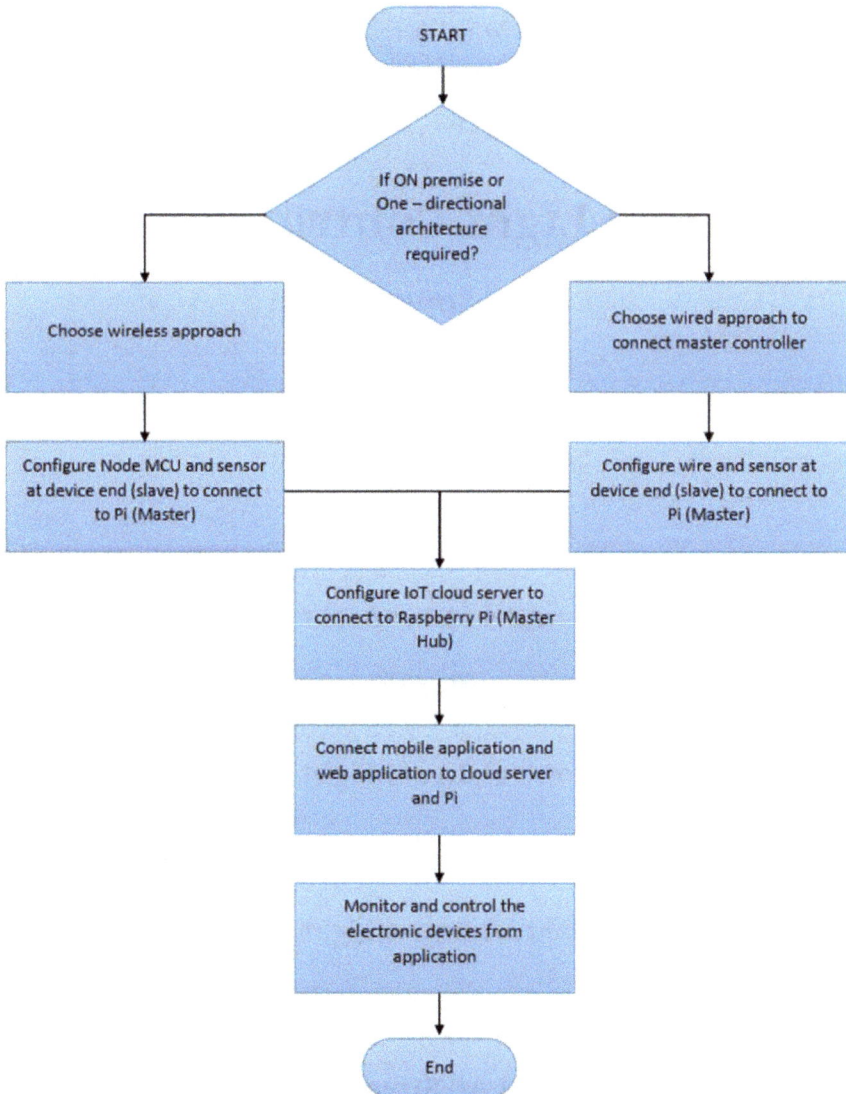

Fig. 5   Flowchart of the algorithm

6 Results Obtained

The above-designed system efficiently works by drastically reducing the consumption of the electric energy and provides complete control on the appliances. The front-end view of the application is as shown in Fig. 6. So, the user can control the lights using a mobile application. When there is a defect in some streetlights then those respective street lights can be made OFF in order to save the electric energy.

Database showing status of each light is as shown in Fig. 7.

Street Light Control Panel

Control Street Light 1:

ON   OFF

Control Street Light 2:

ON   OFF

Control Street Light 3:

ON   OFF

Control Street Light 4:

ON   OFF

Control Street Light 5:

ON   OFF

Control All Street Lights

Turn ON All Street Lights   Turn OFF All Street Lights

Fig. 6 Front view of an application for controlling all streetlights

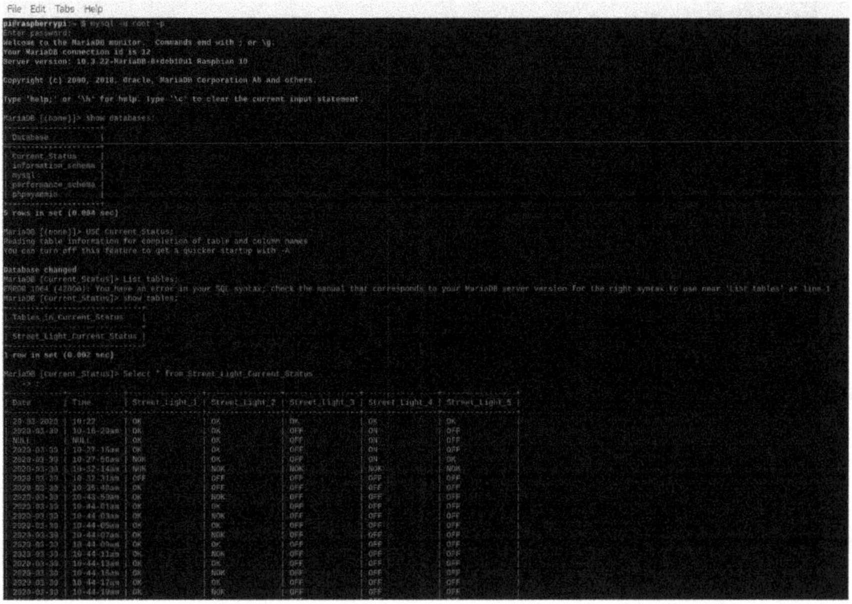

Fig. 7 Table in the database showing the status of each streetlight

User can access the following database to know the status of the lights, and he can control the flow of electric energy in the defective paths to avoid energy wastage or leakage. He will also have prior knowledge about defective lights to go and correct them, and this also saves his/her time to manage the streetlight system.

7 Conclusion

The designed system efficiently works for both indoor and outdoor lighting. It will considerably reduce electric energy consumption by providing central control over the appliances. The result of IoT is to establish a smart society, where all possible systems can be connected to the Internet and can be accessed at users' convenience, which reduces risk, time, etc. This work demonstrates streetlights controlled by users via connecting to the Internet where he/she can operate the entire streetlight system at a single time or can access the streetlights individually. By doing this, the users can identify which streetlight is faulty or not working and stop the electricity connection to that transmission line to reduce power consumption.

8 Future Scope

8.1 Monitoring the Vehicles for Traffic Rules Violation

In case of traffic violations like jumping the signal when it is RED, the message will be issued/sent to the respective users' phones.

8.2 Monitoring the Vehicle for Over-Speeding

Monitoring the vehicles for over-speeding and if found, sending the fine which is to be paid by the vehicle's user via a text message.

References

1. Gokhale P, Bhat O, Bhat S (2018) Introduction to IoT. Int Adv Res J Sci Eng Tech 5(1):41–44
2. Yosuf BA, Musa M, Elgorashi T, Elmirghani JM (2019) Impact of distributed processing on power consumption for iot based surveillance applications. In: 2019 21st international conference on transparent optical networks (ICTON). IEEE pp 1–5
3. Xheladini A, Saygili SD, Dikbiyik F (2017) An iot-based smart exam application. In: IEEE EUROCON 2017-17th international conference on smart technologies. IEEE, pp 513–518
4. Shahzad G, Yang H, Ahmad AW, Lee C (2016) Energy-efficient intelligent street lighting system using traffic-adaptive control. IEEE Sens J 16(13):5397–5405
5. Gupta AK, Johari R (2019) Iot based electrical device surveillance and control system. In: 2019 4th international conference on internet of things: smart innovation and usages (IoT-SIU). IEEE, pp 1–5
6. Murthy KS, Herur P, Adithya B, Lokesh H (2018) Iot-based light intensity controller. In: 2018 international conference on inventive research in computing applications (ICIRCA). IEEE, pp 455–460
7. Abinaya B, Gurupriya S, Pooja M (2017) Iot based smart and adaptive lighting in streetlights. In: 2017 2nd international conference on computing and communications technologies (ICCCT). IEEE, pp 195–198
8. Li F, Chen D, Song X, Chen Y (2009) LEDs: a promising energy-saving light source for road lighting. In: 2009 Asia-Pacific power and energy engineering conference, Wuhan, pp 1–3. Doi: 10.1109/APPEEC.2009.4918460
9. Al-Smadi AM, Salah ST, Al-Moomani AA, Al-Bataineh MS (2019) Street lighting energy-saving system. In: 2019 16th international multi-conference on systems, signals and devices (SSD), Istanbul, Turkey, pp 763–766. https://doi.org/10.1109/ssd.2019.8893160
10. Vinogradov KM, Moskvichev AV (2019) Energy saving at modernization of street lighting. In: 2019 international science and technology conference "EastConf", Vladivostok, Russia, pp 1–5
11. Liu C, Wang Q, Zhang F (2010) Design and development of city street-lighting energy-saving system. In: 2010 second Pacific-Asia conference on circuits, communications and system, Beijing, pp 178–182. https://doi.org/10.1109/paccs.2010.5627060

Chapter 68
Comprehensive Analysis of Classification Techniques Based on Artificial Immune System and Artificial Neural Network Algorithms

Kirti Bala Bahekar⬥

1 Introduction

Data mining is a tool for turning large amounts of data into managed information in the form of patterns, relationships among the largest massive data and generates a comprehensive structure for further decision making. These methods are used in analyzing and summarizing data from different viewpoints and transforming it into useful information [1]. Classification is an essential feature to separate large datasets into classes for rule generation, decision making, pattern recognition, dimensionality reduction, data mining, etc. Innovative computing techniques are inspired by nature like the human body has various systems to execute essential activities, such as genetic algorithms, neural networks, artificial immune systems, etc. The artificial immune system (AIS) and artificial neural network (ANN) emerged as an important tool for classification.

2 Literature Review

Brownlee Jason [2] states that various modern computation techniques are bio-inspired like genetic algorithm, artificial neural network, artificial immune system, swarm optimization, and hybrid models to solve many real-world problems. The human neural system includes three stages (see Fig. 1): receptors, a neural network, and effectors. The receptors receive the signals through internal or by the external world; then, it passes the information in a form of electrical impulses into the neurons [3]. Arbib and Michael [4] state the neural network processes incoming signals, responsible for taking appropriate decisions to generate outputs. The effector

K. B. Bahekar (✉)
Barkatullah University, Bhopal, India
e-mail: kkirt06@gmail.com

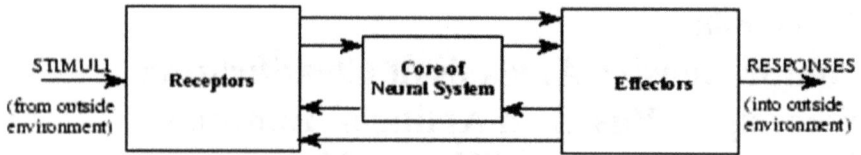

Fig. 1 Three stages of biological neural system [3]

act according to the instruction or decision was taken by the neural network, and its responses to the outside environment. Figure 1 shows the bidirectional communication between the stages for feedback.

The neuron is the central element of the neural network. As shown in Fig. 2, a neuron mainly consists of three parts: dendrites, soma (body), and axon. Dendrites are the treelike structure that receives the signal from surrounding neurons, where each string is connected to one nearing neuron. The axon is a thin cylinder that is responsible for transmitting the signals from one neuron to others. During last axon contact with the dendrites is made through a synapse. The inter-neuronal signal at the synapse is generally a chemical diffusion, but sometimes it is an electrical impulse. A neuron fires an electrical impulse only if a certain conditions are fulfilled [4].

Dasgupta and Nino [5] have defined immune system theory as a well-organized response system. According to them, the immune system is responsible for classifying incoming objects encounter as self or non-self. Non-self is unidentified, like bacteria, viruses, and cancer or tumor cells. Previously known elements present in the health system are identified as self, and they should maintain as it is in the body. Observation of the antigens patterns on the cell surface of bacteria and body cells, classification of self/non-self is performed.

Castro and Timmis [6] as per Fig. 3 presented; the detailed anatomy of the immune system states that the tissues and organs which participate in the immune process are distributed throughout the body. Such organs are called lymphoid organs, they

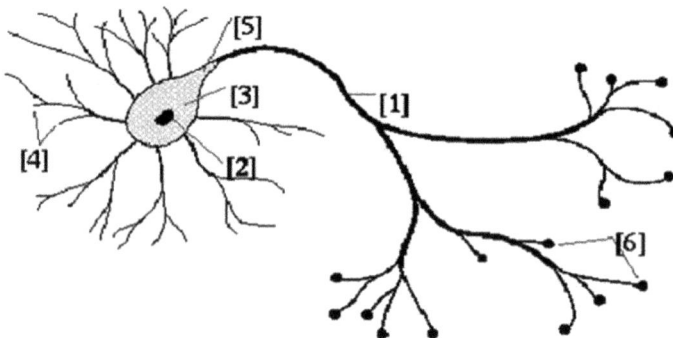

1.Axon 2. Nucleus 3.Soma (Body) 4. Dendrite 5. Axon Hillock 6. Terminals (Synapses)

Fig. 2 A biological neuron [3]

Fig. 3 Anatomy of the immune system [6]

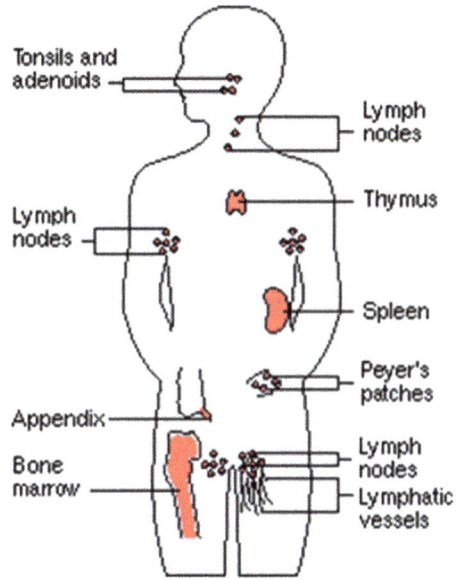

are related to the production, growing, and development of lymphocytes. Human organs as shown in Fig. 3, like tonsils and adenoids, lymphatic vessels, bone marrow, lymph nodes, thymus, spleen, and Peyer's patches play a major role in the biological immune system. The lymphocyte is majorly classified into two types B-cells and T-cells; they both are responsible for identifying and replying to the antigenic patterns. Antigenic material is bent by receptors in the shape space allow for binding, requires high affinity between them, which is the indication of the strong bond among them. Euclidean distance, hamming distance, etc. are the methods generally used to calculate affinity.

ANN teaches the system based on machine learning, where machines are programmed to learn by themselves and generate a model for further processing; it supports two learning methodologies, i.e., supervised learning and unsupervised learning. ANN consists input points also known as input nodes as in Fig. 4; output obtained by these nodes are passed to the hidden layer nodes, and result generated by

Fig. 4 Artificial neural network [3]

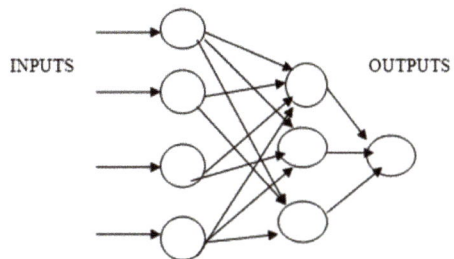

hidden layer is acts as input to the output layer nodes, produces the required output as final result of the model. Here, output produced by each node is called as node value and links are associated with some weight [7].

Dasgupta et al. [6] and Stepney et al. [8] produces biological concepts motivate researchers in the field of science and engineering to map natural concepts into computational practices like genetic algorithm, swam analysis, artificial immune system. Castro de et al. [9] explain the fields of AIS research are concerned with feature extraction (data clustering), and change or anomaly identification, such as intrusion detection systems. Forest et al. [10] have presented AIS as a method for performing tasks like computer security, anomaly detection and fault diagnosis, data mining and retrieval, pattern recognition, adaptive control, chemical pattern recognition, robotics, optimization problems, web mining, fault tolerance, autonomous systems, and engineering design optimization. Different computational models are the negative selection algorithm, clonal selection, immune network model, and danger theory.

3 Proposed Methodology

Different classifier technique is applied to the dataset for the detection of diabetes used to identify the class: tested positive or negative. Positive tested means that the patient has diabetes, and negatively tested means that the person does not have his sugar level high the required parameter. This proposed method of diabetes detection is applied to the various Artificial Immune Recognition System (AIRS) algorithms and Artificial Neural Network (ANN) algorithms, and a comparative analysis are done for appropriate algorithms for further research purposes. These algorithms have major two steps

1. Learning or training phase: where the trained model is generated using the training dataset.
2. Classification phase or testing; model previously generated is used in testing the remaining dataset, where output obtained by the model is compared with the given output and errors and accuracy are calculated.

Further comparison statistics are provided for the selection of suitable techniques, for the implemented dataset. A comparison of the algorithms is performed by comparing several parameters like accuracy, precision, true positive rate, false positive rate, recall values, etc.

Table 1 Dataset used in the experiment

Dataset name	No. of attributes	No. of instances	No. of class values
Pima_diabetes	9	768	2

4 Experiment Analysis

In the proposed method, various Artificial Neural Network (ANN) algorithm and Artificial Immune Recognition System (AIRS) algorithms are will be used for classification of standard dataset, and performance comparison is performed between them.

4.1 Dataset Taken

Real dataset is taken from an online UCI machine learning repository or by Keggale.com; it provides datasets in .arff (Attribute-Relation File Format) or .cvs format (comma-separated values). The Pima_daibetes dataset (Table 1) is chosen and used for the performance analysis of various bio-inspired classifiers.

4.2 Evaluation Setup

Weka (Waikato Environment for knowledge analysis) a well-known tool for data mining and machine learning tool written in the Java programming language. The Weka classalgo 3.6.4 provides classifier algorithms based on artificial immune system algorithms which involve artificial immune recognition algorithms like AIRS1, AIRS2, and parallel AIRS; it also provides clonal selection algorithm like CLONALG, CSCA algorithm, and artificial neural networks algorithms like learning vector quantization, single-layer perceptron, and multi-layer perceptron.

4.3 Experimental Results

In proposed analysis artificial, immune and artificial neural network inspired algorithms listed below are considered as classifiers AIRS1 (Artificial immune recognition system v-1), AIRS2, CLONALG, CSCA (Clonal selection classification algorithm), Immunos1, Immunos2, Immunos99, backpropogation, perceptron.

Every above classifiers a re-analyzed on the two test modes are considered

i. Tenfold cross-validation
ii. Full training set

Fig. 5 Precision analysis

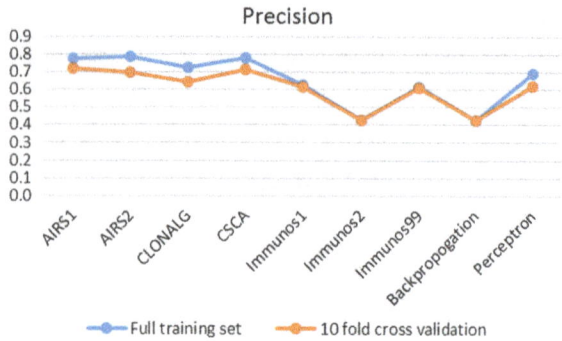

Fig. 6 Accuracy analysis

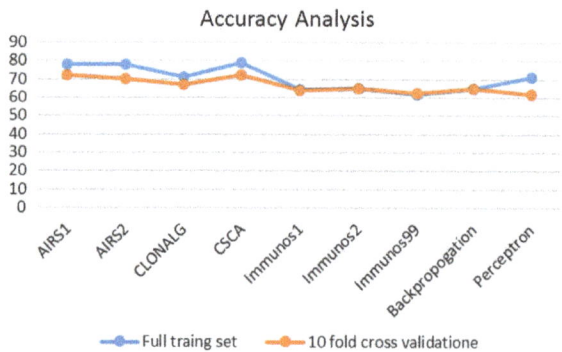

The classifiers are analyzed on the real dataset Pima_diabetes, comparing all the algorithms on the bases of precision we obtain the results shown in Fig. 5.

The precision analysis Figs. 5 and 6: shows that Clonal Selection Classification Algorithm (CSCA) works well in the full training set. Above chart also specifies that algorithms such as immunos1, immunos2, immunos99, and backpropogation works exactly same in both the modes, i.e., tenfold cross-validation and full training set.

The analysis is also done based on the accuracy of the algorithms in both the test modes, shown in Fig. 6. CSCA algorithm gives the highest accuracy in tenfold cross-validation as well as in full training set mode. It is also observed that AIRS1 also has the same performance in the tenfold cross-validation mode. In general, by the experimental results, it can be stated that full training set modes present more accuracy than the tenfold cross-validation.

A comparison of correctly classified instances and incorrectly classified instances is also performed in the experiment. In Fig. 7, the comparative representation of the algorithms is exhibited. From executing the results obtained from the experiment, it is overrated that CSCA algorithm classifies the instances more accurate than all other algorithms. In general, it can be stated that all the algorithms produce the correct classified instances in the range of 500–600 out of 768 instances.

The detailed experimental results on tenfold cross-validation mode produced by all the AIS and ANN algorithms are presented in Table 2. The results show that the

Fig. 7 Correctly/incorrectly classified instances in tenfold cross-validation and full training set

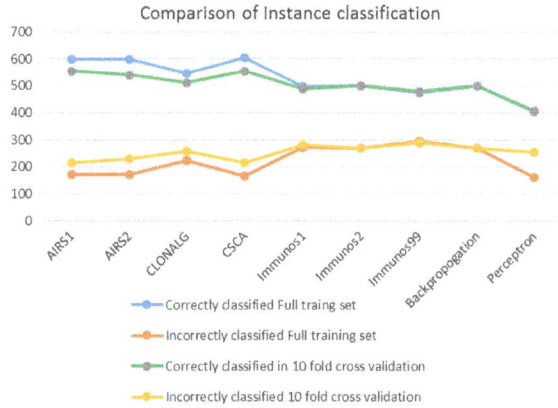

Comparison of Instance classification

— Correctly classified Full training set
— Incorrectly classified Full training set
— Correctly classified in 10 fold cross validation
— Incorrectly classified 10 fold cross validation

Table 2 Tenfold cross-validation

Algorithms	TP rate	FP rate	Precision	Recall	F-measure	ROC area	Correctly classified	Incorrectly classified
AIRS1	**0.720**	**0.337**	**0.720**	**0.720**	**0.720**	**0.692**	553	215
AIRS2	0.702	0.376	0.697	0.702	0.699	0.630	539	229
CLONALG	0.667	0.511	0.641	0.667	0.633	0.578	512	256
CSCA	**0.721**	**0.371**	**0.713**	**0.721**	**0.715**	**0.675**	**554**	**214**
Immunos1	0.637	0.498	0.616	0.637	0.621	0.570	489	299
Immunos2	0.651	0.651	0.424	0.651	0.513	0.500	500	268
Immunos99	0.625	0.499	0.608	0.625	0.613	0.563	480	288
Backpropogation	0.651	0.651	0.424	0.651	0.513	0.500	500	268
Perceptron	0.616	0.453	0.621	0.616	0.619	0.570	408	254

true positive rate, and recall rate of AIRS and CSCA algorithm is higher and competitive; the false positive rate of AIRS1 is low. F-measure and ROC areas are higher in the AIRS1 algorithms. The instance classification ratio is higher in the CSCA algorithm; therefore, it can be mentioned that the performance of this algorithm can be considered better.

From the experimental results obtained by the full training set mode shown in Table 3, it is observed that CSCA algorithm has highest true positive rate, precision, recall, f-measures, and it also provide large number of correctly classified instances AIRS1 algorithm performs well in false positive rate which is lowest among all the algorithms; it also has higher ROC area but CSCA algorithm is more capable of correctly classifying the instances with respect to all other algorithms. Hence, from the above accuracy table, it can be suggested that CSCA algorithm can be treated as the better algorithm in the listed algorithms.

Table 3 Full training set

Algorithms	TP rate	FP rate	Precision	Recall	F-measure	ROC area	Correctly classified	Incorrectly classified
AIRS1	**0.779**	**0.281**	**0.776**	**0.779**	**0.777**	**0.749**	**598**	**170**
AIRS2	0.779	0.248	0.783	0.779	0.780	0.765	598	170
CLONALG	0.711	0.302	0.729	0.711	0.716	0.704	546	222
CSCA	**0.786**	**0.305**	**0.782**	**0.786**	**0.780**	**0.741**	**604**	**164**
Immunos1	0.645	0.488	0.625	0.645	0.629	0.578	495	293
Immunos2	0.651	0.651	0.424	0.651	0.513	0.500	500	268
Immunos99	0.615	0.463	0.615	0.615	0.615	0.576	472	296
Backpropogation	0.651	0.651	0.424	0.651	0.513	0.500	500	268
Perceptron	0.713	0.681	0.692	0.713	0.611	0.536	403	162

5 Conclusion

In the digital era, big data mining is the crucial task, where various pattern recognition, classification, clustering procedure are required frequently. Better methods can be used to perform results more accurately and with less errors. In this paper, artificial immune classifier package tools provided by WEKA are used to compare various human inspired algorithms for classification. Artificial neural network algorithms based on single-layer perceptron and multi-layer perceptron are compared with the artificial immune system algorithms like AIRS1, AIRS2, CLONALG, CSCA, Immunos1, Immunos2, and Immunous99. The analysis done on the bases of precision and accuracy, a comparison of correctly classified instances, and incorrectly classified instances is done for two modes, i.e., tenfold cross-validation and full training set.

The experiment will help in choosing the best analytical algorithm for various datasets in future for better results evaluation and performance of the classification. In future research, many more parameters can be taken in consideration, and more datasets can be used for experiments to evaluate better performance. Various data mining tools can also be compared in future research activities.

References

1. Watkins A, Lois B (2002) A new classifier based on resource limited artificial immune system 'IEEE 0-7803-7282-4 02
2. Brownlee J (2005) Artificial Immune recognition system (AIRS) a review and analysis
3. http://osp.mans.edu.eg/rehan/ann/2_2%20Biological%20Neural%20Networks.htm
4. Arbib MA (1987) Brains, machines, and mathematics (2nd edn). Springer, New York, NY
5. Dasgupta D, Nino LF (2009) Immunological computation theory and application. CRC Press
6. Dasgupta D, Ji Gonzalez Z (2003) Artificial immune system [AIS] research in the last five years. Evol Comput CEC'03. The 2003 Congress, 1:123–130

7. Saravanan K, Sasithra S (2014) Review on classification based on artificial neural networks. Int J Ambient Syst Appl 2(4):11–18
8. Stepney S; Robert SE, Timmis J, Tyrrell AM (2004) Towards a conceptual framework for artificial immune systems. Int Conf Artif Immune Syst LNCS 3239:52–64
9. Castro de LN, von Zuben F (2000b) An evolutionary immune network for data clustering. In: SBRN. Brazil, pp 187–204
10. Forrest AS, Allen PL, Cherukuri R (1994) Self-nonself discrimination in a computer. In: Proceedings 1994 IEEE computer society symposium research in security and privacy, pp 202–292
11. Burnet F (1959) The clonal selection theory of acquired immunity. Cambridge University Press
12. Blake CL, Merz CJ UCI repository of machine learning databases. Dept. of Information and Computer Sciences, University of California, Irvine
13. Castro de LN, von Zuben F (2000a) The clonal selection algorithm with engineering. In: Proceedings of genetic and evolutionary computation. Las Vegas, USA, pp 36–37
14. Castro de LN, Timmis J (2002a) An artificial immune network for multimodal optimisation. In: Congress of evolutionary computation. part of the world congress on computational intelligence. Honolulu, HI, pp 699–704
15. Data Analysis'. Biosystems 55(1/3), pp 143–150
16. Hofmweyr Forest S, Somayaji SA (1998) Computer immunology. In: Proceedings of the twelfth systems administration conference [LISA'98], pp 283–290
17. Jackson Jacob T, Gunsch GH, Claypoole RL, Lamont GB (2003) Novel steganography detection using an artificial immune system approach. CEC IEEE, pp 139–145
18. Jerne N (1974) 'Towards a network theory of the immune system'. Annals of immunology (Inst Pasteur) 125C:373–389
19. Jitha RT, Sivadasan ET (2015) A survey paper on various reversible data hiding techniques in encrypted images. In: International advance computing conference [IACC], 2015 IEEE international, pp 1039–1043
20. Prasasd Babu MS, Somesh K (2015) Artificial immune recognition systems in medical diagnosis. IEEE, pp 882–887
21. Tarakanov AO (2012) Information security with formal immune networks. In: Int J Comput Sci Inf Tech 3(05):54133–5136, ISSN-0975-9646
22. Timmis J, Neal M (2001) A resource limited artificial immune system. Knowl Syst 14(3/4):121–130
23. Timmis J, Neal M, Hunt J (2000) An artificial immune system for data analysis. BIOSYSTEMS 55(1/3):143–150
24. Duch W (2002a) Datasets used for classification: comparison of results. http://www.phys. uni.torun.pl/kmk/projects/datasets.html. Computational Intelligence Laboratory, Department of Informatics, Nicholaus Copernicus University, Torun, Poland
25. Duch W (2002b) Logical rules extracted from data. Computational Intelligence Laboratory, Department of Informatics, Nicholaus Copernicus University, Torun, Poland. http://www.phys. uni.torun.pl/kmk/projects/rules.html
26. Wang W, Tang Z (2009) Improved pattern recognition with complex artificial immune system. Soft Comput 13:1209–1217
27. Bahekar KB, Gupta AK (2018) Artificial immune recognition system-based classification technique. in: Proceedings of international conference on recent advancement on computer and communication. Lecture Notes in Networks and Systems 34:626–635. Springer, Singapore. https://doi.org/10.1007/978-981-10-8198-9_65
28. Shekhar S, Xiong H, Zhou X (2017) Artificial neural network. Springer, Encyclopaedia of GIS
29. Hemlata (2018) Comprehensive analysis of data mining classifiers using weka. Int J Adv Res Compute Sci 9(2):718–723, March–April 2018
30. Reddy S, Sai Prasad K, Mounika A (2017) Classification algorithms on data mining: a study. Int J Comput Intell Res 13(8):2135–2142

Chapter 69
Low Cost IoT-Based Smart Wheelchair for Type-2 Diabetes and Spine-Disorder Patients

Sayanti Dutta, Anusha Chaudhuri, and Arindam Chakraborty

1 Introduction

The smart wheelchair is designed to have self manoeuvrability with the help of user command. This helps the user to independently navigate from one place to another without anyone's help. The first known wheelchair called an invalid's chair was invented in 1595 for Phillip II of Spain by an unknown inventor. A self-propelling chair on a three-wheel chassis was built in 1655 by Stephen Farfler. In 1783, John Dawson of England, designed a wheelchair and named it after the town of bath which served the sick and disabled persons in the early nineteenth century. Wheelchairs with wooden frames made of cane were introduced towards the middle of the nineteenth century. Then, several modifications, such as wire-spoke wheels, rubber tyres along with the push-rims for self-propulsion were introduced in the late nineteenth century. The invention of the folding wheelchair made of tubular steel was one of the most pivotal advancements of the twentieth century. A Canadian inventor named George Klein invented the first electric-powered wheelchair [1] along with his team of engineers while working for the National Research Council of Canada. The cross-frame wheelchair was introduced by Herbert A. Everest and Harry C. Jennings [2] in 1932 which became the standard design for tubular-steel folding chairs. The subsequent developments focussed on increasing reliability, manoeuverability, performance and decreasing the weight of the wheelchair. In the twenty-first century, the incorporation of emerging technologies became an important feature of wheelchair design.

S. Dutta (✉) · A. Chaudhuri (✉) · A. Chakraborty (✉)
Department of Electronics and Communication Engineering, Institute of Engineering & Management, Kolkata, West Bengal, India
e-mail: sayantidutta.dgp@gmail.com

A. Chaudhuri
e-mail: chaudhuri.anusha1@gmail.com

A. Chakraborty
e-mail: arindam.chakraborty@iemcal.com

© The Author(s), under exclusive license to Springer Nature Singapore Pte Ltd. 2021   855
X.-Z. Gao et al. (eds.), *Applications of Artificial Intelligence in Engineering*, Algorithms for Intelligent Systems, https://doi.org/10.1007/978-981-33-4604-8_69

856   S. Dutta et al.

The push-rim-activated power-assisted wheelchair (PAPAW) was one of the new designs introduced in the first part of the century. The motors supplemented the power applied by the user to one or both push-rims during propulsion by employing hybrid of electric and manual wheelchair technologies.

According to the World Health Organisation (WHO), wheelchair is the most commonly used assistive device to improve an individual's functioning, independence and promote well being. About 10% of the global population have disabilities and out of them, 10% need wheelchairs. According to the report of Wheelchair foundation, it is estimated that 10,000,000 or 1% of people require a wheelchair in 34 developed countries whereas in 156 developing countries at least 121,800,000 or 1.85% need a wheelchair. But with the increasing population, there is an additional need for 3500 wheelchairs everyday.

According to an estimate by the United Nations around 10% of the world's population that is approximately 650 million people have a disability and out of this 80% of the population resides in developing countries. There are approximately 56.7 million people living in the United States who had some kind of disability in 2010 according to the US Census Bureau which accounts for 18.7% of the US population. About 12.3 million people aged 6 years and older needed assistance with one or more activities of daily living (ADL). In India, disability constitutes approximately 5% (2% in 2001) of the population out of which at least 8% have locomotion related disabilities. A significant proportion of this population would be benefited from a smart wheelchair to help them with their activities of daily living (ADL). With an unprecedented rate of ageing all over the world, approximately, 2 billion people will be aged 60 and older by 2050. This indicates that the demand for some form of assistive mobility will increase over the next few decades. In [3], Simpson attempts to characterise and quantify the need for smart wheelchairs. It carries out a detailed study of various trends and categories of population who might get benefit from the use of smart wheelchairs and thereby projects a growing market for smart wheelchairs across the globe.

Wheelchairs are required by people for a variety of reasons. People having muscular dystrophy or cerebral palsy or other birth injuries like spina bifida suffer from paralysis. Patients having nervous system disorders typically neuro patients face challenges in pursuing independent mobility. Centres for disease control and prevention estimates cerebral palsy ranging from 1.5 to more than 4 per 1000 live births or children in the range of 2–6 years. Although hand cycles are available which can be used for navigation but it is tiresome to continuously use arms for propulsion. Again for a normal wheelchair, the users have to depend on someone to push the same. The number of aged people requiring wheelchairs is prevalent out of which the number of women depending on wheelchairs is more since calcium deficiency is a big threat to them. From the middle ages onwards, the bone density of women reduces affecting independent mobility in the later years thus making them more dependent on wheelchairs. Moreover, diabetic patients also need wheelchairs since diabetes stems when the body does not produce enough insulin thus making the cells starve from energy. To answer all these issues, wheelchairs have existed for a long time but the idea is to make them smart so that the users can live their life freely without seeking anyone's assistance.

2   Literature Survey

Recently, the focus of research is towards developing an intuitive human-machine interface (HMI) that would minimise the human effort in interacting with the wheelchair. The standard wheelchair are mainly powered by electrical-based power source that is typically some motor control using a joystick and has little computer control. Therefore, researchers today are mainly focusing on computer-controlled wheelchair that integrates both sensors and intelligence in order to reduce the necessity of human involvement. But efforts to build smart wheelchairs lack robustness not only in motion control but also in perception and control techniques. A literary survey was undertaken and some of the approaches are as follows. Some researchers used a vision-based interface where a virtual interface is projected onto a lap tray which is monitored by an overhead camera [4, 5]. Simpson et al. [6, 7] demonstrated the use of voice control of wheelchair. They had developed a modular smart wheelchair component system (SWCS). This was equipped with (analogue joystick, touch-activated switches, pneumatic "Sip n' Puff" switches. The results on testing the prototype demonstrated that the system can provide assistance on wheelchairs using analogue joystick and switch joystick input methods different in front, mid, and rear-wheel drive wheel configurations and different sets of control electronics (MKIVA, Penny +Giles, Curtis). The user's inputs were received from a joystick signal and combining with the status of the sensors navigation occurred.

According to the review paper published by Simpson in 2005, there are many forms for designing a smart wheelchair. Some wheelchairs were developed by modifying commercially existing motorised wheelchairs and some were based on mobile robots to which seats were added. All of these prototypes are sharing the common objectives like easing the usage of wheelchair, collision avoidance, stepping-up distance of travel and reducing the travel time. The Serbian Journal of Electrical Engineering [8] has dealt with controlling wheelchair using head movements. An innovative head motion recognition technique was designed based on processing of accelerometer data. The wheelchair joystick was controlled by the system's mechanical actuator. The prototype consisted of an accelerometer for collecting head motion data, microcontroller for processing the sensor data and a mechanical actuator in order to orient the wheelchair joystick in accordance with the command of the user. Sensor data was processed by a novel algorithm, implemented within the microcontroller. Thus, electric wheelchair joystick position is obtained by the translation of user's head motion. Yanco [9] used eye-tracking or EOG signals for controlling the wheelchair in 1998 which was capable of detecting depth and size of free space, vanishing point, indoor landmarks, and simple motions up to 10 m ahead in its path. Mazo [10, 11] developed an integrated human-machine interface (HMI) that incorporated several types of user inputs like, breath-expulsion, head movement, eye-tracking, face tracking for controlling the wheelchair. According to the Min et al. [4] use force sensors to measure shoulder and head movement which, in-turn, are used for controlling a wheelchair.

So, in these cases, the user has to do some physical movement to make the wheelchair operational. Law et al. [12] developed a cap interface to control a wheelchair using EOG signal from eye-gaze movement and EMG signal from jaw-movement. Another EMG-based control wheelchair control system was demonstrated by Felzer et al. [13] where muscle contractions are translated into appropriate motion commands. Quite recently, researchers have started using brain signals (EEG) to control wheelchairs [14, 15]. Tongue is also used in tongue drive wheelchair as presented in the journal International Journal of Computational Cognition [16]. According to the paper by Mougharbel [4], the sip and puff interface were used for wheelchair navigation. The user did some training test on simulator and drove the wheelchair using sip and puff interface. The driver faced irregular paths. The controller was designed such that fixed radius curvature was performed while turning right and left. The user had to perform multiple operations and go through hard turns. Sometimes, wheelchair also got trapped. The improved controller with optimal adaption parameters was used but this adaption depended on shape of the path.

Our research endeavours to implement a smart wheelchair that can assist users with sensory-motor and cognitive impairments. It can cause physical strain for users to navigate a wheelchair by joystick movement or head movement or through gestures. This paper presents a simple yet smart wheelchair that does not require any complex control sequence thus preventing any confusion of the user. In voice controlled wheelchairs, voice modules are used but this means bearing the cost of extra hardware. Moreover, it consumes time in actually recording the voice commands in the module. This project helps in navigating the same by giving voice commands from a simple voice app. Differential drive mechanism was used for turning sideways. As the mobile phone's Bluetooth is paired with HC-05 Bluetooth module, this will carry the commands from the app to the Arduino. According to the commands, the Arduino adjusts the pin status of L293D motor driver IC and actuates the motors. Obstacles encountered are sensed by infrared sensor and the wheelchair will automatically move away from the same without the user's intervention thus saving the user from accidents.

3   Methodology

Several surveys were conducted to study how smart wheelchairs were implemented. Then, a strategy was formed how to make it more user friendly. Most of the projects talked about using hand movement or head movement to navigate using the wheelchair. But in this project, through a simple voice app, the same can be moved. There will be some preset commands upon which the motion will be controlled. After receiving the text, there will be a speech to text conversion in the mobile app. This message will be then sent to the microcontroller as text over Bluetooth. After that the microcontroller will give the motor drivers signals according to that and motion will be controlled in the same process. HC-SR04 sensors will automatically avoid obstacles. GPS module will record and publish the latitudes and longitudes of the wheelchair position.

4 Proposed Strategy

The strategy of the project is proposed and a prototype is developed.

4.1 Interface for Locomotion

In this project, the goal is to make a smart wheelchair that can be easily available and affordable to everyone. The wheelchair is controlled through a voice app which recognises "Forward", "Backward", "Right", "Left" commands. The mobile phone's Bluetooth is paired with HC-05 Bluetooth module whose receiver and transmitter pins are connected to pins of Arduino. The communication between Arduino and Bluetooth module occurs through software serial (Fig. 1).

The Arduino reads the voice commands sent through Bluetooth and sets the input pins of L293 motor driver IC for rotation of wheels. The basic principle of working of L293D IC is based on the H-bridge and the motor control circuit that allows the voltage to flow in any direction. The voltage will be able to change the direction of rotation of the DC motor in both the directions. Thus, the L293D ICs with H-bridge circuit can be used for perfectly driving a motor.

Fig. 1 Designed prototype of the smart wheel chair

4.2 Accident Avoidance

For obstacle avoidance, IR sensors are used. The IR transmitter emits an infrared signal towards the object. From the surface of the object, a part of the radiation is absorbed and some part of the radiation is reflected back which is received by the IR receiver. Based on the intensity of the signal, the output of the signal is defined. IR sensor send 0 to Arduino if obstacle is present, and accordingly, wheels will be actuated to deviate from obstacle and the IR sensor sends 1 to Arduino if no obstacle is present in the course of the path.

4.3 Position Recording

The global positioning system module named NEO-6 M is used to track the position coordinates of the wheelchair which needs to be recorded and stored in the cloud for further processing. This recorded data will be used to determine the common route that a wheelchair can take in order to reach a destination directly following some set waypoints. The waypoints utilised by the GPS units are commonly expressed in mapping formats such as UTM grid or the latitude/longitude. The pins view the sky for the satellites following the principle of trilateration and give out serial data that are comma separated and follow a data format called NEMEA. The position coordinates will be sent via the ESP8266 Wi-Fi module to the Thing Speak cloud platform. The Wi-Fi module will enable the Arduino to connect to the internet by providing the required TCP/IP stack. Thing Speak is an open-source internet of things (IoT) application and API to store and retrieve data from things using the HTTP and MQTT protocol over the internet or via a local area network. The Thing Speak server is used to visualise the data received from the IoT device that is displayed in the form of graphs on the platform. Thus, Thing Speak can work like platform as a service, and through the API, it can publish information.

4.4 Block Diagram

See Fig. 2.

5 Conclusions

The sole objective of this project is to restore autonomy to disabled people primarily those suffering from spine disorders by making a low cost, user friendly, smart

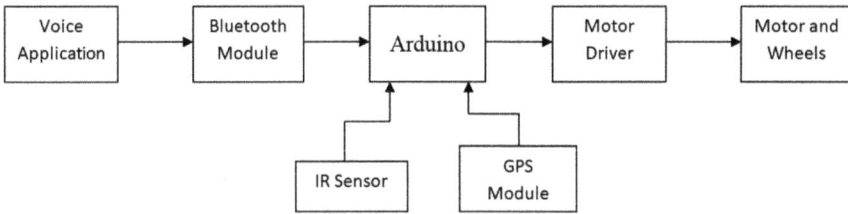

Fig. 2 Working of the smart wheel chair

powered wheelchair. Independent mobility is recognised as an important factor necessary for socio-cognitive development of an individual as it increases vocational and educational opportunities, reduces dependency on care-givers and promotes self-reliance. The wheelchair is controlled through a simple voice app by giving commands like "left" "right", etc. which will be sent via Bluetooth to Arduino, and accordingly, Arduino will manipulate the motor driver IC for actuating the motor. The infrared sensor will automatically avoid obstacles, and the GPS module will track and thereby display the latitudes and longitudes of the relative position of the wheelchair. Thus, the system has been developed to improve the functioning of the wheelchair in order to enhance the comfort of the users.

6 Future Prospects

The future prospects of the projects are adding security features like real-time monitoring, maximum speed limit, bump and wall detection in the terrain. The position of the wheelchair for real-time monitoring can be sent through Google API to Google maps to indicate the place where the user is and then the same can be informed to the family of the patient thereby improving the functional efficiency of the wheelchair. The HC-SR04 ultrasonic sensor can also be used instead of infrared sensor to detect obstacles in dusty and dark environment.

References

1. https://www.thoughtco.com/history-of-the-wheelchair-1992670
2. https://www.britannica.com/topic/history-of-the-wheelchair-1971423
3. Simpson RC, LoPresti EF, Cooper RA (2008) How many people would benefit from a smart wheelchair? J Rehabil Res Dev 45(1):53–72
4. Mougharbel I, El-Hajj R, Ghamlouch H, Monacelli E (2013) Comparative study on different adaptation approaches concerning a sip and puff controller for a powered wheelchair. In: Proceedings of IEEE science and information conference (SAI), London, pp 597–603
5. Matsumoto Y, Ino T, Ogsawara T (2001) Development of intelligent wheelchair system with face and gaze based interface. In: Proceedings of 10th IEEE international workshop on robot and human interactive communication, Bordeaux, Paris, France, pp 262–267

6. Simpson RC, Levine SP (1997) Adaptive shared control of a smart wheelchair operated by voice control. In: IEEE/RSJ International conference on intelligent robots and systems (IROS), Grenoble, pp 622–626

7. Simpson RC, Poirot D, Baxter F (2002) The haphaestus smart wheelchair system. IEEE Transa Neural Syst Rehabil Eng 10(2):118–122

8. Aleksandar P, BrankoD, Diab MO, Nizar A, Al-Hojairi AA, Baalbaki AB, Ghamloush A, El-Harir N (2013) WheelChair control by head motion. Serbian J Electr Eng 10(1):135–151

9. Yanco HA (1998) Wheelesley: a robotic wheelchair system: Indoor navigation and user interface. Assist Tech Artif Intell LNAI 1458:256–268

10. Mazo M (2001) An integral system for assisted mobility. IEEE Robot Autom Mag 8(1):45–56

11. Mazo M, Rodriguez FJ, Lazaro JL, Urena J, Garcia JC, Santiso E, Revenga P, Garcia JJ (1995) Wheelchair for physically disabled people with voice, ultrasonic and infrared sensor control. Auton Robots 2(3):203–224

12. Law CKH, Leung MYY, Xu Y, Tso SK (2002) A cap as interface for wheelchair control. IEEE/RSJ Int Conf Intell Robots Syst (IROS) 2:1439–1444

13. Felzer T, Freisleben B (2002) HaWCoS: the "hands-free" wheelchair control system. In: Proceedings of the fifth international ACM conference on assistive technologies (ASSETS), New York, USA, pp 127–134

14. Rebsamen B, Teo CL, Zeng Q, Ang VMH, Burdet E, Guan C, Zhang H, Laugier C (2007) Controlling a wheelchair indoors using thought. IEEE Intell Syst 22(2):18–24

15. Tanaka K, Matsunaga K, Wang HO (2005) Electroencephalogram-based control of an electric wheelchair. IEEE Trans Rob 21(4):762–766

16. Diab MO, Nizar A, Al-Hojairi AA, Baalbaki AB, Ghamloush A, El-Hariri N, Tongue drive wheelchair. Int J Comput Cogn 10(2). http://www.yangsky.com/ijcc/

17. https://www.nhp.gov.in/disease/non-communicable-disease/disabilities

Chapter 70
Digital Image Forgery Detection Approaches: A Review

Mohassin Ahmad and Farida Khursheed

1 Introduction

Digital image manipulation can be defined as the process of varying the constituents of a picture so as to achieve certain malicious objectives [1, 2]. These kinds of alterations in digital images are well-known as tampering or manipulation or forgery. The concept of photo manipulation is not new; however, they do exist from the decennium. There were numerous circumstances of photo manipulations in history that caused disturbed general public/administrations [3]. There exists wide range of image manipulating gadget like photoshop, paint-slinger, GNU image manipulation program, etc. Few of these are freely offered, and some are paid but easily accessible and inexpensive. Further, images modified by the editing tools are exposed toward some rectification phases and are so realistic that it is not possible for human visual system to distinguish between the authentic and tampered picture with naked eye. This indicates severe susceptibility as well as decreases the trustworthiness of digital images. Henceforth, effective and consistent image manipulation detection approaches need to be developed that can differentiate between genuine and tampered images and also have the ability to localize the tampering. In this research field, a huge number of research publications are found from around the world. A study has been done which exhibits the number of publications per year in the digital image forensics from 2000 to 2019 from two different libraries IEEE (ieeexplore.org) and Elsevier (sciencedirect.com (Fig. 1). This paper provides a survey of blind/passive image manipulation detection approaches. The purpose of digital image manipulation detection approaches is to detect the forgeries in an image. Digital image manipulation detection techniques can be broadly divided into two main classes: Active and passive approaches [4]. This classification of image forgery detection

M. Ahmad (✉) · F. Khursheed
Department of Electronics and Communication, National Institute of Technology, Srinagar, Kashmir, India
e-mail: mohassin_01phd17@nitsri.net

Publications of Image Forgery Detection

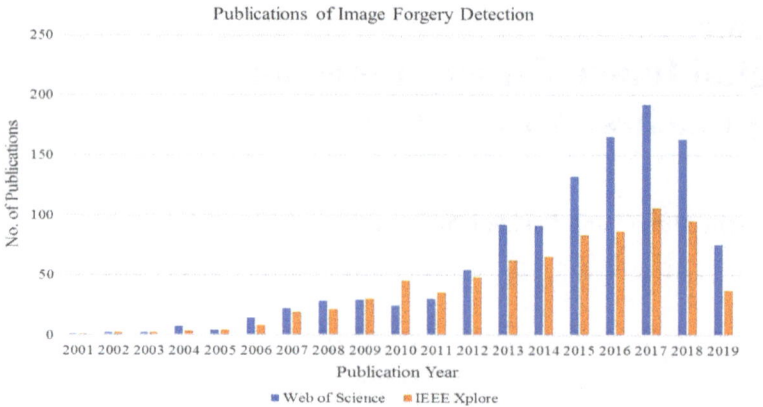

Fig. 1 Number of publications per year in digital image forensics over past 19 years in IEEE (**ieeexplore.org**) and Elsevier (**Sciencedirect.com**) libraries

techniques relies on the fact whether or not the authentic image is available. Each category is further sub-divided

1.1 Active/Intrusive/Non-blind Method

In active approaches, primary knowledge about the image is necessary to the process of verification. Its major concern is with data hiding, wherein specific data is introduced within the image either during image attainment or before the dispersal of the image. The inserted information within the picture is employed to find the basis of that picture or to observe the modification in that picture. That means confirming this encryption validates the authenticity of image. Active/intrusive/non-blind approaches are categorized into two types: digital watermarking and digital signatures. Digital watermarks are inserted in images while image attainment or during dispensation phase and digital signatures insert some subsidiary data, generally obtained from image, by the end of acquisition process. A great deal of research has been occurred in digital signatures and digital watermarking [5, 6]. The major disadvantage of active methods is that they are introduced in images during record by means of special equipment. Therefore, primary knowledge about image is essential.

1.2 Passive/Non-intrusive/Blind Method

In comparison to intrusive approaches, blind tampering detection methods take benefit of the evidences retained by the image dispensation processes achieved in different stages of digital image acquisition and storing [7]. Passive methodologies do not possess past information of the picture. These approaches exploit the existing picture only and a principle that the tampering actions modify the contents of information of the image that can facilitate in tampering detection. Blind methods are basically divided into five basic categories which are pixel based, format based, camera based, physical environment based, and geometry based.

2   Review of Existing Works Regarding the Passive Image Forgery Detection Techniques

This section discusses the several passive approaches of image forgery detection proposed by many researchers respect to the tampering type and changes occurred in an image. Three major types of tampering (CM, IS, and image retouching (IR)) and their detection techniques are presented in this paper. The first two are basically utilized for examining the dependent forgery, and the IR method is employed for revealing of independent manipulation.

2.1   Copy-Move (Cloning) Forgery Detection

It is the widespread and very familiar type of image forgery detection technique [8]. A particular portion of the image is employed to improve/eliminate features. This technique involves replication of some portion of the image and pasting it in other region within the same picture so as to cover definite features in the actual image or duplicate the information that does not exist in the real photo. As the copy-move is done within the same picture, characteristics of the altered region will be similar as that of other areas. An illustration of CM attack is revealed in Fig. 2, where right part is the primary image which having three rockets and left

Fig. 2 Copy-move image forgery [http://latimesblogs.latimes.com/babylonbeyond/2008/07/iran-doctored-m.html]

part is the tampered image with four rockets. Various approaches which are based on either key-point or block based have been developed for detecting copy-move. The input image is split into overlying or non-overlying regions of rigid size or else key-points are extracted by employing feature extractions algorithms which are then matched. It is also important that the extracted image feature must satisfy the process of rotation, scaling, JPEG compressions, and time complication enhancements in image manipulation detection.

Fridrich et al. [9] presented a primary technique for identifying copy-move image tampering in 2003. Firstly, picture is partitioned into segments (16×16 overlying), and then, in order to extract features of these blocks, DCT coefficients are used. These coefficients obtained are sorted lexicographically.

Popescu et al. [10] focused on detecting duplicate image regions. Firstly, PCA is applied on the image of fixed block size ($16 * 16$, $32 * 32$); then, eigenvalues and eigenvectors are computed for each block. Lexicographic sorting is then applied to detect the duplicate regions.

Zandi et al. [11] suggested an adaptive CM forgery detection approach in 2014. The standard deviation is used to evaluate the energy of high frequency coefficients of the blocks. Then, threshold is adapted according to the standard deviation of the block intensity.

Jaberi et al. [12] in 2014 discussed the key-points include mirror reflection invariant feature for forgery detection and Kd-tree for matching similarity. The Random Sample Consensus (RANSAC) process is employed.

Lee et al. [13] in 2015 proposed a method that relies on the histogram of oriented gradient (HOG). The image is partitioned into overlying segments; then, HOG descriptors are obtained from each segment, and Euclidean distance matching is done.

Li et al. [14] proposed one of the hybrid methods which use SIFT algorithm to obtain key-points. The image is first segmented into semantically independent segments by means of Simple Linear Iterative Clustering (SLIC). Then, matching is done by K-nearest neighbor, and EM-based (Expectation-Maximization) algorithm is employed to improve the matching.

Ardizzone et al. [15] built the triangulation on the key-points. The picture is fragmented into triangles after extracting key-points. The triangles are matched by employing a characteristic called Mean Vertex Descriptor (MVD).

Lee [16] in 2015 proposed another method, where Histogram of Oriented Gabor Magnitude (HOGM) is utilized. Image reconstruction has no traces of doctoring as it maintains the surface and structure consistency.

Liang et al. [17] in 2015 proposed a method by employing Central Pixel Mapping (CPM) to explore the dubious segments, greatest zero-connectivity component labeling (GZCL) is utilized to spot altered elements, and ultimately Fragment Splicing detection (FSD) is exploited for discriminating the altered and the original image.

Wo et al. [18] in 2016 targeted larger rotations by exploiting multi-radius rotation invariant Polar Complex Exponential Transform (PCET) features. To decrease the computational time, GPU acceleration is applied

Zhou et al. [19] in 2016 designed the CMFD algorithm based on color information and its histograms. To exploit the color information, the blocks are grouped as per their color resemblance by means of color moments. The extracted features are sent to Compositional Pattern Producing Network (CPPN) in service of deep learning task.

Zhong [20] explored the Radon odd radial harmonic Fourier moments technique (RORHFM) which is robust to several operations like rotation, scaling, conversion, etc. RORHFM employs a circular template to obtain the geometric inherent descriptors.

Beste Ustubioglu et al. [21] in 2016 postulated a method in 2016 which utilizes DCT-phase terms in order to curb the span of feature vectors. Element-by-element equality is used to find similarity between blocks.

Huang et al. [22] proposed a threshold free algorithm in 2017 discovers copy-move tampering in the presence of several attacks like JPEG compression, Gaussian noise, and blurring by employing FFT, SVD, and PCA features which are further matched in a cascading fashion.

Mohd Dilshad et al. [23] in 2017 discussed image projection profiling scheme in which the input picture is transformed into binary then projection profiles both horizontal as well as vertical are computed. These profiles are utilized for tampering detection.

Mohd Dilshad et al. [24] in 2017 used haar wavelet and ring projection transform (RPT) for CM tampering detection. They have considerably improved the time and memory efficiency.

Zhao et al. [25] in 2017 analyzed copy-move forgery segmentation-based swarm intelligence. Initially, swarm intelligence (SI) is chosen to distinguish the ideal parameter for every layer. Mainly, the input image is distributed into flat and coarse layers, and SI algorithm is used to identify the constraints to detect the forgery region by locating SIFT key-points.

Jain and Bansal [26] in 2017 developed an ANT colony optimizations in which features extracted by SIFT and SURF are employed for categorization and detection of forgery.

Yang et al. [27] projected a CM detection technique based on CMFD-SIFT. A SIFT-based detector is used here to detect the key-points. Invariance to mirror transformation is considerably improved.

Jain et al. [28] in 2018 proposed image forgery detection based on Improved Relevance Vector Machine with Glow Worm Swarm Optimization (IRVM-GSO). The extracted feature vectors are arranged lexicographically and identify the similarity matching by Minkowski distance.

Jain et al. [29] in 2018 introduced the Hybrid Extreme Learning Machine with Fuzzy Sigmoid Kernel (HELM-FSK) technique. Active Contour-based Snake (ACS) process is used for effective boundary detection, and Contourlet Transform (CT) is used to obtain the feature vector from object boundary detected area. The similarity matching between vectors are measured by using Distance function with Bat Algorithm (DBA).

2.2   *Image Splicing Detection*

Image splicing is another frequently used way to create the forgery. It involves replacing of one or more regions of an image with portions of another image. Splicing is that method of image manipulation in which the fragments of same or dissimilar images are joined to yield a single complex image (forged image) without any further treatment like blurring and smoothening of borders and edges among diverse segments. Image splicing can produce discrepancies in several ways like the aberrantly sharp transitory at the boundaries, and these discrepancies are utilized to identify the tampering. Figure 3 shows the case of IS, where the pictures of the woman standing in garden, and the original image of pillars is merged into a single picture to create forged image. Combining two separate images which are taken in different time and places can cause to imbalanced in the primary image and doctored image, this mismatch is utilized as parameter for detecting image splicing forgery. Methods which are based on splicing use numerous descriptors like bi-coherence, invariant moments, weber local descriptors, DCT, and DWT coefficients, etc.

Ng and Chang [30] proposed the image splicing detection scheme in which bi-coherence magnitude and phase features are employed. This method has detection accuracy of 70%.

Li et al. [31] used SVM classifier on Hilbert–Huang Transform (HHT)-based and moment features with the detection accuracy of 85.87%.

Carvalho et al. [32] used SVM meta-fusion classifier, which depends on the discrepancies in color of the lighting of picture.

Rao et al. [33] exploited motion blur cues to unveil splicing in image. This method evaluates discrepancies in motion blur in the presence of space-variant muddling circumstances.

Zhao et al. [34] proposed a 2D Markov model which represents the coefficient variance arrangement in two fields, i.e., BDCT and DWT domains. Markov descriptors are one among the leading operative for splicing detection.

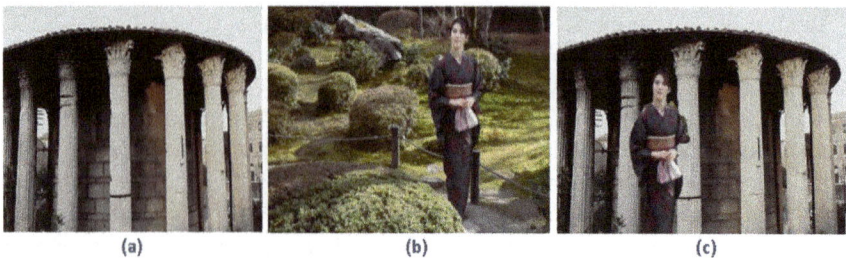

(a)   (b)   (c)

Fig. 3 Example of image splicing forgery, **a** original image of woman standing in a garden, **b** original image of pillars and **c** tampered image with woman and pillar combined. *Source* Image is from CASIA TIDE v1.0 Dataset [64]

Bahrami et al. [35] proposed local blur-type features which are obtained by dividing the image into blocks by means of a Generalized Gaussian Distribution (GGD). Furthermore, a classifier is developed to categorize the segments into distorted or blurred depending on the projected features.

Zhan et al. [36] in 2016 proposed the splicing detection technique designed on PCA minimum eigenvalues. Then, Self-Similarity Pixel Strategy (SSS) is applied to each block for resampling. After this Local Minimum Eigenvalues (LME) of the trial matrix are examined.

Kaur and Gupta [37] in 2016 projected a technique centered on wavelet transform which uses DWT and LBP. To perceive the rapid changes, local sharpness or smoothness is measured by analyzing wavelet coefficients. Low frequency constants are attained by single-level DWT.

Shen et al. [38] in 2016 developed a spliced image forgery detection that depends on Gray Level Co-Occurrence Matrices (GLCM) to detect the texture features. Difference Block Discrete Cosine Transform (DBDCT) arrays are implemented by disintegrating the picture into modules to evaluate the GLCM for extracting the textural features such as mean, standard deviation, and spatial relationship between image pixels.

Tae Hee Park et al. [39] in 2016 proposed another scheme which utilizes the typical function moments for the interspace concurrence matrix in the wavelet field.

Li [40] in 2017 proposed a method of Quaternion Discrete Cosine Transform (QDCT) for IS detection. The quaternion is constructed by exploiting color components and content relating to color from each image block

Salloum et al. [41] in 2017 in order to locate IS forgeries used a Multi-task Fully Convolutional Network (MFCN). Two acquiring tasks were used to comprehend the tag of boundaries and surface of spliced region.

Moghaddasi et al. [42] in 2018 computed the roughness measures of singular value (SV) in order to detect the image splicing with the decomposition of the DCT coefficients.

Shah and El-Alfy [43] in 2018 proposed DCT methods and multi-scale LBP for splicing forgery detection. Multi-scale LBP is applied for texture analysis which generates a binary code for 3×3 block, and DCT is applied to every input module.

Zhang et al. [44] in 2018 postulated a DWT-based methodology to enhance the performance. First step is to apply the block DWT on the basis picture. Then, Markov features are constructed, and a descriptor choosing way SVM-RFE is used.

Jalab et al. [45] in order to capture splicing has implemented Approximated Machado Fractional Entropy (AMFE) of DWT. DWT has been used as the means to disintegrate the prime input into sub-bands of differing frequencies, and then, AMFE is used as a new texture descriptor.

Subramaniam et al. [46] basically employed conformable focus measures and focus measure operators as a combo in obtaining Redundant Discrete Wavelet Transform (RDWT) coefficients.

2.3 Image Retouching Detection

Image retouching is basically an image forgery tool which finds its application in prof-itable and artistic applications. Retouching is essentially a process which is mostly used to improve or diminish the important image features. Retouching encompasses a great deal of actions, e.g., color improvement, glamor correcting, skin retouching, picture renovation, picture cartooning, etc. Retouching is additionally done to form a considerable combination of two pictures which need the geometric transformations like spin, resizing, or enlarging of any of the pictures.

Stamm and Liu [47] explored an algorithm to explain a technique that detects global modifications and also advises procedures for histogram equalization. A comparable design initiated on the probabilistic method of pixels is described in [48] which detect the enrichment in contrast.

Cao et al. [49] established a technique for gamma modification of detection. This method is predicated on approximation of histogram features that are obtained from the highest gap structures. This method is efficient for both local and global gamma adjustments.

Li et al. [50] developed another scheme for forgery detection which depends on the Bi-Laplacian filtering. Here, whole focus is on looking for the comparing segments on the idea of a KD tree algorithm.

Cao et al. [51] developed two innovative procedures to perceive the contrast enrichment concerned with image forgery. Here, emphasis is on the recognition of overall contrast improvement done to JPEG pictures. The histogram peak/gap artifacts experienced through the JPEG compression and pixel values are analyzed and differentiated by obtaining the zero-height gaps.

Chierchia [52] explored the techniques which supported the photo-response non-uniformity (PRNU). Here, device noise is utilized to perceive the forgeries by discovering the lack of the camera PRNU. Basically, a Markov features take pronouncements together on the complete picture instead for every individual pixel.

Ding et al. [53] used edge perpendicular binary coding to identify the distinct features of the texture amendment produced by the unsharp masking.

Zhu et al. [64] method is based on overshoot artifact metric to detect image sharpening operation. At first, canny function is utilized to perceive the edges; then, image edge points are classified using non-subsampled contourlet transform (NSCT).

David and Fernando [55] method is based on Random Matrix Theory (RMT) for the detection of resampling. RMT models the actions of eigenvalues and singular values of arbitrary matrices. RMT has very low computational complexity.

2.4 Deep Learning-Based Detection Methods

Deep learning with the development of techniques in artificial intelligence has evolved promisingly for the image forensics. Training samples are used to learn the features in various deep learning configurations, e.g., Convolutional Neural Network (CNN) and deep residual system. With well-designed and well outperform traditional approaches, median filtering has been preserving the edge statistics of the picture, e.g., eliminating arithmetical clues of blocking artifacts leftover by the JPEG compression or eradicating rectilinear relationships among contiguous elements for hiding the proof of resampling.

Chen et al. [56] proposed the method which uses CNN so as to enhance the performance and to automatize the job of feature abstraction. As soon as the features are extracted, they are directed to two fully connected layers, and after this, the result is provided to a softmax operator in next fully connected level.

Bayar and Stamm [57] explored another method that has a new convolutional layer which will compel the CNN architecture to acquire forgery detection features. Additionally, this technique does not have the requirement of pre-extraction of descriptors or pre-processing.

Wang and Zhang [58] discussed that JPEG compression is one more issue in picture tampering and proposed the technique that describes a set of remarkable features which remain unpretentious by recompression. To fascinate and categorize these features, 1D CNN model is developed.

Zhang et al. [59] suggested another method consisting of two phases to study features. In the first phase of learning, Stacked AutoEncoders (SAE) prototype is exploited to acquire the composite descriptors. For the second stage, the appropriate statistics of image areas is combined to obtain perfect forgery recognition.

Yu et al. [60] presented another technique which utilizes convolutional neural networks that are based on countering anti-forensic approaches. This model encompasses many levels, e.g., convolution, pooling, and fully connected layers

Choi et al. [61] considered numerous post-processing tasks which are employed to hide the clues of the tampering's. The projected structure has the ability to perceive altogether mixtures of distorting, median filtering, and Gaussian noise.

Kuznetsov [62] in 2019 presented an approach for detection of image splicing which utilizes VCC-16 CNN architecture. It selects picture scraps as input and attains grouping outcomes for a patch: genuine or tampered.

Muzaffer et al. [63] in 2019 presented another approach which uses AlexNet model to obtain feature vectors from image segments. Then, resemblance between feature vectors is inspected.

3 Datasets Available for Image Forgery Detection

To evaluate the performance and validate the results of different forgery detection methods, benchmarked image forgery datasets are required. There are a few freely accessible datasets for copy-move forgery, image splicing forgery, and image retouching forgery available. A concise description of accessible datasets is given Table 5.

4 Conclusion

Various tampering detection methods have been projected and implemented in the recent times. Here, a brief review of image manipulation, classification, and its detection strategies has been discussed. Passive or blind techniques do not need any previous input regarding image which is under scanner, thereby possess significant advantage over its counterpart. Also, there is no need of special equipment to insert the cypher into the image during image acquisition. A number of techniques established till today are efficiently enough for perceiving the tampering but are insufficient to localize the forged area. We see a lot of problems with the currently offered techniques. Firstly, most of the schemes need human analysis and hence are not automatic. Second is the issue of localization of the forged region. Next is the issue of robustness to image processing tasks such as JPEG compression, blurring, and scaling. Use of an exact detection procedure is sensible, because an image tampering predictor might not be capable to distinguish which tampering method is used to alter the picture. Hence, there is the compulsion of a forgery detection method that would be capable enough to detect any kind of image tampering.

5 Compliance with Ethical Standards

Funding: There is funding for this study.

Data availability: No new data were created during the study.

Conflict of interest: Author Mohassin Ahmad declares that he has no conflict of interest. Author Farida khursheed declares that he has no conflict of interest.

Acknowledgments We would like to thank the anonymous reviewers for their insightful comments and suggestions.

Appendix

Tables 1, 2, 3, and 4 summarize copy-move, IS, retouching detection, and deep learning-based detection methods, respectively, along with their classification/matching methods, employed techniques. Table 5 gives the description of available datasets that are there for image forgery detection.

Table 1 Comparative study of copy-move forgery detection techniques

Paper serial	Feature extraction technique	Classifier/matching	Details(pros/cons)	Dataset
[9]	DCT of 16 × 16 overlapping blocks	Lexicographical sorting	Non-robust to rotation and scaling are; computational complexity is high	Real-life images
[10]	PCA	Lexicographical sorting	Computational complexity and cost are low; robust against jpeg compression quality level 50; detection accuracy is 50–100%	–
[11]	Standard Deviation (SD)	Adaptive thresholding	Higher performance in matching step and detection phase; less false matches	SBU-CM161
[12]	MIFT	Kd-tree, RANSAC	The accuracy rate is high; Does not work well for flat surface duplicated regions	CASIA v2.0
[13]	HOG	Euclidean distance	perceives multiple occurrences of copy-move: not invariant to rotation and scaling	CoMoFoD
[14]	SLIC segmentation, SIFT	K-nearest neighbor, EM	Results in good performance; precision is 86%; high computational complexity	CMFD, MICC -F600
[15]	Triangulation onto extracted key-points	MVD	Faster; higher the number of triangles the worse is performance	CMFDA
[16]	HOGM	Euclidean distance	Effective for high-resolution images; correct detection ratio = 98.8%	CoMoFoD, CMFDA
[17]	CPM, Hash values	GZCL	Precision is 93%	Literature databases
[18]	PCET	Radius ratio	Mismatch rate is higher for a blurred JPEG compressed image; Lower the computational time	Image manipulation dataset (IMD),Kodak true color suite
[19]	Color moments	CPNN	It is not robust to rotation	CoMoFoD

(continued)

Table 1 (continued)

Paper serial	Feature extraction technique	Classifier/matching	Details(pros/cons)	Dataset
[20]	RORHFM	Lexicographical sorting, Pearson Phase correlation	Robust against various operations like rotation, scaling.; High computational complexity, cost	100 images from Internet
[21]	DCT-phase term	Element-by-element equality	Accuracy is 96%; robust against various post-processing operations	CoMoFoD and Kodak databases
[22]	FFT, SVD, PCA	Euclidean distances	Accuracy is 98%; false detection is 8%	CASIA v1.0
[23]	Binary horizontal and vertical projection profiles	Region similarity	Accuracy is 88%; execution time is less	50 images of various sizes
[24]	Ring projection and haar wavelet transform	Lexicographically sorting	Accuracy is 90.75%; execution time is improved	Realistic database of 100 images
[25]	SI with segmentation and SIFT	SD thresholding	Reduces false positive rate; Effective to identify the small or smooth cloned regions	48 CMF images in the database
[26]	ANT colony optimization with SIFT and SURF	SVM	Better performance	CMFDA
[27]	CMFD-SIFT	AHC	True positive rate is 95.88%; Robust to scaling, rotation, flipping, blurring, etc	CoMoFoD
[28]	BWT-SVD	IRVM with GSO	Efficient performance; precision is 96.87%	CoMoFoD
[29]	CT	DBA, HELM-FSK	Precision is 97%	CoMoFoD

Table 2 Comparison of image splicing detection techniques

Paper serial	Feature extraction technique	Classifier/matching	Details(pros/cons)	Dataset
[30]	Bicoherence features	SVM	Detection accuracy is 70%	CISDE
[31]	HHT and moment feature	SVM	Detection accuracy is 85.87%	CISDE

(continued)

Table 2 (continued)

Paper serial	Feature extraction technique	Classifier/matching	Details(pros/cons)	Dataset
[32]	Inconsistencies in the color of the illumination of images	SVM meta-fusion	85–86% detection accuracy	Dataset of 200 images taken from Internet
[33]	Inconsistencies in motion blur	PSF	Space-variant blurring situations detection	Synthetic and Canon 60D real images
[34]	Markov-based features	SVM	Accuracy = 93.36%. High complexity	CISDE
[35]	Local blur kernels	Expectation-maximization	localization in blurred images only; accuracy = 92.8%; robust to image resizing, splicing boundary blurring	Dataset of 1200 tampered images
[36]	PCA, local minimum eigenvalues	Threshold clustering	High detection accuracy on pixel level	Columbia
[37]	DWT, LBP	SVM	Computational complexity is low; Invariant to monotonic illumination variations; Accuracy = 94.09%	CASIA V1, V2, Columbia compressed and uncompressed
[38]	Textural features-based GLCM	SVM	Reduced dimensionality and computational complexity; Accuracy is 97–98%	CASIA v1.0, CASIA v2.0
[39]	Moments for the inter-scale co-occurrence matrix	SVM with RBF kernel	Accuracy is 95.3–95.6%	CASIA1 and CASIA2
[40]	QDCT	Primal SVM	Accuracy is more than 92.38%	CASIA v1.0, CASIA v2.0, Columbia
[41]	Edge enhancement	multi-task FCN (MFCN)	Achieves finer localization; Robust to post-processing operations	CASIA v1.0, CASIA v2.0, Columbia
[42]	DCT, SVD, and statistical features	SVM	Accuracy is 96.97-99.30%; Low turn-around time	CASIA v1, DVMM v1, and DVMM v2.

(continued)

Table 2 (continued)

Paper serial	Feature extraction technique	Classifier/matching	Details(pros/cons)	Dataset
[43]	DCT coefficients and multi-scale LBP features	SVM	Accuracy is more than 97.3%; Area under ROC curve is 0.99	CASIA v.1.0 and v2.0.
[44]	Markov features with block DWT	SVM-RFE	Reduces computational complexity; Detection rate is 89.88%	DVMM, IFS-TC
[45]	AMFE with DWT	SVM	Detection accuracy is 99.5%	CASIA v2
[46]	CFMs, FMOs with RDWT	SVM	Accuracy is 97.40–98.10%	IFS-TC, CASIA

Table 3 Summarization of retouching detection approaches

Paper serial	Feature extraction technique	Classifier/matching	Details (pros/cons)	Dataset
[47]	Contrast enhancement and histogram equalization	Thresholding classifier	Accuracy is 99%	341 images captured using different digital cameras
[49]	Histogram estimation	Threshold classifier	Effective for both global and local gamma correction	UCID
[50]	Bi-Laplacian filtering	KD-tree	Works well for uncompressed and compressed high-resolution image	–
[51]	Histogram peak/gap artifacts	Thresholding classifier	Detection accuracy is 100%	BOSS UCID
[52]	PNRU	Markov random field	Better performance	Canon EOS 450D, Canon IXUS 95IS, Nikon D200

(continued)

Table 3 (continued)

[53]	Rotation invariant LBP	SVM	Detection accuracy is 90%	UCID
[54]	Multi-resolution Overshoot artifact	NSCT	Detection accuracy is 92%	UCID spite
[55]	Asymptotic eigenvalue Distribution RTM	–	Detection accuracy is 0.0066 s/image, and computational complexity is low	Dresden image database of a total of 1317 raw images

Table 4   Comparative study of deep learning-based detection techniques

Paper serial	Feature extraction technique	Classifier/matching	Detection accuracy	Type of forgery	Dataset
[56]	Median filter residual	CNN	85.14%	Median filtering, Cut and Paste	–
[57]	Prediction error filters	CNN	99.10%	Gaussian blurring, median filtering, AWGN, resampling	Collected from different cameras
[58]	DCT coefficients	CNN	–	Double JPEG compression	UCID, Dresden image database
[59]	Daubechies wavelet	Stacked AutoEncoders (SAE)	91.09%	Splicing, copy-move	Columbia, CASIA V1, CASIA V2
[60]	Anti-forensic features	CNN cross-entropy system	96.9%	JPEG antiforensic, median filtering,	BOSSbase, UCID
[61]	Deep learning based	CNN	87.31%,	Gaussian blurring, median filtering, gamma correction,	BOSS RAW database, Dresden
[62]	Border distortions	VGG-16 CNN	96.4% - 97.8%	Splicing	CASIA V2
[63]	256-dimensional feature vectors, PCA	AlexNet CNN	F-measure is 0.93	Copy-move detection and localization	GRIP

878   M. Ahmad and F. Khursheed

Table 5 Description of various available datasets related to forgery detection

S. No.	Dataset	Description
Dataset of copy-move forgery		
1	CMFDA	Contains a total of 48 images with the resolution of 420 × 300, 3888 ×2592 including forged images (obtained by JPEG compression, rotation, and scaling operation)
2	CoMoFoD [66]	Contains total of 260 images with the resolution of 512 × 512, 3000 × 2000 including forged images (obtained by translation, rotation, scale, distortion, or a combination of them)
3	MICC-F2000 [67]	Contains a total of 1300 images with the resolution of 2018 × 536 including forged images (obtained by translation, rotation, scale)
4	MICC-F220 [68]	Contains a total of 220 images with the resolution of 722 × 480, 800 × 600 including forged images
5	MICC-F600 [68]	Contains a total of 600 images with the resolution of 800 × 533, 3888 × 2592 that are randomly chosen from MICC-F2000 and SATS-130 datasets
6	MICC-F8multi [68]	Contains multiple images with the resolution of 800×532, 2048×1536 and the tampered images with realistic multiple cloning
7	SBU-CM161 [11]	Contains a total of 240 images with the resolution of 800 × 580 including forged images (obtained by rotation, scaling, and compression)
8	CPH [69]	Contains a total of 216 images with the resolution of 845 × 634, 296 × 972 including forged images (obtained by resizing, rotation, scaling, compression, illumination matching)
9	CMFDdb_grip	Contains a total of 160 images, 80 forged images, and 80 pristine images in PNG format with a resolution of 1024 x 768
Dataset of splicing forgery		
1	CISDE [70]	Contains a total of 1845 images with the resolution of 128 × 128 and contains gray images which are in PNG format. A total of 933 forged images and 912 original images.
2	CUISDE [71]	Contains a total of 361 images with the resolution of 757 × 568, 1152 × 768, contains colored images in TIFF format. A total of180 forged images and 181 authentic images.
3	CASIA v1.0 [72]	Contains a total of 1725 with the resolution of 324 × 256, contains colored images in JPEG format, A total of 925 forged images and 800 authentic images
4	CASIA v2.0 [73]	Contains a total of 12614 images with the resolution of 240 × 160, 900 × 600 A total of 5123 forged images and 7491 original images. This set also contains uncompressed images and JPEG images with different Q factors

(continued)

Table 5 (continued)

S. No.	Dataset	Description
5	DVMM	Contains a total of 1845 images with the resolution of 128 x 128, provided by Laboratory of Columbia University and consists of 933 authentic and 912 spliced gray images
6	IFS-TC	Contains a total of 2200 images with the resolution of 1024 × 575, 1024× 768. It consists of 1050 pristine and 1150 forged color images

Dataset of retouching forgery

1	SCUT-FBP [74]	Contains a total of 500 images with the resolution of 384 × 512, contains images of 500 female faces along with their corresponding score based on attractiveness given by 70 different observers.
2	BOSS public dataset [75]	Contains a total of 800 images with the resolution of 2000 × 3008, 5212 × 3468, contains images which are unaltered in raw format
3	UCID [76]	Contains a total of 1338 images with the resolution of 384 × 512, contains images in TIFF format which are uncompressed

References

1. Farid Hany (2009) Image forgery detection. IEEE Sign Process Mag 26(2):16–25
2. Reis G (2004) Digital image integrity. Adobe digital photography white papers
3. Conner K, Farid H (2015) Photo tampering throughout history. Four and Six. Accessed 9 Feb 2015
4. Zhang Z et al (2008) A survey on passive-blind image forgery by doctor method detection. In: 2008 international conference on machine learning and cybernetics 6, IEEE
5. Parashar Nishtha, Tiwari Nirupama, Dubey Deepika (2016) A survey of digital image tampering techniques. Int J Sign Process Image Process Patt Recogn 9(2):415–420
6. Doke KK, Patil SM (2016) Digital signature scheme for image. Int J Comput Appl 49(16)
7. Qazi Tet al (2013) Survey on blind image forgery detection. IET Image Process 7(7):660–670
8. Thajeel SA, Ghazali S (2014) A survey of copy-move forgery detection techniques. J Theor Appl Inf Tech 70(1)
9. Fridrich AJ, David Soukal B, Jan Lukáš A (2003) Detection of copy-move forgery in digital images. In: Proceedings of digital forensic research workshop
10. Popescu AC, Farid H (2004) Exposing digital forgeries by detecting duplicated image regions. Department of Computer Science, Dartmouth College, Tech. Rep. TR2004-515 (2004), pp 1–11
11. Zandi M, Mahmoudi-Aznaveh A, Mansouri A (2014) Adaptive matching for copy-move forgery detection. In: 2014 IEEE international workshop on information forensics and security (WIFS). IEEE
12. Jaberi M et al (2014) Accurate and robust localization of duplicated region in copy–move image forgery. Mach Vision Appl 25(2):451–475
13. Lee J-C, Chang C-P, Chen W-K (2015) Detection of copy–move image forgery using histogram of orientated gradients. Inf Sci 321:250–262
14. Li J et al (2015) Segmentation-based image copy-move forgery detection scheme. IEEE Trans Inf Forens Secur 3(10):507–518
15. Ardizzone Edoardo, Bruno Alessandro, Mazzola Giuseppe (2015) Copy–move forgery detection by matching triangles of keypoints. IEEE Trans Inf Forens Secur 10(10):2084–2094

16. Lee Jen-Chun (2015) Copy-move image forgery detection based on Gabor magnitude. J Vis Commun Image Represent 31:320–334
17. Liang Z et al (2015) An efficient forgery detection algorithm for object removal by exemplar-based image inpainting. J Visual Commun Image Represent 30:75–85
18. Wo Y et al (2016) Copy–move forgery detection based on multi-radius PCET. IET Image Process 11(2):99–108
19. Zhou, Haoyu, et al. "Digital image modification detection using color information and its histograms." *Forensic science international* 266 (2016): 379–388
20. Zhong Junliu, Gan Yanfen, Xie Shuai (2016) Radon odd radial harmonic Fourier moments in detecting cloned forgery image. Chaos, Solitons Fractals 89:115–129
21. Ustubioglu B et al (2016) A new copy move forgery detection technique with automatic threshold determination. AEU-Int J Electron Commun 70(8):1076–1087
22. Huang D-Y et al () Robustness of copy-move forgery detection under high JPEG compression artifacts. Multim Tools Appl 76(1):1509–1530
23. Ansari MD, Satya PG, Mohd W (2017) An approach for identification of copy-move image forgery based on projection profiling. Pertanika J Sci Tech 25(2):507–518
24. Ansari MD, Satya PG (2017) Copy-move image forgery detection using ring projection and Modi_ed fast discrete haar wavelet transform. Int J Electr Eng Inf 9(3):542–552
25. Zhao F et al (2017) Image forgery detection using segmentation and swarm intelligent algorithm. Wuhan Univ J Natur Sci 22(2):141–148
26. Jain N, Er Sushil B (2017) An optimization technique to detect the forgery in digital images by using ANT colony optimization
27. Yang B et al (2018) A copy-move forgery detection method based on CMFD-SIFT. Multimedia Tools and Applications77(1):837–855
28. Jain NK, Neeraj Kumar R, Amit M (2008) An efficient image forgery detection using biorthogonal wavelet transform and improved relevance vector machine. Wirel Person Commun Int J 101(4):1983–2008
29. Jain N, Rathore N, Mishra A (2018) A hybrid extreme learning machine with fuzzy sigmoid kernel based efficient image forgery detection for both copy move and spliced image. Cluster Comput J Springer Publication-US, ISSN: 1386-7857 (print version)
30. Ng T-T, Chang S-F (2004) A model for image splicing. In: 2004 International conference on image processing, 2004. ICIP'04. vol 2. IEEE
31. Li X, Tao J, Li X (2010) Image splicing detection based on moment features and Hilbert-Huang Transform. In: 2010 IEEE international conference on information theory and information security. IEEE
32. Carvalho De, José Tiago et al (2013) Exposing digital image forgeries by illumination color classification. IEEE Trans Inf Forens Secur 8(7):1182–1194
33. RaoMP, Rajagopalan AN, Seetharaman G (2014) Harnessing motion blurs to unveil splicing. IEEE Trans Inf Forens Secur 9(4):583–595
34. Zhao X et al (2015) Passive image-splicing detection by a 2-D noncausal markov model. IEEE Trans Circuits Syst Video Tech 25(2):185–199
35. Bahrami K et al (2015) Blurred image splicing localization by exposing blur type inconsistency IEEE Trans Inf Forens Secur 10(5):999–1009
36. Zhan L, Zhu Y, Mo Z (2016) An image splicing detection method based on PCA minimum eigenvalues. J Inf Hid Multim Sign Process 7(3)
37. Kaur Mandeep, Gupta Savita (2016) A passive blind approach for image splicing detection based on DWT and LBP histograms (International Symposium on Security in Computing and Communication). Springer, Singapore
38. Shen Xuanjing, Shi Zenan, Chen Haipeng (2016) Splicing image forgery detection using textural features based on the grey level co-occurrence matrices. IET Image Proc 11(1):44–53
39. Park TH et al (2016) Image splicing detection based on inter-scale 2D joint characteristic function moments in wavelet domain. EURASIP J Image Video Process 1:30
40. Li C et al (2017) Image splicing detection based on Markov features in QDCT domain. Neurocomputing 228:29–36

41. Salloum R, Yuzhuo R, Jay Kuo C-C (2018) Image splicing localization using a multi-task fully convolutional network (MFCN). J Visual Commun Image Repres 51:201–209
42. Moghaddasi Z, Jalab HA, Noor RM (2019) Image splicing forgery detection based on low-dimensional singular value decomposition of discrete cosine transform coefficients. Neural Comput Appl 31(11):7867–7877
43. Shah A, El-Alfy E (2018) Image splicing forgery detection using DCT coefficients with multi-scale lbp. In: 2018 international conference on computing sciences and engineering (ICCSE). IEEE
44. Zhang Q et al (2018) Digital image splicing detection based on Markov features in block DWT domain. Multim Tools Appl 77(23):31239–31260
45. Jalab HA et al (2019) New texture descriptor based on modified fractional entropy for digital image splicing forgery detection. Entropy 21(4):371
46. Subramaniam T et al (2019) Improved image splicing forgery detection by combination of conformable focus measures and focus measure operators applied on obtained redundant discrete wavelet transform coefficients. Symmetry 11(11):1392
47. Stamm M, Ray Liu KJ (2008) Blind forensics of contrast enhancement in digital images. In: 2008 15th IEEE international conference on image processing. IEEE
48. Stamm MC, Ray Liu KJ (2010) Forensic estimation and reconstruction of a contrast enhancement mapping. In: 2010 IEEE international conference on acoustics, speech and signal processing. IEEE
49. Gang C, Zhao Y, Ni R (2010) Forensic estimation of gamma correction in digital images. In: 2010 IEEE international conference on image processing, IEEE
50. Li X-F, Shen X-J, Chen H-P (2011) Blind identification algorithm for retouched images based on Bi-Laplacian. Jisuanji Yingyong/J Comput Appl 31(1):239–242
51. Cao, Gang, et al. "Contrast enhancement-based forensics in digital images." *IEEE* Trans *information forensics and security* 9.3 (2014): 515–525
52. Chierchia G et al (2014) A Bayesian-MRF approach for PRNU-based image forgery detection. IEEE Trans Inf Forens Secur 4:554–567
53. Ding F et al (2015) Edge perpendicular binary coding for USM sharpening detection. IEEE Sign Process Lett 22:327–331
54. Zhu Nan, Deng Cheng, Gao Xinbo (2016) Image sharpening detection based on multiresolution overshoot artifact analysis. Multimedia Tools Appl 76(15):16563–16580
55. Vázquez-Padín David, Pérez-González Fernando, Comesana-Alfaro Pedro (2017) A random matrix approach to the forensic analysis of upscaled images. IEEE Trans Inf Forensics Secur 12(9):2115–2130
56. Chen J et al (2015) Median filtering forensics based on convolutional neural networks. IEEE Sign Process Lett 22(11):1849–1853
57. Belhassen B, Stamm MC (2016) A deep learning approach to universal image manipulation detection using a new convolutional layer. In: Proceedings of the 4th ACM workshop on information hiding and multimedia security
58. Wang Qing, Zhang Rong (2016) Double JPEG compression forensics based on a convolutional neural network. EURASIP J Inf Secur 2016(1):23
59. Zhang Y et al (2016) Image region forgery detection: a deep learning approach. SG-CRC 1–11
60. Yu J et al (2016) A multi-purpose image counter-anti-forensic method using convolutional neural networks. In: International workshop on digital watermarking. Springer, Cham
61. Choi H-Y et al (2017) Detecting composite image manipulation based on deep neural networks. In: 2017 international conference on systems, signals and image processing (IWSSIP). IEEE
62. Kuznetsov A (2019) Digital image forgery detection using deep learning approach. J Phys Conf Ser 1368(3) IOP Publishing
63. Muzaffer G, Ulutas G (2019) A new deep learning-based method to detection of copy-move forgery in digital images. In: 2019 scientific meeting on electrical-electronics & biomedical engineering and computer science (EBBT). IEEE
64. The dataset mentioned in the paper is freely available to reader at the address [online]. Available: http://forensics.idealtest.org/casiav1/

65. Bharti CN, Purvi T (2016) A survey of image forgery detection techniques. In: 2016 international conference on wireless communications, signal processing and networking (WiSPNET). IEEE
66. Tralic D et al (2013) CoMoFoD—new database for copy-move forgery detection. In: Proceedings ELMAR-2013. IEEE
67. Amerini I et al (2013) Copy-move forgery detection and localization by means of robust clustering with J-Linkage. Sign Process Image Commun 28(6):659–669
68. Amerini I et al (2011) A sift-based forensic method for copy–move attack detection and transformation recovery. IEEE Trans Inf Forens Secur 6(3):1099–1110
69. Silva E et al (2015) Going deeper into copy-move forgery detection: exploring image telltales via multi-scale analysis and voting processes. J Visu Commun Image Repres 29:16–32
70. Columbia DVMM Research Lab (2004) Image splicing detection evaluation dataset. www.ee.columbia.edu/dvmm/researchProjects/AuthenticationWatermarking/Blind
71. Hsu Y-F, Chang S-F (2006) Detecting image splicing using geometry invariants and camera characteristics consistency. In: 2006 IEEE international conference on multimedia and expo. IEEE
72. Dong J, Wei W (2011) CASIA tampered image detection evaluation database
73. Dong J, Wang W CASIA2 tampered image detection evaluation (TIDE) database
74. Xie D et al (2015) Scut-fbp: A benchmark dataset for facial beauty perception. In: 2015 IEEE international conference on systems, man, and cybernetics. IEEE
75. Bas Patrick, Filler Tomáš, Pevný Tomáš (2011) Break our steganographic system": the ins and outs of organizing BOSS. International workshop on information hiding. Springer, Berlin, Heidelberg
76. Schaefer G, Michal S (2003) UCID: an uncompressed color image database. Storage and retrieval methods and applications for multimedia 2004, vol 5307. International Society for Optics and Photonics

Chapter 71
A Framework Based on Latent Neighbourhood and Tensor Based Method for Recommender System

Shital Gondaliya and Kiran Amin

1 Introduction

Recommendation systems are intended to help people make decisions. These system are commonly used on online platforms for video, music and product purchases through service providers such as Netflix, Pandora and Amazon [1]. Massive amount of data are available on social websites, therefore finding the suitable item is a challenging issue. But most users of different domains might have different view and understating for the same item or tag [2]. Collaborating filtering (CF) is the main technique broadly used in making recommendations. A large number of CF based recommendation methods have been developed in recent years [3]. Though, a major task in CF in selecting an appropriate set of users and using them in the rating prediction process. In other words, an important step in CF is to identify relevant users most similar to the target user [4].

Similarity, the size and complexity of these systems can unfortunately lead to information overload and reduce utility [5]. Therefore, in order to efficiently process data whose volume can be very large, many models presented for RSs, clustering the existing data [6]. Consequently, how to utilize rich information in social media to enhance recommendation models has become a hot issue of great interest to both academia and industries [7]. The item recommendation technique for the large number of data increase the computation time of the system [8]. In the real world, items are usually with multiple features and semantic information is latent, hence, a more capable method to extract semantic information is needed [9]. Many research proposed the different method for the efficient item recommendation process and

S. Gondaliya (✉)
Madhuben and Bhanubhai Patel Institute of Technology, Gujarat, India
e-mail: sngondaliya@mbict.ac.in

K. Amin
UV Patel College of Engineering & Technology, Gujarat, India
e-mail: kiran.amin@ganpatuniversity.ac.in

these are majorly affected by the cold start problem [10]. In this research, the Latent Neigbhborhood-Higher Order Singular Value Decomposition (LN-HOSVD) framework is developed to solve the cold start and data sparsity problem which provide the efficient recommendation to the user. To solve the cold start and data sparsity problem, the proposed framework combines the neighbour similarity model for user and item. The experiment verifies that the proposed framework has the higher performance in the recommendation system.

The organization of the paper is as follows. The related works are reviewed in Sect. 2 with their limitations, the proposed model is discussed in Sect. 3, experimental design and results are shown in Sect. 4 and conclusions are drawn in Sect. 5.

2   Literature Review

The user profiling technique helps to provide the recommendation item for the user based on their interest. The recent research on item recommendation methods are surveyed in this section to understand the issue in current recommendation technique.

Symeonidis et al. [5] used the two similarity method for the tags i.e. ontology of Wordnet and Term Frequency—Inverse Document Frequency (TF-IDF) methods. The clustering methods are used to minimize the size of the tensor dimension and its data sparsity. The tag cluster dimension is used instead of the tag dimension in the tensor due to its less size and low noise. The method has been tested on the two real time dataset and this shows that the spectral clustering method has the higher performance than other clustering method. The cold start problem is need to be solved for the better performance of the recommendation.

Taneja and Aroa [6] proposed a method such as Cross Domain Multi-Dimensional Tensor Factorization (CD-MDTF) for the item recommendation. The knowledge is extracted from the source domain and mix with the target domain and based on the expert system to provide the recommendation. Two real-world datasets are used to evaluate the performance of the system namely MovieLens and Book-Crossing. This method deals with the problem of the data sparsity and cold start. The sparsity has been reduced up to 16% and the cold start has been reduced up to 25% respectively. The temporal data can be used in the method to increase the performance of the system.

Barjasteh et al. [7] proposed a method named as DecRec to solve the problem of the cold start by exploiting the side information in the system. The complete rating sub-matrix is generated by the negating the cold start user and item. The knowledge is extracted from the existing rating for the cold start user/information based on side information. The cold start problem is considerably reduced and also minimized the problem of the data sparsity. The error value of the method in the large dataset is need to be reduced.

Zhou et al. [8] developed a method based on the Singular Value Decomposition (SVD) and increment with the approximate value of SVD, named as Incremental ApproSVD. The two datasets such as MoiveLens and Flixster are used to measure the performance of the incremental ApproSVD method. The error value of the ApproSVD is considerably reduced and computation time is high. The cold start problem is required to increase the efficiency of the method.

Park et al. [9] proposed the Reserve Collaborative Filer (RCF) method based on the k-Nearest Neighbor (kNN) graph. The process of the k finding method is reversed and it finds the similarity between rated items. The algorithm is fast due to the less prediction is made by the method is reduced without affecting the accuracy of the recommendation. The cold start and data sparsity problem are required to increase the efficiency of the method.

2.1 Problem Statement

From the survey of the recent researches in the item recommendation, it is clear that the cold start problem need to be solved for the effective performance. The existing methods are inefficient in the process of the recommendation to the new users and user with low data. The efficiency of the item recommendation system is required to improve.

By combing the neighbourhood value of the user and items in the latent factor, the recommendation can be provided for the new users. The latent factor is used to combine the neighbourhood similarity for the user and item. The HOSVD is applied for the system to provide the recommendation to the user.

3 Framework of Recommendation Model

The recommendation model based on the user profile and the social tag is used to give recommendation item for the user. The different techniques were applied for the recommendation model for efficient system. Still, these methods are suffering from the cold start and data sparsity problem. This paper aims to provide the framework for the recommendation system to solve the cold start and sparsity problem and increases the performance of the system. In the proposed Latent Neighborhood-Higher Order Singular Value Decomposition method, the user neighbourhood model is measured. Based on the neighbourhood value, Latent Factor Model is developed that has the combined and provide the recommendation for the user. Then, the latent factor model is applied in the HOSVD to provide the effective recommendation for the user. The proposed framework is depicted in Fig. 1.

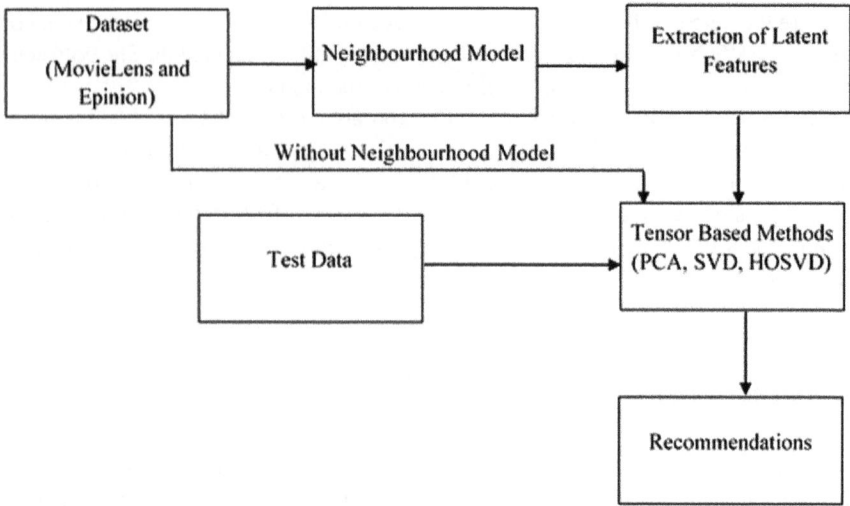

Fig. 1 The proposed framework

3.1 Neighbourhood Model

Normally memory based collaborative filtering models exploit the k-nearest neighbor approach to identify candidate items [10]. This approach uses a neighbourhood of similar users to identify and recommend items to users. Neighborhoods are created by for each user finding users within a certain similarity. Each user's neighbourhood contains the k users' whose similarity is highest, an abstract outline of the algorithm is shown in Algorithm 1. Recommendations are then produced by iterating over all items rated by the user's neighbors which have not been seen by the user, and approximating the preference of each item by averaging all neighbor's rating values for each item multiplied with each neighbor's similarity value. Ultimately, the highly rated items of the most similar neighbors are recommended. The similarity function for neighbor identification can be freely selected.

Common similarity measures used for neighbourhood creation are cosine similarity and pearson product-moment correlation coefficient, when working with data containing ratings. The formula for pearson correlation is shown as follows:

$$p(u, v) = \frac{\sum i \in I_{uv}(r_{ui} - \ddot{r}_u)(r_{vi} - \ddot{r}_v)}{\sqrt{\sum i \in I_{uv}(r_{ui} - \ddot{r}_u)^2}\sqrt{\sum i \in I_{uv}(r_{vi} - \ddot{r}_v)^2}} \tag{1}$$

where u and v are the two users, Iuv the intersection of user u's and v's rated items, and r the rating score. The algorithm to find the k-nearest neighbour is shown as follows.

Algorithm: The k-nearest neighbour algorithm
Input: set U of users
Input: number k of neighbors
Output: k neighbors for each user u $\in$ U
foreach u $\in$ U
foreach u' $\in$ U \ {u} do
$s_{u,u'} = similarity(u, u')$;
Select k neighbors u' $\neq$ u for u with largest $s_{u,u'}$

3.2 Latent Factor Model

The proposed method is based on latent factor model that generates a latent factor vector for each item and user. The basic model based on collaborative filtering algorithm supposes that each rating with which a user evaluates an item mainly consists of the user latent factor and the item latent factor [11]. The model can be described as $\hat{r}_{ij} = \mu + \beta_i + \beta_j + u_i + v_j^T$, where μ is global offset parameter, presenting the average rating of all items from all users, β_i and β_j are user and item bias respectively, μ_i and v_j are K-dimensional user and item factor vector respectively. The basic latent factor recommendation model approximates the parameters by solving the following optimization problem given in the following equation.

3.3 Higher Order Singular Value Decomposition (HOSVD)

When using a social tagging system, a user u tags an item i with a tag t, in order to be able to retrieve information items easily [3]. Thus, the tagging system gathers a collection of usage data, which can be denoted by a set of triplets $\{u, t, i\}$. As users used 5 tags to annotate 3 different items. The series of arrows, which have the same label (explanation) between a user and an item represents that a user used a tag for annotating the specific item. Thus, the annotated numbers on the arrow lines provide the correspondence between three types of the objects. For example, user U1 tagged item I1 with tag "FIAT" denoted as T1. The remaining tags are "BMW" denoted as T2, "JAGUAR" denoted as T3, "CAT" denoted as T4 and "DOG" denoted as T5. That can infer that users U1 and U2 have common interests on cars, while user U3 is interested in animals. A 3-order tensor $A \in R\ 3 \times 3 \times 3$ can be constructed from the usage data. We use the occurrence frequency (denoted as weight) of each triplet user, item and tag as the elements of tensor A and 3-D visual representation of the tensor. HOSVD is applied on the 3-order tensor constructed from these usage data. It uses as input the usage data of a tensor A and outputs the reconstructed tensor $\hat{A}$. $\hat{A}$ measures the relations among the users, items and tags. The elements of $\hat{A}$ can be denoted by quadruplet $\{u, i, t, w\}$, where w measures the likeliness that user u will tag item i with tag t. Hence, items can be recommended to u as per their weights related with $\{u, t\}$ pair.

In this paper, the high order singular value decomposition of three dimensional tensor A is expressed as in Eq. (2).

$$A = S \times U(1) \times U(2) \times U(3) \tag{2}$$

In the formula, the core tensor S R is an orthogonal tensor with the same dimension as tensor A, which determines the interaction between the entity user, item and tag. The mathematical expression of S is given in Eq. (3)

$$\hat{S} = A \times U_{c1}^{(1)^T} \times U_{c2}^{(2)^T} \times U_{c3}^{(3)^T} \tag{3}$$

Finally, due to the large amount of noise in the tensor data, the tensor A does not have low rank and it is necessary to construct the approximate tensor A of the tensor A through the HOSVD. The mathematical expression for reconstructing the tensor A is given in Eq. (4).

$$\hat{A} \approx \hat{S} \times U_{c1}^{1} \times U_{c2}^{2} \times U_{c3}^{3} \tag{4}$$

The recommended items are used to measure the error value of the proposed framework system. In the next section, the various error values are analysed to evaluate the proposed system in the recommendation.

4   Experimental Results

This section gives the validation of the proposed LN-HOSVD framework for the recommendation system. The error values are calculated from the proposed LN-HOSVD framework for the different neighbour value and cluster value.

4.1   Benchmark Datasets

MovieLens: MovileLens is the web based recommendation system and this recommend the user for the movies based on their preference [12]. MovieLens (1 M dataset) contains more than 11 millions ratings for about 8500 movies. This dataset is common preference for the researchers for evaluating their proposed recommendation system. The dataset is in the form of matrix consists of the 943 rows or users and 1682 columns or movies. The rating scale is between 1 and 5 i.e. from dislike to the strong suggestions.

Epinions: This dataset contains the review of the users for the items categories [13]. The dataset contains the categories, trust values between users, categories hierarchy and user expertise on the categories. The dataset contains 49,290 user rated for item 664, 824 reviews and 487,181 issues trusted statements.

4.2   Experimental Platform

The experiment were conducted in the system consist of the i5 processor, 8 GB of RAM and 500 GB HDD. The proposed LN-HOSVD framework is evaluated in the language of python 3.7 with the compiler of Anaconda and in the tool of Jupyter Lab.

4.3   Evaluation Metrics

Error metrics such as Root Mean Square Error (RMSE), Mean Absolute Error (MAE) and Mean Square Error (MSE) are used to evaluate the performance of the proposed system. The metrics such as precision and recall are also used to analyse the efficiency of the proposed LN-HOSVD framework system. The equation for the MSE, RMSE and MAE are given in Eqs. 5, 6 and 7 respectively and the equation for the precision and recall measures are given in Eqs. 8 and 9.

$$\text{MSE} = \sum_{i=1}^{N} \left(r_{ui} - \widehat{r_{ui}}\right)^2 \tag{5}$$

$$\text{RMSE} = \sqrt{\text{MSE}} \tag{6}$$

$$\text{MAE} = \frac{\sum_{i=1}^{N} \left|r_{ui} - \widehat{r_{ui}}\right|}{N} \tag{7}$$

$$\text{Precision} = \frac{TP}{TP + FP} \tag{8}$$

$$\text{Recall} = \frac{TP}{TP + FN} \tag{9}$$

Table 1 Performance of the methods without neighborhood values

Method	RMSE	MAE	MSE	Precision	Recall	F-score
SVD	0.6332	0.2497	0.401	94	85	89
PCA	0.702	0.2478	0.4928	90	42	58
IPCA	0.7024	0.2479	0.4934	89	42	58
HOSVD	0.6129	0.2316	0.3757	95	90	93

4.4 Performance Evaluation

The proposed framework is evaluated using datasets like MovieLens and Epinion and also compared with the existing similar methods such as Principal Component Analysis (PCA), Singular Value Decomposition (SVD).

MovieLens Dataset. The different methods are processed on this dataset to analyse the efficiency of the recommender system without neighbourhood model. The metrics such as RMSE, MSE, SMAE, precision and recall are measured, shown in Table 1. All these methods are processed in the same environment and in the same comparative analysis. PCA and SVD techniques are used in the recommender system and they have low efficiency due to cold start and data sparsity problem.

Table 2 provides the effectiveness of the various methods with neighbourhood model using various metrics. As these combined approach provides better recommendation to the new user as well existing user. The higher performance is achieved through HOSVD because due to its factor decomposition. Hence, neighbourhood model solves the problem of cold start and data sparsity. The HOSVD method has the advantage of the shrinking the tensor at each step and minimize the computation. The RMSE value of the LN-HOSVD is achieved as 0.6122 and the RMSE value of conventional HOSVD is achieved as 0.6129.

The various methods such as PCA, IPCA, SVD and HOSVD are processed in the same scenario and measured their performance. The metric such as precision and recall are measured from the various methods. This shows that the HOSVD method has the higher performance compared to the other methods. The cold start problem is not solved by these methods for new users.

Epinion Datasets. The Epinion dataset are used as secondary dataset to measure the performance of the developed LN-HOSVD framework. The different methods

Table 2 Performance of the methods with neighborhood values

Method	RMSE	MAE	MSE	Precision	Recall	F-score
LN-SVD	0.6503	0.2212	0.423	95	87	90
LN-PCA	0.7247	0.3925	0.5252	88	31	46
LN-IPCA	0.7246	0.3919	0.5251	87	31	46
LN-HOSVD	0.6122	0.2312	0.3747	98.9	93.15	95.94

Table 3 The error values of the various methods without neighborhood values

Method	RMSE	MAE	MSE
SVD	0.147	0.0097	0.0216
PCA	0.1471	0.0109	0.0216
IPCA	0.1471	0.0103	0.0216
HOSVD	0.147	0.0061	0.0216

Table 4 The error values of the various methods with neighborhood values

Method	RMSE	MAE	MSE
LN-SVD	0.147	0.011	0.021
LN-PCA	0.651	0.642	0.424
LN-IPCA	0.651	0.643	0.424
LN-HOSVD	0.147	0.006	0.021

such as SVD, PCA, IPCA and HOSVD methods are used measure the performance. The different error values are measured for the performance analysis in the system as shown in Table 3. The RMSE value of HOSVD is nearly equal to other methods such as SVD, PCA and IPCA. The HOSVD has the low error value compared to the other technique in Epinions dataset. The methods are not capable for handling the cold start and data sparsity problem.

The neighbourhood values are used to solve the problem of the cold start and data sparsity. The neighbourhood values are measured and provided as input to the various methods for analysis the performance as shown in Table 4. This shows that the cold start problem has been solved and the error values are reduced. The methods with Latent Neighborhood value has the higher performance than the traditional methods. The MSE value of LN-HOSVD method has 0.147, while MSE value of HOSVD is 0.0216.

This shows that the method with the neighbourhood values are having less error value for most of the methods. The neighbour values combined with PCA and IPCA methods have high error values. The proposed LN-HOSVD has the lower error value compared to all other techniques due to measure of neighbourhood values and the factorization method.

5 Conclusions

Recommendation system is common in the online shopping and movie websites to provide recommendation to the user. Existing methods of recommendation systems suffer from the cold start and data sparsity problem. These problems need to be solved for the effective recommendation systems. In this research, new framework is developed such as LN-HOSVD system for the effective recommendation and to solve the cold start problem. The neighbour similarity for the user and item is measured and these data are combined to solve the cold start problem. The experiment verifies that the proposed LN-HOSVD framework has higher the performance in the recommendation system. The proposed LN-HOSVD has the precision value of 98.9% while the existing method has the precision value of 95%. In the future work, the various features are selected from the user profile to provide the accurate item recommendation for the user.

References

1. Chaney AJB, Stewart BM, Engelhardt BE (2018) How algorithmic confounding in recommendation systems increases homogeneity and decreases utility. In: Proceedings of the 12th ACM conference on recommender systems, RecSys'18, New York, NY, USA, pp 224–232
2. Rawashdeh M, Shorfuzzaman M, Artoli AM, Hossain MS, Ghoneim A (2017) Mining tag-clouds to improve social media recommendation. Multim Tools Appl 76(20):21157–21170
3. Tang X, Xu Y, Geva S (2015) Integrating tensor factorization with neighborhood for item recommendation in multidimenstional context. In: IEEE/WIC/ACM international conference on web intelligence and intelligent agent technology, pp 377–384
4. Parvin H, Moradi P, Esmaeili S (2019) TCFACO: trust-aware collaborative filtering method based on ant colony optimization. Expert Syst Appl pp 152–168
5. Symeonidis P (2015) ClustHOSVD: item recommendation by combining semantically enhanced tag clustering with tensor HOSVD. IEEE Trans Syst Man, Cybern Syst, pp 1–12
6. Taneja A, Arora A (2018) Cross domain recommendation using multidimensional tensor factorization. Expert Syst Appl 92:304–316
7. Barjasteh I, Forsati R, Ross D, Esfahanian A-H, Radha H (2016) Cold-start recommendation with provable guarantees: a decoupled approach. IEEE Trans Knowl Data Eng, pp 1462–1474
8. Zhou Xun, He Jing, Huang Guangyan, Zhang Yanchun (2015) SVD-based incremental approaches for recommender systems. J Comput Syst Sci 81(4):717–733
9. Park Y, Park S, Jung W, Goo Lee S (2015) Reversed cf: a fast collaborative filtering algorithm using a k-nearest neighbor graph. Expert Syst Appl 42(8):4022–4028
10. Said A, Fields B, Jain BJ, Albayrak S (2013) User-centric evaluation of a k-furthest neighbor collaborative filtering recommender algorithm. In: Proceedings of the 2013 conference on computer supported cooperative work, ACM, New York, NY, USA, pp 1399–1408
11. Qiu L, Gao S, Cheng W, Guo J (2016) Aspect-based latent factor model by integrating ratings and reviews for recommender system. Knowl Syst 110:233–243
12. Harper FM, Konstan JA (2015) The movielens datasets: history and context. ACM Trans Interact Intell Syst
13. Massa P, Avesani P (2007) Trust-aware recommender systems. In: ACM recommender systems conference (RecSys), USA

Chapter 72
An Optimal Demand Response Strategy Using Gray Wolf Optimization

Ankit Kumar Sharma, Akash Saxena, Dheeraj Kumar Palwalia, and Bhanu Pratap Soni

1 Introduction

To encounter the total load demand in the power system, there are two heterogeneous strategies: either expanding the electricity generation which fulfills the newly generated demand side needs or applying managerial policies that provide virtual resources and its management. The first strategy is employed to fulfill the added energy requirement which emerged due to economic modernization. To avoid such financial investments, the second strategy used broadly with the aim of penny-pinching using necessary managerial policies [1]. This implementation of managerial measures made a role in emerging the concept of demand side management (DSM). The term demand side management (DSM) was first given by Clark W. Gellings in the 1980s and concept of DSM has been found in [2–4].

For the future smart grid, DSM is an important function to manage the energy [5, 6] and also support the various functions in smart grid such as controlling the electricity market, decentralize energy management, and electric vehicle. By controlling the energy demand, also can control the peak, can alter the demand curve and can increase the sustainability of the grid by suppressing the energy cost and carbon emission level.

A. K. Sharma (✉) · A. Saxena
Department of EE, Swami Keshvanand Institute of Technology,
Management & Gramothan, Jaipur, India
e-mail: ankit.krishnaa@gmail.com

A. Saxena
e-mail: aakash.saxena@hotmail.com

D. K. Palwalia
Department of EE, UCE, Rajasthan Technical University, Kota, India
e-mail: dheerajpalwalia@gmail.com

A. K. Sharma · B. P. Soni
Department of EE, University of Engineering & Management, Jaipur, India
e-mail: bpsoni.ee@gmail.com; er.bpsoni2011@gmail.com

An efficient DSM can strongly avoid the new construction of generation, transmission and distribution infrastructure or capacity. DR programs can be divided into two main fractions: incentive-based programs (IBP) and price-based programs (PBP) [7, 8]. IBP is moreover divided as reliability DR and economic DR programs. Reliability DR programs include capacity market, ancillary services, and emergency DR. When electricity rates not follow the flat rate and fluctuate hourly, then called price-based programs. DR programs have certain benefits in terms of market side, consumer side, and also in terms of authenticity and merchandise requirements. The profits of DR programs are related to consumers and also centric for merchandise side [9].

In the context of demand response, to flatten the energy utilization curve, utility provides the financial rebate to consumers and load curve during peak and off-peak time can be altered by using six DR methods [1, 10]: peak clipping, valley stuffing technique, strategic conservation, load shifting and flexibility in load shape, load building technique. This is helpful to reduce the cost of network expansion and power generation. To reduce electricity cost and peak demand effectively, there is a need of special DSM program. But such programs have a limitation that these have the ability to lowering the peak demand by 20–42%.

The non-smooth and nonlinear equality and inequality constraints make the DR problem highly complex task for the traditional approaches to find a solution. Since the algorithms based on stochastic processes are non-deterministic and approximate in their nature, the resultant solution may fall under local or global minimum. The gray wolf optimizer (GWO) [11] is a metaheuristic algorithm that is inspired by the hunting behavior and social dominance in the gray wolf pack. The application of GWO for different types of problems across different domains has been fairly successful in the past few years. The application of GWO and its variants in solving different complex problems has been promising in comparison with the existing another approaches.

Due to the uncertainty in energy cost and unrecognized consumption pattern, optimal scheduling of home appliances with user comfort really becomes a difficult task. To solve such problem, Waseem et al. [12] proposed a home appliance scheduling model based on combination of GWO and CSO algorithm. Tikhamarine et al. [13] proposed an efficient hybrid system for forecasting of monthly stream flow by combination of GWO and AI. For prediction of carbon trading in china, Lu et al. [14] used six machine learning models. The prediction models include Arps decline model, which is optimized by GWO. Bouzary et al. [15] used GWO for artificial and optimal cloud manufacturing process. Sankaranarayanan et al. [16] proposed a hybridized gray wolf optimization using static Kalman Bucy process to suppress the effect of noise in parameter estimation process for water distribution system. For solving the problem of energy storage system placement and for micro-grids, Di Miao and Sarmistha Hossain [17] proposed improved gray wolf optimization algorithm. Amir Seifi et al. [18] presented a combination of the GWO and SSA to get optimality in residential applications with renewable resources and storage system. Arther Jain et al. [19] proposed hybrid model of GWO for MPPT and management of power for solar PV generation system. Ahmed Fathy and Hegazy Rezk [20] used GWO for electrical parameter extraction for supercapacitor. Mohseni et al. [21] used

GWO in a comparison of metaheuristics for the optimal capacity planning in micro-grid. Lu et al. [22] predicted the power of offshore wind farm using multiobjectives GWO. Mahian et al. [23] optimized the size and performance of combine heat and power system with storage of energy for residential premises. Fu et al. [24] used GWO with other swarm optimizers for short-term wind speed forecasting. Akash Saxena et al. proposed new variants of GWO as β-chaotic GWO [25] and IGWO [26]. In addition An opposition-based variant of GWO is proposed for energy bidding problem in [28]. To solve multi-objective optimization problems, Liu [27] applied gray wolf optimizer. Applications of WOA and BBO optimization algorithms are in solving DR problem in [29]. Idea about the concept of optimization and evolutionary mechanism used in optimization is given into [30, 31].

In this paper, the DSM program implemented with a large number of shift able and controllable devices with novel load patterns. In the same regard, this problem considered as an optimization problem and to handle such type of complex problem, metaheuristics algorithms have been used. Load minimization and reducing the operational cost problem have been solved using optimization algorithms, in this paper. This paper has certain benefactions as follows.

  (i) In this paper, strategic conservation and load shifting methods have been used. To test these DSM techniques, residential and commercial sectors have been used as test cases.
 (ii) Controlling of switching of various devices of different classes of given test smart grid has been yield by using the metaheuristic, optimization technique as named gray wolf optimization (GWO).
(iii) Calculated the reduction in peak demand and operational cost and compared the results yield from gray wolf optimization (GWO) with whale optimization algorithm (WOA) and moth–flame optimization (MFO).
(iv) Proved the efficacy of GWO over WOA and MFO.

The remaining portions of the paper have been arranged in the given way: Sect. 2 presents the DR problem formulation, Sect. 3 outlines the GWO algorithm, detail about the test system and obtained simulations results have been presented in Sects. 4 and 5 respectively. At last, the important findings of the research work given in form of conclusion.

2 Problem Formulation

For controlling future smart grid, an advanced DSM technique named as strategic conservation and load shifting methods has been used in this paper. The goal of used DSM techniques should be minimized the peak demand and energy cost at any instant. Future smart grid engineer continuously trying to design the optimize load curve with potency to fulfill the main aim of the DSM program.

There are so many optimization techniques in the past literatures, which may be used to solve complex DR problem [32]. DSM techniques always tried to bring the load curve near to the objective curve as possible, and optimization algorithms are try to overlap of both to get the better results [33]. Mathematical formulation for strategic conservation and load shifting technique is given below:

$$\text{Objective function:} \quad \min \sum_{i=1}^{n} (C\,\text{Load}(t) - \text{objective}\,(t))^2 \tag{1}$$

where t is time period, objective(t) stands for objectives value at time t, and term $C\,\text{Load}(t)$ represents the exact utilization at given time t.

At any instance for strategic conservation, exact utilization, $C\,\text{Load}(t)$ can be drawn by equation (2). Range of on–off status of any appliance is ranging from i to N, specific class of any device is given by j to D. Z stands for summation of devices in a specific class.

$$C\,\text{Load}(t) = \sum_{i=1}^{N}\sum_{j=1}^{D} Z_i Z_j + \sum_{i=1}^{N}\sum_{j=1}^{D} (1 - Z_i)\,(Z - Z_j) \tag{2}$$

At any instance for load shifting, exact utilization, $C\,\text{Load}(t)$ is given by equation (3). Forecasted load(t) is forecast load at time t. Connected load (t) and disconnected load (t) are represent amount of load connected and disconnected, respectively.

$$C\,\text{Load}\,(t) = \text{Forecasted load}\,(t) + \text{Connected load}\,(t) - \text{Disconnected load}\,(t) \tag{3}$$

3   Grey Wolf Optimizer

To address the complicated optimization problems, gray wolf optimizer (GWO) proposed by author in [11]. Gray wolves live in a pack of 5 to 12 members and pack is divided into four types: first is Alpha, which is also called decision-maker or dominant wolves; second is beta which is subsidiary wolves; omega the space goat or child minders and last one is delta. Footsteps of the algorithm are given below.

3.1   Prey Encircling

Grey wolves encircle the prey during the hunt and mathematical modeling for encircling given as:

$$\vec{D} = \left| \vec{C} * \vec{X}_p(t) - \vec{X}(t) \right| \tag{4}$$

$$\vec{X}(t+1) = \vec{X}_p(t) - \vec{A} * \vec{D} \tag{5}$$

where t is current iteration; $\vec{X}_p(t)$ is position vector for prey; $\vec{X}(t+1)$ is position update of gray wolf; $\vec{A}$ and $\vec{C}$ are the coefficient. These coefficients are defined by: $\vec{A} = 2\vec{a}\,\vec{r}_1 - \vec{a}$ and $\vec{C} = 2\vec{r}_2$.

3.2   Prey Hunting

Precise identification of prey is not possible in given search space, so three best solutions are given to justify the hierarchy (α, β & δ). The visualization of this fact is represented by three mathematical equations (6), (7) and (8).

$$\vec{D}_\alpha = \left| \vec{C}_1 * \vec{X}_\alpha - \vec{X} \right|; \;\; \vec{D}_\beta = \left| \vec{C}_2 * \vec{X}_\beta - \vec{X} \right|; \;\; \vec{D}_\delta = \left| \vec{C}_3 * \vec{X}_\delta - \vec{X} \right| \tag{6}$$

$$\vec{X}_1 = \vec{X}_\alpha - \vec{A}_1 * (\vec{D}_\alpha); \;\; \vec{X}_2 = \vec{X}_\beta - \vec{A}_2 * (\vec{D}_\beta); \;\; \vec{X}_3 = \vec{X}_\delta - \vec{A}_3 * (\vec{D}_\delta) \tag{7}$$

$$\vec{X}(t+1) = \frac{(\vec{X}_1 + \vec{X}_2 + \vec{X}_3)}{3} \tag{8}$$

The equations represent update in position according with alpha, and $\vec{D}_\alpha$, $\vec{D}_\beta$ and $\vec{D}_\delta$ are the distance of the prey and positions of α, β & δ wolves are $\vec{X}_1$, $\vec{X}_2$ and $\vec{X}_3$, respectively.

3.3   Attacking Prey

In this phase, exploitation is handled by linear decrement in $\vec{a}$. This linear decrement enables the gray wolves to attack the prey, while it stops moving. Fluctuations in $\vec{A}$ are also depending on $\vec{a}$, that means if the higher value of $\vec{a}$ causes more fluctuations. Figure 1 shows the flowchart and pseudo-code of gray wolf optimization algorithm, respectively.

```
Input: Problem Size, Population size
Output: P_g_best
Start
        Initialize the population of grey wolves X_i (i = 1, 2, ... , n)
        Initialize a, A, and C according to combined objective function
        Calculate the fitness values of search agents and grade the agents.
        (X_α = the best solution within the search agent, X_β = the second best solution
        within the search agent, and X_δ = the third best solution within the search
        agent) t = 0
        While (t < maximum number of iterations)
                For each search agent
                        Update the position of the current search according to the equation
                End for
        Update value of a, A, and C according to combined objective function
        Again Calculate the fitness values of all search agents and grade them
        Again update the position of X_α, X_β, and X_δ
        t = t+1
        Store the best solution value.
        End while
End
```

Fig. 1 Pseudocode of GWO

Due to the following reasons, GWO has drawn a lot of interest among the researchers:

(i) GWO has very simple architecture; this architecture permits to researchers to form new variants.

(ii) GWO is very less control parameters derivative free algorithm, and this features make GWO more flexible and user friendly.

(iii) GWO can handle discontinuous and stochastic functions, and due to this, it can be implemented in real world problems.

4 Test System

Proposed demand side management technique is tested on residential and commercial consumers' area to prove the effectiveness of proposed technique. Smart grid network operates on 410 V and has interconnected resistance, reactance, and maximum power limit, 0.0031 pu, 0.011 pu, and 500 kVA, respectively. Length of interconnected links is 2 KM and 3 KM for residential and commercial area, respectively. The main objective of this arrangement is to reduce the electricity bills of consumers of given areas and to fulfill the same, objective curve always chosen to inversely proportional to price of electricity. Details of energy cost for both areas are given in Table 1.

Details about the used devices in both areas are given in below section.

Table 1 Prices of energy with forecasted loads [33]

Hour	Time slab	Price of electricity (Cent/KWh)	Forecasted load (KW)	
			Residential microgrid	Commercial microgrid
1	8–9 h	12	12.2	15.4
2	9–10 h	9.19	11.9	19.2
3	10–11 h	12.3	11.9	24.1
4	11–12 h	20.7	13.5	26
5	12–13 h	26.8	13.7	27.9
6	13–14 h	27.4	12.7	27.9
7	14–15 h	13.8	12.4	27.9
8	15–16 h	17.3	11.4	26.5
9	16–17 h	16.4	11.1	26
10	17–18 h	9.83	15.9	27.9
11	18–19 h	8.63	20.3	30.3
12	19–20 h	8.87	22.2	25
13	20–21 h	8.35	22.7	21.6
14	21–22 h	16.4	20.9	18.3
15	22–23 h	16.2	17.4	15.4
16	23–24 h	8.87	12.7	9.62
17	24–1 h	8.65	7.93	6.73
18	1–2 h	8.11	6.87	6.25
19	2–3 h	8.25	6.08	6.25
20	3–4 h	8.1	5.81	6.73
21	4–5 h	8.14	4.49	7.22
22	5–6 h	8.13	4.49	7.22
23	6–7 h	8.34	6.87	7.22
24	7–8 h	9.35	8.99	11.1

4.1 Residential Area

Devices used in residential consumers' area have highest probability of low power rating and short time of operation. Table 2 has the details of hourly consumption of load with 14 types of total 2605 shift-able devices.

Table 2 Devices data in residential area [33]

Device type	Per hour load of device (KW)			Total number of devices
	I h	II h	III h	
Cloth dryer	1.2	–	–	189
Dish washer	0.7	–	–	288
Washing machine	0.5	0.4	–	268
Oven	1.3	–	–	279
Iron	1	–	–	340
Vacuum cleaner	0.4	–	–	158
Fan	0.2	0.2	0.2	289
Kettle	2	–	–	406
Toaster	0.9	–	–	48
Rice cooker	0.85	–	–	59
Hair dryer	1.5	–	–	58
Blender	0.3	–	–	66
Frying pen	1.1	–	–	101
Coffee maker	0.8	–	–	56
Total				2605

4.2 Commercial Area

Devices used in commercial area have highest probability of high power rating and more time of operation. Table 3 has the details of hourly consumption of load with 08 types of total 810 controllable devices.

Table 3 Devices data in a commercial area [33]

Device type	Per hour load of device (KW)			Total number of devices
	I h	II h	III h	
Water dispensing machines	2.5	–	–	156
Hair dryer machines	3.5	–	–	118
Electric kettle	3	2.5	–	123
Electric oven	5	–	–	77
Coffee machine	2	2	–	99
Electric fan	3.5	3	–	95
A/C machine	4	3.5	3	56
Rooms lights	2	1.75	1.5	87
Total				810

5 Results and Discussion

Used DSM technique has sufficient potential to duplicate load curve with objective curve. Tables and figures also indicate the same. GWO handled this problem of minimization very well. To validate the proposed technique, two test boats have been used.

Test Boat 1: Strategic Conservation

Test Boat 2: Load Shifting

5.1 Test Boat 1: Strategic Conservation

To maintain the desired load shape, efficient and smart appliances are used to suppress the load demand. This method of demand response is only applicable on residential and commercial consumers' area because industrial sector need continuous power supply.

Figure 2 depicts the simulation curves of power consumption for residential area and commercial area got from strategic conservation technique.

Figure 3 shows the comparative results for strategic conservation. Figures 3a and 3b show the reduction in operational cost and peak demand, respectively, for strategic conservation.

Table 4 shows the comparison of operational cost reduction and peak demand reduction in both areas. For residential area, curtailment in cost of operation using WOA and MFO is given by 21.80% and 23.77%, respectively. Here, GWO proved its efficacy over WOA and MFO by giving 23.85% curtailment in operational cost. For the commercial area, curtailment in operational cost using WOA and MFO is given by 15.94% and 15.35%, respectively. Here, GWO again proved its efficacy over WOA and MFO by giving 16.75% curtailment in operational cost. When a number of devices have been used then proposed DSM technique gives the better results.

Table 4 also shows the comparison of peak demand reduction in both areas. Several benefits provided when effective path is selected for DSM. Reduction in peak demand is the one of the example of it. For residential area, peak reduction using WOA and MFO is given by 48.31% and 48.87%, respectively. Here, GWO proved its efficacy over WOA and MFO by giving 50.02% in peak reduction. For commercial area, peak reduction using WOA and MFO is given by 15.92% and 21.71%, respectively. Here, GWO again proved its efficacy over WOA and MFO by giving 28.70% peak reduction. Peak reduction results in increase in grid stability and lower the utility cost.

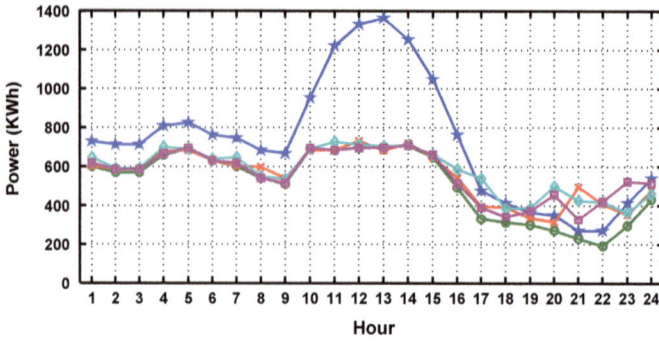

(a) For residential consumers

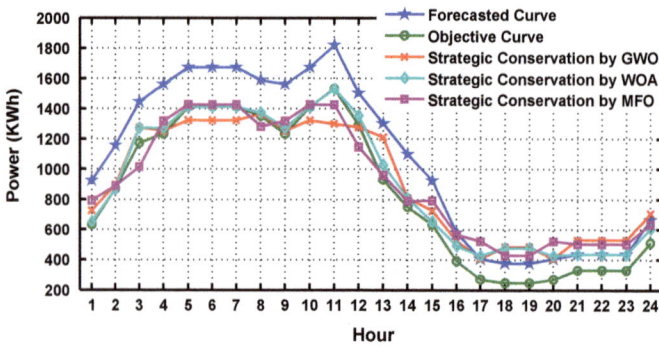

(b) For commercial consumers

Fig. 2 Strategic conservation results

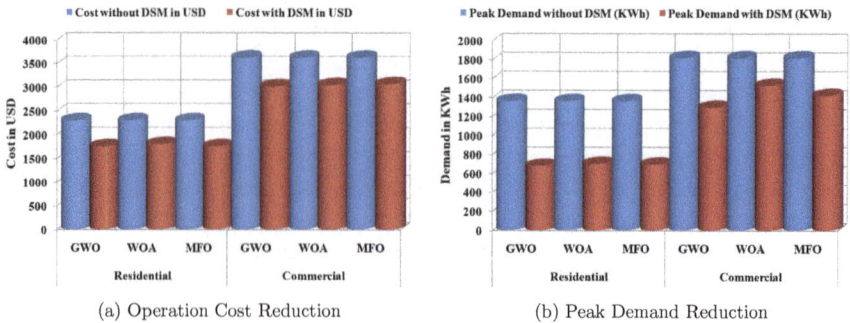

(a) Operation Cost Reduction

(b) Peak Demand Reduction

Fig. 3 Comparative results for strategic conservation

Table 4 Operational cost and peak demand reduction due to strategic conservation

Area	Algorithm	Without DSM	With DSM	Reduction (%)
Operational cost reduction (USD)				
Residential	GWO	2302.87928	1753.689557	23.84795973
	WOA	2302.87928	1800.682662	21.8073358
	MFA	2302.87928	1755.392044	23.77403111
Commercial	GWO	3626.6396	3019.056319	16.75334052
	WOA	3626.6396	3048.512873	15.94111327
	MFA	3626.6396	3069.877037	15.35202349
Peak demand reduction (KWh)				
Residential	GWO	1363.6	681.5215763	50.02041828
	WOA	1363.6	704.8094914	48.3125923
	MFA	1363.6	697.0887916	48.87879205
Commercial	GWO	1818.2	1296.282189	28.70519254
	WOA	1818.2	1528.620662	15.92670431
	MFA	1818.2	1423.308821	21.71879765

5.2   Test Boat 2: Load Shifting

Load shifting technique is used frequently to shift the load from peak time to off-peak time.

Table 5 encountered reductions in operational cost for residential consumers by using WOA and MFO is 6.71% and 9.40%, respectively. GWO proved its iron over

Table 5 Operational cost and peak demand reduction due to load shifting

Area	Algorithm	Without DSM	With DSM	Reduction (%)
Operational cost reduction (USD)				
Residential	GWO	2302.879	2013.285	12.5753
	WOA	2302.879	2148.147	6.719088
	MFA	2302.879	2086.242	9.407254
Commercial	GWO	3626.64	3424.867	5.563623
	WOA	3626.64	3605.217	0.590704
	MFA	3626.64	3465.726	4.437002
Peak demand reduction (KWh)				
Residential	GWO	1363.6	706.3656	48.19848
	WOA	1363.6	765	43.8985
	MFA	1363.6	725.482	46.79657
Commercial	GWO	1818.2	1309.251	27.99194
	WOA	1818.2	1528.621	15.9267
	MFA	1818.2	1481.126	18.5389

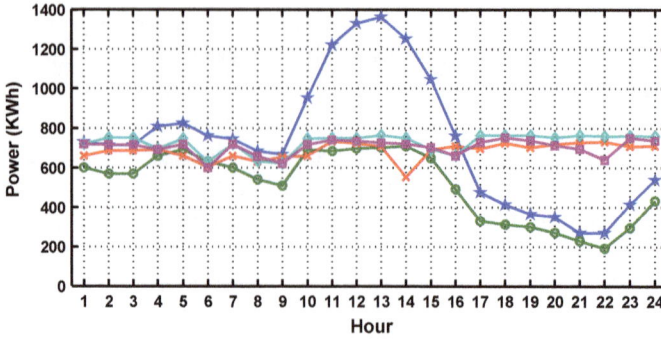

(a) For residential consumers

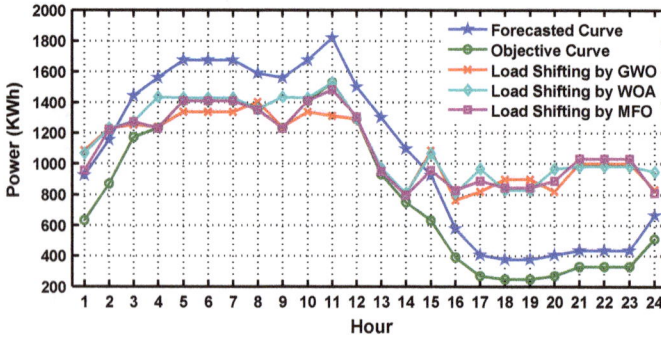

(b) For commercial consumers

Fig. 4 Load shifting results

WOA and MFO by giving reduction in cost by 12.57%. For commercial consumers, reduction in operational cost by using WOA and MFO is 0.59% and 4.43%, respectively. GWO proved its iron over WOA and MFO by giving reduction in cost by 5.56%. WOA gives less than 1% reduction only.

Figure 4a, b shows the results for residential consumers and commercial users by load shifting, respectively. For residential and commercial areas, peak demand is given by Table 5. For residential area, used DSM technique lowered the peak load by using WOA and MFO is given by 43.89% and 46.79%, respectively. GWO again gives better results by 48.19% reduction.

For commercial area, used DSM technique lowered the peak load by using WOA and MFO is given by 15.92% and 18.53%, respectively. GWO again gives better results by 27.99% reduction. Figure 5a, b shows the operational cost and peak reduction, respectively, for load shifting.

Figure 6 depicts the convergence analysis for given problem in four segments: segment (A) shows the convergence analysis of strategic conservation for residential consumers; segment (B) shows the convergence analysis of strategic conservation for commercial consumers; segment (C) shows the convergence analysis of load shifting

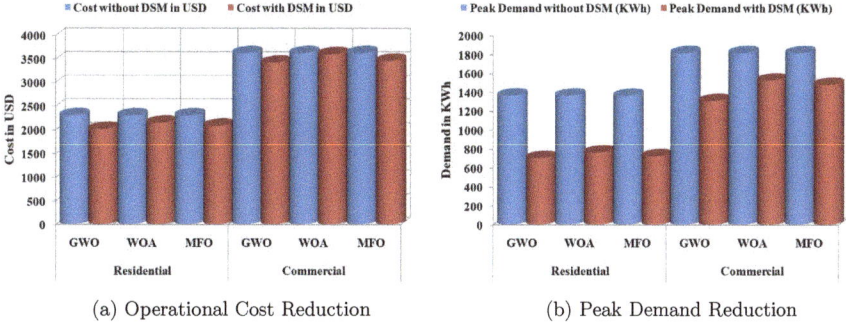

(a) Operational Cost Reduction

(b) Peak Demand Reduction

Fig. 5 Comparative results for load shifting

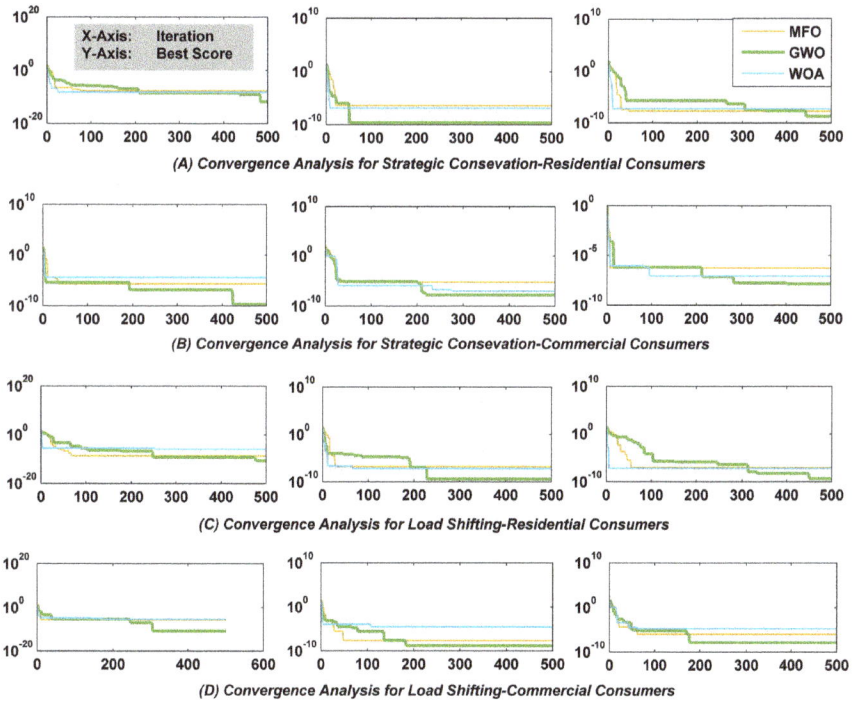

(A) Convergence Analysis for Strategic Consevation-Residential Consumers

(B) Convergence Analysis for Strategic Consevation-Commercial Consumers

(C) Convergence Analysis for Load Shifting-Residential Consumers

(D) Convergence Analysis for Load Shifting-Commercial Consumers

Fig. 6 Convergence analysis of DSM problem

for residential consumers; and segment (D) shows the convergence analysis of load shifting for commercial consumers. These all segments of convergence proved the efficacy of GWO over WOA and MFO.

6   Conclusion

This paper is a meaningful comparison of optimization algorithms, namely GWO, WOA, and MFO. On the basis of obtained results, it is manifested that DSM is merely necessitated for implementation of successive smart grid. Following are the major conclusion drawn from this study:

(i) The energy demand curves give much useful information about the amount of consumption of electricity and energy cost.

(ii) Further, a comparison of the performance of GWO has been conducted for providing demand side management strategies such as strategic conversion and load shifting strategies for residential and commercial consumers' area.

(iii) After comparison, we observe that positive implications exist in the results such as a reduction in peak demand and reduction cost. GWO provides optimal results when compared with contemporary optimizers.

Application of proposed variant on more challenging problems like industrial consumers, protein structure prediction, and model order reduction is kept for future.

References

1. Meyabadi AF, Deihimi MH (2017) A review of demand-side management: Reconsidering theoretical framework. Renew Sustain Energy Rev 80:367–379
2. Gellings CW (1985) The concept of demand-side management for electric utilities. Proc IEEE 73(10):1468–1470
3. Gellings CW (1985) The special section on demand-side management for electric utilities. Proc IEEE 73(10):1443–1444
4. Gellings CW, Barron W, Betley FM, England WA, Preiss LL, Jones DE (1986) Integrating demand-side management into utility planning. IEEE Trans Power Syst 1(3):81–87
5. Rahman S (1993) An efficient load model for analyzing demand side management impacts. IEEE Trans Power Syst 8(3):1219–1226
6. Cohen AI, Wang CC (1988) An optimization method for load management scheduling. IEEE Trans Power Syst 3(2):612–618
7. Albadi MH, El-Saadany EF (2008) A summary of demand response in electricity markets. Electr Power Syst Res 78(11):1989–1996
8. Albadi MH, El-Saadany EF (2007) Demand response in electricity markets: An overview. In: 2007 IEEE power engineering society general meeting. IEEE, New York, pp 1–5
9. Goel L, Wu Q, Wang P, (2006) Reliability enhancement of a deregulated power system considering demand response. In: IEEE power engineering society general meeting. IEEE, New York, p 6
10. Wu Z, Xia X (2017) A portfolio approach of demand side management. IFAC-PapersOnLine 50(1):171–176

11. Mirjalili S, Mirjalili SM, Lewis A (2014) Grey wolf optimizer. Adv Eng Software 69:46–61
12. Waseem M, Lin Z, Liu S, Sajjad IA, Aziz T (2020) Optimal GWCSO-based home appliances scheduling for demand response considering end-users comfort. Electr Power Syst Res 187:106477
13. Tikhamarine Y, Souag-Gamane D, Ahmed AN, Kisi O, El-Shafie A (2020) Improving artificial intelligence models accuracy for monthly streamflow forecasting using grey Wolf optimization (GWO) algorithm. J Hydrol 582:124435
14. Lu H, Ma X, Huang K, Azimi M (2020) Carbon trading volume and price forecasting in China using multiple machine learning models. J Clean Prod 249:119386
15. Bouzary H, Frank Chen F, Shahin M (2019) Optimal composition of tasks in cloud manufacturing platform: a novel hybrid GWO-GA approach. Proc Manuf 34:961–968
16. Sankaranarayanan S, Sivakumaran N, Radhakrishnan TK, Swaminathan G (2020) Dynamic soft sensor based parameters and demand curve estimation for water distribution system: theoretical and experimental cross validation. Control Eng Practice 102:104544
17. Miao D, Hossain S (2020) Improved gray wolf optimization algorithm for solving placement and sizing of electrical energy storage system in micro-grids. ISA Trans
18. Seifi A, Moradi MH, Abedini M, Jahangiri A (2020) An optimal programming among renewable energy resources and storage devices for responsive load integration in residential applications using hybrid of grey wolf and shark smell algorithms. J Energy Storage 27:101126
19. Jain AA, Justus Rabi B, Darly SS (2020) Application of QOCGWO-RFA for maximum power point tracking (MPPT) and power flow management of solar PV generation system. Int J Hydrogen Energy 45(7):4122–4136
20. Fathy A, Rezk H (2020) Robust electrical parameter extraction methodology based on Interior Search Optimization Algorithm applied to supercapacitor. ISA Trans
21. Mohseni S, Brent AC, Burmester D (2020) A comparison of metaheuristics for the optimal capacity planning of an isolated, battery-less, hydrogen-based micro-grid. Appl Energy 259:114224
22. Lu H, Ma X, Huang K, Azimi M (2020) Prediction of offshore wind farm power using a novel two-stage model combining kernel-based nonlinear extension of the Arps decline model with a multi-objective grey wolf optimizer. Renew Sustain Energy Rev 127:109856
23. Mahian O, Javidmehr M, Kasaeian A, Mohasseb S, Panahi M (2020) Optimal sizing and performance assessment of a hybrid combined heat and power system with energy storage for residential buildings. Energy Convers Manage 211:112751
24. Fu W, Wang K, Tan J, Zhang K (2020) A composite framework coupling multiple feature selection, compound prediction models and novel hybrid swarm optimizer-based synchronization optimization strategy for multi-step ahead short-term wind speed forecasting. Energy Convers Manage 205:112461
25. Saxena A, Kumar R, Das S (2019) β-chaotic map enabled grey wolf optimizer. Appl Soft Comput 75:84–105
26. Saxena A, Soni BP, Kumar R, Gupta V (2018) Intelligent grey wolf optimizer–development and application for strategic bidding in uniform price spot energy market. Appl Soft Comput 69:1–13
27. Liu J, Yang Z, Li D (2020) A multiple search strategies based grey wolf optimizer for solving multi-objective optimization problems. Expert Syst Appl 145:113134
28. Sharma P, (2018) An intelligent energy bidding strategy based on opposition theory enabled grey wolf optimizer. In: International conference on power, instrumentation, control and computing (PICC). IEEE, New York
29. Sharma AK, Saxena A (2019) A demand side management control strategy using Whale optimization algorithm. SN Appl Sci 1(8):870
30. Heidari H, Hagh MT (2019) Optimal reconfiguration of solar photovoltaic arrays using a fast parallelized particle swarm optimization in confront of partial shading. Int J Eng 32(8):1177–1185
31. Malik OP (2013) Evolution of power systems into smarter networks. J Control, Automat Electr Syst 24(1–2):139–147

32. Attia HA (2010) Mathematical formulation of the demand side management (DSM) problem and its optimal solution. Cairo University Egypt
33. Logenthiran T, Srinivasan D, Shun TZ (2012) Demand side management in smart grid using heuristic optimization. IEEE Trans Smart Grid 3(3):1244–1252

Chapter 73
Electricity Bill Saving Using Solar Rooftop PV Cell for a Residential Complex in Our Locality—A Case Study

Rehan Aziz, Nirban Chakraborty, and Raubins Kumar

1 Introduction

With increase in electricity demand and awareness of the harmful effects of non-renewable energy sources on the environment, focus has shifted to other sources. This has been in process for more than 30 years now, but its only in the last 10 years that the governments have made it a priority. We can see in countries like Germany, over 50% of countries energy demands are met by Solar [1] and target of 100% from renewable sources by 2050. Similarly, various countries and organizations has dedicated their resources towards it.

In our country, India, similar approach has been taken. As per Ministry of New and Renewable Energy, Govt of India [2], the share of renewable energy out of total installed generation capacity if 23.39% which amounts to 86.3 Gigawatt (GW). Out of this the share of Solar is around 34 GW or Gigawatt. We have a target of total renewable energy of 175 GW till 2022. And because of this target, there is enormous support from public as well as private organizations when it comes to accessing and installing rooftop Photovoltaics (PV's). When we talk about solar, we have to evaluate the overall benefit in terms of cost. Solar irradiation in a location depends on various factors, and it is necessary to calculate it for finding the proper location for installation of solar plants. But when we are talking about Rooftop PV's the calculation is a bit different because of limited application. According to reports by Mercom India [3], the rooftop solar panels capacity has reached 4.4 GW and is growing each day because of efforts by different states and organizations. There are increased subsidies, availability and extension of rules, so as to encourage organizations to focus on them. Geographical impact is also a vital issue for Power generation with the help of rooftop solar PV [4]. Different optimization techniques

R. Aziz · N. Chakraborty (✉) · R. Kumar
Electrical and Electronics Department, Institute of Engineering and Management, Kolkata, West Bengal, India
e-mail: nirban.chakraborty@iemcal.com

has also been adopted for better result [5]. In this paper we mainly focus our case study towards the cost benefit effect of using the solar panel in an existing complex in our local area. We have calculated the total demand of one of the flat as well as the whole building and also find out the monthly electric bill of that flat using the existing tariff plan. After finding out the total consumption we have also shown the peak hours and choose a span of 3 h and replace the power supply during that 3 h with the proposed rooftop solar panel, as a result of the we see a sufficient amount of cost saving in the monthly electricity bill for the whole building as well as for per family also.

Rest of the paper is comprised with in Sect. 3 we have discuss about the model of the proposed complex of our locality and also shown the total energy consumption, In Sect. 4 we have discussed about the total electricity bill for the complex as per the existing tariff of the supplier, In Sect. 5 we have shown the solar calculation and cost saving for the consumers and in Sect. 6 we concluded the paper and also discus about the future scope of this work.

2   Literature Review

Work on estimation of solar irradiation started early in 1980's [6] and by now we have data for every region [7, 8]. Early works on solar energy was focused on estimating the best areas [9] for establishing solar farms. Then it shifted to standalone solar based products like solar LED [10] and other products. With the attention on non-conventional source, advancement in rooftop solar has been made. The study ranges from taking the test cases as a whole city [11–21] or a smaller area [22] and estimating the solar potential by taking the necessary parameters. Some of the previous study has worked on the cost evaluation [23] but the cost evaluation for customer if they install a rooftop for their buildings has not been surveyed much. Here we attempt to cover this portion.

3   Model of the Building and Total Energy Consumption

3.1   Model of the Complex

When we are on the subject of Rooftop PV's, most of the efforts are focused on big organizations and/or government enterprises. Here we are focusing on the benefits in terms of monthly cost reductions when we use rooftop PV's in our residential complexes. We undertook a case study of a 5 story 3 BHK 3 apartments per floor residential building in central Kolkata. We first did a manual calculation based on the ideal usage of different loads used in the building and also shown the consumption of the other loads of the complex (like. Elevators, water pump, Cab Exhaust Fan,

Cab Lighting, Machine Room Cooling) and then shown the consumption of the load for the each flat as well as the total energy consumption of the total building.

3.2 Equipment Table and Energy Rating

After observing the test building in the month of may, and taking into consideration the appliances (LED, Dim Light, fan, Mixer, Refrigerator, Chimney, Microwave, Fan, Television, Washing Machine, Exhaust, Geyser and AC) used taking 4 or 5 star rated models, as per BEE [24] and their duration, we calculated the net energy consumption for a flat per day, and with that as a reference, energy consumption of the whole building. The following tables shows the energy consumption for different rooms (Tables 1, 2, 3, 4, 5, 6).

From the above tables we have seen the different consumption for different appliances, we now calculate the total consumption for each flat and also find out the total energy consumption for the whole building.

Table 1 Master bedroom

S. NO	Equipment	Rating	Quantity	Hours	Energy Consumption/day (KWh)
1.	LED	8 W	6	8	0.384
2.	Fan	48 W	1	10	0.48
3.	Television	80 W	1	4	0.32
4.	AC	1.73KW	1	8	13.84
5.	Lamp	8 W	1	6	0.048

Table 2 Kitchen

S. NO	Equipment	Rating (W)	Quantity	Hours	Energy Consumption/day (KWh)
1.	LED	8	2	6	0.096
2.	Mixer	500	1	1	0.5
3.	Refrigerator	207	1	24	5.0
4.	Chimney	180	1	2	0.36
5.	Microwave	800	1	1	0.8

Table 3 Dining + hall room

S. NO	Equipment	Rating (W)	Quantity	Hours	Energy Consumption/day (KWh)
1.	LED + DLight	8	5 + 1	6	0.288
2.	Fan	48	1	10	0.48
3.	Television	80	1	4	0.32
4.	Washing Machine	370	1	2	0.74

Table 4 Washrooms

S. NO	Equipment	Rating	Quantity	Hours	Energy Consumption/day
1.	LED	8 W	2 + 2	6	0.192KWh
2.	Exhaust	48 W	1	10	0.48KWh
3.	Geyser	–	1	–	–

Table 5 Bedroom 2 and 3

S. NO	Equipment	Rating	Quantity	Hours	Energy Consumption/day (KWh)
1.	LED	8 W	4 + 4	8	0.512
2.	Fan	48 W	1 + 1	10	0.96
3.	Lamp	8 W	1	6	0.048
4.	AC	1KW	1	8	8.0

Table 6 Others

S. NO	Equipment	Rating	Quantity	Hours	Energy Consumption/day (KWh)
1.	Mobile charger	10 W	8	2	0.16
2.	Laptop charger	90 W	2	4	0.72
3.	LED(Gate + Storeroom)	8 W	1 + 1	6	0.072

$$\text{Total Power Consumption per Day per Flat}$$
$$= 15.072 + 6.756 + 1.8280.762 + 9.52 + 1 = 34.176\,\text{KWh} \tag{1}$$

$$\text{Monthly Power Consumption per Flat} = 34.176 {*} 30 = 1,025.28\,\text{KWh}$$

$$\text{Total Power Consumption for}$$
$$(5\,\text{floors}{*}3\,\text{flat per floor} = 15\,\text{Flat})$$
$$\text{per day } 34.176{*}15 = 512.64\,\text{KWh} \tag{2}$$

Common Energy Consumption per day:

$$\text{LED} = 2\,\text{per Floor}(2{*}5) + 8\,\text{for ground floor, } 12\,\text{hours each} = 1.728\,\text{KWh}$$

$$\text{Fans} = 2,\ 24\,\text{hours each} = 2.304\,\text{KWh} \tag{3}$$

$$\text{Water Pump} = 4.5\,\text{hours daily, } 3.5\text{HP}\ (2611\,\text{W})$$
$$= 4.5{*}2611$$
$$= 11.750\,\text{KWh} \tag{4}$$

$$\text{Cab Exhaust Fan} = 0.28 \text{ KWh} \tag{5}$$

$$\text{Cab Lighting} = 2.52 \text{ KWh} \tag{6}$$

$$\text{Machine Room Cooling} = 1.11 \text{ KWh} \tag{7}$$

$$\text{Elevator} = 17.59 \tag{8}$$

Total $= 21.5$ KWh

Total Energy requirement for building per day(x)

$$= 512.64 + 1.728 + 2.304 + 11.75 + 21.5 = 549.92 \text{ KWh} \tag{9}$$

In the Above graph, Hourly energy consumption per flat per day is given, so that we can visualize the peak hours. For a building, the peak hours would be same and the values will be the value in the table multiplied by 15 and then we can add the common consumption (Fig. 1).

From the above figure we can see that the total energy consumption for each hour and from the figure it is clear that the peak demand hour is between 10 PM–5AM. We have also break up the load demand for each day and monthly basis for each 1 h of operation, shown in the Table 7.

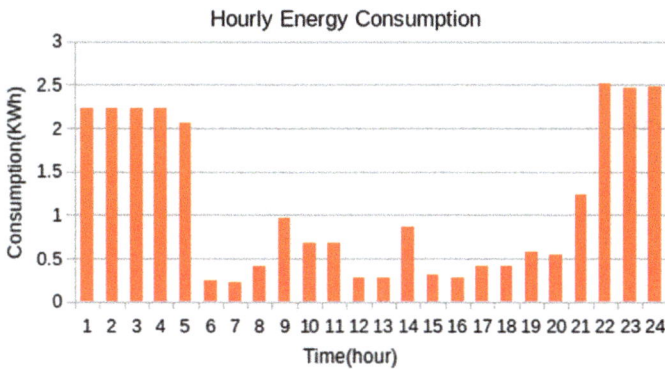

Fig. 1 Hourly energy consumption

Table 7 Peak hours estimation

Duration	Per day	Per month
9–10PM	18.315	549.15
9–11PM	55.83	1674.9
9–12PM	92.865	2785.95
9–1PM	126.060	3781.8

Table 8 Tariff rates

Index	Rate(per unit)
First 25	489 p
Next 35	540 p
Next 40	641 p
Next 50	716 p
Next 50	733 p
Next 100	733 p
Above 300	892 p

4 Electricity Bill as Per the Available Supplier

The current electricity provider at the test building is CESC (RP Sanjiv Goenka Group) [25]. The building falls under Domestic-Urban Category. The tariff [26] here varies as: (Table 8)

Other Charges include:

- Fixed Charges @ 15rupee per KVA per month
- MVCA @ 29p per unit
- Govt. Duty = 10%
- Meter Rent

Electric Bill at the test building for May (11th May–08th June = 199 units(KWh) which is the case we have done the calculations for.

5 Solar Calculations

Various work has been done to calculate the solar irradiation in different locations. As per Government data [27], in West Bengal, India Average solar irradiation in state is 1156.39 W/sq.m and 1 kWp solar rooftop plant will generate on an average over the year kWh of electricity per day (considering 5.5 sunshine hours)

Now we have total available area on terrace = 6000 sq. ft.

$$= 557.42 \text{ sq. m} \tag{10}$$

Cost Evaluation:

If we plan to replace Supply by solar power for the peak hours (refer table) 9–12

Per day solar energy requirement will be = 92.865KWh,

As per calculation available at [27], for our given roof area of 600 sq. Ft. If we use 40% percent of the area to install solar panels, per day electricity generation will be 102.58, which is 9.715KWh more than our requirement. The layout is important, so in order to maximize the efficiency further, we can refer to previous works [28] and personalize them as per the requirement.

If we look into the cost, cost required for installation by MNRE [27] will be without subsidy = 12,26,500 rupees with 30% subsidy = 8,58,550 rupees

Annual saving will be approximately 2,50,000 rupees

If we look into per flat/family savings, its around 1,370 rupees per month.

6 Conclusion

In this paper we have done a real time case study of a Residential complex and showed a saving in the bill for the customer (saving in the electricity bill per family basis of the complex) by applying the rooftop solar PV cell, We saw in the point 3, how usage of rooftop solar during peak hours will help in the short and long term in terms of cost. In the long term, with the proposed capacity ratio of installation of solar plant, the long-term monthly savings will be approximately 20,000 rupees. Apart from these savings, if we look into the environmental impact, carbon dioxide emission mitigated will be approximately 600 tones and will save around 1000 trees [27].

In future we will take weather impact into the consideration of the rainy season and winter season particularly and try to find out the profit in electricity bill from the customer side. In previous studies, refer Sect. 2, similar calculations has been done on a wider scale, like a city and it requires cooperation between various stakeholders. We have kept the focus on a much smaller level, which will only require a clearance from the authorities. We can also incorporate hybrid [29, 30] model to divide the production and also act as backup when the condition is not favorable for solar. Our further work will be focused on this in the test case.

References

1. https://www.investopedia.com/articles/investing/092815/5-countries-produce-most-solar-ene rgy.asp
2. Ministry of New and Renewable Energy(MNRE). https://mnre.gov.in/
3. Mercom India. https://mercomindia.com/category/solar/rooftop-solar/
4. Dutta A, Das S (2020) Adoption of grid-connected solar rooftop systems in the state of Jammu and Kashmir: a stakeholder analysis. Energy Policy 140:111382

5. Awad H, Emtiaz Salim KM, Gül M (2020) Multi-objective design of grid-tied solar photo-voltaics for commercial flat rooftops using particle swarm optimization algorithm. J Build Eng 28:101080
6. Christian G (1989) A two-band model for the calculation of clear sky solar irradiance, illuminance, and photosynthetically active radiation at the earth's surface, Solar Energy 43(5):253–265. ISSN 0038-092X. https://doi.org/10.1016/0038-092X(89)90113-8
7. https://solargis.com/maps-and-gis-data/download/india
8. http://www.mnre.gov.in/sec/solar-resources.htm
9. Sonal S, Vijay N, Sunil L (2017) Investigation of feasibility study of solar farms deployment using hybrid AHP-TOPSIS analysis: case study of India. Renew Sustain Energy Rev 73:496–511. ISSN 1364-0321. https://doi.org/10.1016/j.rser.2017.01.135
10. Yongqing W, Chuncheng H, Suoliang Z, Yali H, Hong W (2009) Design of solar LED street lamp automatic control circuit. In: 2009 International conference on energy and environment technology, Guilin, Guangxi, pp 90–93. https://doi.org/10.1109/iceet.2009.28
11. Thotakura S, Kondamudi SC, Francis Xavier J, Quanjin M, Ramakrishna Reddy G, Gangwar P, Lakshmi Davuluri S (2020) Operational performance of megawatt-scale grid integrated rooftop solar PV system in tropical wet and dry climates of India. Case Stud Therm Eng 18:100602. ISSN 2214-157X. https://doi.org/10.1016/j.csite.2020.100602
12. Phap VM, Huong NTT, Hanh PT, Duy PV, Binh DV (2020) Assessment of rooftop solar power technical potential in Hanoi city Vietnam. J Build Eng 32:101528. ISSN 2352-7102. https://doi.org/10.1016/j.jobe.2020.101528
13. Ko L, Wang J-C, Chen C-Y, Tsai H-Y (2015) Evaluation of the development potential of rooftop solar photovoltaic in Taiwan. Renew Energy 76:582–595. ISSN 0960-1481. https://doi.org/10.1016/j.renene.2014.11.077
14. Yadav SK, Bajpai U (2018) Performance evaluation of a rooftop solar photovoltaic power plant in Northern India. Energy for Sustain Develop 43:130–138. ISSN 0973-0826. https://doi.org/10.1016/j.esd.2018.01.006
15. Dondariya C, Porwal D, Awasthi A, Shukla AK, Sudhakar K, Murali Manohar SR, Bhimte A (2018) Performance simulation of grid-connected rooftop solar PV system for small house-holds: a case study of Ujjain India. Energy Report 4:546–553. ISSN 2352-4847. https://doi.org/10.1016/j.egyr.2018.08.002
16. Byrne J, Taminiau J, Kurdgelashvili L, Kim KN (2015) A review of the solar city concept and methods to assess rooftop solar electric potential, with an illustrative application to the city of Seoul. Renew Sustain Energy Rev 41:830–844. ISSN 1364-0321. https://doi.org/10.1016/j.rser.2014.08.023
17. Hong T, Lee M, Koo C, Jeong K, Kim J (2017) Development of a method for estimating the rooftop solar photovoltaic (PV) potential by analyzing the available rooftop area using Hillshade analysis. Appl Energy 194:320–332. ISSN 0306-2619. https://doi.org/10.1016/j.apenergy.2016.07.001
18. Singh R, Banerjee R (2015) Estimation of rooftop solar photovoltaic potential of a city. Solar Energy 115:589–602. ISSN 0038-092X. https://doi.org/10.1016/j.solener.2015.03.016
19. Gagnon P, Margolis R, Melius J, Phillips C, Elmore R (2016) Rooftop solar photovoltaic technical potential in the united states. a detailed assessment. United States: N. p 2016. Web. https://doi.org/10.2172/1236153
20. Porse E, Fournier E, Cheng D, Hirashiki C, Gustafson H, Federico F, Pincetl S (2020) Net solar generation potential from urban rooftops in Los Angeles. Energy Policy 142:0301–4215 ISSN 0301-4215. https://doi.org/10.1016/j.enpol.2020.111461
21. Mangiante MJ, Whung P-Y, Zhou L, Porter R, Cepada A, Campirano E, Licon D, Lawrence R, Torres M (2020) Economic and technical assessment of rooftop solar photovoltaic potential in Brownsville, Texas, U.S.A. Comput Environ Urban Syst 80:101450. ISSN0198–9715. https://doi.org/10.1016/j.compenvurbsys.2019.101450
22. Vaishnav P et al (2017) Environ Res Lett 12:094015. https://doi.org/10.1088/1748-9326/aa815e
23. Mukherji R, Mathur V, Bhati A, Mukherji M (2020) Assessment of 50 kWp rooftop solar photovoltaic plant at the ICFAI University, Jaipur: a case study. Environ Prog Sustain Energy 39:e13353. https://doi.org/10.1002/ep.13353

24. https://beeindia.gov.in/content/standards-labeling
25. CESC, RP Sanjiv Goenka Group, https://www.cesc.co.in/
26. CESC Tariff, https://www.cesc.co.in/wpcotent/uploads/tariff/TARIFF%20AND%20ASSO CIATED%20CONDITIONS.pdf
27. MNRE, SPIN. https://solarrooftop.gov.in/rooftop_calculator
28. Zhong Q, Tong D (2020) Renewable Energy 150:1e11
29. Performance assessment of hybrid PV-wind systems on high-rise rooftops in the Brussels-Capital Region (2020) Energy and Buildings 224:110137. ISSN 0378–7788. https://doi.org/10.1016/j.enbuild.2020.110137
30. Pettongkam W et al. (2018) Int Energy J 18:331–35

Chapter 74
Low Cost IoT Based Runaway Syndrome Tracking System

Soham Bose and Arindam Chakraborty

1 Introduction

This project is designed to protect the children suffering from Runaway Syndrome. The project aims to track children who are suffering from Runaway Syndrome and help the children's guardian or parents. Whenever the children try to leave the house, the parent or guardian will be notified immediately, and then they can take whatever action is necessary.

The Runaway Syndrome is mostly seen in children when they are supposed to be under some kind of conflict inside their mind which a psychologist calls Stress. Seen predominantly in men, this syndrome involves a preplanned or sudden running away when the child runs away from his home and to a: (1) A relative's place, (2) A friend's place, (3) A place which is very far away and is known to the child, (4) Sometimes in very extreme cases, to a place which is entirely unknown to the child [1].

As noted by experts, these disappearances do not happen for more than a week. When the child runs away and reaches the place where they ran away to, after a certain point they realise that they cannot survive or sustain for long without the financial or emotional support of the parent. When the child realises that there is no point in staying away, they decide to come back. If for some reason they are unable to come back on their own they somehow try to inform their parents about the place where they are hiding [1].

S. Bose (✉) · A. Chakraborty
Department of Electronics and Communication Engineering, Institute of Engineering and Management, Gurukul, Y-12, Block-EP, Sector-V, Salt Lake Electronics Complex, Kolkata 700091, West Bengal, India
e-mail: rickbose.12@gmail.com

A. Chakraborty
e-mail: arindam.chakraborty@iemcal.com

X.-Z. Gao et al. (eds.), *Applications of Artificial Intelligence in Engineering*, Algorithms for Intelligent Systems, https://doi.org/10.1007/978-981-33-4604-8_74

This is precisely what Runaway Syndrome is, and it can be quite dangerous if the child never returns home. There are several reasons as to why Runaway Syndrome might occur in young children, even though maybe it is not possible to eradicate without using professional help. Still, as parents, it is quite possible to prevent it. Runaway Syndrome is serious and should not be taken lightly by any means.

2   Literature Review on Runaway Syndrome

Before we dig into the how and what of the project, we need to understand the motive behind the project. We need to understand what a Runaway Syndrome is and how it impacts society.

2.1   An Insight on Runaway Syndrome

The phenomena of children running away from their home is not a new occurrence.

Research papers from as early as 1932 can be seen which discusses what we now call Runaway Syndrome. There have been many reasons which have been seen in children as to why they run away from their home. We will discuss them soon but first, let us look at some facts.

The following facts are taken from a research paper "Running Away From Home: A Longitudinal Study of Adolescent Risk Factors and Young Adult Outcomes" by Joan S. Tucker, Maria Orlando Edelen, Phyllis L. Ellickson, and David J. Klein. A research was conducted on 4329 youths, 48% were female, and 85% were white. They ranged from grade 9 to age 21. Let us look at the recorded facts: (1) 14% of the children were seen to run away at grade 10 or 11, (2) In grade 9, running away was detected due to a lack of parental support, substance abuse, depression, and troubles in school, (3) Runaways had more depression and drug dependence than non-runaways when they were about 21 years of age, (4) The result of the survey specifically pointed out that substance abuse and depression were significant factors that propelled the children to run away [2].

To put it in numbers, nearly 6–7% of adolescents are known to run away from their homes in any particular year. It is seen that most runaways go as far as 50 miles from their home and then soon enough come back to their parents or guardians. Now running away from their home and living alone for a particular amount of time results in two things: (1) The problems which made the youth run away in the first place are solved, and hence they decide to come back, (2) The second thing which can happen is actually pretty scary for a parent. When a child runs away, he needs to and wants to survive and as a result, may even engage in illicit activities, getting engaged in high-risk behaviour and even exploited by others [3].

As we can see, Runaway Syndrome is not a joke. It is pretty severe and affects a considerable portion of our children.

There are various reasons which have been found, which makes a child leave their home and run away. Children are immature and can be very insecure about themselves, their family or their position in the family. Let us now see a few factors which lead to children running away: (1) If the parents are quarrelling a lot, this could lead to the child feeling very insecure. (2) If the child feels that he or she is unloved, (3) Feeling jealous or insecure when a new child arrives, (4) When the child feels that he will be on the receiving end of a physical punishment or abuse by the father because of some wrongdoing, (5) When the child fears that the parents will stop loving them because of something which they did, (6) When the child fails or does very bad in an examination, (7) Sometimes just out of impulsiveness, (8) Sometimes to help a friend in need [2, 3].

It is seen that the children did not receive proper support from their parents or teacher in school. They were tagged a failure if they did terribly at academics by their teachers. They were also tagged an incompetent child if the child did not perform better than their neighbours by their parents. Hence the children started feeling that their parents are ashamed of them. They started rebelling and found peace and support in substance abuse like Marijuana and Alcohol. The comparisons and not getting enough attention at home made the child depressed and quiet. Some even get bullied at schools which worsen their situations. Since they are immature and insecure, all of these combined makes them give up everything and run away from their homes to find peace and maybe an environment where they will be accepted.

Case Studies The following summary is from "Case Study: Private Investigator Locates Runaway Child" by Robert Mann for a company called Worldwide Intelligence Network which is a licensed private investigator company based in California and New York. They help parents to find their Runaway Child [4].

On one fine day, these investigators received a call from a frantic parent saying that their child ran away. The child who was eleven years old at that time was supposedly upset on his parents and ran away without a word. The parents suspected that the child made an adult friend very recently and probably this very guy was harbouring the child where the child went of his own accord. The parents visited the man's home and asked about their child, and quite obviously he denied having seen or harbouring the child. Then the parents being scared decided to contact the Police. The Police quite frankly told the parents that even if the child is in that man's home, the child went over there of his own accord and hence legally they cannot search the man's home to look for the child. What they can do is ask the man if the Police can search their home, and if allowed by the man, the Police can conduct a search. The only way the Police can get into the man's house and conduct a search is with a warrant which they can get if the man is known to be a sex offender which basically means that the child is in imminent danger. Unfortunately, the man was clean, and hence no warrant could be obtained [4].

Following protocol, the Police visited the man's house and asked him if they could conduct a search. The man gave the police permission, and after intensive searching, the boy was nowhere to be found.

Suddenly the Police received a notification from the parents that one of their neighbours saw their son had gone inside the man's house one evening after the Police conducted the search. The Investigating company assigned an investigator to the case who was a former Police detective and specialised in tracking down runaway children. The Investigator interviewed several neighbours and found out that most of them saw the child around the house of the man, but nobody had seen him enter or leave the man's house. This was enough reason for the Investigator to set up shop and put the man's house under surveillance [4].

On one day the Investigator was about 75% sure that he saw the child leave the man's house from the back door. Following up on this evidence, the Investigation company asked the parents to ask the Police to conduct one more search of the house. The Police agreed and searched the man's house again. Upon searching the second time, the result seemed to be the same as the boy was nowhere to be found. When the Police were about to leave, the lead investigator on the case suddenly noticed an attic door on the second-floor ceiling. When the Police looked up inside the attic, they found the boy curled up into a corner of the attic [4].

The Police could not legally arrest the man because he had no prior criminal record and the boy went there of his own accord. The child was then returned to the parents who were overjoyed having their child back with them. The Investigation company suggested that the parents should counsel the child and find out why he left in the first place and change the circumstances so that the family can be happy and united [4].

As we can see in this case study, we never got to know why the child ran away, but it was evident that the child knew that the parents and the Police were searching for him. Still, he decided to hide and trusted the unknown man over his parents. From this, we can draw a probable conclusion that the child, for some reason, did not want to go back to his parents. This could mean that the parents did something which angered or saddened the child, hence confirming various studies.

As we can see, the Runaway syndrome has deep problematic roots, and if these children are not cured or stopped, they could lead a very traumatic life when they reach adulthood. This is precisely where my system comes in. We will talk about it more in the next section.

Effect of Runaway Syndrome on Rural India. Unlike the consensus of Runaway Syndrome, where the primary victims of this disorder are male kids, in Rural India, it is seen that females tend to run away more. An experiment conducted on ten different women yielded a very concerning result. All of them came from Rural India. It was revealed in the test that almost all of the participants went through some form of domestic violence from a very young age. Whether it be their father or mother, all the girls complained that they were beaten and cursed by their parents [5].

This drove the girls to look for a better life, a better home outside of the house where their parents lived. They did not even want to call it their home because the abuse and neglect they have faced have made a living there intolerable. While looking for a better life, these girls tend to come across a man, who in most cases, is very older than them. They promise the girls a better life and, in exchange, ask them to marry. These girls get wooed and decide to run away from their homes to look for the better life that is promised by the men. In some cases, the girl approaches their parents with the idea of getting married to that man but is vehemently rejected and may be abused again. The fact that she will get married to a man who is not approved by her parents does not concern her at all. She just wants a life away from that place, and to her, any life can be better than that. Hence they run away, with no regard to her own or her parents' future [5].

Now, what follows this is something of a great tragedy. These men are no better than her parents as they abuse and beat the girl. The girl decides to run away again, and if she is successful, she becomes homeless and probably on some street in India.

Now if the girl did not run away in the first place, stayed at her parents' and married a good person, maybe her life would be very different.

3   Precautions Available in the Market

This project was created, keeping in mind the struggles of both parents and children in question. Since you cannot always be with your children, most parents believe that they should either keep someone who will be with their child twenty-four hours a day or to set up a surveillance system. It must be noted that both the cases have their negatives. Let us talk about them and discuss the negatives.

In the case of having a nanny or a security for twenty-four hours, yes it is a great step. The person will be watching over your child as if he or she is the parent and will be watching over the child all the time. If the security feels anything is out of place or anything is bothering him or her, he or she will immediately let you know. A perfect measure, right? Well, not entirely. There are a few disadvantages which need to be discussed as well: (1) The child will feel that his or her's privacy is breached, especially during their teen years. Even though the child probably won't Runaway, chances are due to this, the child will grow to resent the parents, (2) Even the security needs a break. The chances are that if a child has to run away, he or she will when the security is supposed to be in the washroom or taking a call, (3) This way of looking after the child is very expensive.

With these disadvantages in mind, we can safely conclude that having the security to look after a child with Runaway Syndrome is maybe not the best idea.

Now let us see the second option, which is a surveillance system. There is no physical body to look after the child, but that is okay considering how having someone all the time can be a breach of privacy. The surveillance camera can also, at the same time, record the child all the time, and the person sitting on the other side of the screen can monitor the child. Looks pretty good but now let us take a look at a few of the disadvantages: (1) The person monitoring the child, for some reason is not attentive when the child is escaping the house, (2) Even though the camera is recording all the time, if there is no one to see when the child leaves and stops him or her immediately, there is no point of the recording, (3) Would you like to be monitored all the time? A camera on the corner of your room with someone eyeing you all the time? The answer is obviously no, and that is also the child's answer.

Well, now it looks like even a surveillance system is not the way to go. As a parent, you will feel very helpless if you cannot do anything to prevent your child from running away.

Well, thankfully, I have a solution which does not require the child to be monitored at all. With the help of a pressure sensor, if the child tries to make a run at any point, the parents will immediately be notified.

Engineer's touch.

4 Engineer's Touch

As we can see from the surveys and the symptoms, Runaway Syndrome is a psychological disorder that makes children run away from their homes. There are many reasons as to why a child will run away from their homes. Nobody wants to run away from the comfort of one's home but their reason drives them so much that their own safety does not concern them at all. Now the point is to understand all these you either need to be a psychologist or a sociologist. It would help if you had a degree in the understanding of mental health to know why they are running away and how to cure it. As for me, I am an Engineer, so I do not have that degree to fix the child's problem.

It is not possible for me as an Engineer to cure this disorder. Instead with science and technology, I have found a solution.This is exactly where the device comes in. This device cannot cure or tell you why the child is running away but can notify you when the child attempts to escape. This is the solution which I, as an engineer, can bring onto the table for Runaway Syndrome victims.

5 Research Methodology

The basic principle of the project is elementary. For this device, you need to set up piezo sensors under the floor and connect it with the whole system (Figs. 1 and 2)..

As seen in the block diagram, Raspberry Pi is the main processor which is controlling the whole system. All the necessary hardwares are connected to the Raspberry Pi. The Piezo sensor is to sense te movements of the child. Whenever any movement is detected, a message is immediately sent to the IoT website which is a portal made for the parent. The whole system can be even better understood with the picture below [6].

Fig. 1 Proposed flowchart of low cost IoT based runaway syndrome tracking system

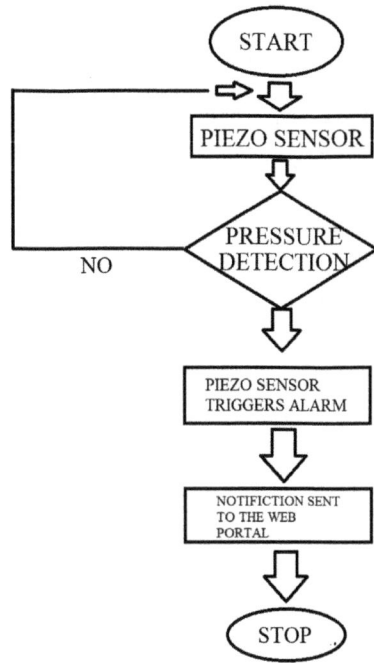

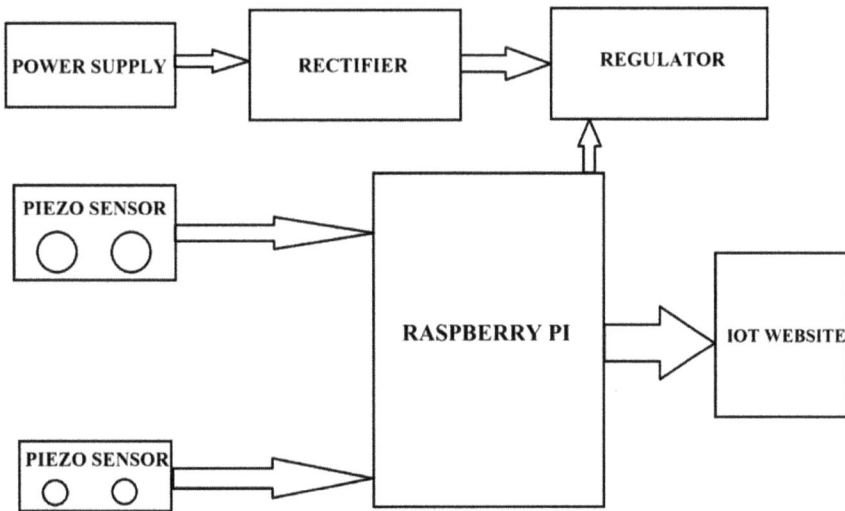

START

PIEZO SENSOR

PRESSURE DETECTION

NO

PIEZO SENSOR TRIGGERS ALARM

NOTIFICTION SENT TO THE WEB PORTAL

STOP

POWER SUPPLY → RECTIFIER → REGULATOR

PIEZO SENSOR → RASPBERRY PI → IOT WEBSITE

PIEZO SENSOR →

Fig. 2   Proposed block diagram of low cost IoT based runaway syndrome tracking system

The parent needs to set up the system outside the child's room and outside their own house, in the courtyard. Hence when the child is thinking about running away and steps out of his or her room, he steps on the tile under which a Piezo sensor is installed. At the same time, the same happens for the system placed outside the house. So if the parent senses two alarms going off instantly, they can be sure that the child is trying to run away.

6   Discussion and Contribution Towards the Society

There are a lot of similar systems that are used for theft detection purposes, but none is made, which helps a troubled child. A runaway syndrome is not a joke, and there is no solution to it other than psychological help. But even with psychological support, it takes time, and in between having such a device is of great use.

This project was made with no intention of making a child feel vulnerable in his or her home where the child is supposed to feel the safest. The purpose of the project is to be the eyes and ears of a parent who cannot always be with their child. A child is supposed to be made to feel safe and nurture him so that their habit of running away goes away slowly. We have seen that the reasons for running away include family problems and how the parents behave in their homes. The parents can put in any such devices and spend money on psychologists, but the child will never recover if the treatment does not start from their home.

People who are psychologically challenged and have symptoms which are similar to Runaway Syndrome can also benefit from this device. Let us take a quick example: (1) Dementia: Patients who have dementia have a habit of forgetting things. They do not run away per se, but sometimes when they go out, they forget their way back home. Having a device that will help the family members know that the person has left the house can be very helpful, (2) PTSD: Patients who have PTSD also suddenly leave their house, which they are not supposed to. They can also be tracked using this system, (3) Anxiety, (4) Depression, (5) Intellectual Disability

As we can see, this system can be used for a variety of purposes. But let's not get sidelined and forget our main concern here, which is to protect our children.

7 Conclusion

This system can help change the future of many children who are suffering from Runaway Syndrome. This system is inexpensive and requires very few items to work with in the first place. The parent just needs some pressure sensors and attach in suitable areas in and around the child's room. That's it then whenever the child gets out of his room, the parents will immediately be notified.

8 Future Scope

This project is by no means perfect. This device has some flaws which can be fixed with future upgrades. Now let us see how this project can be changed and even made perfect: (1) Adding a Camera, (2) Having an option to send videos to the server along with pictures, (3) This device has no memory as of itself. It would be fantastic if all the pictures taken by the device can be stored in a memory somewhere, (4) Adding facial recognition software. This will help the device distinguish between humans and non-humans, (5) Having an advanced facial recognition software which will only alarm the parent if and only if the child has moved. Which means the device will work individually for the child.

These are some of the many additions which can be done on the device. This device is made just for the basic purpose and at a reasonable cost. Adding this feature might increase the price factor, but it will create a fantastic security device for children with Runaway Syndrome.

Acknowledgements The aim of this project was always to contribute something to society. Our children govern as to how the future of our society is framed and hence doing something for their safety, and their future was and still is my aim. Hence I decided to build this device. My parents and also my friends provided their utmost support while I was building this device. I am very proud of this project and would expect this to be used everywhere globally.

References

1. Sreedhar KP (2018) Run away syndrome. [Psychology4all]
2. Edelen MO, Ellickson PL, Klein DJ, Tucker JS (2010) Running away from home: a longitudinal study of adolescent risk factors and young adult outcomes. [NCBI]
3. Edelen MO, Holliday SB, Tucker JS (2016) Family functioning and predictors of runaway behavior among at-risk youth. [NCBI]
4. Mann R (2015) Case study: private investigator locates runaway child. [WINCOR]
5. Pratiksha HR, Stacey PR, Vaishali VR (2010) Damned if they flee, doomed if they don't: narratives of runaway adolescent females from rural India. [Springer]
6. Heath N (2017) What is the raspberry Pi 3? everything you need to know about the tiny, low-cost computer. [ZDNET]

Chapter 75
Application of Robotic Process Automation

**Krishna Kumar, Rachna Shah, Narendra Kumar,
and Ravindra Pratap Singh**

1 Introduction

RPA is used to automate the repetitive tasks which are performed in the industries based on defined rules by using software bots. All the RPA tools can be segregated into four different types of tools which are listed in Table 1.

As each tool has its features, a comparison of companies and methods used for the development is listed in Table 2.

The selection of the RPA tool is based on various criteria like time required for the development, technology used, scalability option, security issues, cost of the tool, etc.

2 Literature Review

To know the development in RPA, an extensive research review has been performed. Huang and Vasarhelyi [1] applied RPA to audit area and proposed an RPA framework for auditors and allows them to focus on tasks that require professional judgment. Syed et al. [2] reviewed the deployment of the RPA organizations. Schörgenhumer and Eitzlmayr [3] proposed a numerical method for the model-based study of the

K. Kumar (✉)
Department of Hydro and Renewable Energy, IIT Roorkee, Roorkee, India
e-mail: kkumar@ah.iitr.ac.in

R. Shah
National Informatics Centre, Dehradun, Uttarakhand, India

N. Kumar
School of Computing, DIT University, Dehradun, Uttarakhand, India

R. P. Singh
Bipin Tripathi Kumaon Institute of Technology, Dwarahat, Uttarakhand, India

Table 1 Type of RPA tools

RPA Tools	Description
Cognitive automation bots	It can handle unstructured data to make decisions
Self-learning tools	It can analyze human activities
Excel	It is a basic automation tool
Software bots	It can interact with other systems

Table 2 Companies and development method

Companies	Method
Another monday	Drag and drop automation process
AntWorks	Bot cloning is possible
Automation edge	Drag and drop automation technique
Automation anywhere	Drag and drop and AI-augmented RPA tool
BluePrism	Enterprise automation with drag and drop automation
Jacada	Desktop automation
Kofax	Unified design environment with built-in analytics
Kryon systems	Strong analytics and deployment efficiency
UiPath	Drag and drop functionality with visual designer
Visual cron	Integration and task scheduling tool

underlying SPH formulation sloshing problems. Modeling of solid wall boundary conditions as well as consistent density field treatment have proved crucial.

Alexovič et al. [4] explained the role of RPA in pharmaceutical industries. A high sample-throughput can be achieved by automating the sample pre-treatment, resulting in high accuracy and low waste output of dangerous solvents and risky biological materials. Schäffer et al. [5] proposed a pragmatic approach for optimizing knowledge-intensive engineering processes, focusing on the incremental, tool-based development of a popular knowledge base. Ma et al. [6] reviewed the RPA system and concluded that it can eliminate the errors in repetitive work. Kommera [7] concluded that the RPA is a software that imitates the steps a human takes to complete repetitive, rule-based tasks. Madakam et al. [8] reviewed the RPA technologies available for different business models. Enriquez et al. [9] examined the current state of RPA to identify potential gaps in both scientific and industrial literature, and concluded that some phases of the RPA lifecycle are already being solved on the market.

Asatiani and Penttinen [10] analyzed the challenges posed by Senior Vice President of the OpusCapita Group, who are looking for ways to market RPA and add value to existing and new customers. Asquith and Horsman [11] analyzed the challenges of tackling increasing caseloads, large volumes of digital data, and maintaining review efficiency. Balasundaram and Venkatagiri [12] concluded that RPA can help to provide better service to employees and managers, ensure compliance of HR processes with standards and regulations. Carden et al. [13] reviewed a large technology company anonymized as TECHSERV as they plan, manage, and implement automation of the robotics processes. Automation of the robotics process has been strategically deployed to automate its business processes.

Chacón-Montero et al. [14] analyzed that an RPA project follows a life-cycle similar to that of a software project. However, a testing environment is not always available in certain contexts. The deployment of the robots in the environment of production thus entails a high risk. Vijaya et al. [15] applied RPA to automate the process of insurance claims, as RPA is already implemented in motor vehicle insurance. Presently an agent dealing with these various types of consumer insurance policy must register each policy dividing it by form and requiring different registration procedures depending on the insurance the consumer is applying.

Automation Anywhere [16] explained that RPA bots are capable of logging into business applications using passwords similar to humans and automating tasks involving any business application, including legacy applications without API. Naveen et al. [17] concluded that the RPA delivers advanced software system robots that take the place of humans whenever complicated processes or routine tasks are controlled by machines. Timo [18] discussed the RPA-Enterprise Architecture partnership in terms of viewpoints and implications. Robotic Process Automation (RPA) is the initial idea of automating business processes through the presentation layer of existing application systems. Issac et al. [19] concluded that RPA can accelerate back-office tasks in commercial industries, remote IT management tasks, and resource conservation in multiple sectors.

Moffitt et al. [20] explained how the Robotic Process Automation (RPA) can interfere with the traditional audit model. RPA is expected to repurpose the role of the auditor with its ability to automate repetitive and manual, rule-based tasks by replacing perfunctory tasks and highlighting higher order-thinking skills which will ultimately contribute to enhanced audit quality. Patil et al. [21] proposed an RPA solution for the education domain for the analysis of student's examination results. Kobayashi et al. [22] explained the communication robot as one of the consumer services, such as an AI. There are still high hurdles for the elderly in the operation of such advanced IT devices, as they have realized high functions seen in the cooperation of network and home appliances.

Kokina and Blanchette [23] explored emerging themes surrounding bot implementation for accounting and finance tasks. Leshob et al. [24] explained the RPA which refers to configuring software-based robots for business processes to do the work previously performed by the organizations. Maalla [25] explained how the existing market-based business process management system replaces manual desktop operations utilizing system development and connects various systems to the business workflow to achieve the effect of improving interaction efficiency. Nam [26] listed factors influencing citizen attitudes regarding job replacement were explored through robotic automation and alternative policy options. An analysis based on regression is used to predict which attitudinal category an individual belongs to what degree he or she supports certain policy options based on their awareness, perceived likelihood, and expectations of positive job replacement outcomes.

Graves et al. [27] installed Laboratory automation systems (LAS) across North America at over twenty-two sites offering automation of many pre-analytical and analytical activities in clinical laboratories. Just a few laboratories for the treatment of hemostasis disorders have automated for the study of citrated whole blood. To have a fast turnaround, the study of coagulation factors in citrated blood requires a significant amount of labor; therefore, automation of this analytical method is attractive. Therefore, they have developed an automated workstation for coagulation, using a systematic approach to design and engineering automation. Firstly, they have used a discrete simulation of events to measure potential performance and define possible bottlenecks for the proposed workcell for coagulation and created a three-dimensional animated workstation computer model to simplify the design of the workstation. Finally, a prototype workcell was constructed using a mobile robot, an articulated robotic arm, and an analytical coagulation system.

Sindhu et al. [28] classified the term RPA which can be split as robotic representing an object capable of being programmed by a machine to execute computer tasks, known as a robot; method meaning the series of actions taken in a sequential order to accomplish a specific purpose; and Automation indicating performing tasks without human interference. Work was undertaken to create a virtual text to voice assistant bot that assists a manager who is responsible for handling item inventory to his varied other jobs. The customized software robot ensures that if the stock in any item's inventory reaches the level of reordering, bots itself sends the message to the supplier by email to reorder the items required. The study shows that RPA can be a better substitute for a human being who can more effectively carry out intense time and necessary activities.

Prasanti et al. [29] explained how a busy manager of the corporate world can engage RPA as an intimate and reliable assistant to more competently and effectively perform repetitive human tasks. With'bot' assuming the human assistant's task of sending routine email responses. Kaelble [30] explored the benefits offered by automation of robotic and desktop processes, giving more and more mundane work to machines that can handle it well and without complaint. It introduces the idea, discusses where it functions best, and describes some of the value that it can bring to your company. Data digitization was an explosive force for productivity and progress, but it also created its new roster of mundane work roles.

Sreeja et al. [31] introduced a user-friendly automation process using the RPA tool. These include record-keeping, questions, estimates, and transactions. Sengupta [32] explained how Software Performance Testing (SPT) can be performed using the RPA tool. SPT is the kind of Non-Functional Testing (NFT) to prove the program to be implemented by the project team/product team is functioning as planned regarding system safety, stability, the robustness of endurance, etc. This check is a very important part of a live device. Several instances have been seen in the software industry history, the live application has been demoted or scrapped due to the system being unable to behave as expected with its features and functionalities when the system is being used simultaneously at high load. Only when we do performance testing can we simulate this type of verification.

Ivančić et al. [33] investigated how the academic community defines RPA, and to what extent it has been investigated in the literature in terms of the state, trends, and application of RPA. Jovanović et al. [34] explained that business processes are part of every business daily routine; they are asked to perform in the best way possible, without any losses. Traditionally, people try to automate processes over time using many techniques that could include machines or mechanical robots. A lot of preformed processes are done using computers in the modern business or e-business, so there is a need to avoid losses caused by bad or slow process performance. The way process automation is done using software that will perform like humans and perform tasks on computers is referred to as automation of robotic processes. It is normally used to automate processes of relatively low to intermediate complexity and high repetitiveness. A business that automates its processes using robotic process automation should gain multiple benefits in terms of cost reduction, improve process efficiency, and significantly reduced the number of process rework tasks.

Romao et al. [35] concluded that inexperienced or untrained models that eventually decrease productivity and increase errors resulting from unfounded or even wrong decisions, as well as a case study in the banking sector that shows some examples of the benefits and risks of BPM solutions using AI-related agents/artifacts. Uskenbayeva et al. [36] presented the findings of the study and the implementation of the RPA methodology. The work is based on the robotic automation approach for administrative business processes. The concept of robotic process automation addressed the benefits of RPA technology and the stages of successful RPA implementation.

Willcocks et al. [37] explained a series of dilemmas facing senior managers, reflecting on the potential application of robotic process automation (RPA) to human resources (HR) operations and global business services (GBS). William and William [38] developed a Corporate Service Providers (CSP) RPA system in Singapore to support their annual compliance processes for their customers and their ad hoc requests. Yatskiv et al. [39] presented a test approach to the program using RPA. It is a business workflow automation approach, in which the program emulates user actions within a graphical user interface to achieve the desired outcome. The summary of the research review is presented in Table 3.

From the above review, we can conclude that RPA can be applied in any area, but it is not applied to date in the energy sector. Therefore there is a scope to identify the areas where RPA can be applied.

3   Application of RPA in Energy Sector

Digitization plays a major role in the development of any industry. The wind and hydropower plants are mostly located in remote areas. Automation is required to

Table 3 Summary of research review

Reference	Work areas
Syed et al. [2], Ma et al. [6], Madakam et al. [8], Enriquez et al. [9], Asquith et al. [11], Carden et al. [13], Naveen et al. [17], Issac et al. [19], Moffitt et al. [20], Kokina et al. [23], Leshob et al. [24], Sindhu et al. [28], Prasanti et al. [29], Sreeja et al. [31], Ivančić et al. [33], Jovanović et al. [34], Yatskiv et al. [39]	Review papers
Balasundaram et al. [12], Willcocks et al. [37]	HR process
Huang et al. [1]	For account auditors
Schörgenhumer et al. [3]	A numerical method for the model-based study
Alexovič et al. [4]	Pharmaceutical industry
Schäffer et al. [5]	Knowledge management
Kommera et al. [7]	Office work automation
Asatiani et al. [10]	Challenges of OpusCapita group
Chacón-Montero et al. [14]	Software testing
Vijaya et al. [15]	Insurance claims
Automation Anywhere [16]	Loan processing
Timo et al. [18]	Enterprise architecture for RPA
Patil et al. [21]	Examination result compilation
Kobayashi et al. [22]	Consumer services
Maalla et al. [25]	Enterprise management
Nam et al. [26]	Citizen attitude examination
Graves et al. [27]	Clinical testing
Kaelble et al. [30]	Digitization of data
Sengupta et al. [32]	Software performance testing
Romao et al. [35]	Banking sector
Uskenbayeva et al. [36]	Administrative business process
William et al. [38]	Corporate services

optimize system performance and to increase the efficiency of the plants. Smart sensors installed at the different locations in a power plant generates a large volume of data every day. To generate reports based on the data is an everyday task. In the case of the energy sector, RPA can be utilized for the generation of shift duty charts, generation of tariff reports, generation of load schedule, generation of fault reports, energy generation reports, generation of salary slips of employees, and also to communicate reports automatically to the concern authorities.

4 Conclusions

RPA is a powerful tool to perform repetitive tasks quickly without any error. RPA is presently implemented in the banking, insurance, and pharmaceutical sector. Some other areas are also under development. It can be utilized in many areas to minimize the manpower requirement. Power plants are running continually and generating large volumes of data. To know better insight and prepare daily reports RPA can be utilized.

References

1. Huang F, Vasarhelyi MA (2019) Applying robotic process automation (RPA) in auditing: A framework. Int J Account Inf Syst 35:100433. https://doi.org/10.1016/j.accinf.2019.100433
2. Syed R, Suriadi S, Adams M, Bandara W, Leemans SJJ, Ouyang C et al (2020) Robotic process automation: contemporary themes and challenges. Comput Ind 115:103162. https://doi.org/10.1016/j.compind.2019.103162
3. Schörgenhumer M, Eitzlmayr A (2019) Modeling of liquid sloshing with application in robotics and automation. IFAC-PapersOnLine 52:253–258. https://doi.org/10.1016/j.ifacol.2019.11.683
4. Alexovič M, Dotsikas Y, Bober P, Sabo J (2018) Achievements in robotic automation of solvent extraction and related approaches for bioanalysis of pharmaceuticals. J Chromatogr B: Anal Technol Biomed Life Sci 1092:402–421. https://doi.org/10.1016/j.jchromb.2018.06.037
5. Schäffer E, Mayr A, Huber T, Höflinger T, Einecke M, Franke J (2019) Gradual tool-based optimization of engineering processes aiming at a knowledge-based configuration of robot-based automation solutions. Proc CIRP 81:736–741. https://doi.org/10.1016/j.procir.2019.03.186
6. Ma Y-W, Lin D-P, Chen S-J, Chu H-Y, Chen J-L (2020) System design and development for robotic process automation. 187–189. https://doi.org/10.1109/smartcloud.2019.00038
7. Kommera V (2019) Robotic process automation. Am J Intell Syst 9:49–53. https://doi.org/10.5923/j.ajis.20190902.01
8. Madakam S, Holmukhe RM, Kumar Jaiswal D (2019) The future digital work force: robotic process automation (RPA). J Inf Syst Technol Manag 16:1–17. https://doi.org/10.4301/s1807-1775201916001
9. Enriquez JG, Jimenez-Ramirez A, Dominguez-Mayo FJ, Garcia-Garcia JA (2020) Robotic process automation: a scientific and industrial systematic mapping study. IEEE Access 8:39113–39129. https://doi.org/10.1109/ACCESS.2020.2974934
10. Asatiani A, Penttinen E (2016) Turning robotic process automation into commercial success-case OpusCapita. J Inf Technol Teach Cases 6:67–74. https://doi.org/10.1057/jittc.2016.5
11. Asquith A, Horsman G (2019) Let the robots do it!–taking a look at robotic process automation and its potential application in digital forensics. Forensic Sci Int Rep 1:100007. https://doi.org/10.1016/j.fsir.2019.100007
12. Balasundaram S, Venkatagiri S (2020) A structured approach to implementing robotic process automation in HR. J Phys Conf Ser 1427. https://doi.org/10.1088/1742-6596/1427/1/012008
13. Carden L, Maldonado T, Brace C, Myers M (2019) Robotics process automation at TECH-SERV: an implementation case study. J Inf Technol Teach Cases 9:72–79. https://doi.org/10.1177/2043886919870545
14. Chacón-Montero J, Jiménez-Ramírez A, Enríquez JG (2019) Towards a method for automated testing in robotic process automation projects. In: Proceedings-2019 IEEE/ACM 14th International work automation software testing, AST 2019 2019:42–7. https://doi.org/10.1109/AST.2019.00012

15. Vijaya GME, Sathish, Shivaram, Dini K, Aishwarya (2019) Claim registration using robotic process automation-a survey. Int J Sci Res Eng Dev 2
16. Automation Anywhere (2020) Enterprise-class security for robotic process automation
17. Naveen Reddy KP, Harichandana U, Alekhya T, Rajesh SM (2019) A study of robotic process automation among artificial intelligence. Int J Sci Res Publ 9:8651. https://doi.org/10.29322/ijsrp.9.02.2019.p8651
18. Timo K (2019) Resilience in security and crises through adaptions and transitions. In: Inform 2019 lecture notes informatics, pp 571–584. https://doi.org/10.18420/inf2019
19. Issac R, Muni R, Desai K (2018) Delineated analysis of robotic process automation tools. In: Proceedings 2018 2nd international conference advaced electronics computing communication ICAECC 2018, pp 1–5. https://doi.org/10.1109/ICAECC.2018.8479511
20. Moffitt KC, Rozario AM, Vasarhelyi MA (2018) Robotic process automation for auditing. J Emerg Technol Account 15:1–10. https://doi.org/10.2308/jeta-10589
21. Patil S, Mane V, Patil P (2019) Social innovation in education system by using robotic process automation (RPA). Int J Innov Technol Explore Eng 8:3757–3760. https://doi.org/10.35940/ijitee.K2148.0981119
22. Kobayashi T, Arai K, Imai T, Tanimoto S, Sato H, Kanai A (2019) Communication robot for elderly based on robotic process automation. Proc-Int Comput Softw Appl Conf 2:251–256. https://doi.org/10.1109/COMPSAC.2019.10215
23. Kokina J, Blanchette S (2019) Early evidence of digital labor in accounting: innovation with robotic process automation. Int J Account Inf Syst 35:100431. https://doi.org/10.1016/j.accinf.2019.100431
24. Leshob A, Bourgouin A, Renard L (2018) Towards a process analysis approach to adopt robotic process automation. In: Proceedings-2018 IEEE 15th International conference E-Bus engineering ICEBE 2018, pp 46–53. https://doi.org/10.1109/ICEBE.2018.00018
25. Maalla A (2019) Development prospect and application feasibility analysis of robotic process automation. In: Proceedings 2019 IEEE 4th Advanced information technology electronics automation control conference IAEAC 2019, pp 2714–2717. https://doi.org/10.1109/IAEAC47372.2019.8997983
26. Nam T (2019) Citizen attitudes about job replacement by robotic automation. Futures 109:39–49. https://doi.org/10.1016/j.futures.2019.04.005
27. Graves S, Holman B, Rossetti M, Estey C, Felder R (1998) Robotic automation of coagulation analysis. Clin Chim Acta 278:269–279. https://doi.org/10.1016/S0009-8981(98)00152-1
28. Sindhu PD et al, SPD et al (2019) Robotic process automation—a virtual assistant for inventory management. Int J Robot Res Dev 9:1–14. https://doi.org/10.24247/ijrrdjun20191
29. Prasanti Dasu S, Pradesh Ch Radhakumari A (2018) Robotic process automation-an expert technology assistant to a busy manager. Int J Adv Res 4:172–181
30. Kaelble S (2018) In: Robotic process automation for dummies, NICE Special Edition
31. Sreeja GG, Shanthini M (2018) Robotic process automation for insurance claim registration. Int J Adv Res 6:23–27. https://doi.org/10.21474/ijar01/6821
32. Sengupta T(2019) Robotic process automation in software performance testing workload modeling. Int J Comput Eng Technol 10:25–30. https://doi.org/10.34218/ijcet.10.2.2019.003
33. Ivančić L, Suša Vugec D, Bosilj Vukšić V (2019) Robotic process automation: systematic literature review. Lect Notes Bus Inf Process 361:280–295. https://doi.org/10.1007/978-3-030-30429-4_19
34. Jovanović SZ, Đurić JS, Šibalija TV (2019) Robotic process automation: overview and opportunities. Int J"Adv Qual 46
35. Romao M, Costa J, Costa CJ. Robotic process automation: a case study in the banking industry. In: Iber conference information system technology Cist 2019 June 1–6. https://doi.org/10.23919/CISTI.2019.8760733
36. Uskenbayeva R, Kalpeyeva Z, Satybaldiyeva R, Moldagulova A, Kassymova A (2019) Applying of RPA in administrative processes of public administration. In: Proceedings-21st IEEE conference bus informatics, CBI 2019, vol 2, pp 9–12. https://doi.org/10.1109/CBI.2019.10089

37. Willcocks L, Lacity M, Craig A (2017) Robotic process automation: strategic transformation lever for global business services? J Inf Technol Teach Cases 7:17–28. https://doi.org/10.1057/s41266-016-0016-9
38. William W, William L (2019) Improving corporate secretary productivity using robotic process automation. In: Proceedings 2019 international conference technology applications artificial intelligence TAAI 2019. https://doi.org/10.1109/TAAI48200.2019.8959872
39. Yatskiv S, Voytyuk I, Yatskiv N, Kushnir O, Trufanova Y, Panasyuk V (2019) Improved method of software automation testing based on the robotic process automation technology. In: 2019 9th International conference advanced computing information technology ACIT 2019-Proc 2019, pp 293–296. https://doi.org/10.1109/ACITT.2019.8780038

Author Index

© The Editor(s) (if applicable) and The Author(s), under exclusive license
to Springer Nature Singapore Pte Ltd. 2021
X.-Z. Gao et al. (eds.), *Applications of Artificial Intelligence in Engineering*, Algorithms
for Intelligent Systems, https://doi.org/10.1007/978-981-33-4604-8

9 789813 346062